T0175714

Molecular and Physiological Mechanisms of Muscle Contraction

Jean Emile Morel

Commissariat à l'Energie Atomique
(Direction des Sciences du Vivant)
Ecole Centrale de Paris &
Université (Pierre et Marie Curie)
Paris, France

CRC Press
Taylor & Francis Group
Boca Raton London New York

CRC Press is an imprint of the
Taylor & Francis Group, an **informa** business

CRC Press
Taylor & Francis Group
6000 Broken Sound Parkway NW, Suite 300
Boca Raton, FL 33487-2742

First issued in paperback 2020

© 2016 by Taylor & Francis Group, LLC
CRC Press is an imprint of Taylor & Francis Group, an Informa business

No claim to original U.S. Government works

ISBN-13: 978-1-4987-2502-6 (hbk)
ISBN-13: 978-0-367-65869-4 (pbk)

Visit the Taylor & Francis Web site at
http://www.taylorandfrancis.com

and the CRC Press Web site at
http://www.crcpress.com

To my wife Marie-Hélène, my son Philippe, my former students and coworkers
To my dear brother Yves who died in July 2014 and will never read this book
This monograph was written in Plouescat, Brittany, France, between 2001 and 2014

CONTENTS

PREFACE

Frog and rabbit are the principal animal species studied in this monograph. The passage from one species to another may make it difficult to follow the reasoning, but this passage is necessary because most of the experimental results concerning physiological properties (e.g. isometric tetanic tension and velocity of shortening) were obtained with intact frog fibres, whereas most of the enzymatic results concerning MgATPase activities were obtained with demembranated rabbit fibres.

This book comprises a long line of reasoning, with many interdependent and complex sections. I therefore recommend reading the titles of the most complicated chapters and sections carefully to facilitate understanding. Consulting the Index could also be useful.

In addition to reviewing and reanalysing the results of studies of my own and many other independent groups, I also report unpublished results for experiments with half-fibres (intact fibres split lengthwise) from white skeletal muscles of young adult frogs and with permeabilised fibre bundles from red skeletal muscles of young adult rats. From this many-faceted 'treatise', a hybrid model emerges, combining the swinging cross-bridge/lever-arm processes and lateral swelling mechanisms.

In these new experimental findings, the relative resting force, recorded at pH 7 and 10°C, in half-fibres from white skeletal muscles (*iliofibularis*) of young adult frogs (*Rana pipiens*), held around the slack length, increased very slightly when the bulk ionic strength was lowered from 180 mM to ~40 mM. Between ~40 mM and ~30 mM, the relative resting force increased very rapidly with further decreases in ionic strength, peaking at high levels, between ~30 mM and ~20 mM. Below ~20 mM, the relative resting force decreased sharply. The dependence of the relative resting force on ionic strength, the existence of a maximum and the rapid decrease at very low ionic strengths demonstrate that strong radial repulsive electrostatic forces are exerted between the myofilaments under resting conditions (see below concerning the conversion of radial forces into axial forces). These radial repulsive electrostatic forces are also effective in half-fibres (and all types of fibre, whether intact or demembranated), under isometric tetanic contraction conditions, and present qualitative characteristics similar to those at rest (only some quantitative features differ).

In another set of experiments, myosin heads were cleaved enzymatically (i.e. digested with α-chymotrypsin) from the rest of the thick myosin filaments, in permeabilised fibre bundles from red skeletal muscles (*tibialis anterior*) of young adult rats (Wistar), held around the slack length, in a buffer mimicking the physiological resting medium, at room temperature. Very small, sometimes tiny, but detectable, transitory contractures resulted from these enzymatic cleavages, demonstrating that thin actin and thick myosin filaments are tethered by a small number of 'resting' (weakly bound) cross-bridges, exerting strong radial tethering forces (together with weak radial attractive/compressive forces) that counterbalance the radial repulsive electrostatic forces, under resting conditions.

Based on these two series of experimental observations, a hybrid model of muscle contraction is proposed, in which the radial tethering forces decrease drastically once contraction is triggered, leading to net radial expansive forces between the myofilaments. Under both resting and contracting conditions, the net radial repulsive (expansive) forces are turned into axial forces. Indeed, in the 1970s, based on theoretical and logical reasoning, mechanisms for this conversion were proposed that are valid at rest and during contraction, regardless of volume variations. In this hybrid model, under standard conditions (e.g. 10°C, slack fibre length, pH ~7, ionic strength ~180 mM), part of the isometric tetanic tension (~40%) results from lateral swelling mechanisms (the usual name for mechanisms involving radial expansive forces inducing axial contractile forces) and another part (~60%) results from swinging cross-bridge/lever-arm processes plus, possibly, the impulsive mechanism developed by Elliott and Worthington plus, possibly, the 'step-wise' mechanism developed by Pollack's group, and other models (e.g. ~30% of swinging cross-bridge/lever-arm models, ~10% of impulsive model, ~10% of Pollack's model, and ~10% of other models). In this work, I neglect the 'unconventional' models and use only swinging cross-bridge/lever-arm mechanisms, with a proportion of ~60%.

The experimental findings, summarised here, and the hybrid model developed and discussed in this book provide the pretext for a long critical and constructive review and an analysis concerning many well-known properties of contracting muscle fibres, as well as complex phenomena (including unexplained, forgotten, ignored, even 'mysterious' experimental and semi-empirical results). Most of the 'forgotten' observations were not accounted for by swinging cross-bridge/lever-arm models and were, therefore, rarely taken into account in the many discussions presented in publications concerning muscle contraction and its molecular basis. By contrast, I think that the hybrid model answers many of these awkward questions.

In 2000, Cyranoski published a commentary paper concerning the symposium held in Osaka (Japan) on in vitro motility and its possible link to muscle contraction. The author provided a severe, but lucid, analysis of the approach of Yanagida and his coworkers (see also Chapter 9 in this book for supplementary arguments). Cyranoski also cited Molloy, who claimed, during the symposium, that 'The tightly coupled lever-arm idea is simple, predictive and inherently testable because of its more restrictive nature' (as opposed to 'the loose-coupled thermal ratchet model' defended by Yanagida and his group). Thus, in 2000, the general view, expressed by Molloy, was apparently clearcut. In this monograph, I demonstrate that the situation is much more complex than previously thought by many specialists in muscle contraction and in vitro motility.

The starting point for writing this book was essentially the conclusion of Cyranoski (2000), who cited A.F. Huxley: 'I came here confused about actin and myosin. Now, I am still confused, but at a higher level'.

The need to reopen the question of the universality of swinging cross-bridge/lever-arm theories and to search for innovative ideas, based on new experiments, has been expressed by Bryant et al. (2007), who wrote: 'the basic actomyosin motor has been embellished, altered, and reused many times through the evolution of the myosin superfamily'. More recently, Grazi (2011) wrote: 'With time clever hypotheses may be accepted as "facts" without being supported by solid experimental evidence. In our opinion this happened with muscle contraction where pure suggestions still occupy the scene and delay the progress of the research'. In the last few years, many authors, including Bryant et al. (2007), have focused on myosin VI and discovered unexpected properties of this motor protein, studied in vitro by brilliant techniques. The titles of the articles by Spudich and Sivaramakrishnan (2010) and Sweeney and Houdusse (2010) were particularly 'explosive': 'Myosin VI: an innovative motor that challenged the swinging lever arm hypothesis' and 'Myosin VI rewrites the rules for myosin motors', respectively.

I have made use of the time available to me since my retirement to read as many papers as possible in the domains of muscle contraction and in vitro motility, with the aim of resolving the confusion. I provide in this work my own analysis of the various questions posed in the muscle and motility areas and suggest a synthesis, including the hybrid model. Moreover, to the best of my knowledge, the most recent monographs describing the traditional mechanisms of muscle contraction only are those by Bagshaw (1993), Simmons (1992) and Squire (2011).* They have the same titles and resemble descriptive textbooks, and the many experiments published during the last 30 years or so have never been critically analysed. In any event, this long and complex book will be a useful working tool for specialists in muscle contraction, professors, doctors in medicine, and graduate students.

Some important keywords emerge from this monograph that may help the reader to understand this work: head–head dimers, thick myosin filaments, radial repulsive electrostatic forces, radial tethering forces and translation of radial forces into axial forces.

This monograph was completed between 2008 and 2011, and the bibliography concerns the period before 2008–2011 (the first reference dates back to 1911). There are 1000–1100 references. In the addendum, I propose a supplementary list of about 200–250 references, corresponding to the period between 2008 and 2012–2013, with a few references from 2014 and 2015.

Finally, this book may be seen as an analysis and a synthesis of many experimental, theoretical and semi-empirical studies published over the last century or so. Even the most firmly 'unconvinced' reader will enjoy reading the long reference list.

REFERENCES

Bagshaw, C. R. 1993. *Muscle contraction.* Chapman & Hall, London, UK.

Bryant, Z., D. Altman, and J. A. Spudich. 2007. The power stroke of myosin VI and the basis of reverse directionality. *Proc. Natl. Acad. Sci. USA* 104:772–7.

Cyranoski, D. 2000. Swinging against the tide. *Nature* 408:764–6.

Grazi, E. 2011. Experimental basis of the hypothesis on the mechanism of skeletal muscle contraction. *Muscles Ligaments and Tendons* 1:77–84.

Simmons, R. M. 1992. *Muscular contraction.* Cambridge Univ. Press, UK.

Spudich, J. A., and S. Sivaramakrishnan. 2010. Myosin VI: An innovative motor that challenged the swinging lever arm hypothesis. *Nat. Rev. Mol. Cell Biol.* 11:128–37.

Squire, J. M. 2011. *Muscle contraction.* eLS. Wiley & Sons Ltd., Chichester, UK.

Sweeney, H. L., and A. Houdusse. 2010. Myosin VI rewrites the roles for myosin motors. *Cell* 131:573–82.

---

* After the acceptance of this monograph, Rall published a book on muscle contraction (2014; see Addendum). These two works are clearly complementary, with little or no overlap.

ACKNOWLEDGEMENTS

This work was supported by grants from the Association Française contre les Myopathies, the Commissariat à l'Energie Atomique (Direction des Sciences du Vivant) and the Ministère de l'Enseignement Supérieur et de la Recherche. I thank Dr. Michel Villaz for teaching me about the preparation and handling of frog half-fibres (split fibres). I also thank Dr. Renée Ventura-Clapier for teaching me about the preparation and handling of permeabilised bundles of rat fibres. I am greatly indebted to all my students for their help in building the home-made apparatuses used to record forces and for long and constructive technical and scientific discussions, over more than 15 years. I thank Nathalie Guillo for her help in preparing cardiac myosin from hamster. I would also like to thank Sir Andrew Huxley (deceased in 2012), Gerald Elliott (deceased in 2013) and Roger Woledge (deceased in 2015) for many fruitful and stimulating discussions and Gerald Pollack for encouragement. I am indebted to the editors of *Acta Protozoologica* (Poland) for publishing, about 35 years ago, a preliminary abstract describing the role of the myosin heads and MgATP in the structure of the thick myosin filaments, resulting in the generation of radial expansive forces during contraction (Morel, J. E., and M. P. Gingold. 1979. Stability of a resting muscle, mechanism of muscular contraction, and a possible role of the two heads of myosin. *Acta Protozool.* 18, 179). I thank Dr. Valérie Bardot and her collaborators, particularly Angélique Idmbarek, for their valuable assistance with some problems that I had with the documentation process. I also thank Dr. Julie Sappa for improving my English (although I am sure some faults remain despite her best efforts) and helping with various practical problems.

AUTHOR

Jean Emile Morel was born in western France in 1940. In 1958, after passing his baccalaureate, he embarked on the demanding preparation for the prestigious 'Grandes Ecoles' and was accepted at the Ecole Centrale de Paris. This course led to postgraduate training in radiochemistry and thermodynamics and a PhD in physical chemistry. At that point, he began working as a researcher in biophysics and made the key decision to apply his knowledge of the mathematical, physical and chemical sciences to the field of biology. After obtaining a second PhD, this time in biophysics, he became a researcher at the Commissariat à l'Energie Atomique at Saclay, where he began to focus on the complex problem of muscle contraction. He remained at Saclay from 1980 to 2004, becoming professor of bioengineering, biophysics and cell physiology at the Ecole Centrale de Paris and Université Pierre et Marie Curie (Paris) and joint director of the DEA course of the Molecular Biophysics Doctoral School of Université Pierre et Marie Curie (Paris). In 1993, he founded the Laboratory of Biology at the Ecole Centrale de Paris. Throughout this period, he applied his critical thinking and knowledge of mathematics and physics to the question of how muscles contract, calling into question the established dogmas, trying to resolve the persistent inconsistencies of the accepted models and developing his own alternative models. Since his retirement in 2004, Professor Morel has devoted much of his time to reviewing the existing data and trying to resolve the conundrums of this field. He presents here the results of his analysis: an extensive critical review of the literature, including his own publications, focusing on the more difficult issues often neglected by the advocates of the traditional swinging cross-bridge/lever-arm models and other approaches, together with a new hybrid model, based on these findings, which fits the data and resolves many of the problems raised or left unresolved by previous models.

1997, 2004; Cooke et al. 1994; Duke 1999, 2000; Eisenberg and Hill 1978, 1985; Eisenberg et al. 1980; Fisher et al. 1995a; Geeves 1991; Geeves and Holmes 1999, 2005; Goldman and Huxley 1994; Hill 1968a,b, 1970, 1974, 1975, 1977; Hill and White 1968a,b; Hill et al. 1975; Holmes 1997; Holmes and Geeves 2000; Huxley 1973b,c; A.F. Huxley 1988, 2000; Huxley and Kress 1985; Huxley and Tideswell 1996, 1997; Irving 1987; Julian et al. 1978a; Linari et al. 2009; Ma and Zahalak 1991; Martyn et al. 2002; Mijailovich et al. 1996; Pate and Cooke 1989; Piazzesi and Lombardi 1995; Piazzesi et al. 2002b; Rayment et al. 1993a; Schoenberg 1980a,b, 1985; Smith and Mijailovich 2008; Smith et al. 2008; Squire 1983). A comment should be made regarding the remarkable experimental work and its interpretation (in terms of the mechanisms of muscle contraction, in particular) presented by Rayment et al. (1993a,b). Indeed, the authors resolved the atomic structure of the myosin subfragment-1 (S1 or head), using crystals of extensively methylated S1. Unfortunately, Phan et al. (1994) demonstrated that 'methylation... causes a complete loss of in vitro motility of actin filaments over methylated HMM [heavy meromyosin, i.e. myosin subfragment containing the two heads, S1, plus the S2 part of myosin; see Figure 5.1 for definitions]... It is concluded that these relatively mild but numerous and important changes impair the function of methylated S1'. Thus, the promising work of Rayment et al. (1993a,b), which was seen as making a major contribution to our understanding of the myosin head, the head–actin interface and the molecular swinging cross-bridge/lever-arm mechanisms of muscle contraction, unfortunately seems to have been essentially a 'non-event' in terms of the mechanisms of muscle contraction.

Buonocore et al. (2004) proposed a hybrid model combining the swinging cross-bridge/lever-arm and biased Brownian motion concepts (the biased Brownian motion theories of muscle contraction were developed by the group of Yanagida; see, for instance, Kitamura et al. 2005; Yanagida et al. 2000a,b, 2007). Grazi and Di Bona (2006) proposed an unconventional model taking into account both swinging cross-bridge/lever-arm processes and the viscous-like frictional forces, already studied and discussed by Elliott and Worthington (2001). As pointed out by Elliott and Worthington (2001), the viscous forces are discounted in all traditional models but should be

taken into account, particularly when the myofilaments slide past each other (isotonic conditions). The two very different hybrid models suggested by Buonocore et al. (2004) and Grazi and Di Bona (2006) differ considerably from that presented and discussed in this monograph. Indeed, the hybrid model presented and discussed here combines the swinging cross-bridge/lever-arm and lateral swelling theories, under isometric tetanic contraction conditions, but does not ignore the viscous forces under isotonic contraction conditions (see p. **281** in Section 8.11).

Many discussions and controversies regarding traditional approaches to muscle contraction have been published over the years. For example, Hoyle (1983) entitled one of the sections of his book 'Why do muscle scientists "lose" knowledge?' Considering in vitro motility as a model of muscle contraction and studying various factors influencing the movement of F-actin filaments propelled by HMM in vitro (see first column on this page for definition of HMM), Homsher et al. (1992) concluded that 'the results of motility assays must be cautiously interpreted'. In a meeting review, Alberts and Miake-Lye (1992) claimed that 'in no case is it understood how [chemo-mechanical] transduction happens' and that 'the problem with the rotating cross-bridge hypothesis is that the major conformational change that it predicts for myosin heads during the power stroke has simply not been observed'. Taylor (1993) rephrased this problem more precisely: 'the fundamental problem with the rotating crossbridge model has been the failure to obtain convincing evidence for a large-scale change in the structure that could account for a movement of the crossbridge of 5 to 10 nm in the direction of motion'. However, Taylor (1993) remained optimistic and proposed that the problem would be resolved by taking into account both the crystallisation of myosin heads, with interpretation of the resulting experimental data (e.g. Rayment et al. 1993a,b), and studies of in vitro motility. Nonetheless, Huxley (1996) expressed doubts about swinging cross-bridge/lever-arm models, as they stood in the mid-1990s: 'The challenge to really understand the mechanism remains'. Analysing experimental data concerning stiffness of muscle fibres, Goldman and Huxley (1994) wrote: 'The studies raise many questions and prod us to reinterpret earlier experiments... We need explicit structural models that explain the energetic discrepancies in both fibres and in vitro data'. A.F. Huxley (2000) was also dubious, stating that '... there is always a possibility—indeed, a probability—that our present concepts are seriously incomplete or even wrong'. Traditional cross-bridge models have been improved in recent years, but the main bases of swinging cross-bridge/lever-arm theories have not markedly changed and I think that the doubts of Alberts and Miake-Lye (1992), Goldman and Huxley (1994), Homsher et al. (1992), Huxley (1996), A.F. Huxley (2000) and Taylor (1993) remain topical. At this point, I believe that one of the major problems to be resolved concerns whether the main features of the usual models should still be considered 'unimpeachable'. The old and recent swinging cross-bridge/lever-arm models are mostly constructed from 'conventional cross-bridge models with one-to-one coupling between the mechanical and ATPase cycles' (Linari et al. 1998). These authors cited many works providing apparently convincing experimental evidence in favor of this traditional view of tight coupling. However, they pointed out that some previous assertions require revision, taking into account, for example, the marked compliance (opposite of stiffness) of the thin actin filaments. More generally, Linari et al. (1998) also raised the issue of the compliance of other structures present in a unit cell and even gave estimates for the compliance of cross-bridges, thick myosin filaments and thin actin filaments. Thus, since the end of the 1990s, swinging cross-bridge/lever-arm theories have become increasingly complex, because, for example, the various compliances must be taken into account and may interfere with interpretation of the experimental results, even potentially blurring the main feature of the swinging cross-bridge/lever-arm mechanisms. Huxley (2000) himself came to a similar conclusion. In this book, many other problems raised by the conventional approaches are analysed and discussed.

In a short abstract published in the *Scientific American*, Yanagida (2001) criticised the swinging cross-bridge/lever-arm theories and claimed that this kind of model 'is still popular because it posits that muscle contraction is, like the operation of ordinary motors, an easy-to-understand, deterministic process'. This is the 'eternal' viewpoint of Yanagida and his coworkers, as well as many other Japanese authors (e.g. Esaki et al.

2003, 2007; Kitamura et al. 1999, 2001, 2005; Oosawa 2000; Oosawa and Hayashi 1986; Shimokawa et al. 2003; Takezawa et al. 1998; Wakabayashi et al. 2001; Yanagida 2007; Yanagida et al. 2000a,b, 2007). This severe 'Japanese view' is not entirely new and the first cross-bridge models were nicknamed 'oar' theories, in the 1970s and 1980s, because of their 'anthropomorphic' aspect. Starting, in many instances, from the same general view as expressed by Yanagida (2001), but with very different concepts, many alternative models have been proposed since the beginning of the 1970s, most taking into account experimental data that were, and are still, generally forgotten (ignored?) (e.g. Brugman et al. 1984; Dragomir et al. 1976; Elliott 1974; Elliott and Worthington 1994, 1997, 2001, 2006; Elliott et al. 1970; Gray and Gonda 1977a,b; Harrington 1971, 1979; Iwazumi 1970, 1979, 1989; Iwazumi and Noble 1989; Jarosh 2000, 2008; Lampinen and Noponen 2005; Levy et al. 1979; McClare 1972a,b; Mitsui 1999; Mitsui and Chiba 1996; Morel 1975; Morel and Gingold 1979b; Morel and Pinset-Härström 1975a,b; Morel et al. 1976; Muñiz et al. 1996; Nielsen 2002; Noble and Pollack 1977, 1978; Oplatka 1972, 1989, 1994, 1997, 2005; Oplatka and Tirosh 1973; Oplatka et al. 1974, 1977; Pollack 1984, 1986, 1990, 1995, 1996; Pollack et al. 1988, 2005; Schutt and Lindberg 1992, 1993; Tirosh 1984; Tirosh and Oplatka 1982; Tsong et al. 1979; Ueno and Harrington 1981, 1986a,b; Wang and Oster 2002). As highlighted above, most of these models have been largely discounted, but, when taken into account, many criticisms have been raised against these unconventional theories and, particularly, against the lateral swelling models proposed to account for axial contraction. In most of the lateral swelling models, the radial repulsive electrostatic forces between the negatively charged myofilaments play a central role in generating axial contractile forces. In this context, it was suggested by April (1969), April et al. (1968, 1972) and Edman and Anderson (1968), using the 'external osmotic pressure technique' on intact fibres (with their sarcolemma), that increases in the internal ionic strength are associated with decreases in the tension-generating capacity of muscle. However, in these 'old' papers, which present experimental results that are a priori in favour of strong electrostatic forces (depending on ionic strength), the complex mechanisms proposed by the authors to explain the relationship between external osmotic pressure, internal ionic strength, and axial contractile force in intact fibres appeared to be largely speculative.

The first lateral swelling models were presented by Elliott et al. (1970) and Ullrick (1967; this pioneering model did not involve radial repulsive electrostatic forces between myofilaments, but solely the elasticity of the Z discs and their possible role during contraction). These models were based on the isovolumic behaviour of intact muscle fibres under contraction conditions, leading to an automatic translation of lateral swelling into axial shortening. The constant volume relationship was first suggested by Huxley (1953) and demonstrated by stretching intact muscles or fibres from crayfish and frog, using traditional x-ray diffraction (e.g. April et al. 1971; Brandt et al. 1967; Elliott et al. 1963, 1965, 1967). However, April and Wong (1976) and Matsubara and Elliott (1972), still using traditional x-ray diffraction on mechanically skinned fibres from crayfish and frog, respectively, interpreted their experimental results as demonstrating that the constant volume relationship does not hold, when skinned fibres are stretched. From these two experimental studies, it was directly and certainly too hastily inferred by most specialists in muscle contraction that lateral swelling models cannot work, because radial repulsive electrostatic forces cannot be translated into axial contractile forces. From the experiments performed by April and Wong (1976) and Matsubara and Elliott (1972), the lateral swelling processes would therefore have canceled out. Nonetheless, Morel (1985a) and Morel and Merah (1997) demonstrated that x-ray diffraction experiments performed on demembranated fibres should be interpreted with great caution (the complex behavior of demembranated fibres is discussed in Sections 3.8, 4.4.2.1 and 8.7). However, based on x-ray diffraction from synchrotron radiation applied to intact frog fibres, Cecchi et al. (1990) claimed that, when the fibres pass from rest to isometric contraction, 'the myofilament lattice does not maintain a constant volume during changes in force' and that their 'observations demonstrate the existence of a previously undetected radial component of the force generated by a cycling cross-bridge. At sarcomere lengths of 2.05 to 2.2 micrometers, the radial force compresses the myofilament lattice'. This compressive force is weak, and, scrutinising the rather complex experimental study and discussion

of these authors, I do not consider these conclusions to be clearcut (see also p. **157** in Section 4.4.2.5.4 for a brief analysis). More recently, Yagi et al. (2004) studied rat papillary muscles by x-ray diffraction from synchrotron radiation and laser light diffraction and concluded that 'the cell volume decreased by about 15% [corresponding to a decrease of ~7% in myofilament spacing] when the sarcomere length was shortened from 2.3 micro m to 1.8 micro m'. One problem with this result is that the thin actin filaments (~0.96–0.97 μm long; see p. **24** in Section 3.2) start to intermingle at sarcomere lengths between ~2 × (0.96–0.97) ~1.92–1.94 μm and ~1.80 μm, possibly inducing unexpected properties of the myofilament lattice. Nonetheless, the findings of these two independent groups could disprove the 'lateral swelling approach', but their conclusions are at odds with the observations presented in the next paragraph and the various experimental and semi-empirical results presented in this book, favouring the existence of strong lateral expansive forces in contracting muscles, regardless of changes in volume.

The theoretical and logical arguments presented by Dragomir et al. (1976), Elliott (1974) and Morel et al. (1976) demonstrated that isovolumic behavior is not a prerequisite for the conversion of lateral swelling into axial shortening during contraction. Nevertheless, lateral swelling theories were 'definitively discounted/ignored', as also highlighted in the preceding paragraph. However, some forgotten (ignored?) experimental findings may generate new interest in lateral swelling models. For example, using intact frog fibres in normal Ringer solution (osmolarity estimated at 245 mOsM by Millman 1986, 1998), I calculated from Figure 8 in the review by Millman (1998) small increases in myofilament lattice spacing, amounting to ~3%–4%, when the fibres passed from resting to isometric contraction conditions (using traditional x-ray diffraction techniques). In his review, Millman (1998) commented on his own findings and those of other authors, obtained with the same biological material, also showing a small detectable increase in filament spacing during tension rise in a tetanus, using x-ray diffraction from synchrotron radiation (e.g. Bagni et al. 1994b; Griffiths et al. 1993). Millman (1998) proposed various explanations but did not take into account a possible significant active role of radial repulsive (expansive) forces, which are considered to play only a passive role in the stability of the myofilament lattice. Using a stereomicroscope coupled with a mechanical apparatus on single intact frog fibres, Neering et al. (1991) showed that, on shifting from rest to isometric tetanic contraction, there was a non-uniform increase in cross-sectional area of between ~1% and ~40% (increase in fibre diameter of between ~0.5% and ~18%), with respect to the same fibre at rest, and an average increase in diameter, along the length of the fibre, of ~10% (estimated from Figures 2 and 3 in the paper by Neering et al. 1991, but not given by the authors). The authors tried to account rationally for their experimental observations, including the non-uniform increase in cross-sectional area in particular. However, they did not take into account a possible contribution of repulsive forces between the thick myosin and thin actin filaments, but I think that they unwittingly gave a good experimental argument for lateral swelling processes. Millman (1998) suggested that '… there is generally not a direct proportionality between fibre diameter and filament lattice under all conditions'. The ~10% difference in the width of frog intact fibre and the ~3%–4% difference in the lattice spacing of frog intact fibre when shifting from resting to isometric tetanic contraction conditions are consistent with this citation from Millman (1998). Further arguments are presented in the next paragraph.

Comparing the two series of experimental results obtained for intact frog fibres passing from rest to isometric tetanic contraction, presented in the preceding paragraph (Millman 1998; Neering et al. 1991), the increase in width when isometric tetanic contraction occurs would be approximately proportional to the cubic power of the lattice spacing $[(1.03-1.04)^3 \sim 1.09-1.12 \sim 1.10]$. Otherwise, Millman (1998) did not dismiss the possibility that the proportionality relationship may hold under particular conditions. In this context, Kawai et al. (1993) used a Na skinning solution at 0°C (chemical skinning) that 'dramatically improved the performance of muscle preparations' and, following the osmotic compression of their chemically skinned fibres from rabbit psoas muscle, they found that the decrease in fibre width was approximately proportional to the decrease in lattice spacing, during relaxation or after rigor induction. Thus, in this special case, the relative decrease in fibre diameter is similar to that in lattice spacing. Nonetheless, the behaviour

of demembranated fibres from rabbit psoas muscle should be interpreted with caution (see Sections 3.8, 4.4.2.1 and 8.7). Very few experiments have been performed on frog demembranated fibres, and we may suggest that frog and rabbit demembranated fibres behave differently.

Thus, the situation is confusing and further experimental, semi-empirical and theoretical results should be presented to support 'lateral swelling hypotheses'. In particular, a demonstration of the existence of strong radial repulsive electrostatic forces between myofilaments and their translation into axial forces during contraction is required. This monograph, and the experimental part especially, provides such a demonstration. Discussions and critical analyses of various problems should also provide further support for both the presence of these forces and their active role in muscle contraction.

As early as 1975–1979, it was suggested that the lateral swelling of fibres during isometric contraction might result from a drastic decrease in radial attractive forces once contraction is triggered (Morel 1975; Morel and Gingold 1979b; Morel and Pinset-Härström 1975b; Morel et al. 1976). This may automatically lead to a net increase in radial repulsive forces, converted into axial contractile forces, independently of possible volume variations (see pp. **4–5** and **102–103** in Section 3.10). New experimental results, presented in this book, show that radial repulsive electrostatic forces are effective, both at rest and under contraction conditions. It is also demonstrated that there are strong radial tethering forces between the thick myosin and thin actin filaments at rest. Consequently, the main features of the 1975–1979 model are maintained in this work. At rest, it is demonstrated that there is, indeed, a balance between radial repulsive electrostatic forces, radial tethering forces and weak radial attractive/compressive forces, whereas, during isometric contraction, the radial tethering forces decrease considerably. This results in the automatic appearance of net radial repulsive (expansive) forces between myofilaments, which are translated into axial contractile forces. The 1975–1979 model is extensively revisited and considerably improved in this monograph, in the light of new experimental data and many published results cited in the preceding paragraphs and obtained by my own and other independent groups. Moreover, in addition to these experimental data, many simple theoretical, phenomenological and semi-empirical approaches demonstrate that self-consistent conclusions can be drawn. The new experimental data show that muscle contraction is partly attributed to the lateral swelling mechanisms, with swinging cross-bridge/lever-arm mechanisms also playing a major role. A complete hybrid model of muscle contraction is proposed, with strong predictive and explanatory power. The notion of hybrid model was first put forward more than 15 years ago (see legend to Figure 5 in the paper by Morel et al. 1998a). The experiments and most of the discussions concern isometrically held intact fibres, half-fibres, mechanically or chemically skinned or permeabilised fibres, isolated myofibrils and intact unit cells (comparative studies of these various biological materials are highly informative; see Sections 3.7, 3.8, 4.4.2.1 and 8.7). Although the experimental data and discussions concern mostly isometric conditions, some major problems raised by the active shortening (isotonic contraction) are also discussed (see Sections 3.4.3.1.1, 3.4.3.1.2, 3.4.4 and 8.11).

The hybrid model addresses many previously forgotten (ignored?) issues relevant to in vivo/in situ conditions but irrelevant to in vitro conditions (see Chapter 9 for a circumstantial discussion). Indeed, the conditions prevailing in vivo/in situ and in vitro are extremely different. For example, in vertebrate skeletal muscle fibres, the myosin molecules are inserted into thick myosin filaments, arranged in a double hexagonal array (the thin actin filaments constitute the second series of myofilaments), whereas myosin or its isolated heads or synthetic myosin filaments are free in vitro. As recalled in Chapter 9, isolated myosin heads can generate movement in vitro. In this context, why does myosin have two heads and what are their roles in vivo/in situ? This is an old question, first posed by A.F. Huxley (1974): 'what is the significance of the fact that each myosin molecule has two heads?' This problem has been studied, from a structural viewpoint, by Morel and Garrigos (1982b) and Offer and Elliott (1978), using indirect reasoning, based, however, on experimental data. For example, Morel et al. (1999) provided experimental evidence to support the hypothesis of Morel and Garrigos (1982b). Briefly, the two independent groups of Morel and Offer gave structural reasons for the existence of two heads. In my group, we demonstrated that the two heads are intimately involved in the structure

of the thick myosin filaments, whereas Offer and Elliott (1978) suggested that the two heads can bind to two different thin actin filaments. The major question, addressed by Huxley (1974) and others, principally concerns the mechanical roles of the two heads. For instance, Reedy (2000) clearly summarised this eternal view: 'what does the second myosin head do in skeletal myosin molecule? This is particularly relevant because myosin in contracting muscle appears to use only one head in a crossbridge'. Huxley and Kress (1985) and Huxley and Tideswell (1997) proposed models for the possible mechanical roles of each head in a given myosin molecule in vivo/in situ. In Section 5.1.1, particularly in Figure 5.1, and on pp. **170–172** and **174–176**, the roles of the two heads are identified from the hypotheses of Morel and Garrigos (1982b) and Offer and Elliott (1978) and the experimental demonstration of Morel et al. (1999). These roles are both structural and functional and, more generally, the two heads play a key role in the hybrid model proposed and analysed in this monograph.

From comparative experimental results, Tyska et al. (1999) deduced that myosin with two heads produces greater force and motion than a single head (S1) in in vitro assays. Based mostly on in vitro experiments, Albet-Torres et al. (2009), Duke (2000), Esaki et al. (2007), Jung et al. (2008) and Li and Ikebe (2003) demonstrated that myosin heads act cooperatively in the motion of myosin along a thin actin filament, with cooperation also occurring between both heads of a single myosin molecule. Cooperativity is demonstrated, in this book, to occur in muscle fibres too, under isometric and isotonic conditions, notably in the hybrid model. However, this cooperativity is not strictly identical to that observed in vitro.

Hill (1978) published a theoretical study on the binding of S1 (isolated myosin head) and HMM (see definition on p. **2**) to F-actin. From this letter to *Nature*, it can be inferred that these two subfragments do not behave similarly and that the presence of the two heads in HMM, assumed to bind to F-actin via the intermediary of either only one head or both heads, results in different binding characteristics. An experimental and semi-empirical study supports many conclusions drawn by Hill (1978) and demonstrates cooperativity between the two heads of HMM in binding to the thin actin filaments (Conibear and Geeves 1998).

Murai et al. (1995), building on the unconventional work of Tonomura and his group, posed questions similar to those posed by Huxley (1974) and Reedy (2000), and stated that: 'it remains unknown why myosin has two heads'. From their experimental studies, concerning MgATPase activity (particularly the Pi burst), performed on myosin heads (S1) in vitro, Murai et al. (1995) suggested 'the existence of two kinds of head in the myosin molecule'. These two families of heads differ enzymatically. This type of enzymatic difference was reviewed by Inoue et al. (1979) but has been largely discarded by many specialists, mostly because of the severe controversy between Taylor and Tonomura, in the 1970s. In the review published by Taylor (1979), the two heads were 'definitively' considered to be identical, from an enzymatic viewpoint. Ten years after the assertion of Taylor (1979), Tesi et al. (1989) demonstrated the existence, in vitro, of two enzymatically different heads, M and M'. According to these authors, who studied essentially the initial transitory phenomena, there are two different sites for MgATP on the two different myosin heads: M, where MgATP is bound and hydrolysed, and M', where MgATP is trapped transitorily without hydrolysis. According to Iorga et al. (2004), '... it could be that both M and M' bind actin and that ATP dissociates both actoM and actoM': actoM with hydrolysis of the MgATP, actoM' without hydrolysis'. As far as I know, the conclusions of Iorga et al. (2004) and Tesi et al. (1989) have not yet been discounted and can be seen as definitive. However, the 'mechanical' consequences of the enzymatic differences were not clearly evoked. Nonetheless, Murai et al. (1995) suggested that 'the two heads of myosin may play different roles in the sliding movement of myosin heads on the thin filament during muscle contraction'.

Regardless of the possible 'enzymatic differences' in vitro, the problem of the mechanical roles of the two heads remains unresolved, despite the possible mechanisms proposed by Huxley and Kress (1985), Huxley and Tideswell (1997) and Esaki et al. (2007). At this point, it should be recalled that Oplatka and his coworkers, in the 1970s and 1980s, demonstrated that movements can be detected upon interaction of the enzymatically active part of myosin (head, S1) with F-actin filaments in vitro. Unconventional models of muscle contraction have been deduced from these observations (e.g. Hochberg et al.

1977; Oplatka 1989, 1994; Oplatka and Tirosh 1973; Oplatka et al. 1974; Tirosh 1984; Tirosh and Oplatka 1982). Moreover, Cooke and Franks (1978) found that, in vitro, threads of single-headed myosin and thin actin filaments (F-actin) can generate tension, and Harada et al. (1987) demonstrated that single-headed myosin can slide along F-actin. Thus, as isolated soluble myosin heads and single-headed myosin can generate movement in vitro, another major question remains: what is the role of the thick myosin filaments in vivo/in situ? In this context, it should be recalled that, in the 1980s, much experimental work was carried out on in vitro motility generated by the interaction of S1 and single-headed myosin with F-actin, without any reference to the pioneering work of Oplatka and his group (see also the similar opinion of Oplatka himself, 2005). All the authors of these studies claimed that in vitro motility and muscle contraction obey the same biological and physical laws. As an illustrative example, Bagshaw (1987) stated that 'S2 [see Figure 5.1 concerning this subfragment] and the second head of myosin are probably useless in muscle contraction'. To be deterministic, Morel and Bachouchi (1988a) noted that 'if this were the case, they [S2 plus the second head] would have disappeared during the course of evolution'. Small (1988) was also dubious about the assertion of Bagshaw (1987) and claimed that 'those who were impressed by the elegantly simple demonstration that single myosin heads bound to a substrate can alone translocate actin filaments may doubt that the rest of the molecule is necessary for motility'. In Chapter 9, a comparison of in vitro motility and muscle contraction is presented and it is concluded that these two processes do not obey the same biological and physical laws.

Other problems, of a physiological nature, are also very important. How can we account for unexplained heat in frog muscles? Where does the negative delayed heat observed in the case of short tetani and twitches in frog muscles come from? What is the origin of temperature-induced contracture in muscle fibres and whole muscles within the body? I have tried to answer these and other problems, by providing as complete a monograph as possible, dealing with many old and recent experimental results that remain unexplained, even somewhat 'mysterious' (see Chapter 8).

Elliott (2007) claimed that 'there are still unknowns in the current view of the contractile event. Modern [and older] work has recently been reviewed by Geeves and Holmes (2005), who present what might be called the majority viewpoint'. As recalled at the beginning of this introduction, this majority view is not a 'universal viewpoint' and I agree with Elliott (2007), who also wrote that 'The discussion continues'. The opinion of Elliott merits careful consideration. One of the many reasons for publishing this book is to provide an alternative to recent 'traditional' reviews, including that presented by Geeves and Holmes (2005), which mostly considers limited aspects of the swinging cross-bridge/lever-arm theories from a purely molecular viewpoint, ignoring many other aspects, such as those recalled in the preceding paragraph, together with many available physiological data that have not yet been taken into account by most authors. Another major reason for publishing this monograph is to propose a new approach, undoubtedly with its own flaws and lacunae, but nonetheless contributing to a better understanding of muscle contraction. In this context, an important experimental study was published by Martin-Fernandez et al. (1994), potentially invalidating the traditional approach, but was ignored by most investigators. Indeed, the authors studied both isometric and isotonic contractions of frog whole muscles, at 8°C, with x-ray diffraction techniques from synchrotron radiation, and stated: 'we conclude either that the required information is not available in our patterns or that an alternative hypothesis for contraction has to be developed'. This comment is, at least partly, consistent with this monograph, which presents such an alternative model. However, using a similar physical technique, but working on single frog fibres, at 4°C, Piazzesi et al. (1999) challenged the conclusions of Martin-Fernandez et al. (1994). Indeed, Piazzesi et al. (1999) concluded from their experimental studies that 'the myosin head conformation changes synchronously with force development, at least within the 5 ms time resolution of these measurements'. This inference clearly supports the traditional view, according to which there is a 'one-to-one coupling between the mechanical and the ATPase cycles' (Linari et al. 1998). However, the experiments of Martin-Fernandez et al. (1994) and those of Linari et al. (1998) and Piazzesi et al. (1999) were performed neither on the same biological material (whole frog muscles vs. single intact frog fibres, respectively) nor at the same temperature (8°C vs. 4°C). Differences in temperature may account for, at least, some of the discrepancies between the results obtained by the two independent groups. In fact, it is demonstrated throughout this book that temperature is an extremely important parameter, for both MgATPase activity and isometric tetanic tension (see Sections 3.4.3.2 and 8.8, respectively). Moreover, Piazzesi et al. (1999) strongly suggested that it would be risky to compare whole muscles and intact fibres. This opinion is consistent with my own experimental and semi-empirical findings, according to which there are major differences between intact unit cells lying in the centre of intact fibres and whole intact fibres, particularly in terms of the isometric tetanic tensions developed by these two types of biological materials (see Section 3.7). A comparison of the conclusions of Martin-Fernandez et al. (1994) with those of Linari et al. (1998) and Piazzesi et al. (1999) again demonstrates that the situation is confusing.

My aim, in this monograph, is not to discuss salient recent and detailed experimental data interpreted on the basis of the traditional approach to muscle contraction (swinging cross-bridge/lever-arm models) or to choose between the conclusions drawn by independent 'conventional' groups, although I give my own opinion in this area. Instead, I focus on the development of a hybrid model, on the basis of new experimental observations presented here and new phenomenological/semi-empirical lines of reasoning. This hybrid model has a good predictive and explanatory power (see Chapter 8). Moreover, as an illustrative example, introducing the hybrid model into the rather divergent interpretations presented by Martin-Fernandez et al. (1994) and Piazzesi et al. (1999), for example (see the preceding paragraph), might make it possible to account entirely for some recent experimental data that do not fit the 'mould' of swinging cross-bridge/lever-arm theories.

Many generalist books, monographs, reviews, minireviews, comments, criticisms, news and views, perspectives and reflections concerning muscle (from biochemical, biophysical, enzymological, physiological and structural points of view) and the possible mechanisms of muscle contraction, in vitro motility and cell motility have been published since the 1970s (e.g. Alberts et al. 2007; Bagshaw 1987, 1993; Bárány 1996; Barclay et al. 2010; Barman and Travers 1985; Block 1996; Borejdo et al. 2006; Bottinelli and Reggiani 2000; Bray 2000; Brenner 1987, 1990; Brenner and Eisenberg 1987; Cooke 1986, 1990, 1995, 1997, 2004; Craig and Woodhead 2006; Curtin and Woledge 1978; Dijkstra et al. 1973; Duke 1999, 2000; Eisenberg and Greene 1980; Elliott 2007; Ferenczi et al. 2005; Geeves 1991; Geeves and Holmes 1999, 2005; Geeves et al. 2005; Gergely and Seidel 1983; Goldman and Huxley 1994; Goody 2003; Gregorio et al. 1999; Harrington and Rodgers 1986; Herzog et al. 2008; Hibberd and Trentham 1986; Hill 1968a,b, 1970c; Holmes 1996, 1997; Holmes and Geeves 2000; Houdusse and Sweeney 2001; Howard 1997, 2001; Hoyle 1969, 1970, 1983; A.F. Huxley 1974, 1980a, 1988, 1998, 2000; Huxley 1971, 1973b,c, 1975, 1980b, 1990, 1996, 2004; Irving 1985, 1991, 1995; Irving and Goldman 1999; Julian et al. 1978a; Kawai 2003; Koubassova and Tsaturyan 2011; Kuhn 1981; Lehninger 2008; Maciver 1996; Mehta 2001; Mehta and Spudich 1998; Mehta et al. 1999; Molloy 2005; Morel and D'hahan 2000; Morel and Merah 1992; Morel and Pinset-Härström 1975a,b; Offer 1974; Offer and Ranatunga 2010; Oplatka 1994, 1997, 2005; Pollack 1983, 1984, 1986, 1988, 1990, 1995, 1996; Pollack et al. 1988, 2005; Pollard 2000; Reedy 2000; Reggiani et al. 2000; Sandow 1970; Sellers 2004; Simmons 1983, 1991, 1992a,b, 1996; Simmons and Jewell 1974; Simmons et al. 1993; Sleep and Smith 1981; Small 1988; Smith et al. 2005, 2008; Spudich 1994, 2001, 2011a,b; Spudich and Sivaramakrishnan 2010; Spudich et al. 1995; Squire 1983, 1989, 1994, 1997, 2011; Sweeney and Houdusse 2010a; Taylor 1972, 1979, 1989, 1993; Thomas et al. 1995; Titus 1993; Trentham 1977; Trentham et al. 1976; Vale and Milligan 2000; Vol'kenstein 1970; Woledge 1971, 1988; Woledge et al. 1985, 2009; Wray et al. 1988). Most authors favour the swinging cross-bridge/lever-arm processes, both in vivo/in situ and in vitro, but there are several bones of contention. For instance, Pollack (1988) presented severe criticisms against the swinging cross-bridge/lever-arm theories. Pollack (1990) has also published a monograph in which unconventional ideas are developed, concerning possible molecular mechanisms of muscle contraction and in vitro motility (particularly the stepwise process for muscle contraction). Jontes

(1995) has proposed a 'calmer' analysis of many mechanisms accounting for muscle contraction.

In this book, I demonstrate that, taking into account experimental data from my group (including new experimental results presented and discussed here) and from many other independent groups, alternative approaches can be proposed that are different from the mechanisms suggested by Elliott and Worthington, by Pollack and his collaborators, and, more generally, by many independent specialists in mechanisms of muscle contraction. Regardless of the choice of the best model, I also take the opportunity to analyse and dismiss a number of pointless discussions and so-called well-established concepts. The experimental work of Bagni et al. (1990a), concerning the relationship between myofilament spacing and force generation in intact frog fibres, using normal and hypertonic or hypotonic Ringer solutions, at 10°C–12°C, put a premium on this view. Indeed, these authors claimed that '… the separation distance [between the myofilaments] influences the force generating mechanism… Even if this effect is not sufficient to challenge the idea of cross-bridges acting as independent force generators, it should be considered in models of the force generation mechanism'. The old notion of cycling cross-bridges acting independently was called into question, even dismissed, by Bachouchi and Morel (1989a), Morel (1984a) and Morel and Merah (1995), who demonstrated that the kinetic characteristics of the cycling cross-bridges (constants of attachment, f, and detachment, g), necessarily depend on sarcomere length, that is, also on myofilament spacing. From the 'mitigated opinion' of Bagni et al. (1990a) and the assertions of Bachouchi and Morel (1989a), Morel (1984a) and Morel and Merah (1995), it appears clear that swinging cross-bridge/lever-arm theories should be revisited. This is done throughout this book, because the 'well-established' notion of cross-bridges acting independently is no longer valid, and this view is strongly, probably definitively, supported in this work.

When reading the many papers cited in the reference list, I noted that the experimental results and their interpretation are extremely muddled and frequently self-contradictory. I have devoted a great deal of work, over the last 12–13 years, to trying to unravel this problem and hope that I have succeeded in this monograph. This notion of complexity was clearly put forward by Pollack

CHAPTER TWO

Materials and Methods

2.1 PRELIMINARY REMARKS ON THE SIMPLE STATISTICAL TESTS EMPLOYED

I have observed a lack of rigor in the use of SE (standard error) and SD (standard deviation) between and within papers. Thus, it is useful to recall briefly the simplest definitions of these two statistical tests (see handbooks on statistics; the precise definitions are rather complex, but the simplified formulae presented here correspond to traditional usage). The first definition is $SD = \left[\sum (x_i - x_{mean})^2 / n \right]^{1/2}$, where n is the number of values measured, x_i is each value measured ($i = 1, 2, ..., n$) and $x_{mean} = \left[\sum x_i \right] / n$, the mean of the n values. The second definition is $SE = SD/n^{1/2}$. In this book, as in most of the papers that I have read, I frequently use SE, except when the authors cited give only the SD and do not give the number n of measurements and in cases in which the use of SD appears to be more appropriate than the use of SE. Indeed, a 'forgotten' advantage of SD is that it is little dependent on n, whereas SE decreases with increasing n and is even close to zero when n is very high, making it hard to obtain a clear notion of the scattering of the experimental points. In this context, I think that SE is potentially misleading, and certainly a much more restrictive statistical tool than SD. Throughout this book, I use these two statistical tests. In some instances, I average SE and SD and use the unconventional test (SE + SD)/2. Finally, it should be stressed that the number n of measurements is an established denomination and should not be mistaken for the proportion n of attached cross-bridges under contraction conditions, which is also an established denomination.

2.2 HALF-FIBRES (SPLIT FIBRES) FROM YOUNG ADULT FROGS (*RANA PIPIENS*)

Half-fibres (split fibres) were prepared from the *iliofibularis* muscle of young adult *Rana pipiens* frogs (length ~ 10–11 cm), using an unconventional and powerful technique first described by Endo et al. (1970) and later improved by Horiuti (1986), Villaz et al. (1987) and Vivaudou et al. (1991), with further improvement introduced here. All dissections were performed in a cold room equipped with a dehumidifier (temperature, 4.0°C ± 0.5°C; relative humidity, 25% ± 5%). All muscles were dissected under a stereomicroscope, in Ringer solution containing 116 mM NaCl, 1.65 mM KCl, 1.80 mM $CaCl_2$, 2.15 mM Na_2HPO_4 and 0.85 mM KH_2PO_4 (pH ~ 7.2 at 4°C, but very little dependent on temperature) (see Horiuti 1986; Villaz et al. 1987). A single intact fibre was then isolated, under the stereomicroscope, in a relaxing buffer similar to that used by Villaz et al. (1987), consisting of 108.5 mM KMS, 5.4 mM Na_2ATP, 10.5 mM Mg(MS)$_2$, 10 mM EGTA and 10 mM PIPES, brought to pH ~ 7.1 at 4°C with KOH, so that the pH was ~7.0 at the working temperature (10°C). Methanesulphonate (MS$^-$) was used for studies of half-fibres (see the next paragraph, pp. **13–14** and many sections concerned with new experimental data presented in this monograph), because of the advantages of this biological material. The dissection process took ~15–20 min. The intact fibre was carefully split lengthwise under the stereomicroscope (see Vivaudou et al. 1991, particularly their Figure 1 and corresponding comments, for precise details on the technique used to split the fibre). This splitting was carried out in the relaxing buffer, described above, and took ~15 min. Binding to the force transducer (AM 801E, Ackers, Horton, Norway) in the trough and measurements of lengths and apparent diameters of the half-fibres (see pp. **12–13**) took ~10 min. Forces were therefore recorded after ~15 min + 10 min ~ 25 min. Force measurements took ~25 min. Thus, the total duration of handling of the half-fibres was ~50 min.

The half-fibres contained half the sarcolemma, but the internal compartment was exposed to the external medium. The sarcoplasmic reticulum was functionally intact, as demonstrated by Villaz et al. (1987, 1989) and confirmed in Section 4.2.2 (see Figure 4.1 and the arrows C corresponding to injection of 30 mM caffeine, and corresponding comments). Using electron microscopy, Asayama et al. (1983) showed that, in mechanically skinned fibres (traditional techniques), the sarcoplasmic reticulum was 'markedly swollen and possibly fragmented', except when the buffers contained 50 mM sucrose. However, the authors used relaxing buffers containing propionate as the major anion. Andrews et al. (1991) demonstrated experimentally that this anion is highly deleterious, as is also Cl$^-$ (in the experimental part of this monograph, care is taken not to use these two anions). Thus, the experimental results obtained by Asayama et al. (1983) probably cannot be extrapolated to half-fibres studied in the presence of MS$^-$ as the major anion. In any event, neither I nor Villaz et al. (1987, 1989) were able to verify by electron microscopy whether, in half-fibres and in the presence of MS$^-$ as the major anion, the sarcoplasmic reticulum was adulterated. However, this kind of behaviour is highly unlikely for half-fibres in the presence of MS$^-$ (see above and Section 3.3, in which it is demonstrated that half-fibres develop the same isometric tetanic tension as intact fibres). This conclusion is clearcut, particularly as Andrews et al. (1991) demonstrated that MS$^-$ is the least deleterious anion and has many other advantages. Thus, in half-fibres in the presence of MS$^-$, the sarcoplasmic reticulum is structurally and functionally intact (see the experimental part of this book, concerning split fibres; see also pp. **221–222** in Section 7.2 for some quantitative data, regarding the absence of

any detectable deleterious effects of MS$^-$, by contrast to Cl$^-$, for example).

The intact sarcoplasmic reticulum around each myofibril in the half-fibres, prepared and used in the presence of MS$^-$ as the major anion, is important for the maintenance of resting conditions at any ionic strength (see pp. **113–115** in Section 4.2.1). Half the sarcolemma and MS$^-$ contribute to the 'similarity' of half-fibres and intact fibres (see pp. **13–14**). It should be pointed out that split fibres are rarely used nowadays by specialists in muscle and muscle contraction, despite their inherent potencies (see the preceding paragraph). The half-fibres were used by Villaz and his group, mostly for studying the sarcoplasmic reticulum. It would have been interesting to measure a possible increase (or decrease?) in lattice spacing (e.g. d_{10}) of the half-fibres when shifting from rest to contraction. As highlighted by Millman (1998), measuring fibre diameter would lead to conclusions concerning the behaviour of the myofilament lattice that could be justified to various extents from the quantitative viewpoint (see also pp. **4–5** in the Introduction). No x-ray apparatus was available in my laboratory and I used a stereomicroscope for observations of possible marked lateral swelling (or shrinkage?).

Each half-fibre was prepared immediately before use. Preliminary experiments were performed before isometric forces were recorded. Only 14 'suitable' half-fibres, from a total of 30, were used for this purpose. The remaining 30 − 14 = 16 half-fibres were rejected because of 'unsuitable' dissection (a high proportion of unsuitable half-fibres as a result of the many problems raised by the difficult preparation). Four of the 14 suitable half-fibres were used for calcium measurements with fura 2 (see p. **15** and p. **113** in Section 4.2.1) and only three half-fibres were used for measurements of isometric forces (mostly resting forces; see Section 4.2.2, Figure 4.2 and corresponding comments). The useable lengths of the remaining 14 − 4 (four half-fibres were used for calcium measurements with fura 2; see above) = 10 half-fibres were measured under a binocular microscope (after binding to the force recorder), and their apparent diameters were determined under the stereomicroscope. Both series of measurements were performed in a relaxing solution similar to that used for lengthwise splitting (see p. **11**), after placing the half-fibre in the trough (the forces were recorded for only three half-fibres, arbitrarily chosen from the 10 selected; see above). The useable length of the 10 half-fibres was ~3.0 ± (0.2 or 0.6) mm (mean ± SE or SD; n = 10) and their 'apparent' diameter was ~85 ± (10 or 32) μm (mean ± SE or SD; n = 10). It can be deduced from the experiments of Edman (1979), performed at a mean sarcomere length of ~2.25 μm, that the diameter of intact fibres from young adult frogs *Rana temporaria* was ~130 ± (8 or 25) μm (mean ± SE or SD; n = 10), corresponding to ~123 ± (8 or 15) μm at the sarcomere length of ~2.5 ± 0.2 μm used here (see p. **14**) (isovolumic behaviour of intact fibres; e.g. April et al. 1971; Brandt et al. 1967; Elliott et al. 1963, 1965, 1967). Half this diameter is therefore ~62 ± (4 or 8) μm (mean ± SE or SD), lower than that obtained here for half-fibres from young adult frogs *Rana pipiens* (~85 ± (10 or 32) μm; see above), but of the same order of magnitude and statistically consistent, if SD is used (see Section 2.1 for precise details concerning the advantage of using SD). Thus, the 10 half-fibres selected had cross-sectional areas of ~5.67 ± 1.34 × 10^{-5} cm^2 (mean ± SE for the diameter). The technique of Blinks (1965), widely used to measure cross-sectional area, was not employed here, because this technique would have led to values difficult to interpret in the case of half-fibres. The volume of the 10 half-fibres was therefore ~1.70 ± 0.38 × 10^{-5} cm^3 (only the SEs for the lengths and diameters are taken into account in this estimate). The apparent diameters, cross-sectional areas and volumes correspond to resting half-fibres in relaxing buffers with bulk ionic strengths exceeding ~40 mM (the compositions of the various relaxing buffers are given on p. **118** in Section 4.2.2). Indeed, for bulk ionic strengths above ~40 mM, no variations in the apparent diameters of the three half-fibres used to record resting forces could be detected, whereas below ~40 mM, the resting half-fibres swelled considerably (see pp. **165–167** in Section 4.4.2.8 for a discussion of the phenomena occurring at low ionic strengths).

Unlike mechanically or chemically skinned or permeabilised fibres in the presence of the anions generally used (e.g. Cl^-, acetate, propionate and sulphate), half-fibres in the presence of MS^- as the major anion (see below, in this section, for precise details) were used over very long periods (~50 min; see p. **12**), with no detectable impairment. Villaz et al. (1989) observed the same kind of behaviour on the same type of biological material, except that they employed *Rana esculenta*, rather

than *Rana pipiens*. They assessed the calcium release induced by caffeine, by measuring the various characteristics of the caffeine-induced transitory contraction, and noted that 'one single [split] fibre could be challenged many times (up to 61) with each assay lasting about 6 min'. Thus, a full set of experiments could last up to ~6 × 60 ~ 360 min (~6 h) in the experiments of Villaz et al. (1989). In the study described here, the duration of the experiments on each half-fibre could not exceed ~1 h. Part of the difference in the maximal duration of experiments (~1 h vs. ~6 h) may be attributed to the use of two different species of frog (see pp. **49–50** in Section 3.4.3.2 for a discussion of some problems relating to different experimental results obtained with different species of frog). The difference in the two durations is also probably partly attributed to the difference in the two types of experiment: caffeine-induced contraction (Villaz et al. 1989), also called 'contracture' by Horiuti (1986), versus resting tension studied in the experimental part of this monograph (see, Sections 4.2.2 and 4.4.2.8).

The unusually long period over which half-fibres can be studied probably results from both the presence of half the sarcolemma and the use of MS^- as the major anion. In this context, many uncontrolled phenomena occur in traditionally mechanically skinned frog or rabbit fibres studied in the usual buffers: disruption/damage/impairment, in the regular arrangement of the myofilament lattice, including disorder in the peripheral myofibrils and at the periphery of each myofibril (e.g. Ford and Surdyk 1978) (furthermore, x-ray diffraction data for such impaired lattices are misinterpreted; see Morel 1985a; Morel and Merah 1997). The same kind of lattice disorder/irregularities occur in chemically skinned frog fibres (e.g. Magid and Reedy 1980). Moreover, in chemically skinned rabbit fibres, the dissolution of myosin, actin and other proteins and their release into the bathing medium were also observed and quantified (see p. **221** in Section 7.2). This protein dissolution phenomenon was not studied in mechanically skinned frog fibres, but it probably also occurs in this biological material. Owing to the many advantages of using half-fibres in the presence of MS^-, as highlighted at many places in this monograph, only three half-fibres were used for studies of resting forces and, occasionally, active forces (see Section 4.2.2, including Figures 4.1 and 4.2). As an illustrative example, a single half-fibre, used for ~25 min (see p. **12**), displayed no significant loss of resting force (no more than ~4%–6% over ~25 min). Many experimental points could therefore be obtained with a single half-fibre. For durations of up to ~25 min, no impairment of the half-fibres was detected under the stereomicroscope and resting tensions remained unaltered, within the limits of experimental error. However, beyond ~30–40 min, the three half-fibres were gradually destroyed, as shown by examination under the stereomicroscope, and there was a gradual decrease in cross-sectional areas, also detected under the stereomicroscope.

MS^- was used as the major anion, rather than Cl^-, which is used in experimental studies performed by many independent groups. The first reason for using MS^- was to avoid the Cl^--induced release of Ca^{2+} from the sarcoplasmic reticulum (e.g. Allard and Rougier 1994; Endo et al. 1970; Ford and Podolsky 1970; Sukharova et al. 1994), resulting in limited, uncontrolled contractions. This Cl^--induced release of Ca^{2+} was observed but not clearly quantified by Endo et al. (1970) and Ford and Podolsky (1970) and was later reinvestigated with greater precision by Asayama et al. (1983). The use of MS^- is therefore essential when resting forces are recorded, as in most of the experiments presented in this book. The second reason for using MS^- is to replace Cl^- with a benign anion (chemically, biochemically and physiologically almost neutral). Indeed, Cl^- and many of the other anions often used (e.g. acetate, propionate or sulphate) are not benign and changes in their concentration affect several parameters, at least in chemically skinned and permeabilised fibres. In this context, in a remarkable comparative experimental study, Andrews et al. (1991) demonstrated, on chemically skinned fibres from rabbit psoas muscle, that MS^- was the least deleterious anion in studies of various properties of skinned fibres bathed in buffers containing various major anions. The authors found that isometric tetanic tensions were maximal in the presence of MS^- and much higher than with any other anion studied (e.g. Cl^- or propionate). Moreover, in Section 3.3, it is shown that the three half-fibres studied here, in the presence of MS^-, developed isometric tetanic tensions similar to those recorded in intact fibres, which is not the case for other skinned fibres in the presence of other anions (see Section 3.8.1). This demonstrates that the resting

and contracting half-fibres in the presence of MS^- were not swollen, except at low and very low bulk ionic strengths (see pp. **165–167** in Section 4.4.2.8 for a discussion). This implies indirectly that, with the 'half-fibres plus MS^- material', the sarcoplasmic reticulum was not 'enormously swollen', by contrast to mechanically skinned fibres (Asayama et al. 1983). Such a swelling of the sarcoplasmic reticulum provides a non-negligible contribution to the overall lateral swelling upon total or partial demembranation (see Section 8.7) and to a decrease in isometric tetanic tension, for purely geometric reasons. Thus, half-fibres in the presence of MS^- can be used as models of intact fibres. Many other advantages of this biological material are described at many places in this book.

Six different fresh stock solutions were prepared and stored for very short periods, at ~4°C, in the cold room (see p. **11**), as described by Horiuti (1986): 1 M KMS, 0.1 M $Mg(MS)_2$, 0.1 M $Ca(MS)_2$, 0.1 M EGTA (brought to pH ~ 7 with KOH), 0.2 M PIPES (brought to pH ~ 7 with KOH) and 0.05 M Na_2ATP. The various chemical compounds were purchased from Sigma France. The fresh stock solutions were mixed, immediately before use, in appropriate quantities to produce the various relaxing and contracting solutions. The final pH was adjusted with KOH, at ~4°C, in the cold room, to give pH 7.0 at the 'experimental' temperature (10°C). Solution compositions were calculated from the stability constants given by Horiuti (1986), with a simple routine.

The half-fibres were attached to the force transducer in a small trough (0.4 ml). The solution in the trough could be changed within ~300 ms, by injecting the new medium from a syringe and removing the overflow by aspiration. The whole apparatus, trough and solution reservoirs were kept at 10.0°C $\pm$ 0.2°C (homemade apparatus). The half-fibres were slightly stretched, as in most independent groups, and their length was adjusted to give sarcomere lengths of 2.5 $\pm$ 0.2 μm, determined by diffraction (He–Ne laser beam).

The calcium contamination of freshly distilled water is of major importance when resting forces are recorded. This contamination was estimated by atomic spectroscopy absorbance and was found to be ~5 μM (mean value). Freshly distilled water containing more than ~7–8 μM calcium was not used. Calcium contamination of the various stock solutions was estimated by the same method. As pointed out by Horiuti (1986), much of the contaminating calcium stems from ATP and this contamination adds to the ~5 μM from freshly distilled water. Indeed, studies of the various stock solutions, other than that containing 0.1 M Ca(MS)$_2$, of course, showed that the stock solution containing 0.05 M (5 × 10^{-2} M) Na$_2$ATP contained concentrations of ~50 μM Ca^{2+} (5 × 10^{-5} M; mean value). Stock solutions of Na$_2$ATP containing more than ~70 μM Ca^{2+} were discarded. The value of ~50 μM for Ca^{2+} contamination gives a Ca/ATP ratio of ~5 × 10^{-5} M/5 × 10^{-2} M ~ 10^{-3}, lower than the maximal value of ~5 × 10^{-3} reported by Horiuti (1986), due simply to improvements in the techniques of ATP purification by the Sigma Company. Two concentrations of ATP were used in the experiments presented in Section 4.2.2: 0.4 mM and 2 mM (see, in Section 4.2.2, p. **118**). Using a Ca/ATP ratio of ~10^{-3} (given above), the estimates of the contaminating calcium concentration stemming from ATP were therefore ~0.4 mM × 10^{-3} ~ 0.4 μM at 0.4 mM ATP and ~2 mM × 10^{-3} ~ 2 μM at 2 mM ATP, respectively.

The other stock solutions contained only trace amounts of Ca^{2+}, frequently undetectable, corresponding to Ca/Y ratios not exceeding ~10^{-6}, where Y is the concentration of any chemical compound mentioned on p. **14**. All stock solutions with a Ca/Y ratio greater than ~1.5 × 10^{-6} were discarded. The stock solution with Y = 1 M KMS contained a maximum of ~1 μM Ca^{2+} and the maximal bulk ionic strength used here (180 mM), corresponding to ~160 mM KMS, resulted in a mean concentration of contaminating Ca^{2+} of only ~0.160 M × 10^{-6} ~ 0.16 μM. Adding the contaminations of calcium stemming from Na$_2$ATP (see end of the preceding paragraph) and distilled water (see beginning of the preceding paragraph), we obtain the maximal concentrations of Ca^{2+} of ~5 μM + 0.4 μM + 0.16 μM ~ 5.56 μM and ~5 μM + 2 μM + 0.16 μM ~ 7.16 μM, respectively (for the two concentrations of ATP used; see end of the preceding paragraph). At low concentrations of KMS, the concentration of contaminating Ca^{2+} is very low and can be neglected, giving maximal concentrations of Ca^{2+} of ~5 μM + 0.4 μM ~ 5.40 μM or ~5 μM + 2 μM ~ 7.00 μM. Thus, regardless of the composition of the buffers, the two reference concentrations of contaminating Ca^{2+} can be taken as ~ (5.40 + 5.56)/2 ~ 5.48 μM ~ 5.5 μM and ~ (7.00 + 7.16)/2 ~ 7.08 μM ~ 7.1 μM. EGTA was present at a concentration of 0.5 mM or

1.0 mM in the various bathing buffers used here and surrounding the half-fibres (see throughout Section 4.2). Thus, the Ca^{2+} concentrations within resting half-fibres were necessarily much lower than ~5.5 μM (at 0.4 mM ATP) or ~7.1 μM (at 2.0 mM ATP) (see Section 4.2.1, for calculations, estimates and conclusions concerning the concentrations of Ca^{2+} in the various buffers).

As repeatedly pointed out above, in this section, the sarcoplasmic reticulum is functional in the three half-fibres used in the experimental part of this monograph (see Figure 4.1 and corresponding comments) and it certainly pumps, at least partly, the remaining trace amounts of Ca^{2+}. In this context, it is interesting to estimate the capacity of the sarcoplasmic reticulum to accumulate Ca^{2+}. For skeletal muscle fibres from young adult rabbits, the sarcoplasmic reticulum can accumulate ~10 mM Ca^{2+} in vivo or in vesicles studied in vitro (Philippe Champeil, personal communication). This value is assumed to be valid for young adult frogs too. Regardless of the calculations presented in Section 4.2.1, experimental estimates of the concentrations of Ca^{2+} in the sarcoplasmic medium of resting half-fibres were made, with fura 2 (purchased from Sigma France) used at a concentration of 2 μM. This chemical marker and technique were first described by Grynkiewicz et al. (1995) and have been successfully used by my group (D'hahan et al. 1997). The experiments were performed in a Hitachi F-2000 spectrofluorometer, regulated at 15.0°C ± 0.1°C (the lowest temperature that can be used without problems of water condensation at room temperature). The buffer (with a volume of 0.5 ml, corresponding to the minimal useable value in the spectrofluorometer) was introduced into the fluorescence cuvette and gently stirred. In parallel, one half-fibre, with an apparent diameter of ~85 ± 10 μm, was prepared and the two ends were cut off to obtain a length of ~3 mm (see p. **12**). This half-fibre was prepared immediately before use and rapidly introduced into the cuvette. Four half-fibres were prepared and studied at four different ionic strengths (see p. **113** in Section 4.2.1). A problem when measuring free Ca^{2+}, even with fura 2, is that it is impossible to detect Ca^{2+} at concentrations lower than ~10 nM (pCa ~ 8). At this pCa, partial activation can occur and the major point relating to the possible presence of trace amounts of calcium under any of the conditions studied here (particularly at low and very low ionic strengths) and under supposedly resting conditions is discussed in Section 4.2.1, to allow definitive conclusions to be drawn.

2.3 PERMEABILISED FIBRE BUNDLES FROM YOUNG ADULT RATS (WISTAR)

Bundles of ~4–6 fibres were prepared from small red muscles from the legs (*tibialis anterior*) of young adult rats (Wistar, weight ~ 400 g). Rat muscles were used because the experiments described in Section 4.3.1 (digestion with α-chymotrypsin) proved impossible with bundles of fibres from young adult frogs *Rana pipiens* (species used for studies of half-fibres; see Section 2.2). Bundles of fibres were used because they develop more force than single fibres. This is necessary to record very small transitory forces (see Section 4.3.1). All the dissection experiments were performed in the dehumidified cold room (~4°C; see p. **11** in Section 2.2). Muscles were dissected in Ringer solution (see composition on p. **11** in Section 2.2) and many bundles were isolated. The isolated bundles of fibres were incubated at room temperature (20°C–26°C) in a relaxing medium (see composition in the next paragraph) containing 1% Triton X-100, for ~1 h, and used immediately after this process. Using this soft permeabilisation technique, the sarcolemma of each fibre was permeabilised and the sarcoplasmic reticulum was almost certainly completely removed (e.g. Aldoroty and April 1984; Aldoroty et al. 1987), but the mitochondria remained intact (Pierre Vignais, personal communication). As observed under a binocular microscope, the useable length of the 24 bundles used (see legend to Figure 4.3, concerning the number of 24) was ~3.0 ± (0.1 or 0.5) mm (mean ± SE or SD; n = 24). The diameter of these 24 bundles, measured under a stereomicroscope, was ~185 ± (7 or 34) μm (mean ± SE or SD; n = 24), corresponding to a mean cross-sectional area of ~2.7 ± (0.2 or 1.1) × 10^{-4} cm^2 (mean ± SE or SD; n = 24). This value is valid for fibre bundles previously incubated as described above and therefore include the unavoidable initial lateral swelling occurring during the permeabilisation process (see Section 8.7) and other complex phenomena also occurring during this process (see pp. **97–98** in Sections 3.8.1 and 3.8.2). The volume of these bundles was therefore ~8.1 ± 4.6 × 10^{-5} cm^3, if the SDs on the length and the diameter of the bundles are taken

into account (see Section 2.1 for precise details on the advantage of using SD rather than SE).

The freshly prepared relaxing medium consisted, in all experiments, of 3 mM Mg(MS)$_2$, 3 mM Na$_2$ATP, 10 mM EGTA, 0.3 mM DTT, 12 mM creatine phosphate and 30 mM imidazole, pH 7.0 (at room temperature), ionic strength ~100 mM (all the chemical compounds were purchased from Sigma France). Atomic absorbance spectroscopy experiments were performed on the relaxing medium used here and showed no detectable traces of Ca^{2+} to be present, consistent with the following estimate. On p. **14** in Section 2.2, it is shown that the mean concentration of contaminating calcium from freshly distilled water was ~5 μM. On p. **15** in Section 2.2, it is also shown that Ca/ATP ~ 10^{-3} and, in the absence of 10 mM EGTA, the concentration of contaminating Ca^{2+}, mostly from 3 mM Na$_2$ATP, would have been ~3 mM × 10^{-3} ~ 3 μM, giving a total concentration of contaminating calcium of ~5 μM + 3 μM ~8 μM. Thus, in the presence of 10 mM EGTA, the concentration of contaminating calcium could not have exceeded ~$10^{-9.5}$ M (pCa ~ 9.5; the equilibrium constant is taken as $10^{6.4}$ M^{-1}; see p. **124** in Section 4.3.2), confirming that the solutions used were actually relaxing buffers (pCa ~ 9.5; Gordon et al. 1973 and Reuben et al. 1971 considered values of pCa ~ 8.7 and pCa > 9, respectively, to correspond to full resting conditions, in buffers mimicking the physiological sarcoplasmic medium, which is approximately the case here). Bulk ionic strength was deliberately fixed at ~100 mM, rather than the reference ionic strength of 180 mM in half-fibres (see, for instance, in Section 4.2.2, pp. **117** and **118** and Figure 4.2), to ensure that the proportion of weakly bound cross-bridges was higher than at 180 mM, because this proportion increases with decreasing ionic strength (see pp. **165–167** in Section 4.4.2.8 for a circumstantial discussion).

As for the three half-fibres selected in Section 2.2 (see p. **14**), the sarcomeres were ~2.5 ± 0.2 μm long, as shown by diffraction (He–Ne laser beam). It was necessary to use techniques different from those employed for half-fibres (see p. **14** in Section 2.2). The apparatus was different and all experiments were performed manually. The cuvette had a volume of 2.5 ml (homemade apparatus). Moreover, the experiments were performed at room temperature, without regulation, to try to increase the resting tensions (temperature of ~20°C–26°C vs. 10°C for the half-fibres; see p. **14**

Morel

in Section 2.2). The buffer in the cuvette was gently stirred to accelerate diffusion of α-chymotrypsin (see the beginning of the first paragraph) into the fibre bundles. By contrast to the situation for the half-fibres (described in Section 2.2 and studied, for example, in Section 3.3), an initial slow swelling of the permeabilised fibre bundles and other uncontrolled phenomena were expected during the ~1 h of incubation (see the first paragraph). These phenomena were not quantified and only the cross-sectional areas corresponding to the end of incubation were taken into account. The transitory increase in force recorded during the in situ digestion of myosin heads with α-chymotrypsin (see Section 4.3.1, especially Figure 4.3 and its legend) should therefore be compared with the forces developed by the 'swollen' permeabilised bundles of fibres immediately before digestion (see pp. **127–128** in Section 4.3.2). The force transducer was the same as that used for frog half-fibres (see p. **11** in Section 2.2).

2.4 PREPARATION OF SYNTHETIC THICK MYOSIN FILAMENTS (MYOSIN FROM THE BACK AND LEGS OF YOUNG ADULT NEW ZEALAND WHITE RABBITS). LIGHT SCATTERING EXPERIMENTS

Myosin from skeletal muscles (back and legs) of young adult New Zealand white rabbits (4 months old, ~2.5–3.0 kg) was prepared as described by Grussaute et al. (1995 and references therein). The reasons for choosing young adult rabbits were explained in the paper by Morel et al. (1999), who demonstrated that natural thick myosin filaments from old rabbits (weighing ~5 kg or more) have 'anomalous' properties (two-stranded filaments and rather low specific MgATPase activity per myosin head, with respect to the three-stranded filaments and rather high specific MgATPase activity per myosin head for young adult rabbits). In this context, to obtain the best experimental results, Schiereck (1982) and Linari et al. (2007) also used young rabbits, weighing ~2.0–2.5 kg and ~2.7–4.2 kg, respectively. Synthetic thick myosin filaments were prepared by the slow dilution technique (Morel et al. 1999; Pinset-Härström 1985; Pinset-Härström and Truffy 1979), at ~0°C (on ice in the dehumidified cold room, regulated at ~4°C; see p. **11** in Section 2.2), in the presence of 1.5 mM MgATP. It has been shown by my

group that, in the presence of 0.5 mM MgATP and at ~0°C, the synthetic thick myosin filaments are two-stranded with half the heads inserted into the core of these filaments (F-filaments; Morel et al. 1999). A similar behaviour of synthetic thick myosin filaments was described by Pinset-Härström (1985), who found that these filaments can fray into two subfilaments only. In the presence of 4 mM MgATP and at ~0°C, we obtained synthetic 'antifilaments' (AF-filaments; Morel et al. 1999), in which most of the heads were inserted into the filament core. The intermediate concentration of 1.5 mM MgATP used here probably corresponds to a mixture of ~60%–70% (e.g. ~65%) F-filaments and ~30%–40% (e.g. ~35%) AF-filaments, although no electron microscopy observations were made (no apparatus available). The use of 1.5 mM MgATP, instead of 0.5 mM, was based on the need to avoid too large a decrease in the concentration of MgATP because of a possible rapid cleavage on the 'external' heads, between the end of synthetic thick myosin filament preparation and the end of the light scattering experiments, particularly at temperatures higher than ~35°C (see below, in this section). The definition of the external heads is given on pp. **170–171** in Section 5.1.1; see also Figure 5.1. Finally, the presence of a limited proportion of antifilaments does not modify the interpretation of the experimental results obtained here.

Essential preliminary experiments were performed to locate the phenomena occurring when the temperature increased to ~40°C–41°C (see Section 5.2 for precise details) and to check whether temperature-dependent filament–filament interactions could interfere with other phenomena. Indeed, experimental studies have shown that filament–filament interactions occur in vitro between myosin minifilaments, via head–head junctions (Podlubnaya et al. 1987). In vivo/in situ, for highly stretched fibres in the absence of an overlap between the thick myosin and thin actin filaments, neighbouring thick myosin filaments are naturally cross-linked (Baatsen et al. 1988; Magid et al. 1984; Suzuki and Pollack 1986). As previously suggested by Suzuki and Pollack (1986), Pollack (1990) proposed that the thick-to-thick cross-links described by his group and by Magid et al. (1984) were 'almost certainly built of myosin S-1 [head]. Possibly, each link is a dimer consisting of S-1 heads projecting toward one another from adjacent thick filaments'. In my group, such filament–filament interactions of external heads were observed in vitro, at 20°C (the only temperature used by Morel et al. 1999). These interactions occur between pure F-filaments and also between pure AF-filaments (Morel et al. 1999, in these filaments, there are some external heads which can interact) but with different features. Both types of filaments were present in this study (see the preceding paragraph), and crossed interactions (of F- and AF-filaments) are almost certain to occur. Moreover, experimental studies, performed on isolated heads (S1), obtained by digesting myosin synthetic filaments with α-chymotrypsin (see Margossian and Lowey 1982 for the principle of the technique) have shown that these heads can dimerise in the presence of MgATP (e.g. Bachouchi et al. 1985; Grussaute et al. 1995; Morel and Garrigos 1982a; Morel et al. 1998a; in this last paper, see the 'supporting information available', for a long list of precautions that must be taken to prepare native S1 able to dimerise). In my group, we also found that when the temperature was gradually increased from ~18°C to ~25°C, the S1 dimers entirely dissociated at ~21°C–22°C, in the presence of 0.15 mM MgATP (Morel and Guillo 2001). Thus, the filament–filament interactions, via the external heads, observed at 20°C (see above) would disappear with increasing temperature, with inevitable consequences for the intensity of light scattered at various temperatures.

In the light of the comments and discussions presented in the preceding paragraph, it was necessary to perform preliminary experiments. Four filament suspensions, prepared as recalled on p. **17**, were studied in their dilution buffers. A volume of 2.5 ml of each suspension was introduced into a fluorescence cuvette (see below), previously regulated at 30.0°C ± 0.1°C. Immediately after injection into the cuvette, the temperature of the suspension was ~6°C–8°C. The cuvette contained 2 mg ml^{-1} myosin (MW ~ 470 kDa, as determined by Morel and Garrigos (1982a), giving a concentration of ~4.26 µM). After injection of filament suspensions into the cuvette, the temperature in the cuvette was increased very rapidly, at a rate of ~+18°C min^{-1} (in these experiments, the phenomena induced by increasing temperature were studied; the case of decreasing temperature is presented on pp. **20–21**). This rate was chosen to prevent denaturation, which occurred at ~40°C (see Figure 5.2). Lower rates (e.g. ~+10°C min^{-1}) resulted in the systematic adulteration of filaments between ~30°C and

~35°C. For purely technical reasons, it was impossible to reach rates higher than ~+18°C min^{-1}. The experimental technique used to increase the temperature in the cuvette so rapidly was merely to adjust very quickly (in less than ~5–10 s) the temperature setting of the thermostat to ~95°C–97°C and to record simultaneously the temperature in the cuvette and the intensity of the scattered light, on a homemade recorder specially equipped to report directly variations in the intensity of the scattered light with temperature in the fluorescence cuvette. The suspension was gently stirred to prevent rapid sedimentation of the synthetic myosin filaments and was studied in a Hitachi F-2000 spectrofluorometer. The intensity of the light (wavelength, 500 nm; arbitrary units) scattered at 90° was studied as a function of temperature. The total time elapsed from the end of the formation of the synthetic myosin filaments (at ~0°C, in the cold room; see p. **17**) to attainment of the maximal temperature in the cuvette (~40°C–41°C) was ~2.5 min (~150 s) in all the scattering experiments.

The intensity of the scattered light was fairly constant, at first, when the temperature gradually increased from ~6°C–8°C to ~27°C. It decreased significantly thereafter, between ~27°C and ~29°C, and was constant beyond ~29°C. The relative amplitude of the decrease, between ~27°C and ~29°C, was ~30%. From Equations 5.1 and 5.2 and the corresponding comments, there appears to be a rather complicated relationship between the molecular weight of a particle and the intensity of the light scattered, even for small particles. It is also recalled that Equations 5.1 and 5.2 are valid solely for small particles, but that they can be used, as a first approximation, for large and very large asymmetrical particles, as here. Equation 5.1 (which is valid over a wider range of temperature than Equation 5.2) shows that the intensity of the light scattered should decrease with decreasing 'apparent' molecular weight of the scattering particle (simple calculations; not shown). Morel et al. (1999) strongly suggested that, at ~20°C and for MgATP concentrations of the same order of magnitude as that used here (see p. **17**), heaps of ~4–6 filaments interact by forming head–head dimers between few external heads from different filaments. Thus, the ~30% decrease in the intensity of the scattered light is certainly related to a decrease in the apparent molecular weight of the scattering objects, resulting from dissociation of the external head–head dimers and, therefore, disappearance of the heaps of filaments, leading to the presence of only isolated filaments above ~29°C. Otherwise, the experimental findings of Morel and Guillo (2001) demonstrated that free isolated myosin head (S1) dimers, in solution, dissociate at ~21°C–22°C (see also the second column on this page). Here, the extruding myosin heads belong to complex structures: the myosin molecules are arranged in synthetic thick myosin filaments, almost certainly providing an explanation for the temperature transition being shifted to higher values than those reported by Morel and Guillo (2001) for S1 in solution (~27°C–29°C vs. ~21°C–22°C, respectively). In any event, beyond ~29°C, the intensity of the scattered light stabilised at a new level, corresponding to 'non-interacting' filaments only. Thus, beyond ~29°C, there are no problems of filament–filament interactions, and the phenomena described in Section 5.2 can therefore be interpreted in a straightforward way.

In the experiments described in the preceding paragraph, too large a decrease in MgATP concentration from the end of the preparation of synthetic thick myosin filament suspensions to the end of the light scattering experiments may occur, owing to hydrolysis by the myosin heads. This is a major problem, because a large decrease in MgATP concentration may result in partial or total dissociation of the internal head–head dimers buried within the synthetic thick myosin filament core (see Section 5.1.1 for a circumstantial discussion of the internal heads and related problems). Indeed, Morel and Guillo (2001) and Morel et al. (1998a) showed that, in the absence of MgATP, only the head monomer is present in vitro. Such a depletion of MgATP would therefore lead to a misinterpretation of the experimental results obtained in Section 5.2. Morel et al. (1999) presented an experimental study of the synthetic thick myosin filaments from young adult rabbit skeletal muscles. The various experimental studies included measurement of the initial rates v_0 of MgATP splitting, after the transitory periods (H^+ and Pi bursts), under steady-state conditions, at 20°C, for MgATP concentrations between ~20 µM and ~1000 µM. Regardless of the interpretation proposed by Morel et al. (1999), at 1500 µM MgATP (i.e. the 1.5 mM used here), it can be deduced from their Figure 2 (extrapolation) and the corresponding equations that, for both F- and AF-filaments, $v_0 \sim 0.20$ µM MgATP s^{-1}. Moreover, this Figure 2 also demonstrates that decreasing MgATP concentration results in an increase in v_0, followed by a flat maximum, and then a decrease for low MgATP concentrations. For the F-filaments, the value of v_0 corresponding to the flat maximum is ~1.1 µM MgATP s^{-1} (at MgATP ~ 400 µM) and, for the AF-filaments, the value of v_0 corresponding to the flat maximum is ~0.7 µM MgATP s^{-1} (at MgATP ~ 300 µM). On p. **21**, it is suggested that there are ~65% F-filaments and ~35% AF-filaments in the filament suspensions. Thus, a rough estimate of the value of v_0 corresponding to the flat maximum for the suspension of synthetic thick myosin filaments used here would be ~ $(1.1 \times 0.65 + 0.7 \times 0.35)$ µM MgATP $s^{-1} \sim 0.96$ µM MgATP s^{-1}. The concentration of MgATP necessarily decreases during scattering experiments and a maximal estimate for v_0 in the course of these experiments would be $v_{0,max} \sim (0.96 + 0.20)/2$ µM MgATP $s^{-1} \sim 0.58$ µM MgATP s^{-1}, and the 'true uniform' value to be used should be $v_0^* < 0.58$ µM MgATP s^{-1} (at 20°C). In this estimate, the concentration of MgATP at the end of each scattering experiment is assumed to be much greater than the ~300–400 µM (see above), ascertained experimentally below, in this section.

We now need to estimate the Q_{10} for v_0. In Figure 2 of the paper by Morel and Guillo (2001), the rate of MgATP splitting by S1 is characterised by kat, another kinetic parameter, probably with a dependence on temperature similar to that of v_0 (see also the next paragraph). In buffers mimicking physiological conditions (pH 7; ionic strength ~100 mM), it can be deduced from this Figure 2 (Morel and Guillo 2001) that, between ~18°C and ~21°C–22°C, $Q_{10} \sim 2.7$ for kat values corresponding to both the monomer and the dimer. Between ~21°C–22°C and ~25°C, $Q_{10} \sim 4.2$, for the monomer only, because, in these conditions, the dimer is entirely dissociated, as recalled in the first column on this page (see Morel and Guillo 2001, for precise details; the estimates of Q_{10} were not presented by the authors and are especially calculated here). A careful study of the rate of MgATP breakdown in permeabilised fibres from rabbit psoas muscle, under relaxing conditions, gives $Q_{10} \sim 2.5$ within the range ~7°C–25°C, and $Q_{10} \sim 9.7$ within the range ~25°C–35°C (Hilber et al. 2001). The origin of this behaviour (very high value of Q_{10} at high temperatures) in resting skinned fibres is, in principle, not known, although Hilber et al. (2001) suggested a mechanism based on the fact that 'some active force may have been generated in relaxing solution at temperatures above 25°C'. In any event, I strongly suggest that there is a correlation between the results obtained in situ by Hilber et al. (2001) and in vitro by Morel and Guillo (2001), as discussed in the next paragraph, where the value of $Q_{10} \sim 9.7$ is analysed and criticised.

As highlighted by Ma and Taylor (1994), there are no major differences between the kinetic characteristics of MgATPase activity in acto-S1 systems in vitro and myofibrils (see the Appendix 2.I for some comments and references on myosin and PGK enzymology), and we can see, in the preceding paragraph, that this inference is also valid when comparing free S1 and permeabilised fibres, supporting the assumption that v_0 and kat depend similarly on temperature (see also the preceding paragraph). Indeed, the experimental values of Q_{10} are similar for the two biological materials, as are the two parameters used in this and the preceding paragraphs (v_0 and kat) and we can choose, as a first approximation, $Q_{10} \sim (2.7 + 2.5)/2 \sim 2.6$ within the $\sim 7°C-20°C$ range, extended to the $\sim 0°C-20°C$ range, $\sim (4.2 + 2.5)/2 \sim 3.3$ between $\sim 20°C$ and $\sim 25°C$ and $\sim (4.2 + 2.5 + 9.7)/3 \sim 5.5$ between $\sim 25°C$ and $\sim 35°C$ (extended to the $\sim 25°C-39°C$ range; 39°C is the body temperature of the rabbit) and also to the range $\sim 25°C-40°C$ ($\sim 40°C$ is the maximal temperature reached in the present experiments; between $\sim 40°C$ and $\sim 42°C$, the filaments are adulterated; see Figure 5.2) (the four values of $Q_{10} \sim 2.5$, 2.7, ~ 4.2 and ~ 9.7 are given in the preceding paragraph). Thus, from the true uniform value $v_0^* < 0.58$ μM MgATP s^{-1} at $\sim 20°C$ (see p. **19**), we deduce that $v_0^* < 0.09$ μM MgATP s^{-1} at $\sim 0°C$, $v_0^* < 0.22$ μM MgATP s^{-1} at $\sim 10°C$, $v_0^* < 0.36$ μM MgATP s^{-1} at $\sim 15°C$, $v_0^* < 1.05$ μM s^{-1} at 25°C and $v_0^* < 17.54$ μM MgATP s^{-1} at $\sim 40°C$.

As pointed out on p. **18**, the total time taken to pass from $\sim 0°C$ to $\sim 40°C$ is ~ 2.5 min (~ 150 s). A rough estimate of v_0^*, between $\sim 0°C$ and $\sim 40°C$, would be $< (0.09 + 0.22 + 0.36 + 0.58 + 1.05 + 17.54)/6 \sim 3.31$ μM MgATP s^{-1}, regardless of temperature. Thus, when passing from $\sim 0°C$ to $\sim 40°C$, the utilisation of MgATP by the filament suspensions is almost certainly <3.31 μM s^{-1} × 150 s ~ 496 μM (~ 0.5 mM). The concentration of MgATP therefore decreased, at most, from 1.5 mM to ~ 1.0 mM. At this stage, it should be stressed that the value of $Q_{10} \sim 9.7$ above 25°C, reported by Hilber et al. (2001) (see the preceding paragraph), is almost certainly unsuitable,

because of inevitable adulteration/fragility of the permeabilised fibres at temperatures within the 25°C–35°C range (see the preceding paragraph concerning this temperature range). The problem of adulteration/fragility is strongly suggested on p. **52** in Section 3.4.3.2, p. **200** in Section 6.3.2, p. **202** in Section 6.3.3, p. **208** in Section 6.3.5 and p. **238** in Section 8.5. For the sake of simplicity, I propose to extend the values of Q_{10} obtained between 20°C and 25°C to the range 25°C–40°C, ignoring adulteration. Thus, the value of Q_{10} becomes $\sim (3.3 + 4.2)/2 \sim 3.8$ between 25°C and 40°C (rather than ~ 5.5 as used in the preceding paragraph; the value of ~ 3.3 is valid between $\sim 20°C$ and $\sim 25°C$ and that of ~ 4.2 is valid between $\sim 21°C-22°C$ and $\sim 25°C$; see the first column on this page concerning the values of ~ 3.3 and ~ 4.2). Thus, the value of 17.54 μM MgATP s^{-1} at 40°C (see the end of the preceding paragraph) should be replaced by 8.37 μM MgATP s^{-1} and a more suitable estimate of v_0^*, between $\sim 0°C$ and 40°C, would be $< (0.09 + 0.22 + 0.36 + 0.58 + 1.05 + 8.37)/6 \sim 1.78$ μM MgATP s^{-1}, corresponding to ~ 1.78 μM s^{-1} × 150 s ~ 267 μM MgATP (~ 0.3 mM). There is therefore a limited maximal decrease in MgATP concentration, which would pass from ~ 1.5 mM to ~ 1.2 mM, with no effect on the conclusions drawn below and in Section 5.2. Indeed, it is recalled on p. **17** that, at 0.5 mM MgATP, F-filaments (in which the head–head dimers are buried within the core) were obtained by Morel et al. (1999) and these F-filaments remain stable for at least ~ 10 min at room temperature (unpublished observations). The lowest estimate for MgATP concentration of ~ 1.2 mM, and even ~ 1.0 mM (see above), is well above 0.5 mM, and there are, in the experiments performed here, neither problems of dissociation of the internal head–head dimers nor problems of stability of the synthetic filaments, owing to MgATP depletion.

From the experiments and calculations presented above, in this section, it appears that (i) the limited depletion of MgATP does not affect the structure and stability of the synthetic thick myosin filaments and (ii) filament–filament interactions do not occur beyond $\sim 29°C$. Thus, above $\sim 37°C$, nonambiguous and extremely interesting phenomena were observed, which are described and discussed in Section 5.2 (see particularly Figure 5.2, its legend and Appendix 5.II).

When the temperature was decreased below $\sim 39°C$, other filament suspensions were used, to check whether the phenomena observed on increasing temperature were reversible (see Section 5.2 and Figure 5.2, its legend and Appendix 5.II). The temperature in the cuvette was previously fixed at 39.0°C ± 0.1°C (maximal temperature that can be used to introduce the filament suspensions without instantaneous denaturation). The filament suspensions reached this temperature ~30 s after injection into the cuvette. The temperature of the thermostat setting was then very rapidly adjusted to ~1°C–2°C (in less than ~10 s). The temperature in the cuvette remained at ~39°C for only ~10 s and then rapidly decreased at a rate of ~−18°C min^{-1} and the experiments were stopped when the temperature reached ~36°C, corresponding to a duration of ~10 s for the decrease in temperature from ~39°C to ~36°C. Thus, the total duration of the experiments was ~ 30 s + 10 s + 10 s ~ 50 s. The true uniform value of $v_0^* < 8.37$ μM MgATP s^{-1} at ~40°C (see p. **20**, for definition of the term 'true uniform value', and p. **20**, for the value of 8.37 μM s^{-1}) becomes <7.46 μM MgATP s^{-1} at ~39°C and <4.92 μM MgATP s^{-1} at ~36°C (Q_{10} ~ 3.8 between ~25°C and ~40°C; see p. **20**). A maximal estimate, between ~36°C and ~39°C, would therefore be ~ (7.46 + 4.92)/2 ~ 6.19 μM MgATP s^{-1}. Thus, MgATP utilisation during the experiments performed here would be <6.19 μM s^{-1} × 50 s ~ 310 μM MgATP (~0.3 mM) and the minimal concentration of MgATP at the end of the experiment would be ~1.2 mM, leading to the same conclusions as those drawn on p. **20**; that is, the cleavage of MgATP during the scattering experiments does not compete with the phenomena described in Section 5.2, shown in Figure 5.2, and commented on in its legend, and in Appendix 5.II.

APPENDIX 2.I

Some References and Comments on Myosin and PGK Enzymology

There are old concepts and data concerning the enzymology of myosin and its subfragments (particularly the isolated heads S1 in vitro). For example, we can cite a list of 20 major contributions in this area: Bagshaw and Trentham (1973, 1974),
Bagshaw et al. (1974, 1975), Biosca et al. (1984), Burke et al. (1974), Geeves and Trentham (1982), Lymn (1978), Lymn and Taylor (1970, 1971), Taylor (1972, 1979, 1989), Taylor and Weeds (1977), Taylor et al. (1970), Trentham (1977), Trentham et al. (1972, 1976), Webb and Trentham (1981) and White and Thorson (1973). Crystallisation of S1 (e.g. Dominguez et al. 1998; Fisher et al. 1995a,b; Houdusse et al. 1999, 2000; Rayment et al. 1993a,b; Winkelman et al. 1985) has improved our knowledge of the enzymatic site. Many authors (cited at many places in this monograph) have studied the enzymatic–mechanical coupling in situ (essentially in demembranated fibres and isolated myofibrils) and also the enzymatic site on myosin heads, in vitro or in situ (e.g. Berger et al. 2001; Geeves 1991; Geeves and Holmes 1999, 2005; Geeves et al. 2005; Huang et al. 2005; Nyitrai et al. 2006; Pate et al. 1997; Smith and Geeves 1995; Yount et al. 1995). In the studies by Brenner et al. (1982, 1984), Goody (2003), Huxley and Kress (1985), Iorga et al. (2004), Kraft et al. (1995), Regnier et al. (1995), Stehle and Brenner (2000) and Xu et al. (2006a), for example, the weakly bound state of the cross-bridges (first observed at low ionic strengths) was taken as an intermediary state in the cross-bridge enzymatic cycle in contracting muscle. As stated in Section 5.1.1, particularly on pp. **170–172** and pp. **174–176**, I believe that this hypothesis is risky and that another approach can be proposed. In any event, I think that there are many unknowns in the enzymatic cycle of myosin in interaction with actin and MgATP. The situation is very different for the enzyme phosphoglycerate kinase (PGK, a transphosphorylase, which obviously forms a ternary complex during the enzymatic cycle). PGK is certainly a well-studied enzyme. Many types of data, particularly from crystals of PGK, are available for this enzyme (e.g. Auerbach et al. 1997; Banks et al. 1979; Blake and Evans 1974; Davies et al. 1994; Flachner et al. 2004; Geerlof et al. 1997, 2005; Harlos et al. 1992; Kováry and Vas 2004; Kováry et al. 2002; Merli et al. 2002; Szabó et al. 2008; Szilágyi et al. 2001; Varga et al. 2001, 2005, 2006). To my mind, much work on the enzymatic, dimerisation, actin binding sites of myosin will be required to obtain results as interesting as those for PGK.

electrical charge densities in the A- and I-bands in demembranated muscle fibres, it has been clearly established that the thick myosin and thin actin filaments are highly negatively charged, under resting conditions and in rigor (e.g. Aldoroty and April 1984; Aldoroty et al. 1985, 1987; April and Aldoroty 1986; Bartels and Elliott 1980, 1985; Bartels et al. 1993; Chichibu 1961; Collins and Edwards 1971; Elliott 1973, 1980; Elliott and Bartels 1982; Elliott et al. 1980, 1984, 1986; Hinke 1980; Naylor 1982; Naylor et al. 1985; Pemrick and Edwards 1974; Regini and Elliott 2001; Scordilis et al. 1975; Stephenson et al. 1981). In this context, Miller and Woodhead-Galloway (1971) presented a simple and very useful analytical expression of the repulsive electrostatic forces between two parallel neighbouring charged cylinders with different radii, valid at any ionic strength. They pointed out that their approach, despite minor simplifications, is consistent with more complex theoretical expressions. Other available formulae are valid for a lattice composed of filaments with identical radii, and most theoretical conclusions can be drawn only numerically (e.g. Elliott 1968; Morel 1985b; Morel and Gingold 1979a), a serious drawback. I therefore use, in this monograph, only the analytical expression of Miller and Woodhead-Galloway (1971), leading to self-consistent results.

In intact fibres from frog muscles, Edman and Anderson (1968) and Gordon et al. (1966b) suggested that the slack length corresponds to sarcomere lengths of ~2.0–2.2 μm and to maximal overlap between the thick myosin and thin actin filaments. Also for frog, He et al. (1997) gave a 'slack' sarcomere length of ~1.9–2.1 μm, whereas Bagni et al. (1990b) proposed lengths of ~1.96–2.24 μm. For rabbit, He et al. (1997) suggested values of ~2.2–2.4 μm, and Woledge et al. (1985) gave values of ~2.3–2.5 μm. From the data recalled by Wang et al. (1999) for rabbit psoas muscle, maximal overlap corresponds to ~2.4 μm. Averaging these 13 values leads to a 'universal' slack sarcomere length $s_0 \sim 2.18 \pm (0.05$ or $0.18)$ μm (mean $\pm$ SE or SD, n = 13). This reference sarcomere length $s_0 \sim 2.18$ μm, corresponding to a 'composite animal species', is systematically taken into account throughout this book (for intact fibres, half-fibres, demembranated fibres, isolated myofibrils and intact unit cells) and I have chosen to call this universal reference length the 'slack sarcomere length', for simplification. This length is slightly shorter than the ~2.5 $\pm$ 0.2

μm used here for experimental studies (see pp. **14** and **16** in Sections 2.2 and 2.3, respectively), but, using the SD, we obtain ~2.18 $\pm$ 0.18 ~ 2.00–2.36 μm, statistically consistent with ~2.5 $\pm$ 0.2 ~ 2.3–2.7 μm (see Section 2.1 concerning the advantage of using SD). Owing to the unavoidable uncertainties in the experimental data discussed in this monograph and the inevitable shortcomings in the semi-empirical reasoning, this slight difference between the two series of sarcomere lengths is not taken into account and no subsequent inconsistencies emerged from this simplification. Thus, the various results and phenomena studied here correspond to the 'composite slack sarcomere length' $s_0 \sim 2.18$ μm, unless otherwise specified.

From Equation 2 in the paper by Miller and Woodhead-Galloway (1971), we can deduce that the repulsive electrostatic force, in an intact unit cell (see Section 3.1 for precisions on this notion of intact unit cell), between two neighbouring thin actin filaments and between two thick myosin filaments from two nearest neighbouring intact unit cells is negligible, with respect to that exerted between a thick myosin filament and a single neighbouring thin actin filament, at the reference sarcomere length $s_0 \sim 2.18$ μm (estimated in the preceding paragraph). These characteristics of the repulsive electrostatic forces are demonstrated below. At physiological bulk ionic strength I = 180 mM (i.e. vicinal ionic strength I* ~ 182 mM; see Section 3.9 for precise details about I*), the Debye–Hückel screening length is λ ~ 0.50 nm, the 'true' radius of a thick myosin filament is R_m ~ 16.6 nm, and the radius of a thin actin filament is R_a ~ 4.3 nm (see pp. **31–34** concerning the values of λ, R_m and R_a). In intact fibres (also intact unit cells) animal species, at s_0 ~ 2.18 μm, the centre-to-centre distance between the myofilaments is c_0 ~ 25.3 nm (see p. **145** in Section 4.4.2.5.1). The closest surface-to-surface distance between two neighbouring thin actin filaments is therefore ~25.3 − (2 × 4.3) ~ 16.7 nm, whereas that between a thick myosin and a thin actin filament is s* ~ 25.3 − 4.3 − 16.6 ~ 4.4 nm. On p. **32**, a comparable value of ~1.8 nm is obtained, from an entirely different approach, and a mean value of ~ (4.4 + 1.8)/2 ~ 3.1 nm can be selected (a supplementary definition of s* is given on p. **29**; the value of s* ~ 3.1 nm is retained in this book). From Equation 2 in the paper by Miller and Woodhead-Galloway (1971), with the

Debye–Hückel screening length $\lambda \sim 0.50$ nm (see above), we deduce that the strength of the repulsive electrostatic force between two neighbouring thin actin filaments is only $\sim 10^{-10}\%$ that corresponding to a thick myosin filament and a single thin actin filament in the same intact unit cell, which is negligible. It can also be demonstrated that the surface-to-surface distance between two thick myosin filaments from two neighbouring intact unit cells is $\sim 43.8 - (2 \times 16.6) \sim 10.6$ nm (43.8 nm $= 3^{1/2} \times 25.3$ nm, in the double hexagonal array of thin actin and thick myosin filaments) and that the repulsive electrostatic force represents only $\sim 3 \times 10^{-7}\%$ that between a thick myosin filament and a single thin actin filament in the same intact unit cell, which is also negligible. Thus, only the repulsive electrostatic forces between a thick myosin filament and a single thin actin filament within the same intact unit cell need be taken into account, at least around the slack length. Using a different approach, Millman and Nickel (1980) drew a similar conclusion and claimed that 'at shorter sarcomere lengths, thick–thin filament [electrostatic] forces dominate. But, as the sarcomere length increases to point of non-overlap between thick and thin filaments, thick–thick filament forces take over' (see pp. **25–26** for more precise details concerning the repulsive electrostatic forces at zero overlap).

A case of particular interest is that of stretched fibres, when the overlap between the two types of myofilaments disappears (below, in this section, and at many places in this monograph, 'zero overlap' or 'non-overlap' state). The zero overlap state corresponds to a sarcomere length of ~ 3.65 μm in *semitendinosus* muscle fibres from frog (Gordon et al. 1966a,b), ~ 3.77 μm in a composite animal species (estimated by Morel 1984a, from experimental data for various animal species and types of muscle available in the 1980s; this value includes the value of ~ 3.81 μm, corresponding to rabbit skeletal muscle, recently recalled by Coomber et al. 2011). Studying mechanically skinned frog fibres by electron microscopy (ultra thin sections) and using cutting-edge techniques, Trombitás et al. (1993) obtained values for the length of the myofilaments in *semitendinosus* and *tibialis anterior* muscle fibres of ~ 1.62 μm and ~ 1.61 μm for the thick myosin filaments, respectively, and $\sim 1.92/2 \sim 0.96$ μm and $\sim 1.94/2 \sim 0.97$ μm for the thin actin filaments, respectively. These two series of values correspond to ~ 3.54 μm and ~ 3.55 μm for the

zero overlap states, respectively. Concerning rabbit psoas muscle, Wang et al. (1999) recalled that the length of a thick myosin filament is ~ 1.64 μm and that of a thin actin filament is ~ 1.12 μm, giving ~ 3.88 μm for the zero overlap state. Averaging the five available values, the non-overlap state for a 'composite animal species and type of muscle' begins at a sarcomere length of $\sim (3.65 + 3.77 + 3.54 + 3.55 + 3.88)/5 \sim 3.68$ μm. This value is taken as the reference, throughout this book, for a composite animal species. In intact fibres, isovolumic behavior holds at rest and under contraction conditions (e.g. April et al. 1971; Brandt et al. 1967; Elliott et al. 1963, 1965, 1967; see pp. **4–5** in Chapter 1 for comments in this domain). Thus, when an intact fibre (particularly from frog) is stretched from full overlap (sarcomere length $s_0 \sim 2.18$ μm; see p. **24**) to non-overlap (sarcomere length of ~ 3.68 μm), its volume is constant and its cross-sectional area decreases by a factor of ~ 0.59. Again on pp. **4–5** in the Introduction, it is recalled that the relationship between fibre width and myofilament lattice spacing is somewhat complex. However, there is no major difference between width and lattice spacing, and the relative myofilament lattice spacing at zero overlap can be approximated by a factor of $\sim 0.59^{1/2} \sim 0.77$.

We need to consider possible increases or decreases, with increasing sarcomere length, in the surface electrical potentials $\Psi_{a,x}$ and $\Psi_{m,x}$ on the thin actin and thick myosin filaments, respectively (see definition of these potentials and the related parameters in the text below Equation 3.1 and also in the text between Equations 1 and 2 in the paper by Miller and Woodhead-Galloway 1971). Working on mechanically skinned fibres from the carpodite extensor muscle of crayfish (*Orconectes*), at rest, Aldoroty et al. (1987) studied the effect of osmotic compression, that is, of myofilament lattice spacing, on the number of fixed negative electrical charges on the thin actin and thick myosin filaments. They demonstrated that, within the range of relative reduction of the lattice spacing (between 1 and ~ 0.77; see the end of the preceding paragraph) this number was independent of compression, within the limits of experimental error (see their Figure 3 and their Table IV). In Section 4.4.2.2, it is recalled that there is no straightforward relationship between the number of fixed negative electrical charges and bound anions and surface electrical potentials, but it is highly unlikely that these potentials vary significantly with increasing osmotic compression, that is, when the myofilaments are brought closer together, whereas the fixed negative electrical charges remain unchanged (see above). As shown in Figure 2 in the paper by Aldoroty et al. (1987), Donnan potential is also approximately constant, within the limits of experimental error, when the relative lattice spacing decreases from 1 (full overlap) to ~0.77 (non-overlap; see above). Taking into account only the results of Aldoroty et al. (1987), obtained with compressed fibres, would be misleading, because the many problems raised by stretching an intact fibre are complex. For example, Aldoroty et al. (1985, 1987) found that increasing sarcomere length itself induces changes in the density of the fixed negative electrical charges, but these changes are limited and statistically similar to changes resulting from decreases in filament spacing (see Section 4.4.2.2). Moreover, on pp. **251–254** in Section 8.9, it is demonstrated that many structural disruptions/distortions of cross-bridge organisation occur with increasing sarcomere length, and with decreasing myofilament spacing. It is even recalled, on p. **254** in Section 8.9, that the stretching of skinned fibres leads to a gradual deformation of the hexagonal myofilament array (at full overlap) to the tetragonal (square) structure (characteristic of the thin actin filaments anchored to the Z discs). On pp. **258–259** in Section 8.9, it is also recalled that, during contraction, Ca^{2+} activation and binding to the myofilaments depend on sarcomere length. These complex and intermingled phenomena almost certainly result in changes of various magnitudes in the numbers of fixed negative electrical charges and bound anions when the fibre is stretched. However, as recalled above, there is no straightforward relationship between the density of fixed negative electrical charges and bound anions and surface potentials. Regardless of this 'electrical' problem, the mathematical function 'tanh' appearing in the expression of Q_x (x = a under contraction conditions; x = r at rest), given in the text below Equation 3.1, is not very sensitive to the values of surface electrical potentials, for absolute values as high as ~100/131 mV found semi-empirically in intact unit cells, at the slack length of the fibre (see pp. **161–162** in Section 4.4.2.6) and also probably at any sarcomere length. Thus, regardless of the many changes in the structure of the myofilaments and interfilament medium (particularly the

main features of the interfilament water and the cross-bridges present in this medium; see above and pp. **27–33**) resulting from stretching of the fibre (and the subsequent decrease in filament spacing), several semi-empirical and mathematical arguments indicate that Q_x displays very little dependence on myofilament lattice spacing and sarcomere length under given conditions (rest or contraction). Taking this assumption into account leads to self-consistent conclusions throughout this work and therefore provides an a posteriori justification of the reasoning presented in this paragraph.

The reasoning presented in this and the following paragraphs is valid at rest or under contraction conditions (Q_r and Q_a are different, as suggested on pp. **161–162** in Section 4.4.2.6, but they do not depend on the amount of overlap; see the end of the preceding paragraph). From the value of ~25.3 nm for the centre-to-centre distance of thick–thin and thin–thin filaments in an intact unit cell at full overlap (see p. **24**) and the corrective factor of ~0.77 at zero overlap (see the end of the preceding paragraph), the centre-to-centre distance at non-overlap is found to be ~0.77 × 25.3 nm ~ 19.5 nm. Assuming that the value of R_a ~ 4.3 nm for thin actin filaments (see p. **34**) is not heavily dependent on the amount of overlap, the closest surface-to-surface spacing between two neighbouring thin actin filaments, at zero overlap, in an intact unit cell, is ~19.5 − (2 × 4.3) ~ 10.9 nm, whereas the closest surface-to-surface spacing between a thick myosin filament and a thin actin filament, at full overlap, in an intact unit cell, is ~3.1 nm (see p. **24**). Note that the notion of an intact unit cell at zero overlap is an abuse of language. Indeed, as recalled in the preceding paragraph, at non-overlap, there are distortions and deformations in both the organisation of the cross-bridges and the structure of the unit cells (see also below). Nonetheless, for simplification, I use the notion of intact unit cell regardless of the amount of overlap. Applying Equation 2 from the paper by Miller and Woodhead-Galloway (1971), with the Debye–Hückel screening length λ ~ 0.50 nm (see p. **32**), the strength of the repulsive electrostatic force between two neighbouring thin actin filaments at zero overlap is only ~2 × 10^{-5}% that of the repulsive electrostatic force between a thick myosin filament and a single thin filament in an intact unit cell, at maximal overlap. This 'residual' force is therefore negligible. At this point, we may consider the true radius of a thick

Morel

myosin filament to be strictly independent of the amount of overlap ($R_m \sim 16.6$ nm; see pp. **33–34** for a detailed discussion of the values of R_m). Thus, the sum of the two true radii is $\sim 2 \times 16.6 \sim 33.2$ nm, very close to the centre-to-centre distance between two neighbouring thick myosin filaments at zero overlap. Indeed, the centre-to-centre distance between thick myosin and thin actin filaments and between thin actin filaments in an intact unit cell, at full overlap, is $c_0 \sim 25.3$ nm (see p. **24**), corresponding to $3^{1/2}c_0 \sim 43.8$ nm between two neighbouring thick myosin filaments and to $\sim 0.77 \times 43.8 \sim 33.7$ nm at non-overlap (the coefficient of ~ 0.77 is recalled above). Starting from the nondependence of R_m on the amount of overlap, there would be an unavoidable interpenetration of the envelopes surrounding each thick myosin filament (see pp. **27–33** for discussions concerning this envelope). Owing to this interpenetration and the subsequent major structural changes, an intact unit cell would be gradually transformed into an altered unit cell when sarcomere length increases, that is, when the myofilament spacing decreases. It would therefore be impossible, in a state of non-overlap, to obtain a suitable estimate of the repulsive electrostatic force between two thick myosin filaments from two neighbouring unit cells. This intermediate conclusion is largely related to the unrealistic assumption concerning the strict nondependence of R_m on the overlap, that is, on myofilament spacing. This problem of the dependence of R_m on the overlap/filament spacing is discussed in the next paragraph and below, in this section, based mostly on experimental and semi-empirical data obtained by several independent groups.

There are several complex questions regarding the value of R_m and its probable dependence on the amount of overlap (see this and the following paragraphs). As recalled in the preceding paragraph, the thick myosin filaments are brought closer together at the sarcomere length of ~ 3.68 μm (corresponding to zero overlap; see p. **25**). This results in almost certain disruption of the envelope in highly stretched fibres, as suggested in the preceding paragraph. Changes in the envelope, themselves resulting from changes in the surrounding medium, are also predicted from experimental and semi-empirical studies performed at room temperature ($\sim 20°C$–$22°C$) on synthetic thick myosin filament suspensions by Grazi and Cintio (2001). The authors strongly suggested that

major distortions of these filaments occur (particularly of the cross-bridges) when the surrounding medium is changed, via 'significant osmotic changes'. In intact fibres, for example, external osmotic alterations induce changes in interfilament spacing (in hypertonic conditions, the fibres shrink, decreasing this spacing, and vice versa in hypotonic conditions). When the sarcomere length is changed (passive stretching or active shortening), in intact fibres, the interfilament distance is also altered, because the intact fibre behaves isovolumically (see p. **25** for references and comments). Thus, the semi-empirical discussion of Grazi and Cintio (2001), valid for synthetic thick myosin filaments, can be extended to intact fibres. Let us consider R_m to be ~ 14.5 nm at a sarcomere length of ~ 3.68 μm (non-overlap state; see above), instead of ~ 16.6 nm, at full overlap (see pp. **32–33** concerning this value of R_m), corresponding to a moderate decrease of only $\sim 100 \times (14.5 - 16.6)$ nm/16.6 nm $\sim -13\%$. At zero overlap, the centre-to-centre distance between two neighbouring thick myosin filaments located in the nearest neighbouring intact unit cells is ~ 33.7 nm (see the preceding paragraph) and the closest surface-to-surface spacing between two neighbouring thick myosin filaments, in nearest neighbouring unit cells, can then be estimated at $\sim 33.7 - (2 \times 14.5) \sim 4.7$ nm, whereas the closest surface-to-surface spacing between a thick myosin filament and thin actin filament at full overlap is $s^* \sim 3.1$ nm (see p. **24**). Applying Equation 2 in the paper by Miller and Woodhead-Galloway (1971) at non-overlap, with the Debye–Hückel screening length assumed to be identical at zero and full overlap ($\lambda \sim 0.50$ nm; see p. **32**), the repulsive electrostatic force between two neighbouring thick myosin filaments is $\sim 5\%$ the value corresponding to the reference repulsive electrostatic force between a thick myosin filament and a single thin actin filament located in an intact unit cell, at full overlap, both at rest and under contraction conditions (see p. **24** and p. **25** for precise details on full overlap and zero overlap, respectively). At this point, as recalled in the preceding paragraph, distortions and deformations occur in stretched fibres, such that intact unit cells at full overlap (slack length) gradually become 'unit cells', different from the intact unit cells defined at the slack length. Thus, the proportion of $\sim 5\%$ is only an order of magnitude. No further discussion is required and the best solution is to consider that the thick-to-thick and thick-to-thin filament repulsive electrostatic forces are very small at non-overlap, with respect to the thick-to-thin filament repulsive electrostatic forces at full overlap (in these conditions, all the other repulsive electrostatic forces are negligible; see p. **25**). This conclusion results in self-consistent discussions and inferences. On pp. **257–258** in Section 8.9, explanations, discussions and precise details are presented regarding the small active tension, recorded at zero overlap by Gordon et al. (1966a) and the approximate 'residual value' of ~5% is taken into account, because it allows a coherent discussion. With the exception of Section 8.9, most of this monograph concerns fibres at the slack length (maximal overlap); hence, the problem of residual repulsive electrostatic forces at zero overlap does not arise.

On pp. **33–34**, the problem of the diameter of the thin actin filaments is analysed, based on the well-documented hypothesis presented by Grazi (1997), and it is clear that these filaments display many phenomena, including changes in diameter with experimental conditions. When the length of a sarcomere is greatly altered, the impact of the experimental conditions, particularly 'the effect of protein osmotic pressure and, in general, of the solvent conditions', invoked by Grazi (1997), gradually becomes different. These conclusions, drawn for thin actin filaments, are almost certainly also valid for thick myosin filaments, consistent with the hypothesis of a limited decrease in R_m from ~16.6 nm, at full overlap, to ~14.5 nm, at zero overlap (see the preceding paragraph). Indeed, concerning the thin actin filaments, the ratios between SE (~0.3 nm) and SD (~1.1 nm) and R_a ~ 4.3 nm are ~7% and ~26%, respectively (see p. **34**). The difference of ~2.1 nm between ~16.6 nm and ~14.5 nm, compared to ~16.6 nm, is ~13%, ranging therefore between ~7% and ~26%. Thus, the value of R_m ~ 14.5 nm is consistent with that of ~16.6 nm, at the experimental and semi-empirical points of view. However, this moderate difference in R_m at full overlap and at zero overlap leads to important conclusions, as demonstrated in this section.

Let us now consider the reference 'physiological' vicinal ionic strength I* ~ 182 mM used in this book (see Section 3.9 concerning the notion of vicinal ionic strength). The elementary repulsive electrostatic force per unit length (expressed in pN μm^{-1}, for example) between a thick myosin filament and a single thin actin filament, in an intact unit cell, at any value of the vicinal ionic strength I*, can be deduced from Equation 2 proposed by Miller and Woodhead-Galloway (1971). After slight modifications and rearrangements, we obtain:

$$F_{e,x} \sim A_x(I^*)Q_x(I^*)(I^*/182)^{3/4}\exp[-H_x(I^*)I^{*1/2}]. \quad (3.1)$$

$A_x(I^*)$ is a coefficient, expressed as a force per unit length (e.g. pN μm^{-1}), depending on the geometry of the myofilament lattice and other parameters (see Miller and Woodhead-Galloway 1971), possibly including vicinal ionic strength I* and the state of the intact unit cell (x = r at rest; x = a under contracting conditions). $Q_x(I^*) = [\tanh(e\Psi_{a,x}/4kT)]$ $[\tanh(e\Psi_{m,x}/4kT)]$, where e is the charge of the electron (1.602×10^{-19} C), k is Boltzmann's constant (1.381×10^{-23} J K^{-1}) and T is absolute temperature in Kelvin ($kT \sim 3.91 \times 10^{-21}$ J ~ 3.91 zJ at 10°C, the reference temperature used in this book; a new definition has been introduced in recent publications: 1 zeptojoule (1 zJ) = 10^{-21} J). $\Psi_{a,x}$ and $\Psi_{m,x}$ are the surface electrical potentials on a thin actin filament and a thick myosin filament, respectively. These potentials may differ under contraction (x = a) and resting (x = r) conditions. Indeed, Ca^{2+} (released from the sarcoplasmic reticulum) and its binding to the myofilaments (particularly the thin actin filaments), the attachment–detachment cycle of the cross-bridges and the subsequent cleavage of MgATP, for example, may result in rearrangements or changes in myofilament structures. Such phenomena may induce changes in the numbers of fixed negative electrical charges and bound anions, and therefore in the surface electrical potentials $\Psi_{a,x}$ and $\Psi_{m,x}$, at a given value of I*. These points are discussed and confirmed at many places in this monograph, such as Section 4.4.2.3 and pp. **161–162** in Section 4.4.2.6. Finally, $H_x(I^*)$ is a coefficient, depending a priori on the state of the intact unit cell (rest or contraction) and on vicinal ionic strength I*. $H_x(I^*)$ is studied on pp. **31–34**.

For frog and rabbit, at the reference sarcomere length s_0 ~ 2.18 μm (see p. **23**), corresponding to the universal slack length of an intact fibre, the closest surface-to-surface spacing between a thin actin filament and the backbone of a thick myosin filament is $s^\dagger$ ~ 11–12 nm (estimated by Bachouchi and Morel 1989a, from traditional x-ray diffraction data available in 1988–1989). From experiments on frog intact fibres, at 10°C–12°C, investigating the relationship between force and sarcomere length, Bagni et al. (1990b) deduced that the maximal force (at maximal overlap) corresponded to a surface-to-surface distance between the thick myosin and thin actin filaments of ~10–12 nm, supporting the estimate of Bachouchi and Morel (1989a). Averaging these four available values gives $s^\dagger \sim$ 11.2 ± (0.4 or 0.8) nm (mean ± SE or SD; n = 4). The universal value of ~11.2 nm is used in this monograph. According to Equation 2 in the paper by Miller and Woodhead-Galloway (1971), $H_x(I^*)$ is, in principle, proportional to $s^\dagger$. However, as highlighted at many places in this section (see pp. **27–33**, concerning cell-associated water and presence of many myosin heads between the thick myosin and thin actin filaments, both at rest and under contraction conditions), the situation is much more complex. As an illustrative example, from the discussion presented by Bachouchi and Morel (1989a), on the basis of geometric and structural data, it appears clear that many myosin heads extrude from the shaft of the thick myosin filaments and are located extremely close to the thin actin filament surface, both at rest and during contraction, regardless of the state of the fibre, at full overlap or stretched/compressed (see also pp. **251–254** in Section 8.9). Studying the effects of osmotic pressure on chemically skinned fibres of the frog, prepared by the old technique of Natori (1954a,b), Tsuchiya (1988) measured fibre diameter under various experimental conditions and interpreted his experimental results as demonstrating the existence of 'passive interactions'. More precisely, combining the results of Elliott and Offer (1978), on the one hand, and his own results and those of Matsubara et al. (1984) (obtained at a sarcomere length of ~2.3 μm in both studies), on the other, Tsuchiya (1988) concluded that 'it is reasonable to postulate that in a compressed fiber, passive interaction is constituted between filaments', even at very low osmotic pressure, in resting and contracting fibres. The geometric and structural conclusions of Bachouchi and Morel (1989a) and the interpretation of experimental data by Tsuchiya (1988) are consistent. At full overlap, the true surface-to-surface spacing is therefore much smaller than $s^\dagger \sim$ 11.2 nm and should be written $s^* = s^\dagger - d$. For simplification, d is assumed to be independent of the state of the muscle fibre (rest or contraction). Owing to specific difficulties in the application of their Equation 2 to some experimental data, Miller and Woodhead-Galloway (1971) themselves suggested the introduction of a layer of strongly bound water molecules surrounding each thick myosin filament, with a thickness d, excluding electrolytes (see their Equation 10), as did Morel (1985b), Morel and Gingold (1979a) and Morel et al. (1976). Morel et al. (1976) concluded their paper, concerning a new model of muscle contraction, by stating that 'interfilament water plays an important role, as it represents about 80% of the myofibrillar matter... any model that does not take into account this interfilament water presents a weak point, inasmuch as it tries to account for motility by reasoning with only about 20% of the myofibrillar matter'. In this context, Ball (2008) considered water to be 'an active constituent in cell biology', consistent with the ideas developed by my group and other independent groups (see Grazi 2008; Oplatka 1994, 1997). However, in the case of muscle, the conclusion of Morel et al. (1976) is incomplete, as demonstrated above, because both bound water layers and many extruding myosin heads contribute to the whole thickness d of the envelope surrounding each thick myosin filament. The value of d is estimated on p. **33** and is demonstrated to be consistent with experimental data obtained by independent groups (see particularly pp. **27–33** for a general discussion).

The notion of major differences between cell-associated water (immobilised and modified water, generally assumed to be unable to dissolve ions to the same extent as bulk water) and bulk water is not new and has been extensively discussed and disputed since the 1960s and 1970s, particularly in muscle (e.g. Belton and Paker 1974; Belton et al. 1972, 1973; Burnell et al. 1981; Chang et al. 1972, 1973; Civan and Shporer 1975; Clark et al. 1982; Clegg 1984a,b, 1986; Cleveland et al. 1976; Cooke and Kuntz 1974; Cooke and Wien 1971; Cope 1967, 1969; Diegel and Pintar 1975; Drost-Hansen 1969, 1971a,b, 2001; Finch and Homer 1974; Finch et al. 1971; Fung 1975; Fung and McGaughy 1974; Fung et al. 1975; Hansen 1971; Hazlewood 1972, 1973, 1975, 1979; Hazlewood et al. 1969, 1971, 1974; Kasturi et al. 1980; Klein and Phelps 1969; Ling 2003; McLaughling and Hinke 1966; Morel 1975, 1976, 1985b; Morel and Gingold 1979a; Morel et al. 1976; Oplatka 1989, 1994, 1997; Palmer et al. 1952; Pézolet et al. 1978; Robinson 1989; Rorschach et al. 1973; Sussman and Chin 1966;

Wiggins 2008; Woessner and Snowden 1973). The most 'prolific author', focusing on cell-associated water and its properties (such as the partial exclusion of various ionic and nonionic solutes) is certainly Ling (e.g. Ling 1965, 1970, 1976, 1977, 1980, 1984, 1988a,b, 1993, 2003; Ling and Hu 1988, 2004; Ling et al. 1973, 1993). Wiggins followed the same train of thought as Ling (e.g. Wiggins 1971, 1972, 1973, 1990, 1996, 2001, 2008; Wiggins and McClement 1987). There has been considerable debate about both the existence and proportion of bound water in muscle, which was estimated, many years ago, to range between ~20% (Belton and Paker 1974; Belton et al. 1972, 1973) and ~70% or ~80%–90% (Hazlewood et al. 1969, 1974), with intermediate estimates of ~30% (Cope 1969) or ~40%–60% (Morel et al. 1976). In the 1970s, using various NMR techniques, Civan and Shporer (1975), Cooke and Wien (1971), Finch and Homer (1974) and Finch et al. (1971) concluded, from their experiments, that muscle water was rapidly exchanged between a small fraction of immobilised water molecules and a large fraction of free water. They therefore took the view that interpretations based on ordering of the intramuscular water were unlikely. Working on very different cells (rat phrenic nerve), Klein and Phelps (1969) also reported evidence against the orientation of water molecules, suggesting that cell-associated water plays a minor role. In a review paper, Cooke and Kuntz (1974) claimed that, in the 1970s, it was almost impossible to draw clear conclusions about the state of water in biological systems. The majority view, especially among muscle specialists, was that water is simply 'nonexistent', not only in muscle but also in any living cell. This remains the majority view among biologists. Nonetheless, the notion of a water layer excluding electrolytes is still topical: Blyakhman et al. (2001) used it to interpret, at least partly, their experimental data. Citing three specialists in cell-associated water (Clegg 1984b; Ling 1984; Wiggins 1990), they wrote: 'water is largely organized around protein filaments, practically as a solid' (a book on strong water adsorption in cells has been published by Pollack 2001; Pollack 2003 went deeply into the role of aqueous interfaces in cells). Wiggins (2008) pointed out that, in the early 1990s, an imaginative leap was made in physics, rather than in biology: coexistence of two types of water in rapidly exchanging microdomains, a low-density water (density of ~0.9 g ml⁻¹) and a high-density water (density of ~1.2 g ml⁻¹). In this area, dominated by physicists and with little input from biologists, Pollack's group, composed of biologists and physicists, has made interesting experimental observations in the last few years (e.g. Chai and Pollack 2010; Yoo et al. 2011; Zheng and Pollack 2003; Zheng et al. 2006). However, the assumption that bound water can be seen as a solid, as strongly suggested by Blyakhman et al. (2001), is probably unwarranted, because this would require a very low dielectric constant of the sarcoplasmic medium at 2–10 (see p. **32**), which is inconsistent with the almost total dissociation of many salts into ions (e.g. NaCl or KCl) and also with the presence of large quantities of soluble ions (see again p. **32**). On p. **125** in Section 4.3.2, it is recalled that Finch et al. (1971) found, in liquid water, that the self-diffusion constant of water in liquid water was similar to most diffusion constants of other simple molecules or ions. Moreover, Finch et al. (1971) pointed out that, at 0°C, the self-diffusion constant of water in ice is ~5 orders of magnitude lower than that in liquid water, also at 0°C. The 'icelike state' concept, first introduced by Ling (e.g. 1965) and frequently raised in the 1960s and 1970s, is therefore questionable. The problem is now much more complex than it was approximately 40–50 years ago and cell-associated water is probably a genuine third state (vapor being the fourth state) that could be called 'living water' (see above, the statement of Wiggins 2008, and her review on water and life). The structure of this third state would allow the 'normal' diffusion of small molecules and ions but would have many other properties different from those of liquid water (e.g. dielectric constant and the subsequent hypothetical dissolution of neutral salts into ions highlighted above; see also p. **32** for precise details and a compromise on this matter). Finally, in many studies cited above, bound water is seen to play only a 'passive' role, whereas an 'active' role has also been attributed to the highly ordered lattice of water associated with proteins (e.g. Clegg 1984a; Cope 1967; Hazlewood 1975; Morel et al. 1976; Robinson 1989; Tigyi et al. 1981; Wiggins 1990, 2008; Wiggins and McClement 1987). Grazi (2008) and Oplatka (1989, 1994, 1997) even suggested that water plays an active and rather complex role in the mechanisms of muscle contraction themselves. More generally, in muscle, the problem is more difficult than in other cells and 'living water' is not the sole partner. For instance, in their discussion, Miller and Woodhead-Galloway (1971) added a complementary assumption to their hypothesis concerning the layer of water excluding ions: they suggested that 'the excluding envelope of myosin is somewhat greater than the myosin diameter indicated by x-ray studies'. They thought that cross-bridges might be involved, but probably only under contraction conditions, when the cross-bridges attach to the thin actin filaments. This conclusion is strongly supported below, in this section, and extended to both resting and contracting conditions.

Haselgrove and Huxley (1973) and Huxley (1980c) interpreted some of their x-ray diffraction data obtained by traditional techniques as showing that up to ~90% of the myosin heads would be in the close vicinity of thin actin filaments in isometrically contracting muscles. These conclusions are consistent with other experimental data and discussions presented on p. **29**, but apply solely to isometric contracting muscles. Furthermore, I consider some interpretations proposed by these authors inadequate. Moreover, these experiments and discussions do not concern resting muscles, going against expectations from the geometric and structural study of Bachouchi and Morel (1989a), experimentally supported by Tsuchiya (1988), as recalled on p. **29**. Owing to the vagueness of the assumptions made by Miller and Woodhead-Galloway (1971) (see the preceding paragraph), some inadequate conclusions drawn by Haselgrove and Huxley (1973) and Huxley (1980c) and the unduly strong assertion of Blyakhman et al. (2001), I take an intermediate position and state that the envelope corresponds partly to bound water, but also (probably mostly) to the many extruding myosin heads present between neighbouring thin actin and thick myosin filaments, at rest and under contraction conditions.

At this point, I would like to recall a controversy and give my own point of view, based on approaches suggested by independent groups, to try to clarify a complex situation. A.F. Huxley (1980) calculated that, under isotonic contraction conditions (active shortening of frog fibres), the viscous-like frictional forces − 'viscous drag forces' or 'hydrodynamic viscous drag' (e.g. Elliott and Worthington 2001, for a careful definition and discussion of these forces) − account for only ~0.01% of the isometric force. Elliott and Worthington (2001) criticised

the calculation of A.F. Huxley (1980) and demonstrated that the true hydrodynamic viscosity of the interfilament medium is much greater than that used by A.F. Huxley (1980), who suggested that this viscosity was only approximately twice that of pure water. Moreover, Worthington and Elliott (1996a,b), commenting on the calculations of A.F. Huxley (1980), claimed that 'the actin–myosin centre-to-centre distance is used as the filament-to-surface distance and therefore the close proximity of the myosin heads to the surface of the actin filament is ignored'. The much smaller filament distance, suggested by Worthington and Elliott (1996a,b) than that suggested by A.F. Huxley (1980), significantly increases the hydrodynamic viscosity of the sarcoplasmic medium and, therefore, the axial frictional forces between the myofilaments. Furthermore, Worthington and Elliott (1996a,b) essentially expressed the same view as myself (see below, in this section), but their view is probably valid only in the case of active shortening, when there is an active sliding of the myofilaments past each other, not at rest or under isometric contraction conditions.

Millman (1998) scrutinised the results of several x-ray diffraction experiments, performed on both relaxed and isometrically contracting fibres from frog (at sarcomere lengths corresponding approximately to fibres at the slack length, i.e. $s_0 \sim 2.18$ μm for the composite animal species; see p. **24**), and deduced that, when the fibre passes from rest to contraction, the myosin heads display a radial shift in position of only ~1 nm toward the thin filaments. Obviously, this very small shift of ~1 nm (i.e. only ~9% of the closest surface-to-surface spacing $s^\dagger \sim 11.2$ nm; see p. **29**) can be interpreted as indicating that the myosin heads extruding from the shaft of the thick myosin filaments are located very close to the thin actin filament surface, both at rest and during isometric contraction. The hypothesis put forward on p. **29**, based on structural data, the discussion based on geometric and structural data presented by Bachouchi and Morel (1989a) and the experimental finding of Tsuchiya (1988) (see p. **29**) are therefore consistent with Millman's interpretation (1998).

Consideration of the envelope surrounding the thick myosin filaments, strongly related to the extruding heads (but not totally; see pp. **29–31** regarding the existence of bound water), is justified, as these heads almost certainly bear a significant proportion of the fixed negative electrical charges and bound anions (see Section 4.4.2.2 for precise details concerning the fixed negative electrical charges and bound anions). Thus, the exponential term in Equation 2 in the paper by Miller and Woodhead-Galloway (1971) should be written $\exp(-s^*/\lambda) = \exp[-(s^\dagger - d)/\lambda]$, where s^* and d are defined on p. **29**. λ is the Debye–Hückel screening length. Miller and Woodhead-Galloway (1971) defined this length, but their formula contains some printing errors and should read $\lambda = (10\varepsilon\varepsilon_0 kT/8\pi e^2 N_A I^*)^{1/2}$ (after some rearrangements), where kT is defined in the text below Equation 3.1; ε is the dielectric constant of the interfilament medium, ε_0 is the dielectric constant of a vacuum (8.85×10^{-12} C^2 N^{-1} m^{-2}), N_A is Avogadro's number (6.022×10^{23}) and the vicinal ionic strength I^* is expressed in millimolar (see Section 3.9 for precise details on I^*). Finally, from Equation 3.1, we deduce that $H_x(J^*)(I^*)^{1/2} = (s^\dagger - d)/\lambda$ and $H_x(I^*) = (s^\dagger - d)(8\pi e^2 N_A/10\varepsilon\varepsilon_0 kT)^{1/2}$ (in this relationship, d can a priori depend on I^*).

The maximal value of I^* used in the experimental part of this monograph was ~182 mM (bulk ionic strength I = 180 mM). This value of bulk ionic strength corresponds approximately to the mean ionic strength in vivo/in situ for any vertebrate skeletal muscle (e.g. Godt and Maughan 1988; Godt and Nosek 1989; Xu et al. 1993). Owing to the existence of bound water (see pp. **29–31**) and the presence of many extruding myosin heads in the interfilament medium, both under contraction and in resting conditions (see pp. **27–33**), the value of the dielectric constant ε is certainly not that of pure water (~78–80). Hazzard and Cusanovich (1986) obtained a value of ~20 near cardiac myosin heads. Palmer et al. (1952 and references therein) measured and discussed the values of the dielectric constant of 'moist clay with water content'. They estimated that this constant was 3 (that for ice) near the solid–liquid boundary, increased exponentially with increasing distance from the boundary, and reached a value of 80 (that for liquid water) at an infinite distance from the boundary. Values of ~2–10 are found for solids (see handbooks on electrostatics). Morel (1985b) and Morel and Gingold (1979a) showed that the usual concepts of colloid stability cannot account for the stability of the myofilament lattice and suggested that this may be because it is necessary (i) to consider the existence of bound water excluding the K^+ ions (see pp. **29–31**) or (ii) to lower the dielectric constant, with respect to that of pure water, although it is not possible to lower it very much. Only this second conclusion is taken into account here. An intermediate uniform value of ~50, between the values of ~78–80, ~20, ~3–80 and ~2–10, can be used for the interfilament medium. Too great a decrease would cause major physical chemical problems concerning the full dissociation of electrolytes into anions and cations and would make it impossible to obtain suitable estimates for ionic strength. Taking values of ~78–80 (pure water) has no major effect on the numerical values and conclusions presented below, in this section. Choosing ε ~ 50 is simply more realistic. At I^* ~ 182 mM, we deduce that λ ~ 0.5 nm, at the reference temperature of 10°C (see the end of the preceding paragraph for the definition of λ). In Equation 3.1 it is assumed that $H_x(I^*)$ depends on vicinal ionic strength I^* and on the state of the fibre (at rest or during contraction). It is shown experimentally (see text above Equation 4.15) that $H_a(I^*)$ is fairly constant under contraction conditions and that H_a ~ 0.26 $mM^{-1/2}$, regardless of the value of I^*, within the limits of experimental error. Moreover, it can reasonably be assumed that $H_r \sim H_a = H$ (H_r at rest), because this assumption generates no inconsistencies in the various calculations and discussions presented in this book. We deduce that $H(182)^{1/2}$ ~ 3.51 ~ $(s^\dagger - d)/0.50$ nm; that is, $s^* = s^\dagger - d$ ~ 1.8 nm (see the end of the preceding paragraph regarding the relationship between H, I^*, $s^\dagger$, d and λ). As highlighted on p. **24**, this value of ~1.8 nm is essentially similar to ~3.1 nm obtained on entirely different bases (the value of s^* ~ 3.1 nm is calculated on p. **24**). On p. **14** in Section 2.2, it is pointed out that the half-fibres used in the experimental part of this monograph were slightly stretched (sarcomere length of ~2.5 ± 0.2 μm). The value of H would therefore be slightly different at the universal slack sarcomere length s_0 ~ 2.18 ± 0.18 μm (see p. **23**; mean ± SD), but no correction is required, because experimental error conceals any possible difference between the slack and slightly stretched half-fibres. A value of ~1 nm for the radial shift of the myosin heads, when the fibre shifts from rest to contraction, was estimated by Millman (1998; see the preceding paragraph). On p. **29** and p. **33**, it is strongly suggested that the studies, calculations, reasoning and conclusions of Bachouchi and Morel (1989a), Millman (1998) and Tsuchiya (1988) are qualitatively (even quantitatively) consistent, despite having very different bases. Indeed, the value of ~3.1 nm recalled here is of the same order of magnitude as the ~1 nm, quantitatively supporting the reasoning presented in this section and, more generally, the hybrid model and its underlying principles, analysed and scrutinised throughout this book. For the composite animal species, at the slack length, we have $s^\dagger \sim 11.2$ nm (see p. **29**), giving $d \sim s^\dagger - s^* \sim 11.2 - 3.1 \sim 8.1$ nm (mean thickness of the envelope surrounding each myosin thick filament, at full overlap). At zero overlap, it is suggested, on p. **27**, that $R_m \sim 14.5$ nm, with $R_m \sim 9.0$ nm $+ d$ (see the beginning of the next paragraph for the definition of R_m), giving $d \sim 5.5$ nm.

At this stage, I am obliged to combine results for frog and rabbit (composite animal species). This does not lead to any inconsistencies, within the limits of experimental uncertainty. The external radius of the backbone of a thick myosin filament from rabbit, and probably also from frog, is $R_b \sim 18.0$ nm$/2 \sim 9.0$ nm (see p. **215** in Section 6.4) and its true radius is therefore $R_m \sim 9.0$ nm $+ d \sim 9.0$ nm $+ 8.1$ nm ~ 17.1 nm, at 10°C (see the end of the preceding paragraph concerning $d \sim 8.1$ nm). Using synchrotron radiation, Malinchik et al. (1997) performed x-ray diffraction experiments on resting and contracting skinned fibres from rabbit psoas muscle. They interpreted their results in a way relevant to the present discussion. They found that the mean distance between the centre of the cross-bridges and the centre of the thick myosin filaments was ~17.5 nm at 4°C and ~13.5 nm at 20°C. Using electron microscopy and working with rabbit myosin natural thick myosin filaments, Ip and Heuser (1983) found an 'apparent' radius (radius of the shaft of the filament plus the mean 'thickness' of the cross-bridges) of ~15.1 nm at room temperature (~20°C–22°C). Using very different and unconventional experimental techniques and reasoning, in which water and osmotic phenomena play major roles, Grazi (2000), Grazi and Cintio (2001) and Schwienbacher et al. (1995) studied suspensions of natural thick myosin filaments from rabbit skeletal muscles, at room temperature (~20°C–22°C), and Grazi and Cintio (2001) deduced that the radius of these filaments was ~13.0 nm. Despite the complex problems relating to certain properties of bound water (see the preceding paragraph, concerning the dielectric constant), this value of ~13.0 nm is comparable to the values obtained

above, taking into account the extruding heads only. Averaging the three values obtained for rabbit at room temperature (~20°C–22°C), we obtain ~13.9 ± (0.5 or 0.9) nm (mean ± SE or SD; n = 3), whereas the single value at 4°C is ~17.5 nm (see above). Assuming, for simplification, that the 'mean distance' (or 'mean true radius') is a linear function of temperature, we obtain (barycentric mean) ~16.2 nm at 10°C, for rabbit. Owing to shortcomings in the reasoning and discussions and to experimental errors, the two values for R_m, at 10°C, of ~17.1 nm for the composite animal species (see the beginning of this paragraph) and ~16.2 nm for rabbit, obtained with very different approaches, are essentially similar, regardless of the important remark of Barman et al. (1998), Stephenson and Williams (1985) and Woledge et al. (1985) concerning possible differences in the behaviour and properties of warm-blooded and cold-blooded species studied at low temperatures (this comment applies essentially to the value of ~17.5 nm at 4°C for rabbit psoas muscle; see above). In any event, there are many consistent qualitative and quantitative arguments supporting the introduction of d (the width of the envelope surrounding the thick myosin filaments) and also its quantitative value and the value of $H \sim 0.26$ mM$^{-1/2}$ (intimately related to d; see p. **32**), deduced from experimental results, valid for crayfish intact and skinned fibres and frog half-fibres (see text above Equation 4.15), thereby supporting the general approach presented in this monograph. The most probable value of R_m, valid at 10°C and at full overlap for crayfish, frog and rabbit (composite animal species), is, therefore, ~(17.1 + 16.2) nm$/2 \sim 16.6$ nm.

Edelman and Padrón (1984) found that the radius of a thin actin filament from frog or rabbit was $R_a \sim 5.0$ nm. In the case of crayfish leg striated muscles, April (1969) obtained a value of ~4.0 nm. Grazi (1997) proposed a hypothesis concerning the best estimate of R_a. He claimed that the thin actin filament cannot be seen as a rigid structure. Instead, he proposed a 'fluttering wing' model that 'predicts a variable diameter as well as a variable orientation of the monomers'. This notion of distortions and conformational changes in the thin actin filaments is not new, having been proposed by Oosawa (1977) for F-actin (in vitro) and by Morel (1990, 1991b) for actively shortening intact fibres or forced lengthening of contracting intact fibres, starting from the phenomenological/

semi-empirical approach repeatedly used in my group over approximately 20 years (first publication by Morel et al. 1976 and the last by Morel and Merah 1995) and improved throughout Section 3.4 in this monograph. Grazi (1997) cited nine values of R_a from independent groups, including the value of ~5.0 nm obtained by Edelman and Padrón (1984) (I do not take into account the old values obtained in the 1950s and 1960s, using the thin sectioning technique and electron microscopy). Adding the value of ~4.0 nm obtained by April (1969), we deduce a mean value of R_a ~ 4.3 ± (0.3 or 1.1) nm (mean ± SE or SD; n = 10). Only the value of R_a ~ 4.3 nm is taken into account in this book. In vivo/in situ, it is unclear whether the thin actin filaments are surrounded by an envelope of bound water, as is the case with thick myosin filaments (see pp. 29–31). For the sake of simplicity, I assume that the thickness of this hypothetical envelope is totally concealed by the experimental uncertainty on the value of R_a, lying between ~3.2 nm and ~5.4 nm, if the SD is taken into account (see Section 2.1 for a brief comment on the advantage of using SD rather than SE). At this stage, there is no point in discussing the problem of the true diameter of the thin actin filaments further. The value of $A_x(I^*)$ in Equation 3.1 is proportional to $[2\pi R_a R_m/(R_a + R_m)]^{1/2}$ (see Equation 2 in the paper by Miller and Woodhead-Galloway 1971 for precise details concerning the various constants appearing in the definition of $A_x(I^*)$). $A_x(I^*)$ would depend on I^* and x solely through R_m, which itself depends on d (see p. 33). However, choosing 'arbitrary and unrealistic' values of d increasing from 3 to 12 nm results in an increase in $A_x(I^*)$ of only ~7%. Thus, $A_x(I^*)$ is fairly constant and we can call this parameter A, regardless of I^* and x. Equation 3.1 can therefore be written in a simpler form:

$$F_{e,x}(I^*) \sim AQ_x(I^*)(I^*/182)^{3/4}\exp(-HI^{*1/2}) \qquad (3.2)$$

where both A and H are constant, regardless of I^* and of the state of the intact unit cell (rest or contraction; the constancy of H is discussed on p. 33). In their theoretical approach, Miller and Woodhead-Galloway (1971) omitted, probably because of printing problems, to multiply the repulsive electrostatic force $F_{e,x}$ by the dielectric constant of a vacuum ($\varepsilon_0 = 8.85 \times 10^{-12}$ C^2 N^{-1} m^{-2}; see handbooks on electrostatics). Applying their various formulae at I^* ~ 182 mM and 10°C,

from Equation 3.2 and using the values of ε, H, R_a and R_m given above, in this section, we deduce the value of A and we obtain $F_{e,a}(182)$ ~ $265Q_a(182)$ pN μm^{-1} for the elementary radial repulsive electrostatic force exerted, under physiological ionic strength and contraction conditions, between a thick myosin filament and a single thin actin filament in an intact unit cell. For 'composite' fibres, the slack length corresponds to s_0 ~ 2.18 μm (see p. 24). The length of the thick myosin filaments for frog and rabbit muscles ranges between ~1.65 μm (Gordon et al. 1966a,b), ~1.62 μm or ~1.61 μm (Trombitás et al. 1993) and ~1.64 μm (Wang et al. 1999) (see p. 25). Averaging these four values gives ~1.63 ± (0.01 or 0.02) μm (mean ± SE or SD; n = 4). The error bar is neglected and only the value of ~1.63 μm is selected. The mean length of the central bare zone on the thick myosin filaments (for frog and rabbit), devoid of myosin heads extruding from the filament core, is ~0.15–0.16 μm (e.g. Craig 1977; Craig and Offer 1976; Gordon et al. 1966a,b; Hanson et al. 1971; Page and Huxley 1963; Sjöström and Squire 1977; Trinick and Elliott 1979; Wang et al. 1999). Many electrical charges are probably located on the myosin heads and the bare zone almost certainly makes a marginal contribution to the repulsive electrostatic forces. A first estimate of the length of the 'fully electrostatically active' overlap zone (at maximal overlap) is therefore ~1.63 − (0.15–0.16) ~ 1.47–1.48 μm. On the other hand, the slack sarcomere length corresponds to s_0 ~ 2.18 μm (see p. 24) and the average length of two thin actin filaments is ~ (2 × 0.96 + 2 × 0.94 + 2 × 1.12)/2 ~ 2.03 μm (see p. 25 concerning the most recent values given by Trombitás et al. 1993 and Wang et al. 1999). Thus, there are no electrostatic interactions between the thick myosin and thin actin filaments beyond a distance ~1.63 − (2.18 − 2.03) μm ~ 1.48 μm. We can choose γ ~ (1.47 + 1.48 + 1.48)/3 ~ 1.48 μm for the 'fully electrostatically active' zone in which the thick myosin and thin actin filaments interact electrostatically. The total radial repulsive electrostatic force per thick myosin filament in electrostatic interaction with a single thin actin filament in an intact unit cell, at maximal overlap γ, is therefore $(\gamma/2)F_{e,a}(182)$ ~ $(1.48/2) \times 265Q_a(182)$ pN ~ $196Q_a(182)$ pN. In some instances, the behaviour of stretched fibres and unit cells is discussed and the notation $\gamma^* < \gamma$ is employed, corresponding to partial overlap. In this event, the notation becomes $(\gamma^*/2)F_{e,a}^*(182)$, where, in principle, $F_{e,a}^*(182) > F_{e,a}(182)$, because the myofilament spacing decreases as the fibres are stretched, whether they are intact (isovolumic behaviour; see p. **25** in Section 3.2 and p. **252** in Section 8.9) or demembranated (working on mechanically skinned frog fibres and using traditional x-ray diffraction, Matsubara and Elliott 1972 and other independent groups found a linear decreasing relationship between myofilament spacing and length of the fibres). However, the behaviour of stretched intact fibres (also stretched unit cells) is complex (see Bachouchi and Morel 1989a and pp. **251–254** in Section 8.9) and the above inequality is not straightforward.

There are four thin actin filaments interacting electrostatically with a single thick myosin filament in an intact unit cell. This value of four thin actin filaments is different from the number of six nearest neighbouring thin actin filaments present in the hexagonal unit cell for defining the traditional lattice spacing d_{10} in the 'crystallographic unit' (e.g. Figure 6 in the paper by Matsubara et al. 1984 and Figure 3 in the review by Millman 1998). On the other hand, the notion of a half-sarcomere is frequently used, because it is related to A.F. Huxley's theory (1957) of side-pieces (cross-bridges) (e.g. Barman et al. 1998; Elliott and Worthington 2001; Huxley 1957; Lionne et al. 1996; Xu et al. 1993), giving the number of 6 thin actin filaments, which is potentially misleading, because several authors take 6 thin actin filaments per thick myosin filament in an intact unit cell, instead of 12 (this number corresponds to all the actin filaments in an intact unit cell, regardless of the neighbouring intact unit cells). Otherwise, each thin actin filament counts as 1/3 (three nearest neighbouring intact unit cells per thin actin filament) and there are $12/3 = 4$ thin actin filaments per intact unit cell. In their discussion, Worthington and Elliott (1996a,b) pointed out that, in crab muscles, the filament lattice is 1:6, with 1 half-thick myosin filament and 6 thin actin filaments per half intact unit cell (i.e. $2 \times 6 = 12$ thin actin filaments per thick myosin filament), whereas in frog and rabbit muscles, the filament lattice is 1:2, that is, 1 half-thick myosin filament and 2 thin actin filaments per half intact unit cell. In other words, for frog and rabbit, there are $2 \times 2 = 4$ thin actin filaments per thick myosin filament, that is, per intact unit cell, exactly the value of 4 suggested above for the number of thin actin filaments interacting electrostatically with a single thick myosin filament. This

value was confirmed by Elliott and Worthington (1997): 'in the half sarcomere 150 myosin molecules, with 300 heads, interact with two actin filaments (defining one actin filament from the Z-line to the end of overlap zone) [this corresponds to 150 heads per thin actin filament]'. As recalled on p. **205** in Section 6.3.5, there are ~600 myosin heads per intact unit cell that are assumed to lie outside the thick myosin filament backbone in the traditional models of these filaments (see pp. **170– 172** in Section 5.1.1 for references concerning the most probable arrangement of the heads). Taking the value of 4 thin actin filaments per intact unit cell (see above), there are therefore ~600/4 ~ 150 heads interacting with a thin actin filament, identical to the value given by Elliott and Worthington (1997). This confirms the choice of four thin actin filaments interacting electrostatically with a single thick myosin filament in an intact unit cell. On p. **85** in Section 3.5 (see text below Equation 3.49), I select $Q_a(182) \sim 0.763$ and the total radial repulsive electrostatic force is $4(\gamma/2)F_{e,a}(182) \sim 4 \times 196 \times 0.763$ pN ~ 598 pN per intact unit cell (the value of $(\gamma/2)F_{e,a}(182) \sim 196Q_a(182)$ is given in the preceding paragraph).

3.3 FULL ISOMETRIC TETANIC TENSION (SHORT TETANI–STEADY STATE) IN INTACT FROG FIBRES AND HALF-FIBRES. SPECIAL CASES OF INTACT FIBRES FROM WHITE MUSCLES OF DOGFISH AND TOADFISH

I take into account here old and more recent experimental results obtained with intact frog fibres, and, on occasion, with whole frog muscles, at the universal reference sarcomere length ($s_0 \sim 2.18$ μm; see p. **24** in Section 3.2), at temperatures between ~1°C and ~12°C, for short tetani, under steady-state conditions for tension (maximal duration of ~1.0–1.5 s; e.g. Altringham and Bottinelli 1985; Bagni et al. 1988b, 1990b; Edman 1966, 1979; Edman and Anderson 1968; Edman and Hwang 1977a; Edman and Reggiani 1987; Elzinga et al. 1989a,b; Granzier and Pollack 1990; Horowitz and Pollack 1993). The necessary corrections for temperature are made below, in this section. On pp. **63–64** in Section 3.4.3.2, values of Q_{10}, for isometric tetanic tension in intact fibres, measured or estimated by several groups, are discussed, and the value of ~1.24 within the 0°C–20°C range of temperature is retained. For temperatures of ~3°C–4°C, Gordon et al. (1966b) reported isometric tetanic tensions of ~2.7 ± (0.1 or 0.3) × 10⁵ N m⁻² (mean ± SE or SD; n = 7), giving, after correction for temperature, ~3.1 ± 0.1 × 10⁵ N m⁻² at 10°C. At ~2.5°C, Edman (1979) obtained tensions of ~2.3 ± (0.1 or 0.2) × 10⁵ N m⁻² (mean ± SE or SD; n = 6), corresponding, at 10°C, to ~2.7 ± 0.1 × 10⁵ N m⁻². At ~1.1°C, Edman and Hwang (1977) obtained values of ~2.2 ± (0.1 or 0.2) × 10⁵ N m⁻² (mean ± SE or SD; n = 6), giving, at 10°C, ~2.6 ± 0.1 × 10⁵ N m⁻². Edman and Reggiani (1987) worked at temperatures between 0.8°C and 3.3°C. From their Figure 3, it can be deduced that the tensions were ~2.5 ± (0.2 or 0.4) × 10⁵ N m⁻² (mean ± SE or SD; n = 4), giving, at 10°C, ~2.9 ± 0.2 × 10⁵ N m⁻². At 4°C, Elzinga et al. (1989a) obtained a value of ~2.5 ± (0.1 or 0.5) × 10⁵ N m⁻² (mean ± SE or SD; n = 23), corresponding to ~2.8 ± 0.1 × 10⁵ N m⁻² at 10°C. At 12°C, Bagni et al. (1990b) reported three values of ~2.0 × 10⁵ N m⁻², ~2.8 × 10⁵ N m⁻² and ~3.0 × 10⁵ N m⁻², giving, at 10°C, ~1.9 × 10⁵ N m⁻², ~2.9 × 10⁵ N m⁻² and ~3.1 × 10⁵ N m⁻², respectively (mean value of ~2.6 × 10⁵ N m⁻²). At 4°C, Cecchi et al. (1990) obtained two values of ~3.0 × 10⁵ N m⁻² and ~3.3 × 10⁵ N m⁻², corresponding, after correction for temperature, to ~3.4 × 10⁵ N m⁻² and ~3.8 × 10⁵ N m⁻² at 10°C (mean value of ~3.6 × 10⁵ N m⁻²). At 2.3–3.0°C, Granzier and Pollack (1990) obtained tensions of ~3.1 ± (0.4 or 0.9) × 10⁵ N m⁻² (mean ± SE or SD; n = 5), giving ~3.6 ± 0.5 × 10⁵ N m⁻² at 10°C. At 4°C, Linari et al. (1998) obtained tensions of ~2.3 ± (0.2 or 0.5) × 10⁵ N m⁻² (mean ± SE or SD; n = 10), giving, at 10°C, ~2.6 ± 0.2 × 10⁵ N cm⁻². At 4°C, Piazzesi et al. (1999) obtained values of ~3.7 ± 0.2 × 10⁵ N m⁻² (mean ± SE; n not specified), giving, at 10°C, ~4.2 ± 0.2 × 10⁵ N m⁻². Working at 0°C or 17°C, Linari et al. (2005) obtained tensions of ~2.1 ± 0.3 × 10⁵ N m⁻² (mean ± SD; n not specified) and ~3.0 ± 0.4 × 10⁵ N m⁻² (mean ± SD; n not specified), respectively, corresponding to similar values of ~2.6 ± 0.4 × 10⁵ N m⁻² at 10°C. At 3°C, Horowitz and Pollack (1993) measured tensions of ~3.8 ± (0.6 or 1.7) × 10⁵ N m⁻² (mean ± SE or SD; n = 7), giving ~4.4 ± 0.7 × 10⁵ N m⁻² at 10°C. Matsubara et al. (1985), working on a whole frog muscle at 0°C, measured a single active tension of ~2.4 × 10⁵ N m⁻², corresponding to ~3.0 × 10⁵ N m⁻² at 10°C. Yagi and Takemori (1995), also working on a whole muscle, at 4°C, found a single active tension of ~2.3 × 10⁵ N m⁻², giving ~2.6 × 10⁵ N m⁻² at

10°C. These two values are statistically similar to those obtained with intact fibres and are taken into account below, in this section.

Working on intact white fibres from toadfish, at 15°C, Rome et al. (1999) recorded tensions of ~2.3 ± 0.2 × 10⁵ N m⁻² (mean ± SE; n = 5 to 8), giving ~2.1 ± 0.2 × 10⁵ N cm⁻² at 10°C (assuming the same Q_{10} ~ 1.24 as for the frog; see the preceding paragraph). The authors obtained values of ~2.1 ± 0.2 × 10⁵ N m⁻² (mean ± SE; n = 5 to 8) for intact red toadfish fibres at 15°C, that is, ~1.9 ± 0.2 × 10⁵ N m⁻² at 10°C. They recalled that Johnston (1983) estimated that the mitochondria contribute up to ~20%–30% of the cross-sectional area of red muscle, whereas these organelles make a negligible contribution in white muscle. Thus, the value of ~1.9 ± 0.2 × 10⁵ N m⁻² for red intact fibres from toadfish corresponds to ~ (1.2–1.3) × (1.9 ± 0.2) × 10⁵ N m⁻² ~ 2.4 ± 0.3 × 10⁵ N m⁻² for white intact fibres. With intact white fibres from dogfish, at 12°C, West et al. (2004) obtained a value of ~2.8 ± (0.1 or 0.4) × 10⁵ N m⁻² (mean ± SE or SD; n = 17), that is, ~2.7 ± 0.1 × 10⁵ N m⁻² at 10°C. The three mean values given by Rome et al. (1999) and West et al. (2004) are statistically similar to those obtained for frog fibres and whole muscles (see above) and we can combine the 17 values (I take only the mean values into account) to obtain a universal average value of T_f^* ~ 3.0 ± (0.1 or 0.6) × 10⁵ N m⁻² (mean ± SE or SD; n = 17).

In Section 4.4.1, it is recalled that the isometric tetanic tension measured at 10°C on the three half-fibres used in the experimental part of this monograph, at physiological ionic strength (I = 180 mM; I* ~ 182 mM; see Section 3.9 for precise details about the vicinal ionic strength I*), in the presence of methanesulphonate (MS⁻) as the major anion, is ~1.2 ± 0.2 × 10⁵ N m⁻² (mean ± SD; n = 6). This value corresponds solely to the electrostatically related axial isometric tetanic tension, accounting for only ~40% of the full isometric tension (see below). The full isometric tension is therefore T_{hf}^* ~ (1/0.40) × (1.2 ± 0.2) ~ 2.5 × (1.2 ± 0.2) ~ 3.0 ± 0.5 × 10⁵ N m². Thus, as pointed out in Section 2.2 and demonstrated in Section 4.4.1, half-fibres in the presence of MS⁻ are extremely interesting, because they develop the same isometric tetanic tension as intact fibres (T_f^* ~ 3.0 ± 0.1 × 10⁵ N m² at 10°C; see the end of the preceding paragraph). This is largely attributed to the use of MS⁻. Indeed, studying

Morel

chemically skinned fibres from rabbit psoas muscle and comparing various anions, Andrews et al. (1991) demonstrated experimentally that 'MS⁻ is to be preferred because it minimizes swelling of the myofilament lattice' (and, therefore, no major increase in cross-sectional area and, for geometric reasons, no detectable change in tension, within the range of experimental error). The most accurate value for T_f^* and T_{hf}^* can therefore be taken as $\sim T_f \sim T_{hf} \sim 3.0 \times 10^5$ N m⁻². On p. **85** in Section 3.5, I select a proportion of ~40% of the full axial contractile tension related to the net radial repulsive (expansive) forces and ~60% related to the swinging cross-bridge/lever-arm processes, consistent with the analysis of Figure 4.2 presented in Section 4.2.2. The conversion of radial repulsive (expansive) forces into axial contractile forces was theoretically and logically predicted by Dragomir et al. (1976), Elliott (1974) and Morel et al. (1976) (see pp. **102–103** in Section 3.10) and unwittingly supported by the experimental results of April and Brandt (1973), Gordon et al. (1973) and Hasan and Unsworth (1985) (see pp. **109–110** in Section 4.1, for a circumstantial discussion) and Neering et al. (1991) (see pp. **4–5** in the Introduction). This conversion is definitively supported by the new experimental results described in this monograph and successfully used throughout this work. Moreover, on the basis of the detailed phenomenological/semi-empirical studies of Morel (1978, 1984a,b, 1990, 1991a,b), Morel and Gingold (1977), Morel and Merah (1995) and Morel et al. (1976), the full isometric tetanic tension in intact unit cells is considered to be much higher than that recorded in intact fibres (e.g. Morel 1991a). Combining experimental data from independent groups and the phenomenological/semi-empirical value, it is shown, on p. **90** in Section 3.7, that the mean value of the isometric tetanic tension is ~7.2 × 10⁵ N m⁻² at 10°C, at the slack length, in an intact unit cell from any animal species, frog, rabbit, crayfish, dogfish and so on (~2.4 times the mean tension of ~3.0 × 10⁵ N m⁻², in intact frog fibres or half-fibres). This apparent discrepancy between the values measured in intact fibres and obtained in intact unit cells is discussed in depth in Section 3.7. On pp. **97–98** in Section 3.8.1, the 'universal value' $T_f \sim T_{hf} \sim 3.0 \times 10^5$ N m⁻² at 10°C is used for intact white fibres of any animal species, for comparison with the average tensions recorded on mechanically and chemically skinned fibres and permeabilised fibres.

3.4 PHENOMENOLOGICAL/SEMI-EMPIRICAL APPROACH TO MICROENERGETICS AND MICROMECHANICS IN AN INTACT UNIT CELL

3.4.1 Nature of the Problems to Be Resolved

The concept of an intact unit cell is used throughout this monograph (see Section 3.1 for precise references). Distinctions are made between intact unit cells lying in the centre of a myofibril, itself located in the centre of an intact fibre, and 'peripheral unit cells' lying in myofibrils located at the periphery of an intact fibre. Given the rather restrictive notion of intact unit cells, it is necessary to distinguish between these 'microscopic' structures, to which most of the calculations and discussions presented in this book apply, and 'macroscopic' structures (intact and demembranated fibres, isolated myofibrils), which are more complex owing, for example, to the presence of additional noncontractile structures in intact fibres (e.g. Millman 1998, for a review). The mechanical behaviour of microscopic intact unit cells, peripheral unit cells and macroscopic intact fibres are compared and discussed extensively in Section 3.7. In Sections 3.8, 4.4.2.1 and 8.7, it is shown that demembranated fibres are extremely complex biological materials, from a structural viewpoint. The phenomenological/semi-empirical approach used in this monograph is valid only for intact unit cells belonging to intact fibres, but many of the experimental data used in most sections concern single intact or demembranated fibres. This combination is unavoidable, because MgATPase activities, for instance, can be directly measured solely on demembranated fibres or isolated myofibrils, whereas isometric tetanic tensions are principally measured in intact fibres, but also, for the last 15–20 years, in demembranated fibres and isolated myofibrils. Obviously, enzymological and mechanical data cannot be obtained directly with microscopic intact unit cells. Nonetheless, it is demonstrated, in all the sections concerned with experimental, phenomenological/semi-empirical and semi-empirical aspects, that the approach proposed throughout this work is self-consistent.

As pointed out on pp. **63–64** in Section 3.4.3.2 and broadly discussed in Section 3.7, there is a 'gap' between the isometric tension in macroscopic structures (intact fibres) and that in microscopic structures (intact unit cells). Thus, all the other features established for intact fibres should be reevaluated in intact unit cells. For this purpose, I use a new approach, deduced from previous studies and experimental data obtained more recently than those employed between 1976 and 1995 by my group. For example, He et al. (1997, 1998a,b, 1999), working on permeabilised frog and rabbit fibres, used powerful techniques to measure Pi release and to deduce the rate of MgATP hydrolysis in these fibres (ATP-caged). They interpreted their 1997 results using the notion of 'turnover', implying a kind of discontinuous dependence of MgATPase activity on time. However, I consider that the experiments performed by He et al. (1997) for short tetani (see Section 3.4.3.2 for essential details concerning this delicate question of short tetani) unquestionably show that the various rates of MgATP splitting are continuous functions of time t (see, e.g. Figures 9 and 10 in their paper), as theoretically supported by Equations 3.10 and 3.11. These authors simultaneously studied the development of tension (rise and stabilisation) and the rate of Pi release over time t and found a rather loose correlation between these two characteristics. After comparable lags in tension rise and Pi release (see Table 1 and Figures 9 and 10 in the paper by He et al. 1997), both tension and Pi release (i.e. MgATPase activity) increase continuously until stable values are reached, probably corresponding to steady states (see, again, Section 3.4.3.2 for essential details concerning this complex problem of steady states).

The situation presented in the preceding paragraph raises many questions concerning, for example, the true rate of MgATP cleavage to be taken into account when calculating isometric tetanic tension, in short tetani under steady-state conditions, and also the possible connection between the development of tension (rise and stabilisation) and variations in the rate of MgATP hydrolysis (transitory and steady) described by He et al. (1997). For this purpose, I present, in Sections 3.4.2 and 3.4.3, a new phenomenological/semi-empirical approach, independent of any mechanical role of the cross-bridges (side-pieces), deduced from the pioneering theoretical and biophysical work of A.F. Huxley (1957). However, I use a major concealed feature of A.F. Huxley's fundamental equation expressing the rate of energy output that A.F. Huxley (1957) unfortunately did not note. The corresponding reasoning

and expression were extensively and successfully used by my group, over approximately 20 years (beginning approximately 40 years ago; see Morel et al. 1976), and are considerably improved in this monograph. The conclusions that can be drawn from this new approach include a 'phenomenological/semi-empirical' expression of the isometric tetanic tension for an intact unit cell, valid for short tetani under steady-state conditions (see Section 3.4.3). It is also definitively demonstrated that both the rate of MgATP splitting and the isometric tension are continuous functions of time t, under transitory conditions, as experimentally shown in Figures 9 and 10 in the paper by He et al. (1997; see the preceding paragraph). I demonstrate, in this monograph, several unexpected properties of contracting (mostly isometrically) intact unit cells and analyse critically some of the major features of the cross-bridge cycle.

3.4.2 Generalisation of the 'Energetic Part' of A.F. Huxley's Two-State Theory of Side-Pieces (Cross-Bridges)

In my group, we demonstrated time and again that the simple two-state side-piece model (applicable to cycling cross-bridges), introduced by A.F. Huxley (1957), can account semi-empirically for many experimental properties of muscle fibres, independently of any mechanical role for cross-bridges (e.g. Sections 8.9 and 8.11 and general conclusion). For our purpose, we took into account only the fundamental equation, as determined by A.F. Huxley (1957), giving the rate of energy output as a function of the shortening velocity of a striated muscle. From this equation, an 'energetic' expression of the contractile force was deduced, valid also under isometric conditions. The 'mechanical' expression of the contractile force proposed by A.F. Huxley (1957) is never taken into account here, as in all the previous publications from my group (see also Sections 8.9 and 8.11 and Chapter 10 on this point, particularly on the lack of consistency between the energetic and mechanical expressions). A slightly modified 'mechanical model' was suggested by A.F. Huxley (1973a), in which cross-bridge attachment takes place in two stages. With the choice of suitable parameters, the additional assumption led to a rate of energy release consistent with that deduced from the experimental work of A.V. Hill (1964a,b), this rate passing through a maximum and declining as the maximal shortening velocity is approached, in the case of whole frog muscles. Barclay (1999) tried to resolve the same problem, but A.F. Huxley (2000) claims that both explanations 'remain speculative'. I do not take into consideration the modifications suggested by A.F. Huxley (1973a) and Barclay (1999), because they were proposed mostly to account for the experimental data of A.V. Hill (1964a,b) obtained on a whole muscle, a biological material very different from an intact unit cell, as demonstrated at many places in this monograph (e.g. Section 8.11). Borejdo (1980), Eisenberg et al. (1980), Hill (1977), Nielsen (2002) and Zahalak (2000) unwittingly tried to mask inconsistencies in the biophysical/theoretical approach to the swinging cross-bridge/lever-arm theories, by suggesting the introduction of multistate models assumed to account better for contraction phenomena at the millisecond timescale. Studying small length perturbations (in particular the C-process), Palmer et al. (2007) rehabilitated the two-state model and this view is supported in this book.

As concerns steady-state 'isometric conditions' (lasting, at most, around 1 s), Morel (1984a) studied the problem of multistate models presented by Eisenberg et al. (1980) and Hill (1977). Let us apply Equations 57 and 58 in the paper by Morel (1984a) and introduce some numerical values given by Eisenberg et al. (1980) and the most recent values of the other parameters used in this monograph. With some mathematical manipulations, the isometric tetanic tension, in intact unit cells (see definition of intact unit cells in Section 3.1 and comments in Section 3.4.1), deduced from the 'energetic part' of A.F. Huxley's (1957) fundamental equation concerning the rate of energy output, but employed in the multistate models, would be $\sim 1.2 \times 10^5$ N m^{-2} at 10°C (the reference temperature used in this book). This value is much lower than the most probable value obtained from various experimental studies and the phenomenological/semi-empirical approach presented in this monograph and taken into account here (T* $\sim 7.2 \times 10^5$ N m^{-2} at 10°C; see p. **90** in Section 3.7). In their quantitative analysis, Eisenberg et al. (1980) claimed that they did not wish to obtain a fit to experimental data but merely to develop the best formalism. At the time, I was disappointed by this assertion, which I found to be rather 'unconventional'. The use of this multistate model under steady-state conditions is not necessarily the best

solution, as highlighted in the preceding paragraph, and may even conceal more fundamental weak points. Furthermore, the two-state model can be employed, even on the millisecond timescale, as demonstrated in several sections of this work. At this point, I maintain my position of taking into account only the simple two-state model, because this leads to self-consistent results and conclusions, as has been the case over approximately 15 years in my group.

It is useful to recall Huxley's first fundamental equation (1957):

$$dn/dx = \partial n/\partial t - (s_0 V/2)\partial n/\partial x = f - (f + g)n \quad (3.3)$$

$n(x,t)$ is the mean proportion (≤ 1) of cross-bridges attached to actin in the thin filaments, with a linear distortion x, at time t. V is the shortening velocity expressed in muscle length s^{-1}. s_0 is a reference sarcomere length, corresponding generally to the slack length of the fibre (see p. **23** in Section 3.2 for precise details on s_0). The factor 1/2 indicates that only a half-sarcomere is taken into account and $s_0 V/2$ is the shortening velocity, expressed, for instance, in µm s^{-1}. This notion of a half-sarcomere is intimately related to the side-piece (cross-bridge) theory, but I maintain it in the model-independent approach presented in this book, to ensure homogeneous reasoning (the notion of half-sarcomere is still used in recent publications). A.F. Huxley (1957) did not clearly define the reference proportion corresponding to the maximal value 1 of $n(x,t)$. However, this problem was later resolved and there is now a consensus that the maximal value of 1 (100%) corresponds to the attachment of all cross-bridges to thin actin filaments in conditions of rigor (see p. **148** in Section 4.4.2.5.2). As pointed out by A.F. Huxley (1957), at time t, $n(x,t)$ is a continuous function of x, provided that a very large number of identical intact unit cells are taken into account. f is the mean constant of attachment of a cross-bridge to a thin actin filament (usually expressed in s^{-1}) and g is the mean constant of detachment from a thin actin filament (also usually expressed in s^{-1}). At the molecular level, the cross-bridges clearly do not all behave similarly and the introduction of this notion of a 'mean constant' is required. Indeed individual actin-binding sites on the thin actin filaments do not exactly face the closest cycling cross-bridge, even at s_0 (e.g. Bachouchi and Morel 1989a), resulting in unavoidable distortions/disruptions (in both the axial and radial directions) of the cross-bridges, as a function of their positions with respect to the corresponding actin-binding sites. Thus, all the kinetic characteristics, including f and g, are different at the molecular level, but, as soon as a very large number of cross-bridges is taken into account, the notion of mean macroscopic values is meaningful. This kind of reasoning is systematic for most of the other parameters, throughout this monograph. After writing this section, I read the minireview by Borejdo et al. (2006), in which the notion of average 'over an entire assembly of observed cross-bridges' is demonstrated. The authors also wrote: 'each cross-bridge has different kinetics, depending on its position relative to the actin binding site'. Thus, the new and older concepts are consistent. In any event, the mean constants f and g depend, a priori, on x and t. For the sake of simplicity, I retain the implicit hypothesis of A.F. Huxley (1957), according to which f and g are independent of the shortening velocity s_0V, because this assumption results in consistent conclusions, within the range of experimental uncertainties. However, on pp. **263–264** in Section 8.11, it is recalled that f and g may depend on s_0V, but strongly suggested that it is not necessary to take this into account. Assuming values of f and g independent of s_0V, the phenomenological/semi-empirical approach predicts the most probable P–V relationship valid for single fibres quite well (reversal of curvature; see again pp. **263–264** in Section 8.11 for a critical analysis of the problem). s_0V is the mean shortening velocity, corresponding to a very large number of identical intact unit cells (see below, in this section). Huxley (1957) unwittingly assumed that s_0V was independent of time t, which is an oversimplification. Indeed, on pp. **61–62** in Section 3.4.3.2, I present a detailed discussion of the dependence of s_0V and s_0V_{max} (corresponding to an unloaded fibre) on time t. On pp. **61–62** in Section 3.4.3.2, I define the notion of a 'pseudo-steady-state isotonic contraction' in which s_0V and s_0V_{max} are independent of time t. In the rest of this monograph, s_0V and s_0V_{max} are considered to be strictly independent of time t. Otherwise, in a brief discussion presented on p. **76** in Section 3.4.5, it is shown that s_0V_{max} (and also certainly s_0V) depends on the position of the unit cell in the fibre. However, it is assumed that ignoring this problem causes no inconsistencies. Thus, the shortening velocities can be considered to be identical in intact fibres and intact unit cells. Finally, the sarcomere length s_0 corresponds,

in all phenomenological/semi-empirical studies presented in this work, to maximal overlap and the resulting maximal isometric tetanic tension (e.g. Gordon et al. 1966b).

At the slack length, the reference universal sarcomere length is $s_0 \sim 2.18$ μm (see p. **23** in Section 3.2). Equation 3.3 is valid solely within the interval of attachment of the cross-bridges $(-h_1, +h_2)$ (see Morel 1984a and Morel et al. 1976 for precise details on this interval, differing from the interval $(0, +h)$ defined by Huxley 1957; our more general definition does not induce any mathematical and biophysical problem; see below, in this section). Outside this active interval, the cross-bridges cannot attach and we have f = 0, but, below $-h_1$, we have 'g' = g*, where g* has a very high value, assumed to be independent of the linear distortion x, as in the model of A.F. Huxley (1957) (g* ≫ g; g is the constant of detachment within the interval of attachment), ensuring rapid dissociation of the cross-bridges from the thin actin filaments (see the next paragraph). As the thin actin filaments are polarised (e.g. arrowhead structure in rigor), no cross-bridges can attach beyond $+h_2$ and we necessarily have $n(+h_2, t) = 0$, regardless of t. This conclusion was also drawn by A.F. Huxley (1957), starting from the same kind of reasoning regarding the polarity of the thin actin filaments, but using a model, in which it was assumed that $h_1 = 0$ (see above). Considering a value of h_1 different from zero is a nonrestricting improvement and also a generalisation of the approach of Huxley (1957), as clearly demonstrated below, in this section. It should be recalled that the use of h_1 has been successful in all the previous publications of my group concerned with phenomenological/semi-empirical approach to many features of contracting fibres (Morel 1978, 1984a,b, 1990, 1991a,b; Morel and Gingold 1977; Morel and Merah 1995; Morel et al. 1976).

In the general case, when s_0V is different from zero (isotonic conditions), that is, when there is active shortening, with a relative sliding of the thick myosin filaments along the thin actin filaments, many cross-bridges are carried away by this sliding, outside the normal interval of attachment cited in the preceding paragraph $(-h_1; +h_2)$. In these conditions, $n'(x,t)$ is the mean proportion of cross-bridges that remain attached below $-h_1$. $n'(x,t)$ can be deduced from Equation 3.3, taking f = 0 (see the preceding paragraph):

Morel

$$\partial n'/\partial t - (s_0 V/2)\partial n'/\partial x = -g^* n' \qquad (3.4)$$

$n'(x,t)$ was considered by A.F. Huxley (1957) to be the fraction of 'resisting' cross-bridges, exerting a negative force opposing the 'driving' force generated by the fraction $n(x,t)$ of active cross-bridges in the interval of attachment $(0, +h_2)$, or, more generally, $(-h_1; +h_2;$ with $h_1 < h_2$; see, e.g. Morel 1984a; Morel et al. 1976). Under the isotonic conditions studied here, according to A.F. Huxley (1957), the net sliding force is the difference between the force exerted by the active cross-bridges and that exerted by the resisting cross-bridges (several problems resulting from this oversimplification are studied in Section 8.11), accounting therefore for a 'pseudo-steady-state isotonic' shortening velocity, $s_0 V$, independent of time t, being reached (the delicate notion of pseudo-steady-state isotonic contraction is discussed in the text below Equation 3.33).

According to A.F. Huxley (1957), the mean turnover rate (usually expressed in s^{-1}) of cross-bridges interacting with the thin actin filaments and resulting in MgATP splitting is given by

$$R(t, s_0 V) = (1/\sigma) \int_{-h_1}^{+h_2} f(1-n)\,dx \qquad (3.5)$$

The turnover rate is different from the rate of MgATP splitting. This difference was not taken into account by A.F. Huxley (1957) and most specialists. This point is commented on and accounted for in this book (see pp. **42–43** in Section 3.4.3.1.1 and pp. **73–74** in Section 3.4.4). σ is the distance between two consecutive sites for the binding of a cross-bridge on the neighbouring thin actin filament. A second improvement to Huxley's biophysical approach (1957) is achieved by considering both f and g to be independent of the distortion x, over the interval of attachment $\Delta h = h_1 + h_2$. This simplifying assumption has proved to be very useful and to give self-consistent conclusions since 1976 (Morel et al. 1976). However, in the most general case, for example, transitory phenomena, f and g depend a priori on time t. The hypothesis that f and g do not depend on x provides satisfactory explanations for many features of contracting fibres and intact unit cells, as recalled and demonstrated at many places in this monograph. Moreover, in the model

of Huxley (1957), $f(x)$ and $g(x)$ are assumed to be zero at $x = 0$, which is conceptually difficult to explain. Combining Equations 3.3 and 3.5, taking into account the assumption that f and g are independent of the linear distortion x and after some mathematical manipulations, we can write $R(t, s_0 V)$ as follows:

$$R(t, s_0 V) = (1/\sigma)(g\nu + \partial n/\partial t) + \nu^*(1/2\sigma)s_0 V \qquad (3.6)$$

with the following definitions of ν and ν^*:

$$\nu = \int_{-h_1}^{+h_2} n\,dx \quad \text{and} \quad \nu^* = \int_{-h_1}^{+h_2} (-\partial n/\partial x)\,dx. \qquad (3.7)$$

Equation 3.6 can be written solely because g is independent of x (see above). The two parameters ν and ν^* depend a priori on time t and the shortening velocity $s_0 V$, and Equation 3.6 is a purely formal definition. For $s_0 V = 0$ (isometric conditions), $R(t, s_0 V)$, given by Equation 3.6, reduces to $R(t, 0) = R_{isom}(t)$, which can be written:

$$R_{isom}(t) = (1/\sigma)(g\nu_0 + \partial n_0/\partial t) \qquad (3.8)$$

ν_0 and n_0 are the values of ν and n when $s_0 V = 0$. Thus, the general expression of $R(t, s_0 V)$, given by Equation 3.6, can also be written:

$$R(t, s_0 V) = R_{isom}(t) + (1/\sigma)[g(\nu - \nu_0) + \partial(n - n_0)/\partial t] + \nu^*(1/2\sigma)s_0 V \qquad (3.9)$$

$R_{isom}(t)$, defined by Equation 3.8, is clearly a continuous function of time t. He et al. (1997) suggested that, for isometrically held permeabilised fibres ($s_0 V = 0$), the rate of MgATP splitting, which is proportional to $R_{isom}(t)$, is approximately biphasic, even triphasic (see p. **50** in Section 3.4.3.2, concerning the notion of successive turnovers; see also Equation 3.42, giving the rate of MgATP breakdown, which can be reduced to the turnover rate of the cross-bridges if $\omega_{isom,\infty} = 1$). The authors attributed this 'polyphasic' behaviour to a first, a second and a third turnover of the cycling myosin heads. In the light of Equations 3.8 and 3.9, this oversimplification appears to have no theoretical basis and the interpretation suggested by He et al. (1997) may be the result of unavoidable experimental uncertainty. Indeed, from Table 1 in their paper, for rabbit at 20°C, and for the second turnover, for example, the SE is ~4.2 s^{-1} (n = 22), which is only ~12% of the mean value of ~35.5 s^{-1}, whereas the SD is ~$22^{1/2} \times 4.2$ s^{-1} ~ 19.7 s^{-1}, which is ~55% of the mean value. In Section 2.1, the relationship between SE and SD is given and it is emphasised that SD is more suitable than SE, particularly when the number n of experimental values is high. At this juncture, in the light of the high value of SD (~55%), it should be stressed that breaking up the MgATPase activities in several turnovers is overly simplified (see also p. **50** in Section 3.4.3.2). In any event, Equations 3.8 and 3.9 are valid under all conditions (transitory states, short tetani under steady-state conditions, pseudo-long-lasting tetani, long-lasting and very long-lasting tetani; all these notions are defined principally in Section 3.4.3.2).

3.4.3 Phenomenological/Semiempirical Approach to MgATPase Activity, Isometric Tension and Characteristics of the Cross-Bridge Cycle, in an Intact Unit Cell

3.4.3.1 MgATPase Activity, Isometric Tension and Related Features, Under Transitory and Steady-State Conditions

3.4.3.1.1 RATE OF MgATP SPLITTING AND ENTHALPY RELEASE AS FUNCTIONS OF SHORTENING VELOCITY, IN AN INTACT UNIT CELL

Equations 3.8 and 3.9 give the turnover rate $R(t,s_0V)$ of the cycling cross-bridges in an intact unit cell, at time t and at any value of the shortening velocity, s_0V, but not the rate $\Pi(t,s_0V)$ of MgATP splitting. Indeed, Morel (1984a,b) claimed that there is no straightforward relationship between the rate $\Pi^{**}(s_0V)$ of MgATP cleavage measured in a whole frog muscle and the turnover rate of the cross-bridges, in an intact unit cell (the superscript ** indicates that the experiments were performed on whole frog muscles; see the next paragraph).

A.F. Huxley (1957) demonstrated that the turnover rate of the side-pieces (cross-bridges) is an increasing function of the shortening velocity. This can also be deduced from Equation 3.44, if $\omega(t,s_0V)$, defined below, in this section, is taken as one (numerical calculations not shown). Morel (1984a,b) stated that $\Pi^{**}(s_0V) = \omega^{**}(s_0V)$ $R(s_0V)$ (I did not distinguish between intact unit cells, intact fibres and whole muscles and did not consider the dependence of $R(s_0V)$ on t; see also

Huxley 1957). The introduction of the parameter $0 \leq \omega^{**}(s_0V) \leq 1$, decreasing with increasing s_0V (Morel 1984a,b), relates to experimental findings obtained more than 30–40 years ago by several independent groups and showing that $\Pi^{**}(s_0V_{\max}) \sim \Pi^{**}(0) = \Pi^{**}_{\text{isom}}$ (i.e. $\Pi^{**}(s_0V_{\max})/\Pi^{**}_{\text{isom}} \sim 1$) within the broad limits of experimental error (see Morel 1984b for a critical analysis of the chemical techniques used). Indeed, it was found that, in whole frog muscles, high-energy phosphate metabolism was similar for isometrically held muscles ($s_0V = 0$) and unloaded muscles shortening at $s_0V_{\max}$ (e.g. Curtin and Woledge 1978; Homsher et al. 1981; Kushmerick and Davies 1969; Kushmerick et al. 1969; Rall et al. 1976; see also Woledge et al. 1985 for a review). More recently, working on intact frog fibres, Piazzesi et al. (1999) proposed a very different explanation for the experimental findings concerning metabolism at $s_0V = 0$ and $s_0V = s_0V_{\max}$, but their discussion was based exclusively on swinging cross-bridge/lever-arm theories. Barclay (1999) and Brokaw (1995) proposed other semi-empirical approaches, valid for amphibian and mammalian skeletal muscles, based on a 'weakly coupled version of the Huxley crossbridge model', assuming rapid detachment of the cross-bridges, without MgATP hydrolysis, resembling approximately the previous hypothesis of Morel (1984a,b) of inactive contacts (see also p. **43** for a brief discussion). However, the approaches of Barclay (1999), Brokaw (1995) and Piazzesi et al. (1999) are based on many questionable choices of various parameters and are highly model dependent and therefore subjective in nature. By contrast, the simple hypothesis, proposed by Morel (1984a,b), regarding $\omega^{**}(s_0V)$, is totally model independent and therefore objective in nature. Thus, I maintain this 'model-independent' assumption in this book.

He et al. (1999), Potma and Stienen (1996), Potma et al. (1994) and Sun et al. (2001) worked on permeabilised or chemically skinned fibres from rabbit psoas muscle and systematically found $\Pi^{*}(s_0V_{\max})$ to be higher than $\Pi^{*}(0) = \Pi^{*}_{\text{isom}}$ (the superscript * indicates that the experiments were performed on permeabilised or chemically skinned rabbit fibres, rather than the whole frog muscles studied in the 1970s and 1980s; see above, in this section). As illustrative examples, He et al. (1999), working on permeabilised fibres from rabbit psoas muscle, found $\Pi^{*}(s_0V_{\max}) \sim (1.5 - 3.0)\Pi^{*}_{\text{isom}}$, Potma and

Stienen (1996) and Potma et al. (1994), working on chemically skinned rabbit fibres, found $\Pi*(s_0 V_{max}) \sim (2.2 - 2.7)\Pi^*_{isom}$, whereas Sun et al. (2001), working on the same biological material, found $\Pi*(s_0 V_{max}) \sim 9.0 \Pi^*_{isom}$. Finally, Barclay (1999) calculated semi-empirically that the ratio is ~5.2 for a single intact frog fibre and ~3.8 for bundles of EDL muscle fibres from mouse. The situation is therefore confusing and these experimental results may be related to at least four factors: (i) the older results were obtained with frog, whereas the more recent data were obtained with rabbit; (ii) the older results were obtained with whole muscles, using debatable chemical techniques (see Morel 1984b for a critical analysis), whereas the more recent values were obtained on chemically skinned or permeabilised fibres, using new techniques (e.g. caged-ATP and laser flash); (iii) the inconsistencies in the experimental results, obtained with permeabilised or chemically skinned fibres from rabbit psoas muscle, would be accounted for by differences in the experimental techniques used for permeabilisation and chemical skinning between groups, as pointed out by Millman (1998) and Sun et al. (2001), who claimed that the various techniques for removing the sarcolemma can result in significant variability in both the qualitative and quantitative behaviours of the fibres; (iv) even in recent studies (see above), it is difficult to define clearly the MgATPase activities corresponding to imperative conditions for short tetani and steady state (see Section 3.4.3.2 regarding the complexity involved in defining these two notions). At this juncture, it is interesting to keep the parameter $\omega**(s_0 V)$ decreasing with increasing $s_0 V$ (see the preceding paragraph). For the sake of simplicity, a single notation $\omega(t, s_0 V)$ is employed, regardless of the animal species (e.g. frog or rabbit), the biological material (whole muscle; intact, skinned, permeabilised fibres; isolated myofibrils; intact unit cells) and the conditions of contraction (transitory or steady state). On pp. **73–74** in Section 3.4.4, this question is considered and arguments are presented, confirming that it is essential to introduce $\omega(s_0 V)$, ensuring both a simple and model-independent approach to active shortening.

By analogy with the semi-empirical relationship $\Pi**(s_0 V) = \omega**(s_0 V) R(s_0 V)$ (see p. **42**), we can write $\Pi(t, s_0 V) = \omega(t, s_0 V) R(t, s_0 V)$ in an intact unit cell, where $R(t, s_0 V)$ is defined by Equations 3.8 and 3.9. The fraction $0 \leq \omega(t, s_0 V) \leq 1$ corresponds to 'active contacts', that is, to cycling cross-bridges

that lead to the hydrolysis of one MgATP molecule per attachment–detachment cycle. By contrast, $[1 - \omega(t, s_0 V)]$ corresponds to 'inactive contacts', that is, to normally cycling cross-bridges, not leading to any MgATP breakdown (Morel 1984a,b; this concept of 'inactive contacts' replaces that of 'accidents' previously and improperly used). By analogy with whole muscles, the proportion of inactive contacts in an intact unit cell is assumed to increase with increasing shortening velocity $s_0 V$; that is, $\omega(t, s_0 V)$ decreases when the myofilaments slide increasingly rapidly past each other. Indeed, upon sliding, many cross-bridges that can attach to thin actin filaments, with difficulty, are unable to complete their full enzymatic cycle, because the duration of attachment is too short. This logical conclusion is confirmed in Section 3.4.4, in which this notion of active and inactive contacts is studied critically. Thus, the rate of MgATP splitting is deduced from Equation 3.9 and is given by

$$\Pi(t, s_0 V) = \omega(t, s_0 V)\Pi_{isom}(t) + \omega(t, s_0 V)(1/\sigma)[g(\nu - \nu_0) + \partial(n - n_0)/\partial t] + \omega(t, s_0 V)\nu*(1/2\sigma)s_0 V \quad (3.10)$$

with (see Equation 3.8)

$$\Pi_{isom}(t) = \omega_{isom}(t) R_{isom}(t) = (1/\sigma)\omega_{isom}(t)(g\nu_0 + \partial n_0/\partial t) \quad (3.11)$$

where $\omega_{isom}(t) = \omega(t, s_0 V = 0)$. Let us call m the total number of cross-bridges (attached + detached, able to enter the attachment–detachment cycle) per unit volume and e_{app} the apparent chemical energy released from the cleavage of one MgATP molecule minus the chemical energy absorbed by Lohmann's reaction (instantaneous rephosphorylation of MgADP to give MgATP; see, for instance, Alberts et al. 2007; Lehninger 2008; Morel and Pinset-Härström 1975a). Introducing $\Pi(t, s_0 V)$, instead of $R(t, s_0 V)$, into the definition given by A.F. Huxley (1957), the rate of energy output per unit volume and per second, in an intact unit cell, resulting from cyclic actin–myosin interactions and the subsequent instantaneous Lohmann reaction is $E^{AM}(t, s_0 V) = me_{app}\Pi(t, s_0 V)$. At this point, let us apply the term enthalpy to any form of chemical energy, for simplification, although this is clearly not the strict meaning of this term. This assimilation is justified for two main reasons: (i) the caloric properties of a muscle are similar to those of water (Woledge et al. 1985) and (ii) there is a wide scattering of the various experimental values of the enthalpy. Thus, the rate of enthalpy output, related solely to actin–myosin interactions, is $\Delta H^{AM}(t,s_0V) \sim \dot{E}^{AM}(t,s_0V) \sim me_{app}\Pi(t,s_0V)$. Under isometric conditions ($s_0V = 0$), the rate of total enthalpy output $\Delta H_{isom}(t)$, in an intact unit cell, results from the cyclic actin–myosin interactions, Lohmann's reaction, and other ATP-consuming structures (e.g. sarcoplasmic reticulum, Na$^+$–K$^+$ pump in the sarcolemma of fibres) and other structures that release enthalpy upon activation of the fibres (e.g. binding of Ca^{2+} to troponin C, in the thin actin filaments, and to calmodulin and parvalbumins). Ca^{2+} binding to troponin occurs in intact unit cells, whereas the ATP-consuming structures are not located in intact unit cells, but the corresponding enthalpies almost certainly 'diffuse', at least partly, throughout the fibres, particularly in the intact unit cells. Calling the additional rate of enthalpy output $\Delta H_{isom}^{\circ}(t)$, we can write $\Delta H_{isom}(t) = \Delta H_{isom}^{AM}(t) + \Delta H_{isom}^{\circ}(t) = [1 + \Delta H_{isom}^{\circ}(t)/\Delta H_{isom}^{AM}(t)]\Delta H_{isom}^{AM}(t)$. $\xi_0(t)$ is a corrective factor, studied and discussed on pp. **197–198** in Section 6.2 (owing to the wide range of error on the estimate of $\xi_{0,\infty}$, under steady-state conditions, it is assumed that $\xi_0(t) \sim \xi_{0,\infty}$; see p. **198** in Section 6.2). More generally, by analogy with these definitions and using the general expression of $\Pi(t,s_0V)$, given by Equation 3.10, we can write the rate of total enthalpy output in an intact unit cell as follows:

$$\Delta H(t,s_0V) = \xi(t,s_0V)me_{app}\Pi(t,s_0V)$$
$$= \Delta H_{isom}(t,s_0V) + \xi(t,s_0V)me_{app}\omega(t,s_0V)$$
$$(1/\sigma)\left[g(\nu - n_0) + \partial(n - n_0)/\partial t\right]$$
$$+ \xi(t,s_0V)me_{app}\omega(t,s_0V)\nu^*(1/2\sigma)s_0V$$

$$(3.12)$$

with $\Delta H_{isom}(t,s_0V) = \xi(t,s_0V)me_{app}\omega(t,s_0V)\Pi_{isom}(t)$, where $\xi(t,s_0V)$ is a 'generalised' corrective factor, which becomes $\xi_0(t)$ under isometric conditions ($s_0V = 0$). Equation 3.12 is purely formal, because $\Delta H_{isom}(t,s_0V)$ depends on s_0V, through both $\xi(t,s_0V)$ and $\omega(t,s_0V)$.

3.4.3.1.2 OLD EXPERIMENTAL RESULTS CONCERNING WHOLE FROG MUSCLES. FENN EFFECT AND HILL'S QUANTITATIVE APPROACH TO THE PHENOMENON

Many experimental results for energy balance have been published, mostly for whole frog muscles at ~0°C–5°C, at the beginning of the 20th century. From the experiments of Fenn (1923, 1924), Fenn and Marsh (1935) and his own results, A.V. Hill (1938), using the notations proposed in this book, but with the same superscript ** as introduced on p. **42** in Section 3.4.3.1.1, to recall that we are dealing with old experimental studies on whole frog muscles, we can write:

$$\Delta H**(V**) \sim \Delta H_{isom}^{**} + (P* + a)V** \qquad (3.13)$$

ΔH_{isom}^{**} and $\Delta H**(V**)$ are the rates of enthalpy output per unit volume, and s^{-1} (e.g. in mJ s^{-1} per gram, assuming that the density of muscle is close to 1; see p. **209** in Section 6.3.5 for justification) for a contracting whole frog muscle. In the experiments of Fenn (1923, 1924), Fenn and Marsh (1935) and A.V. Hill (1938), both transitory (maybe ~5%–10% of the total duration of the tetanus) and steady-state conditions contribute to the 'enthalpy balance' given by Equation 3.13. ΔH_{isom}^{**} corresponds to an isometric tetanus ($V** = 0$). $P*$ is the load lifted per unit volume, in other words, a force per unit volume (for instance in g per unit volume), and $V**$ is the shortening velocity (for instance in mm s^{-1}). $P*V**$ is the mechanical work, in mJ s^{-1} per unit volume, and $aV**$ the 'shortening heat rate', also in mJ s^{-1} per unit volume, first discovered and described by Fenn (1923, 1924) and Fenn and Marsh (1935) and carefully quantified by A.V. Hill (1938a). It was shown experimentally that $a/P_0^* \sim 0.25$, where P_0^* is the load corresponding to the isometric tetanic tension, when $V** = 0$ (Hill 1938a). Hill (1964a,b) later found that a/P_0^* is not constant, but instead depends on the load $P*$ lifted under isotonic conditions ($V**$ different from zero), giving $a/P_0^* \sim 0.16 + 0.18(P*/P_0^*)$. A.V. Hill (1970) commented extensively on and analysed these old experimental results. Worthington and Elliott (1996a,b) discussed the experimental work of Fenn (1923, 1924), Fenn and Marsh (1935) and A.V. Hill (1938, 1964a,b) and gave their own semi-empirical approach to the rate of heat + work production. In this context, Levy et al. (2005) identified the Fenn effect as 'one of the most intriguing features of the skeletal and cardiac muscles' and recalled the discussion presented by Rall (1982), who suggested the existence of feedback control mechanisms. Levy et al. (2005) concluded that the processes underlying the Fenn effect remain unclear. This persistent view is groundless. Indeed, over a period of approximately 20 years, as again explained in this book, my group has shown that the Fenn effect is a straightforward consequence of cyclic attachment to and detachment from thin actin filaments of the cross-bridges, with MgATP breakdown during the cycle, according to the energetic part of the biophysical approach of Huxley (1957). However, simple complementary concepts are introduced: see, for example, p. **43** in Section 3.4.3.1.1 for a short discussion of the active and inactive contacts, leading to an uncoupling of the turnover rate of the cross-bridges and the enzymatic cycle, under isotonic conditions, as demonstrated by Morel (1984a,b) and in this monograph.

Morel (1984a) found, by applying his simplified phenomenological/semi-empirical approach to single intact frog fibres, a coefficient of shortening heat independent of the load lifted, lower than the parameter 'a' defined by Hill (1938a; $a/P_0 \sim 0.25$) (called 1 by Morel 1984a and also here) and given by $1/P_0^* \sim 0.17$. It is recalled, in the preceding paragraph, that Hill (1964a,b) obtained coefficients of between ~0.16 and ~0.34 depending on the load lifted (i.e. on the shortening velocity). On p. **63** in Section 3.4.3.2, I also report values depending on the shortening velocity and ranging between ~0.29 and ~0.69. The values reported by A.V. Hill (1938a, 1964a,b) and Morel (1984a), obtained at ~0°C–5°C, relate to whole frog muscles and single intact frog fibres, respectively, whereas the values of ~0.29 and ~0.69 relate to intact unit cells, also at ~0°C–5°C. Throughout this monograph, especially in Section 3.7, it is demonstrated that a single intact fibre (and a fortiori a whole muscle) is not the sum of intact unit cells, almost certainly accounting for the greater values of ~0.29 and ~0.69 rather than ~0.16 and ~0.34. In any event, as pointed out in Section 3.4.3.1.1 and in the preceding paragraph, most of the energy released from whole muscles, single intact fibres or intact unit cells stems from actin–myosin interactions, directly through cross-bridge cycling (release of chemical energy from MgATP breakdown on the myosin heads, with production of MgADP) and indirectly (instantaneous absorption of chemical energy from Lohmann's reaction, leading to rephosphorylation of MgADP to give MgATP, on the M lines in the sarcomeres, where the specific enzyme, creatine kinase, is located; this location was demonstrated in the particular case of chicken skeletal muscle by Walliman et al. 1977). However, there are also other sources of energy, not directly or indirectly related to actin–myosin interactions, as recalled on p. **44** in Section 3.4.3.1.1. These additional sources of energy are not exactly the same in whole muscles, single intact fibres, mechanically or chemically skinned fibres, permeabilised fibres and intact unit cells (see pp. **197–198** in Section 6.2 for the choice of a 'universal corrective factor' taking into account the various features of all these biological materials). For a whole frog muscle, I define $\Delta H^{**AM}(V^{**})$ and $\Delta H_{isom}^{**\,AM}$ as the rates of total enthalpies released during isotonic and isometric contractions, respectively, corresponding to the enthalpies resulting from actin–myosin interactions. Thus, from the short discussion presented on p. **44** in Section 3.4.3.1.1 and in the preceding paragraph, it can be deduced that $\Delta H^{**}(V^{**}) = \Delta H^{**AM}(V^{**}) + \Delta^{**}(V^{**})$ and $\Delta H_{isom}^{**} = \Delta H_{isom}^{**\,AM} + \Delta_{isom}^{**}$, where $\Delta^{**}(V^{**})$ and Δ_{isom}^{**} correspond to the extra sources of enthalpy. We therefore have

$$\Delta H^{**}(V^{**})\Delta_{isom}^{**} - \Delta H_{isom}^{**} = \Delta H^{**AM}(V^{**}) - \Delta H_{isom}^{**\,AM} + \left[\Delta^{**}(V^{**}) - \Delta H_{isom}^{**}\right].$$ Thus, Equation 3.13 can also be written:

$$\Delta H^{**}(V^{**}) - \Delta H_{isom}^{**} \sim \Delta H^{**AM} - \Delta H_{isom}^{**\,AM}$$
$$+ \left[\Delta^{**}(V^{**}) - \Delta_{isom}^{**}\right] \sim (P^{*} + a)V^{**}$$

$$(3.14)$$

3.4.3.1.3 MICROENERGETICS AND MICRO-MECHANICS IN AN INTACT UNIT CELL. MgATPase ACTIVITY, ISOMETRIC TENSION AND RELATED FEATURES, UNDER TRANSITORY AND STEADY-STATE CONDITIONS

Equation 3.14 is strictly valid for whole frog muscles, but it is also almost certainly valid for whole rabbit muscles. Concerning an intact unit cell, Equation 3.12 can be written in a slightly different form:

$$\Delta H(t,s_0V) - \Delta H_{isom}(t,s_0V) = \xi(t,s_0V)me_{app}\omega(t,s_0V)$$
$$[g(\nu - \nu_0) + \partial(n - n_0)/\partial t] + \xi(t,s_0V)me_{app}\omega(t,s_0V)$$
$$\nu^{*}(1/2\sigma)s_0V \qquad (3.15)$$

In Equations 3.14 and 3.15, the dimensions of the enthalpies and the shortening velocity are similar. For example, s_0V and V^{**} can be expressed in $\mu m\ s^{-1}$ and $mm\ s^{-1}$, respectively (see p. **44** in Section 3.4.3.1.2). There is obviously an impressive analogy between these two equations, and Equation 3.15 can be seen as representing the expression of the 'apparent microscopic enthalpy balance' at time t (the notion 'apparent' is related to the fact that $\Delta H_{isom}(t,s_0V)$ depends, in the most general case, on the shortening velocity s_0V, as already highlighted in Section 3.4.3.1.1; see text below Equation 3.12). Comparing Equations 3.14 and 3.15, we can write:

$$\xi(t,s_0V)me_{app}\omega(t,s_0V)\nu^*(1/2\sigma)s_0 = T(t,s_0V) + l(t,s_0V) \quad (3.16)$$

$T(t,s_0V)$ is the 'apparent transitory microscopic tension' developed by an intact unit cell and $l(t,s_0V)$ is the 'apparent transitory microscopic coefficient of shortening heat', also in the same intact unit cell. Both $T(t,s_0V)$ and $l(t,s_0V)$ are expressed as forces per unit surface area. The term *apparent* reflects the formal nature of Equation 3.16, as pointed out repeatedly in the preceding sections. An interesting problem concerns the tension $T(t,s_0V)$ and the rate of MgATP splitting, $\Pi_{isom}(t)$, for any value of t, including both transitory and steady-state conditions, for short tetani (the complex notions of short tetani and steady-state conditions are studied in Section 3.4.3.2). Indeed, it would be interesting to try to relate the rate $\Pi_{isom}(t)$ to the tension developed $T(t,s_0V)$. From Equation 3.11, we deduce that

$$(1/\sigma) = \Pi_{isom}(t)/[\omega_{isom}(t)(g\nu_0 + \partial n_0/\partial t)] \quad (3.17)$$

Introducing Equation 3.17 into Equation 3.16, we obtain

$$T(t,s_0V) + l(t,s_0V) = \xi(t,s_0V)me_{app}\Pi_{isom}(t)$$
$$[\omega(t,s_0V)/\omega_{isom}(t)]$$
$$\nu^*(1/2)s_0/[g\nu_0 + \partial n_0/\partial t] \quad (3.18)$$

ν^*, $g\nu_0$ and $\partial n_0/\partial t$ depend on time t through $g(t)$ and $n(t,s_0V)$ (see definitions 3.7 and corresponding comments in Section 3.4.2). For an isometric contraction ($s_0V = 0$), we have $\omega(t,s_0V) = \omega_{isom}(t)$ and the proportion $n(x,t)$ of attached cross-bridges is called $n_0(x,t)$ and, from the fundamental Equation 3.3, we deduce that

$$\partial n_0/\partial t = f - (f + g)n_0 \quad (3.19)$$

We deduce

$$g\nu_0 + \partial n_0/\partial t = g\nu_0 + f - (f + g)n_0 \quad (3.20)$$

Combining Equations 3.18 and 3.20 (with $s_0V = 0$) gives

$$T_0(t) + l_0(t) = \xi_0(t)me_{app}\Pi_{ison}(t)$$
$$\nu_0^*(1/2)s_0/[g\nu_0 + f - f(1 - n_0)] \quad (3.21)$$

$T_0(t)$ is the 'microscopic isometric tension' and $l_0(t)$ is the 'microscopic coefficient of shortening heat' in an intact unit cell. $\xi_0(t)$ is the corrective factor $\xi(t,s_0V)$ for $s_0V = 0$. ν_0^* is defined by the relations 3.7 (with $s_0V = 0$). $\Pi_{isom}(t)$ is the rate of MgATP splitting under isometric conditions in an intact unit cell. Equation 3.21 is valid under both transitory and steady-state conditions (see Section 3.4.3.2 for precise details concerning the complex notions of transitory and steady state). On p. **63** in Section 3.4.3.2, it is demonstrated that, under steady-state conditions, the microscopic coefficients of shortening heat are $l_{0,\infty} \sim 0.69 T_{0,\infty}$ and $I_\infty(s_0V_{max}) \sim 0.29 T_{0,\infty}$ ($T_{0,\infty}$ is the microscopic isometric tension under steady-state conditions). It is reasonable to assume, as a rough approximation, that we have $l_0(t) \sim [0.69 + 0.29)/2]T_{0,\infty} \sim 0.49T_{0,\infty}$, regardless of time t. Equation 3.21 therefore gives

$$T_0(t) \sim 0.67\xi_0(t)me_{app}\Pi_{ison}(t)$$
$$\nu_0^*(1/2)s_0/[g\nu_0 + f - f(1 - n_0)] \quad (3.22)$$

From this equation, the isometric tension $T_0(t)$ and the rate of MgATP cleavage $\Pi_{isom}(t)$ seem to be strongly correlated. However, the parameters $\xi_0(t)$, ν_0^*, g, ν_0, f and n_0 also depend on time t, and this correlation is far from proportional. Regardless of the 'molecular' interpretation of the experimental results, a kind of correlation between MgATPase activity and isometric tension was observed experimentally by He et al. (1997), in the case of the rabbit. For instance, they found, in permeabilised rabbit fibres, that the time lags from the photolytic release of ATP to the start of tension rise and Pi release (MgATP splitting) were statistically similar (if the SDs are used; see Section 3.1 concerning the advantage of using SD) and depended similarly on temperature. They also found that there was a similar dependence, over time, of changes in tension (percentage of maximal tension corresponding to the first turnover) and MgATPase activity (rate of Pi release) on temperature, at 5°C, 12°C and 20°C, as shown in their Table 1. A similar kind of correlation was used by Morel (1984a) to explain the origin of the 'creep' in tension described by Gordon et al. (1966a,b) (see p. **250** and p. **251** in Section 8.9, regarding the problem of creep). In Sections 3.4.5 and 3.4.6, it is shown that the use of this apparent correlation between tension and rate of MgATP splitting is not clearcut, at least under steady-state conditions, in short tetani. It is even demonstrated that the coupling between 'isometric' MgATPase activity and isometric tetanic tension is very loose (see pp. **78–80** in Section 3.4.5 and pp. **83–84** in Section 3.4.6).

At this point, let us consider the case of 'pseudo-steady-state isotonic conditions' and the case of steady-state conditions for short isometric tetani. On pp. **62–63** in Section 3.4.3.2, it is demonstrated that active shortening is a complex process, in which s_0V, for example, depends on time t. However, it is possible to define pseudo-steady-state isotonic conditions, in which f and g are assumed to be independent of both t and s_0V, with s_0V strictly independent of t (see particularly p. **62** in Section 3.4.3.2). In these conditions, Equation 3.16 can be written:

$$\xi_\infty(s_0V)m e_{app}\omega_\infty(s_0V)v_\infty^*(1/2\sigma)s_0$$
$$\sim T_\infty(s_0V) + 1_\infty(s_0V) \qquad (3.23)$$

The subscript ∞ is introduced for the sake of homogeneity with the other notations. $T_\infty(s_0V)$ and $l_\infty(s_0V)$ are studied more precisely on pp. **62–64** in Section 3.4.3.2.

We now need to consider the important question of steady-state conditions for short tetani. Under these conditions, the various parameters become independent of time t, and f and g, independent of the linear distortion x (see text between Equations 3.5 and 3.6), are called f_∞ and g_∞. From Equation 3.3, we obtain $n_\infty = f_\infty/(f_\infty + g_\infty)$.

From Equation 3.8, the definition 3.7 of ν, and Equation 3.11, we deduce

$$R_{isom,\infty} = (\Delta h/\sigma)[f_\infty g_\infty/(f_\infty + g_\infty)] \text{ and}$$
$$\Pi_{isom,\infty} = \omega_{isom,\infty}(\Delta h/\sigma)[f_\infty g_\infty/(f_\infty + g_\infty)] \qquad (3.24)$$

$\omega_{isom,\infty}$ is the value of $\omega_{isom}(t)$ under steady-state conditions (t infinity, by contrast to the transitory conditions). Introducing the definition of $n_\infty = f_\infty/(f_\infty$

$+ g_\infty)$ (see above) and through some mathematical rearrangements, Equation 3.24 can also be written:

$$\Pi_{isom,\infty} = \omega_{isom,\infty}(\Delta h/\sigma)n_\infty(1 - n_\infty)(f_\infty + g_\infty) \qquad (3.25)$$

Studying the double-hyberbolic shape of the force–velocity relationship, valid for intact frog fibres, experimentally found by Edman (1988a,b) (see Section 8.11 for detailed comments), Morel (1990) defined a parameter $z_1 = 2\Delta h(f_\infty + g_\infty)/s_0V_{max}$, where s_0V_{max} is the maximal shortening velocity (for an unloaded fibre). However, in the light of the discussion presented on pp. **262–264** in Section 8.11 (see also p. **40** in Section 3.4.2), the double-hyperbolic P–V relationship cannot be retained. Only the parameter $z = 2\Delta h(f_\infty + g_\infty)/s_0V_{max}$ appearing in the papers by Morel (1978, 1984a) is taken into account, because it is consistent with the 'traditional' P–V non-hyperbolic relationship presenting a reversal of curvature at a relative load of ~0.8 (valid for single intact and demembranated fibres from various animal species; see pp. **262–264** in Section 8.11). Note that $z_1 \sim 0.150$ (Morel 1990) and $z \sim 0.157$ (Morel 1984a) are essentially similar and that using z or z_1 leads to similar results. Thus, we can write $(f_\infty + g_\infty) = zs_0V_{max}/2\Delta h$. This semi-empirical expression of $(f_\infty + g_\infty)$ can be introduced into Equation 3.25, from which we deduce the following phenomenological/semi-empirical equation (after some mathematical manipulations):

$$n_\infty(1 - n_\infty) = 2\sigma\Pi_{isom,\infty}/(\omega_{isom,\infty}zs_0V_{max}) \qquad (3.26)$$

This quadratic equation has two obvious analytical solutions, given by

$$n_{\infty,1} = (1/2)\{1 - [1 - 8\sigma\Pi_{isom,\infty}/(\omega_{isom,\infty}zs_0V_{max})]^{1/2}\} \qquad (3.27)$$

$$n_{\infty,2} = (1/2)\{1 + [1 - 8\sigma\Pi_{isom,\infty}/(\omega_{isom,\infty}zs_0V_{max})]^{1/2}\}. \qquad (3.28)$$

It has been shown that z is constant and that its value is ~0.157 (at a mean temperature of ~2.5°C; see Table 1 in the paper by Edman 1979; see also above). Edman (1979) found that $V_{max} \sim 2.5$ muscle length s^{-1}, corresponding to $s_0V_{max} \sim 5.4$ µm s^{-1} ($s_0 \sim 2.18$ µm; see p. **24** in Section 3.2). Note that Edman (1979) defined s_0V_0 as the maximal velocity measured by his technique, but he found that

s_0V_0 was statistically similar to s_0V_{max} measured by more traditional techniques. The estimation of $n_{\infty,1}$ and $n_{\infty,2}$ from Equations 3.27 and 3.28 requires careful definitions and estimates of $\Pi_{isom,\infty}$, σ and $\omega_{isom,\infty}$. This is a very difficult undertaking, particularly for $\Pi_{isom,\infty}$, and several new concepts and definitions should be introduced, progressively and point by point, as in Section 3.4.3.2. In any event, it is obviously impossible to discuss directly the properties of intact unit cells, and many experimental data and semi-empirical reasoning, valid for intact and demembranated fibres, should be extended to intact unit cells. I therefore suggest combining intact unit cells and whole fibres (see, e.g. Equations 3.27 and 3.28 and corresponding comments).

3.4.3.2 Semiempirical Approach to MgATPase Activity, the Proportion of Attached Cross-Bridges and Tension, During Isometric Tetanic Contraction, in an Intact Unit Cell. Survey of Various Problems in these Areas. Combination of Intact Frog and Demembranated Rabbit Fibres, with Some Insights into Isolated Frog and Rabbit Myofibrils

We need here to combine experimental data obtained for frog and rabbit. In this context, I take into account many results for the MgATPase activity in demembranated fibres from rabbit psoas muscle, because few data are available for frog in this domain. This approach is therefore necessary, but the conclusions drawn at the end of this section mostly concern the frog, because many of the fundamental experimental results described in this monograph were obtained with this animal species (e.g. Section 4.2). For the estimation of $n_{\infty,1}$ and $n_{\infty,2}$, given by Equations 3.27 and 3.28, experimental values for intact and chemically skinned or permeabilised fibres should be combined with molecular data, as pointed out in Section 3.4.1. In particular, a clear definition and a good knowledge of the value of $\Pi_{isom,\infty}$ are required (let us recall that $\Pi_{isom,\infty}$ is the rate of MgATP splitting in an intact unit cell, corresponding to steady-state conditions in the case of short tetani; see, e.g. Equation 3.25 and related comments). The importance of choosing the MgATPase activity most precisely 'fitting' a given set of physiological conditions has already been asserted by Morel and Bachouchi (1988), Morel et al. (1993) and Morel

and D'hahan (2000b). Indeed, in my group, we noted that the values of this activity, published over more than three decades or so, obtained in vitro (the exact value of the temperature was frequently 'neglected') and in vivo/in situ (in demembranated fibres, at various temperatures) fell roughly within the ~0.5–100 s^{-1} range (e.g. at ~0°C–20°C for frog or rabbit), and inconsistent conclusions are obviously likely, even for a given biological material, if the rate of MgATP splitting is unwittingly 'randomly selected'. As an illustrative example, Barclay (1999) proposed a 'weakly coupled version of the Huxley crossbridge model' and chose a rate of MgATP breakdown in isometric contraction of only ~1.1 s^{-1}, for frog at 0°C, that is, ~1.9 s^{-1} at 5°C (Q_{10} ~ 3.1; see p. **202** in Section 6.3.3), whereas He et al. (1997; see Table 1) obtained mean values of ~13–15 s^{-1} for frog at 5°C (ratio of ~13–15 s^{-1}/1.9 s^{-1} ~ 7–8). In their discussions, He et al. (1997, 1998a) provided additional experimental arguments supporting this view concerning the spectrum of values of MgATPase activity. Indeed, using the photolytic release of ATP (NPE-caged-ATP technique) and working on rabbit, at 20°C, He et al. (1997) pointed out that the initial MgATPase activity was ~40.6 s^{-1} immediately after the time lag (which lasted ~10 ms; see their Table 1), ~30.6 s^{-1} after ~80 ms, and only ~3.0 s^{-1} after ~600 ms (ratio of ~13–14, between the first and the third values). The authors concluded that the value of ~3.0 s^{-1} corresponds to steady-state conditions. I strongly suggest that the assertion of He et al. (1997) concerning this notion of steady state and the corresponding value of ~3.0 s^{-1} is almost certainly incorrect. Indeed, working on frog at 5°C, He et al. (1997c) found exactly the same behaviour and I conclude that the low value (here ~2.8 s^{-1}; see p. **55**) does not correspond to the steady state (see pp. **53–56** for a detailed discussion). He et al. (1998a), working on rabbit, with the same biophysical technique, but at 7°C, found an initial MgATPase activity of ~4.7 s^{-1} and a final MgATPase activity (under steady-state conditions, according to the authors) of ~1.3 s^{-1}, giving ~17.9 s^{-1} and ~5.0 s^{-1} at 20°C, respectively (ratio of ~3–4) (using the uniform value of Q_{10} ~ 2.8, valid within the range of temperature ~0°C–39°C, as estimated on p. **201** in Section 6.3.3; this estimate is similar to that of ~2.9 proposed by He et al. 1997, within the ~5°C–20°C range of temperature). In the group of He et al. (1997, 1998a), a spectrum of values of MgATPase activity has therefore been obtained. In these conditions, as highlighted below, in this section, the selection of an appropriate and single MgATPase activity, corresponding to a given situation, is essential to avoid fanciful conclusions, when incorrect values of the rate of MgATP cleavage are selected unintentionally and support the author's own bias. However, this is a difficult problem that I try to resolve in this section. The complex nature of this question has already been pointed out and analysed in the discussion of their experimental results by Barman et al. (1998) and He et al. (1997), but these two groups did not, in my opinion, draw clear and definitive conclusions. I present in this section a complementary analysis of this major problem, which has led to many contradictory and self-contradictory reasoning and inferences in past and current studies.

One of the most important factors to be taken into account is the duration of the experiments, as emphasised in the preceding paragraph. This is also the opinion of He et al. (1997) and I frequently focus, in this section, on this duration for the definition of several new concepts. Nevertheless, I do not rule out other specific problems raised by the use of different experimental techniques, as also highlighted by He et al. (1997). He et al. (1997, 1998a) systematically observed rapid, large decreases in MgATPase activities during short tetani, as recalled in the preceding paragraph, and gave plausible explanations for these unexpected features. However, these decreases in MgATPase activity may also result from many as yet not clearly identified phenomena, including an unexpected rapid onset of fatigue (see pp. **53–56** and Chapter 7, for precise details on fatigue). I do not find the many explanations of He et al. (1997c) for the spectrum of MgATP splitting sufficiently convincing I therefore suggest, in this section, a different approach. I take into account mostly the experimental results obtained with the NPE-caged-ATP technique, because most of the data presented by He et al. (1997) were obtained with this technique. My discussion and reasoning are obviously open to criticism, but they are supported by the self-consistency of various conclusions drawn in this book, particularly as concerns the definition of short tetani and steady state.

In their Table 1, He et al. (1997) reported many results for permeabilised fibres from psoas muscle of adult New Zealand white rabbits (47 experiments), but few data for frogs *Rana temporaria* (seven experiments). Before going into detail, it is useful to note that, for a given type of muscle prepared from a given animal species, the notions of short tetanus and steady-state conditions strongly depend on the temperature at which the experiments are performed, as expected from Table 1 in the paper by He et al. (1997). Clearly, many other factors may also affect the behaviour of fibres from a given animal species. These parameters include ionic strength and pH, for skinned or permeabilised fibres, and the technique of skinning and permeabilisation (see point (iii) on p. **43** in Section 3.4.3.1.1). The problem of the definition of short tetanus and steady-state conditions is complex because, even within a given animal genus (that of the frog, for example), the particular species of the animal can play an important role. For instance, Curtin and Woledge (1978) and Woledge et al. (1985), reporting and commenting on experimental results obtained with *semitendinosus* muscles from either kind of frog, *Rana pipiens* or *Rana temporaria*, recalled that this two species had different energetic characteristics: for example, there is noticeably less 'unexplained heat' in *Rana pipiens* than in *Rana temporaria* (see Section 8.4). Without presenting precise quantitative data, Linari et al. (1998) reported that the isometric tetanic tension in intact fibres from *Rana temporaria* is ~1.5 times greater than that from *Rana esculenta*. Taking a reference value of ~1.0 for *Rana esculenta* and averaging these two values, a 'composite frog' would correspond to a factor of ~ (1.0 + 1.5)/2 ~ 1.25. Similar unexpected results were obtained concerning the contribution of thin actin filaments to half-sarcomere compliance in an isometric tetanus. Julian and Morgan (1981) reported a contribution of ~30% in single intact fibres from *Rana temporaria*. In other experiments on single intact fibres from either *Rana temporaria* (Ford et al. 1981) or *Rana esculenta* (Bagni et al. 1990b), this fraction was estimated at less than ~20% (a value of ~10% is taken into account below). Finally, working on *Rana esculenta*, Linari et al. (1998) found a contribution of ~30%. As pointed out by these authors, the various experimental conditions used by the different groups are not the same and this may account, at least partly, for the scattering of the compliance values, but the 'composite compliance' would be

~ (30 + 10 + 30)/3 ~ 23 ± 9% (mean ± SD; n = 3). Thus, the question of the effect of using different species of frog remains. Recently, Offer and Ranatunga (2010) also suggested that cross-bridge and myofilament compliances should be averaged for *Rana esculenta* and *Rana temporaria*. In the light of pp. **96–97** in Section 3.8.1, regarding demembranated fibres, the differences between species of frog lie within the wide range of experimental error. The error bars, ~±20% around the average value of ~1.25, for the 'composite isometric tension', and ~±9% around the mean value of ~23% chosen for the composite compliance, are similar to those found on p. **97** in Section 3.8.1 for isometric tensions measured with 40 demembranated fibres of various origins (e.g. 100 × SE/mean ~ 7%; 100 × SD/mean ~ 43%; 100 × (SD + SE)/2/mean ~ 25%; see Section 2.1 for the definition of this third unconventional statistical test).*

For rabbit, the psoas muscle fibres of adult New Zealand white rabbits are the principal biological material used. As pointed out by Barman et al. (1998), Stephenson and Williams (1985a) and Woledge et al. (1985) and borne out in this work, quantitative comparisons of cold-blooded animals and warm-blooded animals are risky at low temperatures (~0°C–10°C). This is clear from Table 1 presented by Barman et al. (1998) and He et al. (1997), regarding the MgATPase activities of frog and rabbit, at ~4°C–5°C, for example. In this section, I deal mostly with the rabbit, but also try, with caution, to combine experimental data and interpretations for frog and rabbit. In an approach to the definition and estimation of $\Pi_{isom,\infty}$ (defined in Section 3.4.3.1.3 and appearing, for instance, in Equations 3.27 and 3.28), let us consider the concept developed by He et al. (1997), according to which there are several successive and discontinuous turnovers for MgATPase activity, at least at the beginning of contraction (see their Table 1). My own interpretation of the phenomena, including the transitory phenomena, differs from that of the authors, because changes in the rate of MgATP breakdown (e.g. Equation 3.11) and rise in tension (e.g. Equation 3.22) are undoubtedly continuous functions of time t, as also shown by the various experimental curves presented by He et al. (1997c). Breaking up the experimental curves into several discontinuous turnovers is almost certainly 'artificial' but may be justified if the experimental uncertainty is taken into account (see p. **42** in Section 3.4.2 for some numerical data). Nonetheless, I use here the experimental results, presented in Table 1 of the paper by He et al. (1997), as they stand. The authors studied the rates of MgATP splitting and isometric tensions at 5°C, 12°C and 20°C, for rabbit. I tried to fit their experimental data, obtained at 5°C, with an exponential of the form

$$\Pi_{isom}(t) = \Pi^{\circ}_{isom,\infty} + [\Pi_{isom}(0) - \Pi^{\circ}_{isom,\infty}]\exp(-A * t)$$

(3.29)

$\Pi_{isom}(0)$ is the rate of MgATP splitting at time zero, immediately after the time lag (see He et al. 1997, particularly Table 1, for precise details concerning the time lags). $\Pi^{\circ}_{isom,\infty}$ is the rate of MgATP splitting corresponding to the steady state occurring in a pseudo-long-lasting tetanus. These notions of steady state and pseudo-long-lasting tetanus should be explained, and I try, below, in this section, to define them progressively, point by point, from available experimental data. He et al. (1997) obtained three main values for MgATPase activity, corresponding to the three successive turnovers they defined. I use their approach as a 'working tool' to calculate the mean order of magnitude of the apparent rate of MgATP splitting, for each turnover. It is impossible to deduce the order of magnitude of the durations Δt_i (i = 1, 2, 3) of the first, second and third turnovers from the data of He et al. (1997), obtained at 5°C (and other temperatures). The only estimate given by the authors concerns the sum of the durations of the first three turnovers for rabbit permeabilised fibres at 20°C, found to be ~80 ms, that is, a mean estimate of ~27 ms per turnover. On p. **213** in Section 6.4, the value for the first turnover is estimated at $\Delta t_1 \sim t_0 \sim 17 \pm 3$ ms, obtained at 20°C from other data presented in the paper by He et al. (1997), slightly different from ~27 ms, but of the same order of magnitude. Owing to the inevitable experimental uncertainty, I take the same value $\Delta t_i \sim \Delta t$ for the three successive turnovers, at any temperature (between 5°C and 20°C

---

* Evidence for the role of thin actin and thick myosin filament compliance has become increasingly strong over the last 15 years, and the usual swinging cross-bridge/lever-arm models have been improved by introducing this property, which was not anticipated in the 1957, 1965–1969, 1971, and 1973 theories (A.F. Huxley 1957; Huxley 1965, 1969; Huxley and Simmons 1971, 1973). Such improvements were proposed, for example, by Martyn et al. (2002) and Linari et al. (2009).

Morel

for the rabbit; hereafter, Δt is frequently called the 'quantum duration'). After three integrations of $\Pi_{\text{isom}}(t)$, given by Equation 3.29, between 0 and Δt (first turnover), between Δt and $2\Delta t$ (second turnover) and between $2\Delta t$ and $3\Delta t$ (third turnover), we obtain the following definitions of the three corresponding apparent rates:

$$\Pi_{\text{app},1} \sim \Pi_{\text{isom},\infty}^{\circ} + [\Pi_{\text{som}}(0) - \Pi_{\text{isom},\infty}^{\circ}]$$
$$[1 - \exp(-A * \Delta t)]/A * \Delta t \qquad (3.30)$$

$$\Pi_{\text{app},2} \sim \Pi_{\text{isom},\infty}^{\circ} + [\Pi_{\text{isom}}(0) - \Pi_{\text{isom},\infty}^{\circ}]$$
$$\exp(-A * \Delta t)[1 - \exp(-A * \Delta t)]/A * \Delta t$$
$$(3.31)$$

$$\Pi_{\text{app},3} \sim \Pi_{\text{isom},\infty}^{\circ} + [\Pi_{\text{isom}}(0) - \Pi_{\text{isom},\infty}^{\circ}]$$
$$\exp(-2A * \Delta t)[1 - \exp(-A * \Delta t)]/A * \Delta t$$
$$(3.32)$$

He et al. (1997; see their Table 1) obtained the following experimental values for rabbit at 5°C: $\Pi_{\text{app},1}$ ~ 9.0 ± 1.2 s^{-1}, $\Pi_{\text{app},2}$ ~ 7.5 ± 1.1 s^{-1} and $\Pi_{\text{app},3}$ ~ 5.8 ± 0.5 s^{-1} (means ± SEs; n = 7), for the first three turnovers, from the ATP 'released by light' technique (NPE-caged-ATP technique) and lower values, for the Ca^{2+} released-by-light technique (NP-EGTA-caged-Ca^{2+} technique) of only ~5.8 ± 0.3 s^{-1} and ~4.0 ± 0.5 s^{-1} (means ± SEs; n = 4) for the first and second turnovers, respectively (no available value for a possible third turnover). The authors gave no reason for this significant discrepancy between the two series of values (for the second turnover, for example, the ratio of the value for NP-EGTA-caged-Ca^{2+} to that for NPE-caged-ATP is ~0.5). He et al. (1997) performed only four experiments, at a single temperature of 5°C, with the Ca^{2+} technique, whereas they performed 43 experiments with the ATP technique (7 at 5°C, 14 at 12°C and 22 at 20°C). I therefore take into account only the experimental data corresponding to the NPE-caged-ATP technique, from which consistent quantitative conclusions are drawn in this monograph. Using a simple routine and introducing the experimental values of $\Pi_{\text{app},1}$, $\Pi_{\text{app},2}$ and $\Pi_{\text{app},3}$, recalled above, I obtain $A*\Delta t$ ~

0.36 (on p. **213** in Section 6.4, Δt for the rabbit at 5°C is found to be ~143 ms, giving $A*$ ~2.52 s^{-1}), $\Pi_{\text{isom}}(0)$ ~ 11.0 s^{-1} and $\Pi_{\text{isom},\infty}^{\circ}$ ~ 2.0 s^{-1}, giving recalculated rates (to check the semi-empirical approach) of ~9.5 s^{-1}, ~7.2 s^{-1} and ~5.7 s^{-1}, respectively, consistent with the experimental values of He et al. (1997), recalled above, demonstrating that Equations 3.30 through 3.32 are suitable for this purpose. Moreover, the value of $\Pi_{\text{isom},\infty}^{\circ}$ obtained here (~2.0 s^{-1}) is, at least apparently (see the end of this paragraph for this limitation), consistent with values obtained by other authors with a different biological material and with entirely different techniques. Indeed, working on unheld isolated myofibrils from rabbit at 4°C, Barman et al. (1998) obtained a value of ~1.7 s^{-1}. On p. **201** in Section 6.3.3, a uniform value of Q_{10} ~ 2.8 is selected within the temperature range of ~0°C–39°C. Thus, the value of ~1.7 s^{-1}, measured at 4°C, becomes ~1.9 s^{-1} at 5°C, similar to $\Pi_{\text{isom},\infty}^{\circ}$ ~ 2.0 s^{-1} found above for the demembranated rabbit fibres used by He et al. (1997c). On p. **57**, it is demonstrated that the true rate of MgATP splitting, corresponding to a short tetanus under steady-state conditions of MgATPase activity, at 5°C, is ~6.7 ± (0.8 or 2.1) s^{-1} (mean ± SE or SD; n = 7), which is significantly greater than ~2.0 s^{-1} (ratio of ~3–4). This particular point relates to the correspondence of ~6.7 s^{-1} to a short tetanus under steady-state conditions and of ~2.0 s^{-1} to a pseudo-long-lasting tetanus (see text below Equation 3.29 concerning the introduction of the notion of pseudo-long-lasting tetanus). Owing to the various techniques used, the reasoning introduced and the calculations made, the values of ~1.9 s^{-1} and $\Pi_{\text{isom},\infty}^{\circ}$ ~ 2.0 s^{-1}, calculated above, are similar, and the value of ~2.0 s^{-1} for rabbit at 5°C (i.e. ~3.3 s^{-1} at 10°C, the reference temperature used in this book) can be taken as the most suitable value for chemically skinned and permeabilised fibres, for a pseudo-long-lasting tetanus. This unconventional notion of 'pseudo-long-lasting tetanus' corresponds to the state after the steady state and is progressively and carefully discussed below, in this section. Note that the value of ~1.9 s^{-1} for rabbit unheld isolated myofibrils is similar to the value of ~2.0 s^{-1} for rabbit permeabilised fibres, both at 5°C, a conclusion not valid for frog (see pp. **56–57**). Thus, the similarity observed for rabbit is probably coincidental and possibly attributed to the fact that the rabbit (warm-blooded animal) is studied at a low temperature (5°C), that is, noticeably below the normal temperature range of ~20°C–25°C. In any event, it is strongly suggested, on pp. **55–56**, that the pseudo-long-lasting tetani are probably contaminated by fatigue.

Using the NPE-caged-ATP technique, values for MgATPase activity of ~1.5–1.8 s^{-1}, at 19°C, were obtained by Ferenczi et al. (1984b) for glycerinated psoas muscle fibres from adult New Zealand white rabbits weighing more than 5 kg (these rabbits were therefore clearly old; see Morel et al. 1999). According to Figures 5 and 6 in the paper by Ferenczi et al. (1984b), the rate of MgATP splitting was studied for durations of up to ~1.5 s, making it possible, in principle, to obtain directly an experimental value of $\Pi_{\mathrm{isom},\infty}^{\circ}$, corresponding to pseudo-long-lasting tetani (see the preceding paragraph for precise details on this particular state). Indeed, introducing into Equation 3.29 (for simplification), the values, valid for the rabbit at 5°C, of $A^* \sim 2.52$ s^{-1}, $\Pi_{\mathrm{isom}}(0) \sim 11.0$ s^{-1} and $\Pi_{\mathrm{isom},\infty}^{\circ} \sim 2.0\,s^{-1}$ (see the preceding paragraph), we obtain $\Pi_{\mathrm{isom}}(1.5\ s) \sim 2.2$ s^{-1}, similar to $\Pi_{\mathrm{isom},\infty}^{\circ} \sim 2.0\,s^{-1}$ (ratio of ~1.1), within the large range of experimental error and taking into account the oversimplified approach suggested here (see Equations 3.29 through 3.32 and corresponding comments). From the values presented in Table 1 in the paper by He et al. (1997c), and using Equations 3.29 through 3.32 at 20°C, rather than at 5°C, I drew the same conclusion as above at 5°C (for instance, ratio of ~1.2, rather than ~1.1; calculations not shown). Unfortunately, the experimental study of Ferenczi et al. (1984b) is subject to several limitations, even reservations. On p. **201** in Section 6.3.3, it is shown that, for the rabbit, $Q_{10} \sim 2.8$, regardless of temperature, within the range of ~0°C–39°C, giving MgATPase activity of only ~0.58–0.70 s^{-1} at 10°C (~1.5–1.8 s^{-1} at 19°C; see the start of this paragraph), inconsistent with all the recent experimental results (for instance, ~5–6 times lower than the value of ~3.3 s^{-1} at 10°C; see the end of the preceding paragraph). Ferenczi et al. (1984b) wrote: 'any error in the estimate of the amount of myosin per gram protein will affect our estimate of the subfragment 1 concentration'. This is an important problem, as also pointed out by He et al. (1997) and discussed on pp. **206–208** in Section 6.3.5, but Ferenczi et al. (1984b) used a value of ~154 µM, similar to the 'universal experimental concentration' of ~154 µM (see p. **206** in Section

6.3.5). Thus, this is not the most important question. Indeed, Ferenczi et al. (1984b) obtained an isometric tetanic tension, at 19°C, of only ~0.54 × 10^5 N m^{-2}, corresponding to ~0.32 × 10^5 N m^{-2} at 10°C ($Q_{10} \sim 1.8$ between 10°C and 20°C–22°C; see p. **95** in Section 3.8.1). This value for tension is inconsistent with all the other available experimental data for demembranated fibres of various origins (~1.4 ± 0.1 × 10^5 N m^{-2} at 10°C; see p. **97** in Section 3.8.1; the isometric tetanic tension obtained by Ferenczi et al. 1984b, is only ~23% of the true value). Thus, the experimental results obtained by Ferenczi et al. (1984b) are at odds with all the recent and older data. Another reason for not taking into account the results of Ferenczi et al. (1984b) is the use of rabbits that were almost certainly too old (see the beginning of this paragraph and Morel et al. 1999 for precise details concerning 'young adult' and 'old' rabbits), with 'fragile fibres', resulting in inappropriate fibre preparation. These problems of fragility in the experiments performed by Ferenczi et al. (1984b) have also been raised by Linari et al. (1998), albeit under different conditions. Indeed, these authors claimed that they 'preferred α-toxin permeabilisation to protocols that involved the use of glycerol and detergents, because in preliminary experiments detergents were found to produce weakening of the neuromuscular junction region and increase the probability of the fiber breaking while inducing high tension in rigor'. Finally, as recalled in Section 3.4.3.1.1 (see point (iii) on p. **43**), Millman (1998), in his critical analysis of various techniques used to remove the sarcolemma, wrote: '... for repeatable results using glycerol-extracted preparation, it is usually necessary that the extraction process be carefully controlled and the period over which these preparations are stored and used be limited'. Given the likelihood that the experiments of Ferenczi et al. (1984b) were carried out in unfavourable conditions, I conclude that they cannot be taken into account when estimating the rates of MgATP splitting and isometric tensions. Finally, the problem of fragility is also strongly suggested, probably demonstrated, on p. **200** in Section 6.3.2, p. **202** in Section 6.3.3, p. **208** in Section 6.3.5 and p. **238** in Section 8.5.

By contrast, working on mechanically skinned frog fibres at ~0°C–5°C, in surrounding buffers mimicking the sarcoplasmic medium, Ferenczi et al. (1984a) found that, after correction for cross-sectional area, the isometric tetanic tension was ~1.5 × 10⁵ N m⁻² at ~2°C, giving ~3.0 × 10⁵ N m⁻² at 10°C (Q_{10} ~ 2.4 between ~0°C and ~10°C–12°C; see p. **95** in Section 3.8.1). This value of ~3.0 × 10⁵ N m⁻² lies within the wide range of values, recalled on pp. **96–97** in Section 3.8.1, for demembranated fibres and is also coincidentally similar to that recorded in intact frog fibres and half-fibres (see Section 3.3). Thus, the very low isometric tension found by Ferenczi et al. (1984b; see the preceding paragraph) is certainly related to both the use of rabbits that were too old and to the glycerination process used to remove the sarcolemma (see the preceding paragraph), whereas Ferenczi et al. (1984) almost certainly used young adult frogs and traditional mechanical demembranation techniques.

In the light of the results presented in Table 1 in the paper by He et al. (1997), the MgATPase activity, in isometrically contracting permeabilised rabbit fibres, depends strongly on temperature. This relationship is supported, for instance, in Sections 6.3.2 and 6.3.3, for resting and contracting fibres, respectively. For synthetic thick myosin filaments in vitro, temperature plays a key role in interactions involving the myosin heads (see pp. **18–19** in Section 2.4, and Section 5.2, including Figure 5.2, its legend and Appendix II). Thus, temperature plays a key role in terms of myosin structure and enzymological properties. Otherwise, it is well known that the isometric tetanic tension is an increasing function of temperature (see Section 8.8), and, from Table 1 in the paper by He et al. (1997), it appears that the half-time of tension rise decreases strongly with increasing temperature. Temperature must therefore be taken into account in all experiments, reasoning and discussions. In this context, it should be noted that unexpected results are frequently obtained if cold-blooded and warm-blooded animals are not studied at their optimal temperature (e.g. ~0°C–10°C for frog; room temperature or slightly less for rabbit). This point is generally insufficiently highlighted by authors, the notable exception being Barman et al. (1998), Stephenson and Williams (1985) and Woledge et al. (1985).

I focus in this and the next two paragraphs on the values of the quantum duration Δt, corresponding to any turnover of MgATPase activity (Δt is defined on p. **50** and used in Equations 3.30 through 3.32), the total duration Δt^* of tension rise (i.e. duration required to reach full isometric tetanic tension) and the two corresponding time lags after the laser flashes. From Table 1 in the paper by He et al. (1997c), it appears that, for rabbit (very few experiments have been performed on frog, at a single temperature of 5°C), the rate of MgATP breakdown is an increasing function of temperature. It is strongly suggested, on p. **65** in Section 3.4.3.3, that, under steady-state conditions of isometric tetanic contraction, the duration $\tau_{c,\infty}$ of the cycle of a cross-bridge is inversely proportional to the rate $\Pi_{isom,\infty}$ ~ $R_{isom,\infty}$ of MgATP splitting, which means that MgATPase activity increases when this duration is reduced. It is also shown, on p. **65** in Section 3.4.3.3, that this duration $\tau_{c,\infty}$ is statistically similar to Δt. Thus, Δt should decrease with increasing temperature. This conclusion can reasonably be extended to the first complex transitory turnover, the duration of which is also Δt. From Table 1, I also note that the half-time of tension rise decreases with increasing temperature. It appears reasonable to assume that the total duration Δt^* is proportional to this half-time. Δt^*, therefore, decreases with increasing temperature. Thus, both Δt and Δt^* are decreasing functions of temperature and I suggest, as a rough approximation, that Δt^* ~ $\zeta \Delta t$, where ζ is the constant of proportionality. This hypothesis should be supported experimentally, as in most of the following paragraphs, based on the case of frog, retaining the same reasoning and notations as those used in this paragraph for rabbit. In the past, the experiments performed on frog and rabbit were not of the same kind. Intact or mechanically skinned frog fibres have mostly been used to study mechanical properties (e.g. isometric tension, force–velocity relationship and mechanical transients). Chemically skinned and permeabilised fibres from rabbit psoas muscle have mostly been used to study MgATPase activity. Since the 1990s, studies on these two biological materials have become more similar (mostly MgATPase activities, isometric tensions and their relationships). Thus, in an attempt to ensure as broad as possible an approach, I use results obtained for these two animal species and try to obtain conclusions applicable to both frog and rabbit, with only a multiplication factor for the conversion between species (see p. **58**). In this and most of the following paragraphs, I focus on frog. Let us refer to Figure 9 in the paper by He et al. (1997), concerning a single permeabilised frog fibre at 5°C. This figure presents two continuous curves, describing concurrent variations of the quantity of Pi released (i.e. MgATPase activity) and isometric tension against time, for a single short tetanus (duration of ~1 s). In this Figure 9, stabilisation of the isometric tension occurs, at first sight, between ~400 ms and ~450 ms after the slight time lag (after the laser flash) that He et al. (1997) estimated at ~27.3 ± (2.7 or 7.1) ms (mean ± SE or SD; n = 7; see their Table 1), corresponding to a maximal interval of ~20–34 ms. Choosing this maximal interval for the time lag, a period of between ~400 − (20–34) ~ 366–380 ms and ~450 − (20–34) ~ 416–430 ms is therefore required to get beyond the transitory period, that is, to reach a stable value for isometric tetanic tension. Combining these four values (~366 ms, ~380 ms, ~416 ms and ~430 ms) gives $\Delta t^* \sim 398 \pm (13$ or 26) ms (mean ± SE or SD; n = 4): the maximal interval for Δt^* is ~372–424 ms (see the preceding paragraph for the definition of Δt^*).

Obviously, the values of $\Delta t^* \sim 372$–424 ms and ~20–34 ms (for the time lag), at 5°C, for isometric tetanic tension, determined in the preceding paragraph, depend exclusively on the experimental results of He et al. (1997). At this point, it is important to compare them with experimental results obtained at approximately 5°C, with different biological materials, nonetheless all obtained from frog, by independent groups. Martin-Fernandez et al. (1994), using x-ray diffraction from synchrotron radiation, obtained a single value for Δt^* of ~400 ms, at 8°C, for whole frog muscles (this value corresponds to the period starting just after the time lag, after electrical stimulation, also called the 'latent period', estimated at ~20 ms from their Figures 6 through 8, but not given by the authors). Combining the three available values for Δt^* from these two independent groups (~372 ms, ~424 ms and ~400 ms) gives $\Delta t^* \sim 378$–420 ms (maximal interval). Piazzesi et al. (1999) used the same physical technique as Martin-Fernandez et al. (1994), at 4°C, on single intact frog fibres, and gave an estimate of ~7.5 ms for the latent period, but did not estimate the time required to reach full isometric tension. Using the six available values (~20 ms, ~34 ms, ~20 ms, ~20 ms, ~20 ms and ~7.5 ms), the time lag (or latent period) is ~13–28 ms (maximal interval calculated from the mean value and the SD). The half-time of tension rise obtained by He et al. (1997c), on permeabilised frog fibres, at 5°C, was found to be ~56–98 ms (maximal interval; just after the time lag; see Table 1 in the paper by He

et al. 1997; SD is used in this estimate, for the reasons given in Section 2.1), and the half-time of tension rise, estimated from Figures 6 and 7 in the paper by Martin-Fernandez et al. (1994), but not given by the authors, is found to be ~70–80 ms, after the latent period. Combining these six values (~56 ms, ~98 ms, ~70 ms, ~70 ms, ~80 ms and ~80 ms) gives a maximal interval for the half-time of tension rise of ~63–89 ms (maximal interval calculated from the mean value and the SD). Piazzesi et al. (1999) did not give values of Δt^*, but gave the half-time for isometric force development (after the latent period after electrical stimulation) and obtained a maximal interval of ~31.5–35.5 ms. Assuming that the half-time is proportional to Δt^* (this is also assumed on p. 53), and comparing the results of the three independent groups, a good estimate for Δt^*, in the case of Piazzesi et al. (1999), would be ~ (378–420) × (31.5–35.5)/(63–89) ms ~ 132–224 ms (maximal interval). Combining the four available values of Δt^* (~378 ms, ~420 ms, ~132 ms and ~224 ms) obtained at 4°C, 5°C and 8°C (mean value of ~5.7°C), we obtain an estimate of $\Delta t^* \sim 288 \pm (58$ or 116) ms (mean ± SE or SD; n = 4); that is, $\Delta t^* \sim 172$–404 ms ~ 0.17–0.40 s (maximal interval). Adding the time lag of ~13–28 ms, estimated above, gives ~ 185–432 ms ~ 0.18–0.43 s (just after the laser flash or electrical stimulation). Otherwise, considering the rather arbitrary division into two parts of the curve of MgATPase activity against time suggested by He et al. (1997) in their Figure 9, the first two turnovers (presented in their Table 1) are apparently combined by the authors (see their discussion and the next paragraph) and correspond to Δt_{app} ~ $2\Delta t$ ~ 150 − (39–58) ~ 92–111 ms; that is, Δt ~ $\Delta t_{app}/2$ ~ (92–111) ms/2 ~ 46–56 ms (~150 ms is given in the legend to Figure 9 in the paper by He et al. 1997; ~39–58 ms represents the maximal interval for the Pi time lag, deduced from Table 1 in the paper by He et al. 1997c; SD is used for the reasons given in Section 2.1). On p. 53, I suggest writing $\Delta t^* \sim \zeta\Delta t$, and, taking into account only the mean values, I deduce from the calculations presented in this paragraph that $\zeta \sim [(172 + 404)$ ms/2]/[(46 + 56) ms/2] ~ 5–6. Regardless of the many approximations made in this and the two preceding paragraphs, this indicates that there is a correlation between the duration Δt of the turnovers of MgATPase activity and the total duration of the tension rise Δt^* (in the case of the frog) (see p. **53** for the introduction of these two notions). However, the value of ζ is much greater than 1, strongly suggesting that there is extensive uncoupling of isometric MgATPase activity and isometric tetanic tension. This merits further discussion, presented on pp. **78–80** in Section 3.4.5, and pp. **83–84** in Section 3.4.6, from which it is indeed concluded that there is very loose coupling between isometric MgATPase activity and isometric tetanic tension.

Dividing the curve of Pi release against time into two parts (see Figure 9 and its legend in their paper), He et al. (1997) obtained, for the first part, an MgATPase activity of ~16.4 s^{-1} between ~50 ms (time lag in Figure 9 and its legend; value consistent with the maximal interval of ~39–58 ms given at the end of the preceding paragraph) and ~150 ms, that is, a duration of 150 ms − 50 ms ~ 100 ms, consistent with the value of Δt_{app} ~ $2\Delta t$ ~ 92–111 ms (see the end of the preceding paragraph; only this value of Δt_{app} ~ 92–111 ms is retained in this monograph, not the single value of ~100 ms). Thus, this first part corresponds to the first two turnovers, as suggested at the end of the preceding paragraph. Concerning the second part, between ~250 ms and ~400 ms, that is, between ~ 250 ms − 50 ms ~ 200 ms and ~ 400 ms − 50 ms ~ 350 ms (subtracting the time lag of ~50 ms), He et al. (1997) obtained a mean MgATPase activity of only ~2.8 s^{-1}, which they considered to correspond to the steady state. In my opinion, in this second part, the true steady-state conditions for MgATPase activity are almost certainly largely exceeded, probably leading to fatigue in the rate of MgATP breakdown. Thus, steady-state conditions are exceeded somewhere between Δt_{app} ~ 92–111 ms and ~200 ms, and this situation may continue between ~200 ms and ~350 ms (see pp. **210–211** in Section 6.3.5 for further details, and below for a short discussion). If the time lag of ~39–58 ms (see above) is added, these three durations correspond to ~92–111 + 39–58 ~ 131–169 ms, ~ 200 + 39–58 ~ 239–258 ms and ~350 + 39–58 ~ 389–409 ms (maximal intervals), respectively. Owing to the duration of ~100 ms of the first part, statistically consistent with Δt_{app} ~ 92–111 ms (see above), the first value of ~16.4 s^{-1} for the rate of MgATP breakdown certainly corresponds to a mixture of the first two turnovers presented in Table 1 in the paper by He et al. (1997), including the transitory period, corresponding to the first turnover (see Table 1),

and taken as the 'transitory' MgATPase activity throughout this work (see also p. **53** for a short definition of this first complex transitory turnover). This notion of transitory MgATPase activity is incomplete, as demonstrated on pp. **210–211** in Section 6.3.5, where it is strongly suggested that the complex question of the transitory period can be addressed experimentally. I will not, therefore, complicate the discussion further here. In Table 1 presented in the paper by He et al. (1997c), the first 'complex' turnover corresponds to an MgATPase activity of ~15.0 ± 1.5 s^{-1} (mean ± SE; n = 7). The second turnover corresponds to ~13.2 ± 1.3 s^{-1} (mean ± SE; n = 7). If only SE is taken into account, the maximal value for the first complex turnover is ~16.5 s^{-1}, which is consistent with ~16.4 s^{-1}, whereas for the second turnover, the maximal value is only ~14.5 s^{-1}, which is lower than ~16.4 s^{-1}. However, it is shown in Section 2.1 that SD is more suitable than SE in many instances and, using SD, the maximal value for the second turnover now becomes ~ 13.2 + 1.3 × $7^{1/2}$ ~ 16.6 s^{-1}, which is also consistent with ~16.4 s^{-1}. Table 1 and Figure 9 in the paper by He et al. (1997) are therefore self-consistent and the conclusions drawn in this and the preceding paragraphs are appropriate. In conclusion, the time required to go beyond the steady state of MgATPase activity and to reach pseudo-long-lasting tetanus conditions (see the next paragraph for a definition of these special conditions) ranges somewhere between Δt_{app} ~ 92–111 ms and ~200 ms, or ~131–169 ms and ~239–258 ms, and up to ~389–409 ms, if the time lag of ~39–58 ms is added (see above, concerning these values). As shown in this paragraph, below Δt_{app} ~ 92–111 ms (or ~131–169 ms if the time lag is taken into account), there are no signs of fatigue. He et al. (1997) considered a value of ~2.8 s^{-1} to correspond to the steady state. I do not agree with this opinion, as also pointed out above and supported in the next paragraph.

For simplification, regardless of the animal species (e.g. frog or rabbit) and temperature, I systematically omit, in this and the following paragraphs, the first turnover of MgATPase activity. Indeed, it corresponds largely to complex transitory phenomena, as highlighted in several paragraphs of this section (see pp. **53–55** for circumstantial comments) and in Section 3.4.2, where it is demonstrated that all the features of a contracting fibre depend on time t, at least during the first tens, and possibly hundreds, of milliseconds of contraction. For frog at 5°C, the second part of Pi release, defined in Figure 9 of the paper by He et al. (1997) (see the preceding paragraph), corresponds approximately to a pseudo-long-lasting tetanus, that is, to a tetanus that has probably gone beyond the steady-state conditions and is characterised by a value of ~2.8 s^{-1} for the rate of MgATP splitting. Working with the same biological material, Cook et al. (1996) reported, but did not confirm further, a value of ~2.3 s^{-1} at 10°C, for long-lasting tetani (~5 s), corresponding to ~1.3 s^{-1} at 5°C (Q_{10} ~ 3.1; see p. **202** in Section 6.3.3). This value is slightly lower than ~2.8 s^{-1}, corresponding to pseudo-long-lasting tetani. However, the values of ~2.8 s^{-1} and ~1.3 s^{-1} are of the same order of magnitude and the lower value of ~1.3 s^{-1} would be related to the long duration of the tetani (~5 s) and the subsequent possible rapid onset of fatigue, at least regarding MgATP breakdown (see Chapter 7, particularly Sections 7.2 and 7.3, for precise details concerning fatigue, in terms of new concepts based on molecular processes, different from the old ideas of gradual acidification and accumulation of lactic acid). In any event, the low value of ~2.8 s^{-1} possibly (probably?) indicates that fatigue occurs early in the tetanus (within the rather ill-defined range of duration ~92–409 ms; see the end of the preceding paragraph for precise details concerning these two values of ~92 ms and ~409 ms). Barman et al. (1998), working on unheld isolated myofibrils from frog, at 4°C, obtained a value of ~4.6 s^{-1} for the MgATPase activity. Taking into account Q_{10} ~ 3.1 (see above; this value, valid for demembranated fibres, is extended to unheld isolated myofibrils, for simplification), this value of ~4.6 s^{-1} would have been ~5.2 s^{-1} at 5°C, noticeably higher than ~2.8 s^{-1} and ~1.3 s^{-1}. Interpretation of the results obtained by Barman et al. (1998) is not straightforward, as pointed out by the authors themselves (see the next paragraph). The mean value of the MgATPase activity obtained with frog unheld myofibrils (~5.2 s^{-1} at 5°C) is ~2 times that corresponding to pseudo-long-lasting tetani for frog permeabilised fibres (~2.8 s^{-1} at 5°C). In this context, frog unheld myofibrils therefore behave differently from rabbit unheld myofibrils, with respect to permeabilised fibres (see also p. **51**). In any event, the duration of shortening of frog unheld myofibrils is limited (~105 ms at 5°C, deduced from values

given in the paper by Barman et al. 1998; see p. **61**) and fatigue is unlikely to occur with this biological material (the estimate of ~105 ms is of the same order of magnitude as Δt_{app} ~ 92–111 ms, below which there are no signs of fatigue in the permeabilised frog fibre studied in Figure 9 in the paper by He et al. 1997; see the end of the preceding paragraph). Unheld myofibrils display complex behaviour (see the next paragraph), but the finding that the MgATPase activity in this 'non-fatigued' biological material (~5.2 s^{-1}) is ~2 times that corresponding to a pseudo-long-lasting tetanus in a permeabilised fibre (~2.8 s^{-1}) suggests possible contamination by fatigue early in a short tetanus in permeabilised fibres and that the value of ~2.8 s^{-1} corresponds to a mixture of high values recalled in the preceding paragraph (~16.4 s^{-1}) and very low values corresponding to 'pure' fatigue (e.g. ~0.128 s^{-1}; see p. **223** in Section 7.3). For instance, the value of ~2.8 s^{-1} would result from ~16%–17% of ~16.4 s^{-1} and ~83%–84% of ~0.17 s^{-1}. Another circumstantial approach concerning the variation of MgATPase activity over time and the rapid occurrence of fatigue is proposed on pp. **210–211** in Section 6.3.5.

The experiments performed by Barman et al. (1998) may be considered to correspond neither to the pseudo-long-lasting tetani observed in chemically skinned or permeabilised fibres nor to steady-state isometric tetani, but probably to an intermediate state. Indeed, as repeatedly claimed and demonstrated in Sections 3.4.2, 3.4.3.1.1, 3.4.3.1.3, the various features (e.g. MgATPase activity and isometric tension) of a contracting fibre are continuous functions of time and there is necessarily a continuous change from the initial transitory state to a steady-state tetanus, a pseudo-long-lasting tetanus, a long-lasting tetanus and then a very long-lasting tetanus, with the existence of many intermediate states, almost certainly forming a continuum. When trying to compare the rates they obtained for unheld isolated myofibrils with those obtained for chemically skinned or permeabilised fibres under 'steady-state isometric conditions' (with the many problems concerning the definition of these conditions raised in this section), Barman et al. (1998) themselves asked 'Do the ATPase kinetics in isometric fibers and unheld myofibrils agree?' The same type of problem arose 2 years before in the same group: Lionne et al. (1996) claimed: 'from a purely enzymatic point of view, the situation is difficult to understand;

Morel

we have made several attempts at fitting it [their experimental result] to classical ATPase schemes, but without success'. I strongly suggest that it is impossible to reconcile the experimental data obtained by Barman et al. (1998) and Lionne et al. (1996) with unheld myofibrils on the one hand, with those corresponding to isometric conditions with whole fibres on the other, because they correspond to very different physiological states. This problem is very difficult because, as demonstrated in Section 3.4.4, many complex phenomena occur when the myofilaments slide past each other, adding to the difficult questions posed in this section for 'simple' tetanic isometric conditions only. The problem of using unheld myofibrils is universal and several groups have tried to justify their choice, but I and probably other specialists in muscle and muscle contraction are reluctant to use this biological material. The specialists in myofibrils are themselves hesitant and try to justify their choice in many papers: for example, Barman et al. (1998), Candau et al. (2003), Colomo et al. (1997), Herrmann et al. (1992, 1994), Houadjeto et al. (1991), Lionne et al. (1995, 1996, 2002, 2003) and Stehle et al. (2000). By contrast, other authors have studied isometrically contracting myofibrils and recorded isometric forces (e.g. Bartoo et al. 1993; Tesi et al. 2000, 2002). This biological material, under these conditions, probably gives results easier to interpret than the results obtained with unheld myofibrils, in which there is a complex link between isometric MgATPase activity and the maximal velocity of shortening. Unfortunately, to date (as far as I know), no experiments have been carried out on myofibrils, for simultaneous measurements of the isometric force and isometric MgATPase activity.

Let us consider the rabbit again. By contrast to the frog, omitting the first turnover, corresponding to complex transitory phenomena, as suggested on pp. **53–56**, only the second and third turnovers are taken into account. As pointed out in this section, there are no grounds for breaking up MgATPase activity changes over time into several turnovers (see p. **50**, text above Equation 3.29). It is therefore useful to combine the values corresponding to the second and third turnovers (see Table 1 in the paper by He et al. 1997; the error bars correspond to the SEs given by the authors) to obtain a best estimate of the rate of MgATP cleavage, $\Pi_{isom,\infty}$, under steady-state conditions, for short tetani. At 20°C, the mean value of the steady MgATPase activity corresponding to the second and third tetani is $\Pi_{isom,\infty} \sim (35.5 \pm 4.2 + 30.8 \pm 2.5)/2 \sim 33.1 \pm 3.4 \text{ s}^{-1}$ (mean $\pm$ SE; n = 22). The values for the second and third tetani are statistically similar, even if the most restrictive statistical test (SE) is taken into account (see Section 2.1 for precise details). At 12°C, the same type of reasoning and calculation leads to $\Pi_{isom,\infty} \sim (17.1 \pm 1.0 + 15.5 \pm 0.8)/2 \sim 16.3 \pm 0.9 \text{ s}^{-1}$ (mean $\pm$ SE; n = 14) and, here again, the values for the second and third tetani are statistically similar, even using SE. A case of particular interest is that of rabbit at 5°C, for comparison with frog (frequently studied at around this temperature). The mean value of the steady MgATPase activity corresponding to the second and third tetani is $\Pi_{isom,\infty} \sim (7.5 \pm 1.1 + 5.8 \pm 0.5)/2 \sim 6.7 \pm 0.8 \text{ s}^{-1}$ (mean $\pm$ SE; n = 7). The values for the second and third turnovers are of borderline statistical significance, but this is merely attributed to the use of SE as the statistical test. Indeed, using the relationship between SE and SD given in Section 2.1, for the second and third turnovers, the SDs are $\sim 2.9 \text{ s}^{-1}$ and $\sim 1.3 \text{ s}^{-1}$, respectively, and the problem of statistical similarity no longer remains, because the SD test is more appropriate. The value of $\sim 6.7 \pm (0.8 \text{ or } 2.1) \text{ s}^{-1}$ (mean $\pm$ SE or SD; n = 7) is significantly higher than the value of $\Pi^\circ_{isom,\infty} \sim 2.0 \text{ s}^{-1}$ calculated in the paragraph following Equations 3.30–3.32 in this section. This value of $\sim 6.7 \pm (0.8 \text{ or } 2.1) \text{ s}^{-1}$ for rabbit at 5°C corresponds to the steady-state conditions in a short tetanus, whereas the value of $\sim 2.0 \text{ s}^{-1}$ almost certainly corresponds to a pseudo-long-lasting tetanus. Some authors (He et al. 1997, for instance) noted that the concentration of active sites (here myosin heads) is a crucial problem in calculating the MgATPase activities in s^{-1}. Indeed, these activities are measured in μM of Pi released per myosin head per second. Thus, any error in the concentration of heads results in similar errors in MgATPase activities. This problem is too largely forgotten in the 'muscle community'. Fortunately, it is demonstrated in Section 6.3.5, particularly on p. **205**, that the values given above are correct, though the traditional value of 150 μM for the concentration of myosin heads in rabbit psoas muscle is underestimated.

The rates of MgATP splitting for a short tetanus, under steady-state conditions, estimated in the preceding paragraph, are valid for rabbit, whereas the numerical values of z and $s_0 V_{max}$ in Equations 3.27 and 3.28 correspond to frog. This problem of combining experimental results obtained with the two animal species is well known, as repeatedly pointed out in this work (see, for instance, p. **56**). Indeed, most physiological studies have been performed on frog fibres, whereas most biochemical studies have been carried out on rabbit fibres, although much work, over the last 15–20 years, has been devoted to simultaneous studies of MgATPase activity and mechanical performance, particularly in chemically skinned and permeabilised fibres from rabbit psoas muscle (e.g. He et al. 1997, 1998a,b, 1999; Potma and Stienen 1996; Potma et al. 1994; Sun et al. 2001). As an illustrative example of the difference between frog and rabbit, let us consider the experimental results obtained with unheld isolated myofibrils by Barman et al. (1998; Table 1). They show that the rates of MgATP cleavage corresponding to unheld isolated myofibrils from frog and rabbit are different from each other and from those corresponding to chemically skinned and permeabilised fibres (see pp. **56–57** for a detailed discussion). Regardless of these problems, Barman et al. (1998) obtained, at 4°C (owing to the large error bars on the various values obtained at 4°C and 5°C, no corrections are made when passing from 4°C to 5°C), values of ~4.6 ± 0.1 s^{-1} for unheld myofibrils of frog and ~1.7 ± 0.1 s^{-1} for rabbit, corresponding to a factor of ~ (4.6 ± 0.1) s^{-1}/(1.7 ± 0.1) s^{-1} ~ 2.7 ± 0.2 for conversion from rabbit to frog (the authors did not state whether the error bars correspond to SD or SE). According to Table 1 in the paper by He et al. (1997c), for permeabilised rabbit fibres at 5°C, the MgATPase activity corresponding to a short tetanus under steady-state conditions is ~6.7 ± (0.8 or 2.1) s^{-1} (mean ± SE or SD; n = 7; see the preceding paragraph). For permeabilised frog fibres at 5°C, MgATPase activity is apparently biphasic (see Table 1 in the paper by He et al. 1997) and only the value for the second turnover is taken into account, that is, ~13.2 ± (1.3 or 3.4) s^{-1} (mean ± SE or SD; n = 7), corresponding to the steady state and a short tetanus. Taking into account the SEs, the multiplication factor for conversion from rabbit to frog, at 5°C, is therefore ~ (13.2 ± 1.3) s^{-1}/(6.7 ± 0.8) s^{-1} ~ 2.0 ± 0.5. Taking into account the SDs, the multiplication factor is ~ (13.2 ± 3.4) s^{-1}/(6.7 ± 2.1) s^{-1} ~ 2.0 ± 1.0. These two values are lower than the ~2.7 ± 0.2 found above, but statistically similar, regardless of the use of SE or SD. Only the value of ~2.5 is common to the three intervals of uncertainty and is therefore taken into account for converting the MgATPase activity in rabbit to that in frog, at 5°C, and also almost certainly between ~0°C and ~10°C–12°C. Otherwise, on p. **201** in Section 6.3.3, the Q_{10} for rabbit is found to be ~2.8 (using only permeabilised fibres, He et al. 1997c, obtained a similar value: Q_{10} ~ 2.9 between 5°C and 20°C). For frog, Q_{10} ~ 3.1 (see p. **202** in Section 6.3.3) and the value of ~13.2 ± (1.3 or 3.4) s^{-1} (mean ± SE or SD) at 5°C becomes ~9.9 ± (1.0 or 2.5) s^{-1} at ~2.5°C, the mean temperature at which the other parameters were measured by Edman (1979). In Equations 3.27 and 3.28, the best estimate of $\Pi_{isom,\infty}$ for frog to be taken into account is therefore ~9.9 s^{-1} at ~2.5°C.

All the parameters in Equations 3.27 and 3.28 are known, except σ and $\omega_{isom,\infty}$, studied in this and the following paragraphs. As previously suggested by Morel (1984a,b), the probability of inactive actin–myosin contacts existing can be considered negligible for short tetani, under strictly isometric steady-state conditions (no microscopic shortening, zero step distance and zero 'working (power) stroke', in intact unit cells; see p. **67** in Section 3.4.3.3 for precise details). Thus, it is realistic to take $\omega_{isom,\infty}$ ~ 1, resulting in self-consistency in the present approach, as repeatedly borne out in this monograph (e.g. pp. **73–74** in Section 3.4.4). Moreover, in all the publications from my group, self-consistent conclusions were drawn by taking σ ~ 6 nm. This estimate of the spacing between two consecutive actin-binding sites in the thin actin filaments in vivo/in situ is valid under isometric steady-state conditions, for short tetani. Under isotonic conditions, in which active shortening occurs, the problem is much trickier, as suggested semi-empirically by Morel (1990), who found that σ is probably not a universal 'constant' (see, however, reservations concerning this paper on p. **263** in Section 8.11). Changes in σ in high-speed lengthening fibres, under contraction conditions, are also almost certain to occur (Morel 1991b). A similar conclusion, going against the notion of rigid structure for the thin actin filaments, is discussed on pp. **33–34** in Section 3.2 and in the following paragraphs. Under isometric tetanic conditions, a value of σ ~ 36 nm, corresponding approximately to the half-pitch of the double-helical structure of the thin actin filaments (between ~35 nm and ~40 nm; Huxley and Brown 1967), was introduced by Eisenberg et al. (1980) and Hill (e.g. 1974, 1975, 1977) into their biophysical approaches to the swinging cross-bridge/lever-arm theories. Using x-ray diffraction from synchrotron radiation, Linari et al. (2005) and Piazzesi et al. (1999) found, in single intact fibres from frog, at 0°C or 17°C for Linari et al. (2005), and at 4°C for Piazzesi et al. (1999), a half-pitch of ~38 nm, from the long-pitch actin helix (Linari et al. 2005) along the thin actin filaments, at rest, in rigor and at the plateau of isometric tetanic contraction. Barclay (1999), Molloy et al. (1995a,b) and Squire (1981) chose ~38.5 nm for the distance between actin-binding sites, assumed to correspond to the half-pitch of the helix on the thin actin filaments. Averaging the five different values (~35 nm, ~36 nm, ~38 nm, ~38.5 nm and ~40 nm), we obtain a value of ~37.5 ± (0.7 or 1.8) nm (mean ± SE or SD; n = 5), which is marginally different from ~36 nm if SE is taken into account, but statistically similar if SD is taken into account (see Section 2.1 for the advantage of using SD). Owing to various other experimental uncertainties and for self-consistency with the many papers published by my group, there is not need to introduce either the most recent values of ~36.0–38.5 nm, or the mean value of ~37.5 nm. The value of ~36 nm is therefore retained for the half-pitch of the actin helix in the thin actin filaments (see the next paragraph).

In the past, it was systematically assumed that there are 6 actin monomers per apparent pitch on the thin actin filaments. More recently, Steffen et al. (2001) suggested a value of 7 actin monomers per half-repeat. In a short discussion, Linari et al. (2005) suggested that the value of 6 is approximate and, from their experimental values, I deduce that 7 actin monomers would also be valid (see also pp. **59–60** for a short discussion of 'fluctuations' in the thin actin filaments). Taking the value of $\sigma \sim$ 6 nm, recalled in the preceding paragraph, gives ~6 nm × 6 ~ 36 nm for the half-pitch of the thin actin filament, the value retained in the preceding paragraph. There is little point in discussing, in this paragraph, the problem of the number of actin monomers per apparent pitch further (see the next paragraph regarding these two numbers of 6 and 7). Some doubts remain about whether σ should be taken as ~6 nm or ~36 nm. In this context, Steffen et al. (2001) did not clearly choose between 36 nm and 5.5 nm (see the next paragraph concerning this value of 5.5 nm for the diameter of the actin monomer, rather than 6 nm). In any event, the discussion presented by these authors is valid only in vitro, not in vivo/in situ, where the steric constraints are much more drastic (e.g. pp. **39–40** in Section 3.4.2). In my group, we demonstrated that $\sigma \sim$ 36 nm is unrealistic from a phenomenological/semi-empirical viewpoint,

when the approach independent of mechanical roles of the cross-bridges, revisited and improved throughout Section 3.4, is taken into account (e.g. Morel 1984a). Furthermore, Bachouchi and Morel (1989a) demonstrated that $\sigma \sim$ 6 nm is structurally possible and even necessary for consistent results to be obtained. Despite these conclusions, drawn more than 25–30 years ago, the value of ~36 nm has frequently been used, even recently. In this connection, Walker et al. (2000) used this spacing of ~36 nm, implicitly mixing this helical repeat and the working stroke of a myosin head, although their experimental work concerned myosin V, which is taken as 'a model for the rigor state [in skeletal muscle]' (Geeves and Holmes 2005), but which is very different from the myosin II found in skeletal muscle. Further information about myosin V, particularly for the value of ~36 nm, can be found in recent papers (e.g. Cappello 2008; Cappello et al. 2007; Coureux et al., 2003; Dunn and Spudich 2007; Sellers and Veigel 2006; Yildiz et al. 2003). In this work, only myosin II (skeletal muscle myosin) is studied and I would like to point out unequivocally that the introduction of $\sigma \sim$ 36 nm into Equations 3.27 and 3.28 (these equations result from phenomenological/semi-empirical studies independent of the mode of force generation) requires 'calculation' of the square root of a negative term, which is biophysically meaningless and definitively rules out the choice of $\sigma \sim$ 36 nm.

Ishii and Yanagida (2000) and Kitamura et al. (2001) suggested taking into account the distance between pairs of neighbouring actin monomers, as done by my group since 1976, but they took $\sigma \sim$ 5.5 nm, rather than the ~6.0 nm used here and in previous publications from my group. These two values of σ may be considered similar and consistent with the mean diameter of a globular actin monomer, as deduced from traditional x-ray diffraction studies on whole frog skeletal muscles by Huxley and Brown (1967) and Huxley (1973d), who obtained the first estimate of ~5.5 nm, and more recent x-ray diffraction data obtained with synchrotron radiation (e.g. Linari et al. 2005; Piazzesi et al. 1999), from which it can be deduced that the true estimate is either ~38 nm/6 ~ 6.3 nm or 38 nm/7 ~ 5.4 nm, consistent with ~6.0 nm and ~5.5 nm, respectively (see the preceding paragraph concerning the two numbers of 6 and 7). A short, but useful and informative, discussion concerning the actin-based layer lines related to the helical periodicities of the thin actin filaments was presented by Linari et al. (2005). In their discussion, they recalled that the helical structure of the thin actin filaments 'gives rise to layer lines with axial spacing of ca 5.9 nm and 5.1 nm'. They also found that the various axial x-ray reflections depended, to various extents, on the temperature and the state of the fibre (rest, contraction and also, probably, rigor), although they noted that the results for thin actin filaments were unclear. Variations in the state of the fibre (rest or contraction) should be compared with the phenomena of thin actin and thick myosin filament extensibility during contraction, as described by Huxley et al. (1994) and Wakabayashi et al. (1994). Bordas et al. (1999) confirmed that the thin actin filaments are extensible in contracting muscles and Tsaturyan et al. (2005) even showed that 'strong binding of myosin heads [to the thin actin filaments] upon transition from relaxation to low-force rigor' induces distortions in the thin actin filaments, with an increase in actin periodicity. Thus, shifting from rest to contraction or rigor can change the 'geometric features' of the thin actin filament, which is therefore a dynamic structure, as described on pp. **33–34** in Section 3.2 and in the next paragraph.

Based on unconventional but attractive concepts, very different from the traditional ideas, Grazi (1997) suggested that rearrangements of the intermonomer contacts in F-actin filaments (in vitro) can occur, depending on the experimental conditions. Thus, by extrapolation of Grazi's hypothesis (1997), the 'rigid body model of the actin filament' may be called into question, rendering the problem of the exact value of σ, around 6 nm, to some extent irrelevant. This conclusion is strongly justified by the assertion of Morel (1991b) that σ is not a 'universal constant', depending instead on the physiological conditions (e.g. forced lengthening of contracting intact fibres). This inference results from a comparison of various experimental data and the phenomenological/semi-empirical model independent of the mode of force generation introduced by Morel et al. (1976) and repeatedly used ever since (e.g. in this book). Ménétret et al. (1991) provided experimental support for the view that thin actin filaments, at least in vitro (F-actin), are not 'frozen' structures in the presence of isolated myosin heads (S1): 'structural changes in actin filaments occur as a result of interaction with S1.... and raises the question of the involvement of structural changes in the actin filaments in the

contraction mechanism'. Borejdo et al. (2004b) and Shepard et al. (2004) demonstrated that the actin globules in the thin actin filaments change orientation during contraction, consistent with previous findings recalled above in this and the preceding paragraphs, again suggesting that the thin actin filament is not rigid and can be distorted during contraction. The two values given above for σ (~5.5 nm and ~6.0 nm) are not very different and lead to similar conclusions (see the next paragraph). At the supramolecular level, Liu and Pollack (2002) confirmed experimentally that the F-actin filament is not a rigid body, but that it 'is more compliant than previously thought'.

Most of the experimental and theoretical work of Yanagida and his coworkers concerns in vitro motility, whereas my group focused on in vivo/in situ contraction. However, in vitro motility is also dealt with in this monograph (a critical analysis and discussion of in vitro motility is presented in Chapter 9). In vitro, the binding of myosin heads to actin monomer in a thin actin filament is not subject to steric hindrance (see, for instance, pp. **33–34** in Section 3.4.2). This is much less evident in vivo/in situ, when the sarcomeres are at their slack length, or slightly, or significantly stretched, because Bachouchi and Morel (1989b) demonstrated that steric hindrance in the cross-bridges inevitably occurs, at rest and under contraction conditions, with considerable distortions of the myosin molecules and the thick myosin filaments, and also, probably, the actin globules and thin actin filaments (see pp. **251–254** in Section 8.9 for precise details, focusing on the experimental point of view). These various distortions are, therefore, certainly very different in vitro and in vivo/in situ. However, to account for many complex experimental phenomena (described on pp. **251–254** in Section 8.9), it should be assumed that the myosin heads can and must attach to any actin monomer during isometric tetanic contraction, regardless of the relative positions of the myosin heads and actin globules and the various distortions. This leads to the conclusion that a value of ~6 nm can be chosen for σ, at least between 0°C and 10°C, regardless of the physiological state of the fibre (rest, contraction and, probably, rigor). The choice of σ value (within the range of ~5.1 nm and ~5.9–6.0 nm; see p. **59**) has no consequence for the 'numerical' conclusions drawn below, within the limits of experimental uncertainty. For example, using a value of ~6 nm, we can show that the proportion of attached cross-bridges, during isometric contraction (short tetanus and steady-state conditions; see Equation 3.28 and corresponding comments; see also the next paragraph concerning the 'list' of necessary data to be introduced into Equation 3.28) is n_∞ ~0.82 ± (0.03 or 0.06), whereas a value of ~5 nm gives n_∞ ~0.83 ± (0.03 or 0.06). These two values, valid for frog at ~2.5°C, are statistically similar, even if the most restrictive statistical test, SE, is used (see Section 2.1 for precise details). On pp. **64–65**, it is recalled that, like various other parameters considered above, in this section, σ may depend on temperature; hence, its value may differ between experiments, as different temperatures are used by independent groups. However, I think that temperature plays a marginal role in the value of σ, because using a value independent of temperature leads to self-consistent results and conclusions (see below). Finally, the exact value of σ, between ~5 nm and ~6 nm, is unimportant, as all values in this range give a suitable proportion n_∞ of attached cross-bridges (see above). In Section 4.4.2.5.2, I study many experimental values of the proportion of cross-bridges attached during isometric tetani, at various temperatures (between ~0°C and ~20°C–22°C), for various durations (between ~1 s and ~10 s), various animal species, and various types of muscle and fibre. Comparing the experimental and the phenomenological/semi-empirical values obtained above (~0.82 ± [0.03 or 0.06] and ~0.83 ± [0.03 or 0.06]), I define a temperature-independent universal value of n_∞ ~ 0.82, valid for any animal species, at any temperature between 0°C and room temperature (see Section 4.4.2.5.2, particularly p. **152**, for precise details on the calculation of n_∞ ~ 0.82).

Introducing all the experimental and semi-empirical values calculated, estimated and selected above, in this section, and elsewhere, valid for frog at ~2.5°C (z ~ 0.157, s_0V_{max} ~5.4 μm s^{-1} [see the next paragraph for precise details on the definition of the velocity of shortening], $\Pi_{isom,\infty}$ ~9.9 s^{-1}, σ ~ 6 nm, $\omega_{isom,\infty}$ ~ 1) into Equations 3.27 and 3.28, gives $n_{\infty,1}$ ~ 0.18 and $n_{\infty,2} = n_\infty$ ~ 0.82 (see also the end of the preceding paragraph, regarding the universal value of ~0.82). Major problems with many experimental data, collected by independent groups, over more than 45 years, concern the proportion of attached cross-bridges during short tetani and under steady-state conditions. This issue is studied in depth in Section 4.4.2.5.2,

where it is demonstrated that ~0.82 is the only suitable value and that ~0.18 cannot be retained.

Before determining the isometric tetanic tension $T_{0,\infty}$ in an intact unit cell, we should define the notion of pseudo-steady-state isotonic contraction. Edman (1979) and Edman and Hwang (1977), using an unconventional technique, measured, on intact frog fibres, a 'maximal shortening velocity' that they called s_0V_0 (the reference sarcomere length s_0 is introduced in Equation 3.3 and is used throughout this book). They found that the values of s_0V_0 were statistically similar to values of s_0V_{max} obtained by other groups using different techniques. From their experimental data, I deduce that the duration of shortening was ~50 ms at ~2.5°C, that is, ~40 ms at 5°C (see below, concerning the Q_{10} for s_0V_{max}). Barman et al. (1998) performed experiments on unheld isolated myofibrils from frog, shortening at the maximum velocity s_0V_{max} (or s_0V_0). The duration of the period of shortening can be estimated from the velocities and amounts of shortening given by these authors (at 4°C), that is, s_0V_{max} ~ 4.8 μm s^{-1} and shortening distance of ~0.5–0.6 μm, corresponding to a duration of ~0.5–0.6 μm/4.8 μm s^{-1} ~ 0.105–0.125 s ~ 105–125 ms ~ 115 ms. Edman (1979) showed that the Q_{10} for s_0V_{max} (s_0V_0) is ~2.67 ± 0.07 and, extending this value to unheld isolated myofibrils, the value of s_0V_{max} ~ 4.8 μm s^{-1} at 4°C becomes ~5.3 μm s^{-1} at 5°C: the duration of shortening at 5°C would be ~0.5–0.6 μm/5.3 μm s^{-1} ~ 0.095–0.115 s ~ 95–115 ms ~ 105 ms. Piazzesi et al. (2007), working on intact frog fibres at 4°C, and Yamada et al. (1993), working on demembranated psoas rabbit fibres at room temperature (~20°C–22°C), demonstrated that active shortening at s_0V_{max} is a complex process. From the paper by Piazzesi et al. (2007), it is difficult to obtain a suitable estimate of the duration of shortening at s_0V_{max}, but a value of ~20 ms seems to be adequate (valid at ~4°C–5°C). From the paper by Yamada et al. (1993), a value of ~30 ms seems to be suitable (at room temperature, i.e. ~20°C–22°C), corresponding to ~145 ms at 5°C. Although obtained with very different biological materials (unheld isolated myofibrils vs. intact and demembranated fibres, from frog or rabbit), these four scattered estimates can be averaged, giving ~77 ± (25 or 50) ms (mean ± SE or SD; n = 4). The maximal interval for this duration is ~27–127 ms, not too far from the value corresponding to the time lag plus the first complex transitory turnover for isometric MgATPase activity obtained for frog (~13–28 + 46–56 ~ 59–84 ms; see p. **54** for precise details; the interval common to ~27–127 ms and ~59–84 ms is ~59–84 ms). Thus, regardless of the biological material, shortening is of a similar duration to the initial complex transitory phenomena occurring during isometric tetani, and 'steady-state isotonic conditions' for maximal shortening velocity were almost certainly never reached in any experiment. Indeed, s_0V_{max} (or s_0V_0) necessarily depends on time t, as it is unavoidably zero at t = 0 and acceleration should occur during much of the shortening process. This 'intuitive' view is supported experimentally by Yamada et al. (1993), who found that $s_0V_{max} \sim 11$ μm s^{-1} (at room temperature, i.e. ~20°C–22°C), under pseudo-steady-state isotonic conditions (see below for definition of this state). They also found that, 'after a delay of ~20–30 ms, all the fibre segments started to shorten with almost the same time course' and the shortening velocity, over this ~20–30 ms period, was only ~1.9 μm s^{-1} and rapidly reached the asymptotic value $s_0V_{max} \sim 11$ μm s^{-1}. This demonstrates that s_0V_{max} depends on the duration of the shortening experiment, with complex kinetics. More recently, Piazzesi et al. (2007) used a cutting-edge technique for measuring s_0V (and also s_0V_{max} by extrapolation), from which they deduced that, at the molecular level, the definition of s_0V (and s_0V_{max}) and the determination of the force–velocity relationship in skeletal muscle are extremely complex. Thus, there is considerable evidence that the notion of maximal velocity of shortening s_0V_{max}/s_0V_0 is many-faceted. Despite these difficulties, it is possible to assume a pseudo-steady-state isotonic contraction, in which the constants of attachment, f, and detachment, g, are assumed to be independent of both time t and s_0V_{max}, with s_0V_{max} strictly independent of time t, from the beginning to the end of shortening. The complexity of the shortening process and the oversimplification concerning f, g and s_0V_{max} (extended here to $s_0V < s_0V_{max}$) were not taken into account, by any group, in previous experimental and theoretical studies on isotonically contracting unloaded fibres. Fortunately, this unintentional oversight in all the previous papers published by my group and by A.F. Huxley (1957) and all the authors describing biophysical approaches to muscle contraction based on the 1957 theory does not result in self-contradictory

conclusions. Nonetheless, the 'muscle community' should be aware of this biophysical problem occurring under isotonic contraction and, in particular, during isotonic transients.

Despite the complex behaviour during active shortening at any value of s_0V, described in the preceding paragraph, the values of f and g during the pseudo-steady-state isotonic contraction (see the end of the preceding paragraph) are assumed to be identical to those during steady-state isometric tetanic contraction (short tetanus) and are called f_∞ and g_∞ (see also p. **47** in Section 3.4.3.1.3). Equation 3.23 gives the general expression of $T_\infty(s_0V) + l_\infty(s_0V)$. The term v_∞^* in this equation is given by Equation 3.39. Thus, Equation 3.23 becomes, after some rearrangements,

$$T_\infty(s_0V) + l_\infty(s_0V) \sim \xi_\infty(s_0V) \text{me}_{app} \omega_\infty(s_0V)$$
$$[1 - \exp(-zs_0V_{max}/s_0V)](s_0/2\sigma)n_\infty^2 \quad (3.33)$$

When $s_0V = 0$ (isometric tetanic conditions), we deduce that

$$T_{0,\infty} + l_{0,\infty} \sim \xi_{0,\infty} \text{me}_{app} \omega_{isom,\infty}(s_0/2\sigma)n_\infty^2 \quad (3.34)$$

with $\xi_{0,\infty} = \xi_\infty(s_0V = 0)$ and $\omega_{isom,\infty} = \omega_\infty(s_0V = 0)$. For $s_0V = s_0V_{max}$, we necessarily have $T_\infty(s_0V_{max}) = 0$ (unloaded intact fibre/intact unit cell) and Equation 3.33 gives

$$l_\infty(s_0V_{max}) \sim \xi_\infty(s_0V_{max}) \text{me}_{app} \omega_\infty(s_0V_{max})$$
$$[1 - \exp(-z)](s_0/2\sigma)n_\infty^2 \quad (3.35)$$

On pp. **196–198** in Section 6.2, the corrective factors required to pass from the actin–myosin system in vivo to the whole system releasing enthalpy are studied. Taking into account the experimental uncertainty and the existence of other more subtle phenomena, it is shown that the universal corrective factor $\xi_{0,\infty}$ is very close to unity (i.e. no correction required to pass from the enthalpy released from actin–myosin interactions to total enthalpy), in whole muscles, intact and demembranated fibres and isolated myofibrils, extended here to intact unit cells, regardless of the animal species (e.g. frog, rabbit, rat, dogfish or toadfish) and temperature. I assume that the corrective factor $\xi_\infty(s_0V_{max})$ is close to $\xi_{0,\infty}(\xi_\infty(s_0V_{max}) \sim \xi_{isom,\infty})$. The notation $\xi_{isom,\infty}$ is maintained in Equations

3.36 and 3.37 below, for the sake of generality, although the value of this factor is close to unity (see above). Otherwise, let us consider that $l_{0,\infty} \sim \upsilon l_\infty(s_0 V_{max})$. In the case of a whole frog muscle, A.V. Hill (1964a,b) found that the relative coefficient of shortening heat α / P_0^* is ~0.16 when P* = 0 (i.e. maximal shortening velocity $s_0 V_{max}$) and ~0.34 under isometric conditions ($s_0 V = 0$) (see p. **44** in Section 3.4.3.1.2), giving $\upsilon \sim 0.34/0.16 \sim 2.12$, but this coefficient could be different in an intact unit cell. In any event, multiplying $l_\infty(s_0 V_{max})$ by the coefficient υ in Equation 3.35, we obtain $l_{0,\infty}$, given by

$$l_{0,\infty} \sim \upsilon l_\infty(s_0 V_{max}) \sim \upsilon \xi_{0,\infty} m e_{app} \omega_\infty(s_0 V_{max})$$
$$[1-\exp(-z)](s_0/2\sigma)n_\infty^2$$
$$(3.36)$$

Subtracting Equation 3.36 from Equation 3.34 gives

$$T_{0,\infty} \sim \xi_{0,\infty} m e_{app} \{\omega_{isom,\infty} - \upsilon \omega_\infty(s_0 V_{max})$$
$$[1-\exp(-z)]\}(s_0/2\sigma)n_\infty^2$$
$$(3.37)$$

All the parameters in Equations 3.36 and 3.37 are known at ~2.5°C, except $\omega_{isom,\infty}$, $\omega_\infty(s_0 V_{max})$ and υ. However, on pp. **73–74** in Section 3.4.4, it is demonstrated, on experimental and semi-empirical bases, that taking $\omega_{isom,\infty} \sim 1$ and $\omega_\infty(s_0 V_{max}) \sim 0.27$ (for frog, at temperatures of ~0°C–5°C) leads to self-consistent conclusions. On p. **193** in Section 6.2, it is shown that, in vivo/in situ, $e_{app} \sim 6.6 \times 10^{-20}$ J per MgATP molecule. On p. **206** in Section 6.3.5, it is shown that, in a composite animal species, including the frog, the 'density' of cross-bridges is $m \sim 1.16 \ 10^{23} \ m^{-3}$. s_0 is the reference sarcomere length, taken as ~2.18 μm (see p. **23** in Section 3.2). σ is the distance between two consecutive binding sites of a cross-bridge on a thin actin filament ($\sigma \sim 6$ nm; see pp. **58–60** for a detailed discussion of this value). $z \sim 0.157$ at 0°C–5°C (see text below Equations 3.27 and 3.28). n_∞ is calculated on p. **61**, estimated in Section 4.4.2.5.2, and shown to be ~0.82, regardless of temperature. As recalled above, $\xi_{isom,\infty}$ is taken as ~1. Finally, the parameter υ is unknown, but we can write the following relationships (see Equations 3.35 through 3.37):

$$T_{0,\infty} \sim 8.00(1 - 0.027\upsilon) \times 10^5 \ N \ m^{-2};$$
$$l_\infty(s_0 V_{max}) \sim 2.16 \times 10^5 \ N \ m^{-2};$$
$$l_{0,\infty} \sim 2.16\upsilon \times 10^5 \ N \ m^{-2} \qquad (3.38)$$

and also

$$l_{0,\infty}/T_{0,\infty} \sim 0.27\upsilon/(1 - 0.027\upsilon) \ \text{and}$$
$$l_\infty(s_0 V_{max})/T_{0,\infty} \sim 0.27/(1 - 0.027\upsilon) \quad (3.39)$$

At this point, I suggest selecting three values of υ around 2.12 (this value of ~2.12 is valid for a whole muscle; see text between Equations 3.35 and 3.36); that is, $\upsilon = 1, 2, 4$. Thus, at ~2.5°C, choosing $\upsilon = 1$, we obtain the following values: $T_{0,\infty} \sim 7.78 \times 10^5 \ N \ m^{-2}$, $l_{0,\infty}/T_{0,\infty} \sim 0.28$ and $l_\infty(s_0 V_{max})T_{0,\infty} \sim 0.28$. Choosing $\upsilon \sim 2$, we obtain $T_{0,\infty} \sim 7.57 \times 10^5 \ N \ m^{-2}$, $l_{0,\infty}/T_{0,\infty} \sim 0.57$ and $l_\infty(s_0 V_{max})/T_{0,\infty} \sim 0.29$. Finally, choosing $\upsilon \sim 4$, we obtain $T_{0,\infty} \sim 7.14 \times 10^5 \ N \ m^{-2}$, $l_{0,\infty}/T_{0,\infty} \sim 1.21$ and $l_\infty(s_0 V_{max})/T_{0,\infty} \sim 0.30$. I now suggest averaging these three series values, to obtain the best estimates, giving $T_{0,\infty} \sim 7.5 \pm (0.2 \text{ or } 0.3) \times 10^5 \ N \ m^{-2}$ (mean ± SE or SD; n = 3), $l_{0,\infty}/T_{0,\infty} \sim 0.69 \pm (0.22 \text{ or } 0.39)$ (mean ± SE or SD; n = 3) and $l_\infty(s_0 V_{max})/T_{0,\infty} \sim 0.29 \pm 0.01$ (mean ± SD; n = 3). Introducing the mean value of $T_{0,\infty}$ into the expression of $T_{0,\infty}$ given by Equation 3.39, we deduce $\upsilon \sim 2.38$, very close to the value of 2.12 deduced from the experimental studies of Hill (1964a,b) obtained on whole frog muscle (see the first column on this page). In a whole frog muscle, the isometric tension is usually considered to be ~2.5 × 10⁵ N m⁻² and the two relative values of the coefficients of shortening heat are ~0.34 and ~0.16 (see text between Equations 3.37 and 3.38). The three corresponding values in an intact unit cell are greater than those in a whole muscle, but the relative coefficients of shortening heat vary similarly with increasing shortening velocity. The origin of the difference between whole muscle, intact fibre and intact unit cell is briefly discussed in Section 3.4.3.1.2.

For intact frog fibres, at maximal overlap, in a buffer mimicking the physiological extracellular medium and within the 2°C–12°C temperature range, Edman (1979) found the Q_{10} for the measured isometric tetanic tension, T_f, to be ~1.24 (see Section 3.3 for definitions and precise details concerning T_f). $T_{0,\infty}$ and T_f are significantly different at a given temperature (see below), but I assume, as a simplifying hypothesis, that the Q_{10} for $T_{0,\infty}$ is similar to that for T_f, giving $T_{0,\infty}(10°C) \sim 7.5 \times (1.24)^{0.75} \sim 8.8 \times 10^5 \ N \ m^{-2}$. Within the temperature range 0°C–5°C, Homsher (1987) suggested that $Q_{10} \sim 1.30$. Working between 2°C and 17°C, Decostre et al. (2005) found that the average force per attached myosin head increases by ~60% when passing from 2°C to 17°C. The authors used the traditional approach in which the isometric tetanic tension is proportional to the force per head and the Q_{10} for T_f (between 2°C and 17°C) is therefore ~1.37 (value deduced from the ~60% increase; not given by the authors; simple calculations not shown). Working at 0°C or 17°C, Linari et al. (2005) obtained values of isometric tetanic tension from which it can be deduced that the value of Q_{10} is ~1.27 (value not given by the authors). The values of ~1.30, ~1.37 and ~1.27 are similar to that of ~1.24 determined by Edman (1979). They are also consistent with the values recalled in the review by Rall and Woledge (1990): $Q_{10} \sim 1.2$ to ~1.4 within the range 0°C–20°C, and increase of ~40% or more in peak tetanic tension when passing from 0°C to 20°C; that is, $Q_{10} \sim 1.18$ (value not given by the authors). Averaging these seven available values gives $Q_{10} \sim 1.28 \pm (0.03$ or 0.08) (SE or SD; n = 7). Only the value of ~1.24 is retained and extended to the range 0°C–20°C, for two reasons: (i) because it results from a systematic experimental study and (ii) because it is statistically consistent with the average value and the corresponding value of SD (see Section 3.1 concerning the advantage of using SD). This series of results indicates that isometric tension is little dependent on temperature. In Section 3.3, it is demonstrated that, for frog at 10°C, the tensions recorded in intact fibres and half-fibres are $T_f \sim T_{hf} \sim 3.0 \times 10^5$ N m^{-2}, indicating that the tension $T_{0,\infty}$(10°C) calculated here for an intact unit cell (~8.8 × 10^5 N m^{-2}; see above) is ~3 times that recorded in an intact fibre. This sizeable multiplication factor is studied and discussed in detail in Section 3.7. The best estimate of the isometric tetanic tension in an intact unit cell, at 10°C, given on p. **90** in Section 3.7, is $T^* \sim 7.2 \times 10^5$ N m^{-2}, that is, ~2.4 times the tension $T_f \sim T_{hf} \sim 3.0 \times 10^5$ N m^{-2} recorded in a single intact fibre or single half-fibre.

In Section 4.4.2.5.2 (see particularly p. **152**), it is deduced, from various experimental results obtained by independent groups, that the universal proportion n_∞ of attached cross-bridges does not depend on temperature. It is interesting to present here a semi-empirical approach to this problem. Bershitsky et al. (1997) and Piazzesi et al. (2003) have shown that, when temperature increases, the subsequent increase in force (see Section 8.8) is not accompanied by an increase in stiffness of the muscle sarcomeres. Linari et al. (2005) concluded from this that the increase in force observed when the temperature is increased 'is not caused by an increase in the number of myosin cross-bridges interacting with the thin actin filaments'. This assertion is not straightforward, as recalled in Section 5.1.4, where it is pointed out that there is an ill-defined relationship between stiffness and the number of attached cross-bridges. I therefore conclude that there is no straightforward demonstration of possible variations of n_∞ with temperature. Fortunately, this problem can be overcome. Let us consider Equation 3.37, giving a simple relationship between the isometric tension $T_{0,\infty}$ and the proportion n_∞ of bound cross-bridges, in an intact unit cell. On p. **78** in Section 3.4.5, it is demonstrated that the parameter $\varsigma_{0,\infty} = \xi_{0,\infty}\{\omega_{isom,\infty} - \upsilon\omega_\infty(s_0 V_{max})[1 - \exp(-z)]\}n_\infty/[\omega_{isom,\infty}z(1 - n_\infty)]$ is independent of temperature. Thus, we can write Equation 3.37 in a different form (calculations not shown): $T_{0,\infty} = me_{app}s_0\varsigma_{0,\infty}[\omega_{isom,\infty}zn_\infty(1 - n_\infty)/2\sigma]$. In this formula, only the term in brackets may depend on temperature. The dependence of the individual parameters in brackets cannot be predicted. In the preceding paragraph (see text below Equation 3.37) and in Section 3.4.4 (see text between Equations 3.46 and 3.47), it is assumed that $\omega_{isom,\infty} \sim 1$, regardless of temperature, at least within the 0°C–12°C range, and this hypothesis is maintained here. The experimental studies of Linari et al. (2005) suggest that σ may depend on temperature. The definition of z, given in the paragraph between Equations 3.25 and 3.26, suggests that this parameter probably depends on temperature. It is shown, on p. **61**, that the two values of σ taken into account, ~5 nm or ~6 nm, correspond, at ~2.5°C, to the phenomenological/semi-empirical values of $n_\infty \sim 0.83 \pm 0.03$ or $\sim 0.82 \pm 0.03$, respectively (only SEs are taken into account, as an illustrative example). Thus, at ~2.5°C, we have $n_\infty(1 - n_\infty)/\sigma \sim 0.0282 \pm 0.0040$ nm^{-1} or ~ 0.0246 ± 0.0032 nm^{-1}, respectively, and the maximal ratio between these two estimates, related solely to the uncertainty on the exact value of σ, is ~ 0.0282 ± 0.0040 nm^{-1}/(0.0246 ± 0.0032) nm^{-1} ~ 1.146 ± 0.317. This estimate is commented on in the next paragraph. Note that, using SDs, the most suitable statistical tests (see Section 2.1) rather than SEs results in an error significantly greater than ~±0.317.

Morel

On pp. **63–64**, it is reasonably assumed that, between 0°C and 20°C, the Q_{10} for $T_{0,\infty}$ is ~1.24, which is statistically similar to ~1.146 ± 0.317 (maximal value of ~1.463). The most probable value of $\omega_{isom,\infty}$ is ~1 (see the preceding paragraph). At this stage, there is no point trying to deduce possible variations, with temperature, of the individual parameters in the term in bracket $[\omega_{isom,\infty} z n_\infty (1 - n_\infty)/2\sigma]$ presented in the preceding paragraph. Indeed, taking into account only the limited uncertainty on the values of n_∞ (see again the preceding paragraph) results in the uncertainty on the values of $n_\infty(1 - n_\infty)/\sigma$, at ~2.5°C, largely sufficient to blur any conclusion that may be drawn regarding these individual parameters. This results from the weak dependence of isometric tension on temperature (see above for the value of Q_{10}). Thus, the best solution is to retain all the values obtained in the preceding paragraphs, particularly the universal proportion $n_\infty \sim 0.82$, estimated on p. **61**, and in Section 4.4.2.5.2.

3.4.3.3 Some Temporal and Structural Features of the Cross-Bridge Attachment–Detachment Cycle and the Process of MgATP Splitting in an Intact Unit Cell. Case of Young Adult Frogs, with Some Insights into Other Animal Species

Before going into detail, let us give some definitions and recall some problems. In short tetani, under steady-state conditions (see Section 3.4.3.2 for definitions and precise details on these two conditions), it is interesting to estimate the order of magnitude of the duration, $\tau_{c,\infty}$, corresponding to attachment of a cross-bridge, hydrolysis of one MgATP and detachment of the cross-bridge (at least in the case of a simple two-state model, studied in this book). This duration can be defined as $\tau_{c,\infty} \sim 1/R_{isom,\infty}$, where $R_{isom,\infty}$ is the mean turnover rate of the cycling cross-bridges. The working (power) stroke or step size distance of the cross-bridges is a matter of controversy (including the value of the axial displacement of the head for each MgATP molecule hydrolysed and the number of working strokes for each MgATP molecule cleaved). It is even recalled, on p. **66**, that the notions of working stroke or step size distance are meaningless for a purely isometric tetanic contraction with no possibility of myosin head tilting. Below, in this section, estimates of the duration of the cycle of the cross-bridges (attachment, MgATP splitting and detachment) and its

structural characteristics (working stroke and step size distance) are proposed and discussed, together with other related problems.

Under isometric tetanic conditions, $R_{isom,\infty}$ $(2.5°C) \sim \Pi^0_{isom,\infty}(2.5°C) \quad \sim 9.9 \pm (1.0\ or\ 2.5)\,s^{-1}$ for frog (see p. **58** in Section 3.4.3.2). The Q_{10} for $\Pi_{isom,\infty} \sim R_{isom,\infty}$ is ~3.1, regardless of temperature (see p. **202** in Section 6.3.3) and, from the value of $R_{isom,\infty}(2.5°C)$, we deduce that $R_{isom,\infty}(5.7°C)$ $\sim [9.9 \pm (1.0\ or\ 2.5)]\ s^{-1} \times 3.1^{0.32} \sim 14.2 \pm (1.4$ or 3.6) s^{-1}. Thus, $\tau_{isom,\infty}(5.7°C) \sim 1/R_{isom,\infty}(5.7°C)$ ranges between ~56 ms and ~94 ms (maximal interval). Otherwise, again for frog, the value of the quantum duration Δt at 5.7°C, for a single turnover, is found to be ~46–56 ms (see p. **54** in Section 3.4.3.2) and the values of $\Delta t(5.7°C)$ and $\tau_{c,\infty}(5.7°C)$ are therefore statistically consistent, indicating that, at 5.7°C and, almost certainly at any temperature within the range ~0°C–10°C (favorable range for frog, a cold-blooded animal), we have $\tau_{c,\infty} \sim \Delta t$ (in the rest of this work, only the individual values of $\Delta t \sim 46$–56 ms and $\tau_{c,\infty}$ ~ 56–94 ms are considered). Irving (1987) suggested that, for frog fibres contracting within the same range of temperature as used here, the cycling cross-bridges remain bound to the thin actin filaments for only ~2 ms. This estimate corresponds to a very small proportion of the total cycle duration (~100 × 2 ms/(56–94) ms ~ 2%–4%): the duration of binding of ~2 ms can therefore be neglected. Working on whole frog muscles at 0°C, Curtin et al. (1974) presented the first estimate of 'the mean cycle time of a cross-bridge during working and isometric contractions' and found ~340 ms under isometric tetanic contraction conditions. However, they used the value of ~264 μM for the concentration of myosin heads able to enter this cycle. As highlighted on p. **206** in Section 6.3.5, the most realistic concentration is ~193 μM, giving a corrected cycle time of ~ (193 μM/264 μM) × 340 ms ~ 248 ms. Correcting also for temperature and taking Q_{10} ~ 3.1 for the rate of MgATP splitting (see above), this value becomes, at 5.7°C, ~248 ms × $(3.1)^{-0.57}$ ~ 132 ms, which is not too far from $\tau_{c,\infty} \sim 56$–94 ms (see above).

Worthington and Elliott (1996a,b) studied some temporal and structural characteristics of an unconventional approach to muscle contraction (impulsive model, first introduced by Worthington 1962, 1964). They defined an 'average time interval [t] between ATP splitting on sites along the actin filament'. They pointed out that their 'analysis is model-independent', but this assertion is very different from the model-independent approach presented throughout this Section 3.4. Regardless of temperature, but presumably within the ~0°C–10°C range (for frog, a cold-blooded animal), under steady-state isometric conditions and for short tetani, the authors suggested a value of ~2.5 ms for the 'time interval' t (see Figure 3 in the paper by Worthington and Elliott 1996a). They also defined an 'impulse-time' Δt (not to be confused with the quantum duration Δt defined in the text between Equations 3.29 and 3.30, 3.31 and 3.32), corresponding to the 'time interval of the force-generating event', which they found to be ~0.5 ms under both isometric tetanic and isotonic conditions. Commenting later on the value of ~2.5 ms for t, Worthington and Elliott (2003) chose to replace this value with an average value of ~1 ms, but they gave no clear estimate of Δt. I therefore suggest retaining the value of ~0.5 ms. All these values are two or three orders of magnitude lower than those deduced from recent experiments and estimates (~46–56 ms and ~56–94 ms; see the preceding paragraph). Worthington and Elliott (1996a,b) also found step size distances, corresponding to the splitting of one MgATP molecule (at s_0V_{max}; zero load), of only ~1.3 nm for dogfish, ~1.7 nm for frog and ~2.8 nm for tortoise, and Worthington and Elliott (2003) confirmed these values and obtained refined values ranging between ~1.4 nm and ~4.5 nm for a series of animal species and different types of muscle, at different temperatures (see p. **67** for precise details).

All the values recalled in the preceding paragraph are much lower than the traditional values, particularly for step size distances. In this context, many authors used various biological materials (whole muscles, single intact fibres, skinned or permeabilised fibres and isolated myofibrils) and various experimental techniques and conditions, resulting in considerable scattering of the reported step size distances, but these distances were systematically greater than the most recent values of ~1.4–4.5 nm found by Worthington and Elliott (2003). The many values proposed by Worthington and Elliott are, therefore, not taken into account here. The semi-empirical value estimated by Barclay (1996a,b, 1999, 2003) was ~8.7 nm, and this value is retained below. Six years later, Barclay (2005) suggested that a value of ~1.3

nm, not considered below, would be more suitable than a value of ~13 nm. Huxley (1957) deduced from the experimental work of Hill (1938a) a semi-empirical value of ~16 nm. Ford et al. (1977) and Huxley and Simmons (1971, 1973) obtained experimental values of ~12 nm. Huxley and Kress (1985) suggested that the cross-bridges bind to the thin actin filaments over a distance of ~12 nm. Irving et al. (1992) and Yamada et al. (1993) obtained an experimental value of ~10 nm. Ma and Taylor (1994) proposed a value of ~19 nm. Pate et al. (1993) deduced, from their experimental data, a value of ~7 nm. Taylor (1989) suggested a value of ~20 nm. White et al. (1995) gave an estimate of ~4.6 nm. A value of ~11 nm was deduced from structural data (Holmes and Geeves 2000). Using x-ray diffraction data from synchrotron radiation, Huxley et al. (2006), Piazzesi et al. (2002a) and Reconditi et al. (2004) obtained values of ~6 nm to ~10 nm, ~6 nm to ~10 nm and ~7.5 nm to ~10 nm, respectively (only the minimal and maximal estimates are considered here). Combining x-ray diffraction data, physiological measurements and a model-dependent (swinging cross-bridge/lever-arm mechanism) semi-empirical reasoning, Piazzesi et al. (2007) suggested values lying between ~5 nm and ~7 nm (only these two values are considered here). These 21 values give a mean value of the step size distance of ~10.2 ± (0.9 or 4.1) nm (mean ± SE or SD; n = 21), lying within the ~5–15 nm range suggested by Ma and Taylor (1994), who pointed out that the actual value was not accurately known in the 1990s. Note that Yanagida et al. (1985) obtained very high values (≥60 nm) and Higuchi and Goldman (1991) obtained a value of ~60 nm. These two series of values are significantly higher than the other available estimates recalled here and are not taken into account. When in vitro motility assays are performed, under conditions assumed to be 'similar' to those of unloaded fibres or unheld isolated myofibrils shortening at s_0V_{max}, step size distance has been found to be highly scattered (see Chapter 9 for a critical discussion of this 'similarity' and several other related problems). For instance, Harada et al. (1990) and Ishijiama et al. (1991) obtained distances ≥100 nm, Molloy et al. (1995a,b) obtained a distance of ~4 nm, Toyoshima et al. (1990) obtained a distance of ~8 nm, Uyeda et al. (1990) obtained distances of ~10–28 nm and Uyeda et al. (1991a,b) obtained distances of ~5–20 nm. As highlighted by Ma and Taylor (1994), 'the problem is to explain why there is a large variation in the values obtained for 'step distance' (reviewed by Burton 1992)'. Morel et al. (1993) analysed this problem, occurring in vitro, on the basis of their 'molecular jet' hypothesis, to account for in vitro motility (see pp. **270–271** and p. **272** in Chapter 9 for a brief discussion of the molecular jet). We pointed out that our conclusions, drawn for the molecular jet hypothesis, were almost certainly valid for other theories applied in vitro (e.g. rotating cross-bridge models). We demonstrated that there was a close correlation between the 'chosen' MgATPase activity and the calculated step size distance. Morel and D'hahan (2000), in Section 4 of their minireview entitled 'What is the true MgATPase activity to be taken into account?', summarised this confusion and gave some numerical examples (i.e. the case of the square root of a negative value, when the true value of the rate of MgATP splitting is taken into account, rather than that measured in vitro by Ishijiama et al. 1991). In Section 3.4.3.2, it is also recalled that, even in chemically skinned and permeabilised fibres from frog or rabbit, there are broad variations in MgATPase activity: for example, for a given fibre, under isometric conditions, this activity is maximal during the first few milliseconds and then rapidly decreases (see, e.g. p. **48** in Section 3.4.3.2), the duration of the tetanus also being an important parameter, with the rapid onset of fatigue affecting the rate of MgATP cleavage.

Taking into account different experimental data, and using a semi-empirical approach different from the one they had previously used (Worthington and Elliott 1996a,b), Worthington and Elliott (2003) showed that the step size distance depends on the animal species. They found a maximal value for the step size distance of ~1.8 nm (for frog), when the fibre is unloaded and shortens at s_0V_{max}. Indeed, they demonstrated that the step size distance depends strongly on the load lifted and, therefore, on the shortening velocity. Starting from observation and from the dependence of step size distance on animal species (between ~1.4 nm and ~2.6 nm; except for tortoise, ~4.5 nm), Worthington and Elliott (2003) claimed that the 'number of 15 nm has been quoted so often, and for over four decades, that anyone not fully conversant with muscle research might take for granted that this number is a fundamental constant'. Obviously, the same could be said of the mean value of ~10.2 nm estimated in the preceding paragraph (using experimental data from many independent groups). Barclay (2005), in line with Worthington and Elliott (2003), claimed that a value of 1.3 nm is better than 13 nm (see also the beginning of the preceding paragraph). A consequence of the semi-empirical approach of Worthington and Elliott (2003) is that, for a purely isometric tetanic contraction, with no possibility of microscopic sliding in an intact unit cell, the step size distance is obviously zero and the notion of working (power) stroke is meaningless, because, intuitively, the myosin heads cannot swing. More precisely, Elliott and Worthington (2001), revisiting their approach to the impulsive model (see Worthington and Elliott 1996a,b), defined the step size distance $z = vt$ (see text between their Equations 1 and 2; the parameter z should not be mistaken for the dimensionless parameter z defined on p. **47** in Section 3.4.3.1.3), where t is 'average time interval' (see pp. **65–66**) and v is the shortening velocity: when $v = 0$, $z = 0$ (pure microscopic isometric contraction), confirming the conclusion recalled above. At this point, it is important to recall the experimental work of Higuchi and Goldman (1995), who found that the 'interaction distance' ranges between ~0 nm and ~8 nm under macroscopic isometric tetanic conditions and increases with shortening velocity (up to ~60 nm or more, at s_0V_{max}). Piazzesi et al. (2003) showed that the working stroke is force related and decreases as the muscle approaches isometric conditions. Reconditi et al. (2004) demonstrated that the working stroke is smaller at higher load, but not zero under macroscopic isometric tetanic conditions. The question of macroscopic vs. microscopic isometric tetanic contraction and that of microscopic movements and oscillations in an intact unit cell (or a limited number of sarcomeres in a whole fibre) are studied and discussed in Section 3.7. All these results are consistent with step size distance decreasing with shortening velocity and becoming zero for a pure isometric tetanic contraction. This and the problem of the rate of MgATP splitting pointed out in the preceding paragraph would account for the large differences between the various estimates of step size distance.

Despite the concordance between the qualitative inferences deduced in the preceding paragraph, I find no straightforward explanation for the much lower orders of magnitude of the temporal and structural features in the approaches of Elliott and Worthington (2001) and Worthington and Elliott (1996a,b, 2003) than in the approach presented in this monograph (regarding the temporal characteristics, independent of the mode of force generation) and in swinging cross-bridge/lever-arm theories (regarding step size distance). Obviously, the views of Elliott and Worthington (2001) and Worthington and Elliott (1996a,b, 2003) are very different from the traditional views and also that of the phenomenological/semi-empirical model-independent approach presented in this book. However, no definitive explanation can yet be found for the apparently contradictory conclusions drawn from the model-independent approaches of Elliott and Worthington (2001) and Worthington and Elliott (1996a,b, 2003) and alternative models. The discrepancies may be related to differences between intact unit cells, isolated myofibrils, skinned fibres and whole intact fibres, as highlighted in Section 3.7 and, more generally, throughout this monograph. Nonetheless, Elliott and Worthington (2001) follow this thought and distinguish between a 'macroscopic model', valid for whole muscle, and a 'microscopic model', valid for half-sarcomere. Regardless of this specific question, another problem relates to the choice of the rate of MgATP splitting, under isometric and isotonic conditions. Indeed, in Section 3.4.3.2, this issue is shown to be complex, even during isometric tetanic contraction, and it is shown that, under isotonic conditions at s_0V_{max} (for unloaded chemically skinned and permeabilised fibres), MgATPase activity is not very different from that at $s_0V = 0$ (isometric conditions) (see the next paragraph), making the situation even more difficult. In this book, most of these problems are resolved and the general conclusions drawn can be used in future swinging cross-bridge/lever-arm models and future impulsive models (built on the initial work of Worthington 1962, 1964, and further developments of Elliott and Worthington 1997, 2001, and Worthington and Elliott 1996a,b, 2003, 2005).

Bagshaw (1993) proposed an approximate formula for the rate $\Pi(s_0V)$ of MgATP breakdown, which can be written $\Pi(s_0V) \sim s_0V \times n^*(s_0V)/$ (myosin step size), where $n^*(s_0V)$ is the mean proportion of attached cross-bridges at the shortening velocity s_0V (see p. **71** in Section 3.4.4 for definition of this notion of 'mean proportion of attached cross-bridges' $n^*(s_0V_{max})$, in the particular case $s_0V = s_0V_{max}$). The above formula is valid, in particular, for single intact unloaded fibres. For frog, the only available experimental data give $\Pi^{**}(s_0V_{max})/\Pi^{**}_{isom} \sim 1.0$, an old value valid for whole frog muscles at $\sim 0°C$ (see p. **42** in Section 3.4.3.1.1). Experiments have more recently been performed, at 12°C, on permeabilised rabbit fibres, with results that should be interpreted as $\Pi^*(s_0V_{max})/\Pi^*_{isom,\infty} \sim 1.4$ (see p. **75** in Section 3.4.4). Thus, a mean estimate would be $\Pi(s_0V_{max})/\Pi_{isom,\infty} \sim 1.2 \pm 0.2$, regardless of temperature, animal species and biological material (however, no experimental data are available for unheld isolated myofibrils, as highlighted below). I suggest that this ratio of $\sim 1.2 \pm 0.2$ is also valid in intact unit cells. The velocity of shortening s_0V_{max} was measured at several temperatures and the reference temperature is taken here as 4°C. The values of s_0V_{max}, recalled below, correspond mostly to frog and rabbit fibres and myofibrils. As recalled on p. **76** in Section 3.4.5, the Q_{10} for s_0V_{max} in intact frog fibres is ~ 2.67, between 2°C and 12°C and, owing to the large scattering of the experimental results, this value can be extended to any biological material, at any temperature between $\sim 0°C$ and $\sim 20°C–22°C$ (room temperature). Thus, the value $s_0V_{max} \sim 5.4 \, \mu m \, s^{-1}$, measured at $\sim 2.5°C$ (see text below Equations 3.27 and 3.28), becomes $\sim 6.3 \, \mu m \, s^{-1}$ at 4°C. Elzinga et al. (1989b), working on intact frog fibres at $\sim 0°C$, found $s_0V_{max} \sim 4.26 \, \mu m \, s^{-1}$, giving $\sim 6.3 \, \mu m \, s^{-1}$ at 4°C. Woledge et al. (1985) proposed $s_0V_{max} \sim 2.8 \, \mu m \, s^{-1}$ for whole frog muscles at 0°C, giving a value of $\sim 4.1 \, \mu m \, s^{-1}$ at 4°C. Piazzesi et al. (2007) presented a complex experimental study and semi-empirical analysis, in which they tried to find molecular mechanisms accounting for the force–velocity relationship, using intact frog fibres at 4°C. They presented experimental results obtained by x-ray diffraction from synchrotron radiation and mechanical measurements, and semi-empirical approaches based on the traditional concepts (swinging cross-bridge/lever-arm models). However, I have identified several flaws, in their study. For example, extrapolating the two curves presented in their Figure 1C to zero load (maximal shortening velocity), I obtained two values of s_0V_{max}, that is, $\sim 20 \, \mu m \, s^{-1} \times 2 \sim 40 \, \mu m \, s^{-1}$ (for their phase 4) and $\sim 2.0 \, \mu m \, s^{-1} \times 2 \sim 4.0 \, \mu m \, s^{-1}$ (for their phase 2), whereas extrapolating their Figure 3A to zero load, I estimate s_0V_{max} at $\sim 2.8 \, \mu m \, s^{-1} \times 2 \sim 5.6 \, \mu m \, s^{-1}$. Working on mechanically skinned frog fibres at $\sim 0°C–5°C$, in bathing buffers mimicking the sarcoplasmic medium, Ferenczi et al. (1984a) found $s_0V_{max} \sim 4.5 \, \mu m \, s^{-1}$, which is also approximately valid at 4°C. With unheld isolated frog myofibrils at 4°C,

Morel

Barman et al. (1998) found $s_0 V_{max} \sim 4.8$ µm s^{-1} and reported a value of ~7.2 µm s^{-1} at 5°C–7°C, that is, ~5.0 µm s^{-1} at 4°C. With unheld isolated myofibrils from rabbit psoas muscle studied at 4°C, Lionne et al. (1996) found that $s_0 V_{max} \sim 0.6$ µm s^{-1}. Also working on isolated myofibrils from rabbit psoas muscle, but at 20°C, Ma and Taylor (1994) found that $s_0 V_{max} \sim 12$ µm s^{-1}, giving ~2.5 µm s^{-1} at 4°C. Working on chemically skinned fibres from psoas rabbit muscle at 10°C, Cooke and Bialek (1979) found that $s_0 V_{max} \sim 4.6$ µm s^{-1}, corresponding to ~2.5 µm s^{-1} at 4°C. With the same biological material, Higuchi and Goldman (1991) found, at room temperature (~20°C–22°C), that $s_0 V_{max} \sim 5.5$ µm s^{-1}, giving ~1.0 µm s^{-1} at 4°C. Higuchi and Goldman (1995) obtained, under the same temperature conditions, three different values of $s_0 V_{max}$: ~13.6 µm s^{-1}, ~9.6 µm s^{-1} and ~5.4 µm s^{-1}, giving, at 4°C, ~2.3 µm s^{-1}, ~1.6 µm s^{-1} and ~0.9 µm s^{-1}, respectively. Under the same conditions, Yamada et al. (1993) found that $s_0 V_{max} \sim 11$ µm s^{-1}, that is, ~1.9 µm s^{-1} at 4°C. Brenner (1980) found, on this biological material, at 2°C, that $s_0 V_{max} \sim 1.14$ µm s^{-1}, equivalent to ~1.4 µm s^{-1} at 4°C. Working on intact fibres from dogfish muscle at 12°C, Curtin and Woledge (1988) found that $s_0 V_{max} \sim 8.0$ µm s^{-1}, giving $s_0 V_{max} \sim 3.6$ µm s^{-1} at 4°C, consistent with the other values given above, for frog and rabbit. There is, therefore, a broad scattering of the experimental results, giving values of $s_0 V_{max}$ ranging from ~0.6 µm s^{-1} to ~40 µm s^{-1}. The value of ~40 µm s^{-1} is clearly inconsistent with all the other values and I eliminate it in the calculations below. However, a large scattering remains (from ~0.6 µm s^{-1} to ~6.3 µm s^{-1}). I therefore suggest, as throughout this monograph, that all these corrected data should be averaged, giving a value, valid at 4°C, for frog, rabbit and dogfish (intact fibres, chemically skinned and permeabilised fibres and unheld isolated myofibrils). I obtain $s_0 V_{max} \sim 3.3 \pm (0.4$ or $1.8)$ µm s^{-1} (mean $\pm$ SE or SD; n = 18). This average value should not be confused with that of ~5.4 µm s^{-1} at ~2.5°C for intact frog fibres (see p. **60** in Section 3.4.3.2), because all the other data on pp. **60–61** in Section 3.4.3.2 are valid solely for frog intact fibres, but not for a 'composite biological model'. We also have n*$(s_0 V_{max}) \sim 0.08$ (universal value resulting from a combination of data for frog and rabbit at temperatures ranging between 0°C and 12°C; see p. **72** in Section 3.4.4). The problem of the value of the size of the myosin step is complex, as discussed on pp. **65–68**. However, the best solution is to take into account the value of $\sim10.2 \pm (0.9$ or $4.1)$ nm (mean $\pm$ SE or SD; n = 21), the mean of the most frequently used values (see p. **66**), regardless of the animal species and type of biological material used. No experimental data are available for $\Pi(s_0 V_{max})/\Pi_{isom,\infty}$ in the case of unheld isolated myofibrils studied, for instance, by Barman et al. (1998), Lionne et al. (1996) and Ma and Taylor (1994), and, at the time, it is probably impossible to obtain estimates of this ratio (see Section 3.4.3.2 for a long, critical and detailed discussion of the complex problems related to measurements of MgATPase activity, particularly for unheld isolated myofibrils shortening at $s_0 V_{max}$). Thus, it can be assumed that $\Pi(s_0 V_{max})/\Pi_{isom,\infty}$ is also $\sim1.2 \pm 0.2$ for unheld isolated frog myofibrils at 4°C (see p. **68**). This ratio can, therefore, be considered universal, regardless of the biological material used. Using the most restrictive statistical test, SE (see Section 2.1 for precise details), and taking into account n*$(s_0 V_{max}) \sim 0.08$ (see above), $\Pi(s_0 V_{max})/\Pi_{isom,\infty} \sim 1.2 \pm 0.2$, $s_0 V_{max} \sim 3.3 \pm 0.4$ µm s^{-1} (see above) and $\sim10.2 \pm 0.9$ nm for the myosin step size (see above), we deduce, from Bagshaw's approximate formula, $\Pi_{isom,\infty} \sim s_0 V_{max} \times$ n*$(s_0 V_{max})]/[(1.2 \pm 0.2) \times$ (myosin step size)] ~ 14.9 s^{-1} to 31.8 s^{-1} (maximal interval, deduced from the mean values and the SDs), at ~4°C. This temperature is lower than the ~5.7°C referred to below. In principle, no correction for temperature is possible, because the Q_{10} for a composite animal species is not known. However, the Q_{10} for rabbit is ~2.8 (see p. **201** in Section 6.3.3) and that for frog is ~3.1 (see p. **202** in Section 6.3.3): a value of $\sim(2.8 + 3.1)/2 \sim 2.9$ can therefore be selected for a composite animal species. Thus, the values of ~14.9 s^{-1} to ~31.8 s^{-1}, at 4°C, become ~17.8 s^{-1} to ~38.2 s^{-1}, respectively, at 5.7°C. Owing to the large interval between ~17.8 s^{-1} and ~38.2 s^{-1} and the accumulation of experimental errors, also highlighted throughout this book, these rough estimates are statistically consistent (provided SD is used; see Section 2.1 for the advantages of SD over SE) with the value of $\sim14.2 \pm (1.4$ or $3.6)$ s^{-1} (for permeabilised frog fibres at ~5.7°C), recalled on p. **65**. For permeabilised rabbit fibres at 5°C, we have $\Pi_{isom,\infty} \sim 6.7 \pm (0.8$ or $2.1)$ s^{-1} (see p. **57** in Section 3.4.3.2), that is, $\sim7.2 \pm (0.9$ or $2.3)$ s^{-1} at ~5.7°C ($Q_{10} \sim 2.8$; see above). The maximal estimate is therefore ~9.5 s^{-1}, slightly lower than the two lowest estimates of

~17.8 s^{-1} and ~14.2 s^{-1} − 3.6 s^{-1} ~ 10.6 s^{-1} (see above). This may be due to the fact that most of the estimates given above, in this section, relate to frog at low temperature (favourable conditions for a cold-blooded animal) and that the value of ~9.5 s^{-1} relate to rabbit at low temperature (unfavourable conditions for a warm-blooded animal). In any event, the difference between ~9.5 s^{-1} and ~10.6 s^{-1} and ~17.8 s^{-1} is not significant (see above, the problem of the accumulation of experimental errors). Finally, using systematically SDs rather than SEs (see above) for all the parameters in Bagshaw's formula, all the MgATPase activities have a common interval (calculations not presented).

Many shortcomings and experimental errors inevitably appear in the theoretical and semi-empirical reasoning presented here, elsewhere, and by other independent groups. For example, there are problems with the step size distance, which is zero under purely isometric conditions, increases with s_0V and is maximal at s_0V_{max}, with the 'optimal' value for $n^*(s_0V_{max})$ (mean number of attached cross-bridges at maximal shortening velocity; see p. **71** in Section 3.4.4), and problems with $\Pi_{isom,\infty}$ (isometric MgATPase activity, in short tetani, under steady-state conditions) and $\tau_{c,\infty}$ (duration corresponding to attachment of a cross-bridge, hydrolysis of one MgATP molecule and finally detachment of the cross-bridge, under isometric tetanic contraction conditions). Despite these and probably other difficulties, the values estimated in this section are consistent with the available experimental data, and this should again be seen as a demonstration of the self-consistency of the model-independent phenomenological/semi-empirical approach presented in this work.

3.4.4 Active Shortening: Proportion of Attached Cross-Bridges and MgATPase Activity as Functions of Shortening Velocity. Mixture of Frog and Rabbit

Let us consider a pseudo-steady-state isotonic contraction, characterised by a constant shortening velocity s_0V, independent of time. This notion of pseudo-steady-state isotonic contraction is an oversimplification (see pp. **61–62** in Section 3.4.3.2 for a discussion), but is used here, because it gives a simple approach and also because it has unwittingly been employed in all experimental and theoretical studies published to date. Taking into account the condition $n(+h_2) = 0$ (see p. **40** in

Section 3.4.2) and introducing $z = 2\Delta h(f_\infty + g_\infty)/s_0V_{max}$ (see text between Equations 3.25 and 3.26), Equation 2.3 can be solved, giving

$$n(x,s_0V) = n_\infty\{1 - \exp[-(h_2 - x)zs_0V_{max}/\Delta hs_0V]\}$$
(3.40)

This equation is valid only within the interval of attachment $(-h_1; +h_2)$. Below $-h_1$, we have $f_\infty = 0$ and g_∞ becomes $g_\infty^* \gg g_\infty$ (see p. **40** in Section 3.4.2). In these conditions, I use Equation 3.4 with $\partial n'/\partial t = 0$ (pseudo-steady-state isotonic conditions) and $n(-h_1,s_0V) = n'(-h_1,s_0V)$, to ensure the continuity of the two relationships giving the values of the proportion of attached cross-bridges at $x = -h_1$. We therefore obtain

$$n'(x,s_0V) = n_\infty[1 - \exp(-zs_0V_{max}/s_0V)]$$
$$\exp[2g_\infty^*(h_1 + x)/s_0V]$$

(3.41)

Equations 3.40 and 3.41 are comparable to those obtained by A.F. Huxley (1957), in slightly different forms, although he considered f and g to be linearly dependent on x (with these two parameters being zero at x = 0, which is probably a weak point in the model, with no major consequences), but with g* independent of x, as also assumed on p. **40** in Section 3.4.2 and here. In Equation 3.41, we have $x < -h_1$ and the term $(h_1 + x)$ is negative. Under isometric conditions $(s_0V = 0)$, we obviously have $n'(x,0) = 0$, regardless of x. Under isotonic conditions, when s_0V is not zero, $n'(x,s_0V)$ has definite values that rapidly reach zero, even for absolute values of x only slightly below h_1, because of the very high value of g_∞^*. A.F. Huxley (1957) demonstrated theoretically that $n(x,s_0V)$ rapidly decreases with increasing s_0V, regardless of the value of x, which is also predicted from Equation 3.41 (calculations not shown). The mean value of $n(x,s_0V)$ over the interval of attachment $(-h_1, +h_2)$ is $n_{mean}(s_0V)$. According to Equation 3.41, over the interval $(-\infty; -h_1)$ and at any value of s_0V different from zero, $n'(x,s_0V)$ and its mean value $n'_{mean}(s_0V)$ are no longer zero, but the value of $n'_{mean}(s_0V)$ cannot be easily predicted. However, it can be demonstrated that, for any values of f_∞, g_∞ and $g_\infty^* \gg g_\infty$, the sum of $n_{mean}(s_0V)$ and $n'_{mean}(s_0V)$ rapidly decreases with increasing s_0V (see also Barclay 1999; Brokaw 1995; A.F. Huxley 1957). For $s_0V = s_0V_{max}$ (zero load), it can be demonstrated

Morel

from Equations 3.40 and 3.41, as for the approach of A.F. Huxley (1957), that the sum of the two mean proportions $n_{mean}(s_0V_{max})$ and $n'_{mean}(s_0V_{max})$ and the approximate mean proportion $n*(s_0V_{max})$ ~ $[n_{mean}(s_0V_{max}) + n'_{mean}(s_0V_{max})]/2$ are considerably lower than the 'universal proportion' n_∞ ~ 0.82, under isometric tetanic conditions ($s_0V = 0$) (see Section 4.4.2.5.2 concerning n_∞ ~ 0.82). More precisely, Barclay (1999), applying a model derived from that of A.F. Huxley (1957) and using his own values for the constants of attachment and detachment in the case of frog at 0°C (see his Table 2), suggested, in his Figure 2C, that the fraction of attached cross-bridges at s_0V_{max} is ~0.40, giving $n*(s_0V_{max})$ ~ 0.40 × 0.82 ~ 0.33 (value not proposed by the author). From his model, Barclay (1999) also proposes, in his Figure 2C, a fraction at s_0V_{max} of ~0.20 giving $n*(s_0V_{max})$ ~ 0.20 × 0.82 ~ 0.16 (value not proposed by the author). Owing to the reservations raised by Barclay's model (see p. **42** in Section 3.4.3.1.1), the proportions of ~0.33 and ~0.16 are not taken into account here. In an experimental and semi-empirical study, Piazzesi et al. (2007) claimed that 'the force–velocity relationship is primarily a result of a reduction in the number of motors attached to actin in each filament in proportion to the filament load'. This reduction is consistent with all the other experimental and theoretical results. Owing to the considerable scattering of the experimental values obtained by Piazzesi et al. (2007), the approximately linear inverse relationship between the number of motors attached to the thin actin filaments and shortening velocity they report (see, for instance, their Figure 4A) is roughly consistent with Equations 3.40 and 3.41.

On the basis of x-ray diffraction (traditional or synchrotron radiation) performed on intact frog fibres, the theoretical conclusion drawn in the preceding paragraph, that is, $n*(s_0V_{max})$ ~ $[n_{mean}(s_0V_{max}) + n'_{mean}(s_0V_{max})]/2 \ll n_\infty$ ~ 0.82, is supported by the experimental data of Huxley (1979), Huxley et al. (1988) and Griffiths et al. (1993), for example. Using traditional x-ray diffraction, Podolsky et al. (1976) suggested that the proportion of attached cross-bridges does not vary much upon active shortening, but this previous conclusion is no longer valid (see p. **72** for precise details concerning the interpretation of this inference). The majority view, including that described here on a purely phenomenological/semi-empirical basis, is also consistent with experimental results, obtained at 4°C, by traditional x-ray diffraction on whole frog muscle, by Yagi and Takemori (1995), who deduced from their experimental study that a smaller proportion of actin–myosin interactions takes place when the muscle shortens with a smaller load. Quantitatively, these authors used the various values also used by Huxley (1957) and deduced that, at s_0V_{max}, the proportion of attached cross-bridges would be ~0.27 times that under isometric tetanic contraction (~0.82), that is, ~0.27 × 0.82 ~ 0.22 times that in rigor (this reference state corresponds to the proportion of 1 [100%], when all the cross-bridges are attached [see p. **148** in Section 4.4.2.5.2]). In the absence of precise details concerning the various assumptions, parameters chosen and calculations made in the paper by Yagi and Takemori (1995), I think that the value of ~0.22 probably corresponds approximately to the mean value $n*(s_0V_{max})$ ~ $[n_{mean}(s_0V_{max}) + n'_{mean}(s_0V_{max})]/2$ defined in the preceding paragraph. This estimate is significantly higher than the other estimates given below, in this section, because it is based on old values used by Huxley (1957) (see above). However, owing to the scattering of the various experimental values recalled below, in this section, this value of ~0.22 is taken into account.

Using biochemical techniques, He et al. (1999) obtained, at 12°C, a mean proportion of attached cross-bridges of ~0.10, in permeabilised fibres from rabbit psoas muscle shortening at s_0V_{max}. Using biochemical and physiological techniques different from those of He et al. (1999), Sun et al. (2001) estimated that, at 10°C, in chemically skinned fibres from rabbit psoas muscle, the proportion of attached cross-bridges is less than ~0.07 at s_0V_{max} (I take here a value of ~0.04). Using x-ray diffraction from synchrotron radiation, for whole muscles from frog at 8°C, Martin-Fernandez et al. (1994) found proportions of ~0.05–0.10 of attached cross-bridges at s_0V_{max}. Piazzesi et al. (1999), using the same x-ray technique, but applied to single intact fibres from frog at 4°C, rather than whole muscles, observed larger changes in x-ray patterns during rapid shortening than during isometric tetanic contraction but did not estimate the proportion of myosin heads bound to the thin actin filaments during the rapid sliding of the myofilaments past each other. Piazzesi et al. (2007) presented a detailed experimental and semi-empirical study of muscle performance, again on single intact frog fibres at 4°C, and, from their Figure 3D, I estimate the proportion of 'attached motor' (attached myosin heads) to be ~0.05 at 'zero filament load', that is, at s_0V_{max}. For whole frog muscles contracting at 0°C and at s_0V_{max}, Bagshaw (1993) estimated the proportion of attached cross-bridges at ~0.03. Homsher et al. (1981) also performed experiments on whole frog muscles and suggested a value of ~0.02 at 0°C.

According to Equations 3.40 and 3.41, the maximal values of $n(x,s_0V_{max})$ and $n'(x,s_0V_{max})$ correspond to $x = -h_1$ and are both given by $n_\infty[1 - \exp(-z)]$. As recalled in Section 3.4.3.1.3 (see text below Equations 3.27 and 3.28), for single intact frog fibres at ~0°C–5°C, we have $z \sim 0.157$. No value of z can be estimated for skinned fibres or unheld isolated myofibrils from frog or rabbit, between 0°C and 12°C, but using the notion of 'mixing', as employed many times in this monograph, I assume that $z \sim 0.157$ is valid in any case. This gives a maximal expected proportion of cross-bridges attached to actin of ~$0.16n_\infty$, at s_0V_{max}, giving a mean proportion $n^*(s_0V_{max})$ of ~$0.16n_\infty/2 \sim 0.08n_\infty \sim 0.08 \times 0.82 \sim 0.07$ (see p. **71** concerning the value of ~0.82). This is a reasonable estimate, given the various experimental estimates recalled in the next paragraph, for the mean proportion of cross-bridges attached to the thin actin filaments, during the most rapid sliding, at s_0V_{max}, of the myofilaments past each other.

The results, valid for frog or rabbit (experimental, theoretical and phenomenological/semi-empirical values), are therefore ~0.22 as obtained semi-empirically by Yagi and Takemori (1995), ~0.10 as obtained experimentally by He et al. (1999), ~0.04 (based on the value of less than ~0.07 estimated by Sun et al. 2001; see p. **72**), ~0.05–0.10 estimated by Martin-Fernandez et al. (1994), ~0.05 estimated from the work of Piazzesi et al. (2007), ~0.03 calculated by Bagshaw (1993), and ~0.02 deduced from their own experiments by Homsher et al. (1981). These values are consistent both qualitatively and quantitatively. Indeed, averaging the eight estimates gives ~0.076 ± (0.021 or 0.061) (mean ± SE or SD; n = 8) (rounded up to ~0.08), a value similar to the phenomenological/semi-empirical value of ~0.07, obtained independently of the mode of force generation (see the end of the preceding paragraph). Thus, the value of ~0.08 is considered universal and applicable regardless of the animal species, biological material or temperature, at least between 0°C and 12°C (see the preceding paragraphs).

As pointed out by Millman (1998) and recalled on pp. **151–152** in Section 4.4.2.5.2, in x-ray diffraction experiments on isometrically contracting fibres, it is very difficult to determine whether the myosin heads in the vicinity of the thin actin filaments are attached or detached. Yagi and Takemori (1995) also noted that 'a considerable number of myosin heads remain in the vicinity of the thin filaments even during rapid shortening', although 'the number of active cross-bridges does decrease during shortening'. This observation is almost certainly the origin of the misinterpretation by Podolsky et al. (1976) (see p. **71**) of their own experiments. Martin-Fernandez et al. (1994) found, for whole frog muscles shortening at a velocity of $s_0V_{max}/3$, a proportion of bound cross-bridges of ~0.70, whereas Bagshaw (1993) found that this proportion was only ~0.08. Martin-Fernandez et al. (1994) tried to explain this large discrepancy and expressed doubts about the validity of the traditional tilting head models (swinging cross-bridge/lever-arm theories). Yagi et al. (2006), also working on whole frog muscles, at ~4°C–5°C, suggested that 'the number of [attached] cross-bridges during shortening at $0.3T_0$ is ~38% (0.38) of that in the isometric tetanic tension'. According to the hyperbolic relationship, valid for whole frog muscles at ~0°C (Hill 1938, 1964a,b), these conditions of contraction correspond to $s_0V \sim 0.3s_0V_{max} \sim s_0V_{max}/3$. With respect to the rigor conditions, in which 100% of the cross-bridges are attached (see p. **148** in Section 4.4.2.5.2), the proportion suggested by Yagi et al. (2006) gives ~$0.38 \times 0.82 \sim 0.31$ (the value of ~0.82 is recalled on p. **71**). Thus, interpretation of these three series of experiments leads to inconsistent values for this proportion (~0.08, ~0.31 and ~0.70). This may be related to the assertion of Yagi and Takemori (1995), recalled above, that it is very difficult to distinguish, among the cross-bridges in the vicinity of the thin actin filaments, between those that are attached and those that are detached. Fortunately, the approach presented throughout this Section 3.4 is independent of the mode of generation of the contractile force, as repeatedly pointed out in this monograph, and also of specific features (e.g. attached or detached state of the cross-bridges in the vicinity of the thin actin filaments). From Equations 3.40 and 3.41 introducing $z \sim 0.157$ (see p. **47**) and based on exactly the same calculations and reasoning as presented for s_0V_{max}, we obtain a

Morel

proportion of ~0.16 at $s_0 V_{max}/3$, regardless of the animal species, at ~0°C–12°C. This value ranges between ~0.08, estimated by Bagshaw (1993), and ~0.31, estimated by Yagi et al. (2006), on entirely different bases. Thus, it is reasonable to doubt the semi-empirical approach and conclusions of Martin-Fernandez et al. (1994), who suggested a value of ~0.70, but we can be more confident about the estimates given by Bagshaw (1993) and Yagi et al. (2006), and the phenomenological/semi-empirical approach, based entirely on the energetic part of the theoretical and biophysical work of Huxley (1957), revisited and presented in its new formulation in this book (e.g. Sections 3.4.2 and 3.4.3 and this section). Averaging these three 'suitable' values (~0.08, ~0.16 and ~0.31) gives ~0.18 ± (0.06 or 0.10) (mean ± SE or SD; n = 3), statistically similar to ~0.16 deduced from the model-independent approach used in this monograph, regardless of the use of SE or SD. Other concordant theoretical reasoning and experimental results are presented and discussed below, in this section, and, more generally throughout this monograph, in the same approach (e.g. Sections 8.9 and 8.11).

As highlighted at the start of Section 3.4.3.2 (see p. **48**), experimental values obtained for frog and rabbit, with intact fibres, chemically skinned and permeabilised fibres and unheld isolated myofibrils, should be combined with molecular data and the phenomenological/semi-empirical equations. In this context, the value of $\Pi_{isom,\infty}$, valid for an intact unit cell, can be deduced from Equation 3.26, giving the following expression (at this point, no assumption is made regarding the value of $\omega_{isom,\infty}$):

$$\Pi_{isom,\infty} = \omega_{isom,\infty} z s_0 V_{max} n_\infty (1 - n_\infty)/2\sigma \quad (3.42)$$

At this stage, I prefer to avoid introducing numerical values for the various parameters in this equation, for comparison of the phenomenological/semi-empirical model, presented in this Section 3.4, with experimental results obtained for frog and rabbit. I assume below, in this section, that all the numerical values, given above, in this section, as valid for macroscopic structures are also valid for microscopic structures, such as intact unit cells. From definitions 3.7, it can be deduced that, at any value of the shortening velocity $s_0 V$, we have

$$v_\infty^* = \{1 - \exp[-z(s_0 V_{max}/s_0 V)]\} n_\infty^2 \quad (3.43)$$

Thus, taking into account Equation 3.43, Equation 3.10 can be written:

$$\Pi(s_0 V) = \omega(s_0 V)\Pi_{isom,\infty} + \omega(s_0 V)n_\infty\{1 - \exp[-z(s_0 V_{max}/s_0 V)]\}(s_0 V/2\sigma) \quad (3.44)$$

For comparison of $\Pi(s_0 V)$ and $\Pi_{isom,\infty}$, it is useful to introduce into Equation 3.44 the expression of $\Pi_{isom,\infty}$, given by Equation 3.42. After some mathematical manipulations and rearrangements, we obtain

$$\Pi(s_0 V)/\Pi_{isom,\infty} \sim \omega(s_0 V) + [\omega(s_0 V)/\omega_{isom,\infty}] (s_0 V/s_0 V_{max})\{1 - \exp[-(s_0 V_{max}/s_0 V)]\} n_\infty/[z(1 - n_\infty)]\} \quad (3.45)$$

Introducing $z \sim 0.157$ and $n_\infty \sim 0.82$ (these two values are recalled on p. **61**) into Equation 3.45, regardless of temperature (between 0°C and ~20°C–22°C) and the biological material, we deduce

$$\Pi(s_0 V)/\Pi_{isom,\infty} \sim \omega(s_0 V) + 29.0[\omega(s_0 V)/\omega_{isom,\infty}] (s_0 V/s_0 V_{max})\{1 - \exp[-0.157(s_0 V_{max}/s_0 V)]\} \quad (3.46)$$

For short tetani, under isometric steady-state conditions (see Section 3.4.3.2 for precise details concerning these fundamental and complex notions), the probability of inactive contacts existing may be considered negligible (e.g. Morel 1984a,b). This implies that $\omega_{isom,\infty} \sim 1$, which is already assumed on p. **61** and p. **63** in Section 3.4.3.2. Such a simplifying assumption results in self-consistent inferences throughout this monograph. Thus, Equation 3.46 can also be written:

$$\Pi(s_0 V)/\Pi_{isom,\infty} \sim \omega(s_0 V)|1 + 29.0(s_0 V/s_0 V_{max}) \{1 - \exp[-0.157(s_0 V_{max}/s_0 V)]\}| \quad (3.47)$$

For $s_0 V = s_0 V_{max}$, we deduce from Equation 3.47 that $\Pi(s_0 V_{max})/\Pi_{isom,\infty} \sim 4.4\omega(s_0 V_{max}) \sim 1.2 \pm 0.2$ (see p. **68** in Section 3.4.3.3 concerning this last value), from which we deduce that $\omega(s_0 V_{max}) \sim 0.27 \pm 0.05$. This estimate is consistent with previous assertions made by Morel (1984a,b), according to which $\omega(s_0 V)$ is a decreasing function of $s_0 V$, and, therefore, the proportion of inactive contacts, defined on p. **43** in Section 3.4.3.1.1, increases considerably with increasing shortening velocity: $[1 - \omega(s_0 V_{max})] \sim 0.73 \pm 0.05$, whereas $(1 - \omega_{isom,\infty}) \sim 0$. Rejection of this simple concept should result in a calculated value of $\Pi(s_0 V_{max})/\Pi_{isom,\infty} \sim 4.4$, which is much higher than $\sim 1.2 \pm 0.2$ and therefore at odds with the interpretation of the experimental results (see the following paragraphs concerning this notion of interpretation).

Since the mid-1990s, most experimental studies on the rate of MgATP cleavage during active shortening have been carried out on fibres from rabbit psoas muscle, demembranated by various chemical techniques. Using the notations introduced on p. **42** in Section 3.4.3.1.1 (in particular the superscript * for demembranated rabbit fibres), and again assuming that the experimental results are valid in intact unit cells (see the preceding paragraph), it was found that, at 15°C, on chemically skinned fibres stored in 50% glycerol for up to 2 months, $\Pi * (s_0 V_{max}) \sim 2.7 \Pi^*_{isom,\infty}$ (Potma and Stienen 1996) or $\Pi * (s_0 V_{max}) \sim (2.2 - 2.4) \Pi^*_{isom,\infty}$ (Potma et al. 1994). I think that the values of $\Pi^*_{isom,\infty}$ measured by Potma and Stienen (1996) and Potma et al. (1994) are too low (e.g. ~ 2.3 s^{-1} at 15°C, corresponding to only ~ 1.7 s^{-1} at 12°C; $Q_{10} \sim 2.8$; see p. **201** in Section 6.3.3). The temperature used by He et al. (1997c) was 12°C and these authors obtained, for short tetani, values of $\sim 16-18$ s^{-1} with permeabilised fibres from rabbit psoas muscle (see Table 1 in their paper). This major difference between these two independent groups is almost certainly related to the particular techniques used by Potma and Stienen (1996) and Potma et al. (1994), resulting in isometric MgATPase activities similar to those corresponding to long-lasting tetani or even very long-lasting tetani (see Sections 3.4.3.2, 7.2 and 7.3 for definitions of these states, analyses and discussions). Thus, the experimental results of Potma and Stienen (1996) and Potma et al. (1994) are not taken into account in this analysis.

Working on permeabilised fibres from rabbit psoas muscle, at 12°C, at a sarcomere length of ~ 2.7 μm, and determining the rate of release of Pi, He et al. (1999) obtained two possible values for MgATPase activity: ~ 10.3 s^{-1}, immediately after the photolytic release of ATP from NPE-caged-ATP (isometric conditions, before shortening occurs), and ~ 5.1 s^{-1} during the 'isometric phase' before the applied shortening. Unfortunately, no data were presented by the authors concerning the duration of this isometric phase. Thus, the state of the fibres

is unknown (e.g. short tetani, pseudo-long-lasting tetani, long-lasting tetani[?]; see Section 3.4.3.2 for precise details on the various states for MgATPase activity). It is shown, on p. **57** in Section 3.4.3.2, that the best value of isometric MgATPase activity, at 12°C, for short tetani and under steady-state conditions, at sarcomere lengths of $\sim 2.2-2.4$ μm, is $\sim 16.3 \pm$ (0.9 or 3.4) s^{-1}, lying therefore between ~ 12.9 s^{-1} and ~ 19.7 s^{-1} (maximal interval). The value of ~ 10.3 s^{-1} obtained by He et al. (1999) is lower than $\sim 16.3 \pm$ (0.9 or 3.4) s^{-1}, but the value of ~ 5.1 s^{-1}, also obtained by He et al. (1999) (see above), is clearly too low and possibly corresponds to pseudo-long-lasting tetani (see Section 3.4.3.2 for precise details on this notion). At this point, one may ask why the value of ~ 10.3 s^{-1} is lower than $\sim 16.3 \pm$ (0.9 or 3.4) s^{-1}. It should be noted that He et al. (1999) worked at a sarcomere length of ~ 2.7 μm (see above), whereas the reference value of $\sim 16.3 \pm$ (0.9 or 3.4) s^{-1} corresponds to $\sim 2.2-2.4$ μm (see above). Thus, steric hindrance (see Bachouchi and Morel 1989a and pp. **251–254** in Section 8.9) should induce a moderate 'steric' decrease in MgATPase activity at ~ 2.7 μm, with respect to the MgATPase activity at $\sim 2.2-2.4$ μm. Otherwise, at sarcomere lengths of ~ 2.7 μm, there are less cross-bridges interacting with the thin actin filaments (reduced overlap), whereas at $\sim 2.2-2.4$ μm, all the cross-bridges interact with the thin actin filaments (full overlap), resulting in a lower MgATPase activity at ~ 2.7 μm. There are, therefore, steric and 'geometric' origins for the lower MgATPase activity at ~ 2.7 μm. Nonetheless, the value of ~ 10.3 s^{-1} is close to the minimal value of ~ 12.9 s^{-1} (see above) corresponding to maximal overlap (see above), and these two values are almost consistent. Moreover, in the following paragraphs, a sarcomere length of ~ 2.7 μm is systematically used and, for the sake of comparison, I suggest, as a compromise, selecting the average value of $\sim (10.3 + 16.3)/2 \sim 13.3$ s^{-1} for the isometric MgATPase activity at sarcomere lengths of $\sim (2.2-2.4) - 2.7$ μm.

He et al. (1999) studied MgATPase activity, at 12°C, as a function of shortening velocity. They found a hyperbolic increase and a maximal value of ~ 18.5 s^{-1} at maximal shortening velocities >1 muscle length s^{-1} (close to the value of ~ 1.2 muscle length s^{-1}, at 10°C, selected by Sun et al. 2001, with chemically skinned fibres). We therefore have $\Pi*(s_0 V_{max}) \sim 18.5$ s^{-1}. Unfortunately, He et al. (1999) took the value of ~ 5.1 s^{-1} as the reference value for 'the ATPase rate in the isometric fibres, immediately prior to the phase of applied shortening' (see the preceding paragraph). It is demonstrated, in the preceding paragraph, that this is unjustified and that the most suitable value is ~13.3 s^{-1}, if a self-consistent approach to the various experimental results and semi-empirical reasoning is to be achieved. In these conditions, the hyperbolic increase mentioned above is unlikely to occur, because this hyperbolic shape is much more marked at low shortening velocities, that is, when the choice of the true isometric rate of MgATP splitting is crucial. By contrast, at $_{s0}V_{max}$, the choice of this rate is of minor importance, because the duration of active shortening is limited and similar between series of experiments. For example, from the experimental data of He et al. (1999), this duration can be estimated at ~65–75 ms and, in the experiments of Sun et al. (2001), it can be estimated at ~80–85 ms, close to ~65–75 ms (the importance of duration is recalled in the preceding and the following paragraphs, at least in the case of isometric tetanic contraction). Thus, we can retain ~18.5 s^{-1} as the true value of $\Pi^*(_{s0}V_{max})$ at 12°C, giving therefore $\Pi^*(_{s0}V_{max})/\Pi^*_{isom,\infty} \sim 18.5\,s^{-1}/13.3\,s^{-1} \sim 1.4$.

Working on chemically skinned fibres from rabbit psoas muscle, at 10°C, also at a sarcomere length of ~2.7 μm, Sun et al. (2001) did not give clear estimates of the isometric MgATPase activity. They found that the 'rate of MgATP utilisation during shortening clearly increased with increasing shortening velocity, and the relationship was roughly linear' (this conclusion is at odds with the hyperbolic relationship described by He et al. 1999; see the preceding paragraph). Moreover, Sun et al. (2001) found that $\Pi^*(_{s0}V_{max}) \sim 11.3$ s^{-1}, at 10°C, corresponding to ~13.9 s^{-1} at 12°C ($Q_{10} \sim 2.8$, see p. **201** in Section 6.3.3). This value is slightly lower than the ~18.5 s^{-1} obtained by He et al. (1999), at the same sarcomere length (see p. **73**), but this small difference may be accounted for by experimental uncertainty, or the use of different biochemical and physiological techniques in these two studies. Here again, the major problem in trying to find the relationship between MgATPase activity and shortening velocity lies in the measurement of the 'true isometric' MgATPase activity corresponding to short tetani, under steady-state conditions, and imperatively avoiding the rate of MgATP cleavage corresponding to pseudo-long-lasting tetani or even to long-lasting tetani, as described in Section 3.4.3.2. Indeed, as an illustrative example, Hilber et al. (2001), from the same group as Sun

et al. (2001), working on the same biological material, at the same sarcomere length (~2.7 μm) found an isometric rate of MgATP splitting of only ~2.1 s^{-1}, at 10°C, and, taking $Q_{10} \sim 2.8$ (see above), ~2.6 s^{-1} at 12°C, much lower than the realistic value of ~13.3 s^{-1} (see p. **74**). The values of Hilber et al. (2001) and Sun et al. (2001) are not retained below, in this section. The very low value obtained by Hilber et al. (2001) is commented on in the next paragraph.

Assuming that isometric MgATPase activity, at 12°C, is systematically underestimated in the experiments of Hilber et al. (2001), with respect to the findings of He et al. (1999), the value of ~2.6 s^{-1} would be, at best, ~2.6 s^{-1} × 18.5 s^{-1}/13.9 s^{-1} ~ 3.5 s^{-1} (~18.5 s^{-1} and ~13.9 s^{-1} correspond to $_{s0}V_{max}$, measured or calculated at 12°C, in the works of He et al. 1999 and Sun et al. 2001, respectively; see the preceding paragraph). In Section 3.4.3.2 (see text below Equations 3.30 through 3.32) it is shown that, at 5°C, the rate of MgATP cleavage corresponding to pseudo-long-lasting tetani (case of rabbit) is $\Pi^o_{isom,\infty} \sim 2.0\,s^{-1}$, giving ~4.1 s^{-1} at 12°C ($Q_{10} \sim 2.8$; see the preceding paragraph). This value is close to the ~2.6 s^{-1} or ~3.5 s^{-1} found by Hilber et al. (2001), or estimated from their experiments. The authors were therefore almost certainly working under pseudo-long-lasting tetanus conditions, and the rather low experimental value of ~13.9 s^{-1} estimated from the work of Sun et al. (2001), from the same group, was probably 'contaminated' by a certain proportion of pseudo-long-lasting tetani (see above concerning the value of ~13.9 s^{-1}). Thus, I take into account only the results obtained by He et al. (1999) at $_{s0}V_{max}$ (~18.5 s^{-1}) and those obtained under isometric tetanic conditions, from the discussion presented on p. **74** (~13.3 s^{-1}; see the end of the preceding paragraph). This gives $\Pi^*(_{s0}V_{max})/\Pi^*_{isom,\infty} \sim 18.5\,s^{-1}/13.3\,s^{-1} \sim 1.4$, for permeabilised rabbit fibres, at 12°C (see also p. **74**), which is not too far from the old value of $\Pi^{**}(_{s0}V_{max})\Pi^{**}_{isom} \sim 1.0$ obtained with whole frog muscles, at ~0°C–5°C (extended to 12°C), more than 30–40 years ago (see p. **42** in Section 3.4.3.1.1). These two values give a mean value of ~1.2 ± 0.2, consistent with the concept of inactive contacts, the proportion of which increases with shortening velocity (see text below Equation 3.47). The hyperbolic or linear relationships describing variations of $\Pi^*(_{s0}V)$ with $_{s0}V$ obtained by He et al. (1999) and Sun et al. (2001), respectively, are not valid, owing to the use of unsuitable estimates of $\Pi^{*}_{\text{isom},\infty}$ (see above, in this section). Owing to the experimental uncertainty, we can apply the results obtained here to intact unit cells and write $\Pi(s_0V_{\text{max}})/\Pi_{\text{isom},\infty} \sim 1.2 \pm 0.2$ (see p. **73** in Section 3.4.3.3). Again assuming that Equation 3.47 is valid for any animal species (e.g. warm- and cold-blooded animals of various species), for any kind of fibre (intact, chemically skinned and permeabilised) and taking the proportion of active contacts, $\omega(s_0V)$, as a linear function of s_0V, as an illustrative example, Equation 3.47 would give an appropriate relationship between $\Pi(s_0V)$ and s_0V that could be checked in future experimental studies.

3.4.5 Very Loose Coupling between Isometric Tetanic Tension and Isometric MgATPase Activity (Short Tetani–Steady State), in Intact Unit Cells from Intact Frog Fibres, with Some Insights into Demembranated Fibres from Frog and Rabbit

It should be stressed that the term coupling, as used by most investigators in the muscle and in vitro motility domains, is an abuse of language. Indeed, this notion is valid only in irreversible thermodynamics (e.g. Katzir-Katchalsky and Curran 1965; Kedem and Caplan 1965). However, I use it for the sake of simplicity and for consistency with publications concerning contraction and motility.

In this section, I mostly consider intact frog fibres and the corresponding intact unit cells to ensure that the analysis is as accurate as possible. Indeed, a major problem should be raised, concerning the dependence of isometric tension $T_{0,\infty}$ (corresponding to short tetani, under steady-state conditions, in an intact unit cell) on MgATPase activity, $\Pi_{\text{isom},\infty}$, given by Equation 3.42. Combining Equations 3.37 and 3.42, we obtain, after some mathematical manipulations:

$$T_{0,\infty} = \xi_{0,\infty}\text{me}_{\text{app}}s_0(\Pi_{\text{isom},\infty}/s_0V_{\text{max}})\{\omega_{\text{isom},\infty} - \upsilon\omega_\infty(s_0V_{\text{max}})[1 - \exp(-z)]\}n_\infty/[\omega_{\text{isom},\infty}z(1 - n_\infty)]$$

$$(3.48)$$

On p. **63** in Section 3.4.3.2, it is assumed that the Q_{10} for $T_{0,\infty}$ is the same as the Q_{10} for isometric tension T_f recorded in intact frog fibres, that is, $\sim 1.24 \pm (0.01 \text{ or } 0.02)$ (mean $\pm$ SE or SD; n = 6) at 2°C–12°C (Edman 1979), indicating that the isometric tension depends little on temperature. On pp. **245–246** in Section 8.8, a brief discussion of

the Q_{10} for isometric tetanic tension in intact and demembranated fibres is presented. The value of Edman (1979) is the most appropriate for intact frog fibres, within the favourable range of temperature for a cold-blooded animal ($\sim$0°C–12°C). Edman (1979) also found that $Q_{10}(s_0V_{\text{max}}) \sim 2.67 \pm$ (0.07 or 0.17) (mean $\pm$ SE or SD; n = 6) and this value, valid in intact fibres, is again assumed to be valid in intact unit cells. As for isometric tetanic tensions, it is unclear whether s_0V_{max} is similar in intact fibres and intact unit cells. On p. **61** in Section 3.4.3.2, it is recalled that, at $\sim$2.5°C, $s_0V_{\text{max}} \sim 5.4$ μm s^{-1} in intact frog fibres (Edman 1979). The values of s_0V_{max} and $Q_{10}(s_0V_{\text{max}})$ obtained by Edman (1979) for intact frog fibres are extended to intact unit cells. In Section 3.1, key references are given concerning the problem of the location of the unit cells within the fibre. The question of distortions in the various unit cells during active sliding is raised, leading to different values of shortening velocity, depending on this location within an intact fibre (i.e. lower values of s_0V_{max} in peripheral unit cells near the sarcolemma [attributed to frictional phenomena] than in 'free' central unit cells; see also Section 3.7 concerning these two types of unit cells). The simplifying hypotheses, recalled above, concerning the Q_{10} values for isometric tension, $T_{0,\infty}$ and T_f, and maximal shortening velocity, s_0V_{max}, in intact unit cells and intact fibres, are unavoidable and are used throughout the phenomenological/semi-empirical approach presented in this monograph, with no subsequent inconsistencies. Precise details on the definition, calculation and discussions of Q_{10} are given on pp. **201–202** in Section 6.3.3. From p. **202** in Section 6.3.3, we can see that, for frog, regardless of temperature, the dependence of isometric MgATPase activity is characterised by $Q_{10}(\Pi_{\text{isom},\infty}) \sim 3.1$, as deduced from various experiments performed on chemically skinned and permeabilised fibres and unheld isolated myofibrils. From Equation 3.48 and the corresponding comment, it can be deduced that the division of Q_{10} for $\Pi_{\text{isom},\infty}$ by that for the isometric tetanic tension $T_{0,\infty}$ gives $\sim$3.1/(1.24 $\pm$ 0.02) $\sim 2.50 \pm 0.04$, statistically similar to the value of $Q_{10}(s_0V_{\text{max}})$ obtained by Edman (1979) and recalled above; the SDs are used, because they are the best statistical test, as demonstrated in Section 2.1). We can write $\varsigma_{0,\infty} = \xi_{0,\infty}\{\omega_{\text{isom},\infty} - \upsilon\omega_\infty(s_0V_{\text{max}}) [1 - \exp(-z)]\}n_\infty/[\omega_{\text{isom},\infty}z(1 - n_\infty)]$. From the similarity of the two values of $Q_{10}(s_0V_{\text{max}})$ (measured and estimated), it is clear that $\varsigma_{0,\infty}$ is independent of

Morel

temperature, within the range of experimental error. This property is taken into account on p. **78**.

It is difficult to determine whether there is a relationship between the isometric MgATPase activity $\Pi_{\text{isom},\infty}$ and s_0V_{max}. From the experiments performed by Barman et al. (1998) and Lionne et al. (1996) on unheld isolated myofibrils, it is impossible to find a clear relationship between these two parameters, because the MgATPase activities measured by these authors cannot be easily compared with those measured in demembranated fibres contracting isometrically, and resulting in measurements of $\Pi_{\text{isom},\infty}$ (see pp. **56–57** in Section 3.4.3.2 for limitations of the use of unheld isolated myofibrils in this area). However, we can assume that there is a relationship between s_0V_{max} and $\Pi_{\text{isom},\infty}$ and that s_0V_{max} is approximately proportional to $(\Pi_{\text{isom},\infty})^k$ ($k > 0$ is an exponent that should be estimated). This statement is strongly supported below, in this section. We can therefore write $s_0V_{\text{max}} \sim \text{\textcent}(\Pi_{\text{isom},\infty})^k$, where $\text{\textcent}$ is the coefficient of proportionality. This relationship suggests that s_0V_{max} increases with increasing $\Pi_{\text{isom},\infty}$, and vice versa.

It is important to obtain experimental support for the inference concerning the variation of s_0V_{max} with $\Pi_{\text{isom},\infty}$ suggested in the preceding paragraph. In his Figures 11 and 12, Millman (1998) presented many experimental results from independent groups, mostly for intact fibres from various types of frog muscle (only four experimental points for intact toad fibres, in Figure 11). These two figures show that increasing the external osmolarity from ~150 mOsM to ~400–600 mOsM (in vivo external osmolarity estimated at 245 mOsm by Millman 1998 and Millman et al. 1981) leads to significant decreases in both isometric tetanic tension and maximal shortening velocity. It is well known that, in intact fibres, increasing the external osmotic pressure results in an outflow of internal water and a subsequent increase in the internal ionic strength. Increasing this pressure sufficiently (i.e. also increasing ionic strength), the radial repulsive electrostatic forces decrease (see pp. **101–102** in Section 3.10) and the decrease in isometric tetanic tension is expected in the hybrid model, whereas the decrease in s_0V_{max} is more difficult to explain. From Figure 11, it appears that the relative isometric tetanic force is approximately linearly related to osmolarity: [relative isometric force] ~1.5 − 2.3 × 10^{-3}[mOsM], where [mOsM] is the osmolarity, expressed in milliosmoles. This relationship, not given by Millman (1998), is valid within the

~150–600 mOsM range. From Figure 12, it appears that the relative maximal shortening velocity is also approximately linearly related to osmolarity, and we can write, within the ~150–400 mOsM range: [relative s_0V_{max}] ~1.8 − 3.4 × 10^{-3}[mOsM] (this relationship is not given by Millman 1998). Eliminating the osmolarity [mOsM] between these two relationships, we obtain the following relationship, valid within the ~150–400 mOsM range: [relative isometric force] ~0.283 + 0.676[relative s_0V_{max}]. On the other hand, from Equation 3.48, and again using the relationship $s_0V_{\text{max}} \sim \text{\textcent}(\Pi_{\text{isom},\infty})^k$ (see the preceding paragraph), we conclude that $T_{0,\infty}$ is approximately proportional to $(\Pi_{\text{isom},\infty})^{(1-k)}$ and to $(s_0V_{\text{max}})^{(1-k)/k}$. We can therefore also write: [relative $T_{0,\infty}$] = $\bar{e}$[relative s_0V_{max}]$^{(1-k)/k}$, where $\bar{e}$ is the coefficient of proportionality. In the following paragraphs, the relationships between [relative isometric force], [relative $T_{0,\infty}$], [relative s_0V_{max}] and $\Pi_{\text{isom},\infty}$ are compared and important conclusions are drawn.

The [relative $T_{0,\infty}$], corresponding to the relative microscopic isometric tetanic force in an intact unit cell, is almost certainly similar to the relative macroscopic isometric force in a whole intact fibre (see p. **99** in Section 3.7). Thus, we can write, in an intact fibre: [relative isometric force] ~ $\bar{e}'$,[relative s_0V_{max}]$^{(1-k)/k}$, where $\bar{e}'$, is a new constant (see the end of the preceding paragraph for precise details). Comparing this relationship with that given in the preceding paragraph, we can write $\bar{e}'$,[relative s_0V_{max}]$^{(1-k)/k}$ ~ 0.283 + 0.676[relative s_0V_{max}]. In his Figure 12, Millman (1998) presented experimental points for [relative s_0V_{max}] within the ~0.3–1.2 range, from which we deduce that a suitable constant value of k (this constancy results from the experimental error and cannot be mistaken for a mathematical demonstration) can be found: $(1 - k)/k \sim 0.58 \pm 0.01$; that is, $k \sim 0.63 \pm 0.01$ and $1 - k \sim 0.37 \pm 0.01$. Thus, from the preceding paragraph (see the end), we deduce that $T_{0,\infty}$ depends approximately on $(s_0V_{\text{max}})^{(0.58 \pm 0.01)}$ and therefore on $(\Pi_{\text{isom},\infty})^{(0.37 \pm 0.01)}$ (see the end of the preceding paragraph concerning the relationships between $T_{0,\infty}$ and s_0V_{max} and $\Pi_{\text{isom},\infty}$). However, this interpretation of the experimental results in Figures 11 and 12 in the review by Millman (1998), and therefore of the value of the exponents estimated here, is not straightforward, especially for significantly shrunk intact fibres, as demonstrated by Bachouchi and Morel (1989a) and discussed further on pp. **251–254** in Section 8.9 for stretched fibres (problem of steric hindrance of the cross-bridges when the inter-filament spacing is reduced). Moreover, Millman (1998) mixed values obtained for two different animal species (frog and toad) and different types of muscle fibre (e.g. *sartorius*, *semitendinosus* and *tibialis anterior*). These two issues cast doubt on the discussion presented in this paragraph and other approaches should be considered, as described below, in this section. Again using the 'osmotic procedure' discussed above, in this section, it is more appropriate to compare results obtained with a single animal species (frog *Rana temporaria*) and type of fibre (from *semitendinosus* muscle), as was done by Edman and Hwang (1977). To ensure that the experimental conditions resemble those existing in the physiological external medium as closely as possible, I take into account only the values of T_f (isometric tension in intact fibres) and s_0V_{max} within the narrow range of osmolarity of ~180–270 mOsM, around the in vivo extracellular osmolarity (245 mOsM; Millman 1998 and Millman et al. 1981), to try to avoid the problem of steric hindrance highlighted above. From the experiments of Edman and Hwang (1977a), both T_f and s_0V_{max} decrease with increasing external osmolarity, which is qualitatively consistent with Figures 11 and 12 in the review by Millman (1998). Moreover, from Table 3 in the paper by Edman and Hwang (1977a), it can be deduced that T_f is proportional to $\sim(s_0V_{max})^{(0.67 \pm 0.11)}$ (the exponent is the mean $\pm$ SE; n = 6), which again predicts that the isometric tetanic tension T_f and s_0V_{max} above as described above and in the preceding paragraph. The exponent of ~0.67 $\pm$ 0.11 is statistically similar to the value of $(1 - k)/k \sim 0.58 \pm 0.01$ found in the preceding paragraph. From this study, we deduce that $k \sim 0.60 \pm 0.04$ and, therefore, that both T_f and $T_{0,\infty}$ are proportional to $(\Pi_{isom,\infty})^{(1 - k)}$ $\sim (\Pi_{isom,\infty})^{(0.40 \pm 0.04)}$ (see p. **77** concerning the relationship between isometric tension and isometric rate of MgATP splitting). The results obtained here and in the preceding paragraph imply that isometric tetanic tension is loosely coupled to isometric MgATPase activity. As pointed out above and also on pp. **251–254** and in Section 4.4.2.1, the behaviour of intact and demembranated fibres, as functions of osmotic pressure, is complex. Thus, the conclusions drawn in this and the preceding paragraphs, including the value of the exponent found here, should be supported by an entirely different approach. This is done in the next paragraph, where it is demonstrated that the loose coupling is actually a very loose coupling.

Some of the difficulties highlighted at the end of the preceding paragraph are probably related to the other parameters in Equation 3.48 ($\varsigma_{0,\infty}$, z and n_∞), which may depend on external osmolarity. This conclusion is consistent with that presented by Millman (1998), on an entirely different basis, suggesting that the interpretation of experimental results concerning the dependence of T_f and s_0V_{max} on external osmolarity is difficult. Complementary discussions of this subject of external osmolarity are presented in Sections 8.7 and 8.9. Qualitatively, the experimental and phenomenological/semi-empirical results and the discussion presented in this section are consistent and confirm the validity of Equation 3.48. Quantitatively, only the calculations presented below are valid, as they apply to intact fibres from the same species of frog (*Rana temporaria*) and the same types of muscle (*semitendinosus* or *tibialis anterior*, which behave similarly; see Edman 1979), bathed in an isosmotic buffer, with temperature changes limited to a very narrow range (2°C–12°C; Edman 1979). As suggested on p. **77** and experimentally demonstrated in the preceding paragraphs, the isometric tension in an intact unit cell, $T_{0,\infty}$, is approximately proportional to $(\Pi_{isom,\infty})^{(1 - k)}$. In the text below Equation 3.48, it is assumed that the Q_{10} values for T_f and $T_{0,\infty}$ are similar, leading to a series of consistent conclusions. On p. **77**, it is demonstrated that the parameter $\varsigma_{0,\infty}$ is independent of temperature. Thus, using Equation 3.48 and after rather complex mathematical manipulations, we deduce, from the above relationship between $T_{0,\infty}$ and $\Pi_{isom,\infty}$, that $1 - k = \log Q_{10}(T_{0,\infty})/\log Q_{10}(\Pi_{isom,\infty})$ (see p. **201** in Section 6.3.3 for precise details on the mode of calculation of Q_{10}). Introducing the values of Q_{10} given on p. **75**, that is, ~1.24 $\pm$ 0.02 for T_f and $T_{0,\infty}$ and ~3.1 for isometric MgATPase activity, respectively, we deduce that $1 - k \sim 0.190 \pm 0.014$.* Thus, $T_{0,\infty}$ is approximately proportional

---

* On p. **10** in Chapter 1, it is highlighted that the relative uncertainty on MgATPase activity ranges between 1 and ~30–35 or ~13–14, in the case of rabbit fibers. This spectrum is assumed to be valid for frog fibers too. Thus, the corresponding relative isometric tetanic tensions would range between 1 and ~$(30–35)^{0.190} \sim 1.91$–1.96 or ~$(13–14)^{0.190} \sim 1.63$–1.65. It is recalled, on p. **49** in Section 2.4.3.2, that Linari et al. (1998) found that the isometric tetanic tension in *Rana temporaria* is ~1.5 times that in *Rana esculenta* and this confirms that the traditional one-to-one coupling between isometric MgATPase activity and isometric tetanic tension is no longer valid, because the values of ~1.63–1.65/1.91–1.96 are essentially similar to the experimental uncertainty of ~1.5, consistent with very loose coupling.

Morel

to $(\Pi_{isom,\infty})^{(0.190 \pm 0.014)}$. In other words, in an intact unit cell, isometric tetanic tension and isometric MgATPase activity are very loosely coupled, at least for short tetani under steady-state conditions.*

This phenomenological/semi-empirical conclusion is of major importance, because it is totally independent of the mechanical roles of the cross-bridges. Moreover, this inference is valid in a 'composite contractile structure', combining intact unit cells, single isolated myofibrils and intact fibres, bathed in buffers mimicking the physiological media. Indeed, Equation 3.48 is valid for an intact unit cell (this notion of an intact unit cell is introduced and described in Section 3.1 and more precisely defined in Section 3.4.1), whereas the Q_{10} values used above for s_0V_{max} and T_f are those given by Edman (1979), who worked on intact frog fibres. However, the Q_{10} corresponding to $\Pi_{isom,\infty}$ is deduced from experiments on chemically skinned and permeabilised fibres and single unheld myofibrils, with these three types of biological material giving similar values of Q_{10} (see p. **202** in Section 6.3.3). Thus, the main conclusions drawn in this section, concerning the very loose coupling between isometric MgATPase activity and isometric tetanic tension, correspond to genuine phenomena occurring in a 'mean representative unit cell' and can be extrapolated, with confidence, to an intact unit cell, independently of any mechanical role of the cycling cross-bridges, as repeatedly recalled above and at many places in this book. This conclusion is supported in the following paragraphs.

When we performed the experiments, from the mid- to late 1990s, that led to our 1999 paper (Morel et al. 1999), concerning thick myosin filaments studied in vitro and in the absence of F-actin, we were unaware of the paper by Ferenczi et al. (1984a) concerning a very different type of investigation (in particular, studies of various parameters affecting the isometric tetanic tension in demembranated frog fibres). Despite working with different biological materials and using different techniques, the two independent groups reported similar properties of isometric MgATPase activity as a function of MgATP concentration, the other characteristics of the buffers being otherwise similar and mimicking the physiological medium. Ferenczi et al. (1984a) worked on mechanically skinned frog fibres in the presence of Ca^{2+} (contraction conditions) and found that the isometric tetanic tension 'showed a biphasic dependence on MgATP concentration. Tension

increased with MgATP concentration from 1 µM to reach a peak at about 30–100 µM and decreased by about 20% from the value at the peak with further increases in the MgATP concentration to 5 mM (about the physiological concentration)'. It is demonstrated, in the preceding paragraph, that there is a very loose coupling between the isometric tetanic tension and the rate of MgATP splitting, in intact unit cells and intact frog fibres: the tension is proportional to this rate to a power of $\sim 0.190 \pm 0.014$. This conclusion, drawn from the phenomenological/semi-empirical approach presented throughout this section (Section 3.4) and experimental data obtained by independent groups, for various biological materials, including intact fibres and chemically skinned and permeabilised fibres, is almost certainly valid for the mechanically skinned fibres studied by Ferenczi et al. (1984a). Thus, when the rate of MgATP hydrolysis increases, it can be predicted, from the approach presented in the preceding paragraph, that the tension must increase, and vice versa. Furthermore, based on the same approach, when MgATPase activity peaks, so does isometric tetanic tension, consistent with the experimental observations of Ferenczi et al. (1984a), at least qualitatively. This 'prediction' and the experiments of Morel et al. (1999) provide a good quantitative fit to the experimental findings of Ferenczi et al. (1984a), as demonstrated in the next paragraph.

Regardless of the interpretation of Ferenczi et al. (1984a), based on the swinging cross-bridge/lever-arm concepts, I think that the biphasic variations in tension (see the preceding paragraph) demonstrate indirectly that the same qualitative variations in 'in situ isometric actin-activated' MgATPase activity occur. More precisely, I claim that, in the experiments of Ferenczi et al. (1984a), the MgATPase activity is also almost certainly biphasic, initially increasing to a rather flat peak at $\sim 30–100$ µM and then decreasing with further increase in MgATP concentration. This is what we found in my group (Morel et al. 1999) for the 'resting' MgATPase activity of thick myosin filaments from rabbit muscles in the absence of Ca^{2+} and F-actin (in vitro). Mixing the experimental values obtained for various types of synthetic thick myosin filaments and natural filaments, I deduce that, in vitro, the flat peak in the rate of MgATP cleavage occurs at MgATP concentrations of $\sim 200 \pm 100$ µM (mean $\pm$ SD; n = 5), not too far from the values of $\sim 30–100$ µM found by Ferenczi et al. (1984a), in situ, in very different conditions (mechanically skinned fibres under contraction conditions). Moreover, again mixing the various types of thick myosin filaments studied by Morel et al. (1999), I deduce that, at an MgATP concentration of 1 mM (the maximal concentration that we used), the MgATPase activity is $\sim 50 \pm$ (9 or 20)% (mean $\pm$ SE or SD; n = 5), the value at the flat maximum, corresponding to a decrease in isometric tension of only $\sim 13 \pm$ (1 or 3)%: this calculation takes into account the very loose coupling, as recalled in the preceding paragraph, according to which the isometric tetanic tension is proportional to the isometric MgATPase activity to a power of $\sim 0.190 \pm 0.014$ (see p. **78**). Within the wide range of experimental uncertainty and taking into account shortcomings in the phenomenological/semi-empirical reasoning, the maximal value of $\sim 13\% + 3\% \sim 16\%$ at 1 mM MgATP (in vitro) is sufficiently consistent with that of $\sim 20\%$ at 5 mM MgATP (in situ), as found by Ferenczi et al. (1984a) (see the preceding paragraph). The two series of results are therefore qualitatively and quantitatively consistent and mutually supportive. In this context, Ferenczi et al. (1984a) pointed out that the experimental data recalled and commented on above require a detailed treatment of cross-bridge mechanics. It is demonstrated here, and throughout this monograph, that the mechanical functions of the cross-bridges are not involved in these phenomena and that the characteristics of the MgATPase activity and the very loose coupling are basic phenomena of muscle contraction. This conclusion is independent of any molecular model for the generation of axial contractile force, provided only that the cross-bridges cyclically attach to and detach from the thin actin filaments and split MgATP during the cycle (phenomenological/semi-empirical approach described in this section [Section 3.4]; see Subsections 3.4.2 and 3.4.3). Not entirely cogent supplementary experimental arguments are given in the next paragraph (more convincing experimental data are analysed on pp. **83–84** in Section 3.4.6)

The demonstration of very loose coupling, presented in this section (see particularly pp. **78–80**) is valid only for frog. For rabbit and other mammals (i.e. warm-blooded species), it is difficult to infer that the very loose coupling also holds, because (i) there are few, if any, available experimental data concerning isometric tetanic tension in intact rabbit fibres and (ii) it would be risky to interpret experiments performed on rabbit at too low a temperature (e.g. between $\sim 0°C$ and $\sim 10°C$), the optimal temperature certainly being higher (e.g. between $\sim 20°C$ and $\sim 25°C$ in the laboratory and $\sim 39°C$ within the body). Hilber et al. (2001) went some way to addressing the issue. Indeed, they worked on single permeabilised fibres from rabbit psoas muscle and stated that 'the rate of ATP utilisation during isometric contraction had a Q_{10} of 3.6 throughout the temperature range 7–25°C; this was similar to the Q_{10} for isometric force at low temperature (3.5 at 7–10°C) but much larger than that for isometric force at higher temperature (1.3 at 20–25°C)'. Thus, in the most suitable temperature range for rabbit, there is clearly an uncoupling between isometric rate of MgATP splitting ($Q_{10} \sim 3.6$; temperature, 7°C–25°C, i.e. also 20°C–25°C) and isometric tension ($Q_{10} \sim 1.3$; temperature, 20°C–25°C), consistent with loose coupling or very loose coupling in the rabbit. In this context, on p. **78**, it is demonstrated that, for frog, within the optimal temperature range ($\sim 0°C$–10°C), the isometric tension $T_{0,\infty}$ is about proportional to the MgATPase activity $\Pi_{isom,\infty}$ to a power of $\sim 0.190 \pm 0.014$. Assuming that this relationship is also valid for rabbit, in its optimal temperature range ($\sim 20°C$–25°C; interval common to 7°C–25°C and 20°C–25°C; see above), we deduce that the Q_{10} for the isometric tetanic tension would be $\sim 3.6^{(0.190 \pm 0.014)} \sim 1.28 \pm 0.02$, corresponding exactly to the experimental value of ~ 1.30 found by Hilber et al. (2001). This is certainly not coincidental and, to my mind, demonstrates that very loose coupling also occurs in rabbit, provided that the frog and the rabbit are each studied in their optimal temperature range, say, $\sim 0°C$–10°C for frog and $\sim 20°C$–25°C for rabbit. Note that the findings and comments of Barman et al. (1998), Stephenson and Williams (1985a) and Woledge et al. (1985), repeatedly recalled in this book, that it is probably risky to compare cold-blooded and warm-blooded animals at low temperatures, are supported in this paragraph.

Finally, loose coupling is almost certainly a universal behaviour of muscle fibres, cardiac cells and, more generally, molecular machines of living cells (e.g. Barclay 1999; Brokaw 1995; Cooke et al. 1994; Esaki et al. 2007; Harada et al. 1990; Lombardi et al. 1992; Oosawa 2000; Oosawa and Hayashi 1986; Yanagida 1990), although the notion of 'very loose coupling' demonstrated here and that of 'loose coupling' strongly suggested by these authors correspond to very different concepts and biological situations. Indeed, either in vivo/in situ or in vitro, these authors studied contractile systems presenting relative movements of myosin heads along thin actin filaments (e.g. isotonic contraction), whereas I study here solely muscle fibres and intact unit cells, under purely isometric tetanic contraction conditions, with no relative sliding of the myofilaments past each other (see also Section 3.4.6).

3.4.6 Discussion of the Very Loose and Tight Couplings (Short Tetani–Steady State), in Intact Unit Cells from Intact Frog Fibres, with Some Insights into Demembranated Fibres from Rabbit and Dogfish and Intact Fibres from Dogfish

The notion of very loose coupling between isometric tetanic tension and isometric MgATPase activity, deduced independently of the mechanical roles of the cross-bridges (see pp. **78–80** in Section 3.4.5 and pp. **83–84**), clearly goes against the traditional view of intangible tight (one-to-one) coupling, including the close relationship between the mechanical characteristics of cross-bridges and all enzymatic steps of the myosin heads. This was recalled, for instance, by Barclay (1999), who wrote: 'the most straightforward scheme for linking the mechanical and biochemical cycles is that a single ATP molecule is hydrolysed in each mechanical cycle; that is, there is a tight coupling between the mechanical and biochemical cycles'. Burghardt et al. (1983) wrote: '... a particular hypothesis has seemed to rationalise many of the biochemical and physiological data'... [the cross-bridges (more precisely the myosin heads)] are assumed to deliver mechanical impulses to adjacent actin filaments in a cyclical manner. The unitary impulses are assumed to result from the interaction of nucleotides with enzymatic sites on the cross-bridges'. These two independent groups provide their own opinions about these assertions, but these two descriptions are representative of the majority view: isometric MgATPase activity studied in vivo/in situ and actomyosin enzymatic steps studied in vitro, on the one hand, and isometric tetanic tension, measured in vivo/in situ, on the other, are necessarily and unavoidably tightly coupled. For 25–30 years, this notion of tight coupling has been improved, but the main features of this approach remained unmodified. It is still widely thought that there is a one-to-one overall correlation between isometric tetanic tension and isometric MgATPase activity, on the one hand, and the enzymatic steps occurring on the cross-bridges in cycling contact with the thin actin filaments, on the other. This tight coupling results in strict proportionality between isometric tetanic tension and isometric rate of MgATP breakdown, with the various enzymatic steps playing a major role in this relationship (particularly regarding the modulation of active forces as a function of these enzymatic steps under various conditions, including, for instance, the transitory phenomena).

Potma et al. (1994) pointed out that 'there is disagreement in the literature as to whether, for isometric contraction, actomyosin ATPase activity and force are proportional'. The authors claimed that their 'results support the relatively simple concept that myofibrillar ATPase activity and average force (or load) are closely linked' and, more precisely, that their experimental findings demonstrated that 'isometric actomyosin and force are proportional'. However, the MgATPase activities measured in chemically skinned fibres from rabbit psoas muscle by Potma et al. (1994) are not appropriate (see p. **74** in Section 3.4.4 for a brief analysis) and the conclusions drawn by these authors cannot be taken into account. Buonocore et al. (2004), who do not agree with the swinging cross-bridge/lever-arm models, recalled that, according to the eternal explanation, there is a 'one-to-one relation between MgATP hydrolysis and the occurrence of the mechanical event consisting of the power stroke'. This view has also been asserted by all the authors favouring the swinging cross-bridge/lever-arm concept (e.g. Ford et al. 1977, 1981; Goldman and Huxley 1994; Huxley 1965, 1969; Huxley and Simmons 1971; Huxley and Tidesswell 1996; Irving 1991, 1995; Linari and Woledge 1995; Linari et al. 1998; Piazzesi and Lombardi 1995; Piazzesi et al. 2007).

As a means of resolving the many inconsistencies resulting from the notion of tight coupling between isometric tetanic tension and isometric MgATPase activity, it has been suggested that only a small fraction of cross-bridges (≤10%–30%) are attached to the thin actin filaments during contraction (e.g. Barclay 1999; Brokaw 1995; Cooke 1995, 1997; Linari and Woledge 1995; Piazzesi et al. 2007; this problem was also recalled by He et al. 1997; see Section 4.4.2.5.2, particularly p. **150**). This estimate (<20%–30%) is at odds with the universal value of ~82% attached cross-bridges (see p. **152** in Section 4.4.2.5.2). Regardless of slight differences in the assumptions and assertions of the authors cited above, all these opinions favour tight coupling. However, working on skinned muscle fibres from rat and toad, Stephenson et al. (1989) observed a 'dissociation of force from myofibrillar MgATPase', which, I think, may be accounted for by a large uncoupling between isometric tetanic force and isometric MgATPase activity in vivo/in situ, or, in other words, a (very) loose coupling between force and myofibrillar MgATPase activity. The same type of uncoupling is also suggested on pp. **54–55** in Section 3.4.3.2. In any event, the question of tight coupling is at odds with the 'model-independent phenomenological/semi-empirical' demonstration of very loose coupling, unwittingly supported experimentally by independent groups, and detailed on pp. **78–80** in Section 3.4.5 and pp. **83–84**. This model-independent approach is based on the theoretical approach of A.F. Huxley (1957), concerning the energetic aspects of muscle contraction only (see throughout this Section 3.4), and has been successfully used by my group for more than 35 years. The model is also based on very old experimental data (Fenn effect; Fenn 1923, 1924; Fenn and Marsh 1935, and the work of Hill 1938, 1964a,b). These old results have never been called into question and are still considered valid (e.g. Elliott and Worthington 2001; Worthington and Elliott 1996a,b, 2003). This shows that we can be confident in the discussions and conclusions presented here. I would like to finish this paragraph by an intermediate short conclusion drawn by West et al. (2004), who performed many experiments on intact and permeabilised fibres from dogfish white muscles: 'all the results agree in showing that the rate of actomyosin turnover continues to decline after the time, about 0.2 s, when force reaches its steady-state level (their Figure 6 is extremely convincing). In other words, although the force is constant during this time, ATP is not being hydrolysed at a constant rate'. Clearly, this is one of the most recent experimental demonstrations that one-to-one (tight) coupling is an outdated notion. Precise quantitative details are given on pp. **83–84**.

Thus, I think that the tight coupling hypothesis is highly questionable and should be replaced by the notion of very loose coupling, which is at odds with the concept that swinging cross-bridge/lever-arm mechanisms are the only possible explanation for muscle contraction. In this context, I present, in this book, a hybrid model, based on published experimental data and new results presented in Chapter 4, that does not require a tight coupling process.

The notion of very loose coupling in isometrically contracting fibres, demonstrated here by the phenomenological/semi-empirical approach, independently of the mechanical functions of the cycling cross-bridges in contracting muscle, is not entirely new. Indeed, the notion of loose coupling has already been suggested by Barclay (1999), Brokaw (1995), Cooke et al. (1994) and Lombardi et al. (1992), for example. However, as highlighted on pp. **78–80** in Section 3.4.5 and pp. **83–84**, the notion of very loose coupling, demonstrated in this monograph, applies to isometric tetanic contraction in fibres under steady-state conditions, whereas the conclusions of the groups cited above were drawn from experimental studies of actively shortening fibres. As clearly demonstrated in Sections 3.4.4 and 8.11, many complex processes are involved in active shortening that do not appear under isometric conditions ($s_0 V = 0$). The problems become even more complex when the ill-defined 'isometric' MgATPase activities measured by Potma and Stienen (1996) and Potma et al. (1994) (see p. **74** in Section 3.4.4) are taken into account. The notion of loose coupling (more precisely very loose coupling studied on pp. **78–80** in Section 3.4.5, and pp. **83–84**) is also totally different from that defined and discussed by Oosawa (2000), Oosawa and Hayashi (1986) and Yanagida (1990) and from that developed by Yanagida and his coworkers (e.g. Esaki et al. 2003, 2007; Kitamura et al. 2005; Yanagida et al. 2007) to account for both in vitro motility and muscle contraction. In Chapter 9, in vitro motility is studied at length and the question of assuming the same biochemical, biophysical, physiological and structural properties in vivo/in situ and in vitro is criticised. In my opinion, the notions of loose coupling and of tight (one-to-one) coupling resulted from erroneous choices for the rate of MgATP breakdown (e.g. Morel and D'hahan 2000, particularly the section entitled 'What is the true MgATPase activity to be taken into account?') and the obligation felt by some groups to subscribe to the traditional view, according to which the swinging/tilting of the cross-bridges in vivo/in situ and of the myosin heads in vitro is the unquestionable mechanism. This assumption is demonstrated to be incorrect in this monograph. Thus, the phenomenological/semi-empirical approach independent of the mechanical functions of the cross-bridges in vivo/in situ, developed in this section (Section 3.4; see Subsections 3.4.2 and 3.4.3), almost certainly provides the strongest evidence that, under isometric and steady-state conditions, there is very loose coupling between isometric tetanic tension and isometric MgATPase activity. This view is at odds with both the dogma of a tight coupling process and the loose coupling concept developed by Japanese investigators (mostly by Yanagida and his group; references cited above).

I decided to write this book as early as 2001–2002, and its general framework was established at the end of 2003. The 'phenomenon' of very loose coupling between isometric tetanic tension and isometric MgATPase activity, described on pp. **78–80** in Section 3.4.5 and pp. **83–84**, was developed progressively from the model-independent phenomenological/semi-empirical approach described in Sections 3.4.2 and 3.4.3. This new notion is employed in Section 3.4.5, together with the experimental data available, including those of Edman (1979) for intact frog fibres (from *semitendinosus* and *tibialis anterior* muscles), and, to a lesser extent (regarding loose coupling), those of Edman and Hwang (1977) for intact fibres (from *semitendinosus* muscle) and those presented in his Figures 11 and 12 by Millman (1998) also obtained with intact frog fibres (from various muscles) (except for some experimental points, obtained with intact toad fibres). The experimental findings of Ferenczi et al. (1984a) and Hilber et al. (2001), commented on in Section 3.4.5, add further support to these inferences. However, the notion of very loose coupling was indirectly deduced from separate, and possibly disparate, experimental studies of isometric tetanic tension and maximal shortening velocity, the corresponding values of which appear in Equation 3.48, which is itself deduced from the phenomenological/semi-empirical approach presented throughout this section (Section 3.4). Thus, it may be argued that the phenomenological/semi-empirical part of the very loose coupling notion is dubious, because it is based mostly on a straightforward, but unexpected and possibly questionable consequence of A.F. Huxley's theory (1957) (see Section 3.4.2), for which there is little direct experimental evidence, despite the many supporting experimental findings recalled in Section 3.4.5. Fortunately, direct experimental results are recalled below, demonstrating the very loose coupling. After writing the first draft of this section, I read the paper by West et al. (2004). The authors studied intact and permeabilised white fibres from dogfish, at 12°C, and demonstrated experimentally that large decreases in isometric MgATPase activity do not result in any significant decline in isometric tetanic tension. Note that the rate of MgATP splitting was measured directly in permeabilised fibres and estimated indirectly in intact fibres. From Figure 6A in the paper by West et al. (2004), valid for intact fibres, it appears that the isometric tension is constant, within the limits of experimental error, between ~0.3–0.5 s and ~3.5 s (below ~0.3–0.5 s, there is a transitory rise in tension). The duration of ~3.5 s is significantly longer than those used for intact frog fibres: ~1.0–1.5 s in the experiments of Altringham and Bottinelli (1985), Edman and Anderson (1968), Gordon et al. (1966a,b) and Granzier and Pollack (1990), for example. Dogfish at 12°C and frog at ~2.5°C could therefore behave slightly differently. In any event, from Table 4 in the paper by West et al. (2004), it appears that actomyosin MgATPase activity, in intact dogfish fibres, at 0.5 s is ~10.2 ± 1.0 s^{-1} and ~5.9 ± 0.5 s^{-1} at 3.5 s, giving a ratio of ~0.6 ± 0.1. For intact frog fibres, between 2°C and 12°C, the isometric tetanic tension is proportional to the rate of MgATP breakdown to a power of ~0.190 ± 0.014 (see p. **78** in Section 3.4.5). Assuming that this value also applies to intact dogfish fibres at 12°C, the decline in tension would be only ~$(0.6 ± 0.1)^{0.190 ± 0.014}$ ~ 0.91 ± 0.03, a value close to 1, that is, constancy of the tension, within the limits of experimental error (the maximal value is ~0.94 and the minimal value is ~0.88, giving values lower than 1 within the range of ~6%–12%, similar to the experimental error on the tensions estimated at ~±5%–15%; see Table 1 in the paper by West et al. 2004). From Figure 6B in the paper by West et al. (2004), valid for permeabilised dogfish fibres, it appears that the isometric tetanic tension is constant, within the limits of experimental error, between 0.25 s (below ~0.25 s, there is a transitory rise in tension) and, at least, ~0.5 s, whereas, from Table 4 in the paper by West et al. (2004), it appears that the rates of MgATP splitting are ~12.8 ±

0.5 s^{-1} and ~6.2 ± 0.5 s^{-1}, respectively, giving a ratio of ~0.5 ± 0.1. Again taking the exponent of ~0.190 ± 0.014, the decline in tension would be ~0.88 ± 0.03 (the maximal value is ~0.91 and the minimal value is ~0.85, giving values lower that 1 within the range of ~9%–15%, similar to experimental error on the tensions estimated at ~±5%–15%; see Table 1 in the paper by West et al. 2004). This value is statistically similar to ~0.91 ± 0.03, estimated above, and the conclusion is the same. The very loose coupling 'concept' is therefore a reality, unwittingly validated by West et al. (2004), but clearly expressed in the title of their paper (see references) These findings provide a posteriori experimental support for my view. Moreover, the experimental observations of West et al. (2004) provide strong support for the phenomenological/semi-empirical approach described in this monograph, independent of the mode of force generation, and, consequently, for the energetic part of Huxley's theory (1957), improved throughout this section (Section 3.4). By contrast, these observations are again at odds with the tight coupling (one-to-one) concept, according to which isometric tetanic tension is proportional to isometric MgATPase activity.*

3.5 FULL AND ELECTROSTATICALLY RELATED AXIAL CONTRACTILE FORCES IN AN INTACT UNIT CELL FROM FROG (SHORT TETANI–STEADY STATE)

All the calculations presented below, in this section, are valid at 10°C, unless otherwise specified, at maximal overlap, corresponding to the universal sarcomere length s_0 ~ 2.18 μm (see p. **24** in Section 3.2), although the new experiments presented in this monograph were performed at ~2.5 ± 0.2 μm (ratio of only ~+1.15 ± 0.09, much lower than experimental uncertainty on isometric tetanic tensions; see also p. **24** in Section 3.2 for a brief discussion, Section 3.3 and pp. **96–97** in Section 3.8.1), and at physiological bulk ionic

strength (I = 180 mM; I* ~ 182 mM in demembranated fibres; see Section 3.9 for precise details on the vicinal ionic strength I*). As shown on p. **89** in Section 3.7, in these conditions, an intact unit cell from frog develops an isometric tetanic tension T*(10°C) ~ 7.2 × 10^5 N m^{-2} ~ 7.2 × 10^5 pN μm^{-2} during a short tetanus, under steady-state conditions. According to Bachouchi and Morel (1989a), the cross-sectional area of an intact unit cell from frog or rabbit, at the slack sarcomere length (~2.0–2.2 μm), can be estimated at ~1.55 × 10^{-3} μm^2 and ~1.82 × 10^{-3} μm^2, respectively, giving a mean value of ~ (1.55 + 1.82) × 10^{-3}/2 ~ 1.69 × 10^{-3} μm^2. From Table 1 in the review by Millman (1998), concerning various types of muscle from frog and rabbit, at a reference sarcomere length of 2.30 μm, a mean value of ~1.65 × 10^{-3} μm^2 (17 values from independent groups) can be deduced, of the same order of magnitude as ~1.69 × 10^{-3} μm^2 (at sarcomere lengths of ~2.0–2.2 μm; see above). Only the most representative value of ~1.65 ± 0.10 × 10^{-3} μm^{-2} (± SD) is retained. Owing to the scattering of the experimental values, no correction is made concerning the sarcomere lengths of 2.30 μm and s_0 ~ 2.18 μm (see above). The two extreme values of cross-sectional area are therefore ~1.55 × 10^{-3} μm^2 and ~1.75 × 10^{-3} μm^2, for a composite animal species. For the sake of self-consistency in the reasoning and calculations, I select a value of ~1.58 × 10^{-3} μm^2 here. The full axial contractile force is therefore F* ~ 7.2 × 10^5 pN μm^{-2} × 1.58 × 10^{-3} μm^2 ~ 1138 pN per intact unit cell. Comparison of experimental and semi-empirical estimates of the apparent force per myosin head, obtained by various techniques and by independent groups, shows, as described in Section 3.6, show that the values of T* and F*, given above, are the most suitable values at 10°C, for both frog and rabbit, regardless of the arrangement of the myosin heads in the thick myosin filaments. However, the arrangement proposed by Morel et al. (1999) appears to be the most suitable (see p. **88** in Section 3.6 for a brief discussion).

It should be borne in mind that, despite the many negative assertions of most investigators, there are mechanisms for turning net radial repulsive (expansive) forces into axial contractile forces. These processes were demonstrated theoretically and logically by Dragomir et al. (1976), Elliott (1974), Morel et al. (1976), and, although unwittingly, experimentally by Gordon et al. (1973) and Hasan and Unsworth (1985). All these

---

* In Section 3.4.3.2, I refer frequently to the paper by He et al. (1997), and particularly to Figure 9, presenting concurrently the isometric tension rise and MgATPase activity against time, for one permeabilised frog fiber (at 5°C), during a contraction lasting ~1 s. Reading the legend to this figure and examining the two curves, it appears that the changes in these two characteristics of the fibre are almost independent. This probably confirms the very loose coupling.

Morel

points are discussed, for example, in Section 3.10. It is strongly suggested that, in short tetani and under steady-state isometric conditions (see Section 3.4.3.2 for definitions of these two notions), ~40% of the full isometric tetanic tension is related to lateral swelling mechanisms in intact fibres, half-fibres and intact unit cells. The other ~60% of the full tension is generated according to the swinging cross-bridge/lever-arm models. I found that proportions of ~40% and ~60% for the lateral swelling and the swinging cross-bridge/lever-arm processes, respectively, were the most likely, after testing many possibilities. This choice gives self-consistency throughout this monograph (see Section 4.4.1 for a semi-empirical approach). Thus, in an intact unit cell, we obtain $T_e^* \sim 0.4T^* \sim 0.4 \times 7.2 \sim 2.9 \times 10^5 \, \text{N m}^{-2} \sim 2.9 \times 10^5 \, \text{pN } \mu\text{m}^{-2}$, where T* and T_e^* are the full tension and axial tension related to net radial repulsive force, respectively (see the preceding paragraph concerning T*); T_e^* results largely from repulsive electrostatic forces between the myofilaments (~76.0%; see below). The corresponding axial contractile 'electrostatically related' force is $F_e^* \sim 0.4F^* \sim 0.4 \times 1138 \, \text{pN} \sim 455 \, \text{pN}$ pN per intact unit cell (see the preceding paragraph, regarding the value of F* ~ 1138 pN per intact unit cell). Let us call B_a^* the dimensionless coefficient, defined for an intact unit cell, that must be used to pass from the net radial repulsive (expansive) force to the axial contractile force F_e^*. The radial repulsive electrostatic force in an intact unit cell is given by $4(\gamma/2)F_{e,a} \sim 784Q_a$ pN (see, in Section 3.2, Equation 3.2 for the definition of $F_{e,a}$, p. **34** for definition of the overlap length γ, p. **35** for a discussion of the number 4 and pp. **34–35** concerning the value of $4(\gamma/2)F_{e,a}$; the value of the vicinal ionic strength I* ~ 182 mM is omitted here, for the sake of simplicity). The net radial repulsive (expansive) force results from the subtraction from $4(\gamma/2)F_{e,a}$ of the residual radial tethering force NF_{cb}, the radial collapsing force C_a and the radial attractive/compressive force Λ_a, all exerted in an intact unit cell (mostly between the thick myosin and thin actin filaments) during contraction (see definitions and main features of these forces in Sections 4.4.2.4 and 4.4.2.5.2; see Section 4.4.2.5.4 for numerical values). The net radial expansive force per intact unit cell can therefore be written $4(\gamma/2)F_{e,a}^* = 4(\gamma/2)F_{e,a} - \Theta_a$, with $\Theta_a = NF_{cb} + C_a + \Lambda_a$ and $4(\gamma/2)F_{e,a} \sim 784Q_a$ pN per intact unit cell, at full overlap. Thus, we have $F_e^* = B_a^* 4(\gamma/2)$

$F_{e,a}[1 - \Theta_a/4(\gamma/2)F_{e,a}] \sim 455 \, \text{pM}$ per intact unit cell (see above for a definition of the dimensionless coefficient B_a^* and the value of ~455 pN). Let us introduce $\rho_a = \Theta_a/4(\gamma/2)F_{e,a}$, giving

$$F_e^* = B_a^*(1 - \rho_a)4(\gamma/2)F_{e,a} \qquad (3.49)$$

Reintroducing I* ~ 182 mM, for comparison with other vicinal ionic strengths (see Section 4.4.2.7), Equation 3.49 gives $F_e^*(182) \sim B_a^*(182)$ $[1 - \rho(182)]Q_a$ (182) × 784 pN (for the sake of generality, B_a^* is assumed to depend on I*). We deduce that $B_a^*(182)[1 - \rho_a(182)]Q_a$ (182)~ 455 pN/784pN ~0.580. In this relationship, there are three unknown parameters. As a first approximation, let us assume that $B_a^*(182) \sim [1 - \rho_a(182)] \sim$ $Q_a(182)$, giving $B_a^*(182) \sim [1 - \rho_a(182)] \sim Q_a(182)$ ~ 0.834 and $\rho_a(182) \sim 0.166$. Although arbitrarily chosen, these three values are suitable in principle. Indeed, from its definition (see text below Equation 3.1), Q_a must be less than 1, simply because the mathematical function 'tanh' cannot exceed 1. However, the value of $Q_a \sim 0.834$ leads to a mean surface electrical potential on the myofilaments of ~−151 mV, which is certainly too high an absolute value (on pp. **161–162** in Section 4.4.2.6, the electrical potentials are assumed to be similar on the thin actin and thick myosin filaments and the absolute value of this mean electrical potential is discussed). Thus, I choose a more realistic mean surface electrical potential of ~−131 mV, giving $Q_a \sim 0.763$ (the rather high absolute value of ~131 mV is selected for self-consistency of the calculations and reasoning in this book). We deduce that $B_a^*(182)[1 - \rho_a(182)] \sim 0.580/0.763 \sim 0.760$. At this point, I choose $B_a^*(182) \sim 1$, corresponding to a one-to-one translation of the net radial repulsive (expansive) force into an electrostatically related axial force. We deduce that $1 - \rho_a(182) \sim 0.760$ (~76.0%) and $\rho_a(182) \sim 0.240$ (~24.0%). Under contraction conditions, the sum of the residual radial tethering forces, collapsing forces and attractive/compressive forces is smaller than the radial repulsive (expansive) force, implying that $\rho_a(182)$ must be significantly below 1, consistent with the value of ~0.240 suggested here. This is equivalent to the residual radial tethering force, NF_{cb}, being much weaker than the radial tethering force, $N_0F_{cb}^0$, at rest (see pp. **174–176** in Section 5.1.1, concerning the mode of generation of F_{cb}^0, and pp. **158–159** in Section 4.4.2.5.4 and pp. **180–181** in Section 5.1.2, for the inequality $NF_{cb} \ll N_0 F_{cb}^0$). The three values of the dimensionless parameters B_a^*, Q_a and ρ_a proposed here lead to self-consistent results throughout this monograph. Finally, from $Q_a \sim 0.763$, we deduce that $4(\gamma/2)F_{e,a} \sim 784Q_a \sim 598$ pN per intact unit cell and the net radial repulsive (expansive) force is $F_{e,a}^* \sim 4(\gamma/2)F_{e,a}(1-\rho_a) \sim 598 \times 0.760 \sim 454$ pN per intact unit cell. This value is identical to that of the axial contractile electrostatically related force found on p. **85**, particularly because $B_a^*(182) \sim 1$ (see above, the one-to-one conversion process).

3.6 APPARENT AXIAL CONTRACTILE FORCE PER MYOSIN HEAD IN AN INTACT UNIT CELL (SHORT TETANI–STEADY STATE). MIXTURE OF FROG AND RABBIT

In the swinging cross-bridge/lever-arm models of contraction, assuming that the thick myosin filaments consist of myosin molecules with all their heads lying outside the filament core, dividing the full axial contractile force F* in an intact unit cell (see p. **85** in Section 3.5 for the definition and value of F*) by the number of attached cross-bridges in this intact unit cell gives the axial force per rotating myosin head. In the hybrid model of contraction, presented and discussed in this monograph, taking into account the existence of internal B–B′-type head–head dimers within the thick myosin filament core (Morel et al. 1999; see also Sections 5.1.1 and 5.2, and pp. **205–206** in Section 6.3.5), the problem is different and only apparent axial contractile forces can be calculated. For example, the internal B–B′-type head–head dimers play a direct role in generating the radial net expansive force between thick myosin and thin actin filaments, and this radial force is converted into an axial contractile force (see pp. **102–103** in Section 3.10). As there is no straightforward relationship between F* and the internal B–B′-type head–head dimers and their behaviour during contraction in the hybrid model, only an apparent axial contractile force per myosin head can be estimated (see below, in this section).

In young adult frogs and young adult rabbits, it is widely accepted that a thick myosin filament in vivo/in situ contains ~300 myosin molecules, giving ~600 heads per intact unit cell (e.g. Barclay 1999; Cantino and Squire 1986; Huxley and Kress 1985; Kensler and Stewart 1983, 1993; Linari et al. 1998; Malinchik and Lednev 1992;

Offer 1987; Schoenberg 1980a,b; Squire 1972, 1975, 1981; Squire et al. 2005; Stewart and Kensler 1986a; see also pp. **205–206** in Section 6.3.5 for a short discussion of the number of heads per thick myosin filament and other related problems). From the experimental work of Morel et al. (1999) and the hypothesis of Morel and Garrigos (1982b), from Section 5.1.1 (experimentally supported in Section 5.2), and from pp. **205–207** in Section 6.3.5, we deduce that there are ~400 external heads able to interact cyclically with the thin actin filaments and ~200 internal B–B′-type heads bound within the filament core, able to form head–head dimers (see Section 5.1.1, particularly Figure 5.1, its legend and Appendix 5.I). In Section 4.4.2.5.2 (see particularly p. **152**), it is shown that the proportion of ~0.82 for attached cross-bridges in the steady state, in a short tetanus, is a universal value, valid for any animal species traditionally used in the laboratory (mostly frog and rabbit) and at any temperature (between ~0°C and ~20°C–22°C). Thus, at the slack sarcomere length (maximal overlap) and, particularly, at 10°C, there are ~400 × 0.82 ~ 328 of the ~400 external heads bound to the thin actin filaments per intact unit cell at a given moment. From the value for full axial contractile force of F* ~ 1138 pN per intact unit cell, at 10°C (see p. **85** in Section 3.5), we deduce that the elementary apparent axial contractile force is ~1138 pN/328 ~ 3.5 pN per external myosin head. Ignoring the arrangement of the myosin heads described by Morel et al. (1999) and assuming, as is done by most investigators, that all the heads are located outside the filament shaft, we obtain ~600 × 0.82 ~ 492 heads attached to the thin actin filaments per intact unit cell and a corresponding apparent axial contractile force of ~1138 pN/492 ~ 2.3 pN per myosin head. This last value corresponds to the best semi-empirical estimate, valid for both the swinging cross-bridge/lever-arm models and the widely accepted model of the arrangement of the myosin heads in the thick myosin filaments. Regardless of this arrangement and based on older experimental data, Bachouchi and Morel (1989b) proposed axial contractile forces of between ~1.5 pN and ~9.5 pN per myosin head, with a most probable value of ~4.0 pN per myosin head. Morel (1991a) claimed that the 'apparent axial contractile force' is ~8.0 pN per myosin head. Merah and Morel (1993) suggested that values of ~5.0 pN or ~9.0 pN per myosin head would be appropriate (the temperature was not clearly specified by Bachouchi and Morel 1989b, Morel 1991a and Merah and Morel 1993, but corresponds implicitly to those used by Edman 1979 and Edman and Hwang 1977, i.e. ~0°C–5°C). As $Q_{10} \sim 2.4$ for full isometric tetanic axial tension between ~0°C and ~12°C for a 'composite animal and biological material' (see p. **246** in Section 8.8), these eight scattered values become, per myosin head (at 10°C): ~3.5 pN, ~2.3 pN, ~3.0 pN, ~18.8 pN, ~7.9 pN, ~15.8 pN, ~9.9 pN and ~17.8 pN, respectively. These values are used in the next paragraph.

The following estimates, obtained from experimental data of independent groups, are valid for traditionally arranged filaments (i.e. all the myosin heads outside the thick filament core), because it was not possible at the time to take into account the model of thick myosin filaments proposed by Morel and Garrigos (1982b) and supported later by experimental arguments (Morel et al. 1999; see also, in Section 5.1.1, pp. **170–172**, Figure 5.1, its legend and Appendix I). Thus, corrections are made below, and the estimates obtained by the different groups are multiplied by a factor of ~600 heads/400 heads ~ 1.5. However, owing to the large uncertainty on the various published values, both the crude and corrected estimates are taken into account. From semi-empirical considerations concerning frog *sartorius* muscle, Barclay (1999) proposed a value of ~1.7 pN per myosin head, presumably at ~0°C. Between 0°C and 12°C, $Q_{10} \sim 2.4$ (see the end of the preceding paragraph). The value suggested by Barclay (1999) therefore becomes ~4.1 pN per myosin head, at 10°C, or ~4.1 × 1.5 ~ 6.2 pN per myosin head. Working on chemically skinned fibres from rabbit psoas muscle at 5°C, Brenner and Yu (1991) estimated, from their experimental data, that 'While not under osmotic pressure, the radial force of the activated fibre was determined to be 400 pN (single thick filament)$^{-1}$. This is the same order of magnitude as the axial force'. On p. **99** in Section 3.8.2, this assertion concerning the radial force is critically analysed. In any event, we can choose an axial contractile force per thick myosin filament of ~400 pN per thick myosin filament. As pointed out on p. **98** in Section 3.8.1, a multiplication factor of ~2.23 should be used to pass from the axial tensions recorded in a demembranated fibre to the axial tensions that would be recorded in the corresponding intact

fibre. Owing to the wide experimental error, the multiplication factor of ~2.23 can be extended to a 'mean representative central unit cell' (not necessarily intact unit cell) lying in the centre of a demembranated fibre and almost certainly containing one thick myosin filament (see pp. **97–98** in Section 3.8.1 and Section 3.8.2 concerning the high degree of structural complexity of demembranated fibres). As there is one thick myosin filament per mean representative central unit cell, we obtain ~400 × 2.23 ~ 892 pN per unit cell or thick myosin filament (at 5°C). Again, using the most suitable value of $Q_{10} \sim 2.4$ within the ~0°C–12°C temperature range (see the end of preceding paragraph), this axial contractile force would be ~1382 pN per thick filament, at 10°C. The apparent axial contractile force would therefore be ~1382 pN/492 ~ 2.8 pN per myosin head, if the traditional structure of the thick myosin filaments is taken into account (all heads outside the backbone) or ~2.8 × 1.5 ~ 4.2 pN per myosin head. Working on mechanically skinned frog fibres at 4°C, Matsubara et al. (1984) gave an estimate of the axial contractile force of ~500 pN per thick myosin filament, corresponding to ~845 pN per thick myosin filament, at 10°C ($Q_{10} \sim 2.4$; see above). Using the same multiplication factor of ~2.23, the true axial contractile force would be ~1884 pN per thick myosin filament, corresponding to two possible values of ~1884/492 ~ 3.8 pN per myosin head or ~3.8 × 1.5 ~ 5.7 pN per myosin head. Studying the 'contractile force developed by the S1-decorated actin monomer', Grazi (2000) suggested, on a semi-empirical basis, a value of ~4.7 pN per myosin head, presumably at room temperature (~20°C–22°C), that is, ~2.5 pN per myosin head at 10°C (with $Q_{10} \sim 1.8$ between 10°C and room temperature; see p. **246** in Section 8.8) or ~2.5 × 1.5 ~ 3.8 pN per myosin head. Working on intact frog fibres at 4°C, Linari et al. (1998) deduced from their semi-empirical study a value of ~3.6 pN per myosin head, corresponding to ~6.1 pN per myosin head at 10°C ($Q_{10} \sim 2.4$; see above) or ~6.1 × 1.5 ~9.2 pN per myosin head. Oplatka (1972) gave several estimates, ranging from ~2.4 pN to ~4.7 pN per myosin head, presumably at ~0°C–5°C (mean value of ~2.5°C), for intact frog or rabbit fibres. Only these two extreme values are taken into account. They correspond to ~4.6 pN and ~9.1 pN per myosin head, respectively, at 10°C ($Q_{10} \sim 2.4$; see above), or ~4.6 × 1.5 ~ 6.9 pN and ~9.1 × 1.5

~ 13.6 pN per myosin head, respectively. Decostre et al. (2005) suggested a value of ~6.2 pN at 10°C. In a rather complex semi-empirical study, taking into account the swinging cross-bridge/lever-arm models, Piazzesi et al. (2007), working on intact frog fibres, suggested a force of ~6.0 pN per 'myosin motor' (myosin head), at 4°C, that is, ~10.1 pN per myosin head at 10°C (Q_{10} ~ 2.4; see above). However, this estimate was based on a proportion of myosin motors attached to the thin actin filaments of only ~29% and ~588 external heads, rather than ~82% and ~600, respectively. Using these last two values, the true force per myosin head would be ~10.1 × (588/600) × (29%/82%) ~3.5 pN per myosin head. Owing to some flaws in the study of Piazzesi et al. (2007), I do not use the corrective factor of 1.5 (see throughout this paragraph, concerning this factor). Moreover, I am unable to choose between these two estimates (~10.1 and ~3.5 pN per myosin head) and therefore consider both in the next paragraph.

The 25 estimates obtained by independent groups and based on various techniques and biological materials, recalled above, in this section, are broadly scattered, ranging between ~1.9 pN and ~18.8 pN per myosin head. The best solution, systematically used in this book, is therefore to average these estimates to obtain the most suitable value of the apparent force per myosin head. This gives a value of ~7.4 ± (0.9 or 4.7) pN per myosin head (mean ± SE or SD; n = 25), which is approximately statistically similar (provided that SD is taken into account; see Section 2.1) to ~3.5 pN and ~2.3 pN per myosin head, as found on p. **86**, independently of the mode of force generation and the arrangement of the myosin heads in the thick myosin filaments. We can therefore select the most probable value of ~7.4 pN per myosin head, at 10°C, in an intact unit cell from a composite animal and biological material. This value is statistically similar to the ~4.0 pN per myosin head assumed by A.F. Huxley (2000), presumably at ~0°C–5°C (mean value of ~2.5°C), and therefore ~7.7 pN per myosin head, at 10°C (Q_{10} ~ 2.4; see the preceding paragraph). In Chapter 9, the value of ~7.4 pN per myosin head is taken as a reference for three reasons: (i) it is not too far from the values of ~3.5 pN or ~2.3 pN per myosin head found on p. **86**, regardless of the mode of force generation; (ii) it is independent of the arrangement of the myosin heads in the thick myosin filaments; (iii) it is consistent with the spectrum of

experimental results presented above, in this section. Multiplying the value of ~7.5 pN per myosin head by either ~328 or ~492 external heads attached to the thin actin filaments (see p. **86** concerning ~328 and ~492), we deduce that the axial contractile force should lie between ~7.4 pN × 328 ~ 2427 pN and ~7.4 pN × 492 ~3641 pN per intact unit cell, that is, between ~2427 pN per intact unit cell/1.58 × 10^{-3} μm² ~15.4 × 10^5 pN μm⁻² ~ 15.4 × 10^5 N m⁻² and ~3641 pN per intact unit cell/1.58 × 10^{-3} μm² ~23.0 × 10^5 pN μm⁻² ~ 23.0 × 10^5 N m⁻² in an intact unit cell (1.58 × 10^{-3} μm² is the best estimate of the cross-sectional area of an intact unit cell; see p. **84** in Section 3.5). Although rather high, these two values of ~15.3 × 10^5 N m⁻² and ~23.0 × 10^5 N m⁻² can be considered essentially consistent with those deduced from various experimental data assumed to correspond to intact unit cells and discussed on pp. **89–91** in Section 3.7, giving an average tension of ~7.2 × 10^5 N m⁻², and the phenomenological/semi-empirical value of ~8.8 × 10^5 N m⁻² (see, e.g. the first paragraph in Section 3.7). Nonetheless, the value of ~15.3 × 10^5 N m⁻², corresponding to ~328 external heads, is more suitable than that of ~23.0 × 10^5 N m⁻², corresponding to ~492 external heads. This provides indirect, but not entirely convincing, support for the model of three-stranded thick myosin filaments, valid for young adult frogs and young adult rabbits (see, in Section 5.1.1, pp. **170–172**, and Figure 5.1, its legend and Appendix I, and also pp. **205–206** in Section 6.3.5).

3.7 WHY IS THE ISOMETRIC TETANIC TENSION (SHORT TETANI–STEADY STATE) IN AN INTACT FROG FIBRE MUCH SMALLER (APPROXIMATELY 40%) THAN THAT IN AN INTACT UNIT CELL (LOCATED IN THE CENTRE OF THE FIBRE)?

Based on the phenomenological/semi-empirical approach, independent of the mechanical role of the cross-bridges (see Sections 3.4.2 and 3.4.3), it is shown, on p. **63** in Section 3.4.3.2, that, in frog, the isometric tetanic tension is $T_{0,\infty}$ ~ 8.8 × 10^5 N m⁻² in an intact unit cell, at 10°C, at the slack length (s_0 ~ 2.18 μm; see p. **24** in Section 2.2), and at physiological ionic strength (I = 180 mM; I* ~ 182 mM). The isometric tetanic tension in intact fibres (in isosmotic buffers mimicking the external medium) and in half-fibres (in surrounding buffers mimicking the internal medium, but in the presence of MS⁻ as the major anion), at 10°C, is $T_{hf} \sim T_f \sim 3.0 \times 10^5$ N m⁻² (see Section 3.3). Thus, there is a great loss of tension when passing from an intact unit cell to an intact fibre or a half-fibre. However, working on intact frog fibres, Granzier and Pollack (1990) obtained, after correction for temperature, a mean value of ~3.6 × 10⁵ N m⁻² and a maximal value of ~5.3 × 10⁵ N m⁻², at 10°C. Working on the same biological material, Horowitz and Pollack (1993) gave a mean value of ~4.4 × 10⁵ N m⁻² and a maximal value of ~7.2 × 10⁵ N m⁻², also at 10°C. These six values, ranging between ~3.0 × 10⁵ N m⁻² and ~8.8 × 10⁵ N m⁻², raise important questions that are discussed on pp. **89–91**.

Using experimental and semi-empirical methods of stereology, Mobley and Eisenberg (1975) found that the tensions recorded in intact frog fibres should be scaled up by a factor of ~1.2 for comparison with the tensions expected in myofibrils within intact fibres. Merah and Morel (1993) showed, on the basis of purely geometric and logical considerations, that, regardless of the mode of force generation, at ~0°C (also probably within the range ~0°C–10°C), the force recorded in intact frog fibres should be scaled up by a factor of ~1.9 to obtain the force in a single intact myofibril inserted into a whole intact fibre. We considered that 'the myofibrils are obviously arranged in a hexagonally packed structure'. This is probably untrue. Indeed, the myofibrils are almost certainly not perfect solid cylinders and would be stuck together along their whole length. In this event, the factor $2 \times 3^{1/2}/\pi \sim 1.1$ in Equation 7 in the paper by Merah and Morel (1993) would not be taken into account, being simply replaced by 1.0, which would give a corrective factor of only ~1.7, rather than ~1.9. Owing to the many experimental and geometric uncertainties, all three corrective factors (~1.2, ~1.7 and ~1.9) are taken into account in the next paragraph.

In the first paragraph of this section, it is recalled that Horowitz and Pollack (1993) recorded, for intact frog fibres, a maximal isometric tetanic tension of ~7.2 × 10⁵ N m⁻² at 10°C. Suggestions are made below, in this section, to account for the loss of tension when passing from an intact unit cell to an intact fibre, and the reasoning works in both directions. The value of ~7.2 × 10⁵ N m⁻² for an intact fibre results, to my mind, from a concatenation of circumstances and is the same

as that in an intact unit cell (~7.2 × 10⁵ N m⁻²; see below). In these conditions, the corrective factor when passing from an intact fibre to an intact unit cell would be ~1.0. In the general case, for passing from an intact fibre to an intact unit cell, I assume that the corrective factor is ~(1.0 + 1.2 + 1.7 + 1.9)/4 ~ 1.45 ± (0.17 or 0.34) (mean ± SE or SD; n = 4). The reference tension in an intact fibre (and a half-fibre in the presence of MS⁻ as the major anion; see Section 3.3) is ~3.0 × 10⁵ N m⁻², corresponding, in principle, to ~1.45 × 3.0 ~ 4.4 × 10⁵ N m⁻² in an intact unit cell. Taking into account the mean and maximal values of ~3.6 × 10⁵ N m⁻² and ~5.3 × 10⁵ N m⁻² deduced from the paper by Granzier and Pollack (1990) (see the first paragraph of this section), the corrected tensions (valid at 10°C), in an intact unit cell, would be ~1.45 × 3.6 ~ 5.2 × 10⁵ N m⁻² and ~1.45 × 5.3 ~ 7.7 × 10⁵ N m⁻², respectively. Taking into account the mean and maximal values of ~4.4 × 10⁵ N m⁻² and ~7.2 × 10⁵ N m⁻² deduced from the paper by Horowitz and Pollack (1993) (see the first paragraph of this section), the corrected tensions (valid at 10°C), in an intact unit cell, would be ~1.45 × 4.4 ~ 6.4 × 10⁵ N m⁻² and ~1.45 × 7.2 ~ 10.4 × 10⁵ N m⁻², respectively. The scattering of the crude experimental results probably reflects the different techniques used. Moreover, the difference between the theoretical value $T_{0,\infty}$ ~ 8.8 × 10⁵ N m⁻² (see the first paragraph) and the crude experimental values of ~3.0, 3.6, 5.3, 4.4 and 7.2 × 10⁵ N m⁻² (before correction; see the first paragraph) may be attributed to marked and uncontrolled lateral swelling of the myofilament lattice and non-filamentous components after fibre isolation (with purely geometric consequences on the cross-sectional areas and, therefore, on the isometric tetanic tensions). In other words, it is possible that the estimated external osmolarity of the buffer surrounding isolated intact fibres is not the same as that in vivo. It is also possible that other unknown parameters are slightly modified when a single intact fibre is dissected from a whole muscle, owing to the disappearance of possible fibre–fibre interactions (e.g. weak electrostatic repulsive and attractive/compressive forces, or some kind of chemical bond between the sarcolemmas of neighbouring fibres, via lipoproteins and other extrinsic proteins). After writing this short discussion of the preparation of single intact fibres, I read the paper by Elzinga et al. (1989b), in which the authors noted the variability of tensions recorded in intact frog fibres and gave their own interpretation of this feature. I think that their views and mine overlap. From the various results obtained and the various hypotheses suggested by the authors, I note three important points: (i) water content varies between fibres (from 76.2% to 84.5%), supporting my own assumption concerning 'in vivo' and 'in vitro' osmolarities; (ii) the variability seems to be related to the 'age or state of development of the fibre' (this assumption is, in my opinion, similar to the experimental findings concerning young adult and old rabbits; see Morel et al. 1999, and also p. **52** in Section 3.4.3.2, for a discussion of the experimental findings of Ferenczi et al. 1984b, almost certainly obtained with fragile fibres extracted from old rabbits); (iii) 'one might suggest that myofibrils at the periphery of a fibre might produce less force than their counterpart in the fibre core'. This hypothesis is similar to that made on pp. **91–95**, although the authors think that their experimental evidence 'makes the hypothesis of weaker myofibrils near the surface unlikely' (I find the sole experimental argument presented by the authors to exclude this hypothesis unconvincing). In any event, in the absence of definitive arguments in favour of a particular value presented above, in this section, the best way to obtain a suitable estimate of the isometric tetanic tension in an intact unit cell is to average the six available values (obtained after correction for temperature; see above), giving $T^*(10°C) \sim (4.4 + 5.2 + 6.4 + 7.7 + 10.4 + 8.8)/6 \sim 7.1 \pm (0.8$ or 2.1) $\times 10^5$ N m^{-2} (mean $\pm$ SE or SD; n = 6). One may argue that I chose rather arbitrarily the experimental values, but the use of other values gives similar values of $T^*(10°C)$. For instance, averaging the two extreme values of $\sim4.4 \times 10^5$ N m^{-2} and $\sim10.4 \times 10^5$ N m^{-2} gives $\sim(4.4 + 10.4)/2 \sim 7.4 \pm 3.0 \times 10^5$ N m^{-2}, statistically similar to the $\sim7.1 \pm 0.8 \times 10^5$ N m^{-2}. For the sake of self-consistency in the reasoning and calculations, I select a value of $T^*(10°C) \sim 7.2 \times 10^5$ N m^{-2} here. I try below, in this section, to find experimental results and complementary semi-empirical assumptions accounting for the loss of tension when passing from an intact unit cell, where $T^*(10°C) \sim 7.2 \times 10^5$ N m^{-2}, to an intact fibre, where $T_f(10°C) \sim 3.0 \times 10^5$ N m^{-2} (see above). On entirely different bases, Elliott and Worthington (2001), in an analysis of their impulsive model, entitled one of their sections 'The macroscopic and microscopic models

compared'. They demonstrated that the impulse time they defined is $\sim6 \times 10^{-5}$ times smaller in half-sarcomere (microscopic model) than in whole muscle (macroscopic model). Thus, Elliott and Worthington (2001) and I reach comparable qualitative conclusions, regarding major differences between microscopic and macroscopic structures.

Comparison of single isolated myofibrils might provide experimental support for the loss of tension described in the preceding paragraph. Unfortunately, this is not possible (see below). Working at 15°C on single isolated skeletal myofibrils from frog, Colomo et al. (1997) found a tension of $\sim3.7 \times 10^5$ N m^{-2}. There are no systematic experimental data on the dependence of tension on temperature for demembranated biological materials from frog. Let us assume, as a first approximation, that, for frog, the dependence of tension on temperature is similar for mechanically, chemically and permeabilised fibres and single isolated myofibrils. Such data are available only for rabbit and I discuss this problem on pp. **245–246** in Section 8.8. From this discussion, it is concluded that frog and rabbit behave similarly, provided that the experimental results obtained with the two animal species are combined. Using the barycentric mean between 10°C and 15°C, we deduce $Q_{10} \sim 2.2$ (simple calculations not shown). The tensions recorded by Colomo et al. (1997) would therefore be only $\sim2.5 \times 10^5$ N m^{-2} at 10°C. Working at room temperature ($\sim20°C–22°C$) on single isolated myofibrils from rabbit psoas muscle, Bartoo et al. (1993) found scattered tensions between $\sim3.4 \times 10^5$ N m^{-2} and $\sim9.4 \times 10^5$ N m^{-2}, giving values of only $\sim1.8 \times 10^5$ N m^{-2} to $\sim4.9 \times 10^5$ N m^{-2} (mean value of $\sim3.3 \times 10^5$ N m^{-2}) at 10°C ($Q_{10} \sim 1.8$ between 10°C and $\sim20°C–22°C$; see p. **246** in Section 8.8). The mean value of $\sim3.3 \times 10^5$ N m^{-2} is of the same order of magnitude as the $\sim2.5 \times 10^5$ N m^{-2} found above for single isolated myofibrils from frog. Working with the same biological material as Bartoo et al. (1993), but at 5°C, Tesi et al. (2000) found a mean value (averaged over 36 single isolated myofibrils) of $\sim2.6 \times 10^5$ N m^{-2}, giving $\sim4.0 \times 10^5$ N m^{-2} at 10°C (see p. **87** in Section 3.6), which is only slightly higher than for frog, but of the same order of magnitude. Averaging the four available values recalled here, valid for single isolated myofibrils at 10°C ($\sim2.5, \sim1.8, \sim4.9$ and $\sim4.0 \times 10^5$ N m^{-2}), we obtain ~3.3 ± (0.6 or 1.2) × 10^5 N m^{-2} (mean ± SE or SD; n = 4). Thus, single isolated myofibrils from both frog and rabbit develop isometric tetanic tensions similar to that in intact fibres, but much smaller than that developed by intact unit cells, corresponding to intact myofibrils included in intact frog fibres, particularly those in the centre of the fibres (~7.2 × 10^5 N m^{-2} at 10°C; see the preceding paragraph). This discrepancy probably results from the many structural, geometric and biological unknowns in isolated myofibrils, possibly including partial adulteration during the preparation of this biological material, differing from 'intact' myofibrils, located in the centre of an intact fibre (see below, in this section). In this context, Linari et al. (2003) studied energy storage experimentally, during the stretching of active single fibres from frog skeletal muscle. They raised the issue of transverse elastic connections between myofibrils, which are certainly broken during preparation of isolated myofibrils and might account in part for the lower axial tension than expected for myofibrils in vivo/ in situ, owing to mechanical adulterations of the 'free' isolated myofibrils.

At this point, it remains unclear whether the location of the unit cells in the myofibrils and the location of the myofibrils themselves in an intact fibre play an important role: do these various unit cells have different mechanical properties and develop different isometric tetanic tensions? Indeed, during macroscopic isometric tetanic contraction of an intact fibre, the possibility of microscopic oscillations and movements of the various unit cells, modulated by their location within the intact fibre, should be taken into account, with expected consequences on the various tensions, as described below, in this section.

He et al. (1997) addressed the issue of microscopic oscillations and movements and tried to record microscopic sarcomere length changes during their measurements of isometric MgATPase activities, performed on permeabilised fibres. Unfortunately, their setup was not appropriate and they detected no microscopic axial movement of the sarcomeres.

The notion of oscillations, or, more generally, non-uniform motion, during isometric tetanic contraction, is not new, having first been reported many years ago. It is based on macroscopic observations of cardiac cells (e.g. Fabiato and Fabiato 1978; Tameyasu et al. 1985) and

sarcomeres, myofibrils and single intact fibres from skeletal muscle and also from the striated adductor muscle of scallop (e.g. Armstrong et al. 1966; Borejdo 1980; Borejdo and Morales 1977; Goodall 1956; Iwazumi and Pollack 1981; Lorand and Moos 1956; Okamura and Ishiwata 1988; Stephenson and Williams 1982; Tameyasu 1994). Linke et al. (1993) also observed spontaneous oscillations in single isolated cardiac myofibrils. From a more quantitative viewpoint, Horowitz et al. (1992) imposed very small releases on intact frog fibres and found that 'releases as small as 20 nm per sarcomere [i.e. 10 nm per half-sarcomere, approximately the length of a myosin head, estimated at ~12.4–12.5 nm; see p. **109** in Section 4.1] produced a substantial increase in force level'. For unknown reasons, 'the force increase depends on the initial sarcomere length. This increase is insignificant at short sarcomere length, and becomes increasingly pronounced above 2.8 µm'. Thus, the microscopic movements assumed to occur in a sarcomere are observed experimentally, but the conditions in which they can be seen are difficult to define, as pointed out by Horowitz et al. (1992) in their section entitled 'Dependence of force increase on release parameters'. In any event, we deduce that microscopic movements can induce macroscopic increases in isometric force.

The oscillations and movements were generated in experimental conditions of various degrees of reproducibility. One particularly interesting issue concerns the stepwise shortening of single intact fibres or fibre segments at the supramolecular, microscopic and macroscopic levels (e.g. Delay et al. 1981; Granzier and Pollack 1990; Granzier et al. 1987; Jacobson et al. 1983; Pollack 1990; Pollack et al. 1977; Toride and Sugi 1989; Yang et al. 1998). This behaviour was considered by many authors to be artefactual, resulting solely from intersarcomere dynamics and rearrangements during fixed-end isometric contraction (e.g. Altringham et al. 1984; Burton and Huxley 1995; Goldman and Simmons 1984a; Morgan et al. 1982, 1991; Rüdel and Zite-Ferenczi 1979). In more recent papers, Blyakhman et al. (1999, 2001), Pollack et al. (2005) and Yakovenko et al. (2002) have confirmed the stepwise shortening or stretching process in small numbers of activated and unactivated sarcomeres, observing step sizes of only a few nanometers per half-sarcomere, corresponding to the molecular level. The existence of stepwise/stepping phenomena was strongly supported by evidence from an independent group (Borejdo et al. 2004a,b, 2006). Borejdo et al. (2006) used specific labeling and anisotropy of fluorescence techniques and worked with tiny numbers of cross-bridges attached to the thin actin filaments during contraction (e.g. between 1 and 1000 cross-bridges). They demonstrated that, studying a single cross-bridge, the anisotropy changes in a stepwise manner. For sufficiently small numbers of cross-bridges (e.g. 50), the steps are preserved. However, for larger numbers (e.g. 300 or 1000), the anisotropy changes 'smoothly'. Blyakhman et al. (1999) suggested that 'apparently a high degree of cooperativity synchronizes the molecular steps and allows them to be detectable at higher levels of organization'. The stepwise/stepping mechanisms are even observed in studies of in vitro motility on myosin V, which is involved in the transport of intracellular materials, such as vesicles, organelles and proteins, when moving along thin actin filaments (e.g. Cappello 2008; Cappello et al. 2007; Coureux et al. 2003; Mehta 2001; Sellers and Veigel 2006; see also Geeves and Holmes 2005, regarding the use of myosin V for studying certain characteristics of muscle contraction). In my opinion, molecular stepwise/stepping shortening and stretching of muscle are related to the notion of macroscopic and subsequent microscopic movements and oscillations, or vice versa, mentioned above. In this context, it should be stressed that the sarcomere stepping pattern is observed in shear conditions in both directions, that is, not only during shortening but also during stretching (Tourovskaya and Pollack 1998). This again provides support for the hypothesis of microscopic oscillations, because both directions are concerned. Finally, Nagornyak et al. (2004) studied stepwise behaviour under relaxing and contraction conditions and found that 'in the unactivated myofibrils, step size depended on initial sarcomere length, diminishing progressively with increase of initial sarcomere length, whereas in the case of activated sarcomeres, step size was independent of the initial sarcomere length'. Thus, the phenomena observed by Pollack's group are many-faceted.

Some of the experimental results recalled in the preceding paragraphs may be subject to certain limitations. Many experiments were performed on chemically skinned fibres or isolated myofibrils.

The various techniques available for chemical skinning or myofibril preparation would not completely destroy the sarcoplasmic reticulum. This structure would therefore remain, at least partly, functional (see Millman 1998 for a critical analysis of the various techniques used to remove the sarcolemma; to my mind, the problems raised by Millman may also apply to the preparation of isolated myofibrils from chemically skinned fibres). For example, in the isolated myofibrils from skeletal muscles used by Okamura and Ishiwata (1988), the sarcoplasmic reticulum is probably largely destroyed, although this remains unproven. The authors used their SPOC solution (containing, in principle, only trace amounts of Ca^{2+}), that is, a special solution inducing 'spontaneous oscillatory contraction' (SPOC) of sarcomeres, and they noted that, in the SPOC solution, the 'myofibrils are not fully but partially activated'. They concluded that 'the oscillation is a third state of skeletal muscle located in between the contracting and relaxed states'. However, as pointed out above, some of the sarcoplasmic reticulum surrounding the myofibrils would be present and, at least partly, functional in the experiments of Okamura and Ishiwata (1988) and oscillations would be a trivial phenomenon. Indeed, Villaz et al. (1987, 1989), working on frog half-fibres, observed oscillatory isometric forces exclusively related to cyclic variations in the uptake and leakage of Ca^{2+} in the sarcoplasmic reticulum, under partial activation. Although the SPOC solution contains trace amounts of Ca^{2+} (see above), the possible presence of part of the sarcoplasmic reticulum and the subsequent partial Ca-induced activation of the myofibrils would result in oscillations not entirely related to the myofilaments themselves and their interactions, but, at least partly, to the properties of fragments of sarcoplasmic reticulum in the isolated myofibrils. A similar problem is raised by the experimental results of Linke et al. (1993), who observed 'spontaneous sarcomeric oscillations at intermediate activation levels in single isolated cardiac myofibrils'. Here again, the presence of fragments of sarcoplasmic reticulum in the cardiac myofibrils cannot be totally ruled out. This makes the experimental observations of Linke et al. (1993) and Okamura and Ishiwata (1988) difficult to interpret. The experiments of Fabiato and Fabiato (1978) were performed on partially activated skinned cardiac cells and the oscillatory phenomena that they observed would be similar to

Morel

those described by Linke et al. (1993) and Okamura and Ishiwata (1988). However, despite the possible qualifications relating to the experimental data of Villaz et al. (1987, 1989), the experiments and theories of Anazawa et al. (1992), Edman and Curtin (2001), Ishiwata et al. (2011), Sato et al. (2011) and Smith and Stephenson (2009) strongly indicate that spontaneous oscillations/auto-oscillations are genuine phenomena characterising the contractile system, regardless of the presence of fragments of the sarcoplasmic reticulum.

The macroscopic phenomena recalled above, in this section, are also assumed to occur at the microscopic level, that is, in intact unit cells. Macroscopic oscillations and movements in isolated myofibrils or myofibrils inserted into whole fibres may be assumed to be related to some kind of cooperativity between 'oscillating' unit cells. This would result in a large amplification of the microscopic oscillations occurring in unit cells, when passing to myofibrils and whole fibres. This assertion is consistent with the opinions of Blyakhman et al. (1999), concerning cooperativity phenomena (see p. **92**), and with the theoretical conclusion of Borejdo (1980), according to which the macroscopic tension oscillations and movements observed in myofibrils and sarcomeres result from the existence of microscopic oscillations and movements in individual unit cells and even in individual cross-bridges (see also Carlson 1975, who presented the same type of observation). Shu and Shi (2006) considered 'the oscillatory motions in biological motor systems, e.g. spontaneous oscillations of single myofibrils' to be a trivial, even universal, phenomenon. I suggest that the unknown conditions required to generate amplification leading to macroscopic oscillations or fluctuations have not generally been defined. I also suggest that, in myofibrils within intact fibres, microscopic oscillations and movements occurring in central unit cells (called intact unit cells in this monograph) and in unit cells located in the vicinity of the sarcoplasmic reticulum are different (the same is obviously true for unit cells located in the vicinity of the sarcolemma, but fewer unit cells are likely to be concerned). The central unit cells are thought to have noticeable freedom to oscillate and to move, whereas the unit cells close to the sarcoplasmic reticulum and the sarcolemma remain almost motionless, owing to the axial stiffness and elasticity of this membrane-like structure and this membrane, respectively,

and their subsequent damping effects. Both the notions of oscillations and damped oscillations in single intact frog fibres have already been observed, analysed and discussed by Edman and Curtin (2001). After constructing this reasoning, I read the minireview by Borejdo et al. (2006), in which the authors claimed that 'in single-turnover experiments, cross-bridges rotate despite the fact that contraction is isometric'. Regardless of the interpretation of the authors, I think that this sentence is consistent with my own opinion, according to which macroscopic and microscopic/molecular events are probably, to some extent, uncoupled, that is, microscopic (molecular) oscillations/movements can be, at least partly, 'invisible' at the macroscopic level. The notion of 'free' central intact unit cells able to oscillate and to move under macroscopic – apparently strict – isometric tetanic contraction conditions is also supported by the work and opinions of Borejdo et al. (2006).

Regardless of the main features described above, in this section, an interesting theoretical analysis has indicated that the extensibility of the thin actin and thick myosin filaments induces local microscopic shortening (i.e. movement) within isometrically contracting fibres (Mijailovich et al. 1996). Moreover, x-ray diffraction experiments with synchrotron radiation, on whole frog muscles, at 8°C (Martin-Fernandez et al. 1994), and on single intact frog fibres, at 4°C (Piazzesi et al. 1999), have shown that several complex types of movement and oscillation can be detected under isometric conditions. These movements either are transitory and immediately follow stimulation (e.g. 'order–disorder transition during which the register between the filaments is lost'; Martin-Fernandez et al. 1994) or occur throughout the entire duration of the tetanus (in most figures presented by Martin-Fernandez et al. 1994 and Piazzesi et al. 1999, small, but detectable, fluctuations in various parameters, around mean values, are observed). These axial motions and oscillations correspond to the average values obtained for whole muscles or fibres and include, therefore, motionless unit cells, called 'peripheral unit cells' (located at the periphery of myofibrils, themselves located at the periphery of the fibre; see the preceding paragraph), and those that are free to oscillate or to move, called intact unit cells (located in the centre of myofibrils, themselves located in the centre of the fibre; see the preceding paragraph). Thus, the central intact unit cells would display axial motions of much greater magnitude than the peripheral unit cells (possibly approximately 10 times greater in central intact unit cells than in peripheral unit cells). This would be sufficient to induce the shift of attached cross-bridges, in the central intact unit cells, from a stable to a metastable state, as described in the next paragraph.

A probable consequence, already highlighted in the preceding paragraph, of microscopic oscillations and movements is that the central intact unit cells may be able to develop their full axial and radial forces because the cross-bridges attached to thin actin filaments may shift from a stable to a metastable state (see below), via the microscopic oscillations and movements described above, in this section. It is unclear whether the microscopic oscillations and movements result in the instantaneous appearance of this metastable state. Based on the experimental studies of Martin-Fernandez et al. (1994) performed on whole frog muscles at 8°C, the time course of the 'order–disorder transition during which the register between the filaments is lost' (see the preceding paragraph) is ~16 ms, whereas the corresponding half-time for the rise in tension is ~36 ms (see Table 1 in the paper by He et al. 1997, in which this half-time is ~77 ms for permeabilised frog fibres at 5°C and necessarily below ~77 ms at 8°C). There is therefore probably a small temporal dissociation between microscopic events and their translation into tension fluctuations. The time courses recalled here and those recorded by Martin-Fernandez et al. (1994) are markedly shorter than the time of $\Delta t^* \sim 172$–404 ms required to reach the plateau in isometric tetanic tension under the same temperature conditions (see p. **54** in Section 3.4.3.2). Thus, the suggested metastable state would be reached very rapidly and would be fully 'functional' well before the steady-state conditions of isometric tetanic contraction are reached. The existence of two possible states of the attached cross-bridges (stable and metastable) has already been suggested and discussed by Morel (1984a) in his Appendix III entitled 'Can the cross-bridges be in a stable or metastable state?' The peripheral unit cells close to the sarcoplasmic reticulum or the sarcolemma (see p. **91** and p. **93** for a short discussion of the unit cells located in the vicinity of these two structures) would be forced to remain almost motionless, owing to the 'axial'

elasticity or stiffness of this membrane-like structure and this membrane, respectively, resulting in significant damping effects. These unit cells would therefore be unable to develop their full force, because fewer attached cross-bridges would be able to shift to the motion-induced metastable state (see above). Thus, the cycling cross-bridges would be affected differently in the peripheral and central unit cells and in all the intermediate unit cells. There would be a gradient between the peripheral and central unit cells. Many parameters would therefore differ in the various unit cells, with the consequence that both the directly generated tension (swinging cross-bridge/lever-arm mechanisms) and the indirectly generated tension (lateral swelling mechanisms) would differ between unit cells. Microscopic changes in the behaviour and ordering of cross-bridges and thick myosin filaments during macroscopic contraction were strongly suggested by Yagi et al. (1981a,b), supporting the approach presented here. Similar behaviour was confirmed by Brunello et al. (2006), who used cutting-edge techniques, including x-ray diffraction from synchrotron radiation. Finally, Horowitz et al. (1992; see also p. **91**) found that very small imposed releases of only ~20 nm per sarcomere (~1% of the slack sarcomere length) resulted in substantial increases in isometric force, demonstrating the existence of a correlation between microscopic movements and macroscopic increases in axial contractile forces. I suggest that the metastable state of the attached cross-bridges is the key element underlying the phenomena described by Horowitz et al. (1992).

Commenting on the experiments performed by Harry et al. (1990) and the interpretation suggested by Morgan (1990) regarding forced lengthening of frog muscles under active contraction conditions, Morel (1991b) strongly suggested that reversible microscopic modifications of the thick myosin filaments, cross-bridges and thin actin filaments occur. Thus, reversible changes induced by macroscopic or microscopic movements appear to be common to the two types of myofilaments and cross-bridges. Yagi and Takemori (1995) and Yagi et al. (2006) confirmed my phenomenological/ semi-empirical 'predictions' concerning sliding-induced changes in the structure of cross-bridges (Morel 1991b). Such consistent results, obtained with very different approaches, support the phenomenological/semi-empirical reasoning proposed by my group over more than 15 years and developed in this book. These results, together with the other experimental data cited above, in this section, support the notion of molecular oscillations and movements leading to structural and conformational changes in the whole contractile machinery.

On p. **91** and p. **93**, it is suggested that there are major differences between unit cells, depending on their location within the fibre. Thus, regardless of axial contractile force generation, the values of the parameters z and n_∞ in Equations 3.34 to 3.39, giving in particular the phenomenological/semi-empirical expression of isometric tetanic tension, almost certainly depend on the location of the unit cell. On p. **63** in Section 3.4.3.2, isometric tension in an intact unit cell is found to be $\sim 7.5 \times 10^5$ N m^{-2} at $\sim 2.5°$C ($\sim 8.8 \times 10^5$ N m^{-2} at $10°$C), assuming that $z \sim 0.157$ and $n_\infty \sim 0.82$. This high tension and the values of z and n_∞ are valid for intact unit cells (located in the centre of the fibre), but they are almost certainly not valid for all unit cells. This is especially true for peripheral motionless unit cells, which cannot develop their full force (see p. **91** and p. **93**). Taking, at $\sim 2.5°$C, as an illustrative example, $z \sim 0.250$ and $n_\infty \sim 0.70$ for these peripheral, almost motionless unit cells, in which no metastable states of the cross-bridges exist (vs. ~ 0.157 and ~ 0.82, respectively, in central intact unit cells, with similar values of $\sigma \sim 6$ nm and $\omega_{isom,\infty} \sim 1$), and using the same reasoning and calculations as on pp. **62–63** in Section 3.4.3.2, we deduce that, at $10°$C, $T_{0,\infty} \sim 6.0 \times 10^5$ N m^{-2}, rather than the $\sim 8.8 \times 10^5$ N m^{-2} recalled above. The molecular mechanisms leading to the generation of isometric tetanic tension therefore differ between unit cells as a function of the location of the unit cell within the myofibril and also, probably, the location of the myofibril within the whole fibre and the location of the fibre within the whole muscle. This would account, at least partly, for the apparently confusing results obtained concerning axial contractile forces, both in intact fibres (see Section 3.3) and in demembranated fibres. However, other complex reasons contribute to this confusion in demembranated fibres (see pp. **97–98** in Section 3.8.1, and Sections 4.4.2.1 and 8.7 for additional details).

Finally, when molecular aspects of the strength of the various forces are considered, the values that should be used are those valid in an intact unit cell (e.g. $T^* \sim 7.2 \times 10^5$ N m^{-2} at $10°$C; see p. **286**). If relative tensions are taken into account,

the problem of loss of tension when considering successively a central intact unit cell, a peripheral unit cell, a myofibril, a demembranated fibre, a half-fibre and an intact fibre (possibly also a whole muscle) is essentially eliminated, at least within the limits of experimental error. Thus, relative isometric tensions can be used, as they stand, without correction. This is demonstrated in Section 4.2.2 and in Figure 4.2.

3.8 BEHAVIOR AND PROPERTIES OF DEMEMBRANATED FIBRES, REGARDING AXIAL TENSIONS AND RADIAL FORCES DURING ISOMETRIC TETANIC CONTRACTION

3.8.1 Axial Tensions

The strengths of the axial and radial forces (or tensions) recorded under isometric tetanic conditions, in mechanically, chemically skinned and permeabilised fibres, correspond frequently to laterally swollen fibres (no correction for cross-sectional area). This initial 'skinning induced' lateral swelling is a universal phenomenon, systematically observed (see Section 8.7 for a discussion). Moreover, I do not specify below, in this section, the sarcomere lengths used in the experiments cited, but they all range between ~ 1.9 μm and ~ 2.7 μm, and the average value is $\sim 2.34 \pm (0.04$ or $0.25)$ μm (mean SE or SD; $n = 40$). The sarcomere length used in the new experiments presented in this monograph is $\sim 2.5 \pm 0.2$ μm (see p. **14** in Section 2.2 and p. **16** in Section 2.3), whereas the reference sarcomere length used in the theoretical and phenomenological/semi-empirical studies is $s_0 \sim 2.18$ μm (see p. **24** in Section 3.2). Owing to shortcomings in reasoning and wide experimental uncertainty, these three series of values can be considered to be similar and no corrections for sarcomere lengths are made. On p. **246** in Section 8.8, values of Q_{10} are proposed for a 'composite animal species and biological material' (e.g. frog, rabbit, intact or demembranated fibres). These values are $Q_{10} \sim 2.4$ between $\sim 0°$C and $\sim 10°$C–$12°$C, ~ 2.1 between $\sim 12°$C and $\sim 15°$C, ~ 1.5 between $\sim 15°$C and $\sim 20°$C and ~ 1.25 between $\sim 20°$C and $\sim 25°$C. Note that, when passing from room temperature ($\sim 20°$C–$22°$C) to the reference temperature of $10°$C, the value of Q_{10} that should be used is

~1.8 (this value of ~1.8 is assumed to be valid up to ~30°C). Below, in this section, I compare demembranated fibres, from any animal species, at a single temperature of 10°C and, for normalisation of the tensions at 10°C, the values of Q_{10} given above are systematically used. These approximations are justified, as the experimental values for isometric tetanic tension are widely scattered.

As pointed out by Morel (1985a), the general view, in the 1970s and at the beginning of the 1980s, was that the skinning of frog fibres resulted in a systematic loss of axial tension, but the origin of this phenomenon was unclear. This universal behaviour was quantified by Elzinga et al. (1989a), who showed that less 'normalised' force (force corrected for the increase in cross-sectional area upon demembranation; see below, in this section) was produced by the same frog single fibre after chemical skinning (glycerination) than before. A.F. Huxley (2000) stated that demembranated fibres are very interesting, because 'they make it possible to vary the concentration of solutes at will'. However, they 'generally give low values for the tension per unit area..., further, the sarcomeres are less regular' (see also pp. **13–14** in Section 2.2 concerning impairment in demembranated fibres). It should be noted that, in most of the cases studied below, in this section, there is some confusion between the forces recorded in demembranated fibres and tensions (tension = force/cross-sectional area; this is important, because demembranation induces large increases in this area; see below, in this section, and Section 8.7). In the paragraphs below, I present many values for isometric tetanic tension and comment on the mean value obtained, in terms of both impairment and lateral swelling.

Cooke and Bialek (1979) recorded tensions in chemically skinned fibres from rabbit psoas muscle of ~0.8 × 10⁵ N m⁻² at 10°C. Cooke and Pate (1985) obtained, with the same biological material and at the same temperature, a value of ~1.8 × 10⁵ N m⁻². Brenner and Yu (1991), using the same biological material, recorded tensions of ~1.0 × 10⁵ N m⁻² at 5°C, giving ~1.5 × 10⁵ N m⁻² at 10°C. Again, with the same biological material at 5°C, Ford et al. (1991) recorded tensions of ~1.1 × 10⁵ N m⁻², corresponding to ~1.7 × 10⁵ N m⁻² at 10°C. Burghardt et al. (1983) recorded a tension, for the same biological material, at room temperature (~20°C–22°C), of ~2.0 × 10⁵ N m⁻², giving ~1.0 ×

10⁵ N m⁻² at 10°C. For the same biological material, Dantzig and Goldman (1985) obtained a value of ~1.6 × 10⁵ N m⁻² at 20°C, that is, ~0.9 × 10⁵ N m⁻² at 10°C. Again, with the same biological material, Goldman et al. (1987) reported values of ~1.0 × 10⁵ N m⁻² at 10°C, ~1.6 × 10⁵ N m⁻² at 20°C, and a mean value of ~1.7 × 10⁵ N m⁻² at 30°C, giving corrected values of ~0.9 × 10⁵ N m⁻² at 10°C (from the value obtained at 20°C) and ~0.5 × 10⁵ N m⁻² at 10°C (from the value obtained at 30°C). However, from Figure 3B in the paper by Goldman et al. (1987), it appears that the many experimental points at 30°C are highly scattered and, if the SE is taken into account, the maximal value is ~2.0 × 10⁵ N m⁻² (~0.6 × 10⁵ N m⁻² at 10°C) and the minimal value is ~1.4 × 10⁵ N m⁻² (~0.4 × 10⁵ N m⁻² at 10°C), whereas, if the SD is taken into account (n = 27; see Section 2.1 for precise details on use of the SD), the maximal value is ~3.3 × 10⁵ N m⁻² (~1.0 × 10⁵ N m⁻² at 10°C) and the minimal value is ~0.1 × 10⁵ N m⁻² (~0.03 × 10⁵ N m⁻² at 10°C). Given the wide scattering of the values obtained by Goldman et al. (1987) at 30°C (almost certainly attributed to considerable adulteration of the frog fibres), I do not take them into account. Again, with chemically skinned fibres from rabbit psoas muscle, Martyn and Gordon (1992) found ~1.7 × 10⁵ N m⁻² at 10°C. Working on the same biological material, at 15°C, Potma and Stienen (1996) obtained an average value of ~1.4 × 10⁵ N m⁻² (n = 16), giving, at 10°C, ~1.0 × 10⁵ N m⁻². Potma et al. (1994) obtained, under the same conditions, seven values, giving, at 10°C, ~0.7 × 10⁵ N m⁻², ~0.7 × 10⁵ N m⁻², ~0.9 × 10⁵ N m⁻², ~0.9 × 10⁵ N m⁻², ~1.0 × 10⁵ N m⁻², ~1.0 × 10⁵ N m⁻² and ~1.1 × 10⁵ N m⁻². Matsubara et al. (1985) measured tensions for chemically skinned whole small muscles from mouse toe (consisting of only approximately 10 fibres) of ~1.3 × 10⁵ N m⁻², presumably at room temperature (~20°C–22°C; the authors did not specify the temperature, but they performed their x-ray diffraction experiments at room temperature), giving a value of ~0.7 × 10⁵ N m⁻² at 10°C. For mechanically skinned frog fibres at ~20°C–22°C, Hellam and Podolsky (1969) measured tensions of ~1.4 × 10⁵ N cm⁻², giving ~0.8 × 10⁵ N m⁻² at 10°C. At the same temperature (~20°C–22°C), Gordon et al. (1973) obtained an average tension for a series of mechanically skinned frog fibres of ~1.1 × 10⁵ N m⁻², with a maximal value of ~1.8 × 10⁵ N cm⁻², corresponding to ~0.6 × 10⁵ N cm⁻² and ~1.0 × 10⁵ N m⁻² at 10°C, respectively (the number of measurements and the minimal value obtained were not given by the authors). In another series of experiments with the same biological material at the same temperature, the authors also obtained a value of $\sim 1.5 \times 10^5$ N m^{-2}, giving $\sim 0.8 \times 10^5$ N m^{-2} at 10°C. They cited Endo, who published a paper in 1967 (in Japanese), in which he obtained, under the same temperature conditions and with the same biological material, tensions of $\sim 1.5 \times 10^5$ N m^{-2} and $\sim 2.0 \times 10^5$ N m^{-2}, corresponding to $\sim 0.8 \times 10^5$ N m^{-2} and $\sim 1.1 \times 10^5$ N m^{-2}, respectively, at 10°C. Ferenczi et al. (1984a), working on mechanically skinned frog fibres at ~ 2°C, obtained tensions of $\sim 1.5 \times 10^5$ N m^{-2}, that is, $\sim 3.0 \times 10^5$ N m^{-2} at 10°C. Goldman and Simmons (1984a), working on the same biological material, presumably at ~ 4°C (the temperature at which the experiments were performed was not clearly specified), recorded tensions of $\sim 1.3 \times 10^5$ N m^{-2}, $\sim 1.5 \times 10^5$ N m^{-2} and $\sim 1.8 \times 10^5$ N m^{-2}, giving $\sim 2.2 \times 10^5$ N m^{-2}, $\sim 2.5 \times 10^5$ N m^{-2} and $\sim 3.0 \times 10^5$ N m^{-2}, respectively, at 10°C. Using an unconventional technique, Elzinga et al. (1989a) obtained a value of $\sim 1.3 \times 10^5$ N m^{-2} at 4°C, for permeabilised frog fibres, corresponding to $\sim 2.2 \times 10^5$ N m^{-2} at 10°C. He et al. (1997) presented several values for permeabilised frog and rabbit fibres. For frog at 5°C (see their Figure 9), they obtained a value of $\sim 1.3 \times 10^5$ N m^{-2}, corresponding to $\sim 2.0 \times 10^5$ N m^{-2} at 10°C. For rabbit, they obtained slightly different tensions: in their Figure 6, at 20°C, they obtained a value of $\sim 2.2 \times 10^5$ N m^{-2}, that is, $\sim 1.2 \times 10^5$ N m^{-2} at 10°C, and, in their Figure 12, they obtained a value of $\sim 2.0 \times 10^5$ N m^{-2} at 20°C, that is, $\sim 1.1 \times 10^5$ N m^{-2} at 10°C. In other series of experiments (55 measurements) with the same biological material prepared from rabbit psoas muscle, He et al. (1998b, 1999) obtained an average value of $\sim 2.3 \times 10^5$ N m^{-2} at 15°C, that is, $\sim 1.5 \times 10^5$ N m^{-2} at 10°C, and an average value of $\sim 1.9 \times 10^5$ N m^{-2} at 12°C, that is, $\sim 1.7 \times 10^5$ N m^{-2} at 10°C. Again, with chemically skinned fibres from rabbit psoas muscle, Hilber et al. (2001) and Sun et al. (2001) obtained values of $\sim 1.5 \times 10^5$ N m^{-2} and $\sim 1.2 \times 10^5$ N m^{-2}, respectively, at 10°C. Siththanandan et al. (2006), with the same biological material, obtained a value of $\sim 2.1 \times 10^5$ N m^{-2} at 12°C, that is, $\sim 1.8 \times 10^5$ N m^{-2} at 10°C. Working on permeabilised white fibres from toadfish, Rome et al. (1999) obtained a mean value of $\sim 2.4 \times 10^5$ N m^{-2} at 15°C (four measurements), giving $\sim 1.7 \times 10^5$ N m^{-2} at 10°C. Finally, working on permeabilised white fibres from

dogfish, at 12°C, West et al. (2004) reported two slightly different values (depending on the technique used) of $\sim 1.9 \times 10^5$ N m^{-2}, that is, $\sim 1.6 \times 10^5$ N m^{-2} at 10°C and $\sim 2.2 \times 10^5$ N m^{-2}, that is, $\sim 1.8 \times 10^5$ N m^{-2} at 10°C. The values obtained for toadfish and dogfish do not differ significantly from the other values recalled above, and they are therefore taken into account. Averaging the 40 values presented here, obtained for chemically and permeabilised fibres and mechanically skinned fibres from various animal species and various types of muscle, at various temperatures (but corrected to a single temperature of 10°C) gives a mean axial isometric tetanic tension of $\sim 1.4 \pm (0.1 \text{ or } 0.6) \times 10^5$ N m^{-2} (mean $\pm$ SE or SD; n = 40).

Regardless of the animal species and taking into account the mean of the values of the axial tensions for the 40 experiments (see the preceding paragraph), a reasonable estimate for the multiplication factor when passing from an intact fibre to a demembranated fibre is $\sim [1.4 \pm (0.1 \text{ or } 0.6)] \times 10^5$ N m$^{-2}/3.0 \times 10^5$ N m$^{-2} \sim 0.47 \pm (0.03 \text{ or } 0.20)$ (the value of $\sim 3.0 \times 10^5$ N cm^{-2} is the universal value for intact frog fibres and half-fibres [in the presence of MS$^-$; see Section 3.3]). Thus, to convert values from demembranated fibres, prepared from any animal species, to those for intact fibres, a multiplication factor of $1/[0.47 \pm (0.03 \text{ or } 0.20)] \sim 2.13 \pm (0.13 \text{ or } 1.10)$ should be used. The error bars of $\pm 6\%$ or $\pm 52\%$ on the multiplication factor (if the SE or SD is taken into account, respectively) reflect the fact that the tensions in demembranated fibres are greatly scattered, as recalled in the preceding paragraph. These error bars may also partly reflect differences in the temperatures used, with the associated problem of the choice of the most suitable value of Q_{10} (see the first paragraph of this section). Moreover, a maximal contribution of $\pm 2\%$ ($100 \times$ SE/mean) or $\pm 13\%$ ($100 \times$ SD/mean) to total error may be related to the use of sarcomere lengths ranging between ~ 1.9 μm and ~ 2.7 μm (see p. **95**). The value of $\sim 0.47 \pm (0.03 \text{ or } 0.20)$ for converting the tension developed by an intact fibre to that developed by a demembranated fibre provides support for the simplified reasoning presented by Morel (1985a), based on a ratio ≤ 0.50. However, this is probably coincidental. Indeed, Morel (1985a), in his 'simplified model for a myofibril in a skinned fibre', took into account only the possible dissolution of thick myosin filaments at the periphery of each myofibril. It is necessary to also take into account the occurrence of impairment/damage/disorder/disruption recalled on pp. **13–14** in Section 2.2. Otherwise, Morel (1985a) did not consider the systematic increase in cross-sectional area upon skinning, recalled at many places in this section, and studied in Section 8.7, this universal behaviour not having been clearly identified and quantified in the 1970s and at the beginning of the 1980s. This increase in cross-sectional area must lead to a decrease in axial tension, for geometric reasons. For example, Elzinga et al. (1989a), working on permeabilised frog fibres at 4°C, found a mean increase in cross-sectional area of ~50%. Ferenczi et al. (1984a), working on mechanically skinned frog fibres at 0°C–5°C, found a mean increase in cross-sectional area of ~50%. Goldman and Simmons (1986), on the same biological material at 4°C, found increases in fibre diameter of ~10% to ~30%, corresponding to increases in cross-sectional area of ~21% to ~69%. Linari et al. (1998) found a mean increase in cross-sectional area of ~44%, in permeabilised frog fibres at 18°C. Matsubara and Elliott (1972), working on mechanically skinned frog fibres at 4°C, found a mean increase in fibre diameter of ~13%, that is, ~28% in cross-sectional area. Combining these six corrective factors, the most probable value is ~43 ± (6 or 16)% (mean ± SE or SD; n = 6). Thus, the lateral swelling processes result in a corrective factor for cross-sectional area of $1/[0.43 ± (0.06 \text{ or } 0.16)] \sim 2.33$ ± (0.33 or 1.00). The difference between ~2.13 ± (0.14 or 1.10), deduced above, from 40 experimental values, and ~2.33 ± (0.33 or 1.00), deduced here from six experimental values, is ~−0.20 ± (0.47 or 2.10) and ranges between ~−2.30 and ~+1.90 (maximal interval). There is, therefore, a considerable range of uncertainty, due to the 'randomly distributed' lateral swellings and many other complex phenomena (mentioned on pp. **13–14** in Section 2.2 and recalled above). All these phenomena are intermingled and overlap, making it extremely difficult to distinguish between them. In any event, taking into account only the SDs (see Section 2.1 concerning the interest of using SD rather than SE), we can choose the approximate corrective factor of ~(2.13 ± 1.10 + 2.33 ± 1.00)/2, ranging between ~1.18 and ~3.28 (mean value of ~2.23). This 'broad scattering' (or more precisely gap) again confirms the complex behaviour of demembranated fibres, also studied in Sections 3.8.2, 4.4.2.1 and 8.7. These values of ~1.18 and ~3.28 can be considered to constitute 'roughly suitable' multiplication factors for passing from the isometric tetanic tension in a demembranated fibre to that in an intact fibre, regardless of temperature,

animal species and demembranation technique. Owing to this scattering of the experimental values, the multiplication factors of ~1.18 and ~3.28 and the mean value of ~2.23 can also be used for cross-sectional area (see p. **122** in Section 4.3.2 and p. **241** in Section 8.7).

3.8.2 Radial Forces

In Section 3.1, it is recalled that the notion of an intact unit cell can be defined only for an intact fibre. It is shown, on p. **85** in Section 3.5, that there is a net radial repulsive (expansive) force of ~455 pN per intact unit cell under contraction conditions (short tetani–steady state). To the best of my knowledge, no radial expansive force has ever been recorded directly in contracting intact fibres (somewhat conflicting results have been obtained concerning the myofilament spacing and the cross-sectional area when an intact fibre passes from resting to isometric contraction conditions; see pp. **4–5** in the Introduction). In an intact fibre, a radial expansive force is certainly difficult to observe because the sarcolemma and the non-sarcolemmal associated components exert a counterforce opposing the expansive force of ~455 pN, resulting in no net radial force being recorded at the level of the sarcolemma.* However, this does not rule out the

---

* The lateral surface area of the intact unit cell is ~0.29 μm² (see p. **155** in Section 4.4.2.5.3), giving a 'lateral tension' of ~455 pN/0.29 μm² ~ 1570 pN μm⁻². Rapoport (1973) studied the anisotropic elasticity of the sarcolemma. From his Table II (concerning 13 frog fibres, at a mean sarcomere length of ~2.35 μm), it can be deduced that the longitudinal elastic modulus is $E_L \sim 9.10 ± 6.38 \times 10^6$ pN μm⁻², with a longitudinal Poisson ratio $\sigma_L \sim 1.34 ± 0.65$, a circumferential elastic modulus $E_c \sim 0.88 ± 0.57$ 10^6 pN μm⁻² and a circumferential Poisson ratio $\sigma_c \sim 0.18 ± 0.09$. Using his complex equations, in which the longitudinal and circumferential parameters are crosslinked, I found a lateral tension (notion roughly equivalent to rigidity) at the sarcolemma of $\sim 2 ± 1 \times 10^6$ pN μm⁻², considerably greater than ~1570 pN μm⁻² ~ 1.57 × 10³ pN μm⁻². I deduce that the rigidity of the sarcolemma is largely sufficient to counterbalance the net radial tension in an intact contracting fibre. Owing to the complexity of the calculations presented by Rapoport (1973), the conclusion drawn in this footnote may be dubious. It would therefore be an excellent idea to develop experimental techniques for measuring lateral force, at the sarcolemma, or for checking that this force is zero. A brief discussion is presented on pp. **3–4** in Chapter 1, indicating conflicting results concerning small increases or decreases in diameter of intact fibres when passing from resting to contraction conditions (e.g. regarding the presence of expansive or compressive lateral forces, respectively).

existence of a net radial expansive force in an intact unit cell lying in the centre of an intact fibre. This conclusion results from the general approach to the mechanics of solids (see specialist handbooks): a microscopic element (here an intact unit cell) of the solid (here an intact fibre) is assumed to be isolated from the rest of the solid for studies of the strain and stress on this elementary structure. Removing the sarcolemma by mechanical or chemical skinning or permeabilisation should remove the counterforce and 'liberate' the net expansive force. Thus, a contracting demembranated fibre would be an appropriate biological material for recording and studying the expansive force. Unfortunately, the behaviour of these fibres is extremely complex and no clear conclusion can be drawn, as discussed in the following paragraphs. Other major difficulties concerning demembranated fibres are analysed and discussed on pp. **70–72** in Section 3.8.1 and in Sections 4.4.2.1 and 8.7.

Brenner and Yu (1991), using x-ray diffraction from synchrotron radiation, worked on chemically skinned rabbit psoas muscle (sarcomere length of ~2.3–2.4 µm), and Matsubara et al. (1985), using traditional x-ray diffraction, worked on chemically skinned bundles of fibres from mouse toe muscle (sarcomere length of ~2.3–2.4 µm). Brenner and Yu (1991) studied the behaviour of their biological material exclusively with increases or decreases in external osmotic pressure. Matsubara et al. (1985) carried out only some experiments on the role of external osmotic pressure. As recalled in Section 8.7, demembranation inevitably results in lateral swelling of the fibre, which is a complex phenomenon. Increasing the external osmotic pressure allows the demembranated fibres to shrink, making it possible to return to the 'pre-demembranation' width/lattice spacing of the intact fibre from which the demembranated fibre was prepared. Working on contracting demembranated, and therefore initially swollen, fibres, Brenner and Yu (1991), Matsubara et al. (1985) and many other independent groups found sizeable radial forces. At this stage, the situation is confusing. For example, Brenner and Yu (1991) wrote, in their abstract: 'while not under osmotic pressure, the radial force of the activated fibre was determined to be 400 pN (single thick filament)$^{-1}$'. However, the authors defined the radial force between the thick myosin filaments and the thin actin filaments per unit length of thick filaments by their Equation 1,

which includes osmotic pressure, and from which it seems that, when the osmotic pressure is zero, the radial force is also zero. I suggest that there are printing errors or a lack of explanation of the formula of Brenner and Yu (1991). Matsubara et al. (1985) proposed a similar equation, but with a term relating to an increase in osmotic pressure rather than the osmotic pressure itself. They found a radial force of 450 pN per thick myosin filament, but they stated that 'the magnitude of the lateral [radial] force underlying the lattice shrinkage during maximum contraction can be estimated from the osmotic pressure … which causes an equivalent shrinkage'. At this point, I note that Brenner and Yu (1991), Matsubara et al. (1985) and others found that the lattice systematically shrinks when a demembranated fibre was shifted from rest to contraction. As highlighted on pp. **4–5** in the Introduction, the situation is not so clearcut in intact fibres (moderate or more substantial lateral swelling or mild shrinkage may be observed when the intact fibre goes from rest to contraction). There is even greater confusion, as Brenner and Yu (1991) wrote: 'The active radial force was found to be a slightly non-linear function of the lattice spacing, reaching zero at 34 nm. The radial force was compressive at lattice spacing greater than 34 nm and expansive at less than 34 nm'. This behaviour was confirmed and quantified precisely by Xu et al. (1993): using x-ray diffraction from synchrotron radiation, the authors wrote: 'At large separations, the direction of the [active radial] force is toward the centre of the fibre; its magnitude decreases monotonically with decreasing d_{10} until it reaches zero at the equilibrium spacing. The radial force turns expansive if the lattice separation is decreased further'. I believe that many of the problems discussed in this section result from the initial swelling of the demembranated fibres that the authors did not apparently take into account (as recalled above, precise details on the initial swelling induced by demembranation are given in Section 8.7).

At this stage, it should be recalled that demembranation leads to major problems (see pp. **13–14** in Section 2.2, pp. **97–98** in Section 3.8.1, and Sections 4.4.2.1 and 8.7). In this domain, Morel (1985a) presented a 'simplified model for a myofibril in a skinned fibre', from which it was deduced that the measured value of lattice spacing d_{10} (obtained from x-ray diffraction) does not correspond to the traditional arrangement of the thick myosin and thin actin filaments in a perfect crystallographic unit cell (see, e.g. Figure 6 in the paper by Matsubara et al. 1984, for a definition of the crystallographic unit cell), corresponding instead to approximately the distance between neighbouring thin actin filaments (i.e. $2d_{10}/3$). As highlighted on pp. **97–98** in Section 3.8.1, this model is not entirely adequate, because the behaviour of demembranated fibres is much more complex than previously thought. However, I use this model here, as an oversimplification. In these conditions, the critical measured value of 34 nm (at sarcomere lengths of ~2.3–2.4 μm; see the preceding paragraph, concerning the value of 34 nm) would correspond to the distance between neighbouring thin actin filaments, not to the traditional lattice spacing d_{10}, and the true value of d_{10} would be ~$(3/2) \times 34 \sim 51$ nm (see above, concerning the coefficients 2/3 or 3/2). The value of d_{10}, in intact muscles from rabbit psoas, at a sarcomere length of 2.3 μm (in an isosmotic surrounding buffer), is ~40.8 ± (1.0 or 1.8) nm (mean ± SE or SD; n = 3; see Table 1 in the review by Millman 1998). At this point, let us assume that the results obtained with contracting demembranated fibres could be extrapolated to contracting intact muscles and fibres. Taking into account the critical value of ~34 nm, we deduce that in intact muscles and fibres, we would have ~40.8 ± (1.0 or 1.8) nm > 34 nm and the radial forces would be compressive. If the critical value is taken to be ~51 nm, rather than 34 nm, the radial forces in intact muscles and fibres would be expansive, because ~40.8 ± (1.0 or 1.8) nm < 51 nm. These two conclusions are obviously self-contradictory. As pointed out in the first paragraph of this section, no net radial force can be recorded in contracting intact fibres, because of counterforces exerted by the sarcolemma and the associated components. In any event, the self-contradictory conclusions drawn here could apply to intact unit cells from intact fibres (considered as a microscopic contractile element; see the first paragraph of this section). In conclusion, a contracting demembranated fibre cannot be considered as a suitable working tool for testing the existence of net radial repulsive (expansive) forces and checking that their strength is ~455 pN per intact unit cell from an intact fibre (see again the first paragraph concerning this value), owing to the many complex problems involved in interpreting the experimental results. Other drawbacks

resulting from a thorough experimental study of contracting demembranated fibres are considered in Section 4.4.2.1. Nonetheless, there are radial forces in this biological material. Thus, in the near future, improvements in the approach to studying the behaviour of these fibres and the interpretation of the experimental results should lead to more interesting conclusions, supporting the hybrid model, in which the radial expansive forces play a major role. In this context, the use of half-fibres is probably the best solution, because this biological material closely mimics intact fibres (see, e.g. Section 3.3, and p. **129** in Section 4.4.1).

I do not believe that swinging cross-bridge/lever-arm mechanisms can account for the strong radial forces recorded in contracting demembranated fibres and the absence of force at the equilibrium lattice spacing (see, for instance, Bagni et al. 1994b; Brenner and Yu 1991; Goldman and Simmons 1986; Matsubara et al. 1985; Xu et al. 1993). These independent groups considered the strong radial forces to be exerted solely by the attached cross-bridges during isometric tetanic contraction, particularly during their rotation (the various enzymatic states during the cross-bridge cycle are also taken into account in the discussions). This hypothesis is at odds with the reasoning and calculations presented in this monograph. However, I mostly present calculations and estimates concerning intact unit cells, not demembranated fibres. For example, in intact unit cells, the radial forces exerted by the attached cross-bridges during isometric tetanic contraction are compressive (collapsing forces, owing to rotation of the cross-bridges) and correspond to a mean value of only ~73 pN per intact unit cell (see p. **157** in Section 4.4.2.5.4), much lower than the net radial expansive force of ~455 pN per intact unit cell (see p. **98**) (100 × 73 pN/455 pN ~ 16%) and the full axial force of ~1138 pN per intact unit cell found on p. **84** in Section 3.5 (100 × 73 pN/1138 pN ~ 6%–7%). The collapsing forces in intact unit cells are studied quantitatively on p. **157** in Section 4.4.2.5.4. These forces should also exist in isometrically contracting demembranated fibres, not under external osmotic pressure, but their strength is certainly greater than ~73 pN per intact unit cell. Indeed, according to the 'geometric estimations' proposed by Morel and Merah (1997), the collapsing forces should increase with increasing myofilament spacing. In the absence of external osmotic pressure, the demembranated fibres are all swollen and the lattice spacing is greater than in intact fibres and intact unit cells (see Section 8.7). Thus, in contracting demembranated fibres, the collapsing forces are certainly greater than ~73 pN per intact unit cell and may account for a larger proportion of the axial forces than suggested above. Assuming a relative increase of ~20% in the lattice spacing of demembranated fibres over that in intact fibres gives collapsing forces of ~90 pN per intact unit cell. This estimate is based on the paper by Morel and Merah (1997) and represents only ~100 × 90/535 ~ 17% of the full axial force, and ~90 pN per intact unit cell is again significantly lower than the axial force of 535 pN (this value of ~535 pN per unit cell in contracting demembranated fibres is ~47% the value of ~1138 pN per intact unit cell, in intact fibres; see above for the value of ~1138 pN per intact unit cell, and pp. **97–98** in Section 3.8.1 for a discussion of the loss of axial contractile tension in demembranated fibres and the estimate of ~47%; see also Section 4.4.2.1 concerning the impossibility of defining an intact unit cell in a demembranated fibre).

3.9 EXISTENCE OF A VICINAL IONIC STRENGTH IN DEMEMBRANATED FIBRES

The difference between the vicinal ionic strength I^* within demembranated fibres and the bulk ionic strength I of the buffer surrounding these fibres is related to the existence of numerous fixed negative electrical charges and bound anions on the myofilaments in situ, at rest and in rigor (e.g. Aldoroty et al. 1987; Bartels and Elliott 1980, 1985; Collins and Edwards 1971a; Elliott 1980; other references are given in Section 4.4.2.2; see also Regini and Elliott 2001 for a summary concerning the bound anions). The existence of fixed negative electrical charges and bound anions leads to a limited accumulation of contaminating cations, ensuring electroneutrality within demembranated fibres. These contaminating cations come from the bathing medium. As shown in Figure 4.2 and Section 4.2.2, $I^* \sim 2$ mM at $I \sim 0$ (distilled water), under resting conditions. This value is obtained by linear extrapolation of the solid line in this Figure 4.2 (resting tension) between $I \sim 0$ and ~ 10 mM. Owing to the scattering of the recorded relative resting tensions below ~20 mM presented in Figure 4.2, I take

an approximate value $I^* \sim I + 2$ mM, at least in distilled water ($I \sim 0$). I also consider the value of ~2 mM to be independent of the value of I, because this value cannot be estimated for other values of I. It is important to use this value of ~2 mM, particularly at low and very low values of I. For higher values of I (e.g. $I \geq 20–30$ mM), the difference between I and I^* becomes negligible. However, I almost systematically use I^*, rather than I, to ensure consistency in the reasoning and calculations. Finally, I consider the value of ~2 mM, deduced from the values of resting tensions, to apply also to contraction conditions. This is probably not the case, because the number of fixed negative electrical charges and bound anions on the myofilaments is almost certainly different under contraction conditions and at rest (see pp. **161–162** in Section 4.4.2.6). However, for simplification and owing to the scattering of the experimental points, no differences are introduced in the values of vicinal ionic strength between fibres in these two states.

3.10 GENERAL REMARKS ABOUT RADIAL REPULSIVE ELECTROSTATIC FORCES. CONVERSION OF NET RADIAL REPULSIVE (EXPANSIVE) FORCES INTO AXIAL FORCES

In this monograph, it is demonstrated, mostly on the basis of experimental data (see principally Section 4.4), that the radial repulsive electrostatic forces given by Equation 3.2 play a major role in generating a large part of the axial forces, in both resting and contracting fibres. The two electrical surface potentials, $\Psi_{x,r}(I^*)$ and $\Psi_{x,a}(I^*)$, introduced in Section 3.2 (see Equation 3.1 and corresponding comment, and Equation 3.2) are thought to depend on I^*. This is discussed on pp. **161–162** in Section 4.4.2.6, where it is strongly suggested that the electrical surface potentials, at least under isometric contraction conditions (and almost certainly at rest), depend very little on I^*. According to Equation 3.2, regardless of variations of these potentials, that is, of $Q_x(I^*)$ with I^* and with the state of the fibre (rest, $x = r$, or contraction, $x = a$), when I^* reaches infinity, $F_e(I^*)$ and its slope tend asymptotically to zero. If I^* is zero, $F_e(I^*)$ is zero, and if I^* is extremely close to zero, the slope of F_e is given by F_e/I^*, which is proportional to $(I^*)^{-1/4}$, with all the other terms in Equation 3.2 having finite values. Thus, this slope is infinity when

I* reaches very low values. These conclusions, regarding F_e and its slope at I* infinity or zero, are general and do not depend qualitatively on any other parameter. As a direct consequence, F_e must pass through a maximal value, as must the resting and active axial tensions. This is experimentally confirmed in Section 4.2.2, particularly in Figure 4.2, presenting the relative resting tension against bulk ionic strength (indeed, there are mechanisms for converting the radial repulsive forces into axial forces; see the next paragraph). However, both the position and magnitude of this peak in resting tension depend quantitatively on various parameters studied in Section 4.4.2. For axial isometric tetanic tension, the same behaviour, including a maximal value when ionic strength is lowered, must be observed, but, to the best of my knowledge, only two sets of experimental data have been obtained to support this assertion. Using KCl as the major neutral salt (see p. **12** and p. **14** in Section 2.2 for precise details on the major drawbacks concerning the use of Cl^-) in an experimental study of mechanically skinned frog fibres, Gordon et al. (1973) reported, in their Figure 2, isometric tetanic tension as a function of bulk ionic strength I. The experimental curve displays the expected features: it passes through a maximal value at an ionic strength of ~80–120 mM, falls sharply below ~80 mM, and decreases slowly between ~120 mM and ~500 mM (at which it is negligible). Using the 'external osmotic pressure technique', Hasan and Unsworth (1985) studied intact toad fibres as a function of the external osmotic pressure (see their Figure 1). On p. **110** in Section 4.1, a simplified method for passing from external osmotic pressure to ionic strength within intact fibres is suggested. The experimental curve presented by Hasan and Unsworth (1985) again displays the expected features: it passes through a maximal value at an ionic strength of ~75–110 mM, falls sharply below ~75 mM, and decreases slowly between ~110 mM and ~130 mM (maximal value used by the authors). Regardless of the interpretation of their data by Gordon et al. (1973) and Hasan and Unsworth (1985), these experimental results are consistent with the electrostatic repulsive forces being fundamental partners in muscle contraction. These and other experimental findings are discussed and critically analysed on pp. **109–111** in Section 4.1.

The conclusions drawn in the preceding paragraph are discussed in Sections 4.4.2.2 to 4.4.2.8.

It is confirmed that, in resting and contracting fibres (see the preceding paragraph), there are strong net radial repulsive (expansive) forces between myofilaments that can be translated into axial forces and play a major role in the development of axial forces (as demonstrated in Section 4.2.2 and Figure 4.2, for resting conditions, and in several subsections of Section 3, for contraction conditions). Indeed, as pointed out at many places in this monograph, two possible mechanisms for turning radial forces into axial forces, independently of volume variations, were proposed approximately 40 years ago (Dragomir et al. 1976; Elliott 1974, on the one hand, and Morel et al. 1976, on the other). Morel et al. (1976) took into account the elasticity of the myoplasm and myofibrillar lattice and used Hooke's laws to account for the conversion of radial into axial forces. The issue of the elasticity of muscle fibres was also raised by Maughan and Godt (1979), who wrote 'values of bulk moduli of fibres, calculated from the compression experiments and preliminary measurements of Young's modulus from stretch experiments are qualitatively consistent with the idea that skinned fibres behave as nonisotropic elastic bodies'. To my mind, this conclusion, valid for skinned fibres, which do not obey the constant volume relationship (e.g. April and Wong 1976; Matsubara and Elliott 1972), seems to bear out the 'elasticity mechanism' suggested by Morel et al. (1976). The contribution of radial repulsive electrostatic forces to axial forces is strongly supported, even demonstrated, throughout this book, from new experimental results that I and my coworkers have obtained over two decades, and from a careful analysis of many experimental results obtained by many independent groups. The intermediate conclusions drawn in the preceding paragraph and above are not sufficient in themselves. The definitive experimental demonstration of the existence of $F_{e,x}$ in a fibre (here a half-fibre; see Section 2.2), at fixed sarcomere length, and its conversion into axial forces, requires studies of changes in both isometric tetanic tensions ($x = a$) and resting tensions ($x = r$) with I or I* ~ I + 2 mM (see Section 3.9 for precise details on the vicinal ionic strength I*) and with all the other parameters remaining fixed. In this monograph, I make use of this kind of experimental approach in my analysis of many problems and phenomena, based on a hybrid model with good predictive and explanatory power, particularly in cases in which pure swinging cross-bridge/lever-arm models prove inadequate (see mostly Chapter 8). Finally, the conversion of radial forces into axial forces is assumed, in this work, to be a one-to-one process (i.e. the radial forces are entirely converted into axial forces, at rest and under contraction conditions; see, p. **85** in Section 3.5 and pp. **160–161** in Section 4.4.2.6).*

---

* Street (1983) studied the mechanism of 'lateral transmission of tension in frog myofibres' experimentally. Ramaswamy et al. (2011) recently studied the same mechanism in the skeletal muscles of very old rats. These processes may be considered to support the conversion of axial forces into radial forces (and possibly vice versa). However, the titles of these articles are, to my mind, misleading, and the corresponding experimental studies do not directly support the translation of radial forces into axial forces.

April and Brandt (1973), it appears that increasing ionic strength from ~50 mM to ~600 mM is responsible for the decrease in relative isometric tetanic tension.

Millman (1998) collected a series of experimental results, valid for intact fibres of various origins, that had been obtained by independent groups, including his own. His Figure 11 clearly shows that isometric tetanic tension decreases as external osmolarity increases from ~150–200 mOsM to ~500–600 mOsM (in vivo external osmolarity was estimated at 245 mOsM by Millman 1998 and Millman et al. 1981). This increase in external osmolarity leads to a well-known gradual shrinkage of the intact fibres (e.g. April and Brandt 1973; Millman 1998) and, despite several problems (see pp. **106–107**), also leads to an inevitable increase in ionic strength within intact fibres, resulting from the outflow of water. The results presented by Millman (1998) are therefore qualitatively comparable to those presented in the preceding paragraph. April and Brandt (1973) discussed the decrease in isometric tetanic tension as a function of ionic strength. They proposed several hypotheses, including a suggestion that this decrease is related to a significant decrease in steady-state actomyosin MgATPase activity with increasing ionic strength. Indeed, many experimental studies performed in the 1960s and 1970s reported that this activity and its main features, measured in vitro, in contraction conditions, on regulated actomyosin (in the presence of calcium) or desensitised actomyosin, decreased substantially when ionic strength increased, for values between ~0.1–0.2 M and ~1–2 M (e.g. Burke et al. 1974; Danker and Hasselbach 1971; Eisenberg and Moos 1970; Perry 1956; Rizzino et al. 1970; Weber and Herz 1961, 1963). Working on isolated myofibrils from rabbit muscle, Portzehl et al. (1969b) also reported higher levels of MgATPase activity at low ionic strengths of surrounding buffers containing EGTA and no added calcium (resting conditions). Chalovich (1992) pointed out that increasing the ionic strength would decrease the binding affinity between myosin and actin (as shown by biochemical, mechanical and structural studies), which would decrease the isometric tension, by processes of various degrees of complexity. However, with the exception of this 'trivial' effect of ionic strength on MgATPase activity, April and Brandt (1973) did not consider the ionic strength of the sarcoplasmic medium to be of major importance.

This view was shared by Andrews et al. (1991), who worked on permeabilised rabbit fibres from psoas muscle and also found that isometric tetanic tension decreased with increases in ionic strength from 95 mM to 390 mM. However, these two independent groups did not propose clear and definitive explanations for the decrease in tension with increasing ionic strength.

Combining the behaviors of skinned and intact crayfish fibres as functions of ionic strength and external tonicity, respectively, April and Brandt (1973) drew a questionable conclusion from their experimental data: 'it is evident, therefore, that the tension-generating capacity of muscle is not dependent upon interfilament spacing'. This inference appears to be at odds with the dependence of isometric tetanic tension on ionic strength demonstrated by April et al. (1968) and April and Brandt (1973; see their Figure 5), particularly in their studies of osmotically compressed intact fibres. Indeed, as pointed out by Millman (1998), when the external osmotic pressure is increased, the myofilaments are brought closer together and the ionic concentration within the sarcomere increases. The problem of this 'double effect' of osmotic pressure on intact frog fibres is studied in Section 4.4.2.5.1, and it is concluded (see p. **147**) that, for a moderate increase in myofilament lattice of ~+6% (i.e. ~+12% in the cross-sectional areas of the intact unit cells), the net radial expansive force (including the repulsive electrostatic forces as well as the tethering, collapsing and attractive/compressive forces) increases by ~+26.7%, whereas a moderate decrease of ~−6% (i.e. ~−12% in the cross-sectional area of the intact unit cells) induces a decrease in the net radial expansive force of ~−24.9%. There is therefore a relationship between myofilament lattice, internal ionic strength and the net radial expansive force, at least when external osmotic pressure is moderately changed. At this point, it should be recalled that a one-to-one conversion of the radial expansive forces into axial forces is highly likely (see pp. **102–103** in Section 3.10). Otherwise, throughout this monograph, it is claimed that ~40% of the full axial tension results from these radial expansive forces, under standard conditions of temperature, ionic strength/osmolarity and sarcomere length (see Section 4.4.1). Thus, for passing from the increase of ~+26.7% in the net radial expansive force to the 'true' increase in full axial tension, it is necessary (i) to take into

Morel

account the increase of ~+12% in cross-sectional area (see above) and (ii) to also consider the proportion of ~40% recalled above. The increase of ~+26.7% in the net radial force therefore corresponds to a true increase of ~+26.7% × 0.40/1.12 ~ +9.5% in the full axial tension. Similarly, the decrease of ~−24.9% in the net radial expansive force corresponds to a decrease of ~−24.9% × 0.40 × 1.12 ~ −11.2% in the full axial tension. Edman and Anderson (1968), cited on p. **144** in Section 4.4.2.5.1, studied the effects of external tonicity on the sarcomere length–tension relationship in the case of intact isolated frog *semitendinosus* fibres. In their paper, there is a weak point concerning force and tension. Indeed, in their Figure 1, they reported forces (in milligrams), but they called them tensions. This is a frequent 'geometric mistake', even in recent publications, as highlighted at many places in this monograph. The authors found that the increase of ~+6% in myofilament lattice (~+8% in fibre width; see p. **146** in Section 4.4.2.5.1 for comments on ~+8% and ~+6%) corresponds to a relative increase ~+10%–15% in axial tension, whereas the decrease of ~−6% in myofilament lattice (~−8% in fibre width; see p. **146** in Section 4.4.2.5.1) corresponds to a relative decrease of ~−15% − −20% in axial tension. Despite the experimental error and the shortcomings in the reasoning presented here (e.g. superseding ~±8% by ~±6%, maybe for unwarranted reasons) and in Section 4.4.2.5.1, the experimental and calculated values are of the same order of magnitude. The double effect of moderate external compression, recalled at the beginning of this paragraph, results therefore in moderate changes in axial tension, essentially through the radial repulsive electrostatic forces, demonstrated to prevail over the other radial forces under contraction conditions in the hybrid model (see particularly Section 4.4.2.5.4). In skinned fibres, ionic strength has a direct effect on the relative axial tension, as frequently demonstrated in this work (see Sections 4.4.2.7 and 4.4.2.8 and also p. **105**, concerning the results obtained by April and Brandt 1973 for four skinned fibres). It is demonstrated, in Sections 3.8, 4.4.2.1 and 8.7, that the axial tensions of skinned fibres are qualitatively comparable to those of intact fibres, although, quantitatively, skinned fibres appear to develop less axial tension (see Section 3.8.1). However, demembranated fibres display 'singular' structural behaviour and cannot be considered a suitable model for intact fibres. Nonetheless, when studying relative axial tension, appropriate conclusions can be drawn from studies of skinned and intact fibres. When studying structural properties (by x-ray diffraction, for instance), skinned fibres and intact fibres cannot be considered comparable. Part of the confusion that I observed in the article by April and Brandt (1973) results from some unwise combinations of the two types of fibre, particularly when x-ray diffraction techniques are used.

Figure 5 in the paper by April and Brandt (1973) is taken as a reference in the experimental part of this work (see, for instance, Figure 4.2 and corresponding comments in Section 4.2.2). Thus, a preliminary discussion of this article is required. Despite the confusion mentioned at the end of the preceding paragraph and highlighted on p. **106** (tension independent of myofilament lattice), April and Brandt (1973) conceded that myofilament distance varied in their experiments, performed on intact fibres submitted to hypo- and hyperosmotic external pressures: 'the range of thick filament spacings corresponds to thin-to-thick filament spacing of from 260 Å to 390 Å [centre-to-centre distance] in the 6:1 lattice of the crayfish leg muscle'. At this stage, it remains unclear whether the major problem of steric hindrance affected the results obtained by April and Brandt (1973) (see Bachouchi and Morel 1989a and pp. **251–254** in Section 8.9). On p. **213** in Section 6.4, it is shown that the mean radius of the backbone of thick myosin filaments from various animal species is $R_b \sim 180 \text{ Å}/2 \sim 90$ Å. On pp. **33–34** in Section 3.2, it is shown that the radius of a thin actin filament is $R_a \sim 43$ Å. Thus, the centre-to-centre distance of 260 Å corresponds to a closest surface-to-surface spacing between the backbone of a thick myosin filament and the neighbouring thin actin filament of $\sim 260 - (90 + 43) \sim 127$ Å and that of 390 Å to $\sim 390 - (90 + 43) \sim 257$ Å. The maximal chord of the myosin head (also called S1; see Figure 5.1) is, to my mind, the only important feature of a crossbridge, the S2 part (see Figure 5.1) having little to do with the problems raised here. Unfortunately, this maximal chord is not known for crayfish but is probably of the same order of magnitude as for other type II myosin molecules (skeletal muscle myosin). Many values have been published for rabbit skeletal muscle (myosin II from back and legs). For free myosin heads in vitro (S1), I select the following values: ~70–90 Å (Katayama 1989, under contraction-like conditions), ~110–120 Å

(Bachouchi and Morel 1989a calculated these values from various traditional x-ray scattering data obtained in muscles and muscle fibres, available in the 1980s), ~119 Å (Garrigos et al. 1983), ~119 Å (calculated by Morel et al. 1993 from the paper by Kinosita et al. 1984), ~120 Å (Bachouchi et al. 1985; Mendelson 1982; Mendelson and Kretzshmar 1980), ~121 Å (calculated by Morel et al. 1993 from the paper by Yang and Wu 1977), ~120–123 Å (Morel et al. 1993), ~125 Å (Morel and Merah 1997 deduced this value from various values obtained by independent groups, particularly from the radius of gyration obtained by Curmi et al. 1988), ~110–140 Å (Katayama 1989, under rigor-like conditions), ~130–140 Å (Morel et al. 1993 calculated these values from the radius of gyration obtained by Curmi et al. 1988) and ~148 Å (Labbé et al. 1984). For myosin heads attached to F-actin in vitro, under rigor-like conditions, that is, in the arrowhead configuration (mostly studied by electron microscopy and reconstruction), I select the following values: ~115 Å (Taylor and Amos 1970), ~128 Å (Moore et al. 1970; Taylor and Amos 1970; Wakabayashi and Toyoshima 1981), ~120–140 Å (Seymour and O'Brien 1985), ~130 Å (Milligan and Flicker 1987), ~110–150 Å (Toyoshima and Wakabayashi 1985) and ~135 Å (Moore et al. 1970; Taylor and Amos 1970).

In work on myosin heads from whole isolated myosin molecules, mostly carried out by electron microscopy, all the available values are systematically overestimated (see Morel et al. 1993 for a critical analysis of this problem, from which it is concluded that these values should be discounted). In this area, I select the following values: ~170 Å (Heuser 1983), ~190 Å (Elliott and Offer 1978; Walker et al. 1985), ~190–200 Å (Margossian and Slayter 1987), ~200 Å (Yamamoto et al. 1985) and ~210 Å (Takahashi 1978). Averaging these seven values, we obtain ~193 ± (4 or 12) Å (mean ± SE or SD; n = 7). Working on myosin heads from thick filaments, indefinite values were obtained, probably attributed to the ill-defined S2 part in thick filaments (see Figure 5.1 concerning the position of S2). I select the following values: ~100–120 Å (Morel et al. 1993), ~120 Å (Poulsen et al. 1987), >140 Å (Stewart and Kensler 1986b), ~120–160 Å (Suzuki and Pollack 1986), >150 Å (Crowther et al. 1985) and ~160 Å (Stewart et al. 1985). Averaging these eight values gives a very approximate value >134 ± (7 or 21) Å (mean ± SE or SD; n = 8). The values >134 ± (7 or 21) Å and ~193 ±

(4 or 12) Å (see above) are considered mathematically consistent. Note that Garrigos et al. (1992) and Wakabayashi et al. (1992), applying x-ray scattering from synchrotron radiation to myosin heads in solution (S1), obtained unexpectedly high values of ~190 Å and ~167 Å, respectively. In the mid-1980s, a controversy arose concerning the available values, particularly ~120 Å and ~190 Å (see comments by Craig 1985; Craig et al. 1986; Mendelson 1985). Moreover, the higher two values of ~167 Å and ~190 Å for S1 in solution (see the end of the preceding paragraph) were critically analysed and discarded by Morel (1996) and Morel et al. (1993). In my group, we compared various experiments concerning the radius of gyration of the myosin heads in solution (from which the length of these heads is frequently deduced) and concluded that the values of the radius of gyration obtained in many studies cannot be retained, because of unsuitable selection of Guinier's region in small-angle x-ray and neutron scattering experiments (Morel 1996). Only values lying within the range of ~110–130 Å are consistent with the radius of gyration actually measured in the true Guinier region. It should be noted that most authors referring to Guinier's region probably have not read the original work. I advise readers interested in determining the radius of gyration by x-ray or neutron scattering to consult the pioneering article concerned (Guinier 1939). Finally, the most consensual value of the maximal chord of the myosin head from rabbit skeletal muscle that can be chosen is ~120–130 Å (e.g. Morel 1996; Morel and Merah 1997; Morel et al. 1993), which is similar to the mean of the 29 values recalled in the first column on this page for S1, that is, ~121 ± (3 or 17) Å (mean ± SE or SD; n = 29). Other values were obtained with other animal species. The values estimated for S1 with chicken skeletal myosin and from neutron scattering by Curmi et al. (1988) are ~130–140 Å (calculated by Morel et al. 1993 from the radii of gyration, but not given by the authors). Using scallop myosin (from the adductor muscle) and electron microscopy, Flicker et al. (1983) obtained a value of ~180 Å for S1 containing the Ca^{2+} regulatory light chain ('functional' S1, able to generate contraction), whereas Vibert and Craig (1982) obtained a value of ~160 Å, again by electron microscopy, but on acto-S1 in vitro (functional S1, under rigor conditions; reconstruction technique). The mean value is therefore ~170 ± 10 Å and can be seen

as abnormally high (it is not taken into account below). When the Ca^{2+} regulatory light chain is removed, S1 from scallop muscle resembles myosin II from any skeletal muscle. In these conditions, Flicker et al. (1983) and Vibert and Craig (1982) obtained values of ~110 Å and ~120 Å, respectively. Using skeletal avian muscle and crystallised S1, Rayment et al. (1993b) obtained a value of ~165 Å, consistent with the previous estimate (>160 Å) obtained by Winkelman et al. (1985), using the same muscle and technique. Averaging the five values recalled above for animal species different from rabbit (except ~180 Å and ~160 Å for scallop functional S1), we obtain ~133 ± (8 or 19) Å (mean ± SE or SD; n = 5), statistically similar to ~121 ± 3 Å estimated above, even if the SEs are taken into account (see Section 2.1 for precise details). The interval ~124–125 Å is almost common to both series of values. The walking leg of crayfish contains skeletal muscle myosin II, and its maximal chord can reasonably be assumed to be similar to that of traditional skeletal muscle myosin II and would therefore be of a similar order of magnitude: ~124–125 Å, similar to ~127 Å and significantly lower than the ~257 Å mentioned on p. **107**, for the minimal and maximal thin–thick closest surface filament spacing, resulting from the 'osmotic pressure' process used by April and Brandt (1973). Thus, in the experiments performed by these authors on intact crayfish fibres, there were few, if any, problems of steric hindrance and resulting phenomena, such as those described by Bachouchi and Morel (1989a) (see also pp. **251–254** in Section 8.9).

The conclusion that we can draw from the paper by April and Brandt (1973) (which is at odds with the inference of the authors themselves; see p. **106**) and the paper by April et al. (1968) are therefore consistent. Thus, the effect of ionic strength on isometric tetanic tension is clearly, but involuntarily, demonstrated by April and Brandt (1973) and, more precisely, by April et al. (1968). Nonetheless, the trivial role of ionic strength in controlling MgATPase activity, recalled on p. **105**, would blur this conclusion. At this point, it should be highlighted that there is very loose coupling between 'isometric' MgATPase activity and isometric tetanic tension (see pp. **78–80** in Section 3.4.5, and pp. **83–84** in Section 3.4.6, where it demonstrated that isometric tetanic tension is proportional to isometric MgATPase activity to a power of ~0.190 ± 0.014; see p. **78** in

Section 3.4.5). As an illustrative example, if MgATPase activity decreases by a factor of 3, isometric tension decreases by a factor of only ~1.2, which is of the same order of magnitude as the uncertainty on the measurements of the various isometric tetanic tensions. Thus, the trivial effect of ionic strength on MgATPase activity is of little interest, when the isometric tetanic tension is taken into account. Figure 5 in the paper by April and Brandt (1973) is therefore taken as a reference below, in this section, and in the experimental part of this monograph (see Figure 4.2, its legend and corresponding comments), because (i) it gives a straightforward relationship between ionic strength and isometric tetanic tension, regardless of variations in MgATPase activity; (ii) it presents a large number of experimental points (~50–60 points), whereas Gordon et al. (1973) and Hasan and Unsworth (1985) (see the following paragraphs for a discussion of these two papers) present few experimental points; and (iii) it is consistent with the reference experimental points obtained for the half-fibres prepared from frog and used in the experimental part of this monograph (see Figure 4.2, its legend and the corresponding comments).

Gordon et al. (1973) studied mechanically skinned frog fibres (from *semitendinosus* muscle) at room temperature (~20°C–22°C) and in various neutral salts. I focus here on the experimental results obtained in the presence of KCl as the major neutral salt, because Gordon et al. (1973) thought that 'KCl is a representative neutral salt, i.e. there are no specific ion effects', which is actually untrue. For instance, on p. **12** and p. **14** in Section 2.2, it is recalled that Cl^- is a deleterious anion, that Cl^- is not a benign anion, and that, in relaxing conditions, Cl^- would induce the release of Ca^{2+} from the sarcoplasmic reticulum, which is, at least partly, present and functional in the mechanically skinned fibres used by Gordon et al. (1973) (on p. **221** in Section 7.2, other unfavorable characteristics of Cl^- are recalled). For organic neutral salts and anions (tetramethylammonium or tetrapropylammonium), the experimental results obtained by Gordon et al. (1973) were, fortunately, qualitatively similar to those obtained with Cl^-, under both resting and contracting conditions, and the possible 'negative role' of KCl is not evident in these experiments, within the limits of experimental error. The basic ionic strength of the solutions used by Gordon et al. (1973), with no added neutral salt, was ~55 mM. In their Figure 2, the isometric tetanic tension is around zero at an ionic strength of ~500 mM and increases with decreasing ionic strength, peaking at ~80–120 mM and then decreasing strongly between ~80 mM and ~55 mM. I find the author's explanations for this biphasic behaviour insufficiently convincing. Indeed, they suggested only that, at high ionic strengths, the decrease in isometric tetanic tension with increasing ionic strength is probably related to the decline in MgATPase activity, already mentioned on p. **106**, but the rest of this section discounts this viewpoint. Hasan and Unsworth (1985) performed 'osmotic pressure' experiments on intact toad *sartorius* muscle fibres, at ~5°C. They used sucrose for hypertonic solutions and dilution for hypotonic solutions. In their Figure 1, they observed an increase in isometric tetanic tension with decreasing external osmolarity, except for osmolarity values below ~150 mOsM. Indeed, when external osmolarity decreases from ~150 mOsM to ~100 mOsM, the tension remains approximately constant at maximal levels. When osmolarity decreases further, from ~100 mOsM to ~50 mOsM, the tension decreases rapidly. These features are almost certainly related to a significant decrease in ionic strength within the intact fibres, as expected from arguments presented on pp. **101–102** in Section 3.10 (see also the second column on this page concerning a simple relationship between osmolarity and ionic strength), also taking into account the conversion of radial forces into axial forces (see pp. **102–103** in Section 3.10). Hasan and Unsworth (1985) briefly discussed their findings, but did not present a clear analysis of their experimental results. As mentioned above, the explanations provided by Gordon et al. (1973) and Hasan and Unsworth (1985) for the peak in tension and drop at low ionic strengths are insufficiently convincing. Nonetheless, their experimental results are qualitatively consistent with the conclusion drawn on pp. **101–102** in Section 3.10. Indeed, it is demonstrated on these two pages, on purely mathematical bases, that the radial repulsive electrostatic forces (which are fully effective; see Section 3.2) vary in the same qualitative manner as experimentally recalled here for the axial contractile forces. Thus, (i) there are mechanisms for translating radial repulsive forces into axial contractile forces (see pp. **102–103** in Section 3.10) and (ii) the variations of isometric tetanic tension with ionic strength described by Gordon et al. (1973) and with osmotic pressure presented

by Hasan and Unsworth (1985) have the same origin, that is, variations in the radial repulsive electrostatic forces with ionic strength. Variations in isometric MgATPase activity are undoubtedly of marginal importance, as large variations of this activity result in limited variations of tension (see, for instance, p. **109**).

It would have been interesting to select experiments performed on isolated myofibrils, concerning the dependence of isometric tetanic tension on ionic strength. To the best of my knowledge, only Ma and Taylor (1994) studied myofibrils from rabbit psoas muscle at various ionic strengths (~10–100 mM; neutral salt NaCl; the exact ionic strengths were slightly higher than ~10–100 mM, particularly at low values, because the authors did not take into account the other chemical components, PIPES, EGTA, $MgCl_2$, etc.; ~10–100 mM corresponds solely to added NaCl). Unfortunately, Ma and Taylor (1994) studied unheld myofibrils, shortening therefore at maximal velocity. Many authors have recorded isometric tetanic tensions in single isolated myofibrils (e.g. Bartoo et al. 1993; Colomo et al. 1997; Tesi et al. 2000, 2002), but these authors did not study the dependence of isometric tetanic tension on bulk ionic strength. I am unaware of such experimental studies performed by these or other authors.

As recalled in the first column on this page, the isometric tetanic tension in intact fibres from toad *sartorius* muscle reaches a maximum when external osmolarity is between ~150 mOsM and ~100 mOsM. It then decreases rapidly, below ~100 mOsM (experimental results obtained by Hasan and Unsworth 1985). As the physiological external osmolarity is 245 mM (Millman 1998; Millman et al. 1981), these low values of external osmolarity certainly correspond to large swelling of the fibres and, therefore, to low values of ionic strength within intact fibres, resulting simply from the 'massive' inflow of water. Thus, variations of isometric tetanic tension with osmolarity are almost certainly related to subsequent changes in sarcoplasmic ionic strength. Using the rule of three, as an oversimplification, the maximal isometric tension would occur within a range of ionic strength of ~180 mM × (100 mOsM/245 mOsM) ~75 mM and ~180 mM × (150 mOsM/245 mOsM) ~110 mM (180 mM is the in vivo ionic strength corresponding approximately to 245 mOsM; see above). In the first column on this page, it is also recalled that Gordon et al. (1973), working on mechanically skinned fibres from frog *semitendinosus* muscle, observed a maximum between ~80 mM and ~120 mM. More details on the experiments of Gordon et al. (1973) and Hasan and Unsworth (1985) are given on pp. **109–110**. Note that the estimated maximal 'electrostatically related' active tension presented in Figure 4.2 occurs at an ionic strength of ~20–30 mM (see also p. **119** in Section 4.2.2 and p. **165** in Section 4.4.2.8). All these values are sufficiently consistent, owing to shortcomings in reasoning and within the range of experimental error, but, at this point, no clear conclusion can be drawn. However, the rather high values of ~75–110 mM and ~80–120 mM, as opposed to ~20–30 mM, probably indicate a shift toward higher ionic strengths for unclear reasons. The same is observed for resting tensions and plausible explanations are given in Section 4.4.3. In any event, as demonstrated on pp. **78–80** in Section 3.4.5 and pp. **83–84** in Section 3.4.6, there is very loose coupling between isometric MgATPase activity and isometric tetanic tension. This calls into question the assumption that one-to-one coupling between MgATPase activity and isometric tetanic tension (swinging cross-bridge/lever-arm models) provides the only explanation for muscle contraction. In this context, Gordon et al. (1973) found that, above an ionic strength of ~170 mM (their reference 'in vivo' value), the decline of MgATPase activity (measured in vitro on acto-S1) with increasing ionic strength occurred at a lower ionic strength than the decrease in tension in skinned frog fibres. I think, therefore, that this finding of Gordon et al. (1973) is consistent with MgATPase activity and tension being largely uncoupled, again calling into question the tight coupling hypothesis and providing indirect support for the very loose coupling process demonstrated on pp. **78–80** in Section 3.4.5, and pp. **83–84** in Section 3.4.6 on totally different bases.

Partly based on arguments put forward above, in this section, I consider, throughout this book, that, under standard conditions (e.g. around the slack length, low temperature, buffers mimicking the physiological media), ~40% of the relative isometric tetanic tensions recorded by April and Brandt (1973), Gordon et al. (1973) and Hasan and Unsworth (1985) are related to the net radial repulsive (expansive) forces, which are highly dependent on ionic strength, as demonstrated in many sections (see Sections 3.2, 3.5, 4.2.2, 4.4.2.7 and 4.4.2.8), but very little on MgATPase activity

(see pp. **78–80** in Section 3.4.5 and pp. **83–84** in Section 3.4.6, regarding the very loose coupling). The other ~60% of relative isometric tetanic tension is related to swinging cross-bridges/lever-arms and is therefore tightly coupled with MgATPase activity. This hybrid process, described and analysed in this monograph, necessarily results in a 'mixture' of tight coupling between MgATPase activity and isometric tetanic tension (swinging cross-bridge/lever-arm models) and extremely weak coupling between MgATPase activity and isometric tetanic tension (lateral swelling models), resulting in a very loose coupling, demonstrated independently of the mode of force generation on pp. **78–80** in Section 3.4.5 and pp. **83–84** in Section 3.4.6.

April and Brandt (1973) deduced from their experiments that electrostatic forces almost certainly play a minor role in contracting fibres. I do not agree with this view and I demonstrate, throughout this work and in the light of the many experimental data published since 1973, that the experimental data of April and Brandt (1973) are, by contrast, entirely consistent with radial repulsive electrostatic forces playing a major role in the generation of axial contractile forces. However, the situation concerning actively contracting intact and demembranated fibres remains rather puzzling, from an experimental viewpoint, as demonstrated above, in this section. Many problems must be resolved to confirm experimentally the existence of strong radial repulsive (expansive) forces and to determine their mode of action. New experiments have therefore been performed, on resting frog half-fibres and resting permeabilised rat fibre bundles, in which the 'resting' MgATPase activity is very weak. Under contraction conditions, very high levels of this activity, related to the rapidly cycling cross-bridges, occur in half-fibres and other muscle systems. The possible dependence of isometric MgATPase activity on ionic strength (particularly for the swinging cross-bridge/lever-arm part of the hybrid model) would therefore interfere with the effects of ionic strength on the radial repulsive electrostatic forces in contracting fibres. One of the major aims of this monograph is to generate definitive and consistent experimental conclusions, by focusing mostly on resting half-fibres, in which the level of MgATPase activity is very low and cannot mask the response of the half-fibres to variations in ionic strength. The results obtained and their phenomenological/semi-empirical interpretations, presented and discussed throughout Chapter 3, are certainly informative, but it is important to confirm experimentally that radial repulsive electrostatic forces make approximately as large a contribution as the swinging of the cross-bridges to generation of the full isometric axial contractile force. This confirmation is provided in this book (see Section 4.2), but the existence of other forces is also demonstrated, the most important of which are the radial tethering forces, the existence of which is experimentally demonstrated in Section 4.3.

4.2 EXPERIMENTS PERFORMED ON RESTING HALF-FIBRES FROM YOUNG ADULT FROGS

4.2.1 Were the Resting Half-Fibres Actually at Rest at Any Ionic Strength?

Does the solid line in Figure 4.2 represent true resting tensions? To address this question, we need to determine the order of magnitude of the concentration of contaminating Ca^{2+} in the bathing buffers and within the half-fibres, under all the experimental conditions used here, and to show that the solid line in Figure 4.2, particularly at low and very low ionic strengths, cannot be accounted for by trivial Ca-induced contractions.

Using mostly KCl as the major salt (other minor chemical compounds are obviously present in the buffer), Fink et al. (1986) showed that ionic equivalents had a marked effect on the Ca sensitivity of demembranated fibres, in terms of the isometric tetanic forces developed. The definitions of ionic equivalents and ionic strength are recalled by Andrews et al. (1991). Ionic equivalents are given by $I_e = (1/2)\Sigma c_i|z_i|$ (the symbol $|\ |$ corresponds to the absolute value), whereas ionic strength is given by $I = (1/2)\sum c_i z_i^2$, where c_i is the concentration of the ionic species 'i' and z_i is its (positive or negative) electrical charge. Andrews et al. (1991) analysed critically the inference of Fink et al. (1986) in their section entitled 'Ionic strength or ionic equivalents?' and concluded that the interpretation of Fink et al. (1986), based on ionic equivalents, was not clearcut. By contrast, when KMS (MS$^-$ = methanesulphonate) is used as the major salt, at any concentration, ionic strength should be employed in all interpretations of the experimental results (despite the presence of other chemical compounds; see above). From a qualitative viewpoint, ionic strength and ionic equivalents vary similarly with increasing major salt concentration, and, for solutions containing very large quantities of monovalent major salts (e.g. KCl and KMS), ionic strength and ionic equivalents have almost similar values. Thus, for simplification, only the notion of ionic strength is taken into account here, regardless of the state of the fibre (rest or contraction). The results of Fink et al. (1986) can therefore be interpreted as showing that Ca sensitivity increased markedly with decreasing ionic strength. However, these authors did not perform experimental studies at low and very low ionic strengths (the minimal ionic strengths/ionic equivalents used were ~80–100 mM). In any event, it is necessary to demonstrate that half-fibres in the presence of KMS are always at rest, regardless of the total ionic strength (no possibility of Ca-induced contraction with decreasing ionic strength).

In the experimental studies presented in Section 4.2.2, two sets of ATP and EGTA concentrations were used. In the first set of experiments, 0.4 mM ATP was used in the various relaxing buffers (corresponding, before the addition of EGTA, to a mean concentration of contaminating calcium of ~5.5 µM; see p. **15** in Section 2.2), together with 0.5 mM EGTA (see p. **118** in Section 4.2.2). This 'first set' was used for experimental studies on half-fibres (●) and (■) and for values of vicinal ionic strength I* between ~2 mM and ~142 mM (see Section 4.2.2 and Figure 4.2). In the second set of experiments, 2 mM ATP was used in the various relaxing buffers (corresponding, before the addition of EGTA, to a mean concentration of contaminating calcium of ~7.1 µM; see p. **15** in Section 2.2), together with 1.0 mM EGTA (see p. **118** in Section 4.2.2). This 'second set' was used for experimental studies on the half-fibre (▲), for values of I* between ~62 mM and ~182 mM (see Section 4.2.2 and Figure 4.2). Fink et al. (1986) used an apparent equilibrium constant of $10^{6.8}$ M^{-1}, for the EGTA–Ca^{2+} equilibrium, to calculate Ca^{2+} content, regardless of ionic strength, which exceeded ~80–100 mM in their experiments and therefore largely exceeded the lowest value for bulk ionic strength used here (~0; distilled water, corresponding to a vicinal ionic strength I* ~ 2 mM; see Section 3.9). In their experiments, Potma and Stienen (1996) and Potma et al. (1994) took exactly the same value of $10^{6.8}$ M^{-1} for an ionic strength of 200 mM. The value of $10^{6.8}$ M^{-1} is approximately twice that of $10^{6.5}$ M^{-1} (ratio of $10^{0.3} \sim 2$) used by Goldman and Simmons (1984a). At this point, it would be unwarranted to take only these approximate constants into account. Indeed, from the careful experimental work of Harafuji and Ogawa (1980), Horiuti (1986) demonstrated that the equilibrium constant depends on the ionic strength, at a given temperature (see their Table 2 and its legend). In the experimental conditions used in this monograph (pH 7; temperature 10°C), the pK deduced from the paper by Horiuti (1986) is given by pK = 6.77 − [2I*$^{1/2}$/(1 + I*$^{1/2}$) − 0.40 I*], where I* is expressed in molar (in this expression, the nature of the anion is not taken into account, but it is assumed that this relationship is valid in the particular case of MS$^-$, used in the experiments presented and commented on in this work). For instance, at I* $\sim$ 2 mM (distilled water; see Section 3.9), we have pK $\sim$ 6.68, and, at I* $\sim$ 182 mM (physiological conditions; see the next paragraph), pK $\sim$ 6.24 (i.e. an equilibrium constant lower by a factor of $\sim 10^{0.44} \sim 3$ at physiological ionic strength than in distilled water). Variations of pK with I* should therefore be taken into account for the most rigorous calculation of pCa.

The pCa values expressed as functions of I*, for the two sets of ATP and EGTA concentration, should be carefully calculated. As recalled at the beginning of the preceding paragraph, for the first set of ATP concentration (0.4 mM), the concentration of contaminating calcium in the relaxing buffers was $\sim$5.5 µM, before the addition of 0.5 mM EGTA. Thus, between I* $\sim$ 2 mM and $\sim$142 mM and for this set of values of ATP and EGTA, we obtain, using a simple routine, pCa(2 mM) $\sim$8.62, pCa(42 mM) $\sim$8.35, pCa(62 mM) $\sim$8.33 and pCa(142 mM) $\sim$8.22. For the second set of ATP concentration (2 mM), the concentration of contaminating calcium in the relaxing buffers was $\sim$7.1 µM, before the addition of 1 mM EGTA. Thus, between I* $\sim$ 62 mM and $\sim$182 mM, and for this second set of values of ATP and EGTA, we obtain, using the same simple routine, pCa(62 mM) $\sim$8.52, pCa(142 mM) $\sim$8.41 and pCa(182 mM) $\sim$8.37. There is, therefore, a gradual increase in pCa with decreasing I*. Moreover, at I* $\sim$ 62 mM, there are two different values of pCa ($\sim$8.33 and $\sim$8.61), but no detectable difference in the forces recorded (see Figure 4.2), within the limits of experimental error. At I* $\sim$ 142 mM, there are also two values of pCa ($\sim$8.22 and

$\sim$8.37), with no detectable difference appearing in Figure 4.2. Before drawing firm conclusions, direct measurements of Ca^{2+} contents under various conditions of ionic strength are required. As described on p. **15** in Section 2.2, control experiments were performed with fura 2, which was injected, at a concentration of 2 µM, into the fluorescence cuvette containing the chosen buffer. Four buffers were used, with bulk ionic strengths I = 6 mM (close to that of distilled water), I = 25 mM (corresponding approximately to the maximum of the solid line in Figure 4.2), I = 40 mM (corresponding approximately to the breakpoint of the solid line in Figure 4.2) and I = 180 mM (maximal bulk ionic strength used here and corresponding approximately to the physiological ionic strength in intact fibres of vertebrates; e.g. Godt and Maughan 1988; Godt and Nosek 1989; Xu et al. 1993). Each buffer contained a single half-fibre (see p. **15** in Section 2.2). Regardless of the value of I, no traces of Ca^{2+} were detected in the four cases studied, indicating that all the pCa values were greater than $\sim$8 (see p. **15** in Section 2.2), as expected from the values calculated above, ranging between $\sim$8.37 at I* $\sim$ 182 mM and $\sim$8.62 at I* $\sim$ 2 mM. Thus, we are probably dealing with resting conditions, regardless of the values of I*. This is obvious for I* values between $\sim$42 mM and $\sim$182 mM, as the forces recorded are negligible (see solid line in Figure 4.2). In this context, working on mechanically skinned frog fibres, at $\sim$20°C–22°C and for bulk ionic strengths higher than $\sim$55 mM, Gordon et al. (1973) claimed that the pCa of their relaxing solutions was $\sim$8.7. On the other hand, working on mechanically skinned crayfish fibres from the walking leg at 4°C, and for similar ionic strengths, Reuben et al. (1971) considered their fibres to be at rest for pCa > 9. In the experimental study presented in this monograph, relaxation conditions are also probably reached for I* values between $\sim$42 mM and $\sim$2 mM, because the two corresponding values of pCa are $\sim$8.35 and $\sim$8.62 (see above), respectively, both of which are very high. However, at low and very low values of I*, rarely or never used by other independent groups, a careful discussion is required, as the solid line in Figure 4.2 has unexpected features. This discussion is presented below, in this section.

Are the conclusions drawn in the preceding paragraph, regarding resting conditions, irrefutable? The solid line in Figure 4.2 has very interesting characteristics at low and very low ionic strengths (lower than ~42 mM). In these conditions, are pCa values of ~8.35 at $I^* \sim 42$ mM and ~8.62 at $I^* \sim 2$ mM (see the preceding paragraph) low enough to ensure the maintenance of resting conditions? In other words, do the phenomena occurring below ~42 mM actually correspond to Ca-independent contractures or merely to Ca-induced contractions? In all the experiments, the sarcoplasmic reticulum was intact and functional (see p. **12** in Section 2.2), regardless of I^* or I (see arrows C in Figure 4.1 and comments on pp. **117–119** in Section 4.2.2). However, it is unclear whether the presence of a functional sarcoplasmic reticulum is sufficient to give pCa values significantly higher than ~8.35–8.62 and whether it is possible to demonstrate that the half-fibres were at rest at low and very low ionic strengths. The sarcoplasmic reticulum, including the triads and terminal cisternae in which most of the Ca^{2+} is sequestered, occupies ~3.5% of cell volume in the heart (Katz 2006), but, to the best of my knowledge, no such estimate has been made for frog skeletal muscles. Nonetheless, the sarcoplasmic reticulum is known to occupy a higher proportion in skeletal muscle fibres. I therefore propose a value of ~10%. It should be noted that Merah and Morel (1993) presented a geometric rationale for estimating the true isometric tension (developed by the contractile system only) from the 'recorded' isometric tension in intact frog fibres. Using the numerical values proposed in this paper, with a few mathematical manipulations, the relative proportion of the sarcoplasmic reticulum is found to be ~11%, consistent with the estimate given above. I retain here a value of ~10% (~0.10). I take into account the range of uncertainty on the volume of the sarcoplasmic reticulum for 10 half-fibres and therefore $\sim 1.70 \pm 0.38 \times 10^{-5}$ cm^3, given on p. **13** in Section 2.2. Thus, the volume of the sarcoplasmic reticulum is probably $\sim 0.10 \times (1.70 \pm 0.38) \times 10^{-5}$ cm$^3 \sim 1.70 \pm 0.38 \times 10^{-6}$ cm$^3 \sim 1.70 \pm 0.38 \times 10^{-9}$ l. Calcium can accumulate, in the sarcoplasmic reticulum of skeletal muscle, a concentration of ~10 mM ~ 1.0×10^{-2} M (see p. **15** in Section 2.2). This corresponds to $\sim 1.0 \times 10^{-2} \times (1.70 \pm 0.38) \times 10^{-9} \sim 1.70 \pm 0.38 \times 10^{-11}$ mol, which can accumulate in the sarcoplasmic reticulum of the half-fibres used in the experimental part of this work (see Section 4.2.2 and Figure 4.2). As recalled at the beginning of this paragraph, ignoring the sarcoplasmic reticulum, the pCa values in the 0.4-ml trough

are ~8.35 and ~8.62 at low and very low values of I^* (~42 mM and ~2 mM respectively). Thus, immediately before measuring resting forces, the trough (0.4 ml = 4×10^{-4} l; see p. **14** in Section 2.2) contained $\sim (10^{-8.62}$ and $\times 10^{-8.35})$ mol l$^{-1} \times 4 \times 10^{-4}$ l $\sim 0.96 \times 10^{-12}$ and 1.78×10^{-12} mol of free calcium, at $I^* \sim 42$ mM and ~2 mM, respectively. The sarcoplasmic reticulum was functional in all the experiments performed on each half-fibre, even at low and very low ionic strengths, at which strong contractures occurred (see Figures 4.1 and 4.2 and corresponding comments). The loading capacity of the sarcoplasmic reticulum being $\sim 1.70 \pm 0.38 \times 10^{-11}$ mol (see above), the ratio of free calcium concentration in the trough to the full loading capacity of the sarcoplasmic reticulum is therefore $\sim (0.96$ and $1.78) \times 10^{-12}$ mol$/(1.70 \pm 0.38) \times 10^{-11}$ mol ~ 0.046 and ~0.135 (maximal interval), at $I^* \sim 42$ mM and ~2 mM, respectively. These two ratios are markedly lower than the uncertainty on the calcium concentration in the sarcoplasmic reticulum (e.g. $\sim 0.38 \times 10^{-11}$ mol l$^{-1}/1.70 \times 10^{-11}$ mol ~0.236, i.e. ~1.7 and ~5.1 times the two ratios estimated above). Taking into account these values, the sarcoplasmic reticulum would therefore be able to pump most of the external calcium during each run and would almost entirely deplete the bathing buffer and the inner medium of the half-fibres of free calcium. However, it is shown in the next paragraph that other complex problems can occur.

Ca^{2+} is simultaneously taken up by the Ca^{2+} pump located in the sarcoplasmic reticulum (largely via the terminal cisternae, located in the triads, i.e. T-system) and leaked by the sarcoplasmic reticulum in a process different from that involving the Ca^{2+} pump (e.g. Endo 1977; Horiuti 1986). The steady-state Ca^{2+} loading concentration within the sarcoplasmic reticulum therefore results from a balance between these two antagonistic processes. Thus, owing to this leakage of Ca^{2+} toward the sarcoplasmic medium of the half-fibres and the surrounding buffer, it is clearly impossible to deplete the internal and external media of the half-fibres of all their contaminating calcium. As experimentally demonstrated by Endo (1977) and Horiuti (1986), both loading and leakage properties depend on many factors, including, perhaps, ionic strength at pH 7 and 10°C, as used in the experimental part of this monograph (see Endo 1977, for a general, but certainly outdated, review on the sarcoplasmic reticulum). The quantity of free Ca^{2+} in the trough, in the presence of the sarcoplasmic reticulum, may be estimated, for example, at $\sim 0.96 \times 10^{-13}$ and 1.74×10^{-13} mol, rather than the $\sim 0.96 \times 10^{-12}$ and $\sim 1.74 \times 10^{-12}$ mol found in the absence of the sarcoplasmic reticulum (see the preceding paragraph). We therefore deduce that the concentration of Ca^{2+} would be ~ 0.96 and 1.74×10^{-13} mol/4×10^{-4} l $\sim 2.40 \times 10^{-10}$ and 4.43×10^{-10} M, giving pCa values of ~ 9.35 and 9.62, at $I^* \sim 42$ mM and ~ 2 mM, respectively. Nonetheless, the loading capacity of the sarcoplasmic reticulum taken into account here (~ 10 mM; see the preceding paragraph) corresponds to physiological ionic strength (here, $I^* \sim 182$ mM). As far as I know, no experimental data are available concerning the possible dependence of the properties of the sarcoplasmic reticulum on I^*, particularly at low and very low values of I^*, but it is reasonable to suggest that the pCa values of ~ 9.35 (at $I^* \sim 42$ mM) and ~ 9.62 (at $I^* \sim 2$ mM) are not strongly dependent on I^*, within the range ~ 2–182 mM, and to consider these pCa values as corresponding to resting conditions, at any value of the vicinal ionic strength. However, this approach is based on intuition rather than certainty and definitive experimental arguments are provided in the next paragraph.

Even at the very low pCa values of ~ 9.35 (at $I^* \sim 42$ mM) and ~ 9.62 (at $I^* \sim 2$ mM), strongly suggested in the preceding paragraph on the basis of sound arguments, two objections may be raised. Indeed, Brandt and Grundfest (1968), working on crayfish muscle fibres, used the technique of iono-phoretic injections of Ca^{2+} from an intracellular microcapillary and claimed that 'just detectable tensions were observed on extrusion of $\sim 10^{-14}$ M Ca^{2+} (pCa ~ 14)'. However, this very high value may be attributed to artefacts related to the technique used. Indeed, Reuben et al. (1971), in the same group, used traditional techniques and claimed that crayfish fibres are at rest for pCa values >9, at approximately physiological ionic strength. However, for vicinal ionic strengths as low as ~ 42 mM and ~ 2 mM (pCa ~ 9.35 and ~ 9.62, respectively; see above), there may be sufficient free calcium to induce Ca-dependent contraction, because the myofibrillar system would be highly sensitive to trace amounts of calcium, leading to trivial induced contractions at low and very low values of I^*, rather than Ca-independent contractures. Nevertheless, in case of Ca-induced contractions, it would be impossible to account for the extremely rapid increase in resting tension below $I^* \sim 42$ mM, the very high maximum at $I^* \sim 22$–32 mM and the drop below ~ 22 mM (see Figure 4.2). Moreover, Ca-induced axial forces, generated, for example, by the traditional swinging of the cross-bridges, cannot account for the sizeable lateral swelling of the half-fibres (e.g. $\sim 20\%$ in diameter, i.e. $\sim 44\%$ in volume) at bulk ionic strengths of ~ 20–30 mM (see p. **166** in Section 4.4.2.8). Thus, we can definitively conclude that the half-fibres were in states of non–Ca-induced contracture at low and very low vicinal (or bulk) ionic strengths and that the resting half-fibres were actually at rest, at any ionic strength, never in states of Ca-induced contraction, for bulk ionic strengths between ~ 0 (distilled water) and 180 mM ('in vivo-like' conditions).

4.2.2 Dependence of Resting Tension on Ionic Strength

The behaviour of three half-fibres, described in Section 2.2, was studied in various relaxing media. The half-fibres were first rinsed thoroughly with a relaxing buffer at $I = 180$ mM and then studied in relaxing buffers, the compositions of which are given on p. **118**. Figure 4.1 presents typical traces obtained with half-fibre ($\bullet$), also referred to in this section as the 'first series' of experiments or the 'first half-fibre'. Relaxing buffers containing 30 mM caffeine (purchased from Sigma France) were injected into the trough (arrows C) to check that the sarcoplasmic reticulum was intact and functional. Based on the results obtained with half-fibres and presented in Figure 6 of the paper by Villaz et al. (1987), the main features of the traces presented in Figure 4.1 here (presence of 0.5 mM EGTA), corresponding to caffeine-induced contractures, were expected, that is, rapid increase in tension related to the release of Ca^{2+} from the sarcoplasmic reticulum, existence of a shoulder during the decrease and total duration of the return to resting tension of ~ 15–20 s. After each force recording, the half-fibre was kept at resting tension, corresponding to the reference relaxing buffer ($I = 180$ mM), for ~ 30 s, to ensure the maximal reaccumulation within the sarcoplasmic reticulum of the half-fibre of trace amounts of calcium from the surrounding medium and to check that the sarcoplasmic reticulum remained intact and functional throughout the run. This was also done for the other two half-fibres ($\blacksquare$, $\blacktriangle$) used in this experimental study. In the limited number of

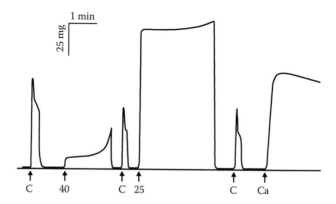

Figure 4.1 A typical set of experiments on the half-fibre (●). Details of the composition of the various buffers used for studying this half-fibre are provided in the main text (see p. **118**). The peak corresponding to the first arrow labelled C (injection of a buffer containing 30 mM caffeine) is higher than the next two peaks 'C'. This phenomenon is almost certainly linked to this first peak corresponding to the first experiment on the half-fibre fixed in the trough. A few initial rearrangements (after dissection and attachment of the half-fibre) affecting the structure and function of the sarcoplasmic reticulum and the overall structure and geometry of the half-fibre might occur, followed by a long 'steady state' extending until the arrow labelled Ca (injection of a buffer with a pCa ~ 4.5).

experiments presented in Figure 4.1 (many more experiments were performed on half-fibre [●]), a continuous accumulation of calcium within the sarcoplasmic reticulum was clearly demonstrated by further injections of caffeine (second and third arrows C), confirming that the sarcoplasmic reticulum was functioning throughout the set of experiments. The 'everlasting' functioning of the sarcoplasmic reticulum, for at least the ~25 min corresponding to a full set of experiments (see p. **12** in Section 2.2), was therefore checked. Beyond this period, the sarcoplasmic reticulum may display fatigue and would accumulate less efficiently Ca^{2+} (see Section 7.4 for a discussion).

Arrow 40 indicates the injection of the same relaxing buffer as used in the preceding paragraph, but with I = 40 mM (I* ~ 42 mM). This fresh buffer had an initial pCa value of ~8.35 (see p. **113** in Section 4.2.1). However, this pCa value increased, probably over a few milliseconds, to ~9.35, owing to the presence of the functional sarcoplasmic reticulum (see p. **115** in Section 4.2.1). The half-fibre was therefore almost instantaneously at rest, within the limits of the maximal duration of buffer injection (~300 ms; see p. **14** in Section 2.2). The initial ATP concentration was only 0.4 mM (see p. **118**), but this was sufficient to maintain the half-fibre at rest at the beginning of the experiment. Indeed, the resting force recorded was stable for ~30–40 s (see Figure 4.1) and corresponded to the resting tension studied and discussed in Section 4.4.2.8. The

resting force then increased, slowly at first, and then more rapidly, as clearly seen in Figure 4.1. This slow increase, followed by a rapid increase, resulted simply from the gradual onset of rigor, owing to the gradual depletion of ATP. Indeed, even at rest, small quantities of MgATP are continuously cleaved by the myosin heads detached from the thin actin filaments and also by the limited number of weakly bound cross-bridges (see pp. **170–171** and **174–176** in Section 5.1.1 for precise details on the 'weakly bound cross-bridges' and their role). Moreover, the Ca^{2+} pump of the sarcoplasmic reticulum also continuously splits ATP (probably MgATP; e.g. Asayama et al. 1983), as does the Na^+–K^+ pump in the half-sarcolemma (present in half-fibres), which also contributes to the gradual depletion of MgATP.

Arrow 25 indicates the injection of a relaxing buffer, similar to the other two buffers used in the preceding paragraphs, but with I = 25 mM. The relative resting force was very high and stable for ~1–2 min. Furthermore, the increase in the recorded force, certainly resulting from the gradual onset of rigor, was slow and reached a plateau (~4–5 min after the beginning of the force recording; not shown). This plateau resulted from addition of the rigor and resting forces and corresponded to an excess of only ~20%–30% with respect to the resting force recorded at the beginning of the experiment.

The arrow Ca indicates the injection of an 'in vivo-like contracting buffer' with pCa ~ 4.5 at

I = 180 mM, to induce full isometric tetanic contraction (see p. **118** concerning the other components of the buffer). The corresponding 'apparent' trace, shown at the right of Figure 4.1, was carefully drawn by eye and was obtained by dividing the true trace recorded by a factor of 2.5. This apparent trace represents ~100%/2.5 ~ 40% of the total recorded isometric tetanic tension (this proportion of ~40% is the best estimate, used throughout this monograph) and corresponds to the axial active tension related to the radial net repulsive (expansive) forces (mostly electrostatic, in a proportion of ~100.0% − 24.0% ~ 76.0%; see p. **85** in Section 3.5). Thus, full relative isometric tetanic tension corresponds to a value of 2.5. The baseline in Figure 4.1 corresponds to a relaxing buffer at I = 180 mM and is taken as the reference zero for all resting tensions. Optimal measurements for the three half-fibres, corresponding to actual resting conditions, were obtained by ensuring that all the resting tensions taken into account were recorded within ~30–40 s of the injection of relaxing buffers and before any detectable onset of rigor (see above, in this section).

Figure 4.2 shows the relative resting tensions plotted against the bulk ionic strength I of the surrounding medium (solid line), together with some results relating to resting tensions obtained by two other independent groups. Note that the solid line was drawn by eye as a best fit to the experimental points, which are significantly scattered for bulk ionic strengths below ~30–40 mM. I also obtained two experimental results (Δ) concerning isometric tetanic tensions at I = 60 mM (see p. **119**). The dotted line corresponds to the best fit of Equation 4.15 to the many experimental points obtained by April and Brandt (1973) (see their Figure 5). All the experimental values obtained by these authors, except for their reference value, are divided by a factor of 2.5, to take into account the ~40% of full isometric tetanic tension related to lateral swelling mechanisms (see the preceding paragraph). The experimental results of April and Brandt (1973) are analysed and discussed on pp. **106–109** in Section 4.1, in which it is shown that the results presented in their Figure 5 can be taken as a reference for the axial contractile forces related to the net radial repulsive (expansive) forces (mostly electrostatic; see the preceding paragraph). Unfortunately, the authors did not record tensions below bulk ionic strengths of ~50 mM and, between bulk ionic strengths of ~0 (distilled water) and ~50 mM, the dotted line is simply the mathematical extrapolation of Equation 4.15. However, this extrapolation gives a 'virtual' maximum at bulk ionic strengths of between ~40 mM and ~50 mM and a 'virtual'

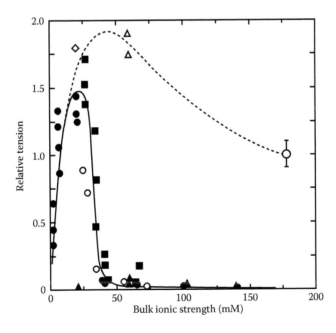

Figure 4.2 Relative resting (solid line; mostly for half-fibres) and isometric tetanic tensions (dotted line; mostly deduced from Equation 4.15) against bulk ionic strength. The way in which the experimental points and the two curves were obtained is described and discussed in the main text (Sections 4.2.2 and 4.4.2.7).

drop below ~30 mM. This is qualitatively consistent with the mathematical conclusions drawn on pp. **102–103** in Section 3.10 and with the experimental results presented in Figure 2 of the paper by Gordon et al. (1973) and in Figure 1 of the paper by Hasan and Unsworth (1985) (see the discussion on pp. **109–111** in Section 4.1). As explained in Section 2.2, only three half-fibres were studied, in various relaxing media (obviously excluding distilled water, which is not a relaxing medium, but a complex rigor plus relaxing medium; see the next paragraph).

The first set of experiments was performed on the first half-fibre (●), with surrounding buffers of the following compositions: 5 mM PIPES, 0.5 mM EGTA, 0.6 mM $Mg(MS)_2$, 0.4 mM Na_2ATP and variable KMS. A concentration of 0.4 mM ATP is sufficient to avoid the onset of rigor for at least ~30–40 s (see Figure 4.1 and corresponding text, particularly pp. **116–117**). For the points at I = 6 mM, the PIPES concentration was only 1 mM. In these conditions, the pH of the surrounding buffer was ~7, but this medium was obviously poorly buffered. However, no major discrepancies were observed with respect to the other experimental points, within the broad limits of experimental error. In freshly distilled water (I ~ 0, pH ~ 6.5), both rigor and resting forces clearly contributed to the observed forces. In distilled water, F-actin (filaments) dissociate in vitro and actin exists in the form of G-actin (soluble globular actin monomers). Fortunately, in situ, the vicinal ionic strength I* of ~ 2 mM within half-fibres bathing in distilled water prevents the thin actin filaments from dissociating (see Section 3.9 for precise details on I*). Thus, Equation 3.2 remains valid and radial repulsive electrostatic forces are exerted, whereas radial tethering forces (see Section 4.3 for precise details on these forces) and other attractive/compressive forces cannot be detected below ~20–25 mM, within the broad limits of experimental error (see Figure 4.2; see also the discussion presented in Section 4.4.2.8). Rigor is induced by the total depletion of ATP. A special method was used to deduce the contribution of resting forces alone, in distilled water, from the sum of the resting and rigor forces necessarily recorded in these conditions. Indeed, using the first half-fibre (●) in the buffer described above, but with only 1 mM PIPES, resting forces were recorded at I = 6 mM, at the start of the experiments. Resting plus rigor forces were then recorded, at the end of

runs, after the total depletion of ATP (~5–6 min after the start of the run). Rigor forces in distilled water and at I = 6 mM were assumed to be similar and this assumption seems to be acceptable. The resting forces in distilled water were estimated by subtracting the rigor forces, recorded at I = 6 mM, from the total forces recorded in distilled water. In all the other experiments, the low concentration of ATP (0.4 mM) was sufficient to keep the half-fibre at rest and to prevent the onset of rigor for at most ~30–40 s (see Figure 4.1 and pp. **116–117**). Regenerating systems (for the rephosphorylation of ADP to generate ATP) were therefore not required to keep the ATP concentration constant. Moreover, the use of such systems invariably increases ionic strength, precluding the study of half-fibres at very low ionic strengths. Full contraction at I = 180 mM was induced by adding ~0.5 mM $Ca(MS)_2$ (pCa ~ 4.5) and only ~40% of the recorded force was taken into account (see again Figure 4.1 and pp. **116–117**). This force is related solely to net radial repulsive (expansive) forces and is frequently called the electrostatically related contribution to the total recorded axial forces in this monograph (the corresponding buffers are called 'in vivo-like contraction buffers').

The second series of experiments was performed on the second half-fibre (■), with surrounding buffers of the following compositions: 9 mM PIPES, with concentrations of the other components identical to those for the first half-fibre (●). Full contraction at I = 180 mM was induced by adding ~0.5 mM $Ca(MS)_2$ (pCa ~ 4.5). The third series of experiments was performed on the third half-fibre (▲), with surrounding buffers of the following compositions: 15 mM PIPES, 1 mM EGTA, 3 mM $Mg(MS)_2$, 2 mM Na_2ATP and variable KMS. The concentration of EGTA was 1 mM and I therefore observed no shoulder during the decrease in tension after the caffeine-induced signal (not shown), as expected from Figure 6 (valid for half-fibres) in the paper by Villaz et al. (1987). Full contraction at I = 180 mM was induced by adding ~1 mM $Ca(MS)_2$ (pCa ~ 4.5). The full forces recorded, corresponding to isometric tetanic contractions, were divided by a factor of 2.5, to assess the contribution of net radial repulsive (expansive) forces alone (see p. **117**). As the concentration of ATP was 2 mM, there were no problems of depletion and the onset of rigor was delayed for several minutes, in the case of resting fibres (~5–6 min; not shown).

The number of fixed negative electrical charges on myosin depends on ATP concentration (Bartels et al. 1993; Deshayes et al. 1993). Fortunately, at 0.4 mM and 2 mM ATP (the two concentrations used here), these charges are similar, within the limits of experimental error (Bartels et al. 1993). Little is known about the bound anions (see Section 4.4.2.2 for a short discussion), but their number is also almost certainly independent of the ATP concentration within the narrow range of 0.4–2.0 mM. Thus, the radial repulsive electrostatic forces do not depend on the ATP concentrations used here and these two different ATP concentrations did not affect the experimental results. The two experimental points (Δ) correspond to the third half-fibre (▲), with pCa ~ 4.5 (contraction conditions), at I = 60 mM. These points again represent ~40% of the measured relative isometric tetanic forces (see p. **117**). They are consistent with the 'electrostatic' curve calculated from Equation 4.15, representing ~40% of the relative isometric tetanic forces (or tensions) deduced from the experimental results of April and Brandt (1973). For reasons given in Section 4.1 (see particularly p. **109**), I select the experiments of April and Brandt (1973) rather than those of Gordon et al. (1973) and Hasan and Unsworth (1985). Forces were normalised, assigning a value of zero to the resting force at I = 180 mM. The value of ~40% full isometric force at pCa ~ 4.5 and I = 180 mM was set to 1 (the vertical bar represents 2 × SD; n = 6). From the values of the relative forces, 'crude values' of the relative resting tensions can be deduced that are not corrected for observed cross-sectional area increases, particularly below I ~ 40 mM (see pp. **165–167** in Section 4.4.2.8 for precise details on the significant lateral swelling below I ~ 40 mM). However, in Section 4.4.1, the resting 'crude tensions' corresponding to I ~ 20–30 mM are corrected to take into account the large increase in cross-sectional area (~+44%), making it possible to draw important conclusions concerning tensions. Points (○) correspond to the experiments of Gordon et al. (1973), performed on resting frog fibres, with the necessary shift in I to low ionic strengths (see Section 4.4.3 for precise details). Point (◇) corresponds to 23 μM MgCl$_2$ and point (◆) corresponds to 1 mM MgCl$_2$. These two points were deduced from the results obtained by Gulati (1983), for resting frog fibres, with the same shift in I as for the experimental results of Gordon et al. (1973), for the same reasons (see again Section 4.4.3).

Based on Figure 4.2, relative resting forces (or 'crude' relative resting tensions) are very low between bulk ionic strengths I = 180 mM (I* ~ 182 mM) and ~40 mM (I* ~ 42 mM) and increase very slightly with decreasing I. A further small decrease in I results in a very rapid increase in resting forces, which peak at high values between ~20 mM and ~30 mM (I* ~ 22–32 mM), later falling sharply with further decreases in I. Resting forces do not reach zero at I ~ 0 (see p. **118** for an estimate of the resting forces in these conditions), and the slope is extremely steep, as predicted from Equation 3.2 (see, in Section 3.10, pp. **101–102** for a simple mathematical explanation and pp. **102–103** concerning the one-to-one process for the conversion of radial forces into axial forces). These experimental and semi-empirical observations confirm that (i) the radial repulsive electrostatic forces are effective (see Sections 3.2 and 3.5), (ii) the vicinal ionic strength I* is not zero when bulk ionic strength I is ~ 0 (distilled water) and (iii) the net radial repulsive (expansive) forces are converted into axial forces (see also pp. **102–103** in Section 3.10). Otherwise, on pp. **109–112** in Section 4.1, consistent conclusions are also drawn from experiments performed by two other independent groups, concerning the role of the radial repulsive electrostatic forces. Finally, in Section 3.9, I take I* ~ I + 2 mM, regardless of the value of I. This choice is confirmed here, at least under resting conditions. However, it is possible, even probable, that the vicinal ionic strength I* is different under resting and contracting conditions (also probably in rigor), owing to the different numbers of fixed negative electrical charges and bound anions (see Section 4.4.2.2 and pp. **160–161** in Section 4.4.2.6; see also Regini and Elliott 2001 and references therein, regarding the question of bound anions). Nonetheless, such differences in I* are not taken into account for the reasons given in Section 3.9.

4.3 EXPERIMENTS PERFORMED ON RESTING PERMEABILISED FIBRE BUNDLES FROM YOUNG ADULT RATS

4.3.1 Detection of Very Small Transitory Contractures and Existence of Radial Tethering Forces

The mode of generation of radial tethering forces at rest is described and discussed on pp. **174–176** in Section 5.1.1. These radial tethering forces are exerted by some weakly bound cross-bridges, called 'active weakly bound cross-bridges' (see pp. **170–172** in Section 5.1.1 concerning the two types of weakly bound cross-bridges, particularly the active weakly bound cross-bridges). The existence of radial tethering forces, in resting conditions, was experimentally investigated by trying to reduce them and observing the consequences of this reduction (see below, in this section). The radial repulsive electrostatic forces are effective, both at rest and under contraction conditions (see Sections 3.2, 3.5, 4.2.2, 4.4.2.6, 4.4.2.7 and 4.4.2.8). Thus, reducing the radial tethering forces should result in favour of the radial repulsive electrostatic forces, resulting in net radial repulsive (expansive) forces, translated into axial contractures, because there are mechanisms of conversion, particularly at rest (see pp. **102–103** in Section 3.10 and p. **119** in Section 4.2.2). There is a great deal of evidence to support the hypothesis of a major role of radial expansive forces (see pp. **109–111** in Section 4.1 and the six sections mentioned above).

The best way to demonstrate the existence of radial tethering forces experimentally is to cleave the weakly bound cross-bridges, particularly the active weakly bound cross-bridges in resting permeabilised fibre bundles, enzymatically (see Section 4.3 for precise experimental details concerning these bundles). I prepared 5 mg of α-chymotrypsin (purchased from Sigma France) in 1 ml of HCl N/1000, immediately before use, as usual (Margossian and Lowey 1982 and references therein). Margossian and Lowey (1982) found that, in vitro, α-chymotrypsin cleaves the myosin molecule included in synthetic thick myosin filaments at the joint between S1 (head) and S2 (soluble part of the myosin tail). The definitions of S1 and S2 are given in Figure 5.1. On p. **175** in Section 5.1.1, the probable mode of action of α-chymotrypsin is presented. This cleavage would lead, in situ, to the de facto detachment from the thin actin filaments of the active weakly bound cross-bridges (and of all other weakly bound cross-bridges). As the sarcoplasmic reticulum is almost certainly completely removed from the fibre bundles (see Section 2.3), there is, in principle, little chance that this structure is digested with α-chymotrypsin to release large quantities of Ca^{2+} into the myofibrillar lattice. This question and that of the very low possibility of digestion

of the mitochondria are studied in Section 4.3.2, where an unequivocal interpretation of the experimental results is presented.

Typical curves are presented in Figure 4.3, which shows the behaviour of bundles of fibres before and after the injection of α-chymotrypsin. Six phases are defined for the 24 fibre bundles studied. In phases 1 and 2, the initial resting forces were recorded. The vertical bar between phases 1 and 2 corresponds to the lifting of the bundles attached to the force transducer and their reintroduction into the cuvette. No artefacts were detected during this process. At the end of phase 2, the bundles were lifted again and 30 µg of α-chymotrypsin was rapidly injected into the cuvette, which contained 2.5 ml of buffer (see Section 2.3). The final concentration of α-chymotrypsin in the cuvette was therefore 30 µg/2.5 ml ~ 12 µg ml^{-1} and its injection lasted ~2–3 s. The bundles were then reintroduced into the cuvette. Phase 3 corresponds, in most cases, to a small increase in resting forces, followed by a decrease (phase 4). In Figure 4.3, it can be seen that (i) the duration of the increase (phase 3) was ~50–70 s (the mean value for the 24 bundles was ~60 s) and that the whole digestion process lasted ~140–170 s (the mean value for the 24 bundles was ~160 s) and (ii) the maximal force recorded for the transitory contracture differed between bundles (the absence of a clear maximum in some cases is discussed in the legend to Figure 4.3). In Figure 4.3, some of the difference in the maximal values reached resulted from differences in bundle diameters (see Section 2.3). The maximal recorded value was ~1.8 ± 0.2 times (mean ± SD; n = 3) the initial resting force (this maximal value was recorded for three fibre bundles). Phase 5 corresponds to the final resting forces. At the end of phase 5, calcium was added, corresponding to a pCa value of ~4.5 (phase 6; contraction conditions). As all the heads had been digested, the cross-bridges were de facto not able to swing and axial contractile forces related to these structures could not be observed. Moreover, the resting radial tethering forces, intimately related to the active weakly bound cross-bridges also disappeared, simply because all the weakly bound cross-bridges had been digested. Owing to the many adulterations of the myofilaments during the digestion process (see below and p. **128** in Section 4.3.2), the numbers of fixed negative electrical charges and bound anions on the

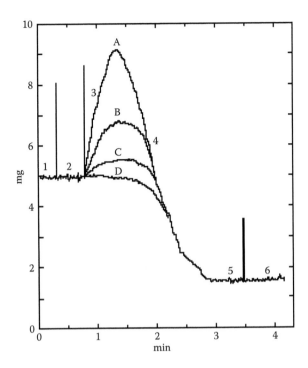

Figure 4.3 Results obtained with 24 bundles of four to six permeabilised fibres. Only four typical 'normalised' traces (see below for precise details) are presented. Forces were recorded before and after injection of α-chymotrypsin at the end of phase 2 (see main text, p. **120**). Seven batches of α-chymotrypsin were used. The forces recorded differed between bundles, partly because of the different diameters of the various bundles (150–250 μm, leading to cross-sectional areas differing by a factor of ~3, when passing from ~150 μm to ~250 μm), but also partly attributed to room temperature not being the same in different series of experiments (between ~20°C and ~26°C; see p. **14** in Section 2.3). However, many uncontrolled parameters also contribute to the scattering of the experimental results (see Section 4.3.2). For the comparison of experimental results between bundles, phases 1 and 2 were normalised as follows: the mean value of all the forces recorded was calculated, giving ~4.95 ± (0.06 or 0.29) mg (mean ± SE or SD; n = 24). The reference force was therefore taken as 5 mg. This process was repeated for phases 5 and 6, giving ~1.47 ± (0.08 or 0.39) mg (mean ± SE or SD; n = 24). The reference force was therefore taken as 1.5 mg. All the traces were carefully drawn by eye, from the true forces recorded and thereafter normalised. Traces A and B correspond to the maximal and minimal values recorded at ~1.4–1.5 min with three batches of α-chymotrypsin (18 bundles used). Traces C and D, presented for information, were obtained with six bundles and correspond to the maximal and minimal values recorded at ~1.4–1.5 min with a single batch, which was probably adulterated, for unknown reasons. Despite the very low minimal force recorded (trace D), single exponential decays would have been recorded only in the total absence of tethering forces, showing that, even with a poor batch, tiny tethering forces can be detected.

myofilaments were certainly considerably altered (almost certainly greatly reduced) at the end of phase 5. However, some of these charges may remain and a small proportion of the radial repulsive electrostatic forces may be exerted, accounting, at least partly, for the residual resting axial forces (~1.5 mg; see Figure 4.3 and its legend). Nevertheless, a large part of the residual resting axial forces is probably related to titin (e.g. Leake et al. 2004), but titin is also present before digestion and, assuming that it is not digested with α-chymotrypsin, its contribution would be similar before and after digestion. The contribution of titin to the various phenomena observed here would therefore be negligible, as the fibre bundles are only very slightly stretched (see Section 2.3). Finally, positive results (detectable transitory axial contractures) were obtained in all the experiments, with small transitory increases in resting axial forces of various magnitudes after the addition of α-chymotrypsin, which digests the weakly bound cross-bridges and then other structures (all heads, thick myosin filaments, etc.; see p. **128** in Section 4.3.2 for a discussion). The addition of α-chymotrypsin therefore resulted in complex phenomena and caused transitory axial contractures, mostly attributed to the disappearance of the radial tethering forces exerted at rest (see Figure 4.3 and Section 4.3.2).

4.3.2 Analysis of the Experimental Data. Complementary Experiments. Discussion

How should we interpret the small transitory axial contractures described in Section 4.3.1 (see Figure 4.3)? On p. **175** in Section 5.1.1, it is strongly suggested that, under resting conditions, the S1–S2 joint is slightly more accessible to α-chymotrypsin for the external A-type myosin heads attached to the thin actin filaments than for the many other external heads stuck to the thick myosin filament backbone (the A-type heads are shown in Figure 5.1; see also pp. **170–172** in Section 5.1.1 for precise details on the A- and F-type heads). As recalled on p. **120** in Section 4.3.1, enzymatic digestion with α-chymotrypsin results in specific cleavage at the S1–S2 joint. This cleavage concerns only a small number of amino acids, maybe approximately 10 (e.g. Garrigos et al. 1983; Mendelson and Wagner 1984). As the external heads, of A- and F-types, attached to the thin actin filaments at rest are slightly more accessible to α-chymotrypsin than heads stuck to the backbone (see above), they are digested first. Digestion therefore results in a rapid decrease in the number of resting A- and F-type heads and a rapid decrease in the total radial tethering forces (see pp. **174–175** in Section 5.1.1, as concerns the relationship between the A-type heads and the resting tethering forces). Thus, the net radial repulsive forces (mostly electrostatic forces) become effective, owing to an imbalance, and this induces axial contracture because of the translation of net radial repulsive forces into axial forces, as demonstrated throughout this monograph, both at rest and under contraction conditions (e.g. pp. **102–103** in Section 3.10 for the conversion and Sections 4.4.2.6, 4.4.2.7 and 4.4.2.8 for precise details on the axial forces). The other heads (stuck to the backbone) are also digested, after a short time lag (see p. **175** in Section 5.1.1, regarding this time lag). This results in the complete digestion of all the heads and of many thick myosin filaments, myofibrils and finally many whole fibres from the bundle (observed under a stereomicroscope). Only transitory axial contractures,

of unpredictable magnitudes, were therefore detected (see p. **128**).

During the digestion process, α-chymotrypsin may have digested troponin or tropomyosin, thereby desensitising actin and accounting for the transitory increase in axial forces, observed in the experiments described in Section 4.3.1, resulting in trivial transitory axial contractions. The possible digestion of troponin or tropomyosin (both purchased from Sigma France) was assessed in vitro, under conditions similar to those used in situ (same buffer, same ionic strength and same temperature; see Section 2.3). The ratios of α-chymotrypsin to troponin and tropomyosin used in situ were calculated as follows. Each thin actin filament is usually assumed to contain ~330 actin molecules. More precise values may be deduced from x-ray experiments with synchrotron radiation (e.g. Linari et al. 2005; Martin-Fernandez et al. 1994; Piazzesi et al. 1999). However, refined calculations are pointless here. It is generally accepted that there are 7 actin molecules per troponin and per tropomyosin molecules in the thin actin filaments. Moreover, there are 4 thin actin filaments per intact unit cell (see p. **35** in Section 3.2). Thus, there are ~330 × 4 ~ 1320 actin monomers and 1320/7 ~ 189 troponin and tropomyosin molecules per intact unit cell. The volume of an intact unit cell for a 'composite animal species' is ~3.44 × 10^{-15} cm^3 ~ 3.44 × 10^{-18} l (see p. **206** in Section 6.3.5). The concentration of these two molecules, in an intact unit cell, is therefore 189/(3.44 × 10^{-18} × 6.022 × 10^{23}) ~ 91 μM (6.022 × 10^{23} is Avogadro's number). However, we are dealing with demembranated fibre bundles, that is, swollen fibres, the mean cross-sectional area and volume of which are ~2.23 times the cross-sectional area and volume of an intact fibre (see p. **98** in Section 3.8.1). This multiplication factor of ~2.23 can reasonably be applied to unit cells belonging to demembranated fibres. Thus, the concentration of troponin and tropomyosin in these unit cells is ~91 μM/2.23 ~ 41 μM. This estimate corresponds to ~2.5 mg ml^{-1} for troponin and ~3.3 mg ml^{-1} for tropomyosin (the molecular weights of troponin and tropomyosin are given in the minireview by Morel and Pinset-Härström 1975a). The α-chymotrypsin was used at a concentration of 12 μg ml^{-1} in permeabilised fibre bundles (see p. **120** in Section 4.3.1), giving ratios of ~1:210 (w/w) for troponin and ~1:275 (w/w) for tropomyosin. These ratios were used for the in vitro digestions described in the next paragraph.

Before performing in vitro digestion experiments, I studied troponin and its three subunits (TnC, TnI and TnT) and tropomyosin and its two identical subunits in non-denaturing conditions in a buffer similar to the 'in situ buffer' and at room temperature (see Section 2.3) and in denaturing conditions (sodium dodecyl sulphate [SDS] added to the in situ buffer; usual preparation of denatured muscle proteins, as described by Margossian and Lowey 1982). Intact and denatured troponin and tropomyosin were observed by high-performance capillary electrophoresis, with absorbance measured at a wavelength of 214 nm, corresponding to peptide bonds (Beckman P/ACE 2100 apparatus; see Morel et al. 1998a, particularly the 'supporting information available', concerning this technique). Under denaturing conditions, three clearly resolved peaks were observed with troponin (corresponding to TnC, TnI and TnT), whereas only one peak was observed with tropomyosin, because both subunits of this protein are identical. Troponin and tropomyosin were 'digested' separately with α-chymotrypsin in the in situ buffer over periods of 60 s, corresponding approximately to the time scale over which the forces increased (phase 3 in Figure 4.3), or 160 s (stabilisation of the forces; phases 5 and 6 in Figure 4.3). Conditions are certainly very different in vitro and in situ, particularly the structure of interfilament water and the presence of many myosin heads in the interfilament medium, commented on and discussed on pp. 27–32 in Section 3.2, and problems of steric constraints, even hindrance, discussed on pp. 251–254 in Section 8.9. In this context, Borejdo et al. (2006) argued that thick myosin and thin actin filaments are crowded in vivo/in situ but highly dispersed in vitro. This view is difficult to refute and is consistent with my own (see also p. 270 in Chapter 9). For these reasons, the two proteins were also treated for 15 min (900 s). The digestion processes were stopped by adding 1% phenyl methyl sulphonyl fluoride (PMSF) as usual (Margossian and Lowey 1982). Treated troponin and tropomyosin were studied in the digestion medium (non-denaturing conditions; in situ buffer described in Section 2.3) or in denaturing buffer (in situ buffer plus SDS; see above). No signs of troponin or tropomyosin digestion were observed, after 60 s, 160 s and 900 s, under either non-denaturing or denaturing conditions (the many traces obtained are not shown).

As recalled on p. 16 in Section 2.3, permeabilisation by Triton X-100 for ~1 h at ~20°C–26°C (room temperature) at least partly destroys the sarcoplasmic reticulum and almost certainly completely removes this structure, as suggested by specialists in muscle fibres and sarcoplasmic reticulum. I did not check whether removal was total here. Let us assume, therefore, as a pessimistic compromise, that ~25% of this structure is intact in the fibre bundles and is entirely digested with α-chymotrypsin. In Section 2.3, it is strongly suggested that frog sarcoplasmic reticulum accumulates ~10 mM (~10×10^{-3} M) calcium. This value can reasonably be extended to the rat, even for red muscles (tibialis anterior; see Section 2.3). The volume of the sarcoplasmic reticulum is taken as ~10% of the volume of the fibre from frog muscle (see p. 114 in Section 4.2.1), also extended to red muscles from rat. As only ~25% of this structure may be functional, the maximal 'potential' concentration of calcium released from this structure within the fibre bundles would be ~10×10^{-3} M $\times 10 \times 10^{-2} \times 25 \times 10^{-2} \sim 2.50 \times 10^{-4}$ M ~ 0.25 mM (10×10^{-2} corresponds to the proportion of ~10% for the intact sarcoplasmic reticulum and 25×10^{-2} to the proportion of ~25% potentially remaining functional after permeabilisation). Although highly unlikely (see Section 2.3), mitochondria in the permeabilised fibre bundles may have been entirely digested, releasing calcium into the fibres. Heart cells contain a large number of mitochondria, owing to the strong aerobic activity of this organ. Indeed, mitochondria account for ~35% of cell volume in the heart (Katz 2006). Red muscles, which also have a high level of aerobic activity, contain ~20%–30% mitochondria (e.g. Johnston 1983). Rabbit skeletal white muscle fibres (from psoas, back or legs) and muscle fibres from frog are known to have much lower levels of aerobic activity than the heart and red muscles and I suggest a proportion of only ~5% mitochondria in this case. Colomo et al. (1997) recalled that cardiac and fast (white) skeletal single isolated myofibrils develop similar average isometric tetanic tensions. However, they also pointed out that cardiac cells develop lower isometric tetanic tensions than vertebrate white skeletal fibres. The higher proportion of mitochondria in the heart than in white skeletal muscle should result in a significant difference in 'mechanically active' cross-sectional areas, accounting for a non-negligible proportion of the lower tension recorded in the heart (here, a loss of, at least, ~35 − 5% ~ 30% is expected). This indirectly justifies the assumption of much lower proportion of mitochondria in skeletal fibres from white muscles than in cardiac cells. The experiments presented in Section 4.3.1 were performed on red muscles (see Section 2.3) and the proportion of mitochondria is taken as ~25%, which is midway between ~20% and ~30% (see Johnston 1983) and less than the ~35% in cardiac cells.

In skeletal muscle fibres, mitochondrial calcium concentration is less than ~1 mM ~ 1×10^{-3} M (Pierre Vignais, personal communication) and the consequence of a putative complete diges- tion of the mitochondria should be evaluated. A mitochondrial calcium concentration of ~0.5 mM (~0.5×10^{-3} M) is taken into account here. The duration of the injection of α-chymotrypsin into the cuvette was ~2–3 s (see p. **120** in Section 4.3.1), but we can assume, for the sake of sim- plicity, that digestion of the mitochondria began instantaneously and lasted a very short time, resulting in the 'instantaneous' release of calcium. Thus, the maximal potential concentration of cal- cium released from the mitochondria within the fibre bundles would be ~0.5×10^{-3} M $\times$ 25 $\times$ 10^{-2} ~ 1.25×10^{-4} M ~ 12.5 mM (25 $\times 10^{-2}$ cor- responds to the proportion of ~25% mitochondria suggested at the end of the preceding paragraph). Adding the maximal putative concentration cor- responding to the 'residual' sarcoplasmic reticu- lum, that is, ~2.50×10^{-4} M (see the preceding paragraph), the maximal putative concentration of calcium corresponding to these two struc- tures would be ~1.25×10^{-4} M + 2.50×10^{-4} M ~ 3.75×10^{-4} M. However, the buffer used here contained 10 mM EGTA (see Section 2.3). Using a value of $10^{6.4}$ M^{-1} for the equilibrium constant, deduced from the work of Harafuji and Ogawa (1980) and Horiuti (1986), with I ~ 100 mM, pH 7.0 and at room temperature (experimental conditions described in Section 2.3), the maxi- mal (but clearly unrealistic) concentration of free calcium released from both the mitochondria and the sarcoplasmic reticulum into the fibre bundle medium would be ~$10^{-7.8}$ M (pCa ~ 7.8), which is sufficient for partial activation of the fibre bun- dles. Nevertheless, once Ca^{2+} is released within the fibres, it must diffuse out of the fibre bundles, because the concentration of Ca^{2+} is much lower

in the bathing buffer (~$10^{-9.5}$ M; pCa ~ 9.5; see p. **17** in Section 2.3). Thus, the concentration of Ca^{2+} within the fibre bundles is certainly lower than $10^{-7.8}$ M and greater than $10^{-9.5}$ M. This problem is studied and discussed in the following paragraphs.

Based on the experimental data of Mannherz (1968) and Martson (1973), Cooke and Pate (1985) took a diffusion constant of ~0.2×10^{-6} cm^2 s^{-1} for the negatively charged MgATP and MgADP molecules (MW ~ 600 Da), within the myo- fibrillar lattice. Kuschmerick and Podolsky (1969) obtained a value of ~1.2×10^{-6} cm^2 s^{-1} for MgATP, also within the myofibrillar lattice. Again, in situ, Yoshizaki et al. (1982) obtained a value of ~2.6×10^{-6} cm^2 s^{-1} for creatine phos- phate (molecular weight slightly less than that of MgATP and MgADP, probably also with a lower negative charge). In situ, for EGTA, which is also negatively charged, with a molecular weight sim- ilar to those of MgATP and MgADP, Elzinga et al. (1989a) obtained a value of ~3.3×10^{-6} cm^2 s^{-1} and Moisescu and Thielececk (1978) obtained a value of ~4.6×10^{-6} cm^2 s^{-1}. Hubley et al. (1996) studied the diffusion constant of MgATP, in vitro, in solutions of physiological ionic strength and pH, and obtained a value of ~3.7×10^{-6} cm^2 s^{-1}, that is, of the same order of magnitude as the scattered values obtained in situ. Averaging these six values gives a value of ~2.6 ± (0.6 or 1.5) $\times$ 10^{-6} cm^2 s^{-1} (mean ± SE or SD; n = 6). MgATP, MgADP, EGTA and creatine phosphate (except when complexed with Mg^{2+}) are all negatively charged, as are the myofilaments (see Section 4.4.2.2), and these anions are probably rapidly ejected from the myofilament lattice, because of strong electrostatic repulsive forces in the myo- filament lattice (principally between myofila- ments). Some of the diffusion constants, recalled here, were measured in situ and would therefore be higher than in a neutral aqueous medium of the same viscosity. Nonetheless, the interfilament medium is extremely complex (see pp. **27–32** in Section 3.2) and its viscosity is much higher than that of pure water (see Elliott and Worthington 2001, for a careful study of viscosity in the myo- fibrillar lattice; see also Grazi and Di Bona 2006, regarding the possible 'active' role of viscosity in muscle contraction). Thus, the diffusion of any molecule or ion, including negative ions, would be slower within a fibre than in a bulk aqueous medium. Considering these two antagonistic phenomena (electrostatically related 'acceleration' and viscosity-related 'deceleration') and taking into account the wide range of uncertainty on the six values recalled above, the best solution is to maintain the mean value of ~2.6 × 10⁻⁶ cm² s⁻¹ for the diffusion constant of MgATP and MgADP within the myofibrillar lattice. However, this estimate is only a rough order of magnitude and a minimal value of ~1.1 × 10⁻⁶ cm² s⁻¹ and a maximal value of ~4.1 × 10⁻⁶ cm² s⁻¹ (taking into account the SD found above; see Section 2.1 for comments on the use of SD rather than SE) would also be suitable. The mean value and these other two possible values are taken into account in the next paragraph.

According to the Stokes–Einstein law for spherical particles in a given neutral liquid medium, the diffusion constant is inversely proportional to the cubic root of the molecular weight (see Tanford 1967). Assuming that this law is also valid in the sarcoplasmic medium, taking a molecular weight of 40 Da for Ca^{2+} and neglecting the positive charge and the shape of this ion, the diffusion constant of Ca^{2+} would be ~ $(600/40)^{1/3}$ × 2.6 × 10⁻⁶ cm² s⁻¹ ~ 6.4 × 10⁻⁶ cm² s⁻¹. Using the other two possible values for the diffusion constant given at the end of the preceding paragraph, we obtain a minimal value of ~2.7 × 10⁻⁶ cm² s⁻¹ and a maximal value of ~10.1 × 10⁻⁶ cm² s⁻¹. From a series of experimental studies in aqueous media and organic solvents (with various viscosities) of approximately 100 inorganic neutral salts and organic compounds, with molecular weights between ~35 Da and ~500 Da, I found, during my PhD studies in the 1960s, that the diffusion constant of inorganic neutral salts and organic molecules was roughly inversely proportional to the square root of molecular weight, regardless of the shape of the compound and the viscosity of the medium. Thus, the diffusion constant for Ca^{2+} would be ~$(600/40)^{1/2}$ × 2.6 × 10⁻⁶ cm² s⁻¹ ~ 10.1 × 10⁻⁶ cm² s⁻¹, with a minimal value of ~4.3 × 10⁻⁶ cm² s⁻¹ and a maximal value of ~15.9 × 10⁻⁶ cm² s⁻¹. Ca^{2+} is positively charged, whereas the myofilaments are highly negatively charged (see the preceding paragraph). This charge difference almost certainly results in attractive electrostatic forces between Ca^{2+} and the myofilaments, and in a lower diffusion constant of Ca^{2+} in the sarcoplasmic medium than in a neutral medium of the same viscosity. Furthermore, the complexity of the interfilament

medium (see the preceding paragraph) almost certainly results in the further slowing down of Ca^{2+} diffusion. Thus, we have here two 'decelerating' phenomena and diffusion of Ca^{2+} in the sarcoplasmic medium is probably slowed down by a moderate factor of 3/4, for example. The six diffusion constants, estimated above, become therefore: ~$(3/4)$ × 2.7 × 10⁻⁶ cm² s⁻¹ ~ 2.0 × 10⁻⁶ cm² s⁻¹, ~$(3/4)$ × 6.4 × 10⁻⁶ cm² s⁻¹ ~ 4.8 × 10⁻⁶ cm² s⁻¹, ~$(3/4)$ × 10.1 × 10⁻⁶ cm² s⁻¹ ~ 7.6 × 10⁻⁶ cm² s⁻¹, ~$(3/4)$ × 4.3 × 10⁻⁶ cm² s⁻¹ ~ 3.2 × 10⁻⁶ cm² s⁻¹, ~$(3/4)$ × 10.1 × 10⁻⁶ cm² s⁻¹ ~ 7.6 × 10⁻⁶ cm² s⁻¹ and ~ $(3/4)$ × 15.9 × 10⁻⁶ cm² s⁻¹ ~ 11.9 × 10⁻⁶ cm² s⁻¹. On p. **30** in Section 3.2, it is strongly suggested that cell-associated water, also called living water, is probably a third state of water, between liquid water and ice. In this living water, the diffusion constants are assumed to be similar to those in 'bulk' liquid water, because (i) there is a lack of experimental data regarding many physical chemical properties of living water and (ii) the various diffusion constants measured or estimated in whole fibres and in bulk water do not differ greatly (see above and the preceding paragraph). Finch et al. (1971) reported three values for the self-diffusion constant of water in liquid water, between 0°C and 25°C, of ~25.0 × 10⁻⁶ cm² s⁻¹, ~25.0 × 10⁻⁶ cm² s⁻¹ and ~11.0 × 10⁻⁶ cm² s⁻¹. These values are close to that of ~22.5 × 10⁻⁶ cm² s⁻¹ (at 24°C) deduced by Yoshizaki et al. (1982) from the data obtained in situ by Mills (1973). Applying either the Stokes–Einstein law or my own experimental formula, recalled above, the expected eight values of the diffusion constant of Ca^{2+} would be ~$(3/4)$ × $(18/40)^{1/3}$ × (25.0, 25.0, 11.0, 22.5) × 10⁻⁶ cm² s⁻¹ ~ (14.4, 14.4, 6.3, 12.9) × 10⁻⁶ cm² s⁻¹ and $(3/4)$ × $(18/40)^{1/2}$ × (25.0, 25.0, 11.0, 22.5) × 10⁻⁶ cm² s⁻¹ ~ (12.6, 12.6, 5.5, 11.4) × 10⁻⁶ cm² s⁻¹. These scattered 6 + 8 = 14 estimates for the diffusion constant of Ca^{2+} in the myofibrillar medium can be averaged, as frequently done in this work, to give a new estimate of ~9.1 ± 2.6 × 10⁻⁶ cm² s⁻¹ (mean ± (SE + SD)/2; n = 14; see Section 2.1 concerning this unconventional statistical test). For the sake of simplicity, only the value of ~9.1 × 10⁻⁶ cm² s⁻¹ is selected below, in this section.

The volume of the cuvette containing the fibre bundles is ~2.5 ml (2.5 cm³) and the mean volume of each fibre bundle is ~8.1 × 10⁻⁵ cm³ (see Section 2.3). Assuming that the concentration of

Ca^{2+} is strictly zero in the surrounding buffer before diffusion out of the fibre bundles, the final concentration of Ca^{2+}, in both the external and the internal media, would have been $\sim 10^{-7.8}$ M $\times$ 8.1×10^{-5} cm^3/2.5 cm$^3 \sim 10^{-12.3}$ M (pCa $\sim$ 12.3; see p. **124** for precise details on the concentration of $\sim 10^{-7.8}$ M, pCa $\sim$ 7.8). The hypothetical pCa value of $\sim$12.3 clearly cannot be reached, because the concentration of contaminating Ca^{2+} in the bathing buffer corresponds to a pCa of $\sim$9.5, which is sufficient to ensure relaxation (see Section 2.3). At this point, a problem should be resolved concerning the kinetics of the process for passing from the highest concentration of Ca^{2+} ($\sim 10^{-7.8}$ M, pCa $\sim$ 7.8, probably corresponding to partial activation) to the lowest concentration ($\sim 10^{-9.5}$ M, pCa $\sim$ 9.5, certainly corresponding to full relaxation). This question is considered in the next paragraph.

Citing a book entitled *Conduction of Heat in Solids*, Elzinga et al. (1989a) suggested that 'the steady-state radial distribution of orthophosphate will be approximated for times τ^* (since the start of contraction) such that diffusion constant $\times$ τ^*/ (radius)^2 > 0.60 [i.e. τ^* > 0.60(radius)^2/diffusion constant]'. Handbooks on matter transfer in liquids state that the duration of transfer of a molecule into a cylinder without a membrane-like barrier and containing a network of very porous material ($\sim$80% liquid, for instance) is $\tau^* \sim 0.50$(radius)^2/diffusion constant. Only this second relationship, comparable to that recalled by Elzinga et al. (1989a), is considered here. It is, in principle, valid for small molecules, ions and inorganic compounds of molecular weight not exceeding $\sim$1–2 kDa, but I also use it below, as a rough approximation, for α-chymotrypsin (MW $\sim$ 22 kDa). As demonstrated in Section 3.2 (see pp. **27–32**), recalled on p. **123** and studied further on pp. **251–254** in Section 8.9, the interfilament lattice is a very complex medium, with a structure of the 'liquid medium' between that of water and ice ('living water' or 'cell-associated water'), with many myosin heads lying between the thick myosin and thin actin filaments, both at rest and under contraction conditions. However, the expression of τ^* is maintained. The mean radius of the fibre bundles is $\sim$ (185) μm/2 $\sim 9.25 \times 10^{-3}$ cm (see Section 2.3). Introducing the mean diffusion constant, estimated on p. **125**, for Ca^{2+} ($\sim$9.1 $\times$ 10^{-6} cm^2 s^{-1}), τ^* would be $\sim 0.50 \times 9.25^2 \times 10^{-6}$ cm^2/9.1 $\times 10^{-6}$ cm^2 s$^{-1} \sim 4.7$ s, which is only $\sim$8%

of the duration of the increase in force ($\sim$60 s; see, in Section 4.3.1, p. **120** and Figure 4.3) and only $\sim$3% of the total duration ($\sim$160 s; see, in Section 4.3.1, p. **120** and Figure 4.3). Thus, the decrease in Ca^{2+} within the fibre bundles (from $\sim 10^{-7.8}$ M to $\sim 10^{-9.5}$ M; see the preceding paragraph) is much more rapid than the changes in transitory forces, even if the fragments of sarcoplasmic reticulum ($\sim$25%; see p. **123**) and the mitochondria are entirely digested. A possible long-lasting increase in the concentration of Ca^{2+} within the fibre bundles can therefore be ruled out. Moreover, the presence of $\sim$25% of the sarcoplasmic reticulum and the digestion of the mitochondria are highly unlikely: we can definitively conclude that Ca-induced contractions should not be taken into account and that the small transitory contractures observed here resulted solely from the digestion process (see the next paragraph) of the active weakly bound cross-bridges present at rest (see the first paragraph in Section 4.3.1 for references concerning these structures), with the subsequent disappearance of the radial tethering forces exerted between the thin actin and thick myosin filaments.

Along the same lines, does α-chymotrypsin (MW $\sim$ 22 kDa) penetrate the fibre bundles rapidly? The diffusion constant of a myosin head monomer (MW $\sim$ 108 kDa) in solution is $\sim 4.43 \times 10^{-7}$ cm^2 s^{-1} and that of the dimer (MW $\sim$ 213 kDa) is $\sim 2.54 \times 10^{-7}$ cm^2 s^{-1} (Morel et al. 1998b). Applying the 'cubic' and 'square' formulae given on p. **125** (i.e. above in the section), for the diffusion constant, the mean value for α-chymotrypsin would be $\sim 7.67 \times 10^{-7}$ cm^2 s^{-1} (deduced from the values for the monomer and the dimer; calculation not shown). Applying the formula given at the beginning of the preceding paragraph, using the values of the radius of the fibre bundles also given in the preceding paragraph, and introducing the estimated diffusion constants, we obtain a rough estimate of τ^* $\sim$ 56 s, which is close to the duration of $\sim$60 s for the increase in force (see the preceding paragraph). This estimate indicates that the diffusion of α-chymotrypsin into the fibre bundles is a major rate-limiting phenomenon in the digestion process. However, the duration of the digestion of the myosin heads should be added to this estimate, although this process is probably very rapid. A more 'physiological' experimental result was reported by He et al. (1997) in their Figure 3 and its legend. Working at 12°C and using a single permeabilised rabbit fibre with a cross-sectional area of ~4.8 × 10⁻⁵ cm², corresponding to the square of the radius of ~1.53 × 10⁻⁵ cm², the authors found that the half-time of diffusion of MDCC-PBP in the permeabilised fibre was ~15 s (see legend to their Figure 3A), corresponding approximately to a full time $\tau^* \sim 2.5 \times 15$ s ~ 37.5 s (the multiplication coefficient of ~2.5 is not given by the authors) for the completion of the diffusion process of MDCC-PBP. MDCC-PBP is a coumarin-labeled A197C mutant of the phosphate-binding protein from *Escherichia coli*, described by Brune et al. (1994). The molecular weight of this protein is ~35 kDa. Applying the formula presented at the beginning of the preceding paragraph for τ^* and using again the 'cubic' and 'square' formulae recalled above, the mean value of the diffusion constant of MDCC-PBP would be ~7.67 × 10⁻⁷ cm² s⁻¹ × (22 kDa/35 kDa)$^{1/2 \text{ or } 1/3}$ ~6.33 × 10⁻⁷ cm² s⁻¹ (calculations not shown; see above concerning the values of 22 kDa and ~7.67 × 10⁻⁷ cm² s⁻¹ for α-chymotrypsin). A mean value of the full time for completion of the diffusion process in the case of fibre bundles and α-chymotrypsin, studied here, would be ~37.5 s × (9.25² × 10⁻⁶ cm²/1.53 × 10⁻⁵ cm²) × 6.33 × 10⁻⁷ cm² s⁻¹/7.67 × 10⁻⁷ cm² s⁻¹ ~ 173 s (see the preceding paragraph and above concerning the values of ~9.25² × 10⁻⁶ cm², ~1.53 × 10⁻⁵ cm² and ~37.5 s). This estimate is greater than the durations of ~60 s and ~160 s, recalled in the preceding paragraph, for the experiments of digestion with α-chymotrypsin in fibre bundles. At this stage, I note that, in their Figure 3B, He et al. (1997) found that, with fibre oscillation, diffusion into the fibre was accelerated by a factor of ~3 (estimate not given by the authors). In the case of fibre bundles, the buffers in the cuvette containing the bundles were stirred to accelerate the diffusion of α-chymotrypsin into the bundles (see Section 2.3). This is, at least to some extent, equivalent to the oscillation in the case of the fibre studied by He et al. (1997), in terms of fluid mechanics. The estimate of ~173 s should therefore be divided by the factor of ~3 and becomes ~58 s, which is reasonably consistent with the observed values of ~60 s and ~160 s recalled in the preceding paragraph. The estimate obtained here for the diffusion of α-chymotrypsin into the fibre bundles (~58 s) again indicates that a large part of the digestion process is controlled by diffusion. This adds complexity to the whole process, together with the other problems raised in this and the preceding paragraphs.

Before performing the digestion experiments described in Section 4.3.1, on fibre bundles, I recorded both cross-sectional areas after initial swelling (inevitable after permeabilisation; see Section 8.7; see also Section 2.3) and isometric tetanic forces on specially prepared fibre bundles (not used for digestion), at room temperature (~20°C–26°C; see Section 2.3). The tensions were ~0.9 ± 0.6 × 10⁵ N m⁻² (mean ± SD; n = 10). On p. **97** in Section 3.8.1, it is shown that, at 10°C, for single demembranated fibres from various types of white muscle of various animal species (mostly frog and rabbit), the isometric tetanic tension is ~1.4 ± 0.6 × 10⁵ N m⁻² (mean ± SD; n = 40), corresponding to values (maximal interval) of ~1.4 × 10⁵ N m⁻² to ~1.5 × 10⁵ N m⁻² at room temperature (~20°C–26°C; in these conditions, $Q_{10} \sim 1.8$; see p. **246** in Section 8.8). The tensions recorded here are therefore significantly lower than those obtained for single demembranated fibres (70%–75% lower, taking into account only the mean values). This is probably attributed to the use of red muscles (as opposed to white muscles), with the presence of more non-contractile structures in the fibres (e.g. mitochondria; see p. **123** for precise details) and, possibly, a proportion of buffer lying between individual fibres in the bundles, thus reducing the active cross-sectional area. Finally, in the hybrid model, at 10°C, ~40% of the isometric tetanic tension is related to lateral swelling processes, corresponding to the translation of the radial net repulsive forces into axial tensions. It is assumed that this proportion is the same at ~20°C–26°C. The electrostatically related axial tensions are therefore ~0.40 × (0.9 ± 0.6) × 10⁵ N m⁻² ~ 0.36 ± 0.24 × 10⁵ N m⁻² ~ 360 ± 240 g cm⁻², before digestion.

The action of α-chymotrypsin leads to a 'geometric' deficit in isometric tetanic tension, owing to a gradual decrease, clearly observed under the stereomicroscope, in the cross-sectional areas of the fibre bundles, which reach sizes of only ~50% of their initial size at the end of the experiments (duration of ~160 s; see p. **120**). In Figure 4.3 and its legend, the ratio between the final normalised force and the initial normalised force is ~1.5 mg/5 mg ~ 0.30 (~30%). This ratio of ~30% is lower than the ~50% decrease in cross-sectional area, demonstrating that geometric phenomena are not sufficient to account for the full loss of force (see the next paragraph). It is demonstrated in the preceding paragraph that ~40% of the tensions recorded before digestion correspond to ~360 ± 240 g cm^{-2} and, therefore, after digestion, to ~(1.5 mg/5 mg) × (360 ± 240) g cm^{-2} ~ 108 ± 72 g cm^{-2}. Thus, as a rough approximation, the reference 'mean' isometric tetanic tension over the total duration of digestion may be taken as ~(360 ± 240 + 108 ± 72)/2 ~ 234 ± 156 g cm^{-2}. The three fibre bundles corresponding to trace A in Figure 4.3 had approximately maximal cross-sectional areas, as deduced from diameter measurements, of ~2.7 × 10^{-4} + 1.1 × 10^{-4} ~ 3.8 × 10^{-4} cm^2 (see Section 2.3; SD is chosen for the sake of homogeneity throughout this reasoning), before digestion with α-chymotrypsin, and, therefore, ~50% × 3.8 × 10^{-4} cm^2 ~ 1.9 × 10^{-4} cm^2 at the end of the experiments. Thus, the average reference cross-sectional area, over the total duration of digestion, can be taken as ~(3.8 × 10^{-4} + 1.9 × 10^{-4}) cm^2/2 ~ 2.8 × 10^{-4} cm^2 and the isometric tetanic force can be estimated at ~234 ± 156 g cm^{-2} × 2.8 × 10^{-4} cm^2 ~ 66 ± 44 mg. From Figure 4.3, it appears that the normalised peak force for curve A corresponds to only ~9 mg. Otherwise, the normalised reference resting force, before digestion, is 5 mg, and that after digestion is 1.5 mg (see legend to Figure 4.3). The approximate resting tension can therefore be taken as ~(5.0 + 1.5)/2 ~ 3 mg. Thus, the peak force shown on curve A in Figure 4.3 is approximately ~9 mg − 3 mg ~ 6 mg, so the maximal expected axial force resulting from radial repulsive electrostatic forces would be approximately ~(66 ± 44) mg/6 mg ~ 11.0 ± 7.3 times the corresponding peak force shown in Figure 4.3. There is therefore a gap between the expected and recorded values. Owing to the complexity of the digestion process, the maximal force, recorded on three fibre bundles (curve A in Figure 4.3), is large enough for a clear conclusion to be drawn (see the next paragraph), but the strength of the transitory forces recorded on the other fibre bundles requires additional comments (see the next paragraph, regarding the many problems raised by the digestion process).

Why is the digestion process so complex? On pp. **126–127**, it is demonstrated that the diffusion of α-chymotrypsin into the fibre bundles makes a major contribution to this complexity. However, other processes, concerning digestion itself, are involved in the delicate problem analysed in this paragraph. We see, in Section 4.3.1, that all the weakly bound cross-bridges, including the active weakly bound cross-bridges (see first paragraph in Section 4.3.1 for references on these structures), are rapidly digested. However, most thick myosin filaments are probably also rapidly disrupted or destroyed, slightly later, with a probable overlap between these two processes. This 'digestion' of the thick myosin filaments certainly results from extensive digestion of all the heads at the S1–S2 joint and also, presumably, of all the tails at the S2–LMM joint (see Figure 5.1 for the definition of these subfragments). The gradual disappearance of many negatively charged thick myosin filaments greatly reduces their contribution to net radial repulsive forces. Furthermore, titin and other structures (e.g. M lines and Z discs) may be, at least partly, digested with α-chymotrypsin. These putative digestions would lead to considerable disorder in the arrangement of the thin actin and remaining thick myosin filaments and a considerable decrease in the radial repulsive electrostatic forces. Thus, the destruction of many thick myosin filaments, disorder in the thin actin and remaining thick myosin filament arrays, and the decrease of ~50% in cross-sectional area (see the preceding paragraph), observed at the end of the digestion process of the fibre bundles, result in (i) the final normalised force recorded being only ~30% the initial normalised force recorded, (ii) only very small transitory contractures being observed and (iii) peak forces being markedly lower than expected, as pointed out in the preceding paragraph. Nevertheless, as shown in Figure 4.3, very small transitory contractures are detected when the radial tethering forces exerted at rest are abolished. In some cases, these contractures are extremely weak or even almost negligible (see Figure 4.3, its legend and the above discussion). However, in all the cases studied, there are contracture forces of various magnitudes, providing experimental support for the existence of radial tethering forces at rest. The most striking observation is that these radial tethering forces are detected despite the many unfavorable phenomena occurring during the digestion process.

Morel

4.4 INTERPRETATION OF THE EXPERIMENTS PERFORMED ON HALF-FIBRES FROM YOUNG ADULT FROGS. COMPARISON WITH PREVIOUS EXPERIMENTAL DATA OBTAINED BY INDEPENDENT GROUPS. COMBINATION OF VARIOUS BIOLOGICAL MATERIALS

4.4.1 Preliminary Remarks on Resting and Active Tensions

As described on pp. **116–117** in Section 4.2.2, Figure 4.1 shows results of experiments performed on the first half-fibre (●). The arrow Ca indicates the injection of a buffer with a bulk ionic strength of 180 mM, containing sufficient calcium to trigger full isometric tetanic contraction (pCa ~ 4.5). The corresponding trace, carefully drawn by eye from the true trace, corresponds to ~40% of this true trace (see p. **117** in Section 4.2.2). Considering this proportion, and taking into account the standardised mean cross-sectional area for the three half-fibres used in this experimental study (~5.67 × 10⁻⁵ cm²; see p. **13** in Section 2.2), the maximal electrostatically related tension is ~1.2 ± 0.2 × 10⁵ N m⁻² (mean ± SD; n = 6; two values were recorded per half-fibre), corresponding to a full actually recorded isometric tetanic tension of ~(1/0.4) × (1.2 ± 0.2) × 10⁵ N m⁻² ~ 3.0 ± 0.5 ×10⁵ N m⁻². This value of ~3.0 ± 0.5 ×10⁵ N m⁻², at 10°C, for the three half-fibres in the presence of MS⁻, corresponding to a bulk ionic strength of 180 mM (approximate in vivo/in situ ionic strength in the sarcoplasmic medium for most vertebrate intact skeletal muscle fibres; e.g. Godt and Maughan 1988; Godt and Nosek 1989; Xu et al. 1993), is identical to that recorded in intact fibres (~3.0 × 10⁵ N m⁻²; see Section 3.3). We can therefore take as a reference the value of ~3.0 × 10⁵ N m⁻². The behaviour of half-fibres in the presence of the benign anion MS⁻, used here as the major anion, confirms that both half-fibres and MS⁻ (see pp. **13–14** in Section 2.2) constitute the most appropriate experimental model for comparing demembranated and intact fibres (see also p. **221** in Section 7.2, and Section 8.7 concerning the unavoidable swelling of traditionally demembranated fibres in standard buffers, by contrast to half-fibres in the presence of MS⁻ as the major anion).

The three half-fibres are only slightly stretched (see p. **14** in Section 2.2) and titin almost certainly plays a minor role in the resting tension (see p. **142** in Section 4.4.2.3 for details about titin). Under resting conditions, at 10°C, and at I ~ 20–30 mM (I* ~ 22–32 mM), all the active weakly bound cross-bridges (see first paragraph of Section 4.3.1 for references on these structures) are detached (see pp. **164–167** in Section 4.4.2.8 for an in-depth discussion of this point). Thus, the resting forces recorded below I* ~ 22–32 mM are related to net repulsive forces between the thick and thin filaments, resulting solely from radial repulsive electrostatic forces minus some weak radial attractive/compressive forces (see Sections 4.4.2.3, 4.4.2.4, 4.4.2.5.3 and 4.4.2.5.4 concerning the radial repulsive electrostatic and attractive/compressive forces), as there are no swinging cross-bridges (resting conditions) and the active weakly bound cross-bridges are all detached, as recalled above. Moreover, on p. **159** in Section 4.4.2.5.4, it is demonstrated that, at physiological bulk ionic strength (180 mM), the weak radial attractive/compressive force Λ_r is only ~1.11% of the radial repulsive electrostatic force $4(\gamma/2)F_{e,r}$. For simplification, this proportion is assumed to be valid at any value of ionic strength and, therefore, negligible. Between I ~20 mM and ~30 mM (I* ~ 22–32 mM), the recorded resting forces are ~160 ± 40 mg (mean ± SD; n = 6). The standardised mean cross-sectional area of the three half-fibres used, in the relaxing buffer with I = 180 mM (I* ~ 182 mM), was ~5.67 × 10⁻⁵ cm² (see the preceding paragraph). At I* ~ 22–32 mM, the half-fibres are considerably swollen (+44% in cross-sectional area; see p. **166** in Section 4.4.2.8) and the resting forces of ~160 ± 40 mg correspond to resting tensions of ~ (160 ± 40) mg/(1.44 × 5.67 × 10⁻⁵ cm²) ~ 20 ± 5 × 10⁵ mg cm⁻² ~ 2.0 ± 0.5 × 10⁵ N m⁻². Ignoring the active weakly bound cross-bridges and, therefore, the resting radial tethering forces, $N_0F_{cb}^0$, at any value of I* and neglecting the radial attractive/compressing forces Λ_r (see above), the radial forces are generated, in these conditions, from the radial repulsive electrostatic forces, $F_{e,r}$, given by Equation 3.2 (with x = r) (or $4(\gamma/2)F_{e,r}$; see above). In Equation 3.2, it is assumed, from experimental data, that $H_r \sim H_a = H \sim 0.26$ mM⁻¹/² (see Section 4.4.2.7, text above Equation 4.15). On p. **163** in Section 4.4.2.7, it is strongly suggested that the parameter Q_a in Equation 3.2 (under contraction conditions) is not very sensitive to I*, and this inference can reasonably be extended to resting conditions, that is, to Q_r. Thus, Equation 3.2 gives $F_{e,r}(182)/F_{e,r}(22–32) \sim 0.50–0.55$. The virtual resting tension, $T_r^*(182)$, that would be recorded in the absence of radial tethering forces and radial attractive/compressive forces is therefore given by $T_r^*(182)/T_r(22–32) \sim F_{e,r}(182)/F_{e,r}(22–32) \sim 0.50–0.55$, from which we deduce that $T_r^*(182) \sim (0.50 – 0.55) \times (2.0 \pm 0.5) \times 10^5$ N m^{-2} $\sim 0.75 – 1.38 \times 10^5$ N m^{-2} (maximal interval). It is recalled, in the preceding paragraph, that the best estimate for total isometric tetanic tension in a half-fibre is $T_{hf}(182) \sim 3.0 \times 10^5$ N m^{-2}. Thus, we deduce that $T_r^*(182)/T_{hf}(182) \sim (0.75 – 1.38) \times 10^5$ N m$^{-2}/3.0 \times 10^5$ N m^{-2} $\sim 0.25 – 0.46$ ($\sim 25\% – 46\%$). Owing to the accumulation of experimental errors and shortcomings in the theoretical approach, these estimated ratios are of the same order of magnitude as the $\sim 40\%$ used throughout this work. Moreover, the wide interval between $\sim 25\%$ and $\sim 45\%$ includes the value of $\sim 40\%$. Thus, it can be considered that $\sim 40\%$ of the contractile force exists before contraction (one-to-one conversion of the radial repulsive forces into axial contractile forces; see pp. **102–103** in Section 3.10). This kind of dichotomy in isometric tetanic tension generation is a key element of this book. The radial repulsive electrostatic forces are counterbalanced by radial tethering and attractive/compressive forces at rest (mostly radial tethering forces; see pp. **174–176** in Section 5.1.1 for precise details on these forces) and are 'liberated' during contraction. This electrostatically related part of the axial isometric tetanic tension is potentially present at rest, and resting conditions would correspond to a kind of active state. The electrostatically related part of contraction may therefore be seen as a 'downhill process', corresponding, at first sight paradoxically, to an inactive state occurring during contraction.

The dotted line in Figure 4.2 was calculated from Equation 4.15 and represents the best fit to the experimental points reported by April and Brandt (1973) in their Figure 5, divided by a factor of 2.5 (see p. **117** in Section 4.2.2 for precise details on this factor). Figure 5 in the paper by these authors corresponds to many contracting intact and some mechanically skinned fibres from the walking leg of crayfish (Orconectes) (see pp. **105–109** in Section 4.1 for a discussion of the paper by April and Brandt 1973). It may be argued that these fibres behave differently from those of frog used here. However, such differences would

disappear when relative tensions are studied (see p. **95** in Section 3.7). This is demonstrated by the two experimental points (Δ) obtained for a single frog half-fibre (see p. **119** in Section 4.2.2), both of which correspond to $\sim 40\%$ of the recorded isometric tetanic tension. These two points are located exactly on the dotted line deduced from the results presented by April and Brandt (1973), via Equation 4.15.

We can assume, as a first approximation, that the difference between vicinal and bulk ionic strengths remains ~ 2 mM, regardless of the bulk ionic strength I, for any type of fibre, at rest or during contraction (see Section 3.9). This assumption is justified, owing to the large scattering in the resting tensions recorded here and the active tensions presented by April and Brandt (1973). As noted in the preceding paragraph, the dotted line in Figure 4.2 corresponds to $\sim 40\%$ of the recorded isometric tetanic tensions (below a bulk ionic strength of ~ 50 mM, the lowest ionic strength used by April and Brandt 1973, the dashed line corresponds to extrapolated values obtained from Equation 4.15). Indeed, there are many consistent experimental and semi-empirical data in this monograph, demonstrating that $\sim 40\%$ full isometric tetanic tension is related to net radial repulsive (expansive) forces. These expansive forces result from subtraction of the collapsing, residual radial tethering forces and radial attractive/compressive forces between the thin actin and thick myosin filaments from the radial repulsive electrostatic forces (see Sections 4.4.2.3, 4.4.2.4, 4.4.2.5.3 and 4.4.2.5.4; pp. **174–176** in Section 5.1.1; and pp. **180–181** in Section 5.1.2 for many precise details on these forces) and their translation into axial contractile forces (see pp. **102–103** in Section 3.10). The experimental and semi-empirical results presented in this section (see the preceding paragraph) confirm that $\sim 40\%$ of the isometric tetanic tension is related to net radial repulsive (expansive) forces (mostly electrostatic), at least at 10°C. This proportion of $\sim 40\%$ is probably not very different at the various temperatures used in the laboratory, between ~ 0°C and ~ 20°C–22°C (room temperature). The major role of the electrostatically related axial isometric tension clearly accounts for the dependence of full relative isometric tension on ionic strength. Furthermore, my own two experimental points (Δ) in Figure 4.2 and the many experimental data reported by April and Brandt (1973) demonstrate that the net radial repulsive (expansive) forces can be turned into axial forces in actively contracting intact fibres, as also demonstrated under resting conditions (see principally Section 4.4.2.8).

4.4.2 Biophysical Interpretation of the Experiments Concerning Resting Forces and Active Forces, during Isometric Tetanic Contraction (Short Tetani–Steady State). Corresponding Relative Tensions

4.4.2.1 Preliminary Remarks on Radial Forces and the Myofilament Lattice in Demembranated Fibres and Intact Unit Cells. Insight into the Singular Behavior of Demembranated Fibres

The main aim of Section 4.4.2 is to analyse, discuss and estimate the various forces in the myofibrillar lattice, at rest and during isometric tetanic contraction. Before going into detail, it is interesting to discuss the question of the radial forces (repulsive, tethering, collapsing and attractive/compressive) occurring in demembranated/permeabilised fibres (the term demembranation is mostly used below, in this section, for the sake of simplicity), under contraction conditions, when the osmotic pressure is changed. Radial forces, recorded on single demembranated fibres, mostly from rabbit psoas muscle, have been attributed, in the swinging cross-bridge/lever-arm theories, exclusively to rotation of the attached cross-bridges and to the various states of these structures (e.g. Bagni et al. 1994b; Brenner and Yu 1991; Goldman and Simmons 1986; Matsubara et al. 1985; Xu et al. 1993). In this section and in Sections 3.8 and 8.7 (see also p. **13** in Section 2.2), it is demonstrated that many-faceted phenomena occur in demembranated fibres. For instance, the systematic lateral swelling of these fibres, triggered by demembranation, leads to several complex phenomena, accounted for by the hybrid model (described in Chapter 5 and discussed in Chapter 8). Otherwise, technical problems probably affect the interpretation of the experiments performed on demembranated fibres. Indeed, in the 1980s and at the beginning of the 1990s, these experiments were carried out with x-ray machines with long and very long exposure times. Xu et al. (1993), for example, using a traditional x-ray source, performed their experiments over periods of many seconds (generally 500 s). Using the intense x-ray source generated by the synchrotron DESY (Hamburg, Germany), Brenner

and Yu (1991) pointed out that 'the exposure time was... generally 5 s'. At many places in this book, I focus on steady-state conditions of plateau isometric tetanic tension, typically corresponding to durations (mostly in intact frog fibres) of ~1.0–1.5 s or slightly less (e.g. Altringham et al. 1984; Bagni et al. 1988b; Edman 1979; Edman and Hwang 1977; Edman and Reggiani 1987; Elzinga et al. 1989a,b; Gordon et al. 1966a,b; Granzier and Pollack 1990; He et al. 1997; Martin-Fernandez et al. 1994; Piazzesi et al. 1999). In this context, it is shown, on p. **54** in Section 3.4.3.2, that the time Δt^* required to reach a stable value of the isometric tetanic tension is ~0.17–0.40 s (this estimate results from a combination of results obtained with permeabilised frog fibres, intact frog fibres and whole frog muscles, at a mean temperature of 5.7°C). To the best of my knowledge, no experimental data are available, in these domains, for permeabilised fibres from rabbit psoas muscles. In any event, combining frog and rabbit, we can assume that some of the problems raised in this section may be attributed to the use of long and very long exposure times in the x-ray experiments of Brenner and Yu (1991; 5 s) and Xu et al. (1993; 500 s) with respect to the various durations of short tetani recalled above (between ~0.17–0.40 s and ~1.0–1.5 s). However, I suggest that the experimental results obtained with demembranated fibres in the 1980s and at the start of the 1990s should be considered genuine. This view is supported by the similar behaviour of all demembranated fibres and the similar conclusions drawn by the authors cited at the beginning of this paragraph.

Xu et al. (1993) recalled that several independent groups found that, in demembranated fibres (these fibres are swollen upon demembranation, before the experiments are carried out, as recalled in the preceding paragraph), the interfilament spacing decreases with the passage from resting conditions to isometric contraction. I think that Xu et al. (1993) and other investigators did not take into account the initial lateral swelling resulting from demembranation and therefore did not work on intact fibres, but on complex biological materials, as demonstrated in this section (see also Sections 3.8 and 8.7). The situation described by Xu et al. (1993) is similar to that regarding the phenomena, first observed by Shapiro et al. (1979) and further studied by Matsubara et al. (1985) and Brenner and Yu (1985, 1991), concerning the shift from rest to contraction, and studied by Brenner et al. (1984), Matsubara et al. (1984), Maughan and Godt (1981b) and Umazume and Kasuga (1984), concerning the shift from rest to rigor. As briefly recalled in the preceding paragraph, working on demembranated fibres leads to complex phenomena, frequently resulting in self-contradictory conclusions. The many problems raised by demembranated fibres include their systematic lateral shrinkage when passing from rest to contraction (see above). By contrast, this phenomenon is not observed in intact fibres, which display mild lateral swelling or a putative tiny shrinkage in a single case (see pp. **4–5** in the Introduction). A discussion should be proposed at this juncture.

It is important to compare the results obtained with demembranated fibres to the behavior of intact unit cells included in intact fibres, because these unit cells represent an intact biological material (structurally, mechanically, biophysically and biochemically), by contrast to demembranated fibres, which are profoundly modified/impaired biological materials. It is useful to write Equation 4.8, valid for an intact unit cell belonging to an intact fibre (from frog, rabbit or any other animal species), at the 'slack length', that is, at full overlap γ, and in the absence of external osmotic pressure, in the form of two equations, as follows:

At rest $(x = r)$: $\sum F_r \sim 4(\gamma/2)F_{e,r} - (N_r F_{cb}^r + \Lambda_r)$

$$(4.1)$$

During contraction $(x = a)$:

$$\sum F_a \sim 4(\gamma/2)F_{e,a} - (N_a F_{cb}^a + C_a + \Lambda_a) \qquad (4.2)$$

At rest, there is no cyclic tilting of the heads of weakly bound cross-bridges and the collapsing force C_r is zero (the notion of collapsing force is defined in the fourth point of Section 4.4.2.4). During contraction, there is a rapid cyclic active rotation of the attached cross-bridges and the weak collapsing force C_a should be taken into account (see p. **157** in Section 4.4.2.5.4 for quantitative estimates). In intact unit cells, resting conditions correspond to equilibrium of the myofilament lattice and the absence of any net radial force. Thus, in Equation 4.1, we necessarily have $\Sigma F_r = 0$. From Equation 4.2, it can be deduced that $\Sigma F_a > 0$, as demonstrated throughout this monograph (net radial expansive force

due mostly to the drastic reduction of the radial tethering force, $N_a F_{cb}^a$). Thus, during contraction, there is a net radial 'driving' force that must result in an increase in the lattice spacing, in an intact unit cell. More precisely, when shifting from rest to contraction (particularly isometric tetanic contraction), in an intact unit cell, the net active radial force is $\Sigma F_a - \Sigma F_r = \Sigma F_a > 0$, corresponding to a radial expansive active force. This is at odds with the 'evidence' and discussions presented by many authors for demembranated fibres, claiming a decrease in the spacing between myofilaments during the transition from the relaxed state to contraction (see the preceding paragraph, particularly the quote from Xu et al. 1993). To my mind, this is further demonstration that intact unit cells and demembranated fibres do not behave similarly. Nonetheless, one may argue that it is demonstrated, on pp. **181–183** in Section 5.1.2 (concerning latency relaxation and elongation in intact fibres), that there is a transitory, very small, 'shrinking' radial force (resulting in a tiny decrease in the lattice spacing that cannot be detected, even with x-ray diffraction methods based on synchrotron radiation; see pp. **181–183** in Section 5.1.2 for a circumstantial discussion), before any sign of active contraction, during the first few milliseconds after electrical stimulus in intact fibres (and also intact unit cells). However, this is not inconsistent with the existence of a radial expansive active force strongly suggested above, under well-identified isometric tetanic contraction conditions and under steady-state conditions. Otherwise, the initial, transitory, slight lateral shrinkage mentioned here, which lasts a few milliseconds, has nothing to do with the lateral shrinkage of demembranated fibres in transition from rest to isometric contraction (see the beginning of the preceding paragraph). Indeed, durations of 5 s and 500 s for x-ray experiments performed on demembranated fibres are recalled in the first paragraph of this section, concerning observations of this transition. Under these conditions, the extremely short latency period is entirely masked, because the time scales of this period, on the one hand, and of the phenomena observed by x-ray diffraction, on the other, are very different. Thus, only experimental results obtained by x-ray diffraction and in well-identified isometric tetanic contraction conditions (after the latency period) can be compared.

It is demonstrated, in the preceding paragraph, that there is an inconsistency between the simultaneous existence of a net radial expansive force in a contracting intact unit cell and a decrease in myofilament lattice spacing, when demembranated fibres shift from relaxation to contraction. At this stage, it should be recalled that Morel (1985a) and Morel and Merah (1997) claimed that skinned fibres cannot be used to draw firm conclusions and, even, that such conclusions are erroneous, concerning lattice spacing. This view is confirmed in Section 3.8, where it is recalled that distortions, disruptions and impairments (even disappearance of thick myosin filaments at the periphery of the myofibrils belonging to demembranated fibres) almost certainly occur. In this context, as already pointed out by Morel (1985a) and Morel and Merah (1997), the use of x-ray diffraction techniques with demembranated fibres would result in misleading conclusions, if, for example, osmotic pressure is steadily increased. Morel and Merah (1997) expressed doubts about whether it was possible to pass from compressive to expansive radial forces in a single demembranated fibre subjected to a steady increase in osmotic pressure (see Brenner and Yu 1991; Xu et al. 1993). We also had reservations about the detailed roles of the cross-bridges, as strongly suggested by these two groups. Taking into account only swinging cross-bridge/lever-arm processes as the sole possible model, most specialists have attributed the phenomena of compression and expansion exclusively to tilting and to the various enzymatic and mechanical properties of the attached cross-bridges, without considering other possible phenomena. Thus, many of the statements of several authors (see, for instance, Xu et al. 1993 and references therein) are probably incorrect, as the alternative approach described in this monograph results in the self-consistent, predictive and explanatory hybrid model, supporting the various assumptions, explanations, reasoning and discussions presented in this work. On pp. **133–137**, the major problem of probable artefacts resulting from the study of demembranated fibres by x-ray diffraction, under conditions of steadily increasing osmotic pressure, is reexamined and discussed.

Despite the major qualifications described above, in this section, some of the experimental results obtained with demembranated fibres from rabbit psoas muscle by Xu et al. (1993) are potentially convincing. Before going into detail, it should be recalled that introducing appropriate amounts of impermeant molecules (e.g. dextran, PVP) into bathing buffers causes the lattice spacing to return to 'pre-demembranation' values (see also pp. **244–245** in Section 8.7). As frequently recalled in this section, drawing general conclusions from studies of swollen demembranated fibres is risky. A good solution would be to perform preliminary experiments on compressed demembranated fibres that have recovered their pre-demembranation lattice spacing. Subsequent complementary osmotic compression would give interesting and, probably, surprising results. Regardless of this major stumbling block, Xu et al. (1993) stated that the decrease in myofilament spacing when passing from rest to contraction suggests that 'force is generated by attached cross-bridges'. One of the major arguments presented by the authors in support of this assertion is that, at non-overlap, lattice spacing is not influenced by physiological state (relaxation, contraction or rigor). The non-overlap state studied by Xu et al. (1993), at 5°C–7°C, corresponded to sarcomere lengths >4.0 μm. Ignoring the inconsistency denounced above, in this section (simultaneous decrease in myofilament lattice spacing and existence of a radial expansive driving force when shifting from rest to contraction; see, for instance, pp. **131–132**), the absence of attached cross-bridges at zero overlap could be seen to support the assumption of Xu et al. (1993). However, the problem is much more complex than thought by the authors and most other specialists in muscle and muscle contraction. Indeed, considering again an intact unit cell belonging to an intact fibre, in the absence of external osmotic pressure (as explained on p. **132**), and taking into account the hybrid model of force generation, it can be deduced from Equation 4.1 that, when $\gamma^* = 0$ (non-overlap state; see pp. **34–35** in Section 3.2 for comments on γ and γ^*), the radial repulsive electrostatic force $4(\gamma^*/2)F_{e,r}$ (at rest) is mathematically zero (see the following paragraphs for reservations concerning this statement). On the other hand, at zero overlap, the number N_r (N_0) of active weakly bound cross-bridges (see pp. **170–172** in Section 5.1.1 for definitions) is obviously also zero. Thus, Equation 4.1 reduces to $\Sigma F_r(\gamma^* = 0) \sim -\Lambda_r(\gamma^* = 0) < 0$: there are solely weak radial attractive/compressive forces, with no possibility of repulsive forces, resulting therefore in a small lateral shrinkage of the myofilament lattice, inconsistent with the myofilament lattice being at equilibrium at rest. The same reasoning can be applied under isometric tetanic contraction conditions. In this case ($\gamma^* = 0$), C_a is necessarily zero, because there are no cycling cross-bridges and obviously no rotation and no collapsing forces (see fourth point in Section 4.4.2.4 and p. **157** in Section 4.4.2.5.4 for precise details on the collapsing forces). Thus, Equation 4.2 reduces to $\Sigma F_a = (\gamma^* = 0) \sim -\Lambda_a(\gamma^* = 0) < 0$: there are solely weak radial attractive/compressive forces, with no possibility of radial repulsive forces, also resulting in a small lateral shrinkage of the myofilament lattice, in contradiction with the lateral swelling part of the hybrid model (see below, in this section, concerning the small lateral expansive force at non-overlap). It is therefore important to investigate further the phenomena occurring at $\gamma^* = 0$.

The very simple reasoning and inferences presented in the preceding paragraph, according to which the radial repulsive electrostatic forces $4(\gamma^*/2)F_{e,r}$ and $4(\gamma^*/2)F_{e,a}$ are zero when there is no overlap ($\gamma^* = 0$), are purely mathematical in nature and not valid in biophysical terms. At this point, let us focus on intact unit cells from frog, because very few quantitative data are available for rabbit psoas muscles. It is recalled, in the preceding paragraph, that Xu et al. (1993) performed their experiments at sarcomere lengths >4.0 μm. On p. **25** in Section 3.2, the non-overlap state ($\gamma^* = 0$) is demonstrated to begin at a sarcomere length of ~3.68 μm. The calculations presented in this paragraph are valid for a sarcomere length of ~3.68 μm, not for sarcomere lengths >4.0 μm. I have checked that the various numerical values are slightly different at ~4.0 μm and ~3.68 μm, but the orders of magnitude are similar and the conclusions are identical (calculations not shown). It is demonstrated, on pp. **27–28** in Section 3.2, that residual radial repulsive electrostatic forces between thick myosin filaments, belonging to neighboring unit cells, exist at zero overlap. I call these residual forces $\Phi_{e,r}(\gamma^* = 0)$, at rest, and $\Phi_{e,a}(\gamma^* = 0)$, during isometric tetanic contraction, per unit cell (see pp. **27–28** in Section 3.2, concerning the notions of 'intact unit cells' and 'unit cells'; for the sake of simplicity, I do not take into account this difference here and I use only the term intact unit cell). Equations 4.1 and 4.2 become $\Sigma F_r(\gamma^* = 0) = \Phi_r(\gamma^* = 0) - \Lambda_r(\gamma^* = 0)$ and

$\Sigma F_a(\gamma^* = 0) = \Phi_a(\gamma = 0) - \Lambda_a(\gamma^* = 0)$, respectively. On p. **27** in Section 3.2, it is strongly suggested that $\Phi_a(\gamma^* = 0)/4(\gamma/2)F_{e,a} = \Phi_r(\gamma^* = 0)/4(\gamma/2)F_{e,r} \sim 0.05$ (~5%). At full overlap, we have $4(\gamma/2)F_{e,a} \sim 598$ pN per intact unit cell and $4(\gamma/2)F_{e,r} \sim 468$ pN per intact unit cell (see p. **142** in Section 4.4.2.3) and, therefore, $\Phi_a(\gamma^* = 0) \sim 0.05 \times 598$ pN ~ 29.9 pN per intact unit cell and $\Phi_r(\gamma^* = 0) \sim 0.05 \times 468$ pN ~ 23.4 pN per intact unit cell. To ensure equilibrium at rest, we necessarily have $\Sigma F_{e,r}(\gamma^* = 0) = \Phi_r(\gamma^* = 0) - \Lambda_r(\gamma^* = 0) \sim 23.4 - \Lambda_r(\gamma^* = 0) = 0$, giving $\Lambda_r(\gamma^* = 0) \sim 23.4$ pN per intact unit cell. On p. **159** in Section 4.4.2.5.4, it is shown that Λ_r at full overlap is ~5.2 pN per intact unit cell: although slightly different, these radial attractive/compressive forces of ~5.2 pN and ~23.4 pN are weak and of the same order of magnitude (most of the difference probably results from the use of a unit cell at zero overlap similar to an intact unit cell at full overlap, which is oversimplified). Also on p. **159** in Section 4.4.2.5.4, it is suggested that Λ_a at full overlap is ~1.2Λ_r, and, assuming the same ratio at zero overlap, we deduce that $\Lambda_a(\gamma^* = 0) \sim 1.2 \times 23.4 \sim 28.1$ pN per intact unit cell. Thus, when shifting from rest to contraction, we have $\Sigma F_a(\gamma^* = 0) - \Sigma F_r(\gamma^* = 0) = \Sigma F_a(\gamma^* = 0) = \Phi_a(\gamma^* = 0) - \Lambda_a(\gamma^* = 0) \sim 29.9 - 28.1 \sim 1.8$ pN per intact unit cell. When shifting from rest to contraction (at zero overlap), there is therefore a very small radial expansive force of ~1.8 pN per intact unit cell, that is, a relative maximal radial expansive force (compared to the radial repulsive electrostatic forces at full overlap) of between ~100 × 1.8 pN/598 pN ~ 0.30% and ~100 × 1.8 pN/468 pN ~ 0.38% (the values of ~598 pN and ~468 pN are recalled above). This tiny radial expansive force leads, in principle, to a limited lateral swelling of the intact unit cell, apparently at odds with the assertion of Xu et al. (1993), recalled at the beginning of the preceding paragraph (no detectable variation in d_{10} when passing from resting to contracting states at zero overlap). However, it is shown, on p. **146** in Section 4.4.2.5.1, that, in intact frog fibres at the slack length (full overlap), a relative increase in the filament lattice spacing of ~+6% (obtained in a hypotonic buffer) results in an increase of ~+20.3% in the radial repulsive electrostatic forces (in the absence of radial tethering forces; the attractive/compressive forces are negligible). Here, we are at zero overlap ($\gamma^* = 0$), rather than full overlap, increasing complexity further (see above and also pp. **25–28** in Section 3.2 and pp. **251–254** in Section 8.9), but I retain, as an oversimplification, the two values of $\sim+20.3\%$ and $\sim+6\%$. Thus, the $\sim0.30\%-0.38\%$ relative maximal radial expansive force when an intact unit cell passes from rest to contraction would correspond to an increase in d_{10} of only $\sim+ (0.30-0.38)\% \times 6\%/20.3\% \sim +0.09\%-0.11\%$. For a composite animal species, the value of d_{10}, at the reference sarcomere length of ~2.18 μm (see p. **24** in Section 3.2), is ~38.5 nm (see p. **145** in Section 4.4.2.5.1), that is, ~29.6 nm at a sarcomere length of ~3.68 μm (approximate constant volume relationship in intact fibres and intact unit cells; see pp. **4–5** in the Introduction for references and discussions). Thus, the expected increase in d_{10} is $\sim(0.09-0.11)\% \times 29.6$ nm $\sim 0.027-0.033$ nm. On pp. **181–183** in Section 5.1.2, it is demonstrated that the detection of a decrease in d_{10} of $\sim0.18-0.19$ nm, during the latency period after the electrical stimulus of an intact frog fibre, is almost certainly impossible. A variation of $\sim0.027-0.083$ nm is also more likely to be undetectable. On the basis of the lateral swelling part of the hybrid model, in an intact unit cell shifting from rest to contraction (at zero overlap), no detectable increase in d_{10} at zero overlap, is therefore expected.

The experiments of Xu et al. (1993) were carried out on demembranated fibres from rabbit, not frog, and performed at 5°C–7°C (as opposed to 10°C, for frog, in this monograph). This may complicate the comparison of intact unit cells from frog with demembranated fibres from rabbit. Otherwise, Xu et al. (1993) studied the case of sarcomere lengths >4.0 μm, whereas the calculations presented in the preceding paragraphs are valid only for a sarcomere length of ~3.68 μm (see comments on p. **134** in the preceding paragraph). However, these are almost certainly minor points, owing to the many approximations made here. In any event, the conclusion drawn above, in this section, based on the lateral swelling part of the hybrid model, valid for frog intact unit cells, is similar to that suggested by Xu et al. (1993) for demembranated rabbit fibres (no detectable variation in d_{10} in the absence of overlap when passing from rest to contraction; see p. **133**). The proposal of Xu et al. (1993) is therefore not the only possibility, because, in the hybrid model, the absence of detectable changes in d_{10} at zero overlap, when passing from rest to contraction, is a genuine property of an intact unit cell.

Still using the 'osmotic compression technique', Xu et al. (1993) and other independent authors drew other important conclusions, from the properties of demembranated fibres from rabbit psoas muscle, on the basis of the swinging cross-bridge/lever-arm theory. In the experiments performed by Xu et al. (1993), the concentration of dextran T500 varied between 0% and 8% (w/v), corresponding to radial compressive forces of between 0 pN and 2400 pN per thick filament, respectively. Steadily increasing the osmotic pressure from 0 pN to 2400 pN per thick myosin filament, Xu et al. (1993) found that the radial forces measured on demembranated fibres, under isometric tetanic contraction conditions, were compressive, zero and then expansive (see their Equation 1). As recalled on p. **133**, Morel (1985a) and Morel and Merah (1997) contested these statements, because x-ray diffraction techniques are not adapted to the study of demembranated fibres. At this stage, I recall that comparing demembranated fibres and intact unit cells permits to demonstrate that these two biological materials behave very differently (see above, in this section). In Section 3.7, it is shown that intact unit cells lying in the centre of myofibrils located themselves in the centre of an intact fibres do not develop the same axial tension as the whole fibre. However, this is only a quantitative difference, resulting solely from the intact unit cells being able to move/oscillate more than the unit cells located at the periphery of the myofibrils or close to the sarcolemma. At the qualitative viewpoint, all the unit cells belonging to an intact fibre are similar, because they all contain thin actin and thick myosin filaments and do not display any distortions, damages or impairments, by contrast to many unit cells belonging to demembranated fibres. Finally, intact fibres in isosmotic buffers do not present the significant initial lateral swelling observed in demembranated fibres, in buffers mimicking the internal medium, and resulting from mechanical or chemical skinning or permeabilisation (inducing significant increases in lattice spacing/width; see Section 8.7). At this point, it is interesting to compare intact and demembranated fibres submitted to external osmotic pressures, particularly when this pressure increases steadily.

If the external osmotic pressure is increased moderately (hypertonic conditions), an intact fibre displays lateral shrinkage owing to an outflow of water (see Millman 1998 for a review). A straightforward consequence of this shrinkage is that the thin actin and thick myosin filaments are brought closer together, and, according to Equation 2 in the paper by Miller and Woodhead-Galloway (1971), the radial repulsive electrostatic force $4(\gamma/2)F_{e,a}$, appearing in Equation 4.2 (under contraction conditions), should, in principle, increase. However, when the intact fibre shrinks, the ionic strength of the sarcoplasmic medium inevitably increases, because the various ionic species remain within the fibre, whereas the water content decreases. It is therefore difficult to predict the strength of the radial repulsive electrostatic force when the intact fibre shrinks, because of the simultaneous appearance of these two antagonistic phenomena (decrease in myofilament spacing and increase in ionic strength; see below for precise details). In the light of Section 5.1.3, the residual radial tethering force $N_a F_{cb}^a$ (or NF_{cb}, using my traditional notation) decreases strongly with decreasing myofilament spacing, that is, increasing external osmotic pressure (even if the increase is moderate). Concerning the other forces, I refer the reader to p. **132** and Equation 4.2. Variations in the weak collapsing force C_a are, in principle, difficult to predict, but, based on the simple geometric and structural approach proposed by Morel and Merah (1997), this force almost certainly decreases when the intact unit cell shrinks, because myofilament spacing decreases. Finally, the force Λ_a also clearly decreases, because the Z discs and the M lines, for example, are less stretched (decrease in the diameter of the intact fibre). Thus, for low and moderate values of external osmotic pressure, the radial tethering plus collapsing and attractive/compressive forces decrease, whereas the variations of the radial electrostatic force remain, in principle, difficult to estimate (see above). Fortunately, this problem is resolved in Section 4.4.2.5.1, on p. **146**, where it is demonstrated that, in intact unit cells belonging to intact fibres, the net radial expansive force ΣF_a decreases steadily with decreasing filament spacing: when the relative filament spacing decreases by 6% (using the osmotic pressure technique), the relative net radial expansive force, ΣF_a, decreases by ~24.9% (see p. **146** in Section 4.4.2.5.1). Thus, increasing moderately the external osmotic pressure, an intact fibre (also an intact unit cell) shrinks, owing to decreases in net radial expansive forces, but there can never be increases in these expansive forces. In any event, the radial forces are always expansive, never compressive when filament spacing decreases, on increasing

external osmotic pressure. This behaviour is very different from that of demembranated fibres, in which steadily increasing the external osmotic pressure, the radial forces can be, at least apparently, expansive, zero or compressive. However, this conclusion is valid only for moderate osmotic pressures. Further increasing external osmotic pressure, the problem becomes more complex, as discussed in the next paragraph. However, the conclusions remain unchanged.

Bachouchi and Morel (1989a), using geometric and structural data, and logical reasoning, provided support for experimental data discussed in a previous paper by Morel (1984a; see Section 5 and 'Note added in proof'). We demonstrated that, at high osmotic pressures, the myofilaments are located very close to each other and that the cross-bridges (mostly the myosin heads) are crushed between the thick myosin and thin actin filaments (see also pp. **251–254** in Section 8.9). These major distortions make it difficult to predict the strengths of the radial repulsive electrostatic and tethering forces (the collapsing and attractive/compressive forces are neglected here), at rest and under contraction conditions. For instance, the crushing of the myosin heads on the thin actin filaments and subsequent massive distortions almost certainly lead to changes in fixed negative electrical charges and bound anions. However, at rest and during contraction, this crushing may mostly result in both the partial disappearance of the radial tethering force and the occurrence of a 'radial counterforce' exerted by the crushed heads on the thin actin filaments, equivalent to a radial repulsive force. There is, therefore, a net radial repulsive force in highly compressed intact unit cells. Thus, regardless of the strength of the osmotic pressure (moderate; see the preceding paragraph; high and very high; discussed here) and the subsequent shrinkage of the intact fibre, only net radial repulsive forces can be predicted, never forces directed toward the centre of the fibre (e.g. Xu et al. 1993, who found compressive, zero or expansive radial forces in demembranated fibres, with steadily increasing osmotic pressure; see also below). Biophysically, intact fibres (also intact unit cells) therefore behave very differently from demembranated fibres, which I think are an inappropriate biological material, if lattice spacing and radial forces are considered (see also Section 3.8.2). Figure 11 in the paper by Millman (1998) strongly supports the conclusions about the existence of radial repulsive forces only, regardless of osmotic pressure, in intact fibres. Indeed, in his Figure 11, Millman (1998) presents many experimental points, from independent groups, concerning the isometric tetanic force developed by intact fibres plotted against external osmolarity, between ~150 mOsM (hypotonic conditions) and ~600 mOsM (hypertonic conditions) (in vivo osmolarity ~245 mOsM; see Millman 1998 and Millman et al. 1981). This range of osmolarity, for intact fibres, certainly overlaps the range of concentration of dextran T500 (0%–8%, see p. **135**) for demembranated fibres and a comparison of these two biological materials is justified. As recalled on p. **77** in Section 3.4.5, from Figure 11 in the review by Millman (1998), the relative isometric tetanic axial force decreases monotonically (approximately linearly) with increasing external osmolarity. Regardless of the exact mechanism of contraction, it is demonstrated, by logical reasoning and from various experimental results presented in this book, that any radial expansive force is translated into an axial contractile force, under contraction conditions (see pp. **102–103** in Section 3.10; throughout this work, it is demonstrated that ~40% full isometric tetanic tension is attributed to the radial expansive forces, at least under standard conditions of contraction). If the conclusions drawn by Xu et al. (1993) and other independent authors are considered universally valid, there should be, in intact fibres, a breakpoint in the isometric tetanic force–osmolarity relationship at which the radial forces are converted from expansive to compressive forces (the breakpoint would correspond to the zero radial force; see below). More precisely, at 0% dextran T500, the myofilament separations are large and Xu et al. (1993) found the radial forces to be compressive, whereas they become expansive when these separations are reduced (e.g. at 8% dextran T500), indicating that significantly osmotically compressed fibres display expansive forces. As a straightforward consequence, there is a critical spacing, called the equilibrium spacing by Xu et al. (1993) and other specialists, corresponding to a radial force of zero (see their Equation 1). This equilibrium spacing depends on the composition of the bathing buffer, the physiological state of the demembranated fibres (rest, contraction, rigor), and the different states of the cross-bridges during contraction, for example. The experimental observations of Xu et al. (1993) are therefore at odds with those for intact fibres (see above, concerning comments on Figure 11 in the review by Millman 1998, and the absence of a breakpoint in the relative force–compression curve for intact fibres) and with the biophysical approach, consistent with this Figure 11, presented in this section. I therefore maintain my position, developed since 1985 (e.g. Morel 1985a; Morel and Merah 1997), according to which a demembranated fibre is a poor biological model for studying structural and biophysical properties, such as the lattice spacing d_{10} measured by x-ray diffraction, radial forces as a function of osmotic pressure and the relationship between the radial forces and d_{10}. Nonetheless, demembranated fibres are essential for studies of many enzymatic characteristics of the actin–myosin system in situ (see Section 3.4.3.2).

4.4.2.2 Fixed Negative Electrical Charges and Bound Anions on the Myofilaments

In vitro, the numbers of fixed negative electrical charges on myosin molecules and F-actin filaments decrease significantly with decreasing ionic strength (Bartels et al. 1993; Deshayes et al. 1993). Indeed, for rabbit myosin + ATP (conditions similar to relaxation), the charge per myosin molecule (necessarily included in thick myosin filaments, but this was not specified by the authors) is ~80e at bulk ionic strength I ~ 80–150 mM and ~50e at I ~ 30–40 mM (Bartels et al. 1993). It has also been suggested, but not confirmed, that the charge per F-actin filament is ~9.5e at I ~ 140 mM and ~6.5e at I ~ 29 mM (Deshayes et al. 1993). Aldoroty et al. (1985, 1987) obtained a series of results for mechanically skinned fibres from the walking leg of the crayfish *Orconectes*. In a state of non-overlap between the thick myosin and thin actin filaments, in relaxing buffers, Aldoroty et al. (1985) found mean linear charge densities of ~6.6 × 10^4 e μm^{-1} along the thick myosin filaments and ~0.7 × 10^4 e μm^{-1} along the thin actin filaments. In their Table IV, Aldoroty et al. (1987) gave a list of values, obtained by independent groups, corresponding to various animal species, under relaxing conditions, for fibres at around the slack length or slightly stretched. Averaging all the values, we obtain linear charge densities of ~6.6 × 10^4 e μm^{-1} along the thick myosin filaments and ~1.1 × 10^4 e μm^{-1} along the thin actin filaments. From the results obtained and reported by Aldoroty et al. (1985, 1987), it therefore appears that the linear charge density along the thick myosin filaments is independent of sarcomere length, whereas this density is slightly dependent on this length along the thin actin filaments (it is slightly higher at the slack length, at full overlap, than in stretched fibres, at non-overlap). Unfortunately, there is no straightforward relationship between the number of fixed electrical charges (plus, probably, bound anions; see below) and the surface electrical potential (e.g. Parsegian 1973; Verwey and Overbeek 1948). However, decreasing this number should also decrease the absolute values of the surface electrical potentials $\Psi_{a,x}$ and $\Psi_{m,x}$ (see definitions in the text below Equation 3.1). Indeed, Millman (1986, 1998) pointed out that, in striated muscle, it has often been shown that decreasing the electrical charges on myofilaments tends to cause the lattice to shrink. The author deduced that this phenomenon resulted from a decrease in the radial repulsive electrostatic forces. According to Equation 3.2, under otherwise given conditions, the radial repulsive electrostatic force $F_{e,x}$ depends on Q_x, that is, on the surface electrical potentials $\Psi_{a,x}$ and $\Psi_{m,x}$ (see text below Equation 3.1 for the definition of Q_x), and the inference of Millman (1986, 1998) is consistent with a decrease in these potentials, which are correlated with a decrease in the number of fixed electrical charges and bound anions. Moreover, $\Psi_{a,x}$ and $\Psi_{m,x}$ are different at rest ($x = r$) and under contraction conditions ($x = a$) (see pp. **161–162** in Section 4.4.2.6). Regini and Elliott (2001, and references therein) assessed the presence of bound anions, the number of which depends on temperature and, probably, on other factors, such as ionic strength and the conformations existing in resting or contracting fibres. These bound anions also certainly contribute to the values of $\Psi_{a,x}$ and $\Psi_{m,x}$. In this context, it is well established that the chloride anion strongly binds to polyelectrolyte gels (Elliott and Hodson 1998) and to myofilaments (Regini and Elliott 2001). As described in Sections 4.2.2 and 4.4.2.8, only MS⁻ was used as the major anion, in the experimental part of this monograph. It is not known whether this anion binds in larger or smaller quantities than the chloride anion. In any event, MS⁻ is the least deleterious anion (Andrews et al. 1991) and may have a binding capacity different from that of chloride.

On p. **32** in Section 3.2, the value of the dielectric constant ε of the interfilament medium is discussed. Cell-associated water (also called living water; notions introduced and discussed on pp. **29–31** in Section 3.2) is the major component (the living water content is probably ~80% of the myofilament lattice; see p. **29** in Section 3.2). However, it can also be deduced, from the structural, geometric and logical study of Bachouchi and Morel (1989a), that this medium is not significantly different at rest and under contraction conditions (see also pp. **31–33** in Section 3.2, for various other experimental data and biophysical approaches). Indeed, both at rest and under contraction conditions, the many myosin heads extruding from the thick myosin filament cores (called 'external' heads; see pp. **170–172** in Section 5.1.1) are located extremely close to the surface of the neighbouring thin actin filaments. The major difference between the two states of a fibre is that, during contraction, the proportion of attached cross-bridges is very large (~82%; see p. **152** in Section 4.4.2.5.2), whereas there are few weakly bound cross-bridges at rest, at least at physiological ionic strength (see, in Section 5.1.1, pp. **170–172** for definitions and pp. **175–176** for precise details, particularly the proportion of only ~9% weakly bound cross-bridges). This last point, concerning ionic strength, should be stressed, because decreasing bulk ionic strength increases the proportion of weakly bound cross-bridges (see pp. **164–165** in Section 4.4.2.8). The composition of the interfilament medium (living water + ions + myosin heads + various soluble proteins and enzymes), at a physiological ionic strength of ~150–200 mM, is therefore essentially similar at rest and during contraction. The dielectric constant of this medium is also almost certainly similar at rest and during contraction ($\varepsilon \sim 50$; see p. **32** in Section 3.2). Finally, differences between resting and contracting conditions, in terms of the radial repulsive electrostatic forces between the thick myosin and thin actin filaments, may be correlated with differences in surface potentials. Reasonable assumptions about these potentials are made on pp. **161–162** in Section 4.4.2.6.

4.4.2.3 Radial Repulsive Forces between Myofilaments, in an Intact Unit Cell

Decreases in vicinal ionic strength I*, for sufficiently high values of I*, result systematically in increases in the radial repulsive electrostatic forces (see pp. **101–102** in Section 3.10), regardless of any other indirect effects. As an example of indirect effects, decreasing ionic strength probably leads to a decrease in the number of fixed negative electrical charges on the myofilaments in vitro (see Section 4.4.2.2) and may also modify the number of anions bound to the myofilaments in situ (Regini and Elliott 2001 and references therein). It has been shown that changes in ionic strength induce structural changes in isolated thick myosin filaments from rabbit skeletal muscles (e.g. Kensler et al. 1994), probably inducing changes in the surface electrical potentials $\Psi_{m,x}$ (on the thick myosin filaments) and, possibly, $\Psi_{a,x}$ (on the thin actin filaments), with consequences for the strength of radial repulsive electrostatic forces, via Q_x, as expected from Equation 3.2 ($x = r$ at rest, and $x = a$ under contraction conditions; the definition of Q_x is given in the text below Equation 3.1). These possibilities were clearly summed up by Coomber et al. (2011): 'The structured protein matrix in skeletal muscle, like other polyelectrolyte protein matrixes, is dependent on the ionic environment and pH, which together will affect the electrical charges on the proteins and thereby govern the structure and function of the proteins (Elliott and Hodson 1998)'. Regardless of possible variations in $\Psi_{m,x}$ and $\Psi_{a,x}$, it is shown, in Sections 3.2 and 3.5 and, more generally, throughout this work, that the radial repulsive electrostatic forces are fully effective in the myofilament lattice. Moreover, there are mechanisms for translating the net radial expansive forces into axial forces, at rest and during contraction (see pp. **102–103** in Section 3.10). This conversion is repeatedly confirmed in this book, both experimentally and semi-empirically. For example, when the ionic strength is lowered, the relative resting tension increases, extremely slowly at first, and then very rapidly from I* ~ 42 mM, peaking at a high level at I* ~ 22–32 mM, and then falling sharply when ionic strength decreases below I* ~ 22–32 mM (see, for instance, pp. **118–119** in Section 4.2.2 and Figure 4.2). These findings qualitatively confirm the existence of strong radial repulsive electrostatic forces, at least at rest (see pp. **101–102** in Section 3.10). Unfortunately, I am unaware of experimental data for bulk ionic strength below ~50 mM (I* ~ 52 mM), concerning isometric tetanic tension (e.g. April and Brandt 1973, Gordon et al. 1973 and Hasan and Unsworth

1985 recorded contractile forces for bulk ionic strengths greater than ~50 mM, ~55 mM and ~50 mM, respectively). However, in the experiments of Gordon et al. (1973) and Hasan and Unsworth (1985), there was a shift toward high ionic strengths, and a sharp decrease in active tension, at low ionic strengths, was observed (see pp. **109–111** in Section 4.1 for a circumstantial discussion). Finally, as pointed out by Andrews et al. (1991) and Millman (1998), a decrease in ionic strength, under contraction conditions, may change protein conformation, probably resulting in unexpected phenomena, not necessarily directly related to the radial repulsive electrostatic forces, such as the 'destabilization of protein structure/function' (Andrews et al. 1991). The question of conformational changes, at rest and under contraction conditions, is studied below, in this section.

Let us assume that all the characteristics of the resting forces, described in the preceding paragraph, when ionic strength is lowered, result solely from rapid conformational changes in the various proteins present in a fibre (e.g. myosin, actin, troponin, tropomyosin, titin, nebulin and desmin). Such forces induced by conformational changes are very different from the radial repulsive electrostatic forces expressed in Equation 3.2 and quantified in Section 3.2, particularly on pp. **25–28**. This hypothesis of forces resulting solely from conformational changes is unrealistic but may be assumed to be valid, at least at rest. It would imply a continuous increase in relative resting tension as a function of bulk ionic strength but not the biphasic behaviour giving the solid line in Figure 4.2 and commented on in the preceding paragraph. Nevertheless, we cannot rule out the existence of such phenomena, under resting and contraction conditions (see below, in this section, for precise details), and they may add their effects to those of the radial repulsive electrostatic forces. Such phenomena, if much stronger than the electrostatic effects, would lead to a continuous increase or decrease in the radial repulsive forces, both at rest and during contraction, inconsistent with the experimental results, as demonstrated here and recalled in the preceding paragraph. Nonetheless, possible I*-dependent changes in protein conformation may be taken into account, at least partly, through continuous changes in the surface electrical potentials $\Psi_{m,x}$ and $\Psi_{a,x}$, themselves resulting from changes in fixed negative electrical charges and bound anions. Within the limits of experimental error (see Figure 4.2) and taking into account the shortcomings of the many consistent phenomenological/semi-empirical and semi-empirical results presented in this monograph, there is no tangible sign of phenomena not accounted for by variations in electrical surface potentials. Thus, the decrease in the number of fixed negative electrical charges, variations in the number of anions bound to the myofilaments (see Section 4.4.2.2) and conformational changes with changes in I* would induce changes in $Q_x(I*)$ only.

In Section 4.4.2.7, it is shown that, under contraction conditions ($x = a$) and probably also at rest ($x = r$), very small variations in $Q_x(I*)$ as a function of I*, can be deduced from experimental results. These extremely limited variations (only a few percent) may be, at least partly, attributed to the phenomena described in this section (e.g. rapid conformational changes). Moreover, it is shown that $Q_a(I*)$ and $Q_r(I*)$ are different (see p. **139**), indicating that major changes in the number of fixed negative electrical charges and bound anions occur, probably induced by conformational changes, when shifting from rest to contraction and vice versa. In this context, between the 1970s and 2000s, Haselgrove (1973, 1975, 1980), Vibert et al. (1972), Wakabayashi et al. (1988, 1991, 1992, 1994, 2001), Yagi and Matsubara (1984, 1989), Yagi and Takemori (1995) and Yagi et al. (1977, 1981a,b, 2005, 2006) detected many changes, based on x-ray diffraction patterns (traditional techniques and synchrotron radiation), when a whole muscle or a single intact fibre (mostly from frog) was shifted from rest to isometric or isotonic contraction conditions. Some of these authors claimed that their data provided evidence of conformational changes in cross-bridges, whereas others thought that tropomyosin movement was responsible for switching on the thin actin filaments, but the situation appears to be more complex (see Bremel and Weber 1972; Bremel et al. 1973; Grabareck et al. 1983; Güth and Potter 1987; Hill et al. 1980; Lehrer and Geeves 1998; Ménétret et al. 1991; Prochniewicz et al. 1996; Shiner and Solaro 1982; Yagi and Matsubara 1989 for important features of thin actin filaments [e.g. cooperativity] and Gordon et al. 2000 for a review on the detailed mechanisms of regulation of the myosin–actin interactions). However, studying in depth the

results and discussions of Haselgrove (1973, 1975, 1980), Haselgrove and Rodger (1980), Haselgrove and Huxley (1973), Huxley (1973b,d, 1979, 1980b), Vibert et al. (1972), Wakabayashi et al. (1988, 1991, 1992, 1994, 2001), Yagi and Matsubara (1984, 1989), Yagi and Takemori (1995) and Yagi et al. (1977, 1981a,b, 2005, 2006), I think that all these experimental data may also be interpreted as indicating that many other structural and conformational changes occur. In this context, numerous rapid conformational changes in the myosin heads, the whole myosin molecule and the thick myosin filaments have recently been described, particularly when demembranated fibres are shifted from rest to contraction conditions (e.g. Gu et al. 2002; Himmel et al. 2002; Houdusse et al. 1999, 2000; Xu et al. 2003, 2006a,b). In the 1970s and 1980s, some very different experiments demonstrated that Ca^{2+} can bind to thick myosin filaments, resulting in various conformational changes in these filaments, particularly during contraction (e.g. Lehman 1978; Morimoto and Harrington 1974). However, in these experiments, the concentrations of Ca^{2+} in the bathing buffers were within the millimolar range, whereas, in intact and demembranated fibres, Ca^{2+} concentration is between ~10 μM and ~30 μM during contraction. Thus, under actual contraction conditions, Ca^{2+} binding levels would be much lower than reported by Lehman (1978) and Morimoto and Harrington (1974), unless the constant for Ca^{2+} binding to the thick myosin filaments is very high, which cannot be ruled out. In such a case, the release of Ca^{2+} from the sarcoplasmic reticulum in the myofilament lattice, when contraction is triggered, would induce additional changes in the conformation of thick myosin filaments.

It is well established that, in skeletal muscle fibres, immediately after the beginning of contraction (some tens of milliseconds after triggering) and during the entire process of contraction, Ca^{2+} almost instantaneously strongly binds to the troponin C in the thin actin filaments, remaining bound throughout contraction. This leads to a series of changes in the conformation and structure of troponins C, M, T and tropomyosin, resulting in the unblocking of myosin binding sites on the actin monomers (this is known as the steric blocking process; see the review by Gordon et al. 2000), probably leading to further conformational changes in thin actin filaments (e.g. shifting of tropomyosin toward the groove between the two strands of the thin actin filaments). Many experimental studies have shown that Ca^{2+} binding to thin actin filaments induces several other phenomena (e.g. cooperative processes; see also the preceding paragraph) resulting in various conformational changes (e.g. Bremel and Weber 1972; Bremel et al. 1973; Grabareck et al. 1983; Hill et al. 1980; Lehrer and Geeves 1998; Shiner and Solaro 1982). More generally, thin actin filaments can display other cooperative phenomena (e.g. Prochniewicz et al. 1996) and structural changes (e.g. Yagi and Matsubara 1989), probably resulting from various Ca^{2+}-induced conformational changes, some of which are briefly reviewed above, in this and the preceding paragraphs. It has even been shown experimentally that the cycling cross-bridges themselves induce conformational changes in the thin actin filaments (Güth and Potter 1987). Ménétret et al. (1991), using an unconventional cryo-electron microscopy technique, also showed that 'structural changes in actin filaments occur as a result of interaction with S1 [isolated myosin head in vitro]'. More generally, Molloy (2005) and Morel and Merah (1992) pointed out that the thin actin filaments play an important role in muscle contraction, in addition to its traditional role as an activator of the MgATPase activity of the myosin heads.

When shifting from resting to contraction conditions, many structural and conformational changes therefore occur, probably inducing changes in the electrical characteristics of both the thick myosin and thin actin filaments and, later, in $Q_a(I^*)$, with respect to $Q_r(I^*)$ (see below, in this and the next paragraphs). In this context, the charge distribution in relaxed demembranated fibres differs from that in contracting fibres (Pemrick and Edwards 1974). There is, therefore, also a difference in surface electrical potentials, resulting in the values of $Q_a(I^*)$ being different from those of $Q_r(I^*)$. Many of the experimental results of Pemrick and Edwards (1974) are difficult to interpret, because of the high degree of scattering of experimental values, but these results are consistent with the notion that the conformation of the proteins (mostly myosin and actin) is different at rest and under contraction conditions. This inference is consistent with the experimental study of Highsmith and Murphy (1992), in which it was demonstrated

that electrostatic changes are induced during force generation. More precisely, during isometric contraction, many rapid cyclic actin–myosin head interactions occur, inducing conformational changes in the thin actin filaments, in addition to those in the thick myosin filaments, described on pp. **139–141**, and the various conformational changes affecting the cross-bridges during MgATP breakdown and studied by many independent groups (e.g. Brunello et al. 2006; Martin-Fernandez et al. 1994; Piazzesi et al. 1999). For this and many other reasons described in several chapters and section of this work, only short tetani (duration $\leq$ 1.0–1.5 s, used by most specialists) can be studied, because, for long- and very long-lasting tetani, MgATPase activity decreases with increasing tetanus duration (see Sections 7.2 and 7.3). During such tetani, the cyclic actin–myosin head interactions simultaneously become less frequent and less numerous (see Section 7.4), with the consequence that the conformational changes described above, in this section, almost certainly depend on the time elapsed from the beginning of contraction. On p. **150** in Section 4.4.2.5.2, it is also recalled that the conformation of the thick myosin filaments depends strongly on the nature of the nucleotide (ATP or ADP.Pi) bound to the myosin heads and, therefore, on the relative amounts of MgATP and MgADP.Pi bound during contraction. As MgATPase activity decreases over time, these relative amounts are also highly dependent on time, as are the resulting conformational changes. Thus, the question of conformational changes during isometric contraction is a complex problem, even for short tetani (see Section 3.4.3.2 for precise details on the notion of short tetani).

Other radial repulsive forces exist in the myofibrillar lattice, including the 'swelling pressure', related to 'Donnan-osmotic effects' referred to here as $D_x(c, I^*)$ per intact unit cell (c is the centre-to-centre distance between the myofilaments in an intact unit cell; see p. **145** in Section 4.4.2.5.1 for precise details about c). One of the first reviews concerning the Donnan equilibrium and its main features in physical chemistry was published by Overbeek (1956). Regini and Elliott (2001, and references therein) stated that these forces exist in both the A- and I-bands, in a sarcomere. However, they also pointed out that the Donnan-osmotic effects are largely discounted by the scientific community, particularly in the muscle field. Thus, the situation remains unclear, with Wiggins et al. (1991) expressing doubts about the 'Donnan membrane equilibrium' and its link to distribution of ions and water in cells. It is interesting to determine whether this swelling pressure, if important in intact and demembranated fibres, is exerted fully in an intact unit cell, or, in other words, whether these osmotic swelling forces are negligible with respect to the radial repulsive electrostatic forces at the microscopic level. In this context, Elliott et al. (1980) clearly showed that Donnan-osmotic effects are just another manifestation of the repulsive (swelling) forces between myofilaments or other filament arrays and that they do not require separate treatment (Gerald Elliott, personal communication). Maughan and Godt (1980) suggested the existence of 'entropic repulsive forces', particularly in demembranated fibres, referred to here as $E_x(c, I^*)$ per intact unit cell. It is not known whether $D_x(c, I^*)$ and $E_x(c, I^*)$ depend on the number of thin actin filaments per intact unit cell (4; see p. **35** in Section 3.2 for discussion of this number) or on the amount of overlap (the maximal value, corresponding to full overlap, is $\gamma \sim$ 1.48 μm; see p. **34** in Section 3.2). At a given value of c, it is assumed that these two forces depend on I^*. This is clearly the case for $D_x(c, I^*)$, if we take into account the works of Elliott et al. (1980) and Millman (1998, and references therein), but it remains unclear for $E_x(c, I^*)$. There may also be correlations between these forces and the other phenomena discussed above, in this section (e.g. conformational changes). A third type of swelling force must be added, stemming from $\Xi_x(s, I^*)$, which is an axial force per intact unit cell dependent on sarcomere length and almost certainly on I^*, through titin (see below). This axial force depends on various elastic components in series. According to Leake et al. (2004), 'titin is responsible for the passive elasticity of the muscle sarcomere'. This is also probably the case for an intact unit cell, because titin is present in any sarcomere and intact unit cell (in a half-sarcomere, it extends from the Z disc to the M line). Titin (also called connectin) was studied by Trinick et al. (1984) and by several independent groups (e.g. Leake et al. 2004; Tskhovrekova et al. 1997). Titin is a giant molecule (MW $\sim$ 3 MDa) and is very long and flexible and contains little α-helix but has a high random coil content. Its elasticity in vitro depends on ionic strength, within the

~15–300 mM range, and on temperature, within the ~10°C–60°C range (Leake et al. 2004). It is not known whether the elastic force $\Xi_x(s, I^*)$ is exclusively passive ($x = r$) and remains unmodified under contraction conditions ($x = a$). This force can be turned into a radial repulsive force and the resulting radial repulsive force is called $\delta_x(c, I^*)\Xi_x(s, I^*)$. $\delta_x(c, I^*)$ is the coefficient, taking into account the mechanisms by which axial forces are translated into radial forces (reverse of the conversion described on pp. **102–103** in Section 3.10). In the experiments presented in this monograph (see most sections of Chapter 4 for experimental data, interpretations and discussions), the sarcomeres were moderately stretched (see p. **14** in Section 2.2). Such moderate stretching is frequent in experiments performed on intact, skinned and permeabilised isometrically held fibres. Thus, the axial force $\Xi_x(s, I^*)$ is probably very weak, although its strength would depend on the state of the fibre (at rest, $x = r$, or under contraction conditions, $x = a$). The force $\delta_x(c, I^*)\Xi_x(s, I^*)$ is, therefore, probably very weak in the experiments presented in this monograph, as is $\Xi_x(s, I^*)$ (see above). The total radial repulsive force per intact unit cell is therefore $4(\gamma/2)$ $F_{e,x}(c, I^*) + D_x(c, I^*) + E_x(c, I^*) + \delta_x(c, I^*)\Xi_x(c, I^*)$. On p. **151** in Section 3.2, it is shown that, at physiological vicinal ionic strength ($I^* \sim$ 182 mM) and at full overlap, we have $4(\gamma/2)F_{e,a}(182)$ $\sim 784Q_a(182)$ pN per intact unit cell. In Section 3.5 (see text below Equation 3.49), it is found that a suitable value of $Q_a(182)$ is $\sim$0.763, giving $4(\gamma/2)F_{e,a}(182) \sim$ 598 pN per intact unit cell. On pp. **161–162** in Section 4.4.2.6, a suitable value of $Q_r(182) \sim$ 0.597 is proposed, giving $4(\gamma/2)F_{e,r}(182)$ $\sim 4 \times 196 \times 0.597 \sim$ 468 pN per intact unit cell (see p. **34** in Section 3.2 concerning the number of 196). Thus, both at rest and under contraction conditions, $4\gamma F_{e,x}(182)$ is large. In these conditions, it is almost certain that $D_x(182) + E_x(182) + \delta_x(c, I^*)\Xi_x(182) \ll 4(\gamma/2)F_{e,x}(182)$ and the sum of these three radial repulsive forces is highly likely of the same order of magnitude as the uncertainty on $Q_x(182)$. For those seeing this hypothesis as not entirely valid, it may be suggested that the major part of the sum $D_x(182) + E_x(182) + \delta_x(c, I^*)\Xi_x(182)$ results in values of surface electrical potentials $\Psi_{m,x}(182)$ and $\Psi_{a,x}(182)$ (i.e. in values of $Q_x(182)$; see definition in the text below Equation 3.1) slightly different from those corresponding only to fixed negative electrical charges and bound anions alone. In any event, it can be concluded that $4(\gamma/2)F_{e,x}(c, I^*)$ is the only effective radial repulsive force to be taken into account (see Sections 4.4.2.5.1 and 4.4.2.5.4 for precise details on this assertion). Indeed, this assumption gives self-consistent conclusions throughout this work and there is no need to complicate the reasoning further.

4.4.2.4 Generalities on the Radial Tethering, Collapsing and Attractive/Compressive Forces between Myofilaments, in an Intact Unit Cell

Let us define and study the radial tethering, collapsing and attractive/compressive forces exerted in an intact unit cell, for comparison with the radial repulsive electrostatic force $4\gamma F_{e,x}(c, I^*)$ studied and discussed in Sections 3.2 and 4.4.2.3. On p. **145** in Section 4.4.2.5.1, c is defined as the centre-to-centre distance between the myofilaments. Generalising the notations proposed in Section 3.5 (see text above Equation 3.49), I suggest calling $\Theta_x(c, I^*)$ the total radial attractive force per intact unit cell. Several forces should be included in $\Theta_x(c, I^*)$. They are described below, in this section.

First, $N_x(c, I^*)$ is the number of myosin heads of the A-type attached to the thin actin filaments per intact unit cell, at a given moment, at rest ($x = r$) or during a short tetanus, under steady-state conditions ($x = a$) (the A-type heads are involved in the generation of radial tethering forces; see, in Section 5.1.1, pp. **170–172** and **174–176**, Figure 5.1, and the corresponding comments, including Appendix I). $N_x(c, I^*)$ is assumed to depend on c and I^*. As there is only one thick myosin filament per intact unit cell, $N_x(c, I^*)$ is also the number of A-type heads per thick myosin filament attached to the four neighbouring thin actin filaments of an intact unit cell (see p. **35** in Section 3.2 concerning the number of 4 for the neighbouring thin actin filaments). $N_a(c, I^*)$ is the mean number of rapidly cycling cross-bridges (during contraction) and $N_r(c, I^*)$ is the mean number of active weakly bound cross-bridges, both corresponding to the A-type heads attached to the thin actin filaments (see pp. **170–172**, in Section 5.1.1 for definition of the active weakly bound cross-bridges). $F_{cb}^x(c, I^*)$ is the mean radial tethering force exerted on a thin actin filament by a single attached cross-bridge (during contraction, $x = a$)

or by a single active weakly bound cross-bridge (at rest, $x = r$). The generation of this force and some of its main features at rest and its decrease during contraction are studied on pp. **174–176** in Section 5.1.1 (see Figure 5.1, its legend and Appendix I) and pp. **180–181** in Section 5.1.2, respectively (see also Section 5.2, including Figure 5.2, its legend and Appendix II). The total radial tethering force per intact unit cell is therefore $N_x(c, I^*)F_{cb}^x(c, I^*)$. It is recalled on pp. **139–141** in Section 4.4.2.3 that many conformational changes occur in the myosin heads, whole myosin molecules and thick myosin and thin actin filaments, depending on various conditions. The full radial tethering force per intact unit cell may therefore vary with the conditions used in the experiments and the state of the intact unit cell. Under given conditions ($x = r$, rest, or $x = a$, contraction), such changes mostly affect $F_{cb}^x(c, I^*)$, but they probably have very little effect on $N_x(c, I^*)$ at given values of c and I^*, provided that the values of c and I^* are around the physiological values (on pp. **164–167** in Section 4.4.2.8, it is demonstrated that the situation is much more complex in more extreme conditions, e.g. low and very low values of I^*).

Second, $F_w(c)$ is the van der Waals–London radial attractive force per unit length (expressed in pN µm^{-1}, for example), in principle independent of I^* (e.g. Elliott 1968; Israelachvili 1974; Miller and Woodhead-Galloway 1971; Parsegian 1973, 1975; Verwey and Overbeek 1948). This force is exerted between a thick myosin filament and a single thin actin filament. In the light of the information provided on pp. **29–31** in Section 3.2, the existence of living water (cell-associated water) in the myofibrillar medium would result in $F_w(c)$ depending on I^*, but I do not consider this possibility here. As for $F_{e,x}(c, I^*)$, in the case of $F_w(c)$, 4 thin actin filaments interact with a thick myosin filament (see p. **35** in Section 3.2 for a discussion of this number of 4). This force is also proportional to the length of the zone of overlap γ (the maximal value of γ is ~1.48 µm at the slack length of the fibre; see p. **34** in Section 3.2). The full van der Waals–London attractive force per intact unit cell is therefore $4(\gamma/2)F_w(c)$.

Third, $\Gamma_x(c, I^*)$ is the sum of various radial attractive/compressive forces per intact unit cell. Such forces, mostly elastic in nature, occur, for example, in the Z discs for thin actin filaments and the M lines for thick myosin filaments or in other extrafibrillar structures (e.g. Magid et al. 1984; Miller and Woodhead-Galloway 1971; Millman 1998). Several experiments have been carried out on the Z discs (e.g. Edwards et al. 1989; Franzini-Armstrong and Porter 1964; Goldstein et al. 1986, 1987, 1988; Rowe 1971). Interesting findings were published by Goldstein et al. (1986, 1987, 1988), who detected variations in the Z discs before, during and after tetanic contraction. It is therefore necessary to distinguish between $x = r$ and $x = a$. Vigoreaux (1994) has reviewed the properties and roles of the Z discs. Other experimental studies concerned the M lines (e.g. Eaton and Pepe 1974; Knappeis and Carlsen 1968; Luther and Squire 1978). Shabarchin and Tsaturyan (2010) have proposed a many-faceted model for the M lines and their roles. Higuchi (1987) demonstrated the existence of elastic components by showing that selective digestion of these structures in single skinned frog fibres resulted in supplementary lateral swelling (see Section 8.7 concerning the 'natural' swelling upon demembranation). It is also probable that the sarcolemma and the associated non-sarcolemmal structures, present in intact fibres and partly present (~50%) in half-fibres (see Section 2.2), compress the whole myofilament lattice, the sarcoplasmic medium and the intact unit cells radially (see pp. **155–156** in Section 4.4.2.5.3 for a discussion). On p. **93** in Section 3.7, the sarcoplasmic reticulum is assumed to play a role comparable to that of the sarcolemma, and this membrane-like structure may be responsible for additional radial compression in all unit cells, particularly in intact unit cells. Other possible weak protein–protein interactions may be taken into account in $\Gamma_x(c, I^*)$.

Fourth, collapsing forces $C_x(c, I^*)$, should be considered. These forces are related exclusively to the rotation of the attached cross-bridges under active contraction but not in resting conditions, when there are few weakly bound cross-bridges and no tilting (no energy available from MgATP splitting; see p. **175** in Section 5.1.1). Consequently, we have $C_r(c, I^*) = 0$. Dragomir et al. (1976), Elliott (1974), Morel (1985a) and Morel and Merah (1997) demonstrated that the attached swinging cross-bridges exert attractive radial forces called collapsing force, $C_a(c, I^*)$, during contraction. Dragomir et al. (1976) described this force clearly: 'to the force producing sliding and therefore shortening of the sarcomere a transverse force is added that tends to bring the myofilaments closer'. The strength of this weak force is estimated and commented on p. **157** in Section 4.4.2.5.4.

4.4.2.5 Complementary Experimental Data and Analyses Concerning the Radial Forces between Myofilaments and the Proportion of Attached Cross-Bridges during Isometric Tetanic Contraction, in an Intact Unit Cell

4.4.2.5.1 PREVIOUS EXPERIMENTS PERFORMED BY INDEPENDENT GROUPS. CONTRADICTORY INTERPRETATIONS. EXPERIMENTAL DATA SUPPORTING THE EXISTENCE OF THE VARIOUS INTERFILAMENT FORCES

Edman and Anderson (1968) studied the effect of external osmolarity on intact contracting frog fibres, at 1°C–2°C, particularly at about the slack length. They found that an increase in external osmolarity (hypertonic conditions) resulted in an outflow of intracellular water and, therefore, in a shrinkage of the fibres (~−8% in fibre width, with respect to the width in an isosmotic buffer), whereas a decrease in osmolarity (hypotonic conditions) resulted in an inflow of external water and, therefore, in a swelling of the fibres (~+8% in fibre width, with respect to the width in an isosmotic buffer). Variations of ~±8% in fibre width result in variations of ~±15%–17% in cross-sectional area, which is a moderate value (for comparison, see p. **98** in Section 3.8.1, p. **119** in Section 4.2.2, p. **166** in Section 4.4.2.8, and Section 8.7). For such a moderate decrease of ~−8% in fibre width, we can consider that the major problem of steric hindrance of the cross-bridges does not occur (see pp. **251–254** in Section 8.9 for precise details). Thus, 'steric alterations' in the radial repulsive electrostatic forces and the radial tethering, collapsing and attractive/compressive forces can be neglected here. Edman and Anderson (1968) were unsure of the reason for the behaviour of their fibres, but they suggested a plausible explanation: 'the nature of this effect is unclear. It may reflect a change in the interaction between the actin and myosin components due to alterations of the ionic composition and the total ionic strength of the intracellular medium'. To my mind, this rather vague conclusion is consistent with my own approach. Indeed, if fibre width decreases (lateral shrinkage under hypertonic conditions) or increases (lateral swelling under hypotonic conditions), the distance between the thin actin and thick myosin filaments decreases or increases, respectively, and the intracellular ionic strength simultaneously increases or decreases, respectively. As concerns the radial repulsive electrostatic force $F_{e,a}(I)$ exerted between a thick myosin filament and a single thin actin filament, increasing myofilament spacing and decreasing ionic strength result in opposite contributions to $F_{e,a}(I)$ (note that the vicinal ionic strength I^* is replaced by the 'bulk' ionic strength I, because, in intact fibres, the notion of vicinal ionic strength is meaningless; see Section 3.9). At this point, we need to try to quantify this inference concerning these two antagonistic phenomena. For this purpose, let us take into account Equation 3.1 (giving the expression of the radial repulsive electrostatic force $F_{e,x}$) and the definitions given on p. **32** in Section 3.2. We deduce a new expression of $F_{e,a}(I)$ (I is expressed in millimolar and 180 mM replaces the vicinal ionic strength $I^* \sim 182$ mM used only for demembranated fibres; 180 mM is the reference bulk ionic strength used in this book for the external buffer, but supposed here to represent the 'bulk' ionic strength within the intact fibre, for the sake of comparison):

$$F_{e,a}(I) \sim A_a(I)Q_a(I)(I/180)^{3/4}$$
$$\exp[-(8\pi e^2 N_A/10\varepsilon\varepsilon_0 kT)^{1/2}(s^\dagger - d)I^{1/2}] \quad (4.3)$$

In an intact unit cell, where the closest surface-to-surface spacing $s^* = s^\dagger - d$ between the thick myosin and thin actin filaments is fixed, the exponential term in brackets in Equation 4.3 is $H_a(I)$ $I^{1/2}$ in Equation 3.1. At a fixed sarcomere length, the volume of an intact unit cell is proportional to c^2, where c is the centre-to-centre distance between the myofilaments (thin–thin and thin–thick filaments distance; isovolumic behaviour of intact fibres; see April et al. 1971; Brandt et al. 1967; Elliott et al. 1963, 1965, 1967; see pp. **4–5** in the Introduction for a brief discussion of this behaviour). Thus, the ionic strength I is necessarily reversibly proportional to c^2, and we have, for purely geometric reasons, $I = 180(c_0/c)^2$, where c_0 is the reference centre-to-centre distance corresponding to s_0 and the reference ionic strength of 180 mM in an isosmotic bathing buffer ($s_0 \sim$ 2.18 μm; see p. **24** in Section 3.2). Several values are presented in Table 1 in the review by Millman (1998), regarding experimental values of the crystallographic spacing d_{10} measured by x-ray diffraction with various animal species and types of muscle (isosmotic buffers; sarcomere length of ~2.3 μm). Correcting for sarcomere length (isovolumic behaviour; simple calculation not shown), we obtain an average value of $d_{10} \sim 38.5 \pm$ (0.6 or 2.5) nm (mean ± SE or SD; n = 17), at $s_0 \sim$ 2.18 μm, from which we deduce $c_0 = (2/3)d_{10} \sim 25.6 \pm$ (0.4

or 1.7) nm (mean ± SE or SD; n = 17). For the sake of self-consistency throughout this section, I select $c_0 \sim$ 25.3 nm, which is compatible with the SD and SE. In this context, this p. **84** refers to Section 3.5, a value of $\sim 1.58 \times 10^{-3}$ μm^2 is selected for the cross-sectional area of an intact unit cell from a composite animal species. This value corresponds to a centre-to-centre distance of ~24.7 nm, only 2% lower than ~25.3 nm, which is negligible. There is a simple relationship between c and $s^\dagger$: $c = s^\dagger + (R_a + R_b)$, where R_a is the radius of a thin actin filament ($R_a \sim 4.3$ nm; see p. **34** in Section 3.2) and R_b is the radius of the thick myosin filament backbone ($R_b \sim 18/2 \sim 9.0$ nm; see p. **215** in Section 6.4 for the value of ~18 nm). We deduce that $s^* = s^\dagger - d = c - (R_a + R_b + d)$ and that $c = s^* + (R_a + R_b + d) = s^* + (R_a + R_m)$, with $R_m = R_b + d$. Comments on the envelope, of thickness d, surrounding each thick myosin filament and on R_m are provided on pp. **27–33** in Section 3.2. On p. **33** in the same section, d is found to be ~8.1 nm. On p. **24**, p. **29** and p. **32** still in Section 3.2, comments on s^* are provided and it is strongly suggested that $s^* \sim 3.1$ nm. These values of d and s^* are valid at the reference sarcomere length $s_0 \sim$2.18 μm (see above). Thus, the reference value of c, at s_0 and under isosmotic buffer conditions, is $c_0 \sim 3.1 + (4.3 + 9.0 + 8.1) \sim 24.5$ nm, statistically consistent with the value of ~25.3 nm selected above. In the text above Equation 3.2, it is demonstrated that $A_a(I)$ is essentially independent of d and, therefore, of c and I (see above: there is a simple mathematical relationship between I and c) and can be replaced by the constant A. $Q_a(I$ or I^*) is shown, on p. **163** in Section 4.4.2.7, to have high values and to depend very little on I or I^*, at least in an intact unit cell, at the reference sarcomere length $s_0 \sim$ 2.18 μm and at the reference centre-to-centre distance $c_0 \sim$ 25.3 nm (see above). When I is altered, the centre-to-centre distance, c, is also 'mathematically' altered (see above), and $Q_a(I)$ may therefore depend indirectly on c. In this context, it is recalled, on pp. **25–26** in Section 3.2, that Q_x ($x = a$, contraction; $x = r$, rest) is practically independent of myofilament lattice spacing. This is recalled in the next paragraph, at least over a limited range of variations in myofilament spacing.

From the discussion presented in the preceding paragraph (see text below Equation 4.3), regarding changes in c mediated by osmotic pressure, we deduce that, at full overlap, corresponding to the slack length of an intact fibre ($s_0 \sim$ 2.18 μm;

see the preceding paragraph), Equation 4.3 can be rewritten, after simple (but fastidious) mathematical manipulations, as follows:

$$F_{e,a}(c) \sim AQ_a(c)(c_0/c)^{3/2}\exp\{-(180 \times 8\pi e^2 N_A/10\varepsilon\varepsilon_0 kT)^{1/2}(c_0/c)[c - (R_a + R_b + d)]\} \quad (4.4)$$

Assuming that $\Delta c = c - c_0$ is a change in c (with $\Delta c < 0$ or $\Delta c > 0$, according to external osmolarity; see the preceding paragraph, concerning c_0), it can be deduced (derivation from the logarithmic function $\ln F_{e,a}$; fastidious calculations not shown) that the relative variation $\Delta F_{e,a}$ in $F_{e,a}$ is given by

$$\Delta F_{e,a}/F_{e,a} \sim \Delta Q_a/Q_a - (\Delta c/c)\{3/2 - (180 \times 8\pi e^2 N_A/\varepsilon\varepsilon_0 kT)^{1/2}c_0[(R_a + R_b + d)/c - (\Delta d/\Delta c)]\} \quad (4.5)$$

The numerical values are $\varepsilon \sim 50$ (see p. **32** in Section 3.2), $k = 1.38 \times 10^{-23}$ J K^{-1}, $T = 283$ K, corresponding to 10°C, the reference temperature used in this monograph (using 1°C–2°C [see the beginning of this section] does not markedly affect the numerical values), $e = 1.602 \times 10^{-19}$ C, $R_a \sim 4.3$ nm and $R_b \sim 9.0$ nm (see p. **145**, concerning R_a and R_b). Thus, from Equation 4.5, we obtain a simple relationship (c, c_0, d and Δd are expressed in nanometres):

$$\Delta F_{e,a}/F_{e,a} \sim \Delta Q_a/Q_a - (\Delta c/c)[1.5 - 6.356c_0[(13.3 + d)/c - \Delta d/\Delta c] \quad (4.6)$$

It is recalled, on p. **144**, that Edman and Anderson (1968) found variations in fibre width of ~±8% when they varied the external osmolarity around intact fibres. Thus, we can write, as an illustrative example, $100 \times \Delta c/c_0 \sim \pm 6\%$ (see Millman 1998 for a discussion of the greater variations in width than in centre-to-centre spacing). Taking $100 \times \Delta c/c_0 = (c - c_0)/c_0 \sim +6\%$ (hypotonic conditions), we obtain $\Delta c = c - c_0 \sim +0.06 \times 25.3$ nm $\sim +1.518$ nm ($c_0 \sim 25.3$ nm is proposed on p. **145**), and $c \sim 25.3 + 1.518 \sim 26.818$ nm, leading to $\Delta c/c \sim +1.518$ nm/26.818 nm $\sim +0.057$ (+5.7%). Thus, Equation 4.6 gives

$$\Delta F_{e,a}/F_{e,a} \sim \Delta Q_a/Q_a - 0.086 + 9.166[(13.3 + d)/26.818 - 0.659\Delta d] \quad (4.7)$$

On p. **163** in Section 4.4.2.7, it is shown that, in demembranated fibres, $Q_a(52) \sim 0.739$ and,

in Section 3.5 (text below Equation 3.49), that $Q_a(182) \sim 0.763$ (52 and 182 are values of vicinal ionic strength, expressed in millimolar), giving $[Q_a(52) - Q_a(182)]/Q_a(182) \sim -0.031$. According to the relationship between I and c (see p. **145**), in intact fibres studied here, we have $\Delta I/180$ mM $\sim -2\Delta c/c_0 \sim -0.12$, that is, $\Delta I \sim -180$ mM $\times 0.12 \sim -21.6$ mM, markedly lower than $52 - 182$ mM ~ -130 mM, and $\Delta Q_a/Q_a \sim -0.031 \times 21.6$ mM/130 mM ~ -0.005, which is very low, as compared with the other terms in Equation 4.7, and confirms the conclusion drawn at the end of the preceding paragraph. Concerning Δd, I define, in the text between Equations 4.3 and 4.4, $c_0 = s_0^* + R_a + R_b + d_0$ and $c = s^* + R_a + R_b + d$, giving $\Delta c = c - c_0 = (s^* + d) - (s_0^* + d_0) \sim 1.518$ nm (see above). On pp. **24** and **32** in Section 3.2, we find $s_0^* \sim 3.1$ nm and $d_0 \sim 8.1$ nm, from which we deduce $s_0^\dagger = (s_0^* + d_0) \sim 11.200$ nm (see p. **29** in Section 3.2, concerning the relationship between s^*, $s^\dagger$ and d) and $(s^* + d) \sim 11.200$ nm + 1.518 nm ~ 12.718 nm. Assuming, for simplification, that $s^*/d \sim s_0^*/d_0 \sim 3.1$ nm/8.1 nm ~ 0.382, we obtain a value of $d \sim 9.203$ nm and $\Delta d = d - d_0 \sim 9.203 - 8.100 \sim 1.103$ nm. Introducing these numerical values into Equation 4.7 gives an estimate of $\Delta F_{e,a}/F_{e,a} \sim +0.203$ (~+20.3%). When the intact fibres studied by Edman and Anderson (1968) are studied in hypertonic buffers, we can choose $100 \times \Delta c/c_0 \sim -6\%$, rather than the ~+6% studied above, and the reasoning can simply be reversed (note that Equation 4.7 is no longer valid; new equation not shown), giving $\Delta F_{e,a}/F_{e,a} \sim -0.189$ (~−18.9%). The conclusions drawn here are promising: the lateral swelling of intact contracting fibres is apparently related to an increase in the radial repulsive electrostatic forces, and the lateral shrinkage is related to a decrease in these forces. Nonetheless, the problem is more complex and can be studied on the basis of the hybrid model, as demonstrated in the next paragraph. In any event, the conclusion is qualitatively the same: there is a straightforward relationship between the behaviour of intact contracting fibres submitted to various external osmotic pressures and the net radial expansive forces).

There are residual radial tethering, collapsing and attractive/compressive forces, under contracting conditions, that can be written, in an intact unit cell (see Equation 4.8 and definition on p. **159** in Section 4.4.2.5.4): $\Theta_a(c, I) \sim N(c, I)$ $F_{cb}(c, I) + C_a(c, I) + \Lambda_a(c, I)$ (I use my traditional notation NF_{cb}, rather than $N_a F_{cb}^a$). At 10°C, under physiological conditions of isometric tetanic contraction (isosmotic bathing medium for an intact fibre), c and I have fixed values, and Θ_a is ~0.240 (~24.0%) times the radial repulsive electrostatic force $4(\gamma/2)F_{e,a}$ (see Section 3.5, text below Equation 3.49 and also p. **159** in Section 4.4.2.5.4). When the external osmotic pressure is changed, it is difficult to anticipate the changes in the radial tethering, collapsing and attractive/compressive forces with changes in c and I (c and I are closely linked, as shown on p. **145**: $I = 180(c_0/c)^2$). In the light of the geometric calculations presented by Morel and Merah (1997), the weak collapsing force $C_a(c, I)$ related solely to rotation of the external myosin heads must increase with increasing c, owing to the subsequent increase in the angle between S2 and the thick myosin filament backbone and vice versa (see, in Section 5.1.1, pp. **170–172** and Figure 5.1, its legend [and Appendix I] and the corresponding comments for precise details on the notion of external and internal heads; see also Figure 5.1 concerning the position of S2). Nonetheless, it is impossible to predict the variations in $C_a(c, I)$ and $\Lambda_a(c, I)$ when I varies, after changes in c. In Section 5.1.3, it is strongly suggested that, at rest, the elementary tethering force F_{cb}^0 increases considerably when c increases very slightly, and vice versa, but there is no reason to assume that the number N_0 of active weakly bound cross-bridges (notion defined on pp. **170–172** in Section 5.1.1) changes with c (at around physiological ionic strength; see pp. **164–167** in Section 4.4.2.8, as concerns the particular case of low and very low ionic strengths). Thus, small changes in c induce considerable changes in the strength of the tethering force $N_0 F_{cb}^0$, via F_{cb}^0. This is also the case for $N(c, I)F_{cb}(c, I)$ (or, more simply, NF_{cb}), via F_{cb} (see Section 5.1.3). Here again, it is impossible to predict the direct role of I (on various conformational changes, for example), but it is almost certain that increasing c results in an increase in $\Theta_a(c, I)$, via, at least, NF_{cb} and C_a. Similarly, we can assume that decreasing c results in a decrease in $\Theta_a(c, I)$. From Equation 4.8, we can write $\Sigma F_a = 4(\gamma/2)F_{e,a} - \Theta_a = 4(\gamma/2)F_{e,a}(1 - \rho_a)$, with $\rho_a = \Theta_a/4(\gamma/2)F_{e,a}$, and $\Delta\Sigma F_a = 4(\gamma/2)\Delta F_{e,a} - \Delta\Theta_a = 4(\gamma/2)\Delta F_{e,a} \times (1 - \Delta\rho_a)$, with $\Delta\rho_a = \Delta\Theta_a/4\gamma\Delta F_{e,a}$. Finally, we obtain $\Delta\Sigma F_a/\Sigma F_a = [4(\gamma/2)\Delta F_{e,a}/4(\gamma/2)F_{e,a}](1 - \Delta\rho_a)/(1 - \rho_a)$. As ρ_a and $\Delta\rho_a$ are, by definition, significantly lower than 1, the sign of $\Delta\Sigma F_a/\Sigma F_a$ is the same as

that of $4(\gamma/2)\Delta F_{e,a}/4(\gamma/2)F_{e,a}$ and the self-consistency of the approach is ensured, as demonstrated in the preceding paragraph. The residual radial tethering forces, the collapsing forces and the radial attractive/compressive forces simply modulate the relative strength of the net expansive force ΣF_a. Taking, for instance, $\Delta\rho_a \sim 0$ and introducing $\rho_a \sim 0.240$ (see the beginning of this paragraph), we obtain $\Delta\Sigma F_a/\Sigma F_a \sim 1.316 \times [4(\gamma/2)\Delta F_{e,a}/4(\gamma/2)F_{e,a}]$. In the preceding paragraph, we found that $100\Delta c/c_0 \sim +6\%$ corresponds to $100 \times 4(\gamma/2)\Delta F_{e,a}/4(\gamma/2)F_{e,a} \sim +20.3\%$, giving therefore $100\Delta\Sigma F_a/\Sigma F_a \sim +26.7\%$. Similarly, $\Delta c/c_0 \sim -6\%$ leads to $100 \times 4(\gamma/2)\Delta F_{e,a}/4(\gamma/2)F_{e,a} \sim -18.9\%$, giving $100\Delta\Sigma F_a/\Sigma F_a \sim -24.9\%$. Thus, the existence of residual tethering forces leads to conclusions consistent with those drawn in the preceding paragraph, at least qualitatively, and the discussion presented here is in favour of the hybrid model, and, more specifically, the part involving lateral swelling. I indicate on p. **144** that the inference of Edman and Anderson (1968) was rather vague. Owing to the complexity of the reasoning presented in this section, the authors were obviously unable, in 1968, to draw clear conclusions, but they were working along the right lines. The conclusions presented here are consistent with the existence, in intact contracting fibres, of both strong radial repulsive electrostatic forces and detectable residual radial tethering forces (the radial collapsing and attractive/compressive forces are weak; see Section 4.4.2.5.4). Finally, it is demonstrated on pp. **106–107** in Section 4.1 that the values of ~+26.7% (hypotonic buffer) and ~−24.9% (hypertonic buffer) are qualitatively and quantitatively consistent with the increase and decrease in full axial tension, respectively, recorded by Edman and Anderson (1968).

Maughan and Godt (1980) studied, both experimentally and semi-empirically, the radial forces in relaxed mechanically skinned frog fibres at ~20°C–22°C (unfavourable conditions for a cold-blooded animal, as pointed out throughout this monograph). The authors found that decreasing ionic strength led to very small decreases in resting fibre width, which they considered to conflict with the expected increase in radial repulsive electrostatic forces. However, they studied only three ionic strengths: 90 mM, 150 mM and 290 mM (major neutral salt KCl). At an ionic strength of 290 mM and at pH ~ 7, synthetic thick myosin filaments in vitro are 'unstable' and generally dissolved, as observed by electron microscopy (Ingrid Pinset-Härström, personal communication). Maughan and Godt (1980) also reported this possibility in skinned fibres, under the same conditions of ionic strength and pH. From their Figure 6, Maughan and Godt (1980) claimed that the 'experiment shows exactly the opposite effects of changes in ionic strength'. This assertion requires some comments. Under the reference external osmotic pressure chosen by the authors and taking 150 mM as the reference ionic strength, I calculate, from this Figure 6, that the fibre width at 90 mM and 290 mM are ~95%–97% and ~104%–106% of that at 150 mM, respectively. From these values, we deduce that no statistical difference in fibre width was observed, within the limits of experimental error, particularly as partial dissolution of thick myosin filaments may occur at 290 mM (see above). This approximately constant value for fibre width is consistent with Figure 4.2. Indeed, between bulk ionic strengths of ~40 mM and 180 mM, I found that the resting axial tensions resulting from the conversion of net radial repulsive (expansive) forces into axial forces were close to zero, indicating that the radial repulsive electrostatic forces and the radial tethering plus radial attractive/compressive forces counterbalance (see pp. **163–165** in Section 4.4.2.8 for further discussions and precise details). Thus, the results obtained by Maughan and Godt (1980) support the arguments developed in this work, despite the 'negative' assertion of the authors, recalled above. In this context, Maughan and Godt (1980) were apparently unaware of the existence of radial tethering forces, although Morel (1975), Morel and Gingold (1979b), Morel and Pinset-Härström (1975b) and Morel et al. (1976) had published, some years previously, theoretical and logical studies concerning the existence of such forces, based on some experimental data available in the 1970s.

4.4.2.5.2 EXPERIMENTAL ESTIMATES OF THE PROPORTION OF ATTACHED CROSS-BRIDGES UNDER ISOMETRIC TETANIC CONTRACTION CONDITIONS (MOSTLY SHORT TETANI–STEADY STATE), BASED ON PUBLISHED DATA. 'UNIVERSAL VALUE' OF THE PROPORTION OF ATTACHED CROSS-BRIDGES

He et al. (1997) stated that tight coupling between isometric MgATPase activity and the mechanical cross-bridge cycle is possible only if a relatively small fraction of cross-bridges are attached to actin (e.g. Cooke 1995; Linari and Woledge 1995). On p. **150**, I propose various reasons for not selecting the small fractions. Moreover, in Section 3.4.5, it is demonstrated that there is a very loose coupling between isometric tetanic tension and isometric MgATPase activity and that the notion of tight coupling is almost certainly outdated (see also Section 3.4.6 for a discussion of very loose and tight couplings). Thus, a high fraction is almost certainly the only possibility. This assertion gives a series of self-consistent conclusions throughout this monograph and is clearly supported in this section.

During isometric tetanic contraction (short tetani and steady-state conditions; see Section 3.4.3.2 for precise details on these two notions) of intact fibres bathed in buffers mimicking the physiological external medium within the body, the mean proportion n_∞ of external heads (major components of the cross-bridges) per intact unit cell, attached to the thin actin filaments, is very high (see pp. **170–172** in Section 5.1.1 for precise details on the external heads). Indeed, on p. **152**, it is shown, from a reasoning independent of the mechanical roles of the cross-bridges and from many experimental results from independent groups, using various experimental techniques, that $n_\infty \sim 0.82$ (~82%) (on p. **172** in Section 5.1.1, it is suggested, on experimental bases, that, at rest, the proportion of weakly bound heads is ~9%). The high proportion of ~82% almost certainly accounts for the existence of a noticeable residual radial tethering force $N_a(I^*)F_{cb}^a(I^*)$, despite the inequality $F_{cb}^a \ll F_{cb}^r$ (see pp. **180–181** in Section 5.1.2) (at 10°C, this residual force represents ~14% of the reference radial tethering force at rest, $N_r(I^*)F_{cb}^r(I^*)$; see pp. **158–159** in Section 4.4.2.5.4 concerning this proportion of ~14%). The percentage of cross-bridges attached to the thin actin filaments during isometric contraction (short tetani and steady state) is $100n_\infty$, and this percentage should be ≤100. A consensual maximal reference value of 100 has been reported in rigor conditions, in acto-myosin–S1 (S1 = head) studied in vitro, in intact and demembranated fibres from rabbit psoas muscle or insect flight muscles (e.g. Cooke and Franks 1980; Craig 1977; Linari et al. 1998; Lovell and Harrington 1981; Lovell et al. 1981; Lymn 1978; Thomas and Cooke 1980). There was controversy concerning the stoichiometry of the binding of S1 to actin in F-actin filaments, studied in vitro. Amos et al. (1982)

and Mornet et al. (1981) found that S1 attached to two actin globules, whereas Grussaute et al. (1995), Iorga et al. (2004), Sutoh (1983) and Tesi et al. (1989) found that S1 attached to only one actin globule. At the beginning of the 1990s, the problem may have been resolved by Andreev and Borejdo (1991) and Andreev et al. (1993), who found that S1 bound to two actin globules. I believe that the choice (1 or 2) depends on the experimental technique used. In any event, the question of the stoichiometry of S1-actin globule binding does not affect the definition of the 100% reference value discussed in this paragraph.

Many x-ray diffraction experiments have been carried out at around the slack length and during isometric tetanic contractions of various durations, mostly on whole muscles or intact fibres from frog and, more recently, on demembranated fibres from rabbit psoas muscle. Other experimental techniques have also been used (e.g. stiffness). Some of the results obtained are presented and discussed below, in this section.

Haselgrove and Huxley (1973), using a traditional x-ray diffraction technique, performed their experiments at 4°C, on whole frog muscles stimulated electrically. During the experiments, oxygen was passed into the Ringer solution and the concentration of ATP was therefore maintained at a constant level. The authors used a large number of successive tetani (duration of 3 s each), separated by release periods of 2 min, and periods of 10 h to 15 h were required to obtain useable patterns. Under these conditions, Huxley and Kress (1985) claimed that Haselgrove and Huxley (1973), Huxley (1980c) and Yagi et al. (1977) found that 'up to 90% of the myosin heads are in the close vicinity of the actin filaments'. However, reading the paper by Haselgrove and Huxley (1973) carefully, I noted that the mean proportion of cross-bridges in the vicinity of the thin actin filaments was only ~50%–60% and probably corresponds to attached cross-bridges. Morel (1984a) demonstrated semi-empirically, independently of the mechanical roles of the cross-bridges, that the proportion of attached cross-bridges inevitably decreases for successive tetani (see also Section 7.4). Moreover, the duration of 3 s for each tetanus does not actually correspond to short tetani as defined in Section 3.4.3.2. In a personal communication, the late Ichiro Matsubara told me that the ratio $I_{1,0}/I_{1,1}$ (used to estimate the number of cross-bridges in the vicinity of the thin actin filaments, from comparison of this ratio during tetanus and at rest) decreases over large numbers of successive tetani. By contrast, Ichiro told me that this ratio does not significantly vary over small numbers of tetani. These unpublished results therefore indicate that the values of ~50%–60% are almost certainly underestimates and I replace them by ~70%–80%. I also take into account the value of ~90% mentioned by Huxley and Kress (1985), from three experimental studies (see above).

Using x-ray diffraction (traditional technique), with single tetani lasting 10 s (i.e. long-lasting tetani, based on the criteria given in Section 3.4.3.2 and working on whole frog muscles, at 0°C, Matsubara et al. (1975) found that '[at least ~80% of] the cycling projections are in the vicinity of the thin filaments when the muscle is producing the maximum isometric tension'. This value of ~80% is discussed below. Matsubara et al. (1975) did not oxygenate their muscles and they pointed out that rigor-like links may be present for a single tetanus of 10 s duration. The case of long- and very long-lasting tetani in frog is studied in Section 7.2 (see p. **222**), where it is shown that, for a duration of 15 s, the maximal cleavage of ~5.2 mM MgATP would occur at 0°C, that is, ~(10 s/15 s) × 5.2 mM ~ 3.5 mM for a duration of 10 s. This 'maximal potential depletion' of MgATP is less than the MgATP concentration in the sarcoplasmic medium (~4–6 mM; see Alberts et al. 2007; Lehninger 2008). Obviously, the MgADP produced by MgATP hydrolysis is immediately rephosphorylated via Lohmann's reaction, but, owing to the long duration of the tetanus (10 s), the partial depletion of MgATP is possible (see, for instance, Morel and Pinset-Härström 1975b for precise details on the metabolism of MgADP). Thus, the question of rigor-like links raised by Matsubara et al. (1975) is posed here, and the result obtained by these authors may be considered only an order of magnitude. However, Matsubara et al. (1975) presented several experimental arguments to justify that their value of ~80% is not 'contaminated by rigor-like links'. This value of ~80% is therefore taken into account, regardless of other problems concerning the steady state of short tetani (analysed and defined in Section 3.4.3.2) being exceeded, including the possible onset of fatigue (see Section 3.4.3.2).

Ford et al. (1981) and Goldman and Simmons (1977) measured the stiffness of intact frog fibres, at ~0°C–5°C. They assumed that stiffness is proportional to the number of attached heads, which is not necessarily the case, as pointed out in Section 5.1.4. At full overlap, taking the stiffness in rigor as corresponding to 100% bound cross-bridges, the two groups found the same proportion of ~70% of heads bound to the thin actin filaments during isometric contraction, in short tetani and under steady-state conditions. Regardless of the discussion presented in Section 5.1.4 (concerning stiffness), I take these proportions into account, because they lie within the wide range of the other experimental values.

Cooke et al. (1982), Corrie et al. (1999), Hopkins et al. (1998) and Ostap et al. (1995) worked at 5°C–10°C on chemically skinned fibres from rabbit psoas muscle and used probes attached to the cross-bridges to monitor their orientations during contraction (labeling technique), comparing the results obtained at rest and in rigor. All these authors concluded that, during the generation of isometric tension, only a very small fraction (~10%–30%) of the cross-bridges are attached to the thin actin filaments, the rest remaining unattached. For example, Cooke et al. (1982) obtained this proportion by comparing rigor, resting and contracting fibres and suggested that, under contraction conditions, ~70%–80% of the heads are in a state similar to that existing at rest and only ~20%–30% are in a state similar to that existing in rigor. This interpretation is unwarranted, because the resting-like position of the cross-bridges, amounting to ~70%–80% of the whole population, does not necessarily imply detachment, and the rigor-like position, representing ~20%–30% of the whole population, does not necessarily imply attachment. Moreover, the problems raised by estimation of the tilting angle in contracting muscle are complex, with a further complexity added by the relationship between this angle and the proportion of cross-bridges attached (see pp. **172–174** in Section 5.1.1). Furthermore, the conclusion drawn by the authors is highly model dependent. Indeed, according to the pioneering, but oversimplified 'oar' model of Huxley (1965, 1969), the cross-bridges attach at the 90° position, corresponding to resting conditions, and, during contraction, they adopt the 45° position, that is, the rigor-like position (arrowhead). Other experimental studies have suggested that, during isometric contraction, only ~12% of the cross-bridges are attached (Hopkins et al. 1998), or

~10%, according to Corrie et al. (1999). Labeling the myosin heads and studying the response at low temperature (~5°C–10°C; unfavourable conditions for a warm-blooded animal species, as highlighted throughout this monograph, particularly in the next paragraph) seems to lead systematically to very low proportions of attached cross-bridges. From the qualifications detailed in this paragraph, and the experimental reasons given in the next paragraph, these very low estimates are not retained.

Using traditional x-ray diffraction, Wray (1987) and Wray et al. (1988) found that, in rabbit muscle, at temperatures below ~15°C–19°C, the cross-bridges lose their helical ordering. By contrast, at higher temperature, they appear to be highly ordered and give x-ray diffraction patterns similar in quality to those for frog. The authors interpreted this transition as reflecting a coupling between nucleotide state and overall conformation (see below). Zoghbi et al. (2004), working on tarantula thick myosin filaments, gave a molecular explanation for the helical order, based on the 'closed' conformation of the myosin head. This type of 'switch-II closed conformational state' was defined by Geeves and Holmes (1999). Xu et al. (1999), working on bundles of fibres from psoas muscle, demonstrated that the 'disordered' conformation observed at low temperature corresponds to most of the myosin heads (M) being in the M.ATP state, whereas the 'ordered' conformation observed at high temperature corresponds to the M.ADP.Pi state (Xu et al. 2006a found that, at 25°C, ~95% of the myosin heads were in the M.ADP.Pi state). Working on isolated thick myosin filaments from rabbit psoas muscle, Kensler et al. (1994) obtained similar results. Moreover, Kensler and Stewart (1983), again working at 25°C, confirmed that isolated rabbit thick myosin filaments were fully preserved and claimed that, when experiments are performed at low temperature (below ~15°C–19°C; see above), the 'increased lability of the cross-bridges order in the rabbit filaments may explain the failure of many probe studies to show the ordered relaxed arrangement of cross-bridges'. Thus, for all these reasons and those presented in the preceding paragraph, the values of ~10%–30% bound cross-bridges during an isometric short tetanus, obtained for rabbit fibres at 5°C–10°C, are not retained.

Burghardt et al. (1983) used a labeling technique to study the proportion of cross-bridges attached to the thin actin filaments during isometric contraction (their labeling technique was different from those used by the authors cited on p. **150**). They worked at room temperature (~20°C–22°C) on chemically skinned single fibres or bundles of two or three fibres from rabbit psoas muscle and obtained a value of ~65% attached heads. Their technique appears to be appropriate, probably because they worked at a temperature corresponding to highly ordered thick myosin filaments (see the preceding paragraph). Nonetheless, the measurements were made within 1 min of the onset of contraction, whereas the experimental data were collected in less than 5 s. It seems likely that the authors carried out several experiments on a given fibre or bundle of fibres, over this period of 1 min. Thus, the notions of short tetani and steady state, described in Section 3.4.3.2, are not fulfilled. The various phenomena concerning fatigue, discussed in Chapter 7, probably also blur the discussion of Burghardt et al. (1983) and it may be suggested that the value of ~65% is an underestimate, owing to the long duration of the experiments and the decrease in the proportion of attached cross-bridges for large numbers of successive tetani (see p. **149**). Thus, I select a value of ~75% rather than ~65%.

Burghardt and Ajtai (1985), working on the same biological material as Burghardt et al. (1983) (at 4–6°C against ~20–22°C) and using a fluorescence anisotropy technique different from that used in 1983 (with durations of 3–4 min against 1 min or less), obtained a proportion of attached cross-bridges >80%. The temperature conditions are a priori unfavourable, but the proportion of bound cross-bridges is comparable to that obtained by Burghardt et al. (1983), for unknown reasons. I select here two values of ~80% and ~90%.

Matsubara et al. (1985), working on chemically skinned toe muscles from mouse, at ~17°C–21°C, concluded from their experimental studies that the proportion of attached cross-bridges during isometric tetani was ~96%. However, each muscle was exposed to x-rays for 5–10 min, to record useable x-ray diffraction patterns (traditional technique). This is a source of concern (see Chapter 7, regarding onset of fatigue in very long-lasting tetani). Nonetheless, the authors checked that tension declined very little during this period. They also checked that their muscles were not in rigor and that there was a close correlation between pCa

and the proportion of attached cross-bridges and between pCa and isometric tension. This interpretation is consistent with Equation 3.37, showing that the isometric tension is proportional to the square of the number of attached cross-bridges, regardless of the mode of generation of the contractile force, provided that the other parameters in Equation 3.37 do not depend on the 'amount of fatigue'. This problem of fatigue cannot be resolved in the experiments of Matsubara et al. (1985), which lasted 5–10 min, but the experimental results can be considered appropriate. The very high proportion of ~96% may be related to the authors working on a type of muscle very different from frog or rabbit muscles. In any event, this proportion is taken into account.

Using x-ray diffraction patterns obtained from synchrotron radiation, Martin-Fernandez et al. (1994) found, at 8°C, for whole frog muscles, a proportion of ~70% or more (two values of ~70% and ~80% are selected here).

Linari et al. (1998) measured and compared the stiffnesses of isolated intact fibres from skeletal muscle of frog, at 4°C, under isometric contraction conditions and in rigor. They also combined these results with those of x-ray experiments (synchrotron radiation). They found a proportion of ~43%. The 'stiffness technique' is highly questionable (see Section 5.1.4), and this estimate is to some extent model dependent (swinging cross-bridge/lever-arm theories), but this proportion is nevertheless retained.

Working on permeabilised muscle fibres from rabbit psoas, using 'stiffness and thermodynamic' techniques, and comparing 'resting and rigor' fibres, Linari et al. (2007) found a proportion of ~33%, regardless of temperature. This proportion is rather low, but it is retained.

Rome et al. (1999) deduced this proportion from measurements of stiffness in permeabilised fibres from white and red muscles of toadfish, under contraction conditions and in rigor. Comparing these two states and taking into account several experimental findings obtained by other groups, the authors calculated proportions of ~61% and ~70% for white and red fibres, respectively, for short tetani (~1 s at most), during the steady-state period (see Section 3.4.3.2 for definitions and discussions of these two notions). In Section 5.1.4, I highlight the risk involved in comparing stiffness measurements for fibres of the same origin, but in two different states (isometric contraction vs. rigor). Nonetheless, the proportions of ~61% and ~70% are similar to the other values obtained by various techniques. Thus, these two values are taken into account.

Combining x-ray diffraction data from synchrotron radiation and mechanical measurements on single intact frog fibres, at 4°C, Piazzesi et al. (2007) suggested a 'molecular explanation' of the force–velocity relationship. In my opinion, there are several flaws in this study and the conclusions are highly model dependent (swinging cross-bridge/lever-arm theories). One of the important conclusions drawn by Piazzesi et al. (2007) is that only ~29% of the cross-bridges (called 'myosin motors' by the authors) are bound during isometric tetanic contraction. On p. **150**, it is recalled that model-dependent values of between ~10% and ~30% have been obtained and are not taken into account here. However, the two series of experiments are very different and cannot be compared. Nonetheless, below, in this section, I do not initially take into account the value of ~29%. However, as it lies approximately midway between ~10%–30% and ~50%–80%, I later do take it into account. These two choices do not result in significant differences in the mean value of the proportion of attached cross-bridges (see below, in this section).

In an experimental study, performed on demembranated rabbit fibres, using temperature methods [T-jump] (described by Bershitsky and Tsaturyan, 2002) and synchrotron radiation, Tsaturyan et al. (2011) found that there are '41–43% [I select ~42%] stereospecifically bound myosin heads at near-physiological temperature [31–34°C]'.

Woledge et al. (2009), studying the frog at 2°C–10°C, suggested a semi-empirical proportion of ~36%. However, they estimated that there were 167,000 myosin heads in a muscle volume of 1.05 μm^3, corresponding to a 'density' of heads of ~264 μM. On pp. **208–209** in Section 6.3.5, three different estimates are proposed: (i) ~262 μM if the thick filaments are assumed to have the traditional structure (three strands; all myosin heads outside the backbone); (ii) ~154 μM, the 'universal empirical concentration', valid for a 'composite adult animal species'; and (iii) ~193 μM, valid for a composite animal species, and corresponding to the external heads being able to enter the rapid cyclic interaction process with the thin actin filaments (see pp. **170–172** in Section 5.1.1 for precise details on the external heads). The only suitable estimate is ~193 μM and the proportion of ~36% becomes ~36% × 264 μM/193 μM ~ 49%, a value taken into account in the next paragraph.

Matsubara (1980) has already highlighted 'scattering' in the proportions obtained by independent groups and critically analysed and discussed the various divergent estimates obtained experimentally, semi-empirically and theoretically. We can conclude again that there is no consensus concerning the proportion of cross-bridges attached under contraction conditions, in the case of short tetani (or, in some instances, long-lasting tetani), mostly because of the use of technologies, temperatures and biological materials that are not necessarily appropriate for obtaining firm conclusions. At this point, it is interesting to average the various proportions, recalled above, in this section, to take into account all the advantages and drawbacks of each technique and to try to 'smooth' the scattering, as is systematically done in this book. Only the results of Cooke et al. (1982), Corrie et al. (1999), Hopkins et al. (1998) and Ostap et al. (1995) are discarded, for reasons presented on p. **150**. In the first calculation, not taking into account the value of ~29% suggested by Piazzesi et al. (2007), we obtain, within the ~0°C–20°C temperature range, an estimate of ~72 ± (4 or 17)% (mean ± SE or SD; n = 20). In the second calculation, the value of ~29% is introduced and we obtain ~68 ± (4 or 17)% (mean ± SE or SD; n = 21). These two proportions are statistically similar, regardless of the choice to use SE or SD. However, it is unclear whether these two series of results are statistically similar to the phenomenological/semi-empirical value of ~82 ± (3 or 6)% obtained on p. **61** in Section 3.4.3.2, for frog at ~2.5°C. Provided that SDs are taken into account rather than SEs (see Section 2.1 concerning the advantage of using SD rather than SE), the interval of ~76%–85% is common to the three series of results and I select the mean value of ~ (76 + 85)%/2 ~ 81% ± 5% (valid between ~0°C and ~20°C; see above). Thus, owing to the qualifications concerning certain techniques used to measure the proportion of cross-bridges attached during isometric contraction (tetani of various durations, at various temperatures, etc.), the 'experimental' proportion of ~81% is statistically consistent with the phenomenological/semi-empirical value of ~82% obtained for short tetani (see above). In this work, I refer to the proportion of ~82% as a universal value, independent of the muscle and the animal species and valid, at least, between ~0°C and ~20°C.

At this stage, it should be noted that most of the estimates cited in this section were obtained from low-angle x-ray diffraction data. In this context, Millman (1998) recalled that this kind of technique can detect cross-bridges in the vicinity of the thin actin filaments, but not necessarily binding. Some values have also been deduced from measurements of stiffness (see the preceding paragraphs). A.F. Huxley (1980a) claimed that stiffness is proportional to the number of attached cross-bridges, but this assertion has proved difficult to demonstrate (see Section 5.1.4). Despite these reservations concerning experimental techniques and their quantitative interpretations, I again consider retention of the 'universal' proportion of ~82% to be justified (see below). One clear consequence, suggested in the preceding paragraph, is that the values of ~10%, 12%, 20% and 30% (similar to the ~18% obtained for the second value of the proportion deduced from Equation 3.27) are at odds with all the other values obtained on various types of muscle. Choosing ~82% is therefore entirely justified. By contrast, the values of ~18% and of ~10%, 12%, 20% and 30% cannot be retained. This inference, according to which only a high proportion of bound cross-bridges for short tetani under steady-state conditions is considered suitable, would be consistent with the experimental observation of He et al. (1997): 'in the early stages of an isometric contraction the fraction of cross-bridges that show maximal ATPase activity appears to be close to 100% … Then, it drops to a lower percentage'. We could deduce that, early in a tetanus, ~100% of the cross-bridges are attached to the thin actin filaments. The proportion of attached cross-bridges would decrease thereafter, with unpredictable kinetics. For short tetani, under steady-state conditions of isometric tetanic tension (stabilisation of the tension at a constant level, after an initial short tension rise) in single intact or permeabilised fibres (durations of ~1.0–1.5 s; e.g. Altringham et al. 1984; Bagni et al. 1988b; Edman 1979; Gordon et al. 1966b; Granzier and Pollack 1990; durations of ~0.3–0.5 s; e.g. Martin-Fernandez et al. 1994; Piazzesi et al. 1999; West et al. 2004), the average proportion of attached cross-bridges, over the duration of short tetani, would be slightly below ~100% but of the same order of magnitude. Caution is required here. Indeed, He et al. (1997) deduced

their estimate of ~100% from a comparison of the MgATPase measured in situ with that measured in vitro. However, on pp. **270** and **272** in Chapter 9, I cite Borejdo et al. (2006) and Grazi (2000), who claim that the preparation of actomyosin systems in vitro requires the disruption of sarcomere structure, resulting in extremely low protein concentrations, very different from the crowding observed in vivo/in situ. Thus, comparing the MgATPase activity of actomyosin in situ and in vitro to deduce the proportion of attached cross-bridges under isometric contraction conditions (see the above quote from He et al. 1997) would be risky. Nonetheless, the conclusion drawn by He et al. (1997) may support the universal value of ~82% given at the end of the preceding paragraph and the value obtained on p. **61** in Section 3.4.3.2 from the phenomenological/semi-empirical approach proposed throughout Section 3.4.

4.4.2.5.3 GENERAL COMMENTS ON THE RADIAL TETHERING AND ATTRACTIVE/COMPRESSIVE FORCES

In his review, Millman (1998) took into account the semi-empirical study of Maughan and Godt (1980), briefly analysed on pp. **148–149** in Section 4.4.2.5.1, but he improved their conclusions. I agree with many of his conclusions, particularly those concerning the major role of the radial repulsive electrostatic forces. However, he drew some conclusions that I consider to be invalid. For instance, he took into account the interpretations of most of the supporters of swinging cross-bridge/lever-arm models, concerning 'additional pressure from the cross-bridges' occurring during contraction or in rigor. This traditional view is analysed and criticised in Section 3.8.2 and is demonstrated to be almost certainly erroneous (see also Section 4.4.2.1 for a better understanding of the problems). I demonstrate, in this book, both experimentally and semi-empirically, that, at rest, some weakly bound cross-bridges (called active weakly bound cross-bridges; see pp. **170–172** in Section 5.1.1) also play a major role in intact fibres or half-fibres, skinned and permeabilised fibres, single isolated myofibrils and intact unit cells. These active weakly bound cross-bridges exert strong radial tethering forces under resting conditions, which decrease considerably under contraction conditions (see pp. **174–176** in Section 5.1.1 and pp. **180–181** in Section 5.1.2, respectively). The radial tethering forces, first envisaged and described solely at rest by Morel (1975), Morel and Gingold (1979b), Morel and Pinset-Härström (1975b) and Morel et al. (1976), are generalised in this monograph to contraction conditions and called $N_x(c, I^*)F_{cb}^x(c, I^*)$ (x = a during contraction; x = r at rest). They should be added to the radial attractive/compressive forces studied by Millman (1998 and references therein), particularly those exerted by the Z discs and the M lines (see third point in Section 4.4.2.4).

It is pointed out that the phenomenological/semi-empirical reasoning presented in this work is self-consistent and the hybrid model described and analysed throughout Chapter 5 has strong experimental, semi-empirical, phenomenological/semi-empirical and theoretical bases, together with high predictive and explanatory power (see Chapter 8). It is also shown that, under contraction conditions, there are collapsing and attractive/compressive forces in the myofilament lattice that add their effects to those of the radial tethering forces (see Section 4.4.2.4). Despite this apparent complexity, the electrostatically related relative axial contractile tension, accounting for ~40% of the full relative axial contractile tension, has a very simple semi-empirical expression (see Equation 4.15). At this point, it is interesting to comment on the attractive/compressive forces introduced and briefly discussed in Section 4.4.2.4. Let us recall that the centre-to-centre distance c is related to the closest surface-to-surface spacing s^*, introduced and discussed on p. **24**, p. **29** and p. **32** in Section 3.2: $c = s^* + (R_a + R_m)$, where R_a is the radius of the thin actin filaments and R_m is the true radius of the thick myosin filaments (see pp. **27–33** and **33–34** in Section 3.2 for precise details on R_m and R_a, respectively).

$F_w(c)$ (see second point in Section 4.4.2.4) may be considered very weak (e.g. Millman 1998 and references therein), owing to the wide closest surface-to-surface spacing, $s^\dagger$, between the thin actin filaments and thick myosin filament backbones. Indeed, the van der Waals–London attractive forces depend on $(s^\dagger)^{-6}$ (e.g. Elliott 1968; Israelachvili 1974; Maughan and Godt 1980; Parsegian 1973, 1975; Verwey and Overbeek 1948). For frog and rabbit, at the universal reference sarcomere length $s_0 \sim 2.18$ μm (see p. **24** in Section 4.2), we have $s^\dagger \sim 11.2 \pm$ (0.4 or 0.8) nm (mean $\pm$ SE or SD; see p. **28** in Section 3.2). The reference sarcomere length used in the experimental part of this monograph was ~2.5 $\pm$ 0.2 μm (see p. **14** in Section 2.2 and p. **16** in Section 2.3). As

isovolumic behaviour holds for intact fibres (April et al. 1971; Brandt et al. 1967; Elliott et al. 1963, 1965, 1967; see pp. **4–5** in the Introduction for some comments on this point), the corresponding mean value of $s^\dagger$ is ~10.4 $\pm$ (1.3 or 1.7) nm in the slightly stretched fibres (calculations not shown), statistically consistent with ~11.2 $\pm$ (0.4 or 0.8) nm. Thus, a value of ~11.2 nm is selected in this work, regardless of the exact sarcomere length, between ~2.18 μm and ~2.5 $\pm$ 0.2 μm. As discussed and clearly shown on pp. **27–33** in Section 3.2, the true radius R_m of the thick myosin filaments is considerably greater than that of the backbone R_b, owing to the presence of large numbers of extruding myosin heads, both at rest and during contraction and, to a less extent, to the existence of living water bound to the thick myosin filaments and, possibly, also to the thin actin filaments (see pp. **29–31** in Section 3.2 concerning living water; see also p. **34** in Section 3.2, regarding the possible non-existence of bound water around the thin actin filaments). In any event, we cannot ignore $F_w(c)$, as assumed, for example, by Maughan and Godt (1980), because the closest surface-to-surface distance s^* between the myofilaments is very small (~3.1 nm; see p. **24**, p. **20** and p. **32** in Section 3.2 for precise details). However, $F_w(c)$ is certainly much weaker than the radial repulsive electrostatic forces and the radial tethering forces at rest and under contraction conditions. Nonetheless, $F_w(c)$ may be similar to the other attractive/compressive forces studied in the next paragraph, at least in intact fibres, half-fibres and intact unit cells, at the slack length, in the absence of other factors (e.g. absence of osmotic compressive forces). One major difference lies in the lack of dependence of $F_w(c)$ on ionic strength (e.g. Elliott 1968; Israelachvili 1974; Miller and Woodhead-Galloway 1971; Parsegian 1973, 1975; Verwey and Overbeek 1948) and on the state of the interfilament medium, because the composition of this medium appears to be similar at rest and during contraction. Indeed, in both cases, the number of external myosin heads in the interfilament medium is similar (see above), the only difference resulting from the large number of rapid cycles of attachment–detachment and the cleavage of large quantities of MgATP during contraction, with most of these external heads not entering these many cycles at rest (see pp. **174–176** in Section 5.1.1 for precise details on these particular points). This 'enzymatic' difference almost certainly has no noticeable effect on $F_w(c)$. The only potential problem is that the characteristics of living water may depend on ionic strength, with putative consequences for the dispersive van der Waals–London force $F_w(c)$, which would depend indirectly on ionic strength in vivo/in situ. However, this assumption is not taken into account here.

Concerning $\Gamma_x(c, I^*)$ (see third point in Section 4.4.2.4), the role of radial elastic compressive forces exerted by the sarcolemma and associated non-sarcolemmal structures on the whole fibre is supported below and on pp. **242–243** in Section 8.7. The sarcoplasmic reticulum surrounding each myofibril is also assumed to exert radial elastic compressive forces (see again third point in Section 4.4.2.4), but no experimental data are available regarding these forces and they are not taken into account quantitatively below, because they can be included in the contribution of the sarcolemma, owing to the broad range of uncertainty concerning the quantitative contribution of the sarcolemma itself. The effects of the sarcolemma on an intact unit cell are difficult to evaluate. Nonetheless, the corresponding compressive forces, although certainly much weaker than the radial tethering forces in a resting intact unit cell (see Sections 5.1.1, 5.1.2, 5.1.3 and 4.4.2.5.4 for discussions of the radial tethering forces), cannot be neglected. Some important aspects of the sarcolemma and the associated non-sarcolemmal components have been studied (e.g. Rapoport 1972, 1973). It is difficult to deduce, from the experimental and semi-empirical study of Rapoport (1973), the value of the lateral pressure relating to the sarcolemma, at rest or under contraction conditions, at about the slack length of an intact fibre ($s_0 \sim 2.18$ μm; see p. **24** in Section 3.2). However, using the Laplace equation for a cylinder (pressure exerted on the myoplasm = surface force/radius of the cylinder) and his semi-empirical approach, Rapoport (1973) found two possible values for the pressure exerted on the myoplasm: $< \sim 0.7 \times 10^4$ dyn cm^{-2} and $< \sim 1.9 \times 10^4$ dyn cm^{-2}. We can therefore use a maximal estimate of $\sim [(0.7 + 1.9)/2] \times 10^4 \sim 1.3 \times 10^4$ dyn cm$^{-2} \sim 1300$ N m^{-2} ~ 1300 pN μm^{-2}. This estimate corresponds to frog fibres at $\sim 22°C$, conditions unfavourable for a cold-blooded animal, as frequently recalled in this monograph. The maximal value of ~ 1300 pN μm^{-2} should therefore be taken with caution. However, the order of magnitude is almost

certainly correct. The cross-sectional area of an intact unit cell is $\sim 1.58 \times 10^{-3}$ μm^2 in a composite animal species (see p. **84** in Section 3.5). At the reference sarcomere length $s_0 \sim 2.18$ μm, (see above), this gives a lateral area of ~ 0.29 μm^2 per intact unit cell (the intact unit cell has a hexagonal cross-section). The maximal radial compressive force is therefore ~ 1300 pN μm$^{-2} \times 0.29$ μm$^2 \sim 377$ pN per intact unit cell, of the same order of magnitude as the net expansive force under contraction conditions (~ 455 pN per intact unit cell, see p. **85** in Section 3.5). This value is undoubtedly largely overestimated. At this point, it is interesting to recall some values of surface forces (at room temperature; see handbooks on surface forces): $\sim 73 \times 10^{-3}$ N m$^{-1} \sim 73 \times 10^3$ pN μm^{-1} for the air–water interface, $\sim 30 \times 10^{-3}$ N m$^{-1} \sim 30 \times 10^3$ pN μm^{-1} for the olive oil–air interface and $\sim 20 \times 10^{-3}$ N m$^{-1} \sim 20 \times 10^3$ pN μm^{-1} for the olive oil–water interface. The surface of the sarcolemma and associated non-sarcolemmal components in contact with the myoplasm (aqueous medium) are mostly hydrophilic and the surface force is certainly very weak, but non-zero: a purely hydrophilic structure in contact with an aqueous medium should display a zero surface force, but a small proportion of the non-sarcolemmal components display hydrophobic zones, rendering the surface force weak but non-zero. Let us consider, as an illustrative example, an entirely hydrophobic structure for the sarcolemma: we can therefore choose the value for the oil–water interface ($\sim 20 \times 10^3$ pN μm^{-1}) and a maximal radius of $\sim (130 + 25)/2 \sim 78$ μm (see p. **13** in Section 2.2). Laplace's equation (see above) gives a pressure on the myoplasm of ~ 256 pN μm^{-2} and ~ 256 pN μm$^{-2} \times 0.29$ μm$^2 \sim 74$ pN per intact unit cell (see above concerning the lateral surface area of ~ 0.29 μm^2). This value is much too high given the attractive/compressive forces $\Lambda_r \sim 5.2$ pN per intact unit cell and $\Lambda_a \sim 6.2$ pN per intact unit cell (see pp. **156** and **158–159** for definition and calculation). Choosing a London–van der Waals force $4(\gamma/2)F_w$ of only ~ 0.5 pN per intact unit cell, the compressive forces Γ_r and Γ_a relating to the sarcolemma are of the order of ~ 4.7–5.7 pN per intact unit cell, corresponding to a surface 'tension' of $\sim (4.7$–$5.7)$ pN per intact unit cell/0.29 μm^2 per intact unit cell ~ 16–20 pN μm^{-2}, much lower than ~ 256 pN μm^{-2} (see above). Otherwise, using the rule of three, the values of ~ 16–20 pN μm^{-2} correspond to surface forces of only $\sim (16$–20 pN μm$^{-2}/$

256 pN μm^{-2}) $\times$ 20 $\times$ 10^3 pN μm^{-1} ~ 1.2–1.6 $\times$ 10^3 pN μm^{-1}. These results are consistent, particularly concerning the low values of ~1.2–1.6 $\times$ 10^3 pN μm^{-1} for the surface forces between the aqueous myoplasm and the surface of the sarcolemma and associated non-sarcolemmal components, which is mostly hydrophilic (only a few zones are probably hydrophobic, and the very low values of ~1.2–1.6 10^3 pN μm^{-1} are therefore much lower than ~20 10^3 pN μm^{-1} (~6%–8%) for the hydrophobic olive oil-water surface; see above). The values of ~5.2–6.2 pN per intact unit cell are only ~100 $\times$ (5.2–6.2) pN/468 pN ~ 1.11%–1.35% of the radial repulsive electrostatic force, at rest, $4(\gamma/2)F_{e,r}$ ~ 468 pN per intact unit cell (see, for instance, pp. **158–159** in Section 4.4.2.5.4 regarding this value). Under resting conditions in vivo/in situ (sarcolemma present), the radial repulsive electrostatic forces exactly counterbalance the radial tethering and attractive/compressive forces. As estimated here, the contribution of the radial compressive force exerted by the sarcolemma (including also, within the range of experimental uncertainty, the contribution of the sarcoplasmic reticulum; see above) is very small but cannot be considered negligible. For example, the total or partial removal of the sarcolemma is expected to induce a tiny relative decrease in the radial attractive/compressive forces. This very small decrease tips the balance in favour of the radial repulsive electrostatic forces, leading to an initial tiny lateral swelling of the myofilament lattice. Although providing a useful starting point for the lateral swelling of fibres upon demembranation, the numerical values given in this paragraph are valid only in intact unit cells. They are not valid in unit cells belonging to demembranated fibres, because these fibres are profoundly different from intact fibres and intact unit cells (see p. **13** in Section 2.2, where it is recalled that demembranated fibres present many signs of impairment). Important consequences of these major differences are studied on pp. **97–98** in Section 3.8.1 and in Sections 3.8.2, 4.4.2.1 and 5.1.3. In any event, from a qualitative viewpoint, the initial slight imbalance can be considered to be the origin of the massive lateral swelling upon demembranation (see pp. **242–243** in Section 8.7), regardless of the nature of the unit cell (intact or impaired).

As recalled on pp. **4–5** in the Introduction, it can be deduced from the experiments of Neering et al. (1991) on isometrically held contracting intact frog fibres that there is a lateral swelling corresponding to a mean, but non-uniform, relative increase in fibre diameter of ~10%, when passing from testing conditions ~10%, to isometric contraction. This swelling is probably related to the net radial repulsive forces exerted during contraction. As pointed out on p. **5** in the Introduction, Millman (1998) claimed that an increase in fibre diameter does not necessarily correspond to the same increase in the centre-to-centre distance measured by x-ray diffraction, because fibre diameter changes more rapidly than myofilament lattice spacing, implying a greater swelling of the extracellular spaces. This argument is supported by Figure 8 in the review by Millman (1998), who showed, by x-ray diffraction (measurement of lattice spacing), under conditions similar to those used by Neering et al. (1991), a mean increase of only ~3%–4% rather than ~10%. Thus, during contraction, the centre-to-centre distance between the myofilaments would become c^* ~ (1.03–1.04)c or ~1.10c at most, slightly or markedly greater, respectively, than c at rest. This may give a value of $\Gamma_a(c^*, I^*)$ slightly or markedly higher than $\Gamma_r(c, I^*)$, probably attributed to slight stretching of Z discs and M lines and an increase in the circumference of the sarcolemma, and also, probably, that of the sarcoplasmic reticulum, with the subsequent increase in the compressive forces exerted by these two structures (see the preceding paragraph for precise details on $\Gamma_x(c, I^*)$).

A complete quantitative study of the radial tethering, collapsing and attractive/compressive forces is proposed in Section 4.4.2.5.4.

4.4.2.5.4 DETAILED AND QUANTITATIVE ANALYSIS OF THE RADIAL TETHERING, COLLAPSING AND ATTRACTIVE/COMPRESSIVE FORCES. GENERAL EXPRESSION OF THE NET RADIAL REPULSIVE (EXPANSIVE) FORCE PER INTACT UNIT CELL, AT REST AND DURING ISOMETRIC TETANIC CONTRACTION (SHORT TETANI–STEADY STATE)

In this section, the vicinal ionic strength I^* is mentioned, because many discussions presented in this work concern demembranated fibres and half-fibres (see Section 3.9 for precise details on I^*). However, some discussions also concern intact fibres and the bulk ionic strength I is also used (see Section 4.4.2.5.1). For the sake of self-consistency of the notations, only I^* is used here.

Adding the radial tethering, collapsing and attractive/compressive forces, the corresponding total radial force, per intact unit cell, under contracting conditions ($x = a$), can be written (using my traditional notations, N for N_a, and F_{cb} for F_{cb}^a): $\Theta_a(c, I^*) = N(c, I^*)F_{cb}(c, I^*) + C_a(c, I^*) + \Lambda_a(c, I^*)$, where $\Lambda_a(c, I^*) = 4(\gamma/2)F_w(c) + \Gamma_a(c, I^*)$ (see Sections 4.4.2.4 and 4.4.2.5.3 concerning the forces $F_w(c)$ and $\Gamma_a(c, I^*)$). Similarly, at rest ($x = r$), we can write $\Theta_r(c, I^*) = N_0(c, I^*)F_{cb}^0(c, I^*) + \Lambda_r(c, I^*)$, where $\Lambda_r(c, I^*) = 4(\gamma/2)F_w(c) + \Gamma_r(c, I^*)$ (there is no collapsing force, because the resting cross-bridges cannot rotate; see pp. **174–176** in Section 5.1.1 for a short discussion).

Starting from simple geometric considerations, structural data and experimental results obtained by independent groups, Morel and Merah (1997) calculated the order of magnitude of the strength of the radial collapsing forces $C_a(c, I^*)$, under contraction conditions, in an intact unit cell belonging to an intact fibre (see Section 3.1 for necessary references). Under physiological conditions (medium surrounding the intact fibre mimicking the physiological milieu, at the slack length), but at temperatures not specified, because of the large uncertainty on the various values, these forces were found to be $C_a \sim 0.3 \times 10^5$ pN μm^{-2}, $C_a \sim 0.5 \times 10^5$ pN μm^{-2} and $C_a \sim 0.7 \times 10^5$ pN μm^{-2}, depending on various geometric and structural assumptions made by Morel and Merah (1997). The 'composite' cross-sectional area of an intact unit cell is $\sim 1.58 \times 10^{-3}$ μm^2 (see p. **84** in Section 3.5) and we obtain $C_a \sim 47$ pN, $C_a \sim 79$ pN and $C_a \sim 111$ pN per intact unit cell, respectively. Bagni et al. (1994b) performed experiments on intact frog fibres, at the slack length, using x-ray diffraction from synchrotron radiation. They employed the 'osmotic pressure' technique (see Section 4.4.2.5.1 for precise details on this technique, in terms of the hybrid model) and detected, under isometric tetanic contraction conditions, a radial compressive force of ~ 74 pN per thick myosin filament, that is, ~ 74 pN per unit cell, including intact unit cells located in the centre of the intact fibre and peripheral unit cells (see Section 3.7 for precise details on these two notions). This small force is certainly a collapsing force, as demonstrated by Morel and Merah (1997). There is a large uncertainty on the estimate of this force, and the experimental value given by Bagni et al. (1994b) is assumed to correspond approximately to intact unit cells. About 35 years ago, Schoenberg (1980b), using a simplified approach

based on the traditional swinging cross-bridge/lever-arm models, claimed that the active radial forces exerted by the attached cross-bridges were compressive and represent $\sim 10\%$ of the active axial forces. In the 1980s, the best estimate of the axial contractile tension was ~ 2.0–2.5×10^5 N $m^{-2} \sim 2.0$–2.5×10^5 pN μm^{-2}. Thus, $\sim 10\%$ represents ~ 0.20–0.25×10^5 pN μm^{-2}, that is, ~ 0.20–0.25×10^5 pN $\mu m^{-2} \times 1.58 \times 10^{-3}$ $\mu m^2 \sim$ 32–40 pN per intact unit cell (the cross-sectional value of $\sim 1.58 \times 10^{-3}$ μm^2 is recalled above). The small 'compressive' force of ~ 32–40 pN per intact unit cell is also, to my mind, a collapsing force. Otherwise, as recalled on p. **4** in the Introduction, Cecchi et al. (1990) presented a complex study in which a compressive 'tension' of $\sim 0.8 \times 10^5$ pN μm^{-2} was detected when an intact frog fibre passes from rest to contraction (see legend to their Figure 2), corresponding to $\sim 0.8 \times 10^5$ pN $\mu m^{-2} \times 1.58 \times 10^{-3}$ $\mu m^2 \sim 126$ pN per intact unit cell. Averaging these seven values gives $\sim 73 \pm$ (12 or 33) (mean $\pm$ SE or SD; n = 7). Owing to the simplified approaches of Morel and Merah (1997) and Schoenberg (1980b), the approximate value estimated by Bagni et al. (1994b) and the complex explanations of Cecchi et al. (1990), and to ensure self-consistency of the various calculations presented in this monograph, I choose a value of $C_a \sim 73$ pN per intact unit cell. On p. **84** in Section 3.5, it is shown that the full axial contractile force, including the directly generated (swinging cross-bridges/lever arms) and the indirectly generated (lateral swelling) forces, is $F^* \sim 1138$ pN per intact unit cell. Using the conversion coefficient $B_a^* \sim 1$ (see p. **85** in Section 3.5), we deduce that the collapsing force corresponds to an axial force of $\sim 1 \times 73$ pN ~ 73 pN per intact unit cell and we have therefore $100 B_a^* C_a / F^* \sim 100 \times 73$ pN/1138 pN $\sim 6\%$–7%. On p. **35** in Section 3.3, it is recalled that Edman (1979), for example, found, at $\sim 2.5°C$, a mean value of the isometric tetanic tension of $\sim 2.3 \times 10^5$ N m^{-2} and an SE of $\sim 0.1 \times 10^5$ N m^{-2} (n = 6). This value of the SE corresponds to a ratio of $\sim 100 \times 0.1 \times 10^5$ N $m^{-2}/2.3 \times 10^5$ N $m^{-2} \sim 4\%$. Using the corresponding value of SD $\sim 0.2 \times 10^5$ N m^{-2} gives a ratio of $\sim 9\%$. Finally, combining SE and SD ((SE + SD)/2; see Section 2.1), we obtain a ratio of $\sim 6\%$–7%. Thus, the relative strength of the 'radial collapsing' force converted into the 'axial collapsing' force $100 B_a C_a^* / F^* \sim 6\%$–7% (see above) is of the same order of magnitude as the experimental uncertainty on the values of isometric tetanic tension, regardless of the statistical test employed. The other conclusion that can be drawn by comparing the strength of the radial collapsing force $C_a \sim 73$ pN per intact unit cell with that of the radial repulsive electrostatic force $4(\gamma/2)F_{e,a} \sim 598$ pN per intact unit cell (see p. **142** in Section 4.4.2.3) is that $100C_a/4(\gamma/2)F_{e,a} \sim 12\%$, which is again of the same order of magnitude as other error bars ($\sim 4\%$, $\sim 6\%$–7% and $\sim 9\%$; see above). These findings and calculations are at odds with what is claimed by many authors, who, in my opinion, misinterpret the experimental data, attributing too great an importance and an unlikely role to strong radial forces resulting solely from rotation of the cross-bridges during isometric tetanic contraction (e.g. Brenner and Yu 1991; Matsubara et al. 1985; Xu et al. 1993; see also Millman 1998 for a review). This involuntary mistake is analysed and discussed in Section 4.4.2.1, on the basis of the hybrid model.

The contribution of half the weakly bound cross-bridges (see discussion on p. **171** in Section 5.1.1, regarding this coefficient of 1/2) to the radial forces, at rest, is given by the radial tethering force $N_0(c, I^*)F^0_{cb}(c, I^*)$ pN per intact unit cell. Under contraction conditions, the residual radial tethering force is given by $N(c, I^*)F_{cb}(c, I^*)$, with $N(c, I^*)F_{cb}(c, I^*) < N_0(c, I^*)F^0_{cb}(c, I^*)$ and, in most cases (particularly at 10°C, the reference temperature in this work), $N(c, I^*)F_{cb}(c, I^*) \ll N_0(c, I^*)F^0_{cb}(c, I^*)$ (see below, in this section, and several sections of this chapter). At this stage, we should examine whether this marked inequality is consistent with the experimental data and various semi-empirical results obtained in this book. From the definitions of Θ_a and Θ_r given in the second paragraph, under physiological conditions (intact fibres or half-fibres at the slack length; $I = 180$ mM/$I^* \sim 182$ mM, the notations I/I^* are omitted below, in this section, unless otherwise specified), it is shown, on p. **85** in Section 3.5, that $\Theta_a = NF_{cb} + C_a + \Lambda_a = \rho_a 4(\gamma/2)F_{e,a}$, with $4(\gamma/2)F_{e,a} \sim 784Q_a \sim 598$ pN per intact unit cell (see the preceding paragraph), and, therefore, $\Theta_a \sim 784\rho_a Q_a \sim 144$ pN per intact unit cell ($\rho_a \sim 0.240$ and $Q_a \sim 0.763$; see text below Equation 3.49), giving $NF_{cb} \sim 144 - (C_a + \Lambda_a)$ pN per intact unit cell. Under resting conditions, there is obviously no swinging of the 'absent attached cross-bridges' (the weakly bound cross-bridges, present at rest and

described in Section 5.1.1 on pp. **170–172** and **174–176** [see also pp. **176–180** for complementary information], are negligible in number and cannot swing, because no chemical energy is available; see pp. **174–176** in Section 5.1.1) and there are no collapsing forces ($C_r = 0$). Under the same conditions of ionic strength and sarcomere length and at the temperatures traditionally used in the laboratory (these points are very important; see Sections 4.4.2.7 and 4.4.2.8, concerning the role of I^*, and many paragraphs of Sections 6.3 and 6.4, as well as Sections 6.5, 6.6 and 8.8, regarding the complex role of temperature), there is a precise balance between the radial repulsive electrostatic force, $4(\gamma/2)F_{e,r}$, and the radial tethering plus attractive/compressive forces, $\Theta_r = N_0 F^0_{cb} + \Lambda_r$. We have therefore $\Theta_r = N_0 F^0_{cb} + \Lambda_r = 4(\gamma/2)F_{e,r} \sim 784Q_r(182) \sim 468$ pN per intact unit cell (see p. **142** in Section 4.4.2.3, concerning the value of $Q_r(182) \sim 0.597$). We deduce that $N_0 F^0_{cb} \sim (468 - \Lambda_r)$ pN per intact unit cell. Thus, we have $NF_{cb}/N_0 F^0_{cb} \sim (144 - C_a - \Lambda_a)/(468 - \Lambda_r)$. Introducing $C_a \sim 73$ pN per intact unit cell, as estimated in the preceding paragraph, we deduce that $NF_{cb}/N_0 F^0_{cb} \sim (71 - \Lambda_a)/(468 - \Lambda_r)$. This ratio is discussed and quantified in the next paragraph.

Millman (1998) considered Λ_r to be the major radial attractive/compressive force counterbalancing the radial repulsive electrostatic force at rest. In the absence of experimental data in 1998, concerning the radial tethering forces, his view was justified. However, the existence of the radial tethering forces is experimentally demonstrated in Section 4.3, and the introduction of these forces into the various calculations leads to a series of self-consistent conclusions in this work. I consider the radial tethering forces to be much stronger than the radial attractive/compressive forces, at least at rest and under physiological conditions of ionic strength ($I = 180$ mM; $I^* \sim 182$ mM) and in the temperature conditions used in the laboratory (between ~ 0°C and ~ 20°C–22°C). Λ_a would be slightly higher than Λ_r, because $\Gamma_a(c^*, I^*)$ is probably higher than $\Gamma_r(c, I^*)$ (see third point in Section 4.4.2.4 for the definition of the force Γ, p. **156** concerning the relationship between Λ and Γ and p. **156** in Section 4.4.2.5.3 concerning the difference between Γ_a and Γ_r). As an illustrative example, let us take $\Lambda_a \sim 7$ pN and $\Lambda_r \sim 5$ pN per intact unit cell, from which we obtain (see the end of the preceding paragraph)

$NF_{cb}/N_0F_{cb}^0 \sim (71-7)/(468-5) \sim 0.138$. Ignoring the likely condition $\Lambda_a > \Lambda_r$ and taking $\Lambda_a \sim \Lambda_r \sim 0$, we obtain $NF_{cb}/N_0F_{cb}^0 \sim 71/468 \sim 0.152$, and choosing $\Lambda_a \sim \Lambda_r \sim 9$ pN per intact unit cell, the ratio is $\sim (71-9)/(468-9) \sim 0.135$. Averaging these three values, we obtain $\sim 0.142 \pm (0.004$ or $0.008)$ (mean $\pm$ SE or SD; n = 3). For the sake of homogeneity with other calculations presented in this monograph, I select $NF_{cb}/N_0F_{cb}^0 \sim 0.14 (\sim 14\%)$. This value is assumed to be valid in the standard experimental conditions used here (10°C; bulk ionic strength I = 180 mM, around the slack length, i.e. $s_0 \sim 2.18$ μm [see p. **24** in Section 3.2]), from which we can definitively conclude that $NF_{cb} \ll N_0F_{cb}^0$, as emphasised at the beginning of the preceding paragraph. However, at bulk ionic strengths lower than ~ 40 mM, other important phenomena occur (see pp. **164–167** in Section 4.4.2.8 for necessary information). Other phenomena also occur when the temperature increases above critical values (temperature-induced contracture; see Section 8.5). I assert implicitly at the beginning of this paragraph that, under 'standard' resting conditions, we have $N_0F_{cb}^0 \gg \Lambda_r$. We need to check that this assumption is consistent with the other calculations presented here. As demonstrated at the beginning of the preceding paragraph and above, the most probable value of $NF_{cb}/N_0F_{cb}^0$ is $\sim (71-\Lambda_a)/(468-\Lambda_r) \sim 0.14$. Λ_a is probably slightly higher than Λ_r, as recalled above, and we can assume, as an illustrative example, that $\Lambda_a \sim 1.2\Lambda_r$. We deduce that $\Lambda_r \sim 5.2$ pN and $\Lambda_a \sim 6.2$ pN per intact unit cell. It is recalled, at the end of the preceding paragraph, that $N_0F_{cb}^0 + \Lambda_r = 4(\gamma/2)F_{e,r} \sim 468$ pN per intact unit cell, giving $N_0F_{cb}^0 \sim 468-5.2 \sim 462.8$ pN per intact unit cell and, therefore, $100\Lambda_r/N_0F_{cb}^0 \sim 100 \times 5.2pN/462.8$ pN $\sim 1.12\%$, that is, $\Lambda_r \sim 0.0112 N_0F_{cb}^0$, giving $N_0F_{cb}^0 \sim 89\Lambda_r \gg \Lambda_r$ and confirming the assumption made above. We also deduce that $100\Lambda_r/4(\gamma/2)F_{e,r} \sim 100 \times 5.2$ pN$/468$ pN $\sim 1.11\%$ and $100\Lambda_a/4(\gamma/2)F_{e,a} \sim 100 \times 6.2$ pN$/598$ pN $\sim 1.04\%$. In other words, the attractive/compressive forces Λ_r and Λ_a represent $\sim 1\%$ of the radial repulsive electrostatic forces, regardless of the state of the intact unit cell (contraction or rest). Although these are merely orders of magnitude, the conditions, stated at the beginning of this paragraph, under which $N_0F_{cb}^0$ largely prevails over Λ_r, are fulfilled and the self-consistency of the various calculations and reasoning is ensured.

Finally, we deduce that $NF_{cb} \sim 0.14 N_0F_{cb}^0 \sim 0.14 \times 462.8 \sim 64.8$pN per intact unit cell. From $NF_{cb} \sim 64.8$ pN per intact unit cell, $C_a \sim 73.0$ pN per intact unit cell (see p. **157**) and $\Lambda_a \sim 6.2$ pN per intact unit cell (see above), we deduce that $\Theta_a \sim NF_{cb} + C_a + \Lambda_a \sim 144$ pN per intact unit cell. On p. **158** in the preceding paragraph, I obviously obtain an identical value (~ 144 pN per intact unit cell). Under isometric tetanic contraction conditions, we have a radial repulsive electrostatic force $4(\gamma/2)F_{e,a} \sim 598$ pN per unit cell (see the preceding paragraph), and therefore, $\Theta_a/4(\gamma/2)F_{e,a} \sim 144$ pN$/598$ pN ~ 0.24 ($\sim 24\%$, obviously identical to the value of $\sim 24\%$ found in Section 3.5, text below Equation 3.49). Finally, multiplying Θ_a by $B_a^* \sim 1$ (coefficient required for the conversion of net radial repulsive forces into axial contractile forces; see Section 3.5, text below Equation 3.49) gives $B_a^*\Theta_a \sim 144$ pN per intact unit cell, that is, $\sim 100 \times 144$ pN$/1138$ pN $\sim 12.7\%$ of the full axial contractile force $F^* \sim 1138$ pN per intact unit cell (see p. **85** in Section 3.5 for precise details on F^*). This non-negligible proportion of $\sim 12.7\%$, at 10°C, corresponding to the conversion of the residual radial tethering, collapsing and attractive/compressive forces into axial contractile forces almost certainly depends on temperature.

Let us call $\Sigma F_x(c, I^*)$ (x = a or r) the net radial repulsive force exerted in an intact unit cell between the myofilaments, at full overlap, γ, that is, at the slack length of an intact unit cell (see Section 3.1 for necessary references concerning the notion of intact unit cells). Considerably improving the pure lateral swelling model suggested by Morel (1975), Morel and Gingold (1979b), Morel and Pinset-Härström (1975b) and Morel et al. (1976), and taking into account the radial repulsive electrostatic, the radial tethering, collapsing and attractive/compressive forces defined and discussed in Sections 4.4.2.3, 4.4.2.4, and in this section, we can write

$$\Sigma F_x(c, I^*) \sim 4(\gamma/2)F_{e,x}(c, I^*) - \Theta_x(c, I^*) \quad (4.8)$$

with

$$\Theta_x(c, I^*) \sim N_x(c, I^*)F_{cb}^X(c, I^*)$$
$$+ C_x(c, I^*) + \Lambda_x(c, I^*)$$

4.4.2.6 General Considerations on the Relative Resting and Isometric Tetanic Tensions (Short Tetani–Steady State) in an Intact Unit Cell, Associated with Net Radial Repulsive (Expansive) Forces between Myofilaments. Related Problems

For short isometric tetani, under steady-state conditions, in vivo/in situ and at full overlap between the thin actin and thick myosin filaments, it is proposed and repeatedly confirmed in this monograph that ~60% of the axial contractile force is related to swinging cross-bridge/lever-arm mechanisms. The other ~40% of this force is related to the net radial repulsive (expansive) force, converted into an axial contractile force (see below concerning the conversion process). In Sections 2.2, 3.3, 3.5, 4.4.1, 7.2 and 7.3, it is shown that half-fibres, in the presence of MS^- as the major anion, behave very differently from traditionally demembranated fibres in the presence of the usual anions (e.g. Cl^- or propionate), instead closely resembling intact fibres. We can simplify the notations as follows: the centre-to-centre distance c is omitted, because I study only isometrically held resting and contracting half-fibres at a fixed sarcomere length (2.5 ± 0.2 μm, close to the reference length $s_0 = 2.18 ± (0.5$ or $0.18)$ μm; see p. **14** in Section 2.2 and p. **24** in Section 3.2, respectively), that is, at a fixed value of c (constant volume behaviour). Cecchi et al. (1990), Millman (1998), Neering et al. (1991), Yagi et al. (2004) and some other authors challenged this view (the constant volume relationship does not hold, particularly when the fibres pass from rest to isometric contraction; see pp. **4–5** in the Introduction for a short discussions), whereas April et al. (1971), Brandt et al. (1967) and Elliott et al. (1963, 1965, 1967) supported this view (constant volume behaviour). Briefly, Cecchi et al. (1990), working on intact frog fibres in physiological buffers, at 4°C, interpreted their experimental results as indicating that there are weak compressive forces and that c decreases slightly when the fibre goes from rest to isometric contraction. By contrast, Millman (1998) and Neering et al. (1991), for example, under the same conditions, observed the opposite phenomenon, that is, c increased slightly or more markedly. Thus, the limited changes in filament spacing c in intact fibres (and intact unit cells), observed by these groups, can be neglected here, as contradictory results have been

reported by Cecchi et al. (1990), on the one hand, and Millman (1998) and Neering et al. (1991), on the other. Moreover, the half-fibres used in the experimental part of this work behave as intact fibres (see Section 3.3) and the conclusions that can be drawn, concerning the approximate constancy of c, are similar in intact fibres and half-fibres. Even when the resting half-fibres displayed significant lateral swelling (see pp. **164–167** in Section 4.4.2.8 for a circumstantial discussion), the increase in c was generally not taken into account (except, e.g. on pp. **129–130** in Section 4.4.1), owing to the scattering of the experimental points, which conceals small and even some marked increases in c (see Section 4.2.2, including Figure 4.2 and the corresponding comments). On p. **85** in Section 3.5, it is assumed that there is a one-to-one conversion of the net radial expansive forces into electrostatically related axial forces ($B_a^* \sim 1$). Thus, under the conditions defined above, we deduce from Equation 4.8:

$$T_a^e(I^*) = (1/\text{IUCCSA})[4(\gamma/2)F_{e,a}(I^*) - \Theta_a(I^*)] \tag{4.9}$$

$T_a^e(I^*)$ is the isometric tetanic tension (expressed in pN μm^{-2}) corresponding to the conversion of the net radial repulsive (expansive) forces into axial contractile forces. The acronym IUCCSA corresponds to the 'intact unit cell cross-sectional area' (expressed in μm^2). Equation 4.9 can also be written:

$$T_a^e(I^*) = (1/\text{IUCCSA})[1 - \rho_a(I^*)]4(\gamma/2)F_{e,a}(I^*) \tag{4.10}$$

with $\rho_a(I^*) = \Theta_a(I^*)/4\gamma F_{e,a}(I^*)$ (see also p. **85** in Section 3.5 concerning this definition). Assuming that, at rest, there is also a one-to-one conversion process ($B_r^* \sim 1$; see below for a brief comment), an equation of the same type as Equations 4.9 and 4.10 can be written:

$$\begin{aligned} T_r(I^*) &= (1/\text{IUCCSA})[4(\gamma/2)F_{e,r}(I^*) - \Theta_r(I^*)] \\ &= (1/\text{UCCSA})[1 - \rho_r(I^*)]4(\gamma/2)F_{e,r}(I^*) \end{aligned} \tag{4.11}$$

with $\rho_r(I^*) = \Theta_r(I^*)/4(\gamma/2)F_{e,r}(I^*)$ (generalisation of the definition of $\rho_a(I^*)$). As recalled on pp. **102–103** in Section 3.10, Dragomir et al. (1976) and

Elliott (1974) proposed similar mechanisms for the translation of radial expansive forces into axial contractile forces, valid solely during contraction, whereas Morel et al. (1976) proposed a mechanism valid both under contraction conditions and at rest. Thus, at rest, only the mechanism proposed by Morel et al. (1976) can apply. Under contraction conditions, all types of mechanism are valid and may be simultaneously involved in the conversion process. Unknown phenomena, different at rest and during contraction, may be involved. It is therefore impossible to predict whether $B_a(I^*)$ and $B_r(I^*)$ are identical and both close to 1. However, as a reasonable assumption, I choose to write $B_r(I^*) \sim B_a(I^*) \sim 1$, regardless of I^* (one-to-one conversion process at rest and under contraction conditions). This simplifying hypothesis results in self-consistent conclusions. Thus, Equations 3.2, 4.10 and 4.11 can be combined to give

$$T_r(I^*)/T_a^e(182) = rel.\ T_r(I^*)/rel.\ T_a^e(182) = G_r(I^*)$$

$$(I^*/182)^{3/4} \exp[H(13.5 - I^{*1/2})][1 - \rho_r(I^*)]$$
$$(4.12)$$

with the definition

$$G_r(I^*) = Q_r(I^*)/[1 - \rho_a(182)]Q_a(182) \quad (4.13)$$

Equation 4.12 and Definition 4.13 are presented in another form on p. **163** in Section 4.4.2.8 (see Equation 4.17). On p. **95** in Section 3.7, it is highlighted that relative tensions are identical in intact unit cells and half-fibres. Thus, Equation 4.12 is valid for both biological materials. In Figure 4.2, at $I = 180$ mM ($I^* \sim 182$ mM), the relative electrostatically related isometric tetanic tension in the three half-fibres is therefore identical to rel. T_a^e (182) in Equation 4.12 and is taken as 1 (reference value). Similarly, the relative electrostatically related resting tension in the three half-fibres is identical to rel. $T_r(I^*)$ in Equation 4.12 and rel. $T_r(182)$ is taken as 0 (reference value). We can deduce from Equation 4.12 that $\rho_r(182) = 1$, which means that the radial tethering and attractive/compressive forces exactly counterbalance the radial repulsive electrostatic forces. $Q_r(182)$ and $Q_a(182)$ are highly unlikely to be similar. There are also probably limited variations in both Q_r and Q_a with I^* (see Section 4.4.2.7). At this stage, we can assume that, when a resting fibre is shifted to isometric tetanic

contraction, or when the vicinal ionic strength in a half-fibre (more generally in a demembranated fibre) is altered, several changes in protein conformation occur, leading to rearrangements in the fixed negative electrical charges, as strongly suggested on p. **141** in Section 4.4.2.3. Indeed, it is pointed out in Section 4.4.2.3 that many conformational changes occur in both the thick myosin and thin actin filaments when contraction is triggered, under physiological conditions of ionic strength and pH. Moreover, the number of anions bound to the myofilaments may also change with state (rest or contraction) and ionic strength (see Section 4.4.2.2 for a brief discussion of the bound anions). The various observations (experimental and semi-empirical) and the various phenomena recalled above certainly alter, to some extent, the values of the electrical surface potentials. All these preliminary inferences are analysed and quantified in the next paragraph.

For simplification, and because we cannot obtain separate estimates of the surface electrical potentials on the two types of myofilament (thin and thick), let us assume that $\Psi_{a,x} \sim \Psi_{m,x} \sim \Psi_{0,x}$, giving $Q_x \sim [\tanh(e\Psi_{0,x}/4kT)]^2$ ($\Psi_{a,x}$ and $\Psi_{m,x}$ are defined in the text below Equation 3.1). In Section 3.5 (see text below Equation 3.49), it is strongly suggested that $Q_a(182) \sim 0.763$, corresponding to $\Psi_{0,a}(182) \sim -131$ mV. This value is comparable to other potentials found in any living cell: values of membrane potential between ~ -70 mV and ~ -100 mV (see Alberts et al. 2007 and Lehninger 2008), a value of ~ -100 mV for the surface potential on a myosin head in a muscle fibre, suggested by Yu et al. (1970), and values for the depth of the potential wells in muscle of $\sim -50/100$ mV proposed by Regini and Elliott (2001). At this point, it would be unreasonable to assume that $Q_r(182)$ is identical to $Q_a(182)$, owing to the various conformational changes occurring during the shift from rest to contraction and vice versa (see text below Equation 4.12 and Definition 4.13). $Q_r(182)$ is also highly unlikely to be higher than $Q_a(182)$, because this would give absolute values of $\Psi_{0,r}(182)$ exceeding ~ 131 mV, which is much too high. Only the inner membrane of mitochondria, during full activation (oxidative phosphorylation and proton translocation), displays very high absolute values of potential of up to $\sim 150/180$ mV, corresponding to hyperpolarisation, according to Mitchell's theory (see the original paper by Mitchell 1974 and the generalist books by Alberts et al. 2007 and Lehninger 2008). It is therefore almost certain that $Q_r(182) < Q_a(182)$. The mean of the five values cited above, with the obvious exception of ~ -131 mV and $\sim -150/180$ mV, is ~ -84 mV, giving an order of magnitude of $\Psi_{0,r}(182)$. For self-consistency in the calculations and conclusions, I select a value of $\Psi_{0,r}(182) \sim -100$ mV here. Such a marked increase of $\sim 100 \times (131 - 100)/100 \sim 31\%$ in the absolute value of surface electrical potential, during the shift from rest to contraction, certainly involves major changes in the properties of the myofilaments (e.g. significant conformational changes, as recalled in the text below Equation 4.12 and Definition 4.13, followed by changes in the number of fixed electrical charges and bound anions). We deduce from the value of $\Psi_{0,r}(182) \sim -100$ mV that $Q_r(182) \sim 0.597$. Finally, the problem of the dependence on I^* of various parameters, including $Q_a(I^*)$ (under contraction conditions; short tetani, steady state), in one half-fibre, is discussed in Section 4.4.2.7.

4.4.2.7 Electrostatically Related Relative Isometric Tetanic Tension (Short Tetani– Steady State) as a Function of Ionic Strength, in Intact (and Some Mechanically Skinned) Crayfish Fibres and in Half-Fibres from Young Adult Frogs

We first need to study the dependence on sarcoplasmic ionic strength I (or vicinal ionic strength I^*) of the relative isometric tetanic tension recorded in intact and skinned fibres from crayfish (Orconectes) walking legs by April and Brandt (1973), and also for half-fibres (see the experimental points at I = 180 mM and the two experimental points (Δ) at I = 60 mM, in Figure 4.2), validating the choice of the experimental data of April and Brandt 1973; see also pp. **106–109** in Section 4.1 for a circumstantial discussion). Only the notation I^* is used for simplification. We also need to interpret the results, based essentially on Equation 3.2 and the various parameters included in this expression. From Equations 3.2 and 4.10, given that $rel.\,T_a^e(182) = 1$ (see p. **161** in Section 4.4.2.6), we deduce, by analogy with Equation 4.12 and Definition 4.13, that

$$T_a^e(I^*)/T_a^e(182) = rel.\,T_a^e(I^*)$$

$$/rel.T_a^e(182) \sim rel.\,T_a^e(I^*) \sim$$

$$K_a(I^*)(I^*/182)^{3/4}\exp[H(13.5 - I^{*1/2})] \quad (4.14)$$

with the following definition:

$$K_a(I^*) = [1 - \rho_a(I^*)]Q_a(I^*)/[1 - \rho_a(182)]Q_a(182) \quad (4.15)$$

April and Brandt (1973) performed their experiments at sarcoplasmic ionic strengths greater than ~ 50 mM ($I^* \sim 52$ mM) for many intact crayfish fibres and four skinned fibres. Thus, below $I \sim 50$ mM, the dotted curve in Figure 4.2 is merely a mathematical extrapolation of Equation 4.16 below. In Section 4.4.2.2, it is recalled that, under resting conditions, the number of fixed negative electrical charges on thick myosin and F-actin filaments, measured in vitro, decreases with decreasing ionic strength. Very little is known about possible variations in the number of bound anions with ionic strength (see Section 4.4.2.2). Nothing is known about the dependence on ionic strength of the other possible phenomena contributing to repulsive forces mentioned on pp. **141–143** in Section 4.4.2.3, including the swelling pressure related to Donnan-osmotic effects $D_a(I^*)$ and the entropic repulsive forces $E_a(I^*)$. $\Xi_r(I^*)$ and almost certainly $\Xi_a(I^*)$ depend on ionic strength (see p. **142** in Section 4.4.2.3). However, we can reasonably assume that the major feature, resulting from these various phenomena, is equivalent to a moderate change in electrical surface potentials with decreasing I^*, leading to limited changes in $Q_a(I^*)$ (see p. **142** in Section 4.4.2.3 and below for confirmation). As an oversimplification, let us take $K_a(I^*) = (I^*/182)^p$, this very simple expression being valid between $I^* \sim 52$ mM and $I^* \sim 182$ mM and obeying the necessary condition $K_a(182) = 1$. The best fit of Equation 4.14 to the experimental points obtained by April and Brandt (1973) was found for $p \sim 0.1$ and $H \sim 0.26$ mM$^{-1/2}$, with a simple routine. Thus, Equation 4.14 can be rewritten (after simple manipulations):

$$rel.\,T_a^e(I^*) \sim 33.5(I^*/182)^{0.85}\exp(-0.26\,I^{*1/2}) \quad (4.16)$$

The dotted line in Figure 4.2 represents the best fit for I > 50 mM ($I^* > 52$ mM) and is a mathematical extrapolation of Equation 4.16 below $I \sim 50$ mM ($I^* \sim 52$ mM), as already highlighted above. My two experimental points (Δ), at I =

60 mM, were obtained with the half-fibre (▲) (see p. **119** in Section 4.2.2). Working on mechanically skinned frog fibres at room temperature (unfavourable conditions for a cold-blooded animal), Gordon et al. (1973) observed, in their Figure 2, a maximum in tension within the ~80–120 mM range and a rapid decrease below ~80 mM (see p. **110** in Section 4.1). These results support qualitatively Equation 4.16 and Figure 4.2. As pointed out on pp. **110–111** in Section 4.1, it can be estimated from Figure 1 in the paper by Hasan and Unsworth (1985) that, for contracting intact toad fibres at ~5°C (favourable conditions for a cold-blooded animal), the isometric tetanic tension passes through a maximum within the range of intracellular ionic strength of ~75–110 mM and decreases rapidly below ~75 mM. Thus, these experimental results, obtained under conditions different from those used here, also support qualitatively Equation 4.16 and Figure 4.2 (see dotted line: maximal extrapolated tension at bulk ionic strength of ~20–30 mM and rapid decrease below ~20 mM). $K_a(I^*)$ is given by Definition 4.15 and, from its semi-empirical expression (see text above Equation 4.16), we deduce that $[1 - \rho_a(I^*)] \, Q_a(I^*)/[1 - \rho_a(182)]Q_a(182) \sim (I^*/182)^{0.1}$. In Section 3.5 (see text below Equation 3.49), it is shown that $[1 - \rho_a(182)]Q_a(182) \sim 0.580$ (because $B_a^* \sim 1$; see p. **85** in Section 4.4.2.7). We deduce that $[1 - \rho_a(I^*)]Q_a(I^*) \sim 0.580(I^*/182)^{0.1}$, and, therefore, at $I^* \sim 52$ mM, $[1 - \rho_a(52)]Q_a(52) \sim 0.513$. As suggested in Section 3.5 (see text below Equation 3.49), the two unknown parameters $[1 - \rho_a(52)]$ and $Q_a(52)$ may be assumed to be identical, giving $[1 - \rho_a(52)] \sim Q_a(52) \sim 0.716$. Thus, it is reasonable to choose $Q_a(52) \sim 0.716$ (giving a suitable value of $\Psi_{0,a}(52) \sim -121$ mV). Moreover, we have $Q_a(182) \sim 0.763$ (giving $\Psi_{0,a}(182) \sim -131$ mV; see p. **85** in Section 3.5). We deduce that $Q_a(I^*)$ is not very sensitive to I^*, at least between ~52 mM and ~182 mM ($100 \times (0.716 - 0.763)/0.763 \sim -5\%$). We also deduce that $[1 - \rho_a(52)] \sim 0.716$ and, therefore, $\rho_a(52) \sim 0.284$, slightly greater than $\rho_a(182) \sim 0.240$ (see text below Equation 3.49), but of the same order of magnitude. We have, by definition, $\rho_a = \Theta_a/4(\gamma/2)F_{e,a}$ (see Section 3.5, text above Equation 3.49) and, from Equation 3.2 and introducing $Q_a(182) \sim 0.763$ and $Q_a(52) \sim 0.716$ (see above), we deduce that $F_{e,a}(52)/F_{e,a}(182) \sim 4(\gamma/2)F_{e,r}(52)/4(\gamma/2)F_{e,r}(182) \sim 1.88$ ($H \sim 0.26$ mM$^{-1/2}$; see text above Equation 4.16). We also deduce that $\Theta_a(52)/\Theta_a(182) \sim 1.88[\rho_a(52)/\rho_a(182)] \sim$

$1.88(0.284/0.240) \sim 2.22$, giving $\Theta_a(182) \sim 2.22 \times 144 \sim 320$ pN per intact unit cell (the value of $\Theta_a(182) \sim 144$ pN per intact unit cell is given on p. **158** in Section 4.4.2.5.4). Thus, the sum of the residual radial tethering, collapsing and attractive/compressive forces Θ_a (see Section 4.4.2.5.4 for quantitative estimates of these forces) would increase with decreasing vicinal ionic strength, at least within the range ~52–182 mM. This conclusion should, however, be considered with caution, because of the many simplifications and approximations made above. Nonetheless, many conclusions drawn in this section lead to self-consistent inferences, are biophysically valid and confirm the various hypotheses and semi-empirical approaches presented in this work.

4.4.2.8 Relative Resting Tension as a Function of Ionic Strength in Three Half-Fibres from Young Adult Frogs

Let us study resting tension within the range of bulk ionic strength between ~0 (distilled water) and 180 mM (solid line in Figure 4.2). For reasons given on pp. **174–176** in Section 5.1.1, the weakly bound cross-bridges cannot swing at rest, and only the net radial repulsive (expansive) forces contribute to the resting tension, because the axial resting forces are probably all negligible in the slightly stretched half-fibres used here (see p. **142** in Section 4.4.2.3, regarding the axial elastic force $\Xi_x(s, I^*)$ [x = a under contraction conditions, x = r at rest], mostly related to titin). Figure 4.2 shows that, for bulk ionic strengths between ~40 mM and 180 mM, the resting tension is almost negligible (the reference zero corresponds to 180 mM) and the relative electrostatically related tension is taken as the reference 1 at 180 mM (see p. **119** and Figure 4.2 in Section 4.2.2 and p. **161** in Section 4.4.2.6 for the definition of these two reference tensions). The relative tension $rel.\, T_a^e(182)$ being 1, Equation 4.12 becomes (see also Definition 4.13 of $G_r(I^*)$)

$$rel.\, T_r(I^*) = G_r(I^*)(I^*/182)^{3/4}[\exp H(13.5 - I^{*1/2})][1 - \rho_r(I^*)] \tag{4.17}$$

For vicinal ionic strength I^* between ~42 mM and ~182 mM (bulk ionic strength I between ~40 mM and 180 mM; see Figure 4.2), the values of $rel.\, T_r(I^*)$ are negligible. We deduce, from Equation 4.17, that $\rho_r(I^*) \sim 1$, and, from the definition of ρ_r given on p. **160** in Section 4.4.2.6, it results that $\Theta_r(I^*) \sim 4(\gamma/2)F_{e,r}(I^*)$: the radial tethering and attractive/compressive forces counterbalance the radial repulsive electrostatic force. In this context, it is important to cite the experimental work of Matsubara and Elliott (1972), who performed traditional x-ray diffraction experiments on resting mechanically skinned frog fibres at the slack length, in the presence of Cl⁻ as the major anion, to determine the lattice spacing, d_{10}, under various conditions. The use of Cl⁻ makes it possibly difficult to interpret the results, as pointed out on p. **14** in Section 2.2. Despite these problems, the results presented by these authors are qualitatively valid (see p. **242** in Section 8.7). They found that lowering the bulk ionic strength from 160 mM to 90 mM resulted in no detectable lateral swelling of the skinned fibres. We can conclude that, in these conditions, there is no net lateral swelling pressure, resulting in negligible resting tensions, owing to the automatic conversion of the net radial expansive forces into axial forces, demonstrated at many places in this monograph (see pp. **102–103** in Section 3.10). The experimental observation of Matsubara and Elliott (1972) is consistent with the experimental findings obtained here, on very different experimental bases, that is, rel. $T_r(I^*) \sim 0$ and $\rho_r(I^*) \sim 1$ for bulk ionic strengths between ~40 mM and 180 mM (see above). This conclusion is the same as that drawn on pp. **147–148** in Section 4.4.2.5.1, regarding the experimental work of Maughan and Godt (1980). These three concurrent series of experimental observations, obtained by independent groups, have the same origin, and we deduce that, for $I^* > 42$ mM, there is no net repulsive (expansive) force and therefore no increase in myofilament lattice spacing or fibre width.

The reasoning presented in the preceding paragraph is valid only for bulk ionic strengths I between ~40 mM and 180 mM (vicinal ionic strengths I^* between ~42 mM and ~182 mM). For bulk ionic strengths below ~40 mM, it is possible that the axial resting forces, represented by $\Xi_r(s, I^*)$, which are negligible at $I^* \sim 182$ mM (see p. **142** in Section 4.4.2.3), are not negligible. However, this is highly unlikely and, moreover, neglecting $\Xi_r(s, I^*)$ does not lead to inconsistencies at any ionic strength. The three most remarkable features of the solid line, representing the relative resting tension rel. $T_r(I)$ in Figure 4.2, are as follows: (i) its extremely rapid increase within a very narrow range of values of I (between ~40 mM and ~30 mM), (ii) its peak at I ~ 20–30 mM and (iii) its drop for very low values of I (below ~20 mM). These characteristics have been qualitatively 'predicted' and supported experimentally by independent groups (working however on isometrically contracting fibres) (see pp. **101–102** in Section 3.10 and pp. **109–111** in Section 4.1): when ionic strength decreases, radial repulsive electrostatic forces must peak and then decrease at low ionic strength, like axial forces (see pp. **102–103** in Section 3.10, concerning the conversion of radial forces into axial forces). These features are discussed and confirmed in the following paragraphs.

As pointed out and discussed in the preceding paragraphs, when I^* decreases from ~182 mM to ~42 mM, rel. $T_r(I^*)$ remains extremely low, and, from Equation 4.17, we deduce that the radial tethering and attractive/compressive force $\Theta_r(I^*)$ counterbalances the radial repulsive electrostatic force $4(\gamma/2)F_{e,r}(I^*)$, that is, $\Theta_r(I^*) \sim 4(\gamma/2)F_{e,r}(I^*)$. This repulsive electrostatic force $4(\gamma/2)F_{e,r}(I^*)$ increases substantially within this range of ionic strength (see Equation 3.2). The force $\Theta_r(I^*)$ therefore necessarily behaves similarly, that is, also increases substantially with decreasing I^*, to maintain the balance between the repulsive electrostatic and tethering–attractive/compressive forces and the subsequent negligible resting axial force (see pp. **102–103** in Section 3.10 concerning the conversion of radial expansive forces into axial forces). From the discussion presented in Section 4.4.2.5.4, concerning the radial tethering–attractive/compressive force $\Theta_r(I^*) = F_{cb}^0(I^*) + \Lambda_r(I^*)$, the major term is the radial tethering force $N_0(I^*)F_{cb}^0(I^*)$ at $I^* \sim 182$ mM ($N_0(182)F_{cb}^0(182) \sim 89\Lambda_r(182)$; see p. **159** in Section 4.4.2.5.4). As a first approximation, we can reasonably assume that this predominance of $N_0(182)F_{cb}^0(182)$ holds regardless of the value of I^*, at least within the range ~42–182 mM. Thus, there is necessarily an increase in $N_0(I^*)F_{cb}^0(I^*)$ when I^* decreases within this range. At this stage, no information is available concerning the contingent variations of F_{cb}^0 with I^*. However, this force F_{cb}^0 (described and analysed in Section 5.1.1, on pp. **174–175**, and in Figure 5.1, its legend and Appendix I) probably depends little on I^*. In these conditions, the only way in which a significant increase in $N_0(I^*)F_{cb}^0(I^*)$ can be assumed is to suggest a sizeable increase in $N_0(I^*)$ with decreasing I^*. This assumption is consistent with several experimental studies, performed by independent groups, in an entirely different context from that studied here. Indeed, it is now well established that the total number of weakly bound cross-bridges and $N_0(I^*)$ (see below concerning the difference between the total number of weakly bound cross-bridges and N_0) steadily increase with decreasing I^*, particularly when the bulk ionic strength I is low (e.g. Brenner et al. 1982, 1984, 1986; Kraft et al. 1995; Squire et al. 1991; Walker et al. 1994; Yu and Brenner 1989). For instance, Brenner et al. (1982, 1984, 1986) and Yu and Brenner (1989), working on chemically skinned fibres from rabbit psoas muscle at 5°C (unfavourable conditions for a warm-blooded animal, as pointed out at many places in this monograph), under resting conditions and at a bulk ionic strength of ~20 mM, found that, at least, ~60% of the cross-bridges are attached to the thin actin filaments in weak binding states (100% correspond to rigor; see p. **148** in Section 4.4.2.5.2; ~82% corresponds to isometric tetanic contraction conditions; see pp. **152–153** in Section 4.4.2.5.2). Working on the same biological material, but at 25°C (favourable conditions for a warm-blooded animal), Kraft et al. (1995) estimated that, under relaxing conditions, ~5% of the weakly bound cross-bridges are attached at a bulk ionic strength of 200 mM, whereas ~20% are attached at 50 mM. As strongly suggested in the next paragraph, in the frog half-fibres studied in the experimental part of this book, at a bulk ionic strength of ~20–30 mM, all the active weakly bound cross-bridges were detached (the number N_0 of active weakly bound cross-bridges = half the total number of weakly bound cross-bridges; see p. **171** in Section 5.1.1 concerning this factor of 1/2). However, frog half-fibres in the presence of MS^- as the major anion cannot be quantitatively compared with chemically skinned fibres from rabbit psoas muscle in the presence of chloride as the major anion or other non-benign anions (see Andrews et al. 1991 for comparative studies of various anions), as demonstrated at many places in this work (see Section 3.3, pp. **96–97** in Section 3.8.1 and p. **221** in Section 7.2). More precisely, it is shown, in Section 3.3, that the half-fibres, in the presence of MS^-, develop the same isometric tetanic tension as intact fibres, whereas there is a loss of ~(100 − 47)% ~ 53% tension in traditionally demembranated fibres in the presence of other anions (see p. **97** in Section 3.8.1). Thus, from a quantitative viewpoint, experimental observations on frog half-fibres, in the presence of

MS^-, cannot, in principle, be quantitatively compared with those obtained by Brenner et al. (1982, 1984, 1986), Kraft et al. (1995), Squire et al. (1991) and Yu and Brenner (1989). However, the various series of experimental results are qualitatively comparable, in terms of changes in $N_0(I^*)$ with changes in I^*, highlighted above. Finally, based on these consistent findings and, as emphasised above, $N_0(I^*)F_{cb}^0(I^*)$ increases with decreasing I^*, counterbalancing the inevitable increase in $4(\gamma/2)F_{e,r}(I^*)$ (see the beginning of this paragraph) and accounting for the subsequent very low resting tension at vicinal ionic strengths between ~42 mM and ~182 mM (see pp. **102–103** in Section 3.10 concerning the mechanisms of conversion of radial forces into axial forces and p. **85** in Section 3.5 regarding the one-to-one process under contraction conditions, extended here to resting conditions).

An important feature of Figure 4.2 is the extremely rapid increase in rel. $T_r(I^*)$, below the critical value of vicinal ionic strength $I^* \sim 42$ mM (I ~ 40 mM). This behaviour requires careful analysis. First, it is important to stress again that any variation in the net radial repulsive (expansive) forces is automatically converted into similar variation in the axial forces, as demonstrated throughout this book (see the end of the preceding paragraph). The 'biophysical' properties of the axial and radial forces are therefore equivalent. In this context, when I^* decreases gradually below ~42 mM, the radial tethering force $N_0(I^*)F_{cb}^0(I^*)$ initially continues to increase, regardless of possible small variations in F_{cb}^0 with I^* (see the preceding paragraph), because $N_0(I^*)$ increases in these conditions, as demonstrated experimentally by independent groups (see the preceding paragraph). The radial repulsive electrostatic force $4(\gamma/2)F_{e,r}(I^*)$ also increases steadily with decreasing I^*, before reaching a maximum at $I^* \sim 22$–32 mM (I ~ 20–30 mM), and then falls sharply at very low values of I^* (see, in Section 4.2.2, Figure 4.2 and p. **119**; see also below for precise details). It is impossible to predict the critical value of I^* corresponding to the extremely rapid increase in resting tension (found here at $I^* \sim 42$ mM [I ~ 40 mM]; see Figure 4.2; see also below for further comments). Nevertheless, I suggest that, at around these values of ionic strength, the radial tethering–attractive/compressive force $\Theta_r(I^*) = N_0(I^*)F_{cb}^0(I^*) + \Lambda_r(I^*)$ (see the preceding paragraph) becomes slightly weaker than the radial repulsive electrostatic force $4(\gamma/2)F_{e,r}(I^*)$. For example, the radial attractive/compressive force $\Lambda_r(I^*)$ (defined on p. **156** in Section 4.4.2.5.4) may decrease very slightly with decreasing I^* (e.g. I^*-dependent conformational changes in the Z discs and the M lines), whereas $N_0(I^*)F_{cb}^0(I^*)$ may steadily increase. Provided that the decrease in $\Lambda_r(I^*)$ is only slightly more substantial than the increase in $N_0(I^*)F_{cb}^0(I^*)$, the tethering–attractive/compressive force $\Theta_r(I^*)$ would slightly decrease and the balance would imperceptibly tip in favour of the radial repulsive electrostatic force $4(\gamma/2)F_{e,r}(I^*)$. A 'snowball effect' would then occur. Indeed, this tiny initial imbalance would result in a very small and experimentally undetectable increase in the closest surface-to-surface spacing between the thin actin and thick myosin filaments. This tiny increase would result in a large increase in F_{cb}^0, because this radial tethering force increases very rapidly with very small increases in myofilament spacing (see Section 5.1.3). This would result in many A-type heads (see pp. **170–172** in Section 5.1.1 and Figure 4.2 for precise details on the A-type heads) being unable to sustain such strong tethering forces and abruptly detaching. This leads to a sudden fall in the number $N_0(I^*)$ of A-type heads, starting, for example, from $I^* \sim 42$ mM ($I \sim 40$ mM) and even resulting in $N_0(I^*) \sim 0$ at $I^* \sim 22$–32 mM ($I \sim 20$–30 mM) and the relative resting tension rel. $T_r(I^*)$ peaking at high levels at $I^* \sim 22$–32 mM ($I \sim 20$–30 mM) and then falling steeply at very low values of I^*. Indeed, in these conditions, $T_r(I^*)$ (and also rel. $T_r(I^*)$) depends almost entirely on the radial repulsive electrostatic force $4(\gamma/2)F_{e,r}(I^*)$, because the A-type heads are all detached ($N_0 \sim 0$ and therefore $\rho_r \sim 0$ in Equation 4.11; the very low radial attractive/compressive force $\Lambda_r(I^*)$ is neglected, because of the large experimental error at low and very low values of I^*). This results in rel. $T_r(I^*)$ presenting the biophysical characteristics of the radial electrostatic repulsive force described on pp. **101–102** in Section 3.10. Although probably impossible to check experimentally, the tiny imbalance and snowball effect lead to consequences providing a simple explanation for the behaviour of the half-fibres, including the general shape of the curve giving the relative resting tension as a function of ionic strength, particularly at low and very low ionic strengths (see Figure 4.2).

The behaviour of half-fibres at low and very low values of I^*, described in the preceding paragraph, resulting in the existence of almost pure lateral swelling (expansive) forces (radial repulsive electrostatic forces), is confirmed experimentally. Indeed, considerable lateral swelling of the resting half-fibres was observed, under a stereomicroscope, with an increase of ~20% in diameter (~44% in cross-sectional area) observed at $I^* \sim 22$–32 mM ($I \sim 20$–30 mM; no systematic measurements were done below these ionic strengths). This observation also provides strong evidence in favour of radial repulsive electrostatic forces exerting and being translated into axial forces, regardless of volume variations. The observed lateral swelling was almost instantaneous, occurring within ~1–2 s, as estimated roughly by eye, after the rapid injection of the buffer into the trough (~300 ms; see p. **14** in Section 2.2). This almost instantaneous 'enormous' lateral swelling should be related to an instantaneous increase in the net radial repulsive (expansive) force, resulting from the abrupt detachment of A-type heads (see the preceding paragraph). It cannot be mistaken for a slow and limited swelling of the extrafibrillar spaces. This observed sizeable lateral swelling leads to an increase in the centre-to-centre distance c between the myofilaments in an intact unit cell, which becomes $c^{**} \sim 1.20c$ at $I^* \sim 22$–32 mM ($I \sim 20$–30 mM). $\Gamma_r(c^{**}, I^*)$, one of the most important components of $\Lambda_r(I^*)$ (see the definition of Γ_r in the third point in Section 4.4.2.4 and the definition of Λ_r on p. **156** in Section 4.4.2.5.4), probably increases markedly at these low values of I^*, because the various elastic restoring forces included in $\Gamma_r(c^{**}, I^*)$ increase when c becomes c^{**} (e.g. increase in the elastic restoring forces exerted by the Z discs and the M lines and increase in the compressive force exerted by half the sarcolemma present in the half-fibres; see also third point in Section 4.4.2.4 and pp. **155–156** in Section 4.4.2.5.3). All these radial 'counterforces' acting against swelling explain why half-fibres cannot swell indefinitely and cannot 'explode' laterally at very low ionic strengths.

A major conclusion can be drawn from all these observations, in resting conditions. Despite marked lateral swelling, the radial expansive forces can be translated into axial forces. This confirms the validity of the lateral swelling models, even for an increase in fibre volume of ~44% (at $I \sim 20$–30 mM or $I^* \sim 22$–32 mM; see the preceding

Morel

paragraph). Thus, isovolumic behaviour is not a prerequisite for the conversion of radial expansive forces into axial forces, despite the thoughts and claims of most specialists in muscle contraction. As predicted by Morel et al. (1976), there must therefore be mechanisms for converting radial expansive forces into axial contracture (at rest). Under contraction conditions, there must also be mechanisms for converting radial expansive forces into axial contractile forces, as theoretically and logically suggested by Dragomir et al. (1976), Elliott (1974) and Morel et al. (1976) (see pp. **102–103** in Section 3.10). Albeit unintentionally, April and Brandt (1973) (see pp. **106–109** in Section 4.1), Gordon et al. (1973), Hasan and Unsworth (1985) (see pp. **109–111** in Section 4.1) and Neering et al. (1991) (see pp. **4–5** in the Introduction) also demonstrated experimentally the existence of processes for converting radial forces into axial forces under contraction conditions.*

4.4.3 Comparison of Experimental Data, Obtained with Mechanically Skinned Frog Fibres by Two Independent Groups, with the Experimental Results Presented Here, Concerning Relative Resting Tension as a Function of Ionic Strength

In the presence of KCl as the major neutral salt and at room temperature (~20°C–22°C), Gordon et al. (1973) showed that mechanically skinned

---

* On p. **166**, it is strongly suggested that the weakly bound cross-bridges detach suddenly when I* decreases below I* ~ 42 mM, resulting in the disappearance of the radial tethering forces, somewhere between ~42 mM and ~22–32 mM (ρ_r ~ 0). This raises questions about the consistency of this conclusion with the experimental data and semi-empirical calculations presented in this book. On p. **129** in Section 4.4.1, $T_{e,a}(182)$ is found to be ~1.2 ± 0.2 × 10^5 N m^{-2}, whereas on p. **129** in the same section, $T_r(22–32)$ is found to be ~205 ± 0.5 × 10^5 N m^{-2}, giving $T_r(22–32)/T_{e,a}(182)$ ~ 1.7–2.1 (maximal interval). Introducing Q_a ~ 0.763 and Q_r ~ 0.597, assumed not to depend on I* (see pp. **161–162** in Section 4.4.2.6 and p. **162** in Section 4.4.2.7, respectively), H ~ 0.26 mM$^{-1/2}$ (see p. **162** in Section 4.4.2.7) in Equation 4.12, and taking into account Definition 3.13, we obtain the semi-empirical expression ~ 1.7. Combining these results with Equation 4.12 and Definition 4.1.3, and adding $\rho_a(182)$ ~ 0.760 (see p. **85** in Section 3.5), we deduce that $\rho_r(22–32)$ ranges between ~−0.3 and +0.2 (maximal interval; mean value of ~0). The value of $\rho_r(22–32)$ ~ 0 is therefore consistent with this wide interval. Owing to the large margin of error, this conclusion can reasonably be extended to $\rho_r(2)$ and $\rho_r(42)$.

frog fibres behaved similarly to the half-fibres studied in this monograph, at least in terms of the shape of the curve of relative resting tension against bulk ionic strength I. Indeed, relative resting tension increased very slightly as I decreased from ~170 mM to ~100 mM and then increased very rapidly as I decreased from ~100 mM to ~55 mM (minimal value used by Gordon et al. 1973). Figure 4.2 shows the experimental points obtained by Gordon et al. (1973) shifted to lower values of ionic strength by taking the breakpoint of the curves as a reference (I ~ 100 mM in the experimental work of Gordon et al. 1973 and I ~ 40 mM here). This shift may be attributed to (i) the use of the non-benign anion Cl$^-$, the deleterious effects of which have been demonstrated in isometrically contracting chemically skinned fibres from rabbit psoas muscle (Andrews et al. 1991; the extension of these results to resting conditions is almost certainly justified), and (ii) the high temperature for studying frog fibres (unfavorable conditions for a cold-blooded animal). Concerning this problem of working temperature, the same shift in relative active tension is observed, regarding the experimental results of Gordon et al. (1973), working on mechanically skinned frog fibres, at room temperature (unfavourable conditions; see above), and those of Hasan and Unsworth (1985) working on intact toad fibres at ~5°C (favourable conditions for a cold-blooded animal species) (see pp. **109–111** in Section 4.1 for precise details). The problem of the shift is therefore complex. All the same, the use of half-fibres in buffers containing MS$^-$ as the major anion also results in differences with respect to mechanically or chemically skinned fibres and permeabilised fibres, both at rest and under contraction conditions (see Section 3.3, pp. **96–97** in Section 3.8.1 and p. **221** in Section 7.2). After the suggested and almost certainly justified correction, the two series of curves are similar and can even be superimposed, except that Gordon et al. (1973) did not explore very low values of ionic strength (they used values of I > 20 mM, after the shift described above; see Figure 4.2 concerning values of I < 20 mM). Nonetheless, they discussed the relative resting tension–ionic strength curve and demonstrated that ionic strength per se was responsible for the observed variations, although the nature of the anion (and also the excessively high temperature for cold-blooded animals; see above) probably affected the curve. They suggested that, at low ionic strengths, the tropomyosin–troponin system is not present or loses its ability to prevent actin–myosin interactions and that rigor complexes are formed, turning on many actin molecules. These possibilities cannot be ruled out entirely, particularly in the presence of Cl⁻ and at room temperature. However, as repeatedly pointed out in this book (see p. **119** in Section 4.2.2 and pp. **165–167** in Section 4.4.2.8), the experimental data presented in Figure 4.2, including the demonstration of a maximum in the resting tension curve at I ~ 20–30 mM, coupled with substantial lateral swelling, demonstrate that radial repulsive electrostatic forces play a considerable role. The phenomena suggested by Gordon et al. (1973) would have led to a gradual and steady increase in relative resting tension with decreasing I, with no detectable lateral swelling, in the absence of any radial expansive force. Finally, commenting on the relative resting tension, the authors suggested that the molecular processes attributed to ionic strength differ for Ca-activated tension and non–Ca-activated resting tension. This conclusion is consistent with what is proposed in the experimental part of this monograph, and I firmly suggest that the Ca-activated tension corresponds to contraction, whereas the non–Ca-activated tension corresponds to contracture, as observed in Figure 4.2, at low and very low ionic strengths.

Gulati (1983) also carried out experiments on resting mechanically skinned frog fibres, with KCl as the major neutral salt, at room temperature (~20°C–22°C, unfavourable conditions). A strong axial force (comparable to full isometric tetanic force, within the large range of experimental error) was recorded in the presence of 23 μM $MgCl_2$ and no axial force was recorded in the presence of 1 mM $MgCl_2$ (the two corresponding experimental points are represented in Figure 4.2). The author attributed this behaviour to the tropomyosin–troponin system being 'on' and contraction occurring at low Mg^{2+} concentration. At high Mg^{2+} concentration, the system would be 'off' and contraction would be abolished. Gulati (1983) also studied the fibres in the presence of KCl as the major salt and worked at room temperature (see the preceding paragraph for remarks regarding these experimental conditions); thus, I applied the same correction, as described in the preceding paragraph, to the results of Gordon et al. (1973), for the same reasons. The corrected value for I, based on Gulati's data, is ~20–25 mM. His experimental system therefore falls approximately into the region in which relative resting tension is extremely unstable and rapidly falls when ionic strength increases slightly (see Figure 4.2). This drop in relative resting tension is observed in Figure 4.2 for higher values of I, ranging from ~30 mM to ~40 mM, but the biological material was not the same and the temperature was inappropriate in the experiments of Gulati (1983), as pointed out above, in this and the preceding paragraphs. Thus, in the experiments of Gulati (1983), the addition of 1 mM $MgCl_2$ increased ionic strength by 3 mM, thereby abruptly abolishing resting tension, as a straightforward consequence of the instability of the resting tension.

CHAPTER FIVE

General Features of the Hybrid Model

5.1 GENERATION OF RADIAL TETHERING FORCES AT REST AND THEIR DECREASE DURING ISOMETRIC CONTRACTION. RELATED PROBLEMS

5.1.1 Thick Myosin Filaments, External and Internal Myosin Heads, Weakly Bound Cross-Bridges at Rest, Tilting of the External Myosin Heads during Contraction, and Generation of Full Radial Tethering Forces at Rest. Various Arguments Supporting the Existence of Radial Tethering Forces, Based on Experimental Data from Several Independent Groups

Before going into detail concerning the generation of radial tethering forces at rest, it should be recalled that Hill (1968a) deduced from his experimental results, obtained with whole frog muscles (at low temperature), that 'in a resting muscle, the crossbridges on the myosin filaments are not entirely inactive, but a very small proportion of them are cross-linked with the actin filaments'. This inference regarding the existence of 'resting' attached cross-bridges has been largely ignored but was later confirmed by several independent groups (see below, in this section).

The way in which the radial tethering forces are generated at rest can be deduced from previous experiments performed by my group and concerning the dimerisation of isolated myosin heads in vitro (S1) (Bachouchi et al. 1985; Grussaute et al. 1995; Morel and Garrigos 1982a,b; Morel and Guillo 2001; Morel et al. 1998b) and the head–head dimerisation of whole myosin molecules in vitro (Morel et al. 1998a). These experimental observations were

supported by Morel et al. (1999), who confirmed the hypothesis of Morel and Garrigos (1982b) and demonstrated that the main features of thick myosin filaments (synthetic and natural) depend strongly on the existence and properties of the head–head dimers.

I would like to recall the 'S1-dimer story', which is probably familiar to many senior specialists in muscle. Until 1976, no experiments supported the possibility of head–head (S1–S1) interactions in solution. For instance, Margossian and Lowey (1973), working on S1 from rabbit skeletal muscle, claimed that 'proteolytic degradation of myosin yields only isolated globules with no tendency toward dimer formation'. In 1976, it was shown, by electron microscopy, that isolated myosin molecules from rabbit skeletal muscle treated with DTNB (bis (5-carboxy-4-nitrophenyl) disulphide) to remove the regulatory light chain (RLC, formerly called LC2) can form head–head dimers (Arthur Elliott and Gerald Offer, personal communication). Schaub et al. (1977) and Kuntz et al. (1980) presented evidence for head–head interactions in cardiac and skeletal myosin. Margossian and Lowey claimed publicly, during the meeting of the Biophysical Society held in the winter of 1977, that a reversible aggregation of isolated globules (S1) might occur in the presence of 0.1 mM Ca^{2+}. Barrett et al. (1978) also reported that a reversible aggregation of S1 occurs in the presence of 0.1 mM Ca^{2+}. Cusanovich and Flamig (1980) and Flamig and Cusanovich (1981), working on cardiac S1, at low ionic strength and in the presence of 10 mM Ca^{2+}, demonstrated that these isolated heads formed reversible polymers. Stewart et al. (1985) described head–head interactions, by electron microscopy, of heads belonging to different myosin molecules included in *Limulus* and scorpion natural thick myosin filaments. Winkelman et al. (1985) were able to obtain crystals of S1 from avian muscle and showed, by electron microscopy and reconstruction techniques, that the elementary structure of the crystals was an S1 dimer (the authors did not stress this point, but this phenomenon is clearly observed in certain figures). Bachouchi et al. (1985, 1986) and Morel and Garrigos (1982a,b) demonstrated that S1 from rabbit skeletal muscle can exist as a rapid reversible monomer–dimer equilibrium in the presence of MgADP or MgATP. Munson et al. (1986) claimed that the S1 dimer does not exist (they called it the 'putative' dimer). Some months after publication of the paper by Bachouchi et al. (1986), concerning the MgATPase activity of the S1 dimer in solution in the absence of F-actin, Margossian and Stafford claimed publicly, during the meeting of the Biophysical Society held in the winter of 1988, that the S1 dimer does not exist, because the diffusion–sedimentation equilibrium ultracentrifugation experiments performed in their laboratory demonstrated an absence of any trace of dimer, although they worked with a buffer similar to that referred to in my group as a 'dimer buffer'. I was not present at the meeting but was aware of the decision of many specialists to reject systematically any further paper on the S1 dimer from my group. I later contacted Margossian and Stafford to obtain precise details about their experiments. After a long period of discussion, we became, in my laboratory, aware that many precautions were necessary, when preparing 'native S1', to retain its full dimerisation capacity. Indeed, we found that the dimerisation site is extremely labile and we published a new paper on S1 dimerisation in 1998 (Morel et al. 1998b), with a long 'supporting information available', in which 15–20 conditions for keeping the dimerisation site intact were given (in particular, it is 'forbidden' to purify S1 by chromatography, but other techniques are required). In this context, Margossian et al. (1993) used purified cardiac S1 and were unable to confirm the results of Bachouchi et al. (1986), according to which the MgATPase activity of the S1 dimer is greater than that of the monomer. However, Margossian et al. (1993) found that their S1 tended to self-associate under particular conditions. We had unwittingly taken all the precautions into account in our initial publications (and also obviously in all subsequent work), whereas Margossian and Stafford and Margossian et al. (1993) had not. Even in this

situation, it could be argued that all the experimental data, concerning dimerisation, came from my group only. Fortunately, Claire et al. (1997), working in an independent group and using an experimental technique very different from those previously used, published a paper in which they confirmed that S1 can form dimers. Moreover, in the 1980s, it was observed, by independent groups, that thick–thick filament interactions occur, in situ and in a state of non-overlap between thick myosin and thin actin filaments, via myosin heads (see p. **17** in Section 2.4), strongly suggesting the existence of head–head dimers between thick myosin filaments belonging to neighbouring unit cells. Unfortunately, these experiments have been ignored by the 'muscle community'. Since 2000, several experiments have been performed on soluble myosin molecules, its soluble subfragment HMM (two heads S1 + S2) and thick myosin filaments that systematically demonstrate intramolecular and cooperative interactions of the two heads of a single myosin molecule, from various animal species and types of muscle (e.g. Albet-Torres et al. 2009; Duke 2000; Esaki et al. 2007; Jung et al. 2008; Li and Ikebe 2003; Zhao et al. 2009). The interpretation of these findings by the authors is complex, but the main conclusion is that head–head interactions of myosin can now be seen as a 'trivial event'.*

To the property of myosin heads to form dimers in solution, we should add the discovery of the various dispositions of the myosin molecules in synthetic and natural thick myosin filaments from rabbit skeletal muscle. Morel et al. (1999) demonstrated that natural thick myosin filaments, extracted from the psoas muscles of young adult rabbits, are three-stranded, and, according to our Figure 14, one strand (the 'third' strand) consists of myosin molecules with both heads lying entirely outside the backbone. The thick myosin filaments from young adult frogs are also three-stranded (see p. **205** in Section 6.3.5)

---

* This footnote was added after the acceptance of this monograph for publication. It has become common practice to ignore what is done by others. In this context, I feel exasperated by the many groups claiming the existence of head–head interactions (see also the recent publications by Oshima et al. (2012) and Pinto et al. (2012); references given in the addendum) and 'omitting' the discoveries of head-head interactions by Schaub et al. (1977) and Kuntz et al. (1980), head-head dimers by Claire et al. (1977, 1997), and head-head dimers by my group, as early as 1982, and confirmed in 1985, 1986, 1998, 1999 and 2001, on isolated heads in solution and whole myosin molecules also in solution or inserted into thick myosin filaments.

and the disposition of the myosin heads is certainly similar to that in young adult rabbits. The myosin molecules belonging to the third strand and their heads are not involved in the generation of the radial tethering forces (see pp. **174–176** for precise details on this generation) but are obviously involved in both MgATPase activity and the rotation of the heads in the swinging cross-bridge/lever-arm part of the hybrid model. I think that the best model for the arrangement of the two external heads of a myosin molecule belonging to the third strand is that proposed by Offer and Elliott (1978), as already mentioned in the legend to Figure 14 in the paper by Morel et al. (1999). According to Offer and Elliott (1978), each head of this myosin molecule can bind to two different neighbouring thin actin filaments. This arrangement of the two external heads is valid in contracting intact unit cells and, almost certainly, in resting intact unit cells (at rest, only some weakly bound cross-bridges are arranged in this way; the notion of 'weakly bound cross-bridges', formerly called 'resting cross-bridges' by my group, is studied below, in this section, particularly on pp. **176–180**). In intact unit cells, there are 600 heads per thick myosin filament, 200 of which are included in the backbone (the B-type heads; see below), whereas there are 400 external heads: a maximum of 200 external heads

can attach to the thin actin filaments, during contraction, as suggested by Offer and Elliott (1978), and 200 external heads are of A-type (corresponding to the 200 B-type heads), whether attached to the thin actin filaments or not (see Figure 5.1, its legend, Appendix 5.I and also pp. **205–206** in Section 6.3.5 for further details). The A-type heads therefore represent $200/400 = 1/2$ the external heads, regardless of the physiological state of the intact unit cell (rest or contraction). Let us call, in a thick myosin filament/intact unit cell, N_0 the number of weakly bound cross-bridges (at rest) and N the number of attached cross-bridges (during contraction), corresponding solely to A-type heads. The A-type heads plus the corresponding S2 part of the myosin molecules (called 'active weakly bound cross-bridges' when attached to the thin actin filaments, at rest; see Figure 5.1 concerning the A-type heads and the S2 subfragment) are involved in the lateral swelling part of the hybrid model of muscle contraction, whereas the 400 external heads are involved in actin-activated MgATPase activity and other mechanical processes of contraction (see below). The heads arranged according to the model of Offer and Elliott (1978) are called here 'external free heads' (F-type heads). Indeed, the F-type heads are not subject to structural and mechanical constraints, by contrast to the A-type heads, which are involved in

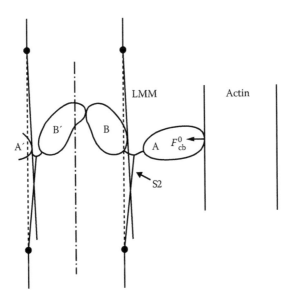

LMM   Actin

A'   B'   B   A F_{cb}^0

S2

Figure 5.1 Generation of the mean elementary radial tethering force, F_{cb}^0, exerted by an 'active weakly bound cross-bridge' (A-type head + S2) at rest (see main text for detailed explanations, particularly pp. **170–172** and **174–176**). Supplementary comments are given in Appendix 5.I. (Slightly modified from Figure 5c of Morel, J. E. et al., Native myosin from young adult rabbit skeletal muscle: Isoenzymes and states of aggregation. *Biochemistry* 37:5457–5463, 1998a. With permission.)

a more tightly constrained system (the external A-type heads are coupled to the internal B-type heads). As an illustrative example, the F-type heads certainly play a major role in the swinging cross-bridge/lever-arm part of the hybrid model of contraction, whereas the A-type heads may be involved in the swinging cross-bridge/lever-arm process, but almost certainly with difficulty. Otherwise, the F-type heads can be involved in the rather complex (external) head–(external) head interactions described by the several groups mentioned on p. **170**. The arrangement of the F-, A- and B-type heads is not consistent with the traditional descriptions repeatedly asserted over four decades or so, according to which the ~600 heads per thick myosin filament lie outside the filament backbone (e.g. Cantino and Squire 1986; Craig and Woodhead 2006; Malinchik and Lednev 1992; Offer 1987; Pepe 1971; Squire 1972, 1975, 1981; Squire et al. 2005). The structure of the thick myosin filaments proposed by Morel et al. (1999), expanded in this paragraph, is an essential element in this work.

In intact unit cells from resting intact frog and rabbit fibres, surrounded by buffers mimicking the extracellular physiological medium, only a small proportions of the A- and corresponding B-type heads are involved in the arrangement described in Figure 5.1, its legend and Appendix 5.I, and below, in this section. At this point, a preliminary remark is required to clarify the discussion. On p. **205** in Section 6.3.5 and in the preceding paragraph, it is recalled that the thick myosin filaments are three-stranded, in both young adult frogs and young adult rabbits. Many 'quantitative consequences' of this arrangement of all the external heads are studied in Section 6.3.5. Unless otherwise specified, I focus mostly on the A-type heads represented in Figure 5.1. The proportion of attached external heads, at rest, would be ~10% of the maximal number of external heads (A- plus F-types) attached in rigor at ~0°C–10°C (the maximum of 100% is discussed on p. **148** in Section 4.4.2.5.2), for frog skeletal muscles (Morel et al. 1976). The estimate of ~10% is consistent with the old value of ~10% suggested by Portzehl et al. (1969a), for isolated myofibrils from rabbit muscle (at room temperature), and the value of ~10% (or slightly less) assumed by Schoenberg (1988), for rabbit and frog fibres (temperature not specified). Finally, Kraft et al. (1995) gave an estimate of ~5% at 25°C, for a bulk ionic strength of 200 mM, in demembranated fibres from rabbit psoas muscle. Thus, these four estimates for the attached external heads at rest (~10%, ~10%, ~10% and ~5%) are all of the same order of magnitude and we can retain an average value of ~9 ± (1 or 2)% (mean ± SE or SD; n = 4) external heads (weakly bound cross-bridges) for young adult frogs at ~0°C–10°C and young adult rabbits at ~20°C–25°C (also, possibly, at 5°C–10°C). This proportion of ~9% also corresponds to resting A-type heads (vs. 100% of A-type heads attached in rigor), or, in other words, to the active weakly bound cross-bridges considered in the preceding paragraph. Note that the definition and general structure of the myosin cross-bridges (corresponding here to the external heads plus the S2 part of the myosin molecule) were suggested by Huxley (1965, 1969), based mostly on biochemical data. Indeed, in the 1960s, it was already known that the globular heads S1 and the S2 parts are hydrophilic and Huxley (1965, 1969) proposed that these two parts of the myosin molecule were located in the aqueous medium between the thin actin and thick myosin filaments, whereas the other part of the tail (LMM) is hydrophobic and was included in the thick myosin filament core. Huxley (1965, 1969) introduced the notion of hinges (or joint) between LMM and S2 and between S2 and S1 (solid circles in Figure 5.1). This hypothesis concerning the existence of two hinges was supported by electron microscopy observations (Elliott and Offer 1978), on whole isolated myosin molecules. Rimm et al. (1989) and Harris et al. (2003) gave other precise details on the location of the head–tail junction and some of its main features. Many experimental studies have been and are still performed on S1, but few have focused on S2, although this subfragment (from human cardiac myosin) has been studied by Blankenfeldt et al. (2006), from structural and functional viewpoints.

According to his basic assumptions, described at the end of the preceding paragraph and the old observation of the arrowhead arrangement of the actomyosin complex under rigor conditions (absence of ATP), Huxley (1965, 1969) proposed the most popular model of muscle contraction. In this oversimplified 'oar' model, when contraction is triggered, most of the external heads (including the A- and F-type heads; see Figure 5.1 and pp. **170–172** for precise detail on the A- and F-type heads) attach to the neighbouring thin actin filament at the 90° position. These heads then undergo a rotation and, at the end of the working (power) stroke, they adopt the 45° position observed in rigor conditions (arrowhead structure), resulting in a change in their axial orientation of $90° - 45° = 45°$. This hypothesis, first expressed by Huxley (1965, 1969) in his swinging cross-bridge/lever-arm model, can obviously also be applied to the traditional part of the hybrid model presented, described and discussed in this work. Using traditional x-ray diffraction, Miller and Tregear (1970), working on asynchronous insect flight muscles (IFM; these muscles have a quasi-crystalline arrangement of thick myosin and thin actin filaments and were therefore easy to observe by traditional x-ray diffraction), confirmed that the cross-bridges attach to the thin actin filaments under contraction conditions. Moreover, the change in the axial orientation of the attached cross-bridges was experimentally described by Reedy et al. (1965), also working on IFM. They showed experimentally, by electron microscopy of 'fixed structures', that the cross-bridges were in the ~90° position upon attachment to the thin actin filaments and in the ~45° position at the end of the working stroke, thereby producing filament sliding under isotonic conditions and force under isometric conditions. However, the situation is actually much more complicated, as discussed by Reedy (2000). There are, indeed, many contentious issues regarding the axial orientation of the myosin heads bound to the thin actin filaments, at rest and under contraction conditions, and also the change in this axial orientation and the main features of the power stroke. For instance, studying muscles other than IFM (i.e. rabbit skeletal muscle), Toyoshima and Wakabayashi (1979) suggested that, in vitro, the tilt angle of a head (S1) in the rigor acto-S1 complex would be only ~15°, rather than ~45°. Heuser and Cooke (1983), also working with rabbit skeletal muscle, showed, in vitro and by electron microscopy, that the tilting angle in the acto-S1 complex under rigor conditions, would range from ~0° to ~20° only and that there would be no major difference between rigor and relaxation, in terms of cross-bridge orientation. They also found, in situ and under contraction conditions, that the myosin heads were at an angle of only ~10°. Bordas et al. (1993) and Martin-Fernandez et al. (1994), using x-ray diffraction from synchrotron radiation, on single frog fibres and whole muscles, respectively, suggested that, at the plateau of isometric tension, the

heads forming an actomyosin complex are not in the orientation suggested by Huxley (1965, 1969). More precisely, Martin-Fernandez et al. (1994) deduced, from their experiments, an axial orientation of the myosin heads relative to the thin actin filament axis of ~65° at rest and ~90° under conditions of isometric contraction, corresponding to a tilting angle of $~65° - 90° ~ -25°$ (this negative value is obviously worrying). The problem of the tilting angle remains an open question, potentially related to both the experimental techniques used and different interpretations of the experimental observations.

The location and amplitude of the orientation changes during contraction have not even clearly been determined. Orientation and structural changes are frequently thought to occur at 'the neck region of the myosin motor domain', more precisely in the RLCs of the myosin heads and in various domains in the heads from myosin molecules (generally from skeletal muscle, but also from other origins). In any event, the debate concerning location and amplitude is contentious (e.g. Adhikari et al. 1997; Allen St. Claire et al. 1996; Baker et al. 1998; Bell et al. 2002; Borejdo et al. 2006; Burghardt et al. 2007; Chaudoir et al. 1999; Cooke 1986, 1995, 1997; Corrie et al. 1999; Hirose et al. 1994; Irving et al. 1995; Ito et al. 2003; Lenart et al. 1996; Ling et al. 1996; Piazzesi et al. 1999, 2007; Rayment et al. 1993b; Sabido-David et al. 1998; Thomas 1987; Vibert and Cohen 1988). In an attempt to try to resolve many problems raised by these aspects of the swinging cross-bridge/lever-arm models, somewhat 'bizarre' notions of 'anti-rigor', 'pre-power stroke', and 'pre- and post-hydrolysis' states, for example, have been introduced (see, for instance, Reedy 2000, mostly concerning IFM). The situation is therefore confusing. Reedy (2000), commenting on IFM fibres, but extending her comments to vertebrate muscle, highlighted the high level of complexity, by pointing out that many authors claim that most cross-bridges are attached at an angle of approximately 90° rather than 45° under contraction conditions, which 'suggests that relatively few bridges are accumulated at the end of their power stroke in a rigor-like conformation'. This remark should be considered in the context of the very low proportion of attached cross-bridges, under the same conditions, estimated by several independent groups (see p. **150** in Section 4.4.2.5.2, as concerns the proportions of ~10%–30%, which should not be retained; see also below, the important reference to Borejdo et al. 2006). Conclusions similar to those drawn by Reedy (2000), regarding the tilting angles and the proportion of attached cross-bridges, were drawn by Bell et al. (2002), who wrote: 'the static rigor distribution peaks at 63°, whereas the active one peaks at 68°. The difference (5°) is small, probably because a relatively small proportion of the cross-bridges contribute to generation of active tension'. These low proportions of 'active' cycling cross-bridges, also recalled by Reedy (2000) (see above), should not be taken into account, because they are at odds with many other independent estimates, the most probable being ~82% (see pp. **152–153** in Section 4.4.2.5.2). Another problem was raised by Borejdo et al. (2006), who suggested, concerning the use of fluorescence polarisation or anisotropy to determine the orientation of cross-bridges in different physiological states, that most measurements have been made on large populations of cross-bridges and that, in these experiments, anisotropy is averaged over a large number of cross-bridges. Regardless of this statement, relating exclusively to a specific technique (fluorescence polarisation or anisotropy), I believe that the many contradictory experimental results and their dubious interpretations, recalled above, in this and the preceding paragraphs, introduce increasing levels of complexity. In this context, this notion of complexity is not new, because, for instance, Huxley and Kress (1985) attributed an active role to the weakly bound cross-bridges in the in vivo/in situ enzymatic cycle and tension generation (see also Brenner et al. 1991; Geeves and Holmes 2005; Goody 2003; Iorga et al. 2004; Kraft et al. 1995; Regnier et al. 1995; Xu et al. 2006a). Huxley and Kress (1985) also proposed breaking up the 120-Å 'attached stroke' and suggested that 'the tension-generating part of the crossbridge stroke is only about 40 Å'. Along the same lines, Laakso et al. (2008) stated that 'the working stroke displacement can be broken up into two substeps'.

The complexity recalled in the preceding paragraph may result from most investigators being constrained by swinging cross-bridge/lever-arm models, which are too rigid. This inference indirectly supports the new approach (hybrid model) presented in this monograph, which is much less rigid. The problems raised in the two preceding paragraphs concern only the external heads

undergoing traditional rotation, leading to ~60% of contraction in the hybrid model. They are of no major interest in the analysis presented in this work, concerning the lateral swelling mechanisms resulting in ~40% of contraction. Moreover, the addition of further freedom to the approach to the complex mechanisms of contraction is appropriate, as demonstrated throughout this book.

Let us now focus on the lateral swelling part of the hybrid model. In an intact unit cell belonging to a resting intact fibre surrounded by a physiological buffer mimicking the extracellular medium, the B–B′-type heads are stuck together to form internal dimers (see Figure 5.1, its legend and Appendix 5.I). The existence of such B–B′-type head–head dimers is strongly supported by various experimental results obtained by my group (see references cited on p. **169** and the 'dimer story' discussed on pp. **169–170**) and by simple and informative new experiments described in Section 5.2 and in Figure 5.2, its legend and Appendix 5.II. To reach the neighbouring thin actin filament, head A must pull the 'dimeric resting' B and B′ heads toward the right-hand side of the diagram. This distorts the strongly bound B and B′ heads, as shown in Figure 5.1, and discussed in the legend to this figure, in Appendix 5.I, and below, in this and the following paragraphs. Moreover, head B is slightly drawn out from the filament core, distorting the close-packing of the LMM parts of the myosin molecules lying along the same axis. This part of the tail comes slightly unstuck from the backbone and lies some nanometres away from the shaft (note that attachment of the A′ head to the thin actin filament, on the left-hand side of the diagram, would lead to considerable distortions and the probability of such a double attachment is negligible). In any event, the B′ head forces this A′ head to compress the filament shaft, resulting in displacement of the corresponding LMM toward the middle axis of the filament, further distorting the thick myosin filament backbone.

The many distortions, described in the preceding paragraph, result in a mean elementary resting radial tethering force F_{cb}^0 exerted by head A on the neighbouring thin actin filament. The notion of 'mean' elementary radial tethering force is related to different arrangements of the individual attached A-type heads, because not all the A-type heads are directly opposite the actin-binding site, inducing different distortions in these heads and, therefore, differences in the strength of each resting radial tethering force. Distortions also occur under contraction conditions, as demonstrated by Bachouchi and Morel (1989a) (see also pp. **251–254** in Section 8.9). Similar distortions, concerning only the various kinetic features of the cross-bridges, during isometric tetanic contraction, were suggested by Hill et al. (1980) and more recently recalled by Borejdo et al. (2006), who wrote: 'each cross-bridge has different kinetics, depending on its position relative to the actin binding site'. For simplification, the term 'mean' is systematically omitted in this monograph. According to Figure 5.1, the S1–S2 joint for the A-type heads is slightly more accessible to digestion with α-chymotrypsin than the corresponding joint for heads stuck to the backbone (see Section 4.3.2 for an illustration of the major consequence of this feature). I also suggest that, at rest, all the A-type heads undergo cyclic attachment to and detachment from the thin actin filaments. In principle, all these cycling resting A-type heads (and also all the cycling resting F-type heads; see pp. **170–172** for precise details on the A- and F-type heads) may split MgATP. Indeed, Morel et al. (1976) pointed out that various arguments support the existence of resting cross-bridges attached to the thin actin filaments, including the results of Portzehl et al. (1969a), who found that the very low MgATPase activity, measured at rest in rabbit myofibril suspensions, was of the actomyosin type. However, under resting conditions, the cyclically attaching and detaching resting heads cannot cleave large amounts of MgATP, because they are too few in number, and the fibre therefore remains at rest (no rotation of the external A- and F-type heads in the swinging cross-bridge/lever-arm part of the hybrid model), by contrast to contraction conditions. This attachment–detachment cycle of the resting heads is the only possible situation and is very different from what occurs in rigor, in which the external heads are immobilised in the 45° position (arrowhead structure, three-dimensional quasi-crystalline arrangement). Indeed, a given external head is unlikely to remain indefinitely attached at rest. If unbroken attachment were to occur, a resting fibre (from frog, for example) could not be passively stretched. Common observation shows that this is clearly not the case. I also suggest that, at the microscopic level, the resting external heads (weakly bound heads) attached to the thin actin filaments probably differ at different time points and that there is therefore a turnover of these heads (almost certainly a very slow turnover). However, considering a very large number of myofibrils (or unit cells) at rest, the mean number of resting external heads attached to the thin actin filaments, including the A-type heads, is statistically constant at a given time, as is the mean tethering force F_{cb}^0. Finally, to my mind, apart from the proportion of attached cross-bridges (~9% at rest and ~82%, during isometric tetanic contraction), there are many analogies between the two states of a living muscle (rest and contraction).

The notion of resting cross-bridges, in skeletal muscle, was first introduced by Hill (1968a) on the basis of experimental data (see first paragraph). This notion was later developed by Morel (1975), Morel and Gingold (1979b), Morel and Pinset-Härström (1975b) and Morel et al. (1976), on theoretical and logical bases, and suggested from the experimental results by Portzehl et al. (1969a) on rabbit myofibrils and by Siegman et al. (1976) on smooth muscles (see also pp. **170–172** for additional details). Matsubara and Millman (1974), in their x-ray diffraction experiments (using traditional techniques) performed on mammalian cardiac muscle, demonstrated that up to ~40%–50% of the cross-bridges were attached 'at rest'. However, it is difficult to define the notions of resting and contracting conditions in cardiac muscle clearly, because, at least in vivo, diastolic and systolic conditions do not actually correspond to pure resting and contracting conditions, respectively, whereas these definitions are much more clearcut in skeletal muscle. As illustrative examples, Kensler (2002, 2005) performed electron microscopy studies on isolated cardiac thick myosin and thin actin filaments, under relaxing conditions, and found that the arrangement of the cross-bridges on the filament surface are similar to those in skeletal thick myosin filaments. However, they also observed that interactions of isolated thick myosin and thin actin filaments, under relaxing conditions, appeared to be markedly more frequent for cardiac than for skeletal filaments, confirming the previous results obtained in situ by Matsubara and Millman (1974). Nonetheless, in vitro, for cardiac proteins (myosin heads, S1, + F-actin), under 'contraction-like' conditions, the actomyosin MgATPase hydrolysis cycle follows the same pathway, with kinetic features similar to those of the skeletal MgATPase cycle (e.g. Hazzard and Cusanovich 1986; Stein and White 1987). Moreover, Xu et al. (2006b), working on relaxed permeabilised rabbit cardiac trabeculae and psoas muscle fibres, showed, by x-ray diffraction from synchrotron radiation, that the traditional $I_{1,1}/I_{1,0}$ ratio was higher in cardiac muscle than in skeletal muscle. They suggested that this was probably attributed to a higher affinity between actin and myosin in cardiac muscle than in skeletal muscle. This interpretation is consistent with the observations of Kensler (2002, 2005) and those of Matsubara and Millman (1974). Finally, on the basis of experimental measurements of chord stiffness by Martyn et al. (2004), Xu et al. (2006b) deduced that their x-ray data probably reflected a larger number of resting cross-bridges in cardiac muscle than in skeletal muscle (compare this qualitative result with the ~9%, in skeletal muscle, suggested on p. **172**, and ~40%–50%, in cardiac muscle, recalled above). Thus, cardiac muscle is more complex than skeletal muscle, although these two types of muscle are similar in many ways (see Section 8.12).

All the assumptions regarding the existence of resting cross-bridges, advanced by my group over more than three decades, were initially discounted, being considered pure speculation by the muscle community. However, since the mid-1980s, many authors have experimentally rediscovered this property of resting muscle fibres, without making reference to previous experimental and logical work. The discounted resting cross-bridges were renamed 'weakly bound' or 'weakly attached' cross-bridges and have been experimentally studied with various mechanical, biochemical and x-ray diffraction techniques (e.g. Brenner 1986; Brenner et al. 1982, 1984, 1986, 1991; Gu et al. 2002; Kraft et al. 1995; Schoenberg 1988; Squire et al. 1991; Walker et al. 1994; Xu et al. 1997, 2002, 2003; Yu and Brenner 1989). Brenner (1986), Brenner et al. (1991), Goody (2003), Iorga et al. (2004), Kraft et al. (1995), Regnier et al. (1995) and Xu et al. (2006a) suggested that the weakly bound cross-bridges represent an intermediate state in the normal enzymatic cycle, during contraction, in swinging cross-bridge/lever-arm theories, although this proposal has remained a matter of discussion (e.g. Xu et al. 1987). Despite the doubts raised by Millman (1998) concerning the existence of weakly bound cross-bridges in all types of muscle (see the next paragraph) and

their role in the enzymatic cycle, this interpretation of the role of the weakly bound cross-bridges in this cycle is still seen as the most probable or even as the only 'reasonable' explanation. Xu et al. (2006a), for instance, at odds with their previous opinions (Xu et al. 1987), wrote: '... it is generally thought that the transition from the weakly bound to strongly bound states, with accompanying conformational changes, underlies the force generation mechanism'. This view was also supported by Goody (2003), who wrote: 'The transition between weakly and strongly bound states of myosin represent the fundamental thermodynamic event in the cross-bridge cycle leading to production of mechanical work'. These statements are entirely different from the view presented in this monograph, but the many consequences of the choice made by my group, since 1975–1979, are self-consistent and I firmly maintain my position. Obviously, this assertion does not rule out the possibility of the weakly bound heads, including the A-type heads (see Figure 5.1) and the F-type heads (see pp. **170–172** for precise details on the A- and F-types of external heads) having many specific properties, including enzymatic activity and subsequent force generation, because, in the hybrid model, the external heads contribute ~60% of the axial contractile force (swinging cross-bridge/lever-arm part). In this book, I use the term weakly bound cross-bridges rather than resting cross-bridges, unless otherwise specified.

An important point, concerning the existence of the weakly bound cross-bridges, should be highlighted. Bagni et al. (1992, 1995), in their mechanical experiments, performed at 4°C, on intact frog fibres, did not confirm the existence of weakly bound cross-bridges, their findings instead tending to rule out this possibility. During the same period, Pollard et al. (1993), working on rabbit actomyosin (more precisely actomyosin subfragment 1), were able to visualise, with a cutting-edge electron microscopy technique, structures that they attributed to weakly bound intermediates in the actomyosin MgATPase cycle. Millman (1998) gave a series of possible reasons for this apparent discrepancy. I therefore need to try to unravel this problem and to draw definitive conclusions. In this context, I think that the 'purely physiological techniques' used by Bagni et al. (1992, 1995) would not have been sufficiently accurate to detect a proportion of only ~9% (proportion suggested on p. **172**, under physiological conditions) and the conclusions of Bagni et al. (1992, 1995) cannot be considered valid, in the light of the many concordant results obtained since the 1990s, definitively demonstrating the presence of weakly bound cross-bridges in the resting fibres of all animal species (cold-blooded and warm-blooded). Further discussions on this and other points are presented in the next paragraph.

The question of the existence and proportion of weakly bound cross-bridges is rather complex, probably because some authors worked on rabbit muscles, whereas others worked on frog muscles, without taking into account sufficiently the temperature at which their experiments were performed. Indeed, as recalled by Barman et al. (1998), Stephenson and Williams (1985) and Woledge et al. (1985), it is risky to compare warm-blooded animals (e.g. rabbit) and cold-blooded animals (e.g. frog) at the same temperature (e.g. between ~0°C and ~10°C). This was also the conclusion of Millman (1998). In any event, it is impossible to predict whether studying rabbit or frog at ~0°C–10°C would lead to higher or lower proportions of weakly bound cross-bridges for these two animal species. The detection of only ~9% weakly bound cross-bridges for frog or rabbit under physiological conditions (see the end of the preceding paragraph) is probably dependent on the experimental techniques used. For instance, the use of x-ray diffraction and stiffness techniques and the comparison of traditionally demembranated fibres and intact fibres lead to different, even contradictory, results (see Sections 3.8, 4.4.2.1, 5.1.4 and 8.7 for discussions of these major points). These problems are important because the proportions of weakly bound cross-bridges are very small (see above, in this and the preceding paragraphs), at least under physiological conditions of ionic strength (see pp. **164–167** in Section 4.4.2.8 regarding the major role of low ionic strength). Nonetheless, combining the many experimental results obtained since the mid-1980s and recalled above, in this section, together with the previous interpretations suggested by Hill (1968a) and Portzehl et al. (1969a) and the logical reasoning presented by Morel (1975), Morel and Gingold (1979b), Morel and Pinset-Härström (1975b) and Morel et al. (1976), we obtain consistent conclusions throughout this monograph. I therefore claim that weakly bound cross-bridges exist in all muscles from all animal

species studied under physiological resting conditions. This is repeatedly demonstrated in this work and is at the heart of the formulation of a self-consistent approach to the molecular mechanisms of muscle contraction (the hybrid model).

Regardless of the various problems analysed above, in this section, it appears clear, from Figure 5.1, its legend, Appendix 5.I and the discussion presented in this section, that there are 'minor' differences between resting and contraction conditions in terms of myosin–actin interactions. Indeed, at first sight, only the angles of the external heads (see pp. **172–174**) and the proportions of these heads bound to the thin actin filaments are different in the two states: ~9% at rest (see the preceding paragraph) and ~82% during contraction (see pp. **152–153** in Section 4.4.2.5.2). This 'qualitative similarity' is probably indirectly supported by the experiments performed, at room temperature, by Blyakhman et al. (1999) on ~20–30 sarcomeres of single bumblebee flight muscle myofibrils, at rest or under contraction conditions, and interpreted by the authors as demonstrating that 'the single sarcomere-shortening trace was consistently stepwise both in activated and relaxed specimens'. The authors also found that the only apparent difference was in the size of the 'active and passive steps'. This quantitative feature may be essentially related to the proportions of attached cross-bridges (~9% at rest vs. ~82% under contraction conditions; see above).

It is interesting to analyse some experimental results indirectly supporting the existence of radial tethering forces. Xu et al. (2006a), working at 25°C, on resting chemically skinned fibre bundles from rabbit psoas muscle (sarcomere length of ~2.4 μm, i.e. approximately the 'reference' length s_0 ~2.18 μm; see p. **24** in Section 3.2), showed, by x-ray diffraction from synchrotron radiation, that 'as ionic strength is lowered from 200 mM to 50 mM...the lattice spacing [d_{10}] decreases from 404 Å to 371 Å, consistent with increased cross-bridge binding'. In other words, when bulk ionic strength decreases strongly, the bundles shrink laterally (by ~100 × (404 − 371)/404 ~ 8%). The authors proposed an interpretation on the basis of the presence of weakly bound cross-bridges, inducing lateral shrinkage. It is demonstrated throughout this work that the radial repulsive electrostatic forces are effective and the interpretation of Xu et al. (2006a) seems, at first sight, to be strange. Indeed, the radial repulsive electrostatic forces must increase when ionic strength decreases, at least for the ionic strengths used by Xu et al. (2006a) (see Equation 3.2, pp. **101–102** in Section 3.10 and Section 4.2.2) and lateral swelling is expected. However, as pointed out by Morel (1985a) and Morel and Merah (1997) and briefly discussed on p. **13** in Section 2.2, caution is required when interpreting x-ray diffraction data for demembranated fibres. More generally, it is demonstrated, in Sections 3.8, 4.4.2.1 and 8.7, that demembranated fibres (except half-fibres) behave very differently from intact fibres (and intact unit cells). Nonetheless, leaving aside these major problems and aside the interpretation of Xu et al. (2006a), their experimental observations are consistent with the hybrid model, including the existence, at rest, of radial repulsive electrostatic forces and radial tethering forces between the thick myosin and thin actin filaments (it is demonstrated at many places in this monograph that these radial forces exist in demembranated fibres as well as in intact fibres and intact unit cells, but their characteristics may differ in these two types of biological material). As it is impossible to present simple calculations for demembranated fibres, given their complexity, only intact unit cells are considered below, in this section, and the conclusions are successfully extended to the results obtained by Xu et al. (2006a) with skinned fibre bundles.

According to Equation 3.2, lowering the bulk ionic strength from 200 mM to 50 mM (see the preceding paragraph concerning the values of 200 mM and 50 mM) results in a significant increase in the elementary radial repulsive electrostatic force $F_{e,r}$ (at rest, $x = r$). Assuming, for the sake of simplicity, that the decrease in ionic strength from 200 mM to 50 mM is almost instantaneous (possibly lasting only a few milliseconds; using the synchrotron radiation, the time resolution of x-ray diffraction experiments is ~1–5 ms; see, for instance, Dickinson et al. 2005; Griffiths et al. 2006; Huxley et al. 1981; Martin-Fernandez et al. 1994; Piazzesi et al. 1999), the radial repulsive electrostatic force passes 'instantaneously' from $F_{e,r}(200)$ to $F_{e,r}(50) > F_{e,r}(200)$. More precisely, applying Equation 3.2 with H ~0.26 mM$^{-1/2}$ (see p. **488** in Section 4.4.2.7), and assuming that $Q_r(I)$ is not very sensitive to I (see p. **163** in Section 4.4.2.7 concerning the very weak dependence of $Q_a(I^*$ or $I)$ on I^* or I, extended here to $Q_r(I^*$ of I), we obtain $F_{e,r}(50) \sim 2.2 F_{e,r}(200)$. The myofilament

lattice would therefore have an 'instantaneous' propensity to swell laterally. However, there are also strong radial tethering forces at rest, described and analysed throughout this monograph (e.g. pp. **174–176** concerning their generation). At this stage, no instantaneous increase in the thick-to-thin filament distance is assumed to occur during the very short initial period (see above). Thus, the elementary radial tethering force $F_{cb}^0(50)$, at an ionic strength of 50 mM, is assumed to be momentarily identical to $F_{cb}^0(200)$, at an ionic strength of 200 mM, because there is no instantaneous increase in thick-to-thin filament spacing (see Section 5.1.3 for precise details on the strong dependence of F_{cb}^0 on thick-to-thin filament distance), unless F_{cb}^0 depends slightly on ionic strength, which seems likely but impossible to prove. The radial attractive/compressive force Λ_r, defined on p. **156** in Section 4.4.2.5.4, is negligible at 200 mM and also almost certainly at 50 mM (see p. **159** in Section 4.4.2.5.4). As recalled on pp. **164–167** in Section 4.4.2.8, when ionic strength is drastically lowered, particularly from 200 mM to 50 mM, the proportion of weakly bound cross-bridges, including the A-type heads (see pp. **170–172** and Figure 5.1 concerning the various types of head), increases instantaneously from ~5% to ~20% (see Kraft et al. 1995 and p. **169** in Section 4.4.2.8, concerning these proportions). The number $N_0(200)$, at 200 mM, of active weakly bound cross-bridges (see pp. **170–172** for definition of this kind of cross-bridge) becomes instantaneously, at 50 mM, $N_0(50) \sim (20\%/5\%)N_0(200) \sim 4N_0(200)$ and the radial tethering force passes instantaneously from $N_0(200)F_{cb}^0$ to $N_0(50)F_{cb}^0 \sim 4N_0(200)F_{cb}^0$ $\left(F_{cb}^0(50) \sim F_{cb}^0(200) \sim F_{cb}^0;\ \text{see above}\right)$. At an ionic strength of 200 mM, the radial repulsive electrostatic forces and the radial tethering forces are in equilibrium (Λ_r is assumed to be zero; see above). We therefore have $4(\gamma/2)F_{e,r}(200) \sim N_0(200)F_{cb}^0$ (see pp. **34–35** in Section 3.2 for precise details on the multiplication factor $4(\gamma/2)$). The instantaneous lowering of ionic strength from 200 mM to 50 mM (see above) results, therefore, in an instantaneous new net radial force, at 50 mM, of $4(\gamma/2)F_{e,r}(50) - N_0(50)\ F_{cb}^0 \sim 4(\gamma/2)$ $F_{e,r}(50) - 4N_0(200)\ F_{cb}^0$. As $4(\gamma/2)F_{e,r}(50) \sim 2.2$ $\times 4(\gamma/2)F_{e,r}(200)$ (see above), the new instantaneous net radial force, at 50 mM, becomes ~ 2.2 $\times 4(\gamma/2)F_{e,r}(200) - 4N_0(200)F_{cb}^0 \sim 2.2 \times 4(\gamma/2)$ $F_{e,r}(200) - 4 \times 4(\gamma/2)F_{e,r}(200) \sim -1.8 \times 4(\gamma/2)$

$F_{e,r}(200) < 0 (N_0(200) F_{cb}^0 \sim 4(\gamma/2) F_{e,r}(200)$; see above): a strong initial radial attractive force is expected. Applying Equation 3.2 and introducing the numerical values of the parameters, we find $4(\gamma/2) F_{e,r}(200) \sim 433$ pN per intact unit cell (obviously close to ~468 pN per intact unit cell, at bulk ionic strength of 180 mM; see, for instance, p. **142** in Section 4.4.2.3), and, finally, $-1.8 \times 4(\gamma/2) F_{e,r}(200) \sim -779$ pN per intact unit cell. There is, therefore, initial instantaneous shrinkage of the intact unit cells, followed by the rapid establishment of a new equilibrium, corresponding to the equality of the new radial repulsive electrostatic and tethering plus attractive/compressive forces, at the new myofilament spacing. The 'instantaneous process' presented here is oversimplified and leads to overestimation of the initial radial attractive force, but the use of a 'continuous process', rather than the instantaneous process, yields similar qualitative conclusions (but moderate radial attractive force; calculations not shown). Thus, regardless of the possible increase in binding of cross-bridges to account for the results of Xu et al. (2006a) (see the start of the preceding paragraph), the lateral swelling part of the hybrid model provides a simple explanation for the experimental observations. The inference of Xu et al. (2006a), according to which the binding of cross-bridges is responsible for the experimental observation, would be valid only in the absence of strong radial forces (repulsive electrostatic and tethering forces), a condition shown to be invalid throughout this monograph. However, the hypothesis of Xu et al. (2006a) could account for part of the phenomenon observed, particularly when the swinging cross-bridge/lever-arm part of the hybrid model is considered. Nonetheless, the consequences of the hypothesis of Xu et al. (2006a) concerning this 'traditional' part of the hybrid model (lateral shrinkage of the fibres upon cross-bridge attachment to thin actin filaments, when passing from a high to a low ionic strength; see the preceding paragraph) are probably concealed by the entirely different phenomena described in this paragraph.

Xu et al. (2006a) recalled that PEG (polyethylene glycol) can increase the formation of the actomyosin complex in solution (e.g. Highsmith et al. 1996; White et al. 1995). For this reason, Xu et al. (2006a) used 5% (w/v) PEG-1000 to study several properties of their demembranated fibre bundles. The authors demonstrated that, at their working temperature (25°C), 5% PEG-1000 mimics, at least qualitatively, a large decrease in ionic strength from 200 mM to, say, ~80–100 mM (this estimate is not given by the authors, but appears to be reasonable, in the light of their various experimental results). In some of their experiments, Xu et al. (2006b) studied their fibre bundles at a sarcomere length of ~4.2 μm (zero overlap), in the presence of 5% PEG-1000. They compared their x-ray diffraction data with those obtained at a sarcomere length of ~2.4 μm (approximately full overlap). They noted that several features of the profiles of intensities along the meridian at non-overlap differed from those at full overlap, and they attributed this to the weakly bound cross-bridges being attached in relaxing conditions, at full overlap, whereas there are obviously no attached weakly bound cross-bridges at zero overlap. In my opinion, these differences also reflected, at least partly, the number of cross-bridges weakly bound, at full overlap, in the presence of 5% PEG-1000 (i.e. ionic strength of ~80–100 mM; see above), being approximately twice that at 200 mM before addition of PEG-1000 (see pp. **164–167** in Section 4.4.2.8 concerning the dependence of the number of weakly bound cross-bridges on ionic strength). In these conditions, the 'full overlap' as defined by Xu et al. (2006a) is a different notion at 0% and 5% PEG-1000 (see the preceding paragraph for precise details on the role of ionic strength, resulting here from the presence of PEG-1000), rendering the comparison of zero overlap with full overlap risky. In this context, Xu et al. (2006a) themselves noted that some of their observations and conclusions were unclear. I suggest here a brief comment on the findings of Xu et al. (2006a), on the basis of the hybrid model. At zero overlap, under resting conditions, the radial repulsive electrostatic force $4(\gamma^*/2) F_{e,r}$ is, in principle, zero, because the length γ^* of the overlap zone (at any sarcomere length) is zero (see pp. **34–35** in Section 3.2 for comments on the factor $4(\gamma/2)$ and, later, $4(\gamma^*/2)$), and the number N_0 of A-type myosin heads attached to the thin actin filaments is obviously also zero (see pp. **170–172** and Figure 5.1 for precise details on the A-type heads and N_0). However, on p. **27** in Section 3.2, it is strongly suggested that, under resting conditions, residual radial repulsive electrostatic forces are exerted at $\gamma^* = 0$, corresponding to ~5% of those at full overlap. On the other hand, it has been shown experimentally that, at non-overlap, head–head interconnections exist in situ between the closest neighbouring thick myosin filaments belonging to neighbouring unit cells (see p. **17** in Section 2.4), with probable residual radial tether-like forces. These forces add to the radial attractive/compressive forces, described in Sections 4.4.2.4, 4.4.2.5.3 and 4.4.2.5.4, and help to counterbalance the residual radial electrostatic forces to establish a stable equilibrium at non-overlap. Thus, at non-overlap, the approach proposed here and, more generally, throughout this monograph, predicts major differences with respect to full overlap (see above, regarding the ambiguity of the definition of full overlap at two different ionic strengths), consistent with the experimental results of Xu et al. (2006a) and their unclear interpretation, based on the presence of weakly bound cross-bridges at full overlap and their absence at zero overlap (see above). Moreover, the discussion presented in this paragraph demonstrates that the situation is much more complex than previously thought by these authors, and this may be responsible for some of the problems of interpretation raised by unclear observations (see above, in this and the preceding paragraphs). To my mind, one of the advantages of the hybrid model, repeatedly stressed in this book, is its capacity to account for many properties of muscle fibres for which the swinging cross-bridge/lever-arm models require a somewhat 'tortuous' reasoning.

5.1.2 Radial Tethering Forces Are Much Weaker during Isometric Tetanic Contraction (Short Tetani–Steady State) Than under Resting Conditions. Latency Relaxation and Elongation. Relaxation after Isometric Tetanic Contraction

During isometric tetanic contraction, the ~9% 'resting A-type heads', studied in Section 5.1.1, inevitably become 'contracting A-type heads', indistinguishable from the other contracting A-type heads (not attached to the thin actin filaments at rest). For short tetani, under steady-state conditions, most of the contracting A-type heads and the other external heads, including F-type heads (see pp. **170–172** in Section 5.1.1 for precise details concerning the A- and F-type heads), are attached to the thin actin filaments at a given time (~82% at any temperature, between ~0°C and room temperature; see pp. **152–153** in Section 4.4.2.5.2). They rapidly and cyclically interact with the thin actin filaments (rapid association and dissociation, e.g. Brenner 1991) and large amounts of chemical energy are available (mostly resulting from MgATP splitting by the external myosin heads minus Lohmann's reaction). The rapidly cycling external heads, including the A- and F-type heads, are generally assumed to act asynchronously and they remain attached to the thin actin filaments for only ~2 ms, the duration of the cycle being $\tau_{c,\infty}$ ~56–94 ms at 5.7°C (see p. **65** in Section 3.4.3.3). This rapid cycling process and the subsequent hydrolysis of large quantities of MgATP result in the dissociation of many internal B- and B′-type heads (see Figure 5.1, its legend and Appendix 5.I for the arrangement of these heads). This dissociation, experimentally demonstrated and commented on in Section 5.2 for synthetic thick myosin filaments (see Figure 5.2, its legend and Appendix 5.II) and quantified in Chapter 6 (e.g. Sections 6.3.5, 6.4, 6.5 and 6.6), automatically leads to a decrease in most of the distortions described on pp. **170–172** in Section 5.1.1 and a subsequent considerable decrease in F_{cb}^0 (called F_{cb} under contraction conditions; see pp. **174–176** concerning the generation of F_{cb}^0 and some properties of this force). In other words, during a short tetanus, under steady-state conditions, there is a number $n*(\theta)$ of dissociations of internal dimers, which increases with temperature θ (see Section 6.5, particularly Equation 6.2). Unfortunately, the relationship between the number of dissociations $n*(\theta)$ and the strength of the residual radial elementary tethering force $F_{cb} \ll F_{cb}^0$ is purely qualitative and cannot be quantified. On the other hand, the number N of contracting A-type heads (under contraction conditions) is much greater than the number N_0 of resting A-type heads (at rest) (see above), but the residual radial tethering force during contraction, NF_{cb}, is much smaller than the radial tethering force at rest, $N_0 F_{cb}^0$ ($NF_{cb} \sim 0.14 N_0 F_{cb}^0$ at 10°C, for a short tetanus under steady-state conditions; see p. **159** in Section 4.4.2.5.4).

Early in isometric contraction, there is a small time lag between the attachment of the external heads (A- and F-type heads defined on pp. **170–172** in Section 5.1.1) to the thin actin filaments and force development (e.g. Bagni et al. 1988a, working on intact frog fibres) and also the full release of chemical energy, via MgATPase activity (Matsubara et al. 1989, working on heart muscles). Harford and Squire (1992), working on fish muscles, found that the cross-bridges shifted to the vicinity of the thin actin filaments ~15 ms before force was developed. He et al. (1997), working on permeabilised rabbit fibres, claimed that 'in the early stages of an isometric contraction the fraction of cross-bridges that show maximal ATPase activity appears to be close to 100%...' (see pp. **152–153** in Section 4.4.2.5.2 concerning the context of this citation). This conclusion drawn by He et al. (1997), from their experimental results, concerns the rise in force (after the short latent period, during which no active force is recorded). The situation, concerning the relationships between the proportion of attached cross-bridges, the rise in force, the initial MgATPase activity and the time lag, is therefore confusing. At this point, I suggest that, during the short latent period (after stimulation and before any active force can be recorded), there is an instantaneous unblocking of the myosin binding sites on the thin actin filaments (this unblocking results from Ca^{2+} binding to troponin C and the subsequent displacement of tropomyosin in the groove of the two-stranded thin actin filaments; see Gordon et al. 2000, for a review; more recent experimental data have also been obtained; see Galinska-Rakoczy et al. 2008). Thus, considering an intact unit cell, I suggest, as a compromise, that a few milliseconds after stimulation, the number N_{00} of attached A-type heads during the latent period very rapidly reaches a mean proportion of ~18%. This rough estimate ranges between ~100% (for a permeabilised rabbit fibre; see above), ~9% at rest (for various biological materials; see p. **172** in Section 5.1.1) and ~82% during isometric tetanic contraction (for various biological materials; see pp. **152–153** in Section 4.4.2.5.2). The number N_0, corresponding to resting conditions and to a proportion of ~9%, would therefore 'almost instantaneously' become $N_{00} \sim (18\%/9\%)N_0 \sim 2N_0$. Owing to the small time lag between the attachment of the heads and the full release of chemical energy (see the start of this paragraph), there is almost certainly no instantaneous effect on the elementary radial tethering force, the strength of which remains $\sim F_{cb}^0$ (value at rest) for a very short period (see Sections 6.3.5, 6.4, 6.5 and 6.6 as concerns the complex, but undeniable, relationship between MgATPase activity, increase in temperature within the fibres and the core of the thick myosin filaments and the radial tethering forces; briefly, when the MgATPase activity increases, this

'internal' temperature increases, and the radial tethering forces decrease). The resting tethering force $N_0 F_{cb}^0$ therefore becomes approximately $N_{00} F_{cb}^0 \sim 2N_0 F_{cb}^0$ during a few milliseconds (see above concerning $N_{00} \sim 2N_0$). This results in a temporary, tiny decrease in myofilament spacing (see below), and it can reasonably be assumed that the radial repulsive electrostatic forces remain unmodified. At rest, before stimulation, the radial repulsive electrostatic forces and the radial tethering plus attractive/compressive forces (see Sections 4.4.2.4, 4.4.2.5.3 and 4.4.2.5.4 for precise details) exactly counterbalance and the resulting net radial force is strictly zero. Neglecting the radial attractive/compressive forces Λ_x ($x = r$ at rest; $x = a$ during contraction) (they represent ~1% of the radial repulsive electrostatic forces and the radial tethering forces; see p. **159** in Section 4.4.2.5.4), we have, at rest, $4(\gamma/2)F_{e,r} \sim N_0 F_{cb}^0$, whereas, during the latency period, the difference between the unmodified radial repulsive electrostatic forces and the increased tethering forces becomes transitorily $4(\gamma/2)F_{e,r} - N_{00} \ F_{cb}^0 \sim 4(\gamma/2)F_{e,r} - 2N_0 \ F_{cb}^0 \sim -4(\gamma/2)F_{e,r}$. There is therefore a transitory 'shrinking' force, after stimulation. Obviously, this calculation gives only a rough estimate of the shrinking force in an intact unit cell. The situation would be more complex in an intact fibre, because intact unit cells and intact fibres do not behave similarly mechanically (see Section 3.7), although their general behaviour is qualitatively similar. In any event, there are mechanisms for converting radial forces into axial forces, at rest and under contraction conditions (see pp. **102–103** in Section 3.10). Thus, an intact unit cell/intact fibre lengthens slightly and the axial resting force decreases slightly before any sign of active contraction: there must be simultaneous latency elongation and relaxation, observed experimentally early in isometric contraction, with a tiny relative lengthening of ~1% the length of the fibre (i.e. a relative decrease of ~0.5% in myofilament spacing) and a very small decrease (a few percent at most) in the resting tension of an intact fibre (e.g. Bartels et al. 1976, 1979; Haugen and Sten-Knudsen 1976; Herbst 1976; Mulieri 1972; Sandow 1944, 1966).

It would be interesting to try to detect experimentally the expected small decrease in myofilament spacing (e.g. measurement of the distance d_{10}; see below). This would be possible with the most recent intense x-ray sources based on synchrotron radiation, with maximal time resolution of the order of ~1–5 ms (e.g. Dickinson et al. 2005; Griffiths et al. 2006; Huxley et al. 1981; Martin-Fernandez et al. 1994; Piazzesi et al. 1999). Indeed, in equatorial x-ray diffraction experiments, it is possible to measure the lattice spacing d_{10} (see Figure 3 in the review by Millman 1998 for the definition of d_{10}). In intact frog fibres at the slack length, d_{10} is ~36–38 nm (see Table 1 in the review by Millman 1998). The ~1% relative increase in intact unit cell/intact fibre length (see the preceding paragraph) corresponds to a relative decrease of ~0.5% in d_{10}, that is, ~0.18–0.19 nm. Martin-Fernandez et al. (1994) distinguished between values of 14.34 nm and 14.56 nm (difference of 0.22 nm) and Piazzesi et al. (1999) distinguished between values of 14.568 nm and 14.587 nm (difference of 0.019 nm) for axial spacings in the myosin heads, these two series of values corresponding to the 'crown' defined by many authors (e.g. Cantino and Squire 1986; Kensler and Stewart 1983, 1993; Morel and Garrigos 1982b; Offer 1987; Squire 1981; Squire et al. 2005; Stewart and Kensler 1986b; see also the legend to Figure 5.1 and Appendix 5.I). These values of 0.22 nm and 0.019 nm are therefore of the same order of magnitude or much smaller than the ~0.18–0.19 nm estimated for variations in d_{10} during the latency period (see above). Thus, experimental studies of latency axial elongation (i.e. the latency lateral shrinkage described in the preceding paragraph) would be possible. However, this estimate of ~0.18–0.19 nm and the available experimental data (0.22 nm and 0.019 nm) do not refer to the same spatial directions (equatorial vs. meridional) and the experimental detection of tiny variations in d_{10} is probably still impossible.

It should be noted that Huxley (1957) gave a plausible explanation for latency relaxation, as did Herbst (1976) and Goldman and Simmons (1986). However, these groups provided no explanation for latency elongation. Yagi (2007) suggested a structural origin for latency relaxation, based principally on experimental results obtained with x-ray diffraction from synchrotron radiation, but did not study latency elongation itself. He thinks that 'detachment of myosin heads that are bound to actin in the resting muscle is the cause of the latency relaxation'. I conclude that the processes underlying both latency relaxation and elongation presumably have several origins and all the available explanations should be taken into account. In any event, these two transitory phenomena

could not be used to distinguish between swinging cross-bridge/lever-arm theories and the hybrid model, although this model provides a single clearcut explanation for both latency elongation and relaxation.

At the end of a short tetanus, most of the 'external' heads (of A- and F-types; see pp. **170–172** in Section 5.1.1) detach from the thin actin filaments and no more chemical energy is available from MgATP splitting. The internal temperature of the fibre and thick myosin filaments decreases (see Sections 6.3.5 and 6.4) and, consequently, the internal B- and B′-type heads gradually reattach to reform internal B–B′ dimers (see in Section 5.1.1, Figure 5.1, its legend and Appendix 5.I). This process of reattachment is experimentally supported in Section 5.2 (see Figure 5.2, its legend and Appendix 5.II), in which it is demonstrated that attachment–detachment of the B–B′-type head–head dimers within synthetic thick myosin filaments is a reversible phenomenon. Thus, during relaxation, the elementary radial tethering force, $F_{cb}(t)$, gradually increases from its steady-state value, F_{cb}, under isometric tetanic contraction to its resting value, F_{cb}^0. The fibre then returns to rest (reappearance of the balance between radial repulsive electrostatic forces, on the one hand, and radial tethering and attractive/compressive forces, on the other). There is no reason to assume that the durations of the period from the beginning of contraction to the achievement of the plateau tension and of relaxation are similar, at least for sarcomere lengths around the slack length of the fibre. Indeed, in the lateral swelling part of the hybrid model, dissociation of the B–B′ internal dimers, at the beginning of contraction, is mostly related to the 'pressure–volume' process, permitting rapid dissociation of the internal dimers. This rapid dissociation is attributed to a rapid and significant increase in temperature within the core of the thick myosin filaments, triggered by the energy released from splitting of MgATP by the external heads and inducing a chain reaction leading to a massive dissociation of the B–B′ internal dimers (see Section 6.4). By contrast, when the production of energy stops, the temperature of the fibres, particularly within the core of the thick myosin filaments, decreases slowly, because transmission of heat is a slow process (see specialist handbooks). Reassociation of the B–B′-type heads should therefore be a slow process. Thus, according to the hybrid model, relaxation must be a slower process than tension rise. The asymmetry of tension rise and relaxation is familiar to all physiologists studying muscle contraction and is exemplified by two old papers in which intact frog fibres are studied at ~1°C–3°C: in Figures 1 and 3 in the paper by Edman and Reggiani (1987) and Figure 1 in the paper by Gordon et al. (1966b), the asymmetry appears to be considerable. Furthermore, studying tension responses to joule temperature jumps (T-jumps from 5°C–9°C up to 40°C) in skinned rabbit muscle fibres, Bershitsky and Tsaturyan (1992) found three-exponential tension transients: 'Phases 1 and 2 had rate constants k_1 = 450–1750 s^{-1} and k_2 = 60–250 s^{-1} respectively, characterising the tension rise, whereas phase 3 had a rate constant k_3 = 5–10 s^{-1} representing tension recovery due to fibre cooling'. Clearly, this result confirms the asymmetry mentioned above, as well as the slow process underlying heat transfer during relaxation. In any event, the new explanation presented above does not rule out the traditional explanations: repumping of Ca^{2+} by the sarcoplasmic reticulum during relaxation being much slower than Ca^{2+} release at the beginning of contraction, and slow rearrangement of the sarcomeres during relaxation.

5.1.3 Some Expected Biophysical Properties of the Radial Tethering Forces. General Comments on the Radial Repulsive Electrostatic and Tethering Forces. Related Problems

Let us consider a 'virtual intact unit cell', at the slack length, in which the surface-to-surface spacing can be virtually changed (not using demembranation or osmotic pressure; see below, in this section), with the ionic strength remaining constant (I = 180 mM, corresponding to the mean ionic strength in vivo/in situ for any vertebrate skeletal muscle; see p. **32** in Section 3.2 for some references). Owing to the mode of generation of the elementary radial tethering force F_{cb}^0 (see in Section 5.1.1, Figure 5.1, its legend, Appendix 5.I and the corresponding comments; see pp. **174–176** regarding the role of the A-type heads), a small increase in surface-to-surface spacing between the thick myosin and thin actin filaments must result in a large increase in the various distortions of the thick myosin filaments and a subsequent large increase in F_{cb}^0, and vice versa. Concerning the elementary radial repulsive electrostatic force $F_{e,r}$ (at rest), it is shown, on p. **32**

of Section 3.2, that it is proportional to the term $\exp(-s^*/\lambda)$, where λ ~0.5 nm is the Debye-Hückel screening length, and s^* ~3.1 nm is the closest surface-to-surface spacing, before any change in this spacing (see p. **32** in Section 3.2 concerning the numerical values of λ and s^*). Increasing s^* therefore decreases $F_{e,r}$. Under standard conditions (e.g. 10°C, slack length, I = 180 mM), the attractive/compressive force Λ_r represents ~1.12% of the radial tethering force $N_0 F_{cb}^0$ (see p. **159** in Section 4.4.2.5.4; a small increase in s^* is reasonably assumed not to change the strength of Λ_r) and the equilibrium of the myofilament lattice, before any change in the surface-to-surface spacing, corresponds to $\Sigma F_r \sim 4(\gamma/2)F_{e,r} - 1.0112 N_0 F_{cb}^0 = 0$, that is, $4(\gamma/2)F_{e,r} \sim 1.0112 N_0 F_{cb}^0 \sim N_0 F_{cb}^0$ (see Equation 4.1). At this point, we are in the presence of two antagonistic phenomena: when increasing s^*, $F_{e,r}$ decreases, whereas F_{cb}^0 increases. More precisely, it is easy to show that a relative increase in s^* of $\Delta s^*/s^*$ results in a relative decrease in $F_{e,r}$ of $\Delta F_{e,r}/F_{e,r} = -(s^*/\lambda)(\Delta s^*/s^*) \sim -6.2(\Delta s^*/s^*)$ (the values of s^* and λ are given above). Thus, a very small decrease in s^* of 1% induces a noticeable decrease in $F_{e,r}$ of ~6.2%. If we assume that the equilibrium of the myofilament lattice is preserved, then $N_0 F_{cb}^0$ should also decrease, inconsistent with the hypothesis of an increase (see above). It is therefore necessary to leave the problem of equilibrium in the virtual intact unit cell to one side, instead focusing on $N_0 F_{cb}^0$. Applying a very small increase of ~1% in s^*, we can reasonably suggest that F_{cb}^0 increases moderately (e.g. by ~5%) and the attached cross-bridges can sustain this increase: N_0 remains constant. By contrast, if a larger increase in s^* were to be applied (e.g. ~5%), F_{cb}^0 would increase by (5%/1%) × 5% ~ 25% or more. This increase is considerable and few, if any, A-type heads can sustain such 'enormous' radial tethering forces. This leads to an abrupt detachment of most of the A-type heads, resulting in the almost total disappearance of the radial force $N_0 F_{cb}^0$, because N_0 becomes ~0. In these conditions, the dominant force becomes the repulsive electrostatic force $F_{e,r}$, proportional to $\exp(-s^*/\lambda)$. Increasing s^* moderately by ~5% results in $F_{e,r}$ decreasing considerably, by ~6.2 × 5% ~ 31% (see above concerning the coefficient ~6.2). More generally, the virtual intact unit cell would swell indefinitely, because the term $\exp(-s^*/\lambda)$ remains positive, regardless of the value of s^*. However, this term rapidly reaches very low values for large increases in s^*, and the attractive/compressive force

Λ_r increases with increasing myofilament spacing, because the Z discs and the M lines become highly stretched, thus exerting rather strong radial elastic-like restoring forces (the sarcolemma and the sarcoplasmic reticulum also exert strong radial compressive forces). Thus, a new equilibrium is reached between the decreasing radial repulsive electrostatic forces and the increasing attractive/compressive forces, and the virtual intact unit cell does not 'explode'. In any event, it is impossible to predict (i) the amount of lateral swelling required for the abrupt detachment of all the A-type heads and (ii) the new equilibrium between the repulsive and attractive/compressive forces, and, therefore, the new filament spacing, after detachment of all the A-type heads. It is shown, in Section 3.3, that half-fibres behave like intact fibres and, therefore, like intact unit cells. A half-fibre would thus serve, to some extent, as a model for the virtual intact unit cell studied here. In half-fibres, the internal vicinal ionic strength I* can be changed and the behaviour of these half-fibres under resting conditions is studied in Section 4.4.2.8: on pp. **163–167** (see particularly pp. **165–166** regarding the role of Λ_r), it is shown that, for sufficiently low values of I*, the half-fibres swell and most of the A-type heads detach abruptly from the thin actin filaments, resulting in $N_0 \sim 0$. Thus, abrupt detachment is common to half-fibres and a virtual intact unit cell, and the two approaches are mutually consistent.

At this point, it is interesting to consider the complex case of demembranated fibres (this complexity is discussed on p. **13** in Section 2.2 and in Sections 3.8, 4.4.2.1 and 8.7). As recalled in Section 8.7, the demembranated fibres are inevitably swollen and very different from intact fibres and intact unit cells. They are impaired and may even be adulterated. However, as in intact fibres/intact unit cells, in demembranated fibres, there are radial repulsive electrostatic forces, radial tethering and attractive/compressive forces (see p. **179** in Section 5.1.1). The radial tethering forces are therefore exerted, unless all the A-type heads are detached (because of considerable swelling, as described and recalled in the preceding paragraph). Obviously, the strength of the radial tethering forces cannot be predicted, because they depend on the amount of swelling, which is also unpredictable (see Section 8.7). I suggest that, in some unit cells, all the A-type heads are detached, whereas in other unit cells they are not, giving a mean radial tethering force different from zero,

over all the unit cells (the notion of 'unit cell' is used here, rather than that of an 'intact unit cell', because it is demonstrated at many places in this work, including Section 3.8.2, that the unit cells in demembranated fibres are not intact). This average radial tethering force cannot be predicted and almost certainly differs between demembranated fibres. In any event, a new equilibrium is reached at rest, after the removal of the sarcolemma and the subsequent lateral swelling. In this new equilibrium, the new radial repulsive electrostatic forces counterbalance the mean radial tethering and attractive/compressive forces. As for intact fibres/intact unit cells, when isometric contraction is triggered (at a fixed sarcomere length and a fixed myofilament spacing), according to the lateral swelling part of the hybrid model, the new equilibrium of the forces is broken, because the mean radial tethering forces decrease considerably, whereas the radial repulsive electrostatic forces do not change very much (isometric contraction conditions; fixed myofilament spacing; only a limited change in strength when passing from rest to contraction, e.g. p. **139** in Section 4.4.2.3). The radial repulsive electrostatic forces therefore automatically predominate and this induces some of the axial contractile force (see pp. **102–103** in Section 3.10 for precise details on the mechanisms, independent of volume changes, for the conversion of radial forces into axial forces). With moderately osmotically compressed contracting demembranated fibres, the filament spacing decreases moderately and the residual radial tethering force (called NF_{cb} in an intact unit cell) decreases markedly, as for the radial tethering force at rest (called $N_0F_{cb}^0$ in an intact unit cell), because the elementary residual radial tethering force (called F_{cb} in an intact unit cell) decreases markedly with decreasing the filament spacing, as does the elementary radial tethering force at rest (called F_{cb}^0 in an intact unit cell) (see p. **183** concerning the rapid increase/decrease in F_{cb}^0 with increasing/decreasing filament spacing). Conversely, the radial repulsive electrostatic force (called $4(\gamma/2)F_{e,a}$ in an intact unit cell) increases, because the filament spacing decreases (in an intact unit cell, $4(\gamma/2)F_{e,a}$ is proportional to $\exp(-s*/\lambda)$, as is $4(\gamma/2)F_{e,r}$ [see p. **184** for definitions and precise details]). Thus, the net radial expansive force must increase, as it is the difference between an increasing radial repulsive electrostatic force and a decreasing radial tethering force, when the osmotic pressure increases moderately. Moreover, there are mechanisms for converting this net expansive force into an axial force, independently of volume variations (see above). I conclude that a moderate increase in the osmotic pressure should lead to an increase in the axial contractile force. Simultaneously, the cross-sectional area of the demembranated fibre decreases slightly, resulting in a supplementary 'purely geometric' increase in contractile axial tension. When the first experiments on the increase in tension were performed, there was confusion between force and tension and it was difficult to distinguish between the increase in axial contractile force and the trivial increase in cross-sectional area. It is now well established that only the 'normalised' tension should be retained, without the need to take into account the decrease in cross-sectional area. The straightforward consequence of the predicted increase in normalised tension described above is given, for example, by Kawai and Schulman (1985) and Zhao and Kawai (1993), working on inevitably swollen chemically skinned fibres (see the beginning of this paragraph) from contracting rabbit psoas muscle: they found that low levels of osmotic compression increased normalised isometric tetanic tension. This behaviour is systematic, as recalled at many places in this book (see p. **187** in Section 5.1.4, p. **244** in Section 8.7 and p. **252** in Section 8.9).

5.1.4 Major Qualifications Concerning the Proportionality of Stiffness to the Number of Attached Cross-Bridges. Conditions Required to Relate Stiffness and Tension Directly. Some Experimental Arguments, Obtained by Several Independent Groups, in Favour of the Lateral Swelling Part of the Hybrid Model

Goldman and Simmons (1986), for instance, measured the stiffness of contracting and relaxed mechanically skinned frog fibres, which are unavoidably swollen after demembranation (see Section 8.7). I believe that their experimental findings provide further support for some conclusions drawn in Section 5.1.3, concerning the increase in axial contractile forces when the fibre is moderately compressed, as long as stiffness and axial tension are closely related (see below, in this section). Owing to the complexity of the problem, a discussion of stiffness is required. A. F. Huxley (1980a) claimed that the stiffness of

a half-sarcomere is linearly related to the fraction of myosin cross-bridges attached to the thin actin filaments, as also assumed by Ford et al. (1977, 1981) and Goldman and Simmons (1977). In this context of a linear relationship between stiffness (i.e. number of attached cross-bridges) and axial contractile force, it has repeatedly been concluded that the cross-bridges are independent force generators, from the traditional linear length–tension relationship as a function of the amount of overlap between the thick myosin and thin actin filaments found, for example, by Altringham and Bottinelli (1985), Bagni et al. (1988b), Edman (1966), Gordon et al. (1966b), Granzier and Pollack (1990) and Morgan et al. (1991). Morel and Merah (1995) demonstrated that the notion of cross-bridges acting independently is self-contradictory, using the phenomenological/semi-empirical approach independent of the mechanical roles of the cycling cross-bridges developed by my group since 1976 (Morel et al. 1976) and improved throughout Section 3.4. The complex and contentious problem of length–tension relationships is discussed in Section 8.9. Moreover, other tricky problems concerning the relationships between stiffness, the number of attached cross-bridges and axial contractile force, briefly raised in this paragraph, are analysed below, in this section.

Before the demonstration by Morel and Merah (1995) of self-contradiction in the swinging cross-bridge/lever-arm models, it was invariably assumed that stiffness is proportional to tension, via the number of attached independent force-generating cross-bridges. This hypothesis is still frequently accepted and extremely interesting for the 'majority view', because it results in a 'definitive' demonstration of the validity of the swinging cross-bridge/lever-arm models. However, Goldman and Huxley (1994) wrote that 'the X-ray measurements give a strong additional reason why stiffness cannot be used safely as a measure of cross-bridge attachment'. In the particular case of the complex transitory phenomena occurring at the beginning of isometric contraction (see Section 3.4.3.1.3 for a general approach), Bagni et al. (1994a) demonstrated that, during the early tension rise, stiffness develops before cross-bridge attachment, strongly suggesting that stiffness and the fraction of myosin cross-bridges attached to the thin actin filaments are essentially uncoupled (this conclusion was not drawn by the authors). Linari et al. (1998), taking the swinging cross-bridge/lever-arm models as the 'universal' reference, showed that, even under steady-state conditions, the problem is much more difficult than previously thought by A. F. Huxley (1980a) and that there is no straightforward relationship between stiffness and the number of attached cross-bridges and, therefore, tension. This complexity probably also renders the relationship between tension and stiffness more complicated. Based on the experimental work of Brenner and Yu (1991), performed on chemically skinned fibres from rabbit psoas muscle and relating largely to the role of interfilament spacing (which can be altered by external osmotic pressure), it could even be argued that there is no relationship between stiffness and tension. Indeed, when demembranated fibres were shifted from isometric tetanic contraction to rigor, Brenner and Yu (1991) found radial forces to be similar in these two states, whereas radial stiffness was approximately five times stronger in rigor than during contraction. Owing to the inevitable one-to-one translation of radial forces into axial forces (see pp. **102–103** in Section 3.10), these findings could be considered to suggest that no firm conclusion can be drawn regarding the relationship between axial stiffness and axial tension. However, this inference would be too hasty, because it is based solely on radial forces and stiffness, not on axial contractile forces and stiffness. Moreover, Brenner and Yu (1991) worked on demembranated fibres, as in most of the experimental studies performed over a period spanning more than three decades or so. In Section 8.7, it is recalled that demembranation results systematically in marked lateral swelling of the fibre. Thus, a demembranated fibre may be seen as an 'artefactual' fibre, likely to generate erroneous conclusions concerning many physiological and structural features, as pointed out in Sections 3.8, 4.4.2.1 and 8.7.

The results of Brenner and Yu (1991), recalled in the preceding paragraph, were obtained with chemically skinned fibres from rabbit psoas muscle in two very different states (isometric tetanic contraction or rigor), with very different structural properties (see the 'frozen' arrowhead structure of the cross-bridges attached to the thin actin filaments, in rigor). To my mind, only fibres in the same conditions (relaxation, contraction or rigor) should be compared. For example, Goldman and Simmons (1984a), using mechanically skinned frog fibres, found that 'in active contraction,

stiffness was closely related to steady developed tension at sub-saturating calcium concentrations' (i.e. pCa values of ~5.5 or ~5.9, corresponding to ~1/3 and ~1/6 of the maximal active tension measured at pCa ~4.5, respectively). The authors studied subsaturating calcium concentrations for technical reasons. Using the osmotic pressure technique to compress their mechanically skinned frog fibres, Goldman and Simmons (1986) found that the demembranated fibres were less stiff than intact fibres at a given isometric tetanic tension, when they compared their data for demembranated fibres with those previously obtained for intact fibres by Ford et al. (1977). This may simply be caused by the absence of the sarcolemma in demembranated fibres (see third point in Section 4.4.2.4 and pp. **155–156** in Section 4.4.2.5.3). I believe that it is pointless to compare these two different biological materials, because they are under different conditions (absence or presence of the 'elastic' sarcolemma). In any event, in demembranated fibres, an increase in isometric tetanic tension is accompanied by an increase in stiffness and vice versa, as shown in Figure 10 from the paper by Goldman and Simmons (1984a), in which stiffness appears to be essentially proportional to isometric tension, for mechanically skinned frog fibres of similar origins and studied under similar conditions. More generally, the various experimental findings of Goldman and Simmons (1984a, 1986) can be understood as follows: normalised isometric tetanic tension (see definition on p. **570** in Section 5.1.3) should increase as lateral spacing decreases, in a given skinned frog fibre, under given conditions, provided that lateral osmotic compression remains limited; the case of significant osmotic compression is complex and was studied by Bachouchi and Morel (1989a; see also pp. **251–254** in Section 8.9). The increase in normalised tension, with limited compression, supports the existence of strong radial repulsive electrostatic and residual radial tethering forces that increase and decrease, respectively, with decreasing lattice spacing (see also Section 5.1.3 for a circumstantial discussion). Thus, the decrease in lattice spacing tips the balance in favour of the net radial repulsive (expansive) forces and, therefore, in favour of an increase in axial contractile force/normalised tension (see pp. **102–103** in Section 3.10 for the basic models of conversion of radial into axial forces). This increase in axial contractile force, in response to moderate compression, results in a new balance of the radial force and, therefore, a new axial contractile force, stronger than that in the absence of compression. I believe that the straightforward consequence of this approach is supported by several experimental data obtained with demembranated fibres of various origins (frog, rat and rabbit). Ford et al. (1991), Maughan and Godt (1981a) and Millman (1998; see comments on his Figure 13, combining the results of nine independent groups) found that, 'in skinned muscle, as the myofilament lattice or fibre diameter shrinks, isometric force first increases...' (Millman 1998; note that the notions of force and tension are confused in most experimental studies, as already pointed out on p. **185** in Section 5.1.3). Godt and Maughan (1981), working on skinned rabbit fibres, and Wang and Fuchs (1995), working on bovine ventricular muscle, made similar observations, but they attributed this to changes in calcium sensitivity with lattice spacing. In any event, all the experimental observations recalled here strongly support the conclusions drawn above from the existence of effective radial repulsive electrostatic forces and residual radial tethering forces, as well as the observations and conclusions described in Section 5.1.3 (see particularly p. **185**), on p. **244** in Section 8.7 and on p. **252** in Section 8.9. However, calcium can modulate the increase in force with the osmotic compression, essentially through the traditional part of the hybrid model (rotation of the cross-bridges).

To conclude this section, it should be recalled that Martyn and Gordon (1992) demonstrated, with contracting glycerinated fibres from rabbit psoas muscle, that force and stiffness are generally fairly closely related, but that these two parameters may vary differently, according to the particular biochemical conditions in which experiments are carried out. Their experimental results confirm that the relationship between force and stiffness is complex, but, with more traditional buffers than those used by Martyn and Gordon (1992), the main conclusions drawn in the preceding paragraph remain valid. Thus, there is a correlation between axial force and stiffness, provided that the fibres are in similar physiological states (demembranated fibres under contraction conditions, for example) and under limited osmotic compression. Nonetheless, the linear relationship between 'axial' stiffness and the number of attached cross-bridges during contraction is hazardous and almost certainly coincidental.

5.1.5 Important Remark

In this work, I focus principally on the lateral swelling mechanisms of contraction, accounting for ~40% of contraction. Clearly, under full contraction conditions, all the external heads, including the A- and F-type heads (see pp. **170–172** in Section 5.1.1 for precise details) attached to the thin actin filaments and splitting large amounts of MgATP, display the traditional conformational changes resulting in the other ~60% of contraction (swinging cross-bridge/lever-arm mechanisms).

5.2 'RESPIRATION' OF SYNTHETIC THICK MYOSIN FILAMENTS

Synthetic thick myosin filaments were prepared and studied as recalled and described in Section 2.4. The intensity of the scattered light sharply and reversibly increased or decreased at ~37°C–38°C (see Figure 5.2, its legend and Appendix 5.II). Dimeric interactions of synthetic thick myosin filaments via their extruding heads, in the presence of either 0.5 mM or 1 mM MgATP (see p. **17** in Section 2.4), were described by Morel et al. (1999) and recalled on p. **17** in Section 2.4. These interactions occur at temperatures up to ~27°C and entirely disappear above ~29°C (see pp. **18–19** in Section 2.4). The phenomenon, observed here for synthetic thick myosin filaments, at ~37°C–38°C (vs. ~27°C–29°C), cannot therefore be related to this type of filament–filament interaction. Concerning the B–B'-type heads inserted into the filament cores to form dimers (in relaxing conditions) and their principal features and role (see Section 5.1.1, pp. **174–176** and Figure 5.1, its legend and Appendix 5.I), it should be noted that the structural, physical–chemical and electrical conditions are very different from those prevailing in the surrounding medium. In this context, the pressure within the shaft of the filament is ~1.49 × 10^4 times the atmospheric pressure (see p. **215** in Section 6.4). This probably explains why, at atmospheric pressure (in vitro), monomer–dimer transitions occur at ~27°C–29°C, rather than at ~37°C–38°C at the very high pressure found within the backbone. It appears clear, from Figure 5.2, its legend and Appendix 5.II, that there is a rapid reversible increase/decrease in the intensity of the scattered light at ~37°C–38°C, undoubtedly caused by reversible phenomena of dissociation–association of the internal myosin

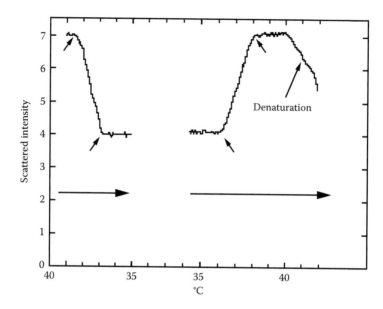

Figure 5.2 'Respiration' of synthetic thick myosin filaments. Below 37°C, on the one hand, and between 38°C and 40°C, on the other, the intensities of scattered light are independent of temperature. This reflects the fact that, below 37°C, all the internal B–B'-type head–head dimers described in Section 5.1.1 (see particularly Figure 5.1) are associated, whereas, above 38°C, all the internal B–B'-type head–head dimers are dissociated (see Section 2.4 for the many precautions that should be taken to avoid misinterpretation of this observation). Supplementary comments are given in Appendix 5.II.

B–B'-type head–head dimers. Nonetheless, this assertion should be supported by a simplified physical–chemical approach. According to Tanford (1967), we have

$$i/i_0 = K_0 M c_p / [1 + 2BMc_p + 3C^2(Mc_p)^2 + \ldots] \quad (5.1)$$

i_0 and i are the intensities of the incident beam and the scattered light, respectively. K_0 is the apparatus constant for a given wavelength. M is the molecular weight of the scattering particle and c_p is its concentration. B and C are the second and third virial coefficients, respectively. The denominator in Equation 5.1 is a corrective factor $(1 + \beta)$ that may be markedly higher than 1, but that can reasonably be assumed to display no marked variations with the experimental conditions used in this study (temperatures between ~36°C and ~40°C). Thus, we deduce from Equation 5.1:

$$i/i_0 \sim K_0 M c_p / (1 + \beta) \quad (5.2)$$

Equations 5.1 and 5.2 are valid only for small spherical or slightly asymmetrical particles, much smaller than the wavelength of the incident light (here, 500 nm) as pointed out by Tanford (1967).

This is obviously not the case for the highly asymmetric synthetic thick myosin filaments (~2 μm long, i.e. ~2000 nm >> 500 nm, but ~18–34 nm in diameter, i.e. <<500 nm; see Morel et al. 1999), but these two equations are used as first approximations. Let i_+ be the intensity of the scattered light at ~40°C and i_- be the intensity of the scattered light at ~36°C. From Equation 5.2, we deduce that $i_+/i_- \sim (M_+/M_-)$. M_+ and M_- are the 'apparent molecular weights' of the filaments at ~40°C and ~36°C, respectively. The mean length of a synthetic thick myosin filament is reasonably assumed to be insensitive to temperature within the narrow range of ~36°C–40°C. The observed changes in intensity therefore result mostly from changes in filament diameter. Morel et al. (1999) suggested that the apparent molecular weight M of a synthetic thick myosin filament may be approximately proportional to the diameter D_b of the backbone and we deduce that $i_+/i_- \sim D_b^+/D_b^-$. However, the assumption of Morel et al. (1999) is not sufficiently convincing and I suggest here that i_+/i_- would be $\sim \left(D_b^+/D_b^-\right)^2$. The best solution is to take $i_+/i_- \sim \left(D_b^+/D_b^-\right)^{1.5}$ as a compromise. i_+/i_- reaches values of up to ~1.8 ± (0.1 or 0.3) (mean ± SE or SD; n = 12) in all the experiments

Morel

performed (12 experiments, one of which is shown in Figure 5.2). Thus, D_b^+/D_b^- reaches values of up to ~1.48 $\pm$ (0.06 or 0.16) and, taking $D_b^- \sim 18$ nm (see p. **215** in Section 6.4), we deduce that $D_b^+ \sim 27 \pm (1$ or 3) nm. This is statistically close to the values of ~30–50 nm, obtained by Pinset-Härström and Truffy (1979) for the backbone of synthetic thick myosin filaments prepared as recalled on p. **17** in Section 2.4, but in the total absence of MgATP and therefore in the absence of internal myosin B–B'-type head–head dimers (see Figure 5.1 concerning the arrangement of the B- and B'-type heads). Thus, according to Figure 5.2, its legend and Appendix 5.II, a gradual temperature increase, within the range ~37°C–38°C, results in synthetic thick myosin filaments swelling considerably (mean increase of ~100 × (27 − 18)/18 ~ 50% in the diameter of the backbone), owing to the total dissociation of internal B–B'-type head–head dimers, and this phenomenon is reversible. This respiration of the synthetic thick filaments provides experimental support for the mode of generation of the radial tethering forces, at rest, and their decrease during contraction, followed by their reappearance during relaxation, presented and discussed on pp. **174–176** in Section 5.1.1 and on p. **183** in Section 5.1.2, respectively. Moreover, the consequences of this experimental finding, studied and discussed throughout Chapter 6, are of major importance for the construction of the self-consistent hybrid model, the lateral swelling part of which accounts for ~40% of the isometric tetanic tension, in an intact unit cell from frog or rabbit muscle, and probably from human muscle.

APPENDIX 5.I

SUPPLEMENTARY COMMENTS ON FIGURE 5.1

A 'double helical' position of the A- and B-type myosin heads was proposed for the thick myosin filaments by Morel and Garrigos (1982b). To test our hypothesis, we built a Perspex model, taking into account all the characteristics deduced from the x-ray diffraction results available in 1982 (with the help of Mrs. Anne-Marie Bardin). For simplification, we considered only a two-stranded dummy filament. The A- and B-type heads were arranged on two helices (strands) with a pitch that was a multiple of 14.3 nm. Indeed, working

on whole muscle, by traditional x-ray diffraction, Huxley and Brown (1967) attributed a value of 14.3 nm to the axial spacing between two successive crowns (see, for instance, Morel et al. 1999; Squire 1981; Squire et al. 2005 for definition of a crown). Huxley and Brown (1967) found that there are three myosin molecules (or crossbridges) per apparent axial repeat of 14.3 nm × 3 = 42.9 nm. Eaton and Pepe (1974), Morel et al. (1979) and Pinset-Härström et al. (1975) also found an axial repeat of ~43.0–43.1 nm for synthetic thick myosin filaments. Koretz (1979), studying synthetic filaments prepared from column-purified myosin, obtained values of ~43.0 nm and ~14.3 nm. More recent and precise values of 14.34 nm or 14.58 nm were obtained, by x-ray diffraction from synchrotron radiation, by Martin-Fernandez et al. (1994) and Piazzesi et al. (1999), for the shifting of skeletal whole muscles or single intact fibres from rest (14.34 nm) to the isometric tetanus plateau (14.58 nm). Using the same x-ray technique, Linari et al. (2005) were able to distinguish, under isometric tetanic contraction conditions, between a value of 14.568 nm at 0°C and 14.587 nm at 17°C. All these values are consistent with the 'old value' of 14.3 nm and the more recent values of 14.568 nm and 14.587 nm, corresponding to axial repeats of 14.568 × 3 ~ 43.704 nm and 14.587 × 3 ~ 43.761 nm, that is, only ~2% higher than ~42.9 nm. Morel and Garrigos (1982b) pointed out that their Perspex model necessarily gave marked distortions of the double helical arrangement of both the external and internal heads (we did not give the order of magnitude of these distortions, but I estimate it at ~±20%–25% around the 'perfect' double-helical position, deduced from x-ray diffraction experiments and described in brief above). The Perspex model was photographed and optical diffraction analysis was carried out on a greatly reduced negative. The diffraction pattern corresponded to a pure helical arrangement. It was therefore concluded that the x-ray diffraction techniques available in the 1960s, 1970s and 1980s gave the mean position and were not sensitive to distortions, even when marked (see above, ±20%–25%). More recent x-ray diffraction techniques would give a better analysis of such distortions and would also provide new explanations for the rather complex phenomena observed in situ by Linari et al (2005), Martin-Fernandez et al. (1994) and Piazzesi et al. (1999) (see above). However, an 'averaging' process may occur in all x-ray diffraction experiments, even those using synchrotron radiation, and this may also be the case for many other techniques (see, for instance, Borejdo et al. (2006) concerning major differences between a single cross-bridge and several cross-bridges; see also p. **40** in Section 3.4.2 for citations from these authors). In this context, it should be noted that analysing the x-ray diffraction pattern obtained by synchrotron radiation techniques from a resting muscle fibre, Piazzesi et al. (1999) pointed out that the layer lines were 'associated with the quasi-helical arrangement of the myosin heads on the surface of the thick filaments'. This sentence unwittingly supports my above assertion that x-ray diffraction experiments lead to 'average', or 'quasi-helical', positions, regardless of the sophistication of the apparatus. Stewart and Kensler (1986b) reported similar findings and noted that natural thick myosin filaments from frog 'are not strictly helical'. Kensler and Stewart (1993) also pointed out that the myosin heads are systematically displaced from an ideally helical arrangement. In any event, I wonder whether the various interpretations of diffraction patterns depend on assumptions made a priori about the position of the myosin heads, as seems to be the case for the inferences proposed by the authors cited above, who are trapped in the traditional model for thick myosin filaments. In other words, would it actually be possible to distinguish, on the basis of x-ray diffraction alone, between the traditional model for thick myosin filaments and the model presented by Morel and Garrigos (1982b) and analysed in this chapter (see Sections 5.1.1 and 5.2). In Section 6.3.5, it is demonstrated that this model for the structure of the thick myosin filaments (with B–B'-type head–head dimers buried within the filament core) is consistent with experimental data of various origins, although this structure caused some controversy (Davis 1988; Morel and Bachouchi 1988c). Finally, Morel and Garrigos (1982b) did not suggest the existence of the radial tethering force F_{cb}^0 (described on pp. **174–176** in Section 5.1.1 and represented in Figure 5.1), although they concluded that the arrangement they described might be interpreted in terms of the mechanisms of contraction themselves.

APPENDIX 5.II

SUPPLEMENTARY COMMENTS ON FIGURE 5.2

The transition occurring between 37°C and 38°C is valid in vitro for synthetic thick myosin filaments. The body temperature of the rabbit is 39°C, and a living muscle in situ would work at the limit of temperature-induced contracture (see Section 8.6). However, as conditions are very different in vitro and in vivo/in situ (as highlighted on pp. **270** and **272** in Chapter 9), the transition may be shifted slightly toward ~43°C–44°C (e.g. ~43.5°C), at least in myofibrils lying, for instance, in the core of arm and leg muscles, but not necessarily elsewhere in the body. Assuming that the temperature transition observed here also holds for humans (body temperature of 37°C), the transition suggested above for the rabbit would occur at ~(37°C/39°C) × 43.5°C ~41.3°C for humans. This plausible value of ~41.3°C would account for contractures and cramps, particularly in myofibrils from the arm and leg muscles (see above), as frequently occurs in severe hyperthermia (high fever owing to influenza, or exhausted athletes, for example). In this context, Davies et al. (1982) demonstrated that the intramuscular temperature in humans can rise to ~39°C after exercise. This is consistent with local temperatures reaching up to ~41.5°C in certain myofibrils, buried in the centre of the muscles. In such cases of hyperthermia, lowering the temperature of the heated muscles is known to be a good solution, making it possible to return to physiological conditions (disappearance of contractures and cramps). This simple explanation of well-known physiological phenomena provides further support for the hybrid model of muscle contraction within the body, although more traditional interpretations remain valid (e.g. occurrence of fatigue; see Chapter 7). I wrote this legend in 2010. I learned, in April 2012, that a drug (MDMA; 3,4 methylenedioxy-N-methyl amphetamine) prepared from ecstasy can lead to severe hyperthermia (up to ~42°C) within the body (particularly in the skeletal and smooth muscles, heart and brain), resulting in irreversible destruction of the brain, severe simultaneous contracture, and inevitable sudden death. This recent observation supports the reasoning put forward here.

CHAPTER SIX

Quantitative Data and Calculations, Concerning Short Tetani under Steady-State Isometric Contraction Conditions, Required for Quantification of the Hybrid Model

6.1 GENERAL ASPECTS OF VARIOUS PROBLEMS

In this chapter, I focus essentially on data for rabbit, referring occasionally to frog. Many experimental studies of MgATPase activity have been performed, by independent groups, on chemically skinned and permeabilised fibres from psoas muscle. Several experimental results obtained in these studies are used in this chapter and at many places in this work. These data concern whole fibres, whether intact, skinned or permeabilised, isolated myofibrils and intact unit cells. The problems of dealing with such a mixture of biological materials are addressed throughout this monograph. Fortunately, consistent conclusions can nonetheless be drawn.

A key part of the lateral swelling mechanism involved in the hybrid model, accounting for ~40% of isometric tetanic contraction (short tetani, steady state), is the process of reversible dissociation–association of internal myosin B–B'-type head dimers, buried within the thick myosin filament core (see Figure 5.1, its legend and Appendix 5.I; see also pp. **170–172** in Section 5.1.1, pp. **180–181** and p. **183** in Section 5.1.2, and Section 5.2, including Figure 5.2, its legend and Appendix 5.II). Many experimental results, obtained by my group and other independent groups, demonstrate systematic reversible temperature-induced transitions in both the myosin head (S1) and the whole myosin monomer–dimer equilibria (myosin either free in solution or inserted within the core of synthetic and natural thick myosin filaments, in vitro and in the absence of F-actin). When contraction is triggered, there is an increase, $(\Delta U)_{t*}$ (expressed in mJ cm^{-3}), in the total energy available in a contracting intact unit cell, from time zero to time t*, given by (see handbooks on thermodynamics)

$$(\Delta U)_{t*} = (\Delta H)_{t*} + (P\Delta V^*)_{t*} \qquad (6.1)$$

In this equation, the subscript t* corresponds to a reference 'virtual' duration of a short tetanus. More details of this complex notion of virtual duration are given on pp. **210–212** in Section 6.3.5.

In Equation 6.1, the enthalpy increase $(\Delta H)_{t*}$ is mostly related to the MgATP breakdown process occurring on the cycling cross-bridges and the subsequent instantaneous Lohmann reaction (rephosphorylation of MgADP, resulting from MgATP cleavage). However, it also concerns ATPase (mostly MgATPase) activities related to non-myofibrillar structures and other processes (e.g. Ca^{2+} pump in the sarcoplasmic reticulum, binding of Ca^{2+} to parvalbumins and Na^{+}–K^{+} pump in the sarcolemma of intact fibres). This enthalpy increase $(\Delta H)_{t*}$ is released outside the thick myosin filament cores. P is a pressure, studied on pp. **214–215** in Section 6.4, relating to the inner part of the thick myosin filament core. $\Delta V^* = V_m^* - V_d^* \sim +0.015$ ml g^{-1} is the change in specific volume for the head monomer $\rightarrow$ dimer transition, determined experimentally on myosin heads (S1) in vitro (S1 from myosin of rabbit back and legs muscles; Morel and Garrigos 1982a) and assumed to be similar within the core of the thick myosin filament. A plus sign for ΔV^* is rarely observed for reversible monomer–dimer equilibria. Moreover, the absolute value of ΔV^* is surprisingly high. These experimental results are probably due to the significant hydration (strongly bound plus loosely bound water) of the myosin heads (Garrigos et al. 1983; Morel and Garrigos 1982a). This suggestion of a relationship between the plus sign and the significant value of ΔV^*, on the one hand, and the very high degree of hydration, on the other, results from long and fruitful discussions with the late Bill Harrington. The energy $(P\Delta V^*)_{t*}$ is released within the thick myosin filament cores, where the B–B′-type head–head dimers are located (see, in Section 5.1.1, pp. **174–176**, Figure 5.1, its legend and Appendix 5.I for precise details on the internal B–B′ dimers).

The numerical values used in this monograph correspond mostly to isolated myosin heads in vitro (S1), isometrically contracting intact fibres, chemically skinned and permeabilised fibres and unheld isolated myofibrils. I combine the corresponding experimental data with many phenomenological/semi-empirical results to obtain quantitative values valid for intact unit cells. This method is successfully used throughout this work. Indeed, no other approach is currently possible, when considering intact unit cells directly. However, I take into account this mixture of results and try to explain the consequences of this choice in every instance.

Finally, the reasoning and calculations presented in the rest of this book are valid under physiological conditions: pH ~7 and ionic strength of ~150–200 mM, unless otherwise specified.

6.2 RELATIONSHIP BETWEEN THE INCREASE IN AVAILABLE ENTHALPY IN THE SARCOPLASMIC MEDIUM AND THE CHEMICAL ENERGY RELEASED FROM MgATP SPLITTING BY THE CYCLING CROSS-BRIDGES AND NON-MYOFIBRILLAR STRUCTURES, AND ABSORBED BY LOHMANN'S REACTION, DURING ISOMETRIC CONTRACTION (SHORT TETANI–STEADY STATE)

In vitro, under conditions mimicking the sarcoplasmic medium of intact frog fibres (i.e. pH,

pMg, ionic strength, various reactant and product concentrations), Woledge and Reilly (1988) measured the enthalpy resulting from the hydrolysis of creatine phosphate. This enthalpy increase corresponds to the difference between the enthalpy released from MgATP breakdown and absorbed by MgADP rephosphorylation via Lohmann's reaction. The authors found an apparent enthalpy $e^*_{app} \sim 5.8 \times 10^{-20}$ J (58 zJ [zeptojoules]) per MgATP molecule, regardless of temperature, within the 0°C–37°C range, almost independently of the ionic strength and reactant concentrations. This value and those given in the following three paragraphs are valid in vitro (on p. **193**, the possible differences between in vitro and in vivo/in situ values are discussed).

Values corresponding, in vitro, to the cleavage of one MgATP molecule (not taking into account Lohmann's reaction) can be recalled: $e^*_1 \sim 8.5 \times 10^{-20}$ J (Morel et al. 1976; Piazzesi et al. 2007) and $e^*_1 \sim 8.3$ or 9.5×10^{-20} J (Benzinger 1969; Morel and Pinset-Härström 1975a). The mean of these six values is $e^*_1 \sim 8.8 \pm (0.2$ or $0.5) \times 10^{-20}$ J per MgATP molecule (mean $\pm$ SE or SD; n = 6). Other values were obtained or suggested by several authors: $e^{**}_1 \sim 8.0 \times 10^{-20}$ J per ATP molecule (not MgATP; see below for a brief discussion) was proposed by Curtin and Woledge (1978), Homsher and Kean (1978), Cappello (2008) and Cappello et al. (2007); $e^{**}_1 \sim 8.3 \times 10^{-20}$ J per ATP molecule was proposed by Kushmerick and Davies (1969); $e^{**}_1 \sim 10 \times 10^{-20}$ J per ATP molecule was suggested by Yu et al. (1970); $e^{**}_1 \sim 22.8 kT \sim 8.9 \times 10^{-20}$ J per ATP molecule (at 10°C) was proposed by Simmons and Hill (1976); $e^{**}_1 \sim 8.3 \times 10^{-20}$ J per ATP molecule was used by He et al. (1998a) and also by Bagshaw (1993) and Oiwa et al. (1991). Grazi and Di Bona (2006) used the value $e^{**}_1 \sim 7.5 \times 10^{-20}$ J per ATP molecule. Averaging the 11 values, valid in vitro, gives $e^{**}_1 \sim 8.3 \pm (0.2$ or $0.6) \times 10^{-20}$ J per ATP molecule (mean $\pm$ SE or SD; n = 11). This value of e^{**}_1 corresponds to one ATP molecule, whereas the value of e^*_1 corresponds to one MgATP molecule. The small difference between $e^*_1 \sim 8.8 \pm (0.2$ or $0.5) \times 10^{-20}$ J per MgATP molecule and $e^{**}_1 \sim 8.3 \pm (0.2$ or $0.6) \times 10^{-20}$ J per ATP molecule is ~0.5 $\pm$ (0.4 or 1.1) $\times$ 10^{-20} J per ATP molecule. This difference is of borderline statistical significance and would take into account the binding of Mg^{2+} to ATP to 'produce' the functional molecule, MgATP, which can be cleaved by the myosin heads, whereas ATP alone cannot. Much of the slight difference between e_1^* and e_1^{**} is attributed to the systematic use of ATP, rather than of MgATP, by most authors. As experimentally demonstrated by Morel and Garrigos (1982a,b), mistaking ADP for MgADP results in opposite phenomena being reported for the myosin head dimer (MgADP promotes dimer formation, whereas ADP dissociates the dimer). The same is true for ATP and MgATP (Morel et al. 1998b), and it is, therefore, imperative to distinguish between ATP and MgATP from both structural and enzymological viewpoints. In any event, the value of $e_1^* \sim 8.8 \times 10^{-20}$ J per MgATP molecule is retained here. In vitro, the energy related to Lohmann's reaction (occurring in vivo in the sarcoplasmic medium of an intact fibre, which contains all the necessary enzymes, including creatine kinase, mostly located in the M line, as shown by Walliman et al. 1977, at least in chicken skeletal muscle and almost certainly in any other animal species), is $e_2^{**} \sim -2.3 \times 10^{-20}$ J per ATP molecule (Curtin and Woledge 1978; Homsher and Kean 1978; Woledge et al. 1985). From the discussion and values presented above, e_2^*: per MgATP molecule would be $\sim (-2.3 \ 10^{-20}$ J per ATP molecule) $\times$ (8.8 10^{-20} J per MgATP molecule/8.3 10^{-20} J per ATP molecule) $\sim -2.4 \ 10^{-20}$ J per MgATP molecule. The apparent energy released, in vitro, from the splitting of one MgATP molecule can therefore be taken as $e_{app}^* = e_1^* + e_2^*$ per MgATP molecule. As highlighted in the first paragraph, another value of $\sim 5.8 \times 10^{-20}$ J per MgATP molecule is available. The mean of these two values is $e_{app}^* = e_1^* + e_2^*$ $\sim 8.8 \times 10^{-20} - 2.4 \times 10^{-20} \sim 6.4 \times 10^{-20}$ J per MgATP molecule.

It would be interesting to estimate the apparent energy released per MgATP molecule in intact fibres containing all the components necessary for contraction (e.g. proteins, enzymes, etc.). Unfortunately, for purely technical reasons, most of the experimental data recalled below, in this section, are valid solely in demembranated fibres. In these biological materials, the chemical compounds required for Lohmann's reaction are absent, due simply to the dissolution of creatine and creatine phosphate in the buffers used during demembranation. In permeabilised fibres from rabbit psoas muscle, Siththanandan et al. (2006) suggested taking into account an energy for MgATP hydrolysis of $e_{app}^1 \sim 5.3 \times 10^{-20}$ J per

MgATP molecule, close to the $e_{app}^2 \sim 5.6 \times 10^{-20}$ J per MgATP molecule chosen by West et al. (2004) for permeabilised white muscle from dogfish. Owing to these rather low values, the two groups almost certainly unwittingly included Lohmann's reaction in their estimates. This inference is the more probable as Worthington and Elliott (1996a,b) clearly selected $e_{app}^3 \sim 5.7 \times 10^{-20}$ J per MgATP molecule, as a result of both MgATP splitting and Lohmann's reaction. Working on intact frog fibres, Linari et al. (1998) suggested a value of $e_{app}^4 \sim 10.0 \times 10^{-20}$ J per MgATP molecule. This value probably takes into account Lohmann's reaction, because the fibres are intact, but the authors confused ATP with MgATP, an error denounced in the preceding paragraph. This value of $\sim 10.0 \times 10^{-20}$ J per MgATP molecule is ~ 2 times the three preceding values ($\sim 5.3 \times 10^{-20}$, $\sim 5.6 \times 10^{-20}$ and $\sim 5.7 \times 10^{-20}$ J per MgATP molecule, for demembranated fibres). Furthermore, the estimate of $\sim 10.0 \times 10^{-20}$ J per MgATP molecule appears to be model dependent (swinging cross-bridge/lever-arm theory) and results from a discussion in which many experimental parameters are introduced. It thus includes an accumulation of experimental uncertainties. However, all the other estimates also have shortcomings and we cannot discard the value proposed by Linari et al. (1998), particularly as the other three values were obtained with different biological systems (demembranated fibres from two different animal species). An average value e_{app}, valid for a 'composite' biological system, that is, both intact and demembranated fibres (of any origin), is therefore $\sim (5.3 + 5.4 + 5.7 + 10.0) \times 10^{-20}/4 \sim 6.6 \pm (1.0$ or 2.0) $\times 10^{-20}$ J per MgATP molecule (mean $\pm$ SE or SD; n = 4). This value of $e_{app} \sim 6.6 \pm 1.0 \times 10^{-20}$ J per MgATP molecule, valid in situ, for intact and demembranated fibres, is statistically comparable to $e_{app}^* \sim 6.4 \ 10^{-20}$ J per MgATP molecule, valid in vitro (see the preceding paragraph). For simplification, only the mean values are used in this work, that is, $e_{app} \sim 6.6 \times 10^{-20}$ J per MgATP molecule (in vivo/in situ in intact and demembranated fibres) and $e_{app}^* \sim 6.4 \ 10^{-20}$ J per MgATP molecule (in vitro; see above).

Morel (1984b) claimed that the energies of both MgATP splitting and MgADP rephosphorylation (Lohmann's reaction) would be different in vivo/in situ and in vitro (the absolute values of these energies are assumed, in principle, to be higher in vivo/in situ than in vitro). The results presented in the preceding paragraph may support this view, but the scattering of the experimental values blurs the conclusion. It should be recalled that this notion of differences between in vivo/in situ and in vitro energies was first raised by Woledge and Canfield (1971) some 13 years before the suggestion of Morel (1984b), but these concordant notions were entirely overlooked by the 'muscle community'. The 'living water' (or cell-associated water) present within muscle fibres (certainly intact fibres and almost certainly also demembranated fibres) and all living cells, described on pp. **29–31** in Section 3.2, is different from bulk water. This difference would be, at least partly, responsible for the different free energy values obtained in vitro and in vivo/in situ. Unfortunately, taking into account the error bars, the difference between the in vitro and in vivo/in situ estimates of e_{app} and e_{app}^* is not significant (see the preceding paragraph). In the absence of definitive arguments, I retain the value of e_{app} $\sim 6.6 \times 10^{-20}$ J per MgATP molecule, valid for all biological materials, including intact unit cells (see also the end of the preceding paragraph).

Regardless of the various numerical values, the energy output, as a result of cyclic actin–myosin interactions, and resulting from both MgATP splitting by the external myosin heads (see pp. **170–172** in Section 5.1.1 for precise details on the external heads) and the instantaneous rephosphorylation of MgADP, via Lohmann's reaction, is $E^{AM} = m\Pi e_{app}$, expressed, for instance, in mJ cm^{-3} s^{-1} (see p. **43** in Section 3.4.3.1.1 concerning the definition of E^{AM}; m is the number of cross-bridges per cubic centimetre, introduced in the text below Equation 3.11, defined, discussed and calculated on pp. **205–207** in Section 6.3.5). I use here simplified versions of the notations given in Section 3.4.3.1.1. For example, Π, the rate of MgATP splitting, replaces the notation $\Pi_{isoim}(t)$ introduced in Equation 3.11 and applied in Sections 3.4.3.1.1 and 3.4.3.1.3. Moreover, $\Pi_{t^*} = \Pi_{isom}(t^*)$ represents the dimensionless rate of 'isometric' MgATP splitting. With these notations, we can write $(E^{AM})_{t^*} = m\Pi_{t^*}e_{app}$, where $(E^{AM})_{t^*}$ is the 'time-independent' energy/enthalpy within the range of time $0–t^*$, expressed in mJ cm^{-3}. This notion of time-independent energy/enthalpy is discussed on pp. **210–211** in Section 6.3.5.

It is difficult to estimate the proportion of the energy output $(E^{AM})_{t^*}$ used to induce movement of the 'side-pieces' (notion introduced by Huxley 1957), which can be seen as the cross-bridge in modern swinging cross-bridge/lever-arm theories. Huxley (2000) remained rather vague in this area and I interpret this as meaning that after attachment of the cross-bridges, the problem of the proportion of chemical energy released from MgATP splitting by each cycling cross-bridge and used to induce conformational changes in the attached cross-bridges is not resolved. It is also difficult to estimate this proportion of chemical energy used to induce both the swinging of the cross-bridges and dissociation of the internal myosin head–head dimers in the hybrid model (see Figure 5.1, its legend, Appendix 5.I and the corresponding comments in Section 5.1.1 concerning the internal dimers). This task is very difficult in the case of isometric contraction, involving no mechanical work, as defined in purely physical terms. Indeed, in this case, only an isometric force is developed, without macroscopic displacement and, therefore, without measurable mechanical work. In a short abstract published in the *Scientific American*, Yanagida (2001) recalled that the swinging cross-bridges can work 'even if the chemical energy is barely more powerful than the noise represented by ambient heat'. He also claimed that 'the molecules [actin + myosin] convert chemical energy—in the form of adenosine triphosphate [ATP]—into kinetic energy with an efficiency of about 50 percent'. Obviously, this last remark, concerning efficiency, only holds when there is an active sliding of the myofilaments past each other during active shortening of the fibre and the value of ~50% merely corresponds to the maximal 'mechanical energy output' (defined, e.g. by PV** in Equation 3.13). This efficiency of ~50% is only an order of magnitude. For instance, Barclay (1996) studied this mechanical efficiency on bundles of fibres from fast-twitch extensor *digitorum longus* (EDL) and slow-twitch *soleus* muscle from mouse, at 25°C, and found maximal yields of ~33%, and ~43% or ~48%, respectively. He et al. (1999) studied contraction yield as a function of relative load, for permeabilised fibres from rabbit psoas muscle, at 12°C. They found a maximal yield of ~35% for a relative load of ~0.5. Sugi et al. (2003) claimed that 'a conservative estimation of the maximum mechanical efficiency at a relative load of 0.5–0.6 was 70%'. However, they used glycerinated psoas muscle fibres from *Oryctolagus cuniculus* L., a rarely studied species of rabbit, with the caged-Ca^{2+} technique, which is certainly less appropriate than the caged-ATP technique used by He et al. (1999) (see text below Equations 3.30, 3.31 and 3.32 for brief remarks on this point). There are, therefore, major concerns relating to the experiments recalled here, giving values of ~33%, ~35%, ~43%, ~48% and ~70%. The main reasons are as follows: (i) it is difficult to compare animals from different species (i.e. rabbit and mouse), (ii) the comparison of two types of muscle from the same animal species (mouse) is unwarranted, (iii) glycerination would give results potentially at odds with most of the other results (see p. **52** in Section 3.4.3.2) and (iv) caged-Ca^{2+} was used by Sugi et al. (2003), raising many problems (see remark above). A careful discussion and analysis of the mechanical efficiency of muscle contraction has been published by Smith et al. (2005). These authors did not necessarily reach the same conclusions as me (i.e. the four points mentioned here). In any event, most of the detailed discussions presented in this monograph relate to isometric conditions, except in Sections 3.4.4 and 7.10, and the problems of mechanical efficiency are not of major interest here.

Let us now focus on the notion of 'ambient heat' recalled in the preceding paragraph, because most sections concern isometric contraction (no active shortening and no 'physical work'). The thermal energy is $kT \sim 3.77 \times 10^{-21}$ J per molecule at ~0°C (temperature at which muscle fibres of cold-blooded animals are frequently studied) and $kT \sim 4.06 \times 10^{-21}$ J per molecule at ~20°C–22°C (room temperature, at which muscle fibres of warm-blooded animals are frequently studied). A mean value of $kT \sim (3.77 + 4.06) \times 10^{-21}/2 \sim 3.92 \times 10^{-21}$ J per molecule can therefore be selected. Moreover, to ensure a margin of error, I suggest that $\sim 5kT \sim 1.96 \times 10^{-20}$ J per molecule is required for a single cross-bridge to change its conformation under isometric tetanic contraction conditions (short tetani; steady-state conditions). Let us assume that one MgATP molecule is split per cycle of attachment–detachment of an external myosin head to and from a thin actin filament (see, in Section 5.1.1, Figure 5.1, its legend, Appendix 5.I and pp. **170–172** regarding this notion of external heads; see also below, concerning conformational changes; all the contacts between the thin actin filaments and the myosin heads are active, leading to $\omega_{isom,\infty} \sim 1$, e.g. text between Equations 3.46 and 3.47). Thus, the maximal energy available per external myosin

head for conformational changes in vivo/in situ, is e_1 (the energy e_2 corresponding to Lohmann's reaction is not absorbed by the external myosin heads, being absorbed instead somewhere in the sarcoplasmic medium, probably on the M lines where creatine kinase is located, as demonstrated by Walliman et al. 1977, at least in chicken skeletal muscles, and also probably in any other animal species). The apparent energy $e_{app} \sim 6.6 \times 10^{-20}$ J per MgATP molecule, in vivo/in situ, is given on pp. **193–194**. Owing to its rather low value and the wide range of uncertainty on the various values, the energy e_2 (in vivo/in situ, in intact unit cells) can be taken as $\sim -2.4 \times 10^{-20}$ J per MgATP molecule (see p. **193**), giving $e_1 \sim e_{app} - e_2 \sim 9.0 \times 10^{-20}$ J ~ 90 zJ per MgATP molecule in vivo/in situ (precise details on these three energies are given on pp. **192–194**). Woledge et al. (2009) mentioned exactly the same value of e_1. I refer the reader again to pp. **170–172** in Section 5.1.1 and to Figure 5.1, its legend and Appendix 5.I for precise details on the notion of external and internal heads. As suggested above, as little as $\sim 5kT \sim 1.96 \times 10^{-20}$ J per molecule is sufficient to ensure the isometric conformational changes, that is, a ratio of $\sim 5kT/e_1 \sim 0.218$ (~21.8%), required to induce isometric tetanic contraction. This ratio corresponds to a single external myosin head changing its conformation and inducing full isometric tetanic force, in the case of a swinging cross-bridge/lever-arm process. In the hybrid model, there are external heads almost similar to the heads in the traditional approach to the structure of the thick myosin filaments, in which all the heads lie outside the thick myosin filament core, as well as internal heads, playing a major role, as described throughout this work. I assume, for simplification, the same order of magnitude for the 'energetic ratio', as estimated above (~21.8%), for these internal heads. This is because the internal heads almost certainly also change their conformation during isometric tetanic contraction, by cooperative, possibly allosteric, phenomena (see p. **170** in Section 5.1.1 for precise details on these notions of cooperativity and head–head interactions), to induce dissociation of the 'internal dimers' (see, in Section 5.1.1, Figure 5.1 and pp. **170–172** and **174–176** concerning the B–B'-type dimers and their role, see also Sections 5.1.2 and 5.2, concerning association–dissociation). The energy released in the sarcoplasmic medium, resulting solely from myosin–actin interactions and not used to induce the various conformational changes, is therefore $\sim e_1 - 5kT \sim e_1 - 0.218e_1 \sim 0.782e_1$, implying that $\sim$78.2% of the energy generated by MgATP cleavage is 'useless' for isometric contraction. Assertions of this kind are discounted below, in this section. In any event, the release of energy, over the duration $0-t^*$, in the sarcoplasmic medium of an intact unit cell is $(E^R)_{t^*} \sim m\Pi_{t^*}(0.782e_1 + e_2) \sim m\Pi_{t^*}e_{app}(0.782e_1 + e_2)/e_{app} \sim 0.703m\Pi_{t^*}e_{app} \sim 0.703(E^{AM})_{t^*}$. Necessary references and definitions relating to the density m of crossbridges and the rate of MgATP splitting Π_{t^*} are given on p. **194**.

Let us call $(\Delta H_0)_{t^*}$ the 'actin–myosin related' enthalpy increase corresponding to the energy $(E^R)_{t^*}$ released into the sarcoplasmic medium (see the end of the preceding paragraph). On p. **43** in Section 3.4.3.1.1, the assimilation of enthalpy to chemical energy is discussed. Under isometric conditions, the value of $\sim$5kT, suggested in the preceding paragraph, is arbitrarily chosen and values of $\sim$3kT or $\sim$8kT would also be suitable, but I retain here $\sim$5kT, which induces no marked change in the semi-empirical reasoning, owing to the scattering of various experimental data. We therefore deduce that $(\Delta H_0)_{t^*} \sim (E^R)_{t^*} \sim 0.703(E^{AM})_{t^*}$ (see the end of the preceding paragraph). The approximate equality between $(\Delta H_0)_{t^*}$ and $(E^R)_{t^*}$, within the wide limits of experimental uncertainty, has been systematically and unwittingly used in all the experimental studies concerning the energy balance in whole frog muscles (for reviews, see Curtin and Woledge 1978; Homsher and Kean 1978; Woledge et al. 1985; in the experiments presented and commented on by these authors, the temperature was $\sim$0°C–5°C and the durations of the tetani were $\sim$1–10 s). In any event, it is unreasonable to consider muscle as a kind of 'thermic machine' that releases most of the chemical energy available from MgATP splitting by the myosin heads ($\sim$70.3%; see above) in the form of heat, at least under isometric tetanic conditions. Thus, assuming that the swinging cross-bridge/lever-arm mechanisms are the only possible models of muscle contraction is unwarranted, at least from an energetic viewpoint. Indeed, it is highly unlikely that the crossbridges in isometrically contracting intact fibres or intact unit cells, consuming little energy in the traditional models, can generate full axial contractile force via the tight coupling mechanism in which there is a one-to-one relationship

between 'actomyosin' MgATPase activity and isometric tetanic tension. This 'one-to-one view' is also called into question, even discounted, on pp. **78–80** in Section 3.4.5 and pp. **83–84** in Section 3.4.6, where it is demonstrated that there is a very loose coupling between the rate of 'isometric' MgATPase splitting and isometric tetanic tension, independently of the mechanical roles of the cross-bridges. Regarding the $\sim$40% of contraction resulting from lateral swelling mechanisms (the hybrid model), the term $(P\Delta V^*)_{t^*}$ (see Equation 6.1) plays a major role. However, there is a close correlation between these two terms, $(P\Delta V^*)_{t^*}$ and $(\Delta H)_{t^*}$: the slight increase in temperature within the sarcoplasmic medium, related to $(\Delta H)_{t^*}$, leads to a chain reaction within the thick myosin filament core related to $(P\Delta V^*)_{t^*}$ (see pp. **212–213** in Section 6.4 for precise details on this chain reaction). Thus, in the hybrid model, the energy/enthalpy available from the cleavage of MgATP is used to generate contraction, via rather complex phenomena, very different from the one-to-one process recalled above. The hybrid model is consistent with the very loose coupling recalled above (see also p. **237** in Section 8.5 and p. **249** in Section 8.8 for a hypothesis, regarding the relationship between the very loose coupling and the two molecular phenomena underlying the hybrid model, i.e. the swinging cross-bridge/lever-arm and lateral swelling processes).

In Equation 6.1, the total enthalpy increase $(\Delta H)_{t^*}$ is assumed to be different from the enthalpy increase $(\Delta H_0)_{t^*} \sim (E^R)_{t^*} \sim 0.703(E^{AM})_{t^*}$ defined and estimated in the preceding paragraph. Indeed, in a whole muscle and an intact fibre, there are other sources of enthalpy: Ca^{2+} pump in the sarcoplasmic reticulum, binding of Ca^{2+} to parvalbumins and calmodulin, Na^+–K^+ pump in the sarcolemma and so on (see Curtin and Woledge 1978, 1979; Homsher 1987; Homsher and Kean 1978; Woledge et al. 1985). Barclay (1996) claimed that the 'noncrossbridge heat is thought to be dominated by energy used to pump Ca^{2+} into the sarcoplasmic reticulum' (see also MacIntosh et al. 2000). The existence of these other sources of enthalpy would account for some of the unexplained heat (see Section 8.4). In demembranated fibres and isolated myofibrils, from frog and rabbit, as studied by Barman et al. (1998), He et al. (1997, 1998a,b, 1999), Lionne et al. (1996), Potma and Stienen (1996) and Potma et al. (1994), the sarcoplasmic reticulum is, at least partly, destroyed. It can thus be considered to be 'at least partly' present (see p. **92** in Section 3.7 for precise details concerning this notion of 'at least partly' employed for the sarcoplasmic reticulum). However, the contribution of the 'residual' sarcoplasmic reticulum to the total enthalpy released in the sarcoplasmic medium can be neglected for these biological materials, within the broad range of experimental uncertainty. Parvalbumins, calmodulin and other soluble proteins are removed by dissolution in the various buffers used during the preparation of this kind of biological material. For such fibres and for isolated myofibrils, we almost certainly have $(\Delta H)_{t*} \sim (\Delta H_0)_{t*} \sim 0.703(E^{AM})_{t*}$. For whole frog muscles, at $\sim 0°C–5°C$, all the additional sources of enthalpy listed here represent $\sim 25\%$ of $(\Delta H_0)_{t**}$ for high values of the duration $t**$ of the tetani ($t**$ is several seconds and is $\gg t* = 1$ s; e.g. Curtin and Woledge 1978, 1979; Homsher and Kean 1978; Woledge et al. 1985; there is almost certainly a contribution of fatigue in these tetani; see Sections 7.3 and 7.4 for precise details on fatigue; this problem of fatigue, concerning MgATPase activity, and, therefore $(\Delta H_0)_{t**}$ and $(\Delta H)_{t**}$, is also raised in Section 3.4.3.2, but ignored here, owing to the wide scattering of the experimental data). Taking into account the $\sim 25\%$ recalled above, we can write $(\Delta H)_{t**} \sim 1.25(\Delta H_0)_{t**}$ $\sim 1.25 \times 0.703(E^{AM})_{t**} \sim 0.879 m\Pi_{t**}e_{app}$. Homsher (1987) estimated that actomyosin ATPase accounts for $\sim 65\%–80\%$ of the ATP consumed by a frog muscle (isometric tetanus; duration of 5 s, at $0°C$) and ignoring the non–ATP-consuming sources of enthalpy, which is justified by the 'error bar', gives $(\Delta H)_{t**} \sim (1.250–1.538) \times 0.703(E^{AM})_{t**} \sim$ $(0.879–1.081) m\Pi_{t**}e_{app}$ (simple calculations not presented). Experimental values are also available for bundles of fibres from mouse (Barclay 1996). For EDL and *soleus*, studied at $25°C$, Barclay (1996) obtained corrective factors of ~ 1.527 and ~ 1.595, the corresponding tetanus durations being 0.4 s and 1 s, respectively. Thus, we can write $(\Delta H)_{t*} \sim 1.527(\Delta H_0)_{t*} \sim 1.527 \times$ $0.703(E^{AM})_{t*} \sim 1.073 m\Pi_{t*}e_{app}$ and $\sim 1.595 \times$ $0.703(E^{AM})_{t*} \sim 1.121 m\Pi_{t*}e_{app}$ (with $t* \ll t**$; see above this inequalty and fatigue, Barclay 1996 suggested that the relative non–cross-bridge enthalpy does not depend on fatigue, i.e. on $t*$ or $t**$ in EDL muscle, whereas it depends markedly on $t*$ or $t**$ in *soleus* muscle, confirming that the problem of the duration of the tetanus is complex, as demonstrated in Section 3.4.3.2). MacIntosh et

al. (2000, and references therein) estimated that $\sim 10\%–40\%$ of the total energy is required for ion pumping in the sarcoplasmic reticulum, regardless of the type of muscle or intact fibre, giving corrective factors of ~ 1.111 and ~ 1.667 (simple calculations not presented) and leading to $(\Delta H)_{t*}$ $\sim 1.111 \times 0.703(E^{AM})_{t*} \sim 0.781 m\Pi_{t*}e_{app}$ and $\sim$ $1.667 \times 0.703(E^{AM})_{t*} \sim 1.172 m\Pi_{t*}e_{app}$, respectively. In mechanically skinned fibres and half-fibres from frog, the sarcoplasmic reticulum is present and functional and the corrective factor when passing from $(\Delta H_0)_{t*}$ to $(\Delta H)_{t**}$ may be ~ 0.879, ~ 0.879, ~ 1.081, ~ 1.073, ~ 1.121, ~ 0.781 or ~ 1.172 (see above), assuming, for the sake of simplicity, that skinned fibres and half-fibres are intermediate between whole muscles and intact fibres from frog and bundles of fibres from mouse: I suggest taking an average value of $\sim (0.879 + 0.879 + 1.081 + 1.073 + 1.121 + 0.781 + 1.172)/7 \sim 0.998$ for these two biological materials. West et al. (2004), working on intact dogfish fibres, at $12°C$, and for tetani lasting 3.5 s, found that the additional sources of enthalpy accounted for $\sim 62 \pm 16\%$ of $(\Delta H_0)_{t**}$. Thus, we deduce that $(\Delta H)_{t**} \sim$ $(1.62 \pm 0.16) \times 0.703(E^{AM})_{t**} \sim (1.139 \pm 0.112)$ $(E^{AM})_{t**}$. There are therefore 12 corrective factors: ~ 0.703 (probably for chemically skinned and permeabilised fibres and almost certainly isolated myofibrils, in which most of the extrafibrillar structures are reasonably assumed to be absent), ~ 0.879, ~ 0.879, ~ 1.081, ~ 1.073, ~ 1.121, ~ 0.781, ~ 1.172 (these seven values are valid for whole muscles and intact fibres from frog or mouse; see above), ~ 0.998 for mechanically skinned frog and rabbit fibres and frog half-fibres (see above), ~ 1.027 and ~ 1.251 (minimal and maximal values for intact dogfish fibres; see above) and ~ 1.139 (mean value for intact dogfish fibres; see above). Owing to the high degree of experimental uncertainty and the various approximations made throughout this monograph, together with the use of four very different animal species (dogfish, frog, mouse and rabbit) and several types of biological materials (whole muscles, permeabilised, chemically and mechanically skinned fibres and half-fibres) and the difference in $t*$ and $t**$ between experiments, we can average the 12 available values mentioned here. We therefore obtain a 'universal corrective factor' $\xi_{0,\infty} \sim 1.007 \pm (0.046$ or $0.159)$ (mean $\pm$ SE or SD; $n = 12$; see the end of the next paragraph for comments on $\xi_{0,\infty}$). This corrective factor is rounded down to $\xi_{0,\infty} \sim 1.00$ and we have therefore $(\Delta H)_{t*} \sim (\Delta H_0)_{t**} \sim m\Pi_{t*}e_{app}$, with $e_{app} \sim 6.6 \times 10^{-20}$ J per MgATP molecule, regardless of the biological material (the necessary numerical values and definitions can be found on pp. **192–194**; see also Equation 6.1 and corresponding comments).

The various corrective factors studied in the preceding paragraph are moderate and are systematically taken as ~1.00 below (see also p. **63** in Section 3.4.3.2), regardless of the animal species; the type of myofibril (isolated or belonging to an intact fibre) or intact fibres, half-fibres, mechanically skinned, chemically skinned, permeabilised fibres; temperature; and duration of the tetanus, including the transitory period, the steady state corresponding to a short tetanus, a pseudo-long-lasting tetanus or a long-lasting tetanus (the definitions of these various states are given in Section 3.4.3.2). Thus, in any case (including that of intact unit cells), we can write $(\Delta H)_{t*} \sim m\Pi_{t*}e_{app}$ (see the end of the preceding paragraph). A corrective factor $\xi_0(t)$ is introduced into Equation 3.21 that takes into account the complex transitory period immediately after the start of stimulation (after the short time lag; see Table 1 in the paper by He et al. 1997), and which tends toward $\xi_{0,\infty}$ (introduced into Equation 3.23), that is, the value corresponding to steady-state conditions. It is demonstrated that the duration Δt of the first turnover (and also the second and third turnovers) of MgATPase activity (including complex transitory phenomena and Pi and H$^+$ bursts; see pp. **208** and **210–211** in Section 6.3.5 for precise details on the Pi and H$^+$ bursts) is ~46–56 ms ~ 0.046–0.056 s for frog at 5°C (see p. **54** in Section 3.4.3.2). Obviously, this duration of ~0.046–0.056 s is negligible, with respect to the durations of the long-lasting tetani, recalled in the preceding paragraph (e.g. ~5 s used in many old experimental studies, or ~3.5 s used by West et al. 2004) and even the duration of short and very short tetani (~1.0–1.5 s, e.g. Bagni et al. 1988b; Edman 1966, 1979; Edman and Anderson 1968; Edman and Hwang 1977; Edman and Reggiani 1987; Elzinga et al. 1989a,b; Gordon et al. 1966a,b; Granzier and Pollack 1990; ~0.4 s and ~1 s, Barclay 1996; ~0.5 s, West et al. 2004). Otherwise, the experimental uncertainty on the value of $\xi_{0,\infty}$ is rather large (see the preceding paragraph; if the (SE + SD)/2 test [defined in Section 2.1] is used, the error bar is ~±10%). Thus, we can write, as a reasonable simplifying assumption, $\xi_0(t) \sim \xi_{0,\infty} \sim 1.00$, that is, a 'universal value'

regardless of the animal species and experimental conditions of isometric contraction (e.g. transitory or steady-state conditions, duration of isometric tetanic contraction, different animal species and different temperature conditions, for example; see the preceding paragraph).

6.3 NUMERICAL VALUES OF THE ENTHALPY INCREASE AND RELATED QUESTIONS REGARDING DEMEMBRANATED RABBIT FIBRES, MYOSIN AND ITS SUBFRAGMENT 1 (S1)

6.3.1 General Remarks Concerning Various Questions Considered in This Section

Let us recall the question repeatedly posed in this work, regarding the validity of using and combining experimental data for isometric tetanic tensions recorded mostly in intact fibres and for actomyosin MgATPase activities routinely measured in demembranated fibres, isolated myofibrils and free myosin molecules or free myosin heads (S1) in vitro. There seems to be no major problem in the use of different types of data, because consistent conclusions are drawn, when taking into account the conclusions presented throughout Section 3.4, making it possible to avoid this issue.

On pp. **174–176**, in Figure 5.1, its legend and Appendix 5.I, in Section 5.1.1, it is shown that, at rest, the internal myosin heads (of B- and B′-types) form dimers within the thick myosin filament core, whereas many internal dimers dissociate during contraction (see Sections 5.1.2 and 5.2, Figure 5.2, its legend and Appendix 5.II). This association–dissociation process of the B- and B′-type heads is 'indefinitely' reversible (see Section 5.2, Figure 5.2, its legend and Appendix 5.II). A major question concerning the internal B- and B′-type heads relates to the way in which the energy released from MgATP splitting during contraction, as a result of cyclic interactions of the external heads with the thin actin filaments (see pp. **170–172** in Section 5.1.1 for precise details on the external heads), is used by the internal B–B′-type head–head dimers for induction of the dissociation phenomenon. I strongly suggest that the internal B–B′-type head–head dimers split MgATP. However, the internal heads are never attached to the thin actin filaments and their MgATPase activity may be seen as corresponding to resting

Morel

conditions, regardless of the state of the fibre (at rest or under contraction conditions). For this reason, it might be assumed that the MgATPase activity of the internal heads, during contraction, is negligible. Nonetheless, this assumption is almost certainly incorrect. Indeed, during contraction, conformational changes occurring in the external heads interacting cyclically and rapidly with the thin actin filaments (e.g. Brenner 1991, concerning rapidity) are transferred to the internal heads, probably resulting in 'indirect activation' of the MgATPase activity of these heads, giving a rate of MgATP cleavage of the same order of magnitude as that of the external heads (see below). Finally, for both young adult frogs and young adult rabbits, the proportion of internal heads is only ~33.3%, whereas that of external heads is ~66.7% (see pp. **204** and **208** in Section 6.3.5). Thus, the contribution of the internal heads to total MgATPase activity during contraction would lie within the limits of experimental uncertainty on the various measurements of MgATP splitting. In this general context, it is recalled, on p. **170** in Section 5.1.1, that several independent groups have demonstrated the existence of head–head interactions and the subsequent occurrence of cooperativity within individual myosin molecules. Therefore, the B-type heads, in interaction with the B–B'-type head–head dimers, may split MgATP at rates comparable to those of the external heads. As a compromise, I assume that the rate of MgATP splitting by the B-type heads lies between ~1/3 and ~1/2 that corresponding to the external heads. This proposed lower 'internal' MgATPase activity does not affect the cooperative phenomena. This particular point of MgATP splitting by the internal heads is analysed and discussed on p. **208** in Section 6.3.5, where the contribution of the internal heads to MgATPase activity in situ/in vivo is quantified.

6.3.2 MgATPase Activity in Resting Chemically Skinned and Permeabilised Fibres, and Myosin Heads (S1) In Vitro (Fibres and Myosin from Rabbit) in the Absence of F-Actin

Little is known about the actomyosin MgATPase activity in resting skinned and permeabilised rabbit fibres. Under physiological conditions, the proportion of weakly bound cross-bridges is small (~9%, at ~0°C–25°C, regardless of the animal species; see p. **172** in Section 5.1.1), and the 'resting' MgATPase activity corresponds mostly to the external myosin heads not bound to the thin actin filaments plus a weak contribution of the weakly bound external heads and, to a lesser extent, the internal B–B'-type head–head dimers (see Figure 5.1 for precise details on the various types of head; see also the end of Section 6.3.1 concerning the possible contribution of the internal heads to overall MgATPase activity). Under resting conditions, between ~0°C and ~20°C, the Q_{10} for MgATPase activity is ~2.6, whereas it is ~3.3 between ~20°C and ~25°C (see pp. **19–20** in Section 2.4). He et al. (1997) measured the actomyosin MgATPase activity in permeabilised fibres from rabbit psoas muscle, at 20°C, under resting conditions, and obtained a value of ~0.11 s^{-1}, whereas Hilber et al. (2001), using the same biological material, obtained a value of ~0.04 s^{-1} at 10°C, corresponding to ~0.04 × 2.6 ~0.10 s^{-1} at 20°C. Myburgh et al. (1995) studied the MgATPase activity in relaxed myofibrils from rabbit psoas muscle and noted that 'steady-state measurements of myofibrillar ATPase rate using conventional methods showed that the rate was ~0.13 s^{-1} at 20°C'. However, using an unconventional technique, the authors obtained a much lower value of only ~0.03 s^{-1} at 25°C, that is, ~0.03 s^{-1} × $(3.3)^{-1/2}$ ~0.017 s^{-1} at 20°C (Q_{10} ~ 3.3 is the value between ~20°C and 25°C; see above). Although the explanations for this discrepancy given by the authors are, to my mind, unclear, I take their two values into account. Finally, Herrmann et al. (1994) reported a value of ~0.07 s^{-1} at 20°C, for suspensions of myofibrils from rabbit psoas muscle. Averaging these five values gives ~0.085 ± (0.018 or 0.039) s^{-1} (mean ± SE or SD; n = 5) at 20°C. Hilber et al. (2001) identified an important property of the MgATPase activity, under relaxing conditions (using demembranated fibres from rabbit), when the temperature increases above ~25°C–30°C and up to ~35°C (see their Figure 4). Indeed, between ~6.5°C and ~25°C, the resting rate of MgATP utilisation is very low, as recalled above. Between ~25°C and ~30°C, this rate increases rapidly, the rate at ~30°C being approximately twice that at ~25°C. Beyond ~30°C, there is a breakpoint and the rate increases very rapidly between ~30°C and ~35°C, the rate at ~35°C being approximately 10 times that at ~20°C. This behaviour of rabbit demembranated fibres is probably related to fragility. The problem of fragility and the subsequent adulteration of demembranated fibres from rabbit psoas muscle is also raised on p. **52** in Section 3.4.3.2, p. **202** in Section 6.3.3, p. **208** in Section 6.3.5, and p. **238** in Section 8.5, and would become crucial at high temperatures (e.g. ~30°C–35°C).

Is the value of ~0.085 s^{-1}, in situ and at 20°C, estimated in the preceding paragraph, consistent with other experimental results obtained for myosin heads (S1) and synthetic and natural thick myosin filaments, in vitro, at 20°C, in the absence of F-actin, for rabbit (back and leg muscles)? Herrmann et al. (1992) obtained a value of ~0.080 s^{-1} for S1, at 20°C. However, Bachouchi et al. (1986) found that the MgATPase activity of S1 monomers, at 20°C and in conditions mimicking physiological resting conditions in the absence of F-actin, was only ~0.015 s^{-1}, whereas that of S1 dimers was ~0.046 s^{-1}. Morel and Guillo (2001) revisited this issue and reported these activities to be ~0.011 s^{-1} and ~0.018 s^{-1}, respectively, in conditions similar to those used by Bachouchi et al. (1986). The reason for these slight discrepancies is unclear. Morel et al. (1999) performed many experiments in vitro and in the absence of F-actin, under approximately physiological conditions and at 20°C, on two-stranded synthetic thick myosin filaments, with one head of each myosin molecule within and the other outside the backbone (F-filaments). At physiological MgATP concentration (~4–6 mM; see, for instance, Alberts et al. 2007; Lehninger 2008), the MgATPase activity was ~0.040 s^{-1}. This value was obtained by extrapolation of the experimental curve, in Figure 2b, giving the rate of MgATP splitting against MgATP concentration, with a cut-off at ~1 mM MgATP, and also deduced from the equation best fitting the curve. Morel et al. (1999) also carried out many experiments, under the same conditions, with natural three-stranded thick myosin filaments from psoas muscles of young adult rabbits (4 months old; see our Figure 3). In the presence of ~5 mM MgATP, using the same extrapolation as described above for myosin F-filaments, we obtain an average enzymatic activity of ~0.015 s^{-1}. I suggest averaging the seven values of MgATPase activity obtained at 20°C and in vitro, with different biological materials and various methods, together with the value obtained by Herrmann et al. (1992; see the beginning of this paragraph), giving therefore ~0.032 ± (0.009 or 0.023) s^{-1} (mean ± SE or SD; n = 7). Thus, the MgATPase

activity in situ and in resting conditions (~0.085 s^{-1}; see the preceding paragraph and start of this paragraph) is ~100 × (0.085 − 0.032)/0.085 ~62% higher than that measured in vitro. This difference could be accounted for by arguing that the presence of thin actin filaments in situ is, at least partly, the source of this 'excess' MgATPase activity (some unblocking of the myosin head binding site on the thin actin filaments, at rest, almost certainly occurs, as highlighted approximately 40 years ago by Morel et al. 1976; in this context of some unblocking, the presence of weakly bound cross-bridges, formerly called resting cross-bridges by my group, is relevant). Another reason could be proposed: in vitro, the myosin heads and the thick myosin filaments are diluted and 'randomly arranged', whereas the myofilaments are more ordered and crowded in demembranated fibres, resulting in possible cooperative effects, with subsequent higher MgATPase activity in situ than in vitro. At this stage, only the value of ~0.085 s^{-1} should be taken into account in resting rabbit fibres at ~20°C. Using, under resting conditions, Q_{10} ~ 2.6 between ~0°C and ~20°C, Q_{10} ~ 3.3 between ~20°C and ~25°C and Q_{10} ~ 3.8 between ~25°C and ~40°C (see pp. **19–20** in Section 2.4) and taking into account the virtual duration t* = 1 s (see pp. **210–211** in Section 6.3.5 for a circumstantial discussion of t*), we obtain the dimensionless values of ~0.020 at ~5°C, ~0.040 at ~12°C, ~0.085 at ~20°C, ~0.154 at ~25°C and ~1.073 at ~39°C (the body temperature of the rabbit) for MgATPase activity at rest. In Section 6.3.3, only short tetani lasting t* = 1 s (i.e. the reference 'virtual time'; see the beginning of Section 6.3.3 for information on t*) are taken into account.

6.3.3 MgATPase Activity and Enthalpy Increase during Isometric Tetanic Contraction (Short Tetani–Steady State), in Chemically Skinned and Permeabilised Fibres from Rabbit, with an Insight into Frog and Rabbit Isolated Myofibrils. Comparison with Resting Conditions

In Section 3.4.3.2, 'isometric' MgATPase activities in permeabilised rabbit fibres from psoas muscle, for short tetani under steady-state conditions, are deduced from the experimental study of He et al. (1997). The notions of short tetanus and steady state, valid for the rates of MgATP splitting, are

Morel

defined point by point in Section 3.4.3.2. For a reference virtual time $t^* = 1$ s (see p. **191** in Section 6.1 for definition and pp. **210–212** in Section 6.3.5 for precise details on this reference virtual time), we obtain the dimensionless value $\Pi_{t^*}(5°C) \sim 6.7$, corresponding to the activation factor $\sim 6.7/0.020 \sim 335$. At 12°C, we have $\Pi_{t^*}(12°C) \sim 16.3$ and the activation factor is $\sim 16.3/0.040 \sim 410$. At 20°C, we have $\Pi_{t^*}(20°C) \sim 33.1$ and the activation factor is $\sim 33.1/0.085 \sim 389$ (see p. **57** in Section 3.4.3.2 concerning the three MgATPase activities $\Pi_{isom,\infty}$, and therefore Π_{t^*}, at 5°C, 12°C and 20°C; see also p. **200** in Section 6.3.2 concerning resting MgATPase activities). He et al. (1997) gave an estimate of ~ 350 at 20°C. It is certainly useful to try to estimate the rate of MgATP splitting at 39°C (the body temperature of the rabbit). In this context, it is necessary to recall that a parameter $P(T)$ measured at the absolute temperature T is given by $P(T) = P^* \exp(-\Delta H^*/kT)$, where P^* is a constant and ΔH^* the energy of activation. The same equation can be written at a reference absolute temperature $T_0 < T$: $P(T_0) = P^* \exp(-\Delta H^*/kT_0)$. Combining these two definitions gives $P(T)/P(T_0) = \exp[(\Delta H^*/kT_0) \times (T - T_0)]$. Alternatively, we also have $P(T)/P(T_0) = (Q_{10})^{(T-T0)/10}$. Equating these two expressions, we deduce that $Q_{10}(T) \sim \exp(10\Delta H^*/kT^2)$, taking $T_0 \sim T$. ΔH^* is expressed in joules per molecule and assumed to be independent of T, at least within the 4°C–20°C range (see Figure 5 in the paper by Barman et al. 1998 for rabbit psoas myofibrils). As an approximation, this non-dependence is assumed to extend up to $\sim 25°C$, and even up to $\sim 39°C$. ΔH^*, as measured by Barman et al. (1998) on unheld rabbit psoas myofibrils, concerns various kinetic parameters, including the rate of MgATP splitting (see their Table 1 for values of 68 ± 2 and 67 ± 2 kJ mol^{-1}). Candau et al. (2003), also working on unheld psoas myofibrils, between 39°C and 42°C, obtained $\sim 71 \pm 5$ kJ mol^{-1}, giving an average value of $\sim (67 + 68 + 71)/3 \sim 69$ kJ mol^{-1}, within the $\sim 4°C–42°C$ temperature range, reasonably extended to $\sim 0°C–42°C$. This mean value of ΔH^* is valid for the MgATPase activity of unheld myofibrils. However, it is shown, on pp. **55–57** in Section 3.4.3.2, that the MgATPase activity of unheld myofibrils does not resemble the 'isometric' MgATPase activity of demembranated fibres (short tetani, steady state). Nonetheless, the energy of activation ΔH^* is reasonably assumed to be similar in both biological materials. This simplification is justified in the next paragraph, where it is shown that Q_{10}, deduced from the ΔH^* measured on unheld myofibrils, is not significantly different from that measured directly on demembranated fibres.

From the value of $\Delta H^* \sim 69$ kJ mol^{-1}, for unheld isolated myofibrils and demembranated fibres from rabbit psoas muscle (see the end of the preceding paragraph), it can be deduced that $Q_{10}(0°C) \sim 3.0$, $Q_{10}(5°C) \sim 2.9$, $Q_{10}(10°C) \sim Q_{10}(12°C) \sim 2.6–2.7$, $Q_{10}(20°C) \sim 2.5$ and $Q_{10}(39°C) \sim 2.3$ (see the preceding paragraph for the relationship between ΔH^* and Q_{10}). He et al. (1997), working on permeabilised fibres from rabbit psoas muscle, estimated that $Q_{10} \sim 2.9$, within the 5°C–20°C temperature range. Hilber et al. (2001), working on the same biological material, found that 'the rate of ATP utilisation during isometric contraction had a Q_{10} of 3.6 throughout the temperature range 7°C–25°C'. Working on demembranated human psoas fibres, between 12°C and 30°C, Stienen et al. (1996) obtained $Q_{10} \sim 2.6–2.8$. Between 20°C and 37°C, Wang and Kawai (2001, and reference therein) proposed $Q_{10} \sim 3.3$ for demembranated rabbit psoas fibres. As the first series of values is calculated indirectly, and essentially for unheld myofibrils, whereas those of Stienen et al. (1996), Wang and Kawai (2001) and Hilber et al. (2001) were directly measured for demembranated fibres, the best solution is to average the 11 available values, giving $Q_{10} \sim 2.84 \pm (0.11$ or $0.35)$ (mean $\pm$ SE or SD; $n = 11$), within the range $\sim 0°C–39°C$, that is, $Q_{10} \sim 2.8$. The dimensionless MgATPase activity Π_{t^*} at 39°C may therefore be $\sim \Pi_{t^*}(20°C) \times (2.8)^{1.9} \sim 33.1 \times 7.0 \sim 230$, corresponding to the activation coefficient of $\sim 230/1.073 \sim 214$ (the value of ~ 1.073 is given on p. **200** in Section 6.3.2). Working on unheld soleus and psoas myofibrils, Candau et al. (2003) found activation coefficients of ~ 213 and ~ 223 at 39°C and 42°C, respectively. Owing to the scattering of the values of the 'active' and resting MgATPase activities, the seven available activation factors (see the preceding paragraph and above) are not essentially different and the mean value (mean $\pm$ SEM or SD; $n = 7$). Again for rabbit, for short tetani, Siththanandan et al. (2006), working on permeabilised fibres, found that 'there was no substantial increase in the steady rate of phosphate release' (i.e. in the 'steady-state' MgATPase activity) when the temperature increased from 12°C to 20°C ($Q_{10} \sim 1$).

They obtained a steady rate of MgATP hydrolysis of ~8.6 s^{-1} at 12°C and 20°C, corresponding to Π_{t*} ~ 8.6 (see their Table 2). This value is significantly lower than the ~16.3 (at 12°C) and ~33.1 (at 20°C) given in the preceding paragraph and deduced from the experimental work of He et al. (1997). This is probably attributed to steady-state conditions being exceeded, for unclear reasons, in the experiments of Siththanandan et al. (2006), such that some pseudo–long-lasting conditions were included in the tetani studied (see Section 3.4.3.2 for precise details on these notions, regarding the various states of contracting fibres). It is difficult to obtain precise estimates of the durations of the isometric tetani in the experiments of Siththanandan et al. (2006), but these durations do not seem to exceed ~0.2–0.4 s, which would correspond to short tetani under steady-state conditions. However, the authors obtained Q_{10} ~ 1.0 (see above), at odds with all the other values mentioned in this monograph. To my mind, this value of Q_{10} is therefore highly questionable. I note that the authors used New Zealand white rabbits weighing more than 5 kg, whereas the rabbits used by He et al. (1997) were lighter, weighing only ~4 kg or slightly more. At this juncture, the problem of weight and ageing, associated with fragility of the fibres, must again be considered (see p. **52** in Section 3.4.3.2, p. **200** in Section 6.3.2, p. **208** in Section 6.3.5 and p. **238** in Section 8.5). Assuming that the rabbits used by Siththanandan et al. (2006) were 'old rabbits' and those used by He et al. (1997) were 'young adult rabbits', the problems raised by the experimental results of Siththanandan et al. (2006) would be resolved, as Morel et al. (1999) demonstrated that natural thick myosin filaments from old rabbit psoas muscle (two-stranded) are markedly less active, enzymatically, than those from young adult rabbit psoas muscle (three-stranded). We can therefore dismiss the values obtained by Siththanandan et al. (2006) and maintain the values obtained by He et al. (1997), together with the six consistent activation coefficients given in the preceding paragraph and at the beginning of this paragraph. Finally, from this and the preceding paragraphs, we can conclude that, regardless of temperature, strong activation occurs, the activation coefficient being fairly insensitive to temperature (~305; see above). Thus, we have, for young adult rabbits (permeabilised fibres from psoas muscle), $\Pi_{t*}(5°C)$ ~ 6.7,

$\Pi_{t*}(12°C)$ ~ 16.3, $\Pi_{t*}(20°C)$ ~ 33.1 and $\Pi_{t*}(39°C)$ ~ 230 (see p. **201** concerning the dimensionless 'contracting' activities).

In the light of Table 1 in the paper by Barman et al. (1998), concerning isolated unheld frog myofibrils, the energy of activation, at 4°C, is ~75 kJ mol^{-1}, versus ~69 kJ mol^{-1} for unheld psoas rabbit myofibrils within the 0°C–42°C temperature range (see p. **201** concerning the value of ~69 kJ mol^{-1}). A multiplication factor of ~1.094 should, therefore, be used to convert the Q_{10} of MgATPase activity of rabbit to that of frog, under isometric tetanic conditions (see p. **201** concerning the relationship between the energy of activation ΔH^* and Q_{10}). This factor of ~1.094 is assumed to be valid for any biological material (e.g. demembranated fibres and unheld isolated myofibrils; see pp. **201–202** for justification of this simplification). Thus, from the value given on p. **201** for rabbit (Q_{10} ~ 2.8 between ~0°C and ~39°C), we deduce a value of Q_{10} ~ 2.8 × 1.094 ~ 3.1 for frog, within the same range of ~0°C–39°C, regardless of the biological material (demembranated fibres or unheld isolated myofibrils). Studying intact frog fibres at ~0°C for ~5 s, Curtin et al. (1986) found that the Q_{10} for the stable maintenance heat rate was ~4.1. The Q_{10} for Π_{t*}, at 0°C, for demembranated fibres and unheld isolated myofibrils studied here, is only ~3.1 (see above) and the difference between ~4.1 and ~3.1 may be related to non-myofibrillar structures present in intact fibres but not in demembranated fibres and isolated myofibrils. Moreover, Woledge et al. (1985) gave an estimate of Q_{10} for the stable maintenance heat rate in a whole muscle of ~5.1. The origin of this difference between ~5.1 and ~4.1 is unknown, but we may assume that, during the preparation of intact fibres, many non-myofibrillar structures are functionally altered, perhaps adulterated (in another context, problems of adulteration related to the dissection processes of single intact fibres are also raised on pp. **89–90** in Section 3.7). Only the value of ~3.1 for the actomyosin MgATPase activity is taken into account, because the values of ~4.1 and ~5.1 relate to the stable maintenance heat rates (there is no a priori one-to-one relationship between the rate of actomyosin MgATP breakdown in demembranated fibres and unheld myofibrils and the stable maintenance heat rate in intact single fibres and whole muscle).

The strong activation of MgATPase activity during the transition from rest to contraction (mean activation coefficient of ~305 regardless of temperature), discussed on p. **201**, confirms that the external myosin heads (see pp. **170–172** in Section 5.1.1 for precise details on these heads) massively, rapidly and cyclically interact with the thin actin filaments during contraction (see Section 3.4.3.3 concerning the duration of the cycle, under isometric tetanic conditions). The contracting external heads later undergo considerable conformational changes, with respect to resting external heads, including the A- and F-type heads (see also pp. **170–172** for definition of the A- and F-type heads and pp. **174–176** in Section 5.1.1 regarding, in particular, the cycle of the external heads and the negligible MgATPase activity at rest). In this context, various conformational changes have been observed for S1 (isolated myosin head in vitro) and thick myosin filaments, with changes in temperature (see comments and references on pp. **246–248** in Section 8.8), and for myosin heads, whole myosin molecules and thick myosin filaments, particularly when demembranated fibres pass from rest to contraction (see comments and references on pp. **139–141** in Section 4.4.2.3). Under contraction conditions, in vivo/in situ, I suggest that the many conformational changes accompanying the cyclic interactions of the external myosin heads, including the A-type heads, with the thin actin filaments, are, at least partly, transferred to the internal head–head dimers (B–B′-type heads; see Figure 5.1), by cooperative (possibly allosteric) phenomena. Such cooperative processes have been observed by several independent groups that have demonstrated that the two heads of a single myosin molecule or its soluble subfragment HMM (two heads S1 + S2) interact physically and cooperate (see p. **170** in Section 5.1.1 for references). These internal conformational changes would lead to various levels of dissociation of the internal B–B′-type head–head dimers, in an intact unit cell. In Section 6.5 (see Equation 6.2), in Sections 6.3.4 and 6.3.5, and also in Section 6.4, it is shown that the number of dissociations of the B–B′-type heads, $n^*(\theta)$, increases with increasing temperature. There is therefore a correlation between temperature, conformational changes and the number of dissociations of the B–B′-type head–head dimers.

6.3.4 Locations and Widths of the Temperature-Induced Reversible Monomer–Dimer Myosin Head Transition In Vitro. Case of Young Adult Rabbits

The hybrid model presented and discussed in this monograph involves a contribution of ~40% from lateral swelling mechanisms, the fundamental theoretical and experimental features of which are described in Sections 5.1.1 (see pp. **170–172** and Figure 5.1, its legend and Appendix 5.I), 5.1.2 (see p. **183**) and 5.2. As discussed below, in this section, and throughout this work, the temperature transition shown in Figure 5.2 (between ~37°C and ~38°C; midpoint ~37.5°C; width ~1°C) is a key point in the development of the hybrid model. For warm-blooded animals (e.g. rabbit, rat and chicken), muscles can work, in the laboratory, from ~10°C (exceptionally 5°C; see He et al. 1997) to ~20°C, and up to ~40°C–41°C (temperature within the body of the chicken), but with different mechanical performance at different temperatures (isometric tension and velocity of shortening, for example). For cold-blooded animals (e.g. frog, toad, dogfish, toadfish and crayfish), muscles work mostly within the ~0°C–12°C temperature range. This raises an important question: can temperature transitions, similar to that described in Section 5.2 and Figure 5.2, its legend and Appendix 5.II, occur at any temperature? We can address this question, using experimental data for myosin and its soluble isolated heads (S1) from rabbit muscle (back and legs). On pp. **18–19** in Section 2.4, it is shown that the external myosin heads of neighbouring synthetic thick myosin filaments interact in the presence of 1.5 mM MgATP, but that these interactions disappear entirely and reversibly between ~27°C and ~29°C (midpoint ~28°C; width ~2°C). Many other experimental observations by my group have shown that reversible monomer–dimer transitions occur in S1 at various temperatures and in the absence of F-actin. (i) Using a 'dimer' buffer similar to that used by Morel and Garrigos (1982a), but in the presence of only 2 mM $MgCl_2$ (absence of ATP), Grussaute et al. (1995) observed and discussed such transitions. In our Figure 4, we found a transition between ~3°C and ~7°C (midpoint ~5°C; width ~4°C). In sedimentation–diffusion ultracentrifugation experiments, we found a transition between ~3°C and ~4°C (midpoint ~3.5°C; width ~1°C; see our Figure 7). (ii) Using dimer buffers systematically containing 4 mM $MgCl_2$ and 0.5 mM or 2 mM AMP–PNP or 2 mM ADP, Morel et al. (1998b) confirmed these results. In our Figure 3, we found a transition between ~5°C and ~6°C (midpoint ~5.5°C; width ~1°C) in the presence of 0.5 mM AMP–PNP, and a transition between ~5°C and ~8°C in the presence of 2 mM AMP–PNP (midpoint ~6.5°C; width ~3°C). In our Figure 4, we found transitions between ~4°C and ~8°C in the presence of 2 mM ADP (midpoint ~6°C; width ~4°C). Sedimentation–diffusion ultracentrifugation experiments on the same biological material showed the transition to occur between ~4°C and ~6°C (midpoint ~5°C; width ~2°C). (iii) Morel and Guillo (2001), studying the enzymatic properties of S1, in the absence of F-actin, observed the same type of transition, in the presence of 0.15 mM MgATP, between ~21°C and ~22°C (midpoint ~21.5°C; width ~1.0°C).

As pointed out in the 'supporting information available' accompanying the paper by Morel et al. (1998b), many factors affect the quantitative characteristics of the dimerisation process and the positions and widths of the temperature-induced monomer–dimer transitions. I do not discuss these complex questions here. I would merely like to recall the importance of the nature of the active molecules and ions (Mg^{2+}, MgATP, MgADP and MgAMP–PNP) and of their concentration in the main features of the reversible monomer–dimer transitions, including their position and width (see also the preceding paragraph). For example, the narrowest widths are observed in the presence of MgATP. Moreover, on the basis of all these observations, recalled in the preceding paragraph and briefly analysed here, I conclude that conformational changes in the myosin heads, resulting in dimer–monomer dissociation–association phenomena in the absence of F-actin, may occur at any temperature, within the range ~0°C–43.5°C (see legend to Figure 5.2 and Appendix 5.II for precise details on this value of ~43.5°C for rabbit). Indeed, it is recalled in the preceding paragraph that the lowest temperature at which dimer dissociation occurs in vitro is ~3°C, whereas the highest temperature, corresponding to monomer–dimer association–dissociation within the core of synthetic thick myosin filaments, is ~37°C–38°C. Furthermore, we deduce from all these values that the mean width, regardless of temperature (between ~3°C and ~37°C–38°C), is ~2.11 ± (0.41 or 1.23)°C (mean ± SE or SD; n = 9). In vivo,

for short tetani, only MgATP is present in the myofibrillar lattice, probably with a slight excess of magnesium (for short tetani, MgADP and Pi levels are negligible, owing to the instantaneous recycling of these metabolites via Lohmann's reaction; e.g. Alberts et al. 2007; Lehninger 2008; Morel and Pinset-Härström 1975a). Moreover, it is recalled above that MgATP is the molecule that regulates the characteristics of the temperature--induced transitions most efficiently (see also Morel et al. 1998b). Three series of experiments were performed on S1 or myosin in the absence of F-actin and in the presence of MgATP. (i) Morel and Guillo (2001) found a width of ~1.0°C. (ii) On pp. **18–19** in Section 2.4, a width of ~2°C is found. (iii) In Section 5.2 and Figure 5.2, a width of ~1°C is found. Combining these three experimental results gives a mean width of ~1.33 ± (0.37 or 0.64)°C (mean ± SE or SD; n = 3). If the most restrictive SE test (see Section 2.1 for precise details) is taken into account, the value of ~2.11 ± 0.41°C leads to the lowest value of ~1.70°C, whereas the value of ~1.33 ± 0.37°C leads to the highest value of ~1.70°C. These two values are similar and I retain the 'consensual' value of ~1.70°C in the next paragraph. As the monomer–dimer equilibrium is reversible, an increase of ~1.70°C corresponds to the total dissociation of the dimers into monomers, whereas a decrease of ~1.70°C corresponds to the total association of the monomers into dimers.

Intact unit cells, isolated myofibrils, chemically or mechanically skinned fibres, permeabilised fibres, half-fibres and intact fibres contain thin actin filaments, and rapid attachment–detachment cycles of the external myosin heads begin as soon as contraction is triggered (after a very short time lag; see Section 3.4.3.3 concerning the temporal characteristics of the cycle). Large quantities of MgATP are hydrolysed and the conformational changes, recalled on p. **203** in Section 6.3.3, occur in both the external heads, leading to their swinging, and the internal heads, leading to the dimer dissociation process (see, in Section 5.1.1, pp. **170–172**, Figure 5.1, its legend and Appendix 5.I for precise details on the external and internal heads). Temperature-induced dimer–monomer dissociation–association processes, resulting from conformational changes, can occur at any temperature, even in the absence of F-actin (in vitro) and are likely to occur if thin actin filaments are present (in vivo/in situ). In other words, in vitro and in vivo/in situ, regardless of temperature of the surrounding medium, monomer–dimer reversible transitions occur (at least between ~0°C and ~20°C–22°C and ~27°C–29°C, in vitro, and at temperatures of up to ~43.5°C in vivo; see legend to Figure 5.2 and Appendix 5.II, concerning this temperature of ~43.5°C for rabbit). At the end of the preceding paragraph, a value of ~1.70°C for the width of the reversible transition is selected, and a value of ~1.42°C is obtained on p. **216** in Section 6.4, based on an entirely different method, for the monomer–dimer reversible transition within the core of the thick myosin filaments. Thus, averaging these two comparable values, an increase of ~ (1.42 + 1.70)°C/2 ~ 1.56°C results in the total dissociation of the 100 B–B'-type head–head internal dimers (see pp. **205–206** in Section 6.3.5 concerning this number of 100), total reassociation being observed if the temperature decreases to the same extent (reversibility of the process, recalled at the end of the preceding paragraph and demonstrated, for instance, in Section 5.2). The value of ~1.56°C can reasonably be assumed to be valid in vivo/in situ. These assertions, based on experimental data and a semi-empirical reasoning, lead to consistent qualitative and quantitative conclusions, providing a justification of the view put forward in this work (see pp. **174–176** in Section 5.1.1, and Section 5.1.2; experimental support is presented in Section 5.2).

6.3.5 Consequence of the Increase in Enthalpy in the Sarcoplasmic Medium, during Isometric Tetanic Contraction (Short Tetani–Steady State), on the Number of Dissociations of Myosin Head–Head Dimers Lying within the Thick Myosin Filament Core. Case of Young Adult Rabbits, with Some Insights into the Case of Young Adult Frogs

In young adult frogs, the thick myosin filaments are three-stranded (e.g. Cantino and Squire 1986; Kensler and Stewart 1983; Stewart and Kensler 1986b). Morel et al. (1999) demonstrated that natural thick myosin filaments from young adult rabbits are also three-stranded (see pp. **170–172** in Section 5.1.1). In all cases, there are ~100 crowns/filament, giving ~3 × 100 ~ 300 myosin molecules and ~2 × 300 ~ 600 heads, that is, exactly the same value of ~600/2 ~ 300 heads per intact half-unit cell proposed by Schoenberg (1980a,b)

and ~600 heads per intact unit cell used by Elliott and Worthington (1997; see p. **35** in Section 3.2) (in the legend to Figure 5.1, including Appendix 5.I, the definition and characteristics of a 'crown' are presented). Huxley and Kress (1985) wrote that there are '49 repeats of the 143 Å axial periodicity [crown] of the myosin heads per half sarcomere and three myosin molecules (six heads) for each 143 Å repeat', giving 49 × 2 × 6 = 588 heads per sarcomere. This value of ~588 heads ('myosin motors') was taken into account by Piazzesi et al. (2007). Barclay (1999) and Linari et al. (1998) used a value of ~580. The values of ~580 and ~588 are sufficiently consistent with ~600 and this last value is therefore taken as the reference value in this book (the mean of the eight values recalled here is ~589 ± (3 or 8) [mean ± SE or SD; n = 8], which can be rounded up to ~600, for self-consistency with previous studies of my group). All these features are also valid for young adult rabbits (e.g. Kensler and Stewart 1983; Kensler et al. 1994; Maw and Rowe 1980; Morel and Garrigos 1982b; Morel et al. 1999; Offer 1987; Squire 1971, 1972, 1973, 1975, 1981; Squire et al. 2005). Morel et al. (1999) provided precise details concerning the notions of 'young adult' and 'old' rabbits and I present here and, more generally throughout this monograph, results and reasoning valid for young adult animals only. Otherwise, as repeatedly pointed out in this work, warm-blooded and cold-blooded animals do not behave similarly at different temperatures. Indeed, it is known that, at room temperature and beyond, the cross-bridge ordering in rabbit skeletal muscle, observed by electron microscopy, is greater than that at low temperature (e.g. ~4°C), whereas the ordering in frog muscle is unaffected by temperature (Kensler and Stewart 1983; Kensler et al. 1994; see also p. **150** in Section 4.4.2.5.2). Similar features have been observed regarding the Q_{10} of isometric tetanic tension in frog and rabbit (see pp. **245–246** in Section 8.8). These ordering–disordering phenomena may strongly affect the efficiency of contraction (e.g. isometric tetanic tension), via the rotation of the cross-bridges in swinging cross-bridge/lever-arm models, but they probably have a lesser effect on the main features of the hybrid model described in the next section (only ~60% of contraction results from rotation of the cross-bridges in this new model).

On the basis of the information provided in the preceding paragraph and in Section 5.1.1, including Figure 5.1, its legend and Appendix 5.I, there are, at rest and at temperatures used for experimentation (~0°C–20°C), 100 B–B′-type head–head dimers buried within the thick myosin filament core. This corresponds to ~200 internal myosin heads of the B-type (~33.3% of the heads), the other ~400 heads lying outside the backbone (~200 are of the A-type and ~200 of the F-type; see pp. **170–172** in Section 5.1.1 for precise details on the A- and F-type heads). During contraction, the F-type external heads interact cyclically and rapidly with the thin actin filaments, splitting large amounts of MgATP and undergoing the traditional swinging during contraction. The A-type heads may also work in the same way, but with greater difficulty (see pp. **170–172** in Section 5.1.1). In any event, this cyclic rotation of ~400 external heads (~66.7% of the heads) is the origin of ~60% of contraction, as claimed and demonstrated in this monograph. In a 'composite animal species', at the reference sarcomere length $s_0 \sim 2.18$ μm (see p. **24** in Section 3.2), the cross-sectional area of an intact unit cell is ~1.58×10^{-3} μm^2 (see p. **84** in Section 3.5), corresponding to a mean volume of ~3.44×10^{-15} cm$^3 \sim 3.44 \times 10^{-18}$ l. Borejdo et al. (2006) described and analysed a new method for observing volumes of the order of attolitres (10^{-18} l) within demembranated fibres. This breakthrough involves labelling myosin, particularly the heads, and observing, for example, a single cross-bridge in a working muscle, with the CTIR (confocal total internal reflection) technique. This new technique may constitute a formidable tool for studying a single unit cell in situ. Throughout this work, I focus on intact unit cells and I hope that these microscopic structures will be made 'quantitatively visible' in the near future, in any of their complex functions, not only in the rotation of individual cross-bridges. Regardless of this important development, the concentration of external heads, in a composite animal species, is m ~$400/3.44 \times 10^{-18}$ l ~ 1.16×10^{20} l^{-1} (~1.16×10^{17} cm^{-3} ~ 1.16×10^{23} m^{-3}), that is, ~$1.16 \times 10^{20}/6.022 \times 10^{23} \sim 193$ μM (6.022×10^{23} is Avogadro's number): 193 μM external myosin heads are able to enter the rapid cycle of attachment–detachment to the thin actin filaments, thereby inducing the splitting of large amounts of MgATP under contraction conditions. It is difficult to determine whether the internal B–B′-type head–head dimers can split detectable amounts of MgATP, because these heads do not interact with the thin actin filaments. However, as highlighted on p. **203** in Section 6.3.3, they almost certainly display an increase in their MgATPase activity during contraction, via cooperative (allosteric?) phenomena involving the highly active external heads (particularly the A-type heads), because the A and B heads belong to the same myosin molecule. The assumption concerning cooperativity is strongly supported by several independent groups (see p. **170** in Section 5.1.1). In any event, the internal MgATPase activity is probably lower than the activity of the external heads (see p. **199** in Section 6.3.1 and p. **207** in Section 6.3.3 for complementary information).

It is interesting to try to estimate, mostly from experimental data, the concentrations of the external and total (external + internal) myosin heads. Using mostly structural data for rabbit, Squire (1971, 1972, 1973, 1975, 1981) and Squire et al. (2005) gave a value of ~260 μM for total myosin head concentration (external + internal, because the author considered the thick myosin filaments to be three-stranded, with all heads lying outside the thick myosin filament shaft). More recently, Woledge et al. (2009), using the same approach as Squire (e.g. 1981), suggested that, in 1.05 μm^3 of muscle, there are 167,000 myosin heads, giving a concentration of ~264 μM (simple calculations not shown). These two values are similar and the best estimate is ~ (260 + 264)/2 ~ 262 μM. The authors below used various methods for estimating the concentration of myosin heads, but the techniques used were mostly chemical and obviously did not take into account the intimate structure of the myosin thick filaments. In other words, the estimates given below correspond, in principle, to total myosin heads. Ferenczi et al. (1984b) used a value of 154 μM for rabbit. In a comparative and critical study of various experimental results obtained by independent groups, He et al. (1997) considered the most probable concentration of total myosin heads to be 180 μM for frog and 150 μM for rabbit. Rome et al. (1999) deduced from their experiments a value of 167 μM for white muscle fibres from toadfish. Bagshaw (1993) gave an estimate of 120 μM for frog. West et al. (2004) took 150 μM for dogfish. Averaging these six values gives a 'universal empirical concentration' of myosin heads for a 'composite young adult animal species' of ~154 ± (7 or 18) μM (mean ± SE or SD; n = 6), with a maximal value of ~ 154 + 18 ~

172 μM. The estimates obtained by Squire (e.g. 1981) and Woledge et al. (2009) (~262 μM) are much greater than ~172 μM. A plausible explanation for this discrepancy is proposed on p. **208**. The semi-empirical value of ~193 μM, calculated in the preceding paragraph, corresponds solely to the external heads. The concentration of external + internal heads can be obtained by multiplying the value of ~193 μM by a factor of ~600 total heads/400 external heads ~1.5 (see the beginning of the preceding paragraph concerning the numbers of ~600 and ~400), giving ~193 μM × 1.5 ~ 289 μM. Clearly, this value is consistent with that obtained by Squire (e.g. 1981) and Woledge et al. (2009) for the external + internal heads (~262 μM; see above). Below, in this section, I select a value of ~193 μM for the concentration of external heads able to interact with the thin actin filaments.

On pp. **96–97** in Section 3.8.1, it is recalled that the isometric tetanic tension is significantly lower in demembranated fibres than in intact fibres. As demonstrated on pp. **97–98** in Section 3.8.1, this problem is complex and related, in many instances, to the x-ray diffraction technique, leading probably to misleading conclusions, which were denounced by Morel (1985a) and Morel and Merah (1997) and ignored by the muscle community. Moreover, these fibres systematically swell upon demembranation (see Section 8.7) adding a 'geometric' loss of tension. Otherwise, as suggested on p. **221** in Section 7.2 on the basis of experimental results obtained by Andrews et al. (1991), for chemically skinned fibres from rabbit psoas muscle, part of the loss of tension also seems to be related to the dissolution of some of the myosin molecules in the surrounding buffer, as already suggested by Morel (1985a). We may assume that, at most, ~10% of the myosin is lost in this way (see, as an illustrative example, p. **221** in Section 7.2). This proportion almost certainly corresponds to the myosin molecules with both heads outside the filament core, because the myosin molecules with one head buried in the core are tightly bound to the backbone. Thus, the concentration of ~193 μM (see the end of the preceding paragraph) would be ~193 μM × 0.90 ~ 174 μM. This value is similar to the maximal value of ~172 μM (see the preceding paragraph), but this is almost certainly coincidental (see p. **208** for a short discussion). At this point, I chose a value of ~193 μM for the concentration of external myosin heads, in the most probable structure

of the thick myosin filaments (see pp. **205–206**). These external heads enter into rapid cyclic interactions with the thin actin filaments, resulting in sizeable MgATPase activity. In a large part of the many discussions presented in this monograph (see Section 3.4.3.2), I focus on the experimental data of He et al. (1997), obtained with rabbit, for which the concentration used by the authors was 150 μM (see the preceding paragraph), which is ~22% lower than ~193 μM, leading to an overestimation of ~22% in the rates of MgATP splitting (expressed in μM μM^{-1} s^{-1} = s^{-1}), a degree of error similar to the experimental errors on MgATP splitting rates (see Table 1 in the paper by He et al. 1997). Otherwise, from this discussion, it appears that, in the hybrid model, we do not need to correct the values of the concentrations used by He et al. (1997) and other authors, to calculate specific MgATPase activities per myosin head. In adopting this approach, I implicitly assume that only the external heads, interacting with the thin actin filaments, have high levels of MgATPase activity, much greater than that of the internal heads. This hypothesis merits a short discussion. Let us call Π_{ext} and Π_{int} the rates of MgATP splitting per head by the external and internal heads, respectively (see Section 6.3.1 for precise details on the 'internal' and 'external' MgATPase activities). The apparent rate per head, regardless of its location, is therefore (barycentric mean) $\Pi = (400\Pi_{ext} + 200\Pi_{int})/600 = (2/3)\Pi_{ext} + (1/3)\Pi_{int}$ (see pp. **205–206** concerning the numbers of 600, 400 and 200). In Section 6.3.1, the problem of internal MgATPase activity is raised and ratios of $\Pi_{int}/\Pi_{ext} \sim 1/2$ or ~ 1/3 are considered a suitable compromise (see p. **199** in Section 6.3.1). We deduce that $\Pi \sim (2/3 + 1/6)\Pi_{ext} \sim 0.83\Pi_{ext}$ or $\sim (2/3 + 1/9)\Pi_{ext} \sim 0.78\Pi_{ext}$ and this underestimation of ~17%–22% is very close to the overestimation of ~22% found above: the uncertainties therefore cancel each other out. Moreover, I maintain that the experimental error renders the internal MgATPase activity 'invisible'.

The most frequently used value of ~150 μM for the 'total' concentration of myosin heads in demembranated rabbit fibres (see p. **206**) are slightly lower than the semi-empirical value of ~193 μM, corresponding to the external heads for the three-stranded thick myosin filaments of young adult rabbits (see pp. **205–206**). From the value of ~193 μM, it is deduced that the concentration of all heads is ~289 μM (see p. **207**). Thus, there is a gap between the values of ~150 μM and ~289 μM. In this context, Morel et al. (1999) demonstrated that, for olds rabbits, the thick filaments from psoas muscle are two-stranded and the value of ~150 μM probably correspond to old rabbits. If the rabbits used had been young adult rabbits, the concentrations would have been ~1.5 × 150 μM ~225 μM (see p. **207** concerning the coefficient of 1.5). This 'corrected' estimate is comparable to ~193 μM. This highlights the issue of the age of the animals used (with the subsequent fragility of the corresponding fibres; see also p. **52** in Section 3.4.3.2, p. **200** in Section 6.3.2, p. **202** in Section 6.3.3 and p. **238** in Section 8.5). Many investigators favour the use of very large rabbits (~5 kg or more, to maximise the amount of myosin obtained and the largest psoas muscle) and these rabbits are necessarily old. The underestimation of the concentration of myosin heads may be attributed, at least partly, to the use of old rabbits.

Owing to the complexity of the problems, experimental measurement of the concentration of myosin heads cannot demonstrate that ~33.3% (~200/600) of the heads are inserted within the backbone of the thick myosin filaments in young adult frogs and young adult rabbits. The presence of ~200 'internal heads' does not depend on the age of the animals, as demonstrated by Morel et al. (1999). Regarding the proportions of ~33.3% B-type internal heads and ~66.7% external heads, in young adult animals, of the A- and F-types (these heads are defined on pp. **170–172** in Section 5.1.1; see pp. **205–206** concerning these proportions), I confirm below these proportions, based on an approach used by Barman et al. (2006), although the authors did not draw the conclusions that I draw here. In their Figure 2, the authors present results obtained by ATP chase and Pi burst, with myosin S1 (head) from rabbit muscle (back and legs), under multiturnover conditions ([ATP] > [S1]), at 15°C, on the 0–300 ms time scale. The 'steady-state rates', expressed in [Pi]/[S1] (mol/mol), are ~0.80 (ATP chase) and ~0.55 (Pi burst). In their Figure 3, the authors present results obtained by Pi burst with S1 under single-turnover conditions ([ATP] < [S1]), at 4°C, in a buffer very different from that corresponding to their Figure 2 and on the 0–110 s scale. Their Figure 3 'reveals a very rapid transient burst phase of Pi (too fast to measure on the time scale used) followed by an exponential rise to the complete hydrolysis of the ATP'. Extrapolation to time 0

gives [Pi]/[S1] (mol/mol) ~ 0.62 (not estimated by the authors). These three values are not very different and are independent of temperature (at least within the 4°C–15°C range) and buffer composition. Averaging them gives ~0.657 ± (0.061 or 0.105) (mean ± SE or SD; n = 3). This value is surprisingly close to the ~66.7% corresponding to the proportion of external heads. However, this may be coincidental and another qualitative experimental argument can be presented, confirming the existence of two types of myosin heads (called M and M′ below). Iorga et al. (2004), working in the same group as Barman et al. (2006), discussed the results of single-turnover transients and demonstrated biphasic kinetics with S1. They commented on this specific behaviour and wrote: 'It could be that both heads M and M′ bind actin and that ATP dissociates both actoM and actoM′: actoM with hydrolysis of the ATP, actoM′ without hydrolysis'. The authors considered their experimental results to be consistent with the existence of states of weak and strong binding of myosin to thin actin filaments during the enzymatic cycle. However, M and M′ could also correspond to the B-type internal heads (see Figure 5.1) and the external heads (including the A- and F-type heads, defined on pp. **170–172** in Section 5.1.1), respectively, or vice versa. It is difficult to determine whether M′ corresponds to B-type heads (internal heads) and M to the external heads, but this is unimportant here. In any event, the existence of M and M′ heads (in vitro) is consistent with the existence of internal and external heads in thick myosin filaments (in vivo/in situ).

We now need to take into account intact unit cells from young adult rabbits, because most of the experimental data regarding MgATPase activity were obtained with demembranated fibres from the psoas muscles of young adult rabbits. Indeed, given the self-consistency of the experimental results recalled below, in this section, it seems likely that young adult animals were used by most of the independent groups referred to in this monograph (see, however, first column on this page, but with no consequence for the values of MgATPase activity, as demonstrated on p. **207**). A problem concerns the linking of results obtained with demembranated rabbit fibres with those for intact fibres (see Sections 3.8, 4.4.2.1 and 8.7). I am unaware of experimental data on intact fibres from young adult rabbits. However, West et al. (2004) performed experimental studies on dogfish white intact and permeabilised fibres. This type of fibre may be intermediate between frog and rabbit fibres. Working at 12°C, the authors found, for tetani lasting between 0.055 s and 0.500 s, that there were no significant differences between intact and permeabilised fibres, concerning the amount of MgATP hydrolysed by actomyosin alone (see their Table 4; only the SEs are presented and the values of MgATPase activity at 0.500 s are apparently different in intact and permeabilised fibres; using the SDs there are no longer statistical differences; see Section 2.1 concerning the advantage of using SD rather than SE; it is recalled, on p. **83** in Section 3.4.6, that the MgATPase activity was measured directly in permeabilised fibres and estimated indirectly in intact fibres). The enthalpy increase related to actin–myosin interactions is therefore similar in intact and permeabilised fibres. Moreover, taking into account the 'universal' multiplication factor $\xi_{0,\infty} \sim 1.00$ for passing from the enthalpy increase resulting from actin–myosin interactions to total enthalpy increase (see pp. **195–198** in Section 6.2 for precise details on the mode of calculation of this multiplication factor), there are no significant differences between intact and permeabilised fibres. Thus, intact unit cells (belonging to intact fibres) and permeabilised fibres behave similarly, in terms of total enthalpy increase. This conclusion can be extended to young frogs and young adult rabbits: the total enthalpy increase, $(\Delta H)_{t*}$, in an intact unit cell, estimated below, in this section, is independent of the nature of the corresponding fibre (intact or permeabilised).

For the composite animal species, including the rabbit studied here, the value of $(\Delta H)_{t*}$ is $\sim m\Pi_{t*}e_{app}$ (see p. **198** in Section 6.2), with $m \sim 1.16 \times 10^{17}$ cm^{-3} (see p. **206**) and $e_{app} \sim 6.6 \times 10^{-20}$ J per MgATP molecule (see pp. **193–194** in Section 6.2), giving $me_{app} \sim 7.66$ mJ cm^{-3}. In Section 6.3.3, it is shown that Π_{t*} depends strongly on temperature (see p. **201** in Section 6.3.3). For a short tetanus of virtual duration $t* = 1$ s, based on the values given for the various values of Π_{t*}, on p. **201** in Section 6.3.3, we obtain $(\Delta H)_{t*}(5°C) \sim 51$ J cm^{-3}, $(\Delta H)_{t*}(12°C) \sim 125$ mJ cm^{-3}, $(\Delta H)_{t*}(20°C) \sim 254$ mJ cm^{-3} and $(\Delta H)_{t*}(39°C) \sim 1762$ mJ cm^{-3}. Using the value of $Q_{10} \sim 2.8$ uniformly applied to the $\sim$0°C–39°C range of temperature (see p. **201** in Section 6.3.3), we obtain $(\Delta H)_{t*}(10°C) \sim 92$ mJ cm^{-3} ($\sim$85 mJ cm^{-3} and $\sim$98 mJ

cm^{-3}, when starting from $(\Delta H)_{t*}(5°C) \sim 51$ mJ cm^{-3} or $(\Delta H)_{t*}(12°C) \sim 125$ mJ cm^{-3}, respectively) and $(\Delta H)_{t*}(0°C) \sim 31$ mJ cm^{-3}. The value of $\sim$51 mJ cm^{-3}, valid for rabbit at 5°C, corresponds to $\sim 51 \times 2.5 \sim 128$ mJ cm^{-3} for frog at 5°C (the multiplication factor of $\sim$2.5 is estimated on p. **58** in Section 3.4.3.2). Similarly, the value of $\sim$31 mJ cm^{-3} (for rabbit at 0°C) corresponds to $\sim 31 \times 2.5 \sim 78$ mJ cm^{-3} for frog at 0°C, the value of $\sim$92 J cm^{-3} (for rabbit at 10°C) corresponds to $\sim 92 \times 2.5 \sim 230$ mJ cm^{-3} for frog at 10°C, and the value of $\sim$125 mJ cm^{-3} (for rabbit at 12°C) corresponds to $\sim 125 \times 2.5 \sim 313$ mJ cm^{-3} for frog at 12°C.

Throughout Section 3.4.3.2, it is demonstrated that the change in MgATPase activity as a function of time is difficult to determine. Indeed, in both frog and rabbit, there is an initial rapid transitory period, corresponding to high rates of MgATP splitting, followed by stabilisation of this rate at a lower value (corresponding to steady-state conditions) and then a slow decrease, corresponding to the pseudo–long-lasting tetanus. However, the value of $\sim$78 mJ cm^{-3} for frog at 0°C (see end of the preceding paragraph) and for a short tetanus lasting a virtual period 0–$t* = 1$ s, corresponding to the rate of MgATP splitting under steady-state conditions (precise details on this complex notion of steady state over the period 0–$t*$ are provided below, in this section), is comparable to experimental data published by Curtin et al. (1986). Indeed, working at $\sim$0°C on single intact fibres from frog muscle, over a period of $\sim$0–(1–5) s, the authors found that, for the virtual period 0–$t* = 1$ s (applying the rule of three), the dimensionless rate of MgATP splitting corresponded to $\sim$84 mJ g^{-1}. Elzinga et al. (1989b) and Truong (1974) reported the density of a frog fibre (also probably of fibres from other animal species) to be $\sim$1.066 g cm^{-3} and $\sim$1.035 g cm^{-3}, respectively, and the mean value of $\sim (1.066 + 1.035)/2 \sim 1.050$ g cm^{-3} is selected here. Thus, the value of $\sim$84 mJ g^{-1} becomes $\sim$84 mJ g$^{-1} \times 1.050$ g cm$^{-3} \sim 88$ mJ cm^{-3} after correction for density. This value of $\sim$88 mJ cm^{-3}, deduced from the experimental work of Curtin et al. (1986), and the estimated value of $\sim$78 mJ cm^{-3} (see above) are sufficiently consistent (only $\sim\pm6\%$ around the mean value of $(78 + 88)/2 \sim 83$ mJ cm^{-3}), providing strong support for the reasoning and calculations presented in this work, in the preceding and following paragraphs in particular. However, in the experiments of Curtin et al. (1986), lasting up to

5 s, there would be a favourable concatenation of circumstances to explain this consistency (these favourable conditions would include myothermic measurements on single frog fibres, which may induce unexpected results, with respect to those obtained with the techniques used in permeabilised frog and rabbit fibres).

On p. **55** in Section 3.4.3.2, it is recalled that He et al. (1997), working on a single permeabilised frog fibre at 5°C (see their Figure 9), found that beyond ~250 ms (or ~200 ms if the time lag of ~50 ms is subtracted), the MgATPase activity is only ~2.8 s^{-1}, against ~16.4 s^{-1} during the first two turnovers, the duration of which is Δt_{app} ~ 92–111 ms (not taking into account the time lag). I strongly suggest that the value of ~2.8 s^{-1} includes a large part of fatigue. The MgATPase activity of ~16.4 s^{-1}, obtained during Δt_{app} ~ 92–111 ms (at a mean temperature of ~5.7°C; see p. **54** in Section 3.4.3.2), after a time lag of ~39–58 ms (see p. **55** in Section 3.4.3.2), is assumed to be approximately valid over the period 0–t* = 1 s, in the fibre used by He et al. (1997) in their Figure 9 (the period 0–t* = 1 s is studied in depth below, in this section). The inclusion or exclusion of the time lag of ~39–58 ms from this period can be considered unimportant, as it is only ~4%–6% of 1 s, much lower than the experimental uncertainty on the various parameters. From Table 1 in the paper by He et al. (1997), the time lags for Pi release (i.e. MgATPase activity), for the rabbit, are ~40–93 ms, at 5°C, ~21–29 ms, at 12°C, and ~0–24 ms, at 20°C (for reasons given in Section 3.1, SDs are calculated from SEs and used here). Here again, the inclusion or exclusion of the time lags can be considered as unimportant, as they range from only ~0% to ~9% of 1 s. However, it is demonstrated, in the following paragraphs, that important phenomena almost certainly occur during the time lag.

When considering MgATPase activities, particularly during the first turnover defined by He et al. (1997), we need to include various complex transitory phenomena, such as the Pi burst observed under both relaxing and contraction conditions (the H$^+$ burst studied, for example, by Morel et al. 1999, under resting conditions in the absence of F-actin, should not be forgotten, as is unfortunately frequently the case in many experimental studies and discussions). Ferenczi (1986) and Ma and Taylor (1994), for instance, described and quantified the Pi burst in resting

and contracting permeabilised fibres and unheld myofibrils from rabbit psoas muscle, respectively. He et al. (1997a), studying isometrically contracting permeabilised rabbit fibres at various temperatures and using various experimental techniques, claimed that 'there was no burst of Pi formation'. However, they observed, in many figures in their paper, a rapid increase of Pi, followed by a rapid decrease, and attributed this to laser flash artefacts. Nonetheless, they noted that 'variability in the fluorescence level immediately following the flash artefact made analysis difficult'. No clear conclusion can therefore be drawn from the paper by He et al. (1997). I am aware of no detailed experiments carried out on demembranated fibres and unheld myofibrils from frog, concerning Pi and H$^+$ bursts. One of the most precise experimental studies concerning the kinetic mechanism of the MgATPase of rabbit unheld myofibrils was carried out by Ma and Taylor (1994) (as pointed out on p. **56** in Section 3.4.3.2, this biological material is almost certainly not contaminated by fatigue). According to their Figure 3B, the steady-state rate at 20°C is ~20 s^{-1} (measured between ~20–25 ms and ~100 ms), whereas the transient rate constant, corresponding to the Pi burst (occurring between 0 and ~20–25 ms, again according to Figure 3B) is ~150 s^{-1}, that is, mean rate of Pi release of ~50–60 s^{-1} between 0 and ~20–25 ms (see Figure 3B for other precise details; the mean rate of ~50–60 s^{-1} is not given by the authors and the calculation is not shown here). It is difficult to know whether the Pi burst represents a large proportion of the transitory MgATPase activity, but I assume, as a rough approximation, that it does. The ratio of ~50–60 s^{-1} to ~20 s^{-1} is ~2.5–3.0, valid for the rabbit at 20°C, and can reasonably be extended to the frog at 5°C, but a value of ~5.0 cannot be ruled out. The mean transitory MgATPase activity would therefore be ~(2.5–3.0) × 16.4 s^{-1} ~ 41.0–49.2 s^{-1} or ~5.0 × 16.4 s^{-1} ~ 82.0 s^{-1}, and even more if the H$^+$ burst is taken into account. The value of ~16.4 s^{-1} is recalled at the end of the preceding paragraph and corresponds to the period between ~39–58 ms (time lag; see the preceding paragraph) and ~150 ms (value given in the legend to Figure 9 in the paper by He et al. 1997). The mean transitory MgATPase activities of ~41.0–49.2 s^{-1}, ~82.0 s^{-1} or more correspond to the Pi and H$^+$ bursts, which almost certainly occur during the time lag of ~39–58 ms after the laser flash (this last value is not very different from ~20–25 ms, corresponding to the Pi burst, found by Ma and Taylor 1994, for a different biological material, at a different temperature; see above). At this point, it should be noted that, for the frog, the time lag for tension is ~13–28 ms (see p. **54** in Section 3.4.3.2), whereas the time lag for MgATPase activity is ~39–58 ms (see above). Thus, there is no common interval between these two time lags, so the tension rise must begin ex nihilo, before any sign of MgATPase activity. This is obviously impossible and, during the time lag for MgATPase activity, there is in fact a true MgATPase activity, corresponding almost certainly to the Pi and H^+ bursts, as claimed above. This confirms the initial values of ~41.0 s^{-1}, ~49.2 s^{-1}, ~82.0 s^{-1} or more, suggested here.

As frequently pointed out at many places in this monograph and recalled, for instance, on p. **55** in Section 3.4.3.2, the first turnover, defined by He et al. (1997) and after the time lag of ~39–58 ms (see the preceding paragraph), corresponds to complex transitory phenomena and is therefore included in the period ~(39–58)–50 ms (see also the preceding paragraph concerning this period). Beyond ~39–58 ms + Δt ~ 39–58 + 46–56 ~85–114 ms (Δt ~ 46–56 ms corresponds to the first complex turnover; see p. **54** in Section 3.4.3.2), the 'normal' steady-state 'isometric' MgATPase activity is reached for at least a period corresponding to Δt ~ 46–56 ms (second turnover). Adding this last value to that of ~85–114 ms gives ~131–170 ms, similar to ~150 ms (on p. **54** in Section 3.4.3.2, I give an estimate of ~131–169 ms, but this slight difference is merely because all the numerical values are rounded off in the two approaches). Beyond these periods of ~131–170 ms or ~150 ms and up to ~(39–58) + 200 ~ 239–258 ms (statistically similar to ~250 ms; see pp. **55–56** in Section 3.4.3.2 concerning the value of ~250 ms), we can assume that there is a gradual slight decrease in MgATPase activity, related to the gradual onset of fatigue (see pp. **55–56** in Section 3.4.3.2, where it is suggested that fatigue occurs somewhere between ~92–111 ms and ~200 ms, or ~131–169 ms and ~239–258 ms if the time lag of ~39–58 ms is added). However, according to the comments, presented on pp. **55–56** in Section 3.4.3.2 about Figure 9 in the paper by He et al. (1997), beyond ~250 ms (the time lag of ~50 ms, valid for Figure 9, consistent with ~39–58 ms, is included in this value), MgATPase activity falls dramatically to ~2.8 s^{-1}. At this stage, the initial

high values of MgATPase activity (~41.0 s^{-1}, ~49.2 s^{-1}, ~82.0 s^{-1} or more, and ~16.4 s^{-1}) and the low value (~2.8 s^{-1}) (all these values are given in the preceding paragraph and above) almost certainly cancel each other out, resulting in a mean steady-state value of ~13.2 ± 3.4 s^{-1} (mean ± SD; n = 7). This value corresponds to the second turnover (see p. **58** in Section 3.4.3.2) and is extended to the period between 0 and $t^* = 1$ s (apparent steady state; see also the next paragraph). This approximation is rendered even more valid by the high SEs and SDs on the values of MgATPase activity (SD represents ~100 × 3.4 s^{-1}/13.2 s^{-1} ~ 26% the mean value of ~13.2 s^{-1}; mixing all the values presented in Table 1 in the paper by He et al. 1997 and the corresponding SDs, the error can reach up to ~55%).

High initial levels of MgATPase activity were also reported by He et al. (1998a), who found an initial rate of MgATP splitting, for permeabilised fibres from rabbit psoas muscle, at 7°C, and at a sarcomere length of ~3.0 μm, ~3–4 times higher than that found a few tens of milliseconds later (see Table 1 in the paper by He et al. 1998a). Thus, the experimental findings, at 5°C and 7°C, for frog and rabbit, respectively, are qualitatively comparable and we can reasonably assume that, in young adult rabbits, the same notion of 'apparent' steady-state MgATPase activity between 0 and $t^* = 1$ s also holds, as in young adult frogs. These notions may not be valid at higher temperatures, but they are retained here, regardless of temperature (between ~0°C and ~43.5°C; see Figure 5.2, its legend and Appendix 5.II concerning this temperature of ~43.5°C in rabbit). Owing to the complexity of the change in MgATPase activity over time, the calculations presented on pp. **210–211** should be considered as providing only appropriate orders of magnitude, supported however a posteriori by the many self-consistent discussions and conclusions drawn in this book.

For young adult rabbits, at 5°C, the ~51 mJ cm^{-3} ~ 0.051 J cm^{-3} of enthalpy released during 1 s (see p. **210**) corresponds to a temperature increase of ~0.0116°C (taking a conversion factor of 4.184 J g^{-1} °C^{-1} and 4.393 J cm^{-3} °C^{-1}; see specialist handbooks on calorimetry concerning the constant 4.184 J g^{-1} °C^{-1}; on p. **209**, the density of a fibre is estimated to be ~1.050 g cm^{-3}). As highlighted on pp. **209–210**, concerning the enthalpy release, the corresponding mean increase with temperature relates to intact or demembranated fibres.

Thus, I use the experimental data available as they stand, that is, for intact or demembranated fibres, extended to intact unit cells. Some of the increase in temperature of ~0.0116°C would be transferred to the medium surrounding the fibre. However, this increase is very low and we can neglect this transfer. Otherwise, the volume of a thick myosin filament is ~4.3% that of an intact unit cell in the frog (see p. **234** in Section 8.4) and this proportion is probably similar for the rabbit. Thus, the ~0.0116°C increase in temperature in the sarcoplasmic medium is almost instantaneously transferred to the inside of the thick myosin filament cores, with a very short time lag of only a few milliseconds, according to general approaches to heat transfer from a large medium to a much smaller medium (see handbooks on heat transfer). The temperature within these cores therefore also increases almost instantaneously by ~0.0116°C. Regardless of temperature of the surrounding medium, we see, on p. **205** in Section 6.3.4, that an increase in temperature of ~1.56°C results in the complete dissociation of the 100 B–B′-type head–head dimers located within the thick myosin filament cores (see Figure 5.1, its legend and Appendix 5.I concerning the dimers, and pp. **180–181** and **183** in Section 5.1.2, and Section 5.2, concerning the dissociation and the reversible association–dissociation processes; see also Section 6.4 for other precise details; the number of 100 internal dimers is recalled on p. **206**). An increase in temperature of ~0.0116°C should lead to the dissociation of a number, n_h, of internal dimers per thick myosin filament and we can write (rule of three) $n_h(5°C) \sim 100 \times 0.0116°C/1.56°C \sim 0.74 \sim 1$. At 0°C, the value of $(\Delta H)_{t*}$ is ~ 31 mJ cm^{-3} (see p. **209**), corresponding to ~0.00705°C and to $n_h(0°C) \sim 100 \times 0.00705°C/1.56°C \sim 0.45$. Owing to the shortcomings of the preceding reasoning and to the accumulation of experimental error, the value of n_h at 0°C is also probably close to 1, as at 5°C. Similarly, at 10°C, ~92 mJ cm^{-3} (see p. **209**) corresponds to an increase in fibre temperature of ~0.0209°C. The difference in temperature between the inside of the fibre and the surrounding medium is again very small and no significant outflow of heat can occur, so there is no subsequent temperature decrease. Thus, we obtain $n_h(10°C) \sim 100 \times 0.0209°C/1.56°C \sim 1.34 \sim 1$. At 20°C, ~254 mJ cm^{-3} (see p. **209**) gives a maximal increase in temperature within the thick myosin filament core of ~0.0578°C, with a possible slight

outflow of heat from the inside of the fibre to the surrounding buffer. An increase of ~0.0500°C can be retained. Thus, we obtain $n_h(20°C) \sim 100 \times 0.0500°C/1.56°C \sim 3.2 \sim 3$. Finally, at the body temperature of the rabbit (39°C), the ~1762 mJ cm^{-3} (see p. **209**) corresponds to ~0.4011°C. This rather high value implies that there is almost certainly a noticeable outflow of heat. Assuming that ~60% of this maximal increase is transferred to the surrounding buffer, a plausible value for the temperature increase within the fibre would be ~0.4011°C × 0.60 ~ 0.2407°C and $n_h(39°C)$ would be ~100 × 0.2407°C/1.56°C ~ 15.4 ~ 15.

The number, n_h, of dissociations of B–B′- type head–head internal dimers, related to $(\Delta H)_{t*}$, and calculated in the preceding paragraph, is an essential parameter, because it is related to both the total number $n*(\theta)$ of dissociations of internal dimers and the temperature θ, via Equation 6.2 and because the strength of the residual radial tethering force NF_{cb} (during contraction) is directly related to the number $n*(\theta)$ (see p. **181** in Section 5.1.2).

6.4 ESTIMATE OF THE ENERGY/ENTHALPY INCREASE RELATED TO MYOSIN HEAD–HEAD DIMER DISSOCIATIONS AND ELECTROSTATIC PRESSURE WITHIN THE THICK MYOSIN FILAMENT CORE, DURING ISOMETRIC TETANIC CONTRACTION (SHORT TETANI–STEADY STATE). CASE OF YOUNG ADULT RABBITS, WITH SOME INSIGHTS INTO THE CASE OF YOUNG ADULT FROGS AND SOME OTHER ANIMAL SPECIES

From pp. **211–212** in Section 6.3.5, we deduce that, regardless of temperature, a limited number of dissociations, n_h, of internal B–B′-type head–head dimers per thick myosin filament occurs, related to $(\Delta H)_{t*}$ (see Section 6.1 for definition and p. **209** in Section 6.3.5 for calculation of $(\Delta H)_{t*}$). This small number of dissociations is certainly insufficient to account for the ~40% full contraction, raising questions about the validity of the lateral swelling part of the mechanisms of contraction, at this stage of the analysis. However, I suggest that this moderate number of dissociations of internal dimers is sufficient to initiate a chain reaction (see also p. **217** in Section 6.5), related to the term $(P\Delta V*)_{t*}$ (see Section 6.1 for definition). For young

Morel

adult rabbits, a short preliminary discussion of the duration of the transitory phenomena occurring in MgATPase activity is required. On p. **54** in Section 3.4.3.2, I estimate that, for frog, at 4°C, 5°C and 8°C (mean temperature of ~5.7°C), the duration of the transitory period, after the time lag of ~39–58 ms, is $\Delta t \sim$ 46–56 ms, corresponding to the first 'complex' turnover (see particularly p. **210** in Section 6.3.5 concerning this notion of first complex turnover; more generally, the complexity of the beginning of MgATP breakdown is discussed on pp. **210–211** in Section 6.3.5). The duration $\Delta t \sim$ 46–56 ms is the same for the other turnovers (see p. **54** in Section 3.4.3.2). For rabbit, nothing is known a priori about Δt, corresponding to the duration of the turnovers for the rate of MgATP splitting, but it is possible to estimate its order of magnitude. Indeed, from Table 1 in the paper by He et al. (1997), it appears that, at 5°C (close to ~5.7°C), the half-time of tension rise for rabbit is $t_{1/2} \sim$ 216 ± 27 ms (mean ± SE; n = 7), whereas it is $t_{1/2} \sim$ 77 ± 8 ms (mean ± SE; n = 7) for frog, again at 5°C. Thus, taking into account only the SEs, I suggest, as a rough approximation, that, for rabbit at 5°C, the duration of the transitory period for MgATPase activity is Δt(rabbit, 5°C) ~ [(216 ± 27) ms/(77 ± 8) ms] × Δt(frog, 5°C) ~ (2.81 ± 0.65) × (46–56) ms ~143 ± 36 ms. Calling t_0(5°C) the duration Δt(rabbit, 5°C) of the transitory period for MgATPase activity, t_0(5°C) represents ~14.3 ± 3.6% of the reference virtual time $t^* = 1$ s. Again from Table 1 in the paper by He et al. (1997) and based on the rule of three, this transitory period, at 12°C, is t_0(12°C) = Δt(rabbit, 12°C) ~ [(84 ± 7) ms/(77 ± 8) ms] × Δt(frog, 5°C) ~ (1.09 ± 0.21) × (46–56) ms ~ 56 ± 12 ms (~5.6 ± 1.2% of $t^* = 1$ s) and, along the same lines, t_0(20°C) ~ 17 ± 3 ms at 20°C (~1.7 ± 0.3% of $t^* = 1$ s). At the reference temperature used throughout this monograph (10°C), the duration of the transitory period is approximately (barycentric mean between the values at 5°C and 12°C) t_0(10°C) ~ 81 ± 19 ms (~8.1 ± 1.9% of $t^* = 1$ s). Regardless of temperature, the durations t_0 of the transitory periods are therefore much lower, even significantly so, at 10°C, 12°C and 20°C, than the reference virtual duration $t^* = 1$ s. Thus, for rabbit, the rate of MgATP splitting can be considered to reach initial high values rapidly (see pp. **210–211** in Section 6.3.5 for a circumstantial discussion of the results obtained with a mixture of permeabilised frog fibres and unheld rabbit myofibrils), leading to a rapid dissociation of the B–B'-type head–head dimers lying within the thick myosin filament core (precise details are given below, in this section). Apparent steady-state MgATPase activity is reached over the interval $0–t^* = 1$ s (see again pp. **210–211** in Section 6.3.5, in the case of the frog). This approach is clearly an oversimplification, because the kinetics of establishment for the various phenomena leading to the notion of apparent steady-state MgATPase activity cannot be predicted (the kinetics of dissociation of the B–B'-type head–head dimers cannot be predicted either). However, this approach leads to self-consistent discussions and conclusions.

An important phenomenon occurs almost certainly during the initial period studied in the preceding paragraph. Indeed, for rabbit, when the internal B–B'-type head–head dimers dissociate rapidly, there is an increase in the specific volume, $\Delta V^* \sim + 0.015$ ml g^{-1} (see p. **191** in Section 6.1), that is, ~+1.89 × 10^{-3} m^3 per mole (MW of a head ~ 126 kDa; Bachouchi et al. 1985). At this stage, a preliminary comment should be made concerning this rather rapid dissociation of the internal B–B'-type head–head dimers in synthetic thick myosin filaments studied in vitro and in the absence of F-actin. In Figure 5.2, its legend and Appendix 5.II, it is demonstrated that the width of the temperature transition (at ~37°C–38°C) for the total dissociation of the internal dimers is ~1°C, whereas the rate of temperature increase is +18°C min^{-1} (see p. **18** in Section 2.4), which would give a maximal duration for the dissociation of ~3.3 s. It could be argued that this very long duration is at odds with the hypothesis of rather rapid dimer dissociation (see the preceding paragraph). However, this estimate of ~3.3 s results purely from the rule of three and is misleading, because it depends solely on technical problems (see p. **18** in Section 2.4). Thus, the estimate of ~3.3 s has nothing to do with the rapid dissociation process. The positive value of ΔV^*, recalled above, is commented on on pp. **191–192** in Section 6.1. Owing to the plus sign of ΔV^*, as soon as one dimer dissociates, there is an unavoidable tiny increase in temperature within the core of the myosin thick filaments (see Equation 6.1 concerning the formal equivalence of the terms $(\Delta H)_{t^*}$ and $(P\Delta V^*)_{t^*}$, and p. **212** in Section 6.3.5 for the relationship between $(\Delta H)_{t^*}$ and the number, n_h, of dimer dissociations and temperature). This initial tiny increase in temperature leads to the dissociation of a second dimer and so on. A chain reaction is initiated, which stops when a 'certain' number of internal dimers have dissociated. It is currently impossible to predict this certain number of dissociations as a function of the temperature of the external medium. For simplification, regardless of the number of dissociations, I suggest that the chain reaction is completed during the initial period, lasting only $\sim 81 \pm 19$ ms at the reference temperature of $10°C$ used in this monograph (see the preceding paragraph). In such conditions, there should be a rapid initial increase in the 'pressure–volume term', which can be written $(P\Delta V^*)_{t0}$ (t_0 is defined, estimated and commented on in the preceding paragraph). The size of this 'almost instantaneous' increase in $(P\Delta V^*)_{t0}$ is considered maximal and depends on the value of P within the thick myosin filament core. The existence of a considerable internal pressure P (see the next paragraph) results mostly from the presence of numerous fixed negative electrical charges on the thick myosin filaments from various animal species (e.g. Aldoroty et al. 1985, 1987; Bartels and Elliott 1980, 1985; Collins and Edwards 1971; Elliott 1980) and also from the presence of bound anions (Regini and Elliott 2001, and references therein). Unfortunately, no estimate of the density of the bound anions is currently available. However, as a compromise, I suggest that part of P, called P_{ba}, is related to the bound anions (see the next paragraph).

The pressure P appearing in Equation 6.1 is mostly electrostatic and is given by $P = P_\sigma + P_{ba}$, where $P_\sigma = \sigma^2/\varepsilon\varepsilon_0$ is the electrostatic pressure related to the fixed negative electrical charges (see handbooks on electrostatics). ε_0 is the dielectric constant of a vacuum (8.85×10^{-12} C^2 N^{-1} m^{-2}; see handbooks on electrostatics). ε is the dielectric constant within the thick myosin filament core (concerning the interfilament medium, a value of ~ 50 is proposed on p. **32** in Section 3.2). Within the thick myosin filament core, the density of proteins (mostly internal myosin heads) is very high (around 11–12 times that in the sarcoplasmic medium; see p. **215**), rendering the inside of the backbone very different from the outside and much more similar to a solid. This major difference between the inside and outside of the thick myosin filaments is exemplified by the temperature transitions corresponding to head–head dimer dissociation

($\sim 27°C$–$29°C$ outside and $\sim 37°C$–$38°C$ inside; see pp. **18–19** in Section 2.4, and Section 5.2, Figure 5.2, its legend and Appendix 5.II, respectively). Values of ε between 2 and 10, valid for solids (see handbooks on electrostatics), are probable and I choose a value of ~ 6 (midway between 2 and 10) for the internal medium of the thick myosin filaments. A reasonable estimate of the pressure P_{ba} related to the bound anions is $P_{ba} \sim 0.5P_\sigma$, giving $P \sim 1.5\sigma^2/\varepsilon\varepsilon_0$. At physiological pH and ionic strength, in fibres at the slack length or slightly stretched, the mean linear density of fixed negative electrical charges on a thick myosin filament in situ, according to a compilation of several experimental studies concerning various animal species and types of muscle (see Aldoroty et al. 1987), is $\sim 6.6 \times 10^4$ e $\mu m^{-1} \sim 1.06 \times 10^{-8}$ C m^{-1} under resting conditions ($e = 1.602 \times 10^{-19}$ C; see handbooks on electrostatics; see also Section 4.4.2.2 for precise details on the linear charge densities). It is strongly suggested, on pp. **161–162** in Section 4.4.2.6, that the absolute values of the surface electrical potentials are higher in contracting fibres than in resting fibres, leading almost certainly to densities of fixed negative electrical charges and bound anions higher under contracting conditions than at rest. Unfortunately, it is currently impossible to measure these densities in contracting fibres. We therefore need to assume that the value of $\sim 1.06 \times 10^{-8}$ C m^{-1} also applies to contracting fibres. This is justified, given that the experimental results obtained by several authors for the density of fixed negative electrical charges on thick myosin filaments are highly scattered, even at rest (see Table IV in the paper by Aldoroty et al. 1987; according to Table V in this paper, the same is true in rigor). Comparing Tables IV and V, it appears that the mean densities of fixed negative electrical charges deduced from several experimental studies are similar at rest and in rigor, within the wide range of experimental error, although these two states are physiologically very different: presence of MgATP at rest, most cross-bridges detached, presence of few weakly bound cross-bridges (see pp. **170–172** in Section 5.1.1 for definition of these cross-bridges), negligible resting tension at the slack length; total absence of MgATP in rigor, all cross-bridges attached, in the 'frozen' arrowhead structure, considerable rigor tension. The numerical calculations presented in this section mostly provide orders of

Morel

magnitude and the value of $\sim1.06 \times 10^{-8}$ C m^{-1} (see above) can be used, regardless of the state of the fibre (rest, contraction, rigor), without inconsistencies.

For rabbit, the first estimates of the diameter of the myosin tail, obtained in the 1960s, were ~1.5 nm (Huxley 1963), ~2.5 nm (Huxley 1969), ~3.0 nm (Perry 1967) and ~2.0 nm (Lowey et al. 1969). In the 1970s, values of ~2.8 nm were obtained by Morel et al. (1979) and Pinset-Härström et al. (1975). More recently, Woodhead et al. (2005), working on natural thick filaments from tarantula leg muscle, obtained a value of ~2.0 nm. Averaging these seven values gives $\sim2.4 \pm (0.2$ or $0.5)$ nm (mean $\pm$ SE or SD; $n = 7$). The value of ~2.4 nm is retained below, because it is consistent with the rather scattered values for rabbit. Again, for rabbit, the external diameter of a thick myosin filament backbone, whether synthetic or natural, is taken as ~18 nm (Morel et al. 1999 provided structural justifications for this value; Pinset-Härström 1985 and Pinset-Härström and Truffy 1978 obtained the same value with a cutting-edge technique for sample preparation before observation by electron microscopy: the Kleinschmidt process). Ip and Heuser (1983) found that the external diameter of the backbone of a thick myosin filament from rabbit was only ~15 nm. This rather low value, not used here, is almost certainly related to the use of traditional electron microscopy, known to shrink structures (e.g. Millman 1998). April (1969) and April and Wong (1976) obtained values of ~18 nm for striated muscle from the crayfish walking leg (*Orconectes*). Woodhead et al. (2005), using a cutting-edge technique (cryoelectron microscopy, followed by three-dimensional reconstruction and atomic fitting), found the diameter of thick myosin filaments to be ~16 nm, for tarantula. The five values of ~18 nm and that of ~16 nm give a mean value for a composite animal species of $\sim17.7 \pm (0.3$ or $0.7)$ nm (mean $\pm$ SE or SD; $n = 6$). In this section, the rabbit is studied and a value of ~18 nm, lying within the range of experimental error, is retained. For rabbit, the length of the myosin tail is $\sim130–135$ nm (e.g. Lowey et al. 1969; Moos et al. 1975; Morel et al. 1979) and the axial repeat on the thick myosin filaments is $\sim42.9–43.8$ nm long (see legend to Figure 5.1 and Appendix 5.I). A similar value of ~43.5 nm was found for natural thick myosin filaments from tarantula (e.g. Crowther et al.

1985; Offer et al. 2000; Woodhead et al. 2005). This gives an approximate value of the overlap of $\sim(130–135)$ nm$/(42.9–43.5–43.8)$ nm ~ 3.0 to $3.2 \sim 3$ between the tails of the myosin molecules lying along the same axis in thick myosin filaments (particularly for the rabbit). From these values, the internal diameter can be estimated at $\sim(18.0 - 2.4 \times 3) \sim 10.8$ nm. Thus, the mean diameter is $\sim(10.8 + 18.0)/2 \sim 14.4$ nm, corresponding to a mean circumference of ~45.2 nm $\sim 4.52 \times 10^{-8}$ m.

From the values presented in the two preceding paragraphs, we deduce that $\sigma \sim 1.06 \times 10^{-8}$ C m$^{-1}/4.52 \times 10^{-8}$ m ~ 0.23 C m^{-2}, giving (see p. **214** for the definition of the pressure P within the thick myosin filaments) P $\sim 1.5\sigma^2/\varepsilon\varepsilon_0 \sim 1.5 \times (0.23)^2/(6 \times 8.85 \times 10^{-12}) \sim 1.49 \times 10^9$ N m^{-2} ($\sim1.49 \times 10^4$ times atmospheric pressure; 6 is the dielectric constant within the filament core; see p. **214**). Taking into account this value of P and that of ΔV^* recalled on p. **213** ($\sim1.89 \times 10^{-3}$ m^3 per mole; the value of $\varepsilon_0 = 8.85 \times 10^{-12}$ C^2 N^{-1} m^{-2} is recalled on p. **214**), we obtain the 'crude value' of P$\Delta V^* \sim 2.82 \times 10^6$ J per mole, independently of kinetic considerations (see below concerning the notion of crude value and kinetics). We now need to estimate the number of moles of internal heads per thick myosin filament. In both frog and rabbit, there are ~100 internal B–B'-type head–head dimers, at rest (i.e. ~200 internal heads; see p. **206** in Section 6.3.5), regardless of the age of the animals (Morel et al. 1999). In a composite animal species, the length of a thick myosin filament is ~1.63 μm $\sim 1.63 \times 10^{-6}$ m (see p. **34** in Section 3.2) and its internal diameter is ~10.8 nm $\sim 1.08 \times 10^{-8}$ m (see the end of the preceding paragraph), giving an internal volume of $\sim[\pi \times (1.08 \times 10^{-8})^2/4] \times 1.63 \times 10^{-6}$ m$^3 \sim 1.49 \times 10^{-22}$ m$^3 \sim 1.49 \times 10^{-16}$ cm^3. The ~200 internal B-type heads correspond to $\sim200/6.022 \times 10^{23} \sim 3.32 \times 10^{-22}$ mol of internal heads per thick myosin filament (6.022×10^{23} is Avogadro's number), giving a concentration of $\sim3.32 \times 10^{-22}$ mol$/1.49 \times 10^{-16}$ cm$^3 \sim 2.23 \times 10^{-6}$ mol cm^{-3} (i.e. ~ 2.23 mM ~ 2230 μM, much greater [ratio of ~ 2230 μM$/193$ μM $\sim 11–12$] than the value of ~193 μM for the concentration of external heads estimated on p. **206** in Section 6.3.5, for a composite animal species). In other words, 1 mol corresponds to ~1 mol$/2.23 \times 10^{-6}$ mol cm$^{-3} \sim 0.45 \times 10^6$ cm^3.

Thus, the maximum of $P\Delta V^*$, called $(P\Delta V^*)_{t0}$ (see pp. **213–214** for a discussion of the value of t_0, and the definition of the maximal value $(P\Delta V^*)_{t0}$, reached almost instantaneously), in a single thick myosin filament is $\sim 2.82 \times 10^6$ J per mole/0.45×10^6 cm^3 per mole ~ 6.27 J cm^{-3}, corresponding to a maximal increase in temperature of ~ 6.27 J cm^{-3}/4.393 J cm^{-3} $^\circ$C^{-1} $\sim 1.42^\circ$C within the thick myosin filament core, for a short tetanus under steady-state conditions, regardless of external temperature (see p. **211** in Section 6.3.5 concerning the constant 4.393 J cm^{-3} $^\circ$C^{-1}; the density of the thick myosin filament is assumed to be similar to that of a whole fibre). This estimate of $\sim 1.42^\circ$C presumably corresponds to an almost instantaneous increase in temperature occurring early in the tetanus (see above regarding this notion of almost instantaneous). The estimated maximal value of $\sim 1.42^\circ$C, corresponding to $(P\Delta V^*)_{t0}$, is close to the experimental value of $\sim 1.70^\circ$C, selected on p. **204** in Section 6.3.4. This similarity in the maximal increase in temperature obtained with very different approaches and experimental results is very interesting and the average value of $\sim (1.42 + 1.70)/2 \sim 1.56^\circ$C is taken into account throughout this chapter. Moreover, the estimate of $(P\Delta V^*)_{t0}$ is very high and is almost certainly sufficient to induce the total dissociation of the ~ 100 internal B–B′-type head–head dimers. However, there is certainly a rapid decrease in the value of $(P\Delta V^*)(t)$, between time t_0 ($\ll t^*$), corresponding to $(P\Delta V^*)_{t0}$, and $t^* = 1$ s (the necessary references concerning t_0 are given above), owing to the unavoidable heat transfer from the thick myosin filament core, representing only $\sim 4.3\%$ the volume of an intact unit cell (see p. **234** in Section 8.4), to the sarcoplasmic medium. This decrease is probably exponential but is impossible to predict either qualitatively or quantitatively. In any event, the increase in temperature within the thick myosin filament core contributes to a large increase in the number of dissociations of internal myosin B–B′-type head–head dimers and to the lateral swelling part of the hybrid model. In Sections 6.5 and 6.6, I suggest a simplified semi-empirical approach to the number of dissociations and to the relative contributions of the 'enthalpy-related' and 'pressure–volume-related' B–B′-type head dissociations within the thick myosin filament core. Finally,

the true value of $(P\Delta V^*)_{t^*}$ is given by the following formal definition:

$$(P\Delta V^*)_{t^*} = (1/t^*) \int_{t_0 \sim 0}^{t^*} (P\Delta V^*)(t)\, \mathrm{d}t \quad (6.2)$$

Unfortunately, this mean value cannot be calculated, because the kinetics of $(P\Delta V^*)(t)$ cannot be estimated as highlighted above and commented on in Section 6.5.

6.5 ESTIMATE OF THE NUMBER OF MYOSIN HEAD–HEAD DIMER DISSOCIATIONS WITHIN THE THICK MYOSIN FILAMENT CORE, DURING ISOMETRIC TETANIC CONTRACTION (SHORT TETANI–STEADY STATE), AS A FUNCTION OF TEMPERATURE. CASE OF YOUNG ADULT RABBITS

Before going into detail, it should be recalled that the number of dissociations of myosin B–B′-type head–head dimers within the thick myosin filament core (see Figure 5.1, its legend and Appendix 5.I in Section 5.1.1 regarding the arrangement of these dimers) is an essential parameter for estimation of the strength of the residual radial tethering force, NF_{cb}, during isometric tetanic contraction (see pp. **158–159** in Section 4.4.2.5.4 and pp. **180–181** in Section 5.1.2 concerning NF_{cb}). More precisely, the maximal strength of the tethering forces per intact unit cell, $N_0 F_{cb}^0$, corresponds to resting conditions (all internal heads bound to form dimers), whereas the minimal strength of these forces, NF_{cb}, corresponds to isometric tetanic contraction conditions and can reach zero, if all the dimers dissociate. As claimed throughout this book, the lateral swelling mechanisms (accounting for $\sim 40\%$ of the isometric tetanic contraction in the hybrid model, at 10°C) strongly depend on the total energy available $(\Delta U)_{t^*}$ given by Equation 6.1. From this equation, this energy, available in intact unit cells during a short tetanus of virtual duration $0-t^* = 1$ s, is the sum of two terms, $(\Delta H)_{t^*}$ and $(P\Delta V^*)_{t^*}$, studied in Sections 6.3.5 (see p. **205**) and 5.4 (see pp. **214–216**), respectively. Each source of energy gives rise to an increase in temperature resulting in dissociation of internal dimers. However, $(\Delta H)_{t^*}$ causes a limited number of internal dimer dissociations, n_h, particularly at low temperatures (see pp. **211–212** in Section 6.3.5). Nonetheless, as emphasised on p. **213** in Section 6.4, this is sufficient to trigger a chain reaction related to $(P\Delta V^*)_{t*}$, which is responsible for most of the dissociations. The number, $n_h(\theta)$, of dissociations at temperature θ (in degrees Celsius), directly related to $(\Delta H)_{t*}$, can be estimated with confidence (see pp. **211–212** in Section 6.3.5). Unfortunately, it is impossible to estimate the number, $n_{pv}(\theta)$, of dissociations related to $(P\Delta V^*)_{t*}$. Indeed, as highlighted at the end of Section 6.4, after the almost instantaneous increase $(P\Delta V^*)_{t0}$ (see p. **213** in Section 6.4, concerning the term almost instantaneous), the inevitable decrease in $(P\Delta V^*)(t)$ over time t (see p. **216** in Section 6.4) cannot be predicted, because this decrease depends on many phenomena relating to the transfer of heat from the core of the thick myosin filaments to the sarcoplasmic medium and then to the bathing buffer. During this decrease, the almost instantaneous dissociations of many internal dimers must be followed by gradual and slower reassociations (heat transfer is a slow process; see specialist handbooks). The real proportion of 'internal dissociations' is therefore not 100% (at $t_0 \sim 0$; see p. **213** in Section 6.4). It is smaller, because it is related to the value of $(P\Delta V^*)_{t*}$ given by Equation 6.2 (see the corresponding comment). At this point, let us call $n^*(\theta)$ the mean number of dissociations at temperature θ, during the period $0–t^* = 1$ s. I have tested several analytical and numerical approaches to $n^*(\theta)$, giving consistent conclusions. As a rough approximation, justified a posteriori below, in this and the following sections, I suggest that the number $n^*(\theta)$ of dissociated internal B–B'-type head–head dimers, during a short tetanus and under steady-state conditions (corresponding to the reference virtual duration $0–t^* = 1$ s, defined in Section 6.1 and justified semi-empirically on pp. **210–211** in Section 6.3.5), is given by

$$n^*(\theta) \sim [100 - n_h(\theta)](\theta°C/43.5°C)^{1/4} + n_h(\theta) \quad (6.3)$$

The total number of internal head–head dimers is 100 (see p. **206** in Section 6.3.5). The value of 43.5°C is discussed and justified in the legend to Figure 5.2 (see also Appendix 5.II). Obviously,

$n_{pv}(\theta)$ corresponds to the difference between $n^*(\theta)$ and $n_h(\theta)$ and is given by:

$$n_{pv}(\theta) \sim [100 - n_h(\theta)](\theta°C/43.5°C)^{1/4} \quad (6.4)$$

In Equations 6.3 and 6.4, a fractional exponent with an integer as a denominator was initially chosen. The value of 1/4 for the exponent was selected, using a simple routine, because it gave the most consistent conclusions. Owing to the rather limited increase in $n_h(\theta)$ with increasing temperature θ of the external medium (see pp. **211–212** in Section 6.3.5), it can be deduced from Equation 6.3 that $n^*(\theta)$ increases substantially with temperature, mostly because of the pressure–volume term, as expected from the purely logical reasoning presented in Section 6.4. If the temperature of the surrounding medium is $\theta = 43.5°C$, Equation 6.3 gives $n^*(43.5°C) \sim 100$ (all internal dimers dissociated). At $\theta = 0°C$, we deduce from Equations 6.3 and 6.4 that $n^*(0°C) = n_h(0°C) \sim 1$ (see p. **212** in Section 6.3.5) and $n_{pv}(0°C) \sim 0$, indicating that the lateral swelling part of the hybrid model may be negligible at 0°C, but this conclusion is based on a rough sketch and should not be taken literally (Equations 6.3 and 6.4 are approximations; e.g. replacing 0°C by 0.2°C leads to $n_{pv}(0.2°C) \sim 26$, rather than ~ 0). At $\sim 1°C–3°C$, that is, at about the working temperature for the experiments performed on frog intact fibres by Edman (1979), Edman and Anderson (1968) and Gordon et al. (1966a,b), $n_h(1°C–3°C)$ is also ~ 1 (see p. **212** in Section 6.3.5, in which it is shown that $n_h(0°C) \sim n_h(5°C) \sim 1$) and we deduce that $n^*(1°C–3°C) \sim 40–52$ and $n_{pv}(1°C–3°C) \sim 39–51$. At the temperature of 10°C, used throughout this work, we have $n_h(10°C) \sim 1$ (see p. **212** in Section 6.3.5) and we obtain, from Equations 6.3 and 6.4, $n^*(10°C) \sim 70$ and $n_{pv}(10°C) \sim 69$. At 20°C, we have $n_h(20°C) \sim 3$ (see p. **212** in Section 6.3.5) and $n^*(20°C) \sim 83$ and $n_{pv}(20°C) \sim 80$. Finally, at $\theta = 39°C$ (body temperature of the rabbit), we have $n_h(39°C) \sim 15$ (see p. **212** in Section 6.3.5) and we obtain, from Equations 6.3 and 6.4, $n^*(39°C) \sim 98$ (only ~ 2 internal B–B'-type head–head dimers remain bound) and $n_{pv}(39°C) \sim 83$. Thus, taking into account the mode of generation of the radial tethering forces at rest, $N_0 F_{cb}^0$, (see, in Section 5.1.1, pp. **174–176** and Figure 5.1, its legend and Appendix 5.I), the residual radial tethering force, $N F_{cb}$, studied on pp. **158–159** in Section 4.4.2.5.4, must decrease rapidly with increasing temperature, as a result of the decrease in the number $n^*(\theta)$ of bound internal B–B′-type head–head dimers, and vice versa. However, the relationship between the number of bound internal dimers and the contribution of the net radial repulsive forces, translated into axial contractile forces (see pp. **102–103** in Section 3.10), cannot be predicted. This problem and its consequences are discussed in Section 8.8 (see p. **248**). The calculations presented in this section are valid only for short tetani, under steady-state conditions. The problems raised by very long-lasting tetani and fatigue are considered in Chapter 7 (see Sections 7.3 and 7.4).

6.6 ESTIMATE OF THE RELATIVE CONTRIBUTIONS OF THE TWO TYPES OF ENERGY-PRODUCING PHENOMENA TO THE GENERATION OF APPROXIMATELY 40% (LATERAL SWELLING PROCESSES) OF THE ISOMETRIC TETANIC TENSION (SHORT TETANI–STEADY STATE), AS A FUNCTION OF TEMPERATURE. CASE OF YOUNG ADULT RABBITS

As suggested and strongly supported throughout this monograph, at 10°C, ~40% of the axial contractile force, under isometric tetanic conditions (short tetani, steady state), at around the 'slack length' of the sarcomeres, results from lateral swelling mechanisms. This '40% contraction' is mostly related to the chain reaction occurring within the thick myosin filaments (see p. **217** in Section 6.4 and also the rest of the section for numerical calculations). Indeed, very little of this 40% contraction is directly related to the increase in enthalpy/energy resulting from the cycling cross-bridges, as demonstrated in the sections above (from Sections 6.3.5 to 6.5). This increase in enthalpy/energy, owing to a high level of MgATP splitting resulting from the rapid cyclic attachment to and detachment from the thin actin filaments of the external myosin heads minus Lohmann's reaction, is used principally to initiate the chain reaction (see above), leading to a high proportion

of internal B–B′-type head–head dimer dissociations (see Figure 5.1, its legend and Appendix 5.I concerning the arrangement of the dimers) and to ~40% axial contraction. For rabbit, the increase in enthalpy is required to trigger the dissociation of only ~1 internal dimer at ~1°C–3°C (and also probably at ~0°C; see comments on Equations 6.3 and 6.4 in Section 6.5) and up to ~15 internal dimers at ~39°C (see Section 6.3.5). The chain reaction is then initiated. The rest of this reaction is closely related to the energy $(P\Delta V^*)_{t^*}$ (see Section 6.4, particularly Equation 6.2, for definition of this parameter), providing the major contribution to the dissociation phenomena observed at any temperature. The ratio $\kappa(\theta) = n_{pv}(\theta)/n_h(\theta)$ can be deduced from Equation 6.4, and we obtain

$$\kappa(\theta) = n_{pv}(\theta)/n_h(\theta) \sim [100/n_h(\theta) - 1](\theta/43.5)^{1/4} \tag{6.5}$$

From the values of $n_h(\theta)$ estimated on p. **212** in Section 6.3.5, at 1°C–3°C, we have $n_h(1°C–3°C) \sim 1$ and the ratio given by Equation 6.5 is $\kappa(1°C–3°C) \sim 39–51$. At 10°C, we have $n_h(10°C) \sim 1$ and $\kappa(10°C) \sim 69$. At 20°C, $n_h(20°C) \sim 3$ and $\kappa(20°C) \sim 27$. Finally, at 39°C, $n_h(39°C) \sim 15$ and $\kappa(39°C) \sim 8$. These estimates provide only rough orders of magnitude. However, they appear reasonable in that, at low temperature, the 'pressure–volume effect' within the thick myosin filaments largely prevails over the 'enthalpy effect' outside these filaments, whereas the contribution of the pressure–volume effect gradually decreases with increasing temperature. At 39°C, the pressure–volume effect contributes only slightly more than the enthalpy effect to the axial contractile force, within the large range of uncertainty (a ratio of ~1, or less, is possible). However, as already highlighted in Section 6.5, at both low and high temperatures, it is impossible to determine the relationship between κ and the net radial repulsive (expansive) force and the subsequent 'electrostatically related' axial contractile force.

CHAPTER SEVEN

Fatigue in One Isometrically Contracting Half-Fibre from a Young Adult Frog, and in Intact and Traditionally Demembranated Fibres of Various Origins

DISCUSSION BASED ESSENTIALLY ON THE HYBRID MODEL

7.1 TRADITIONAL VIEWS AND PUBLISHED EXPERIMENTS CONCERNING FATIGUE, FROM SEVERAL INDEPENDENT GROUPS

West et al. (2004) recalled that, in intact frog fibres, the rates of energy release and MgATP hydrolysis are high at the beginning of an isometric contraction, declining thereafter as contraction continues. These declines in energy release and MgATP splitting may be related to the rapid onset of fatigue, as suggested and discussed in Section 3.4.3.2 and quantified in Sections 7.2 and 7.3. In isolated intact single fibres, whole muscles and muscles within the whole body, it is generally thought that an accumulation of lactic acid, resulting in significant acidification of the internal medium and toxicity, causes fatigue (e.g. decline in isometric tetanic tension, demonstrated in this chapter to be mostly coupled with decrease in MgATPase activity with increasing duration of contraction for very long-lasting tetani; see Sections 7.2 and 7.3 for precise details in this area; see also below). As recalled by Dawson et al. (1978), the 'lactic acid hypothesis', valid for muscles within the body, was put forward as early as 1807 by the Swedish chemist Berzelius. This assumption has gradually become a dogma and is still considered irrefutable by non-specialists and even by many specialists. Dawson et al. (1978, 1980), using phosphorus nuclear magnetic resonance (^{31}P-NMR), studied the problem of fatigue on whole skeletal muscles

from frog stimulated electrically and under strict anoxic conditions. They briefly stimulated the muscles and then waited for several seconds: 'six experiments were done, two using each the following patterns of stimulation: 1 s every 20 s, 1 s every 60 s, or 5 s every 500 s; the experiments lasted 18–92 min'. Using these traditional conditions for inducing fatigue, the authors found that fatigue and some of its main features were correlated, to various degrees, with metabolite levels, including increases in Pi (see also Potma and Stienen 1996) and lactic acid, and with a decrease in pH (acidification) from 7.0 to 6.5. Dawson et al. (1978) claimed that, in fatigued muscles, force development is proportional to the rate at which MgATP is hydrolysed, apparently at odds with what is demonstrated on pp. **78–80** in Section 3.4.5 and pp. **83–84** in Section 3.4.6, concerning very loose coupling between MgATPase activity and isometric tetanic tension, in intact fibres and, almost certainly, also in whole muscles. However, very loose coupling is observed only in short tetani and may not be valid in fatigued fibres. Nonetheless, given the approximations introduced into their equations and the wide scattering of their experimental results, I wonder whether the conclusions drawn by Dawson et al. (1978), particularly for the tight (one-to-one) coupling between isometric force and MgATPase activity, are well founded and of general interest.

Godt and Nosek (1989) performed many experiments on chemically skinned fibres from rabbit psoas muscle and skinned cells of papillary muscle and concluded that 'maximal calcium-activated force and calcium sensitivity were markedly decreased in detergent-skinned fibres from skeletal and cardiac muscle by solutions that mimicked the total milieu changes (assumed to exist in intact fibres) associated with fatigue'. The authors did not study the role of lactic acid, but they confirmed that lowering pH decreases isometric tetanic tension. They also pointed out that they simply added together the individual changes observed in media studied separately. They claimed that their data indicated that, with one exception, the decline in tension with time can be accounted for by the additive effects of the changes in individual solutions. One of the weak points of this experimental study is that buffer conditions, assumed to be valid for the sarcoplasmic medium of fatigued intact fibres, were imposed on their demembranated fibres. Caution is therefore required, because the parameters studied are involuntarily assumed to represent all the possible factors responsible for the onset of fatigue. Although many of the experimental results presented by Dawson et al. (1978, 1980) and Godt and Nosek (1989) are very informative, I therefore consider the interpretation of their experimental data to be far from straightforward. I suggest, in this chapter, entirely different approaches that would, however, be merely complementary to more traditional explanations and hypotheses described, for instance, by Allen et al. (1995), Barclay (1992, 1996), Barclay et al. (1993, 1995), Edman and Lou (1990), Edman and Mattiazzi (1981), Fitts (1994), Nassar-Gentina et al. (1978), Westerblad and Allen (1991, 1992), Westerblad and Lännergren (1994) and Westerblad et al. (1991), to account for various experimental data. The 'molecular' interpretations proposed by these authors are essentially based on the traditional concepts (swinging cross-bridge/lever-arm processes). Nonetheless, from all these studies, several features of the physiology of fatigued fibres were quantified: decrease in isometric tetanic tension, decrease in maximal velocity of shortening, decrease in maximal energy output, changes in the force–velocity relationship and so on. Despite these encouraging results, many problems raised by fatigue remain and the general features of fatigue

are revisited in this chapter, mostly on the basis of the hybrid model.

7.2 BEHAVIOUR OF CHEMICALLY SKINNED AND PERMEABILISED FIBRES AND WHOLE MUSCLES DURING ISOMETRIC TETANI OF VARIOUS DURATIONS. PARTICULAR CASE OF ONE HALF-FIBRE FROM A YOUNG ADULT FROG. MIXTURE OF RABBIT AND FROG

In demembranated fibres and half-fibres, accumulation of lactic acid and other metabolites under contraction conditions is unlikely, because the internal compartment is exposed to the surrounding medium, resulting in the dilution of all these metabolites, including lactic acid, owing to a balance between the production of metabolites and their diffusion out of the demembranated fibres (the rapidity of dilution depends on the molecular weight of the chemical species and the diameter of the biological material, via diffusion processes; see pp. **125–127** in Section 4.3.2 and pp. **225–226** in Section 7.3). For a justification of the results obtained with half-fibres, as opposed to traditionally demembranated fibres, let us consider the many differences between half-fibres, in the presence of methanesulphonate (MS^-) as the major anion (see Section 2.2), and traditionally demembranated fibres, in the presence of acetate, chloride or propionate (frequently used by many independent groups). As demonstrated in Section 3.3, half-fibres, in the presence of MS^-, display isometric tetanic tension identical to that recorded in intact fibres during short tetani and under steady-state conditions. Regardless of the animal species, the technique of skinning (chemical or mechanical), the mode of permeabilisation and the composition of the surrounding buffer (e.g. potassium acetate as a neutral salt for Cooke and Pate 1985; potassium propionate for He et al. 1997, 1998a,b, 1999), it is pointed out, on p. **97** in Section 3.8.1, that demembranated fibres systematically develop only approximately half the isometric tetanic tension recorded in intact fibres, during short tetani, under steady-state conditions, except in the 'anomalous' case, in which the tensions recorded are extremely low (see p. **52** in Section 3.4.3.2). The problems of isometric tetanic tension and MgATP hydrolysis are much more complex for long-lasting and very long-lasting tetani than for short tetani, as demonstrated in Section 7.3. However, in this event, there is a gradual and rather large loss of tension in intact frog fibres (e.g. Edman and Mattiazzi 1981) and traditionally demembranated rabbit fibres (e.g. Cooke and Pate 1985), because of the onset of fatigue. In Figure 4.1 (arrow Ca), corresponding to a single contracting half-fibre in the presence of MS⁻ as the major anion (behaving 'mechanically' like an intact fibre during short tetani and under steady-state conditions, as recalled throughout this monograph; see above, for instance), a key problem is determining whether the loss of tension is related to fatigue only or to other phenomena. Andrews et al. (1991) worked on chemically skinned fibres from rabbit psoas muscle in relaxing buffers at a reference total ionic strength of 390 mM, in the presence of chloride as the major anion (see their Table VII). On pp. **147–148** in Section 4.4.2.5.1, the possibility of partial dissolution, within chemically skinned rabbit fibres, of thick myosin filaments at an ionic strength of 290 mM (neutral salt KCl) is raised, although no such phenomenon can be definitively demonstrated. At 390 mM, the phenomenon should be more pronounced than at 290 mM, and partial or total dissociation is possible. Thus, the quantitative data of Andrews et al. (1991) obtained at 390 mM should be interpreted with caution, as they are likely to result in overestimations of the amount of myosin released into the surrounding buffer (especially in the presence of KCl, a highly deleterious neutral salt). Regardless of this problem, the authors found that, at 390 mM, 0.9% of the myosin and 0.1% of the actin were lost in the surrounding buffer after 1 min, that 4.5% and 0.2%, respectively, were lost in the buffer after 10 min and that 10.0% and 0.4%, respectively, were lost in the buffer after 30 min. For practical reasons, these values were not measured in activating solutions. However, losses would probably have been higher under contraction conditions, owing, for example, to the existence of strong net radial repulsive (expansive) forces between the myofilaments, as repeatedly demonstrated in this book. These strong repulsive forces would lead to the 'ejection' of many myofilaments (naturally covered with a large number of fixed negative electrical charges and bound anions) from the skinned fibre toward the external medium.

Despite the qualifications mentioned in the preceding paragraph, concerning the quantitative data of Andrews et al. (1991), their experimental results are consistent with experiments performed in resting and contracting conditions, showing lattice disorders/disruptions in traditionally demembranated fibres, in the presence of Cl⁻ as the major anion or other anions deleterious to various extents (e.g. Asayama et al. 1983; Ford and Surdyk 1978; Goldman and Simmons 1984b; Magid and Reedy 1980). These 'undesirable effects' show that traditionally demembranated fibres, in the presence of Cl⁻ as the major anion, cannot be used for studies of the molecular events that generate fatigue. The replacement of Cl⁻ or other deleterious anions by MS⁻ results in less complex phenomena. Indeed, Andrews et al. (1991) found, in experiments on chemically skinned fibres from rabbit psoas muscle, in resting conditions, in the presence of MS⁻ (at an ionic strength of 390 mM), that no myosin or actin was lost for durations of up to 10 min and that only 0.1% of the myosin and actin was lost for durations of up to 30 min. The MS⁻ anion is certainly of major importance, and a significant part of the protective effect of MS⁻ is probably attributed to this anion being much less able to dissolve the thick myosin filaments than Cl⁻, even at a concentration of 390 mM. However, this is probably not sufficient in itself to account for the 'mechanical' behaviour of half-fibres in the presence of MS⁻ being similar to that of intact fibres. Both the presence of half the sarcolemma and the use of MS⁻ are likely to have strong protective effects. This is clearly confirmed by the similarity of the isometric tensions recorded in half-fibres in the presence of MS⁻ and intact fibres, at least for short tetani, under steady-state conditions, as recalled on p. **220**. Unfortunately, as pointed out in Section 2.2, half-fibres are rarely, if ever, used in studies of isometric tension, velocity of shortening, mechanical performance and enzymatic features, and there are no available experimental data concerning changes in the composition and structure of half-fibres over time, particularly during very long-lasting tetani. Nevertheless, as a first approximation, I suggest that, for the duration of ~2.5 min (see next paragraph concerning this duration), there is no major difference between isometrically contracting half-fibres in the presence of MS⁻ and intact fibres, except that, in fatigued muscles and single intact fibres, the pH of the internal medium is altered and metabolites can accumulate (see Section 7.1), whereas in half-fibres (and all demembranated fibres), H$^+$, MgATP, Pi and various metabolites are exchanged between the internal and external media. Phenomena other than those anticipated here may also contribute to fatigue (see Section 7.4). However, I consider that the half-fibre studied in this section, during a very long-lasting tetanus, displays no major additional 'undesirable phenomena', over and above those for a short tetanus. This hypothesis is justified a posteriori, because the reasoning leads to self-consistent conclusions.

I refer below to the slow decline in active tension after the injection of an 'in vivo-like contracting buffer' (see p. **117** in Section 4.2.2), in the case of the half-fibre (●). The clear decline is shown in Figure 4.1 (a decline of ~20% over a period of ~2.5 min × 60 s − 20 s ~ 130 s) begins a few seconds (~20 s) after the injection of the 'in vivo-like contracting buffer' (see arrow Ca in Figure 4.1). We need to estimate the quantity of MgATP hydrolysed during the ~150 s period, starting from injection of the buffer. Note that the decline intension is reasonably assumed not to be related to slow changes in structure within the half-fibre for the reasons given in the preceding paragraph. It is instead considered to be solely due to the onset of fatigue. On pp. **195–198** in Section 6.2, it is demonstrated that, for passing from the energy released from the actin-myosin interactions to the enthalpy/energy available in the sarcoplasmic medium, the corrective factor corresponding to the many non-myofibrillar ATPases (present in whole muscles, intact fibres, half-fibres) can be neglected (the 'universal' corrective factor is $\xi_{0,\infty}$ ~1.00). Thus, as a first approximation, we can focus solely on the MgATPase activity due to actin-myosin interactions. This is justified by the large experimental errors and shortcomings in the reasoning. As highlighted on p. **211** in Section 6.3.5, in permeabilised frog fibres at 5°C, the mean steady-state value of the 'isometric' MgATPase activity between 0 and t* = 1 s is ~13.2 ± 3.4 s^{-1} (mean ± SD; n = 7) i.e. ~23.2 ± 6.0 s^{-1} at 10°C (Q_{10} ~ 3.1; see p. **202** in Section 5.3.3; the value of Q_{10} is assumed to be identical in a whole muscle and a demembranated fibre). In isometrically contracting whole frog muscles at 0°C, using chemical techniques analysed by Morel (1984b), Curtin and Woledge (1979) measured chemical changes over a period of 0–1 s from which I deduce here that the mean

rate of MgATP splitting is ~4.7 ± 0.9 s^{-1} (mean ± SD; n = 14) corresponding to ~14.6 ± 2.8 s^{-1} at 10°C (Q_{10} ~ 3.1; see above). This value is statistically similar to ~23.2 ± 6.0 s^{-1} (common interval of ~17.2–17.4 s^{-1}; the value of ~17.3 s^{-1} is selected below, in this section). Another value of the rate of MgATP breakdown was obtained by Curtin and Woledge (1979): ~1.8 ± 0.3 s^{-1} again for the isometric tetanic contraction of whole frog muscles at 0°C, but for a period of 0–15 s after the beginning of contraction.* This value is significantly lower than ~4.7 ± 0.9 s^{-1} for a tetanus lasting only 1 s, also for whole frog muscles at 0°C (see above). The value of ~1.8 s^{-1}, at 0°C, becomes ~5.6 s^{-1} at 10°C (Q_{10} ~ 3.1; see above), about three times lower than the value of ~17.3 s^{-1} (see above). This rapid decrease between 0–1 and 0–15 s.

The 'universal' density of cross-bridges is m ~ 1.16 × 10^{17} cm^{-3} (see p. **206** in Section 5.3.5), and the volume of the half-fibre used here (●) was not measured but is taken as ~1.70 × 10^{-5} cm^3, which is the most probable value (~1.70 ± 0.38 × 10^{-5} cm^3; see p. **13** in Section 1.2). There are, therefore, ~1.16 × 10^{17} cm^{-3} × 1.70 × 10^{-5} cm^3 ~ 1.97 × 10^{12} cross-bridges in the half-fibre. Thus, the number of MgATP molecule split during 15 s, at 0°C, is ~15 s × 1.8 s^{-1} × 1.97 × 10^{12} ~ 5.3 × 10^{13}. The volume of the half-fibre being ~1.70 × 10^{-5} cm^3 ~ 1.70 × 10^{-8} l, this number of ~5.3 × 10^{13} corresponds to a 'virtual depletion' of MgATP of ~5.3 × 10^{13}/6.022 × 10^{23}/1.70 × 10^{-8} ~ 0.52 × 10^{-2} M ~ 5.2 mM (6.022 × 10^{23} is Avogadro's number). This estimate does not correspond to accumulation of MgATP within the half-fibre used here as a model, because of the numerous exchanges between the internal and external media, as recalled on p. **220** and studied on pp. **225–226** in Section 6.3. By contrast, this estimate is valid in a whole muscle and an intact fibre and corresponds to a maximal depletion, in the absence of the Lohmann reaction. This value is taken into account on p. **149** in Section 3.4.2.5.2 to discuss the experimental results obtained by Matsubara et al. (1975).

---

* Still for whole frog muscle under tetanic conditions, at 0°C, for similar durations (>1 s; more precisely ~2 − 5 s), Curtin et al. (1974) and Homsher (1984) reported similar values of the rate of MgATP splitting (~1.0 − 1.5 s^{-1} vs. ~1.8 s^{-1} obtained by Curtin and Woledge (1979) for a duration of 15 s).

Morel

7.3 EXPERIMENTAL DATA OBTAINED BY SEVERAL INDEPENDENT GROUPS, CONCERNING FATIGUED CHEMICALLY SKINNED AND PERMEABILISED FIBRES. SPECIFIC PROBLEMS RAISED BY VERY LONG-LASTING TETANI. CASES OF RABBIT AND FROG, FOCUSING ON ONE HALF-FIBRE FROM A YOUNG ADULT FROG IN PARTICULAR. INTERPRETATION OF THE RESULTS

As a preliminary remark, it should be pointed out that mixing data for frog and rabbit is inevitable, because experimental results are often available for only one or other of these animal species. This is highlighted systematically throughout this book, but general conclusions can nevertheless be reached.

To the best of my knowledge, nothing is known about the MgATPase activity during very long-lasting tetani (durations >15 s; see p. **222** in Section 7.2 concerning the maximal duration of 15 s), such as that presented in Figure 4.1, the duration of which is ~2.5 min. On p. **222** in Section 7.2, it is highlighted that a clear ~20% decline in tension occurs during ~130 s. It is difficult to predict quantitatively the decrease in MgATPase activity beyond 15 s. However, on p. **222** in Section 7.2, it is shown that, for frog at 10°C, the rate of MgATP splitting under isometric tetanic conditions, over the period 0–t* = 1 s, is $\Pi^*_{isom,\infty}$(0–1 s) ~17.3 s^{-1}, whereas, over the period 0–15 s, we have $\Pi^*_{isom,\infty}$(0–15 s) ~5.6 s^{-1}. Owing to this rapid decrease, it is useful to write the instantaneous value of the rate of MgATP splitting $\Pi^*_{isom,\infty}$(t) ~ $\Pi^*_{isom,\infty}$(0)exp(− A**t), where t is a priori different from the duration 0–t of the tetanus and where A** has nothing to do with A* in Equations 3.29 to 3.32. For simplification, and owing to the high degree of uncertainty on the various values, t is taken here as identical to the duration 0–t and we deduce that A** ~0.081 s^{-1} and that $\Pi^*_{isom,\infty}$(0) ~18.6 s^{-1}. For t = 20 s, 60 s, 90 s, we obtain $\Pi^*_{isom,\infty}$(t) ~3.72 s^{-1}, ~0.150 s^{-1}, ~0.012 s^{-1}, respectively. For the maximal duration t ~2.5 min ~150 s, we obtain $\Pi^*_{isom,\infty}$(150 s) ~5.2 × 10^{-6} s^{-1}. The values of ~0.012 s^{-1} and 5.2 × 10^{-6} s^{-1} are considerably underestimated, resulting from the many experimental errors and uncertainties concerning MgATPase activity values for tetani lasting 1 s and 15 s, the oversimplifications and shortcomings in the reasoning and the choice of

a simple exponential decrease suggested here. Nonetheless, it is demonstrated experimentally that the rate of MgATP splitting decreases significantly with increasing tetanus duration (see above concerning the marked difference between durations of 0–1 s and 0–15 s). In any event, the rate of MgATP breakdown, during contraction, cannot be lower than the resting MgATPase activity. As far as I know, there are no available experimental data concerning resting MgATPase activity in demembranated frog fibres. By contrast, such data exist for permeabilised fibres from rabbit psoas muscle: on p. **200** in Section 6.3.2, it is recalled that the rate of MgATP splitting at rest is ~0.020 s^{-1} at 5°C and ~0.040 s^{-1} at 12°C. A simple barycentric mean gives ~0.034 s^{-1} at 10°C. On p. **58** in Section 3.4.3.2, it is demonstrated that a factor of ~2.5 should be used to pass from the MgATPase activity of the rabbit to that of the frog. The resting MgATPase activity, at 10°C, for frog, would therefore be ~2.5 × 0.034 s^{-1} ~0.085 s^{-1}. As an illustrative example, I suggest the same value of MgATPase activity in a contracting frog fibre, at 90 s and 130 s, only 1.5 times the resting activity: ~1.5 × 0.085 s^{-1} ~ 0.128 s^{-1}. As a rough approximation, let us assume that the mean rate of MgATP splitting between 0 s and ~150 s is ~ [$\Pi^*_{isom,\infty}$(0) $\Pi^*_{isom,\infty}$ (1 s) + $\Pi^*_{isom,\infty}$ (15 s) + $\Pi^*_{isom,\infty}$ (20 s) + $\Pi^*_{isom,\infty}$ (60 s) + $\Pi^*_{isom,\infty}$ (90 s) + $\Pi^*_{isom,\infty}$ (150 s)]/6 ~6.5 s^{-1} (see above for the values of the rate of MgATP splitting). Thus, for the total duration of ~150 s, the number of MgATP molecules split is ~1.97 × 10^{12} × 150 s × 6.5 s^{-1} ~1.85 × 10^{15} (1.97 × 10^{12} is the number of cross-bridges in the half-fibre used here; see p. **222** in Section 7.2). The initial concentration of MgATP was 0.4 mM in the 0.4 ml trough (see p. **14** in Section 2.2, and **118** in Section 4.2.2), corresponding to 4 × 10^{-4} M × 4 × 10^{-4} 1 × 6.022 10^{23} ~9.64 10^{16} molecules of MgATP (6.022 10^{23} is Avogadro's number). Thus, the maximal relative quantity of MgATP 'pumped' from the trough by the half-fibre is ~1.85 × 10^{15}/9.64 × (~ 2%). At the beginning of contraction, the MgATP concentration in the trough is 0.4 mM = 400 μM (see above) and the maximal decrease, over ~2.5 min, would be only ~400 μM × 1.92 × 10^{-2} ~7.68 μM. Thus, the maximal concentration of MgATP in the trough does not markedly decrease during ~2.5 min. However, MgATP hydrolysis releases H$^+$ within the half-fibre and the maximal possible increase in the concentration of H$^+$ in the trough is ~7.68 μM (owing to the very rapid diffusion of H$^+$

in aqueous media and, almost certainly, within the half-fibre, no H^+ accumulation within the half-fibre is possible and H^+ is almost instantaneously present in the trough, at identical concentrations in the internal and external media; see specialist handbooks on physical chemistry concerning H^+ diffusion). The value of ~7.68 μM for the concentration of H^+ corresponds to a minimal pH of ~5.1 in a 'hypothetical' unbuffered medium surrounding this half-fibre. Concerning Figure 4.1 analysed here, the bathing solution surrounding the half-fibre (●) contains 5 mM PIPES (see p. **118** in Section 4.2.2) and is therefore poorly buffered and a detectable decrease in pH would be possible. Ignoring all the other chemical compounds in the buffer, except the KOH used to neutralise PIPES (see pp. **11** and **14** in Section 2.2), the pK for the formation of the PIPES-OH complex, at 10°C, is ~$10^{6.9}$ M^{-1} (see Table 2 in the paper by Horiuti 1986), whereas we have $[H^+][OH^-] = K_w$ ~10^{-14} M^2, regardless of temperature, within the range 0–25°C. The concentration of PIPES being 5 mM and, using a simple routine, it can be concluded that the maximal increase of ~7.68 μM in H^+ concentration results in a negligible decrease in pH (less than ~0.01). Thus, there is no detectable acidification in the surrounding buffer or within the half-fibre: the decline in tension is not related to acidification (the possible role of acidification in fatigue is discussed in Section 7.1).

Let us now consider the other two metabolites produced during contraction, MgADP and Pi, under experimental conditions different from those studied in the half-fibre (●). Working at 10°C, Cooke and Pate (1985) found that, in tetani of ill-defined duration (probably ~1 s), the addition of MgADP and Pi to the bathing buffer had complex effects on the isometric tension and shortening velocities of glycerinated fibres from rabbit psoas muscle. To my mind, the authors cannot easily explain these intricate effects on the basis of swinging cross-bridge/lever-arm mechanisms (one-to-one coupling between the rate of MgATP hydrolysis and isometric tension, for short tetani under steady-state conditions, and also, presumably, in fatigued muscles, as recalled in Section 7.1). Moreover, a large part of the problem is probably related to the fact that determination of the 'true' MgATPase activity, even under isometric conditions, is a complex undertaking, as pointed out throughout Section 3.4.3.2. The analysis and conclusions of Cooke and Pate

(1985), based on swinging cross-bridge/lever-arm concepts with dubious values for MgATPase activities, are therefore questionable. Regarding the problem of fatigue, the authors finished their abstract by writing: 'the results obtained here suggest that levels of MgADP in fatigued fibers play no role in these decreases in function [tension and maximal velocity of shortening], but the elevation of both phosphate and H^+ is sufficient to account for much of the decrease in tension.' Despite this apparent clarification and owing to some inconsistencies in the results and interpretations identified by the authors themselves, the last sentence of the discussion in their paper is: 'additional factors in the process of fatigue must be involved.' These two statements are essentially contradictory. The aim of this and the following paragraphs is to try to identify some of the 'mysterious' additional factors, alluded to by Cooke and Pate (1985) to account for fatigue. For this purpose, I first use the experimental results obtained by He et al. (1997) with the same kind of biological material as used by Cooke and Pate (1985), but permeabilised by a special chemical treatment, rather than with glycerine (see Millman 1998 for a critical analysis of the various techniques of chemical demembranation, including glycerination). In swinging cross-bridge/lever-arm theories, there is tight coupling (one-to-one; see above) between MgATPase activity and isometric tension, and the parameters affecting the rate of MgATP splitting must also directly affect isometric tension, through a proportionality relationship (see Sections 3.4.4.3 and 3.4.6 for detailed discussions). However, Cooke and Pate (1985) showed that increasing the [MgADP]/[MgATP] ratio resulted in an increase in isometric tension, which is unexpected, because MgADP is a competitive inhibitor of MgATP at the two myosin enzymatic sites. By contrast, He et al. (1997), working, at 20°C, on only two permeabilised rabbit fibres, apparently obtained the opposite phenomenon for MgATPase activity. More precisely, He et al. (1997) found that, for short tetani lasting up to ~0.4–0.8 s, the presence of 10 mM Pi (first fibre) had no significant effect on MgATPase activity and that 0.5 mM MgADP (second fibre) decreased the rate of MgATP splitting by ~23%. According to Table 1 in the paper by He et al. (1997), the three SEs corresponding to the first three turnovers range between ~6% and ~12% of the mean values. For the comparison of only two experiments with a series of experiments,

Morel

the SE on this series is misleading and it is more appropriate to use the SD (see Section 2.1). For this series of experimental results, we obtain SD values ranging between ~27% and ~56% of the mean values. It can therefore be inferred that the value of ~23% given above is not statistically significant and that the discussion presented by He et al. (1997), based on inhibition by MgADP, is questionable. This is expected, because the concentration of MgATP is 5 mM, whereas that of MgADP is only 0.5 mM, certainly too low a concentration to compete with MgATP, which is well known to bind much more strongly than MgADP to the enzymatic and dimerisation sites (Morel et al. 1998b; see the 'supporting information available'). Thus, the experimental results and discussions presented by Cooke and Pate (1985) and by He et al. (1997) are apparently inconsistent, maybe because these two groups worked at different temperatures, with different biological materials and with very different techniques. These two series of results cannot therefore be used as starting points for improving our knowledge of fatigue. I present, below, in this section, another approach to the putative roles of the MgADP and Pi produced by MgATPase hydrolysis, particularly in the half-fibre (●) referred to in this section. I also present other aspects of the decrease in MgATPase activity for very long-lasting tetani (on p. **223** preliminary calculations of the decrease in this activity and subsequent conclusions are presented). Finally, the gradual onset of fatigue is related to the gradual decrease in MgATP breakdown, mostly via the hybrid model (see below, in this section, particularly pp. **225–227**).

Concerning the half-fibre (●), on p. **223** we found that, from time zero to ~150 s, the maximal number of MgATP molecules split is ~1.85 × 10^{-5}, corresponding to a half-fibre volume of ~1.70 × 10^{-5} cm^3 ~1.70 × 10^{-8} l (see p. **222** in Section 7.2) and to a maximal concentration of ~1.85 × 10^{15}/6.022 10^{23}/1.70 × 10^{-8} ~18.1 × 10^{-2} M ~181 mM of MgATP split and also a maximal production of ~181 mM of MgADP and Pi (6.022 10^{23} is Avogadro's number). There would thus be a maximum of ~181 mM/150 s ~1.207 mM s^{-1} of MgATP split and MgADP and Pi produced. Obviously, there is no massive accumulation of MgADP and Pi within the half-fibre, due to diffusion out into the bathing buffer. Thus, the actual concentration results from the difference between production and outflow. To try to estimate this

actual concentration of MgADP and Pi within the half-fibre, we need to take into account the experimental works of He et al. (1997) and Martson (1973). Working on glycerol-extracted rabbit fibres, Martson (1973) found that the fluxes out of an isometric rabbit skeletal muscle fibre are 0.08 s^{-1} at 1.5°C, for the range of metabolites of molecular mass < 1 kDa. In their work on permeabilised rabbit fibres, He et al. (1997) used this value of 0.08 s^{-1} for NADH. This value, valid for any metabolite of molecular mass < 1 kDa, includes therefore MgADP and Pi. I suggest writing, as a first approximation, Met(t) ~ Met(0)exp(−0.08t), where Met(0) is the concentration of a metabolite (or NADH) at time zero and Met(t) its concentration at time t. He et al. (1997) presented, in their Figure 4, a time course of NADH diffusion out of a permeabilised rabbit fibre, at 20°C, and I deduce, from their curve, corresponding to a fibre held at constant length, Met(t) ~ Met(0)exp(−0.06t), whereas this relationship becomes Met(t) ~ Met(0)exp(−0.25t), when this fibre is mechanically oscillated around the reference length. I believe that this large increase in NADH diffusion out of the fibre, due to mechanical oscillations, results from both a large decrease in the thickness of the boundary layer of water enveloping the fibre and the occurrence of turbulent streaming in the bathing buffer around the fibre (see handbooks on fluid mechanics), these two phenomena interacting and inducing a high rate of NADH dissolution in the surrounding buffer. Regardless of temperature (1.5°C or 20°C) and the chemical compound used (MgADP, Pi, NADH), in the absence of definitive explanations for the kinetics of matter transfer around a fibre surface, I choose an average value, valid for rabbit, of ~ (0.08 + 0.06 + 0.25) s^{-1}/3 ~ 0.13 ± (0.05 or 0.09) s^{-1} (mean ± SE or SD; n = 3).

To the best of my knowledge, there have been no studies of the aspect described in the preceding paragraph, in demembranated frog fibres. For frog half-fibres, at 10°C, the problem may even be more complex, as half the sarcolemma is present, potentially either slowing down or accelerating the outflow of the metabolites present within the half-fibres, with respect to entirely demembranated fibres. Moreover, in Section 3.4.3.2, it is repeatedly shown that entirely demembranated fibres from frog and rabbit do not behave similarly. As it is impossible to predict the value of the time constant for a frog half-fibre, I take the mean time constant calculated at the end of the preceding paragraph for rabbit, i.e. ~0.13 s^{-1}, giving the following approximate relationship: Met(t) ~ Met(0)exp(−0.13t). This choice is justified, particularly as the values of SE and SD on this mean value of ~0.13 s^{-1} are very large for rabbit (± 38% for SE or ± 69% for SD; see last line of the preceding paragraph). Calling Met_0(t) the 'virtual' concentration of MgADP or Pi as a function of time t, regardless of the outflow of these metabolites, we should first study the increase in Met_0(t). Owing to the many uncertainties, we can ignore the possible exponential variation of Met_0(t), suggested by He et al. (1997) for short tetani and we can therefore propose a 'maximal' linear increase in Met_0(t) ~1.207 mM s^{-1} t (see the preceding paragraph regarding the value of ~1.207 mM s^{-1}). We now need to take into account the outflow of MgADP and Pi into the bathing buffer. For this purpose, we can write, as a rough approximation, Met(t) ~ Met_0(t) × t × exp(−t/$\tau^{\bullet}$) ~1.207 mM s^{-1} × t × exp(−t/$\tau^{\bullet}$), where the time constant 1/$\tau^{\bullet}$ has the value of ~0.13 s^{-1} ($\tau^{\bullet}$ ~7.7 s), recalled above. The maximal value of Met(t), within the half-fibre (●) is obtained for t = $\tau^{\bullet}$ ~7.7 s and is ~3.92 mM. This value may be too high (~8–9 times the concentration of MgATP, which is 0.4 mM; see p. **223**). The affinities of MgADP and Pi_i for the enzymatic sites on the myosin heads are much lower than that of MgATP, but this maximal ratio of ~8–9 may give a slightly lower than expected rate of MgATP splitting (competitive inhibition by MgADP and Pi). Regardless of this possible 'inhibition-induced' slowing, a large decrease in the rate of MgATP cleavage beyond ~7.7 s would be observed. For example, at t = 20 s (clear beginning of the decrease in tension; see p. **222** in Section 7.2), Met(20) would be only ~1780 μM. As the affinity of MgADP and Pi is much lower than that of MgATP (here 0.4 mM = 400 μM; see above), marked inhibition by ~1780 μM MgADP and Pi is probably not possible. Finally, it can be inferred from this short discussion that a limited decrease in the rate of MgATP splitting resulting from detectable amounts of MgADP and Pi (up to ~3.42 mM; see above) within the half-fibre would be observed at ~7.7 s. Nevertheless, this putative MgADP and Pi concentrations of ~3.42 mM at ~7.7 s cannot be demonstrated experimentally and is merely related to many mathematical approximations introduced here,

together with complex phenomena occurring during the first ~20 s (before the decline in tension; see above). At ~20 s (and beyond) there is no possible inhibition by MgADP and Pi, owing to their very low concentration within the half-fibre (~1780 μM and less; see above) and the clear decline of ~20% in isometric tetanic tension between ~20 s and ~150 s (see p. **222** in Section 7.2) is certainly not related to such a phenomenon of inhibition by MgADP and Pi. Other processes therefore control the ~20% decline as described below, in this section.

Let us now briefly consider rabbit fibres, in the context of the hybrid model. At temperatures between ~5°C and ~39°C, it is recalled in Section 6.3.5, that, for a short tetanus, there is a rapid increase in enthalpy, $(\Delta H)_{t^*}$, outside the thick myosin filament cores (on pp. **210–211** in Section 6.3.5, a circumstantial discussion of the reference virtual time t* = 1 s is presented). At 10°C, the small increase in temperature, resulting from $(\Delta H)_{t^*}$ ~92 mJ cm^{-3} (see p. **209** in Section 6.3.5), is transferred into the thick myosin filament cores, leading to the almost instantaneous dissociation of internal myosin B–B′-type head–head dimers (see Figure 5.1, its legend, and Appendix I, regarding the arrangement of these dimers) the number of which is n_h ~1 (see p. **212** in Section 6.3.5). Once one dimer dissociates, a chain reaction is initiated (see p. **213** in Section 6.4). Indeed, supplementary energy, related to the term $(P\Delta V^*)_{t^*}$ in Equation 6.1, is released from internal head–head dimer dissociations (see pp. **214–216** in Section 6.4 for a detailed discussion). This supplementary increase in energy occurs within the core of the thick myosin filaments and the subsequent increase in temperature induces the dissociation of many other internal dimers (see Sections 6.4 and 6.5). A value of n*(10°C) ~70 dimer dissociations, for a maximum of ~100, is strongly suggested for rabbit (see text below Equation 6.4 regarding the number of ~70 internal dimers, and p. **208** in Section 6.3.5 and p. **217** in Section 6.5, regarding the number of ~100 internal dimers).

Returning to the frog, experimental data are available for both energetic and mechanical aspects. In Figure 4.1 the isometric tetanic tension was recorded for a very long-lasting tetanus on the half-fibre (●), in the presence of MS$^-$ as the major anion (see Section 2.2 and the experimental part of this work, concerning half-fibres 'bathed' in MS$^-$). The Ca trace in Figure 4.1 represents only

Morel

the electrostatically related trace, but it retains all the characteristics of the whole trace actually recorded (it represents ~40% of the recorded force; this proportion is systematically used in this monograph). The tetanus lasted ~2.5 min ~150 s (see Figure 4.1). Having reached its maximal value, within a few seconds of the injection of the calcium-containing buffer and clearly beginning to decrease at ~20 s (see p. **222** in Section 7.2), relative isometric tension then decreased steadily by ~20% over ~150 s − 20 s ~130 s, corresponding to a relative decrease of ~1.54 × 10^{-3} s^{-1}. On p. **223**, the rate of MgATP splitting is estimated to be ~0.128 s^{-1} at ~150 s. This value corresponds to the mean dimensionless values for enthalpy release $(\Delta H)_{t*}(150)$, at ~150 s, of ~1.16 × 10^{17} cm^{-3} × 0.128 s^{-1} × 6.6 × 10^{-17} mJ per MgATP molecule × 1 s ~0.98 mJ cm^{-3} for the virtual duration $t* = 1$ s (1.16 × 10^{17} cm^{-3} is the 'universal' density m of cross-bridges defined on p. **206** in Section 6.3.5; 6.6 × 10^{-17} mJ is e_{app} defined and calculated in Section 6.2 (see particularly pp. **193–194**); the definition of the virtual duration $0–t* = 1$ s is discussed on pp. **210–211** in Section 6.3.5).

Let us consider the rabbit again. On p. **58** in Section 3.4.3.2, it is shown that a multiplication factor of ~1/2.5 ~0.4 should be used to convert MgATPase activities for frog into values for rabbit, at ~10°C, for short tetani, under steady-state conditions. Owing to the many approximations made above, in this section, I maintain this factor of ~0.4 for long-lasting and very long-lasting tetani (at 10°C). The rate of MgATP splitting of ~0.128 s^{-1} at ~150 s, recalled in the preceding paragraph for frog, therefore becomes ~0.128 s^{-1} × 0.4 ~ 0.051 s^{-1} for rabbit. This low value is comparable to, but sufficiently higher than, the 'resting' rates of ~0.020 s^{-1} at 5°C and ~0.040 s^{-1} at 12°C (i.e. ~0.034 s^{-1} at 10°C; see p. **200** in Section 6.3.2 for precise details), giving a self-consistent approach. Again for rabbit, we have m ~1.16 × 10^{17} cm^{-3} and e_{app} ~6.6 × 10^{-17} mJ (see the end of the preceding paragraph) and we deduce that $(\Delta H)_{t*}(150\ s)$ is approximately ~1.16 × 10^{17} cm^{-3} × 0.051 s^{-1} × 6.6 × 10^{-17} mJ × 1 s ~0.390 mJ cm^{-3}. As recalled on p. **226**, we have $(\Delta H)_{t*}$ ~92 mJ cm^{-3}, giving n_h ~1 enthalpy-related internal dimer dissociations under steady-state conditions corresponding to a short tetanus. As also recalled on p. **226**, this is sufficient to 'trigger' pressure-volume-related internal dimer dissociations (chain reaction), resulting in ~70 dissociations for short tetani. The number of

dissociations $n_h(150\ s)$, at ~150 s, would be only ~1 × 0.390 mJ cm^{-3}/92 mJ cm^{-3} ~0.092 << 1, for rabbit at 10°C. Thus, $n_h(130)$ is necessarily zero and, according to the hybrid model, analysed quantitatively in Chapter 6, 'pressure-volume-related' internal dimer dissociation cannot be triggered. In other words, this process is entirely blocked and the lateral swelling part of the axial contractile force disappears, at least in rabbit psoas muscle.

We now need to return to the frog, because Figure 4.1 concerns this species. For frog, in the same conditions as for rabbit (e.g. 10°C), the number of internal dimer dissociations would be $n_h^* \times 0.98$ mJ cm^{-3} × 1 s / 230 mJ cm^{-3} ~ $0.0042 n_h^*$ where n_h^* is the number of enthalpy-related dissociations for a short tetanus under steady-state conditions (calculated in the first column on this page the value of ~230 mJ cm^{-3} is estimated on p. **209** in Section 6.3.5). Let us assume that n_h^* is similar to n_h ~1 (for rabbit; see the preceding paragraph), therefore giving a number of dissociations of ~1 × 0.042 ~0.042 << 1, for frog at ~130 s. Even taking the very unlikely value of $n_h^* \sim 20$ would give a number of dissociations of ~20 × 0.042 ~0.084, which remains negligible. Thus, here again, the number of dissociations at ~150 s is zero. For both frog and rabbit, at 10°C, the lateral swelling part of the hybrid model is therefore totally inoperative (blocked) and only the swinging cross-bridge/lever-arm mechanisms can, in principle, work, but almost certainly less efficiently than during the first few seconds of contraction (see Section 7.4 for precise details). This blocking builds up gradually, almost certainly beyond ~20 s (see the first column on this page), but it is impossible to predict the kinetics of this process. This oversimplified approach gives a straightforward molecular explanation for the slow decline in active tension observed in Figure 4.1, i.e. for the onset of fatigue, at 10°C, regardless of the contribution to fatigue of the swinging cross-bridge/lever-arm mechanisms, which account for ~60% of contraction in the hybrid model (this contribution of ~60% is valid for short tetani, but would be different for very long-lasting tetani). It would have been interesting to study the problem of fatigue onset for humans at 37°C on the basis of the reasoning presented here, but, to the best of my knowledge, no quantitative data under these conditions are available. However, the phenomena described in this and the preceding paragraphs for frog and rabbit, at 10°C, including the gradual blocking of the lateral swelling mechanisms, almost certainly occur in humans at 37°C.

7.4 OTHER PROCESSES THAT MAY CONTRIBUTE TO FATIGUE

The phenomena described in Sections 7.1 through 7.3 probably account for much of the onset of fatigue and answer the many questions recalled above, in this chapter, and posed by Cooke and Pate (1985) who asserted that: '*additional factors in the process of fatigue must be involved*'. However, it was also suggested, on the basis of a phenomenological/semi-empirical theory independent of the mode of force generation (see Section 3.4 for the updated approach); that both a possible decrease in the activity of the sarcoplasmic reticulum and a decrease in the proportion of attached cross-bridges probably account for another noticeable part of fatigue (see Morel 1984b, for precise details on these two points).

Concerning fatigue of the sarcoplasmic reticulum, West et al. (2004) noted that '*from earlier experiments in which the actomyosin ATPase* has been varied by changing the filament overlap, it has been shown that the rate of energy turnover due to Ca^{2+} pumping does not decrease during a contraction....' This remark concerns tetani of short or moderate durations and the experimental observation, recalled by West et al. (2004), seem to demonstrate that the sarcoplasmic reticulum displays no signs of fatigue. Indeed, as recalled at the start of Section 7.1, fatigue occurs probably rapidly for the myofibrillar MgATPase activity, whereas the MgATPase activity of the sarcoplasmic reticulum shows no signs of fatigue. However, it should be stressed that decreasing the myofilament overlap leads to complex phenomena, as demonstrated on pp. **251–254** in Section 8.9, and conclusions drawn too rapidly are therefore likely to be questionable.

In their discussion, Edman and Lou (1990) suggested that 'the decrease in contractile strength after frequent tetanizations ... is attributable to altered kinetics of cross-bridges function leading to reduced number of active cross-bridges and, most significantly, to reduced force output of the individual bridge'. In their discussion, He et al. (1997) noted that the number of attached cross-bridges probably decreased with increasing duration of the isometric tetanus. These authors stated: 'how this change comes about is unclear'. Edman and Lou (1990) based their discussion on the swinging cross-bridge/lever-arm theories. Obviously, their suggestions are valid for the traditional part of the hybrid model. Nonetheless, Morel (1984b) found a decrease in the number of attached cross-bridges, independently of the mode of generation of the contractile force (see first paragraph).

It is impossible to anticipate the kinetics of all the phenomena mentioned and commented on in this chapter, but these processes may occur early in the tetanus. Indeed, it is shown on p. **222** in Section 7.2 that, at 10°C, for frog, the rate of MgATP breakdown for a tetanus duration of 1 s is ~17.3 s^{-1}, whereas that for a 15 s tetanus is only ~5.6 s^{-1}. There is therefore a rapid decrease in MgATPase activity, even in the first seconds of the tetanus. However, this decrease in the rate of MgATP splitting, by a factor of ~3.1, between 1 s and 15 s (i.e. during 14 s), probably results in a decrease in isometric tetanic tension by a factor of only ~1.24 (the tension at 15 s is ~1/1.24 ~0.81 times that at 1 s). This calculation is based on the phenomenological/semi-empirical and experimental results obtained on pp. **78–80** in Section 3.4.5 and pp. **83–84** in Section 3.4.6, regarding the very loose coupling between MgATPase activity and isometric tension for short tetani, under steady-state conditions (the isometric tetanic tension is proportional to the MgATPase activity to the power of ~0.190; see p. **78** in Section 3.4.5), but not necessarily for long-lasting and very long-lasting tetani. I suggest firstly extending this power of ~0.190 to any duration. The relative loss of ~1 − 0.81 ~0.19 in tension, over 14 s, corresponds to a relative decline of ~12.1 × 10^{-3} s^{-1}, i.e. ~8–9 times greater than the experimental value of ~1.54 × 10^{-3} s^{-1} over ~150 s recalled on p. **227** in Section 6.3. However, the orders of magnitude of these two series of values are sufficiently consistent, because the theoretical values calculated here result from oversimplifications, according to which long-lasting and short tetani obey the same biochemical and biophysical laws (see above concerning the power of ~0.190) and also because the error bars on the rates of MgATP breakdown are probably very large (see Morel 1984b, for a critical analysis of the chemical techniques used to determine the two rates of ~17.3 s^{-1} and ~5.6 s^{-1} recalled above). For example, taking a ratio of ~3.1/2 ~1.55 rather than ~3.1 (see above, concerning the decrease in MgATPase activity between 1 s and 15 s) and replacing the exponent of ~0.190 (see above) by ~0.190/3 ~0.063, the decline in tension during 14 s would be only ~2.19 × 10^{-3} s^{-1} (detailed calculations not shown), sufficiently close to ~1.54×10^{-3} s^{-1}. Thus, I maintain that decreases in MgATPase activity and isometric tension attributable to fatigue may occur early in tetanus, as already suggested in Section 3.4.3.2.

Bianchi and Narayau (1982) found experimentally that the transverse tubules played a role in muscular fatigue in single isolated fibres or whole muscles, with a possible decrease in the efficiency of excitation-contraction coupling. Moreover, the steric-blocking process, recalled and described by Gordon et al. (2000) and related to the troponin-tropomyosin system in the thin actin filaments, may also lead to a gradual blocking of the myosin-binding sites on actin, therefore contributing to fatigue. In Section 4.4.2.3, many rapid conformational changes in muscle proteins (mostly myosin and actin) are described for short tetani. Slow, gradual, reversible, molecular changes may also occur for long-lasting and very long-lasting tetani, with further possible implications for the onset of fatigue and, possibly, reversible changes in the biochemical and biophysical features of the fibres. Studying various types of skeletal muscle, with various creatine kinase contents, Dahlstedt et al. (2000) suggested that a muscle from mouse deficient in this enzyme fatigues more rapidly than a 'normal' muscle, because creatine kinase is involved in Lohmann's reaction and part of MgADP is not rephosphorylated into MgATP and the excess of MgADP would inhibit MgATP binding to myosin heads competitively, thus decreasing MgATP splitting and resulting in less energy being available for contraction.

We must add to the molecular contribution of the lateral swelling part of the hybrid model to fatigue, studied in this chapter, another contribution, related to the swinging cross-bridge/lever-arm part of the hybrid model. In a model-independent phenomenological/semi-empirical approach, Morel (1984b; see also the first paragraph of this section; this model is updated in Section 3.4) studied the case of successive tetani, which is systematically used for experimental studies of muscle fatigue (see Section 7.1). I demonstrated that there should be a gradual decrease in the proportion of attached cross-bridges and a subsequent decrease in tension.

7.5 CONCLUSION

There are many arguments in favour of molecular events accounting for the onset of fatigue. Cooke (2007), studying fatigue, also focused on the factors directly affecting the actomyosin interaction and recalled, for example, that 'the decrease in pH, long thought to be a major factor, is now known to play a more minor role'. Moreover, he concluded that the various factors inhibiting the actomyosin interaction could not fully explain fatigue and that additional modulators of this interaction remained to be discovered. I present some possible additional modulators in this chapter. In any event, the molecular aspects of fatigue are now undeniable and the old 'lactic acid dogma' is questionable, as demonstrated in the discussion presented in this chapter. Concerning this dogma, for muscles within the body, I would like to recall that when lactic acid is produced via the anaerobic processes, it simultaneously leaves the muscle fibres by passive diffusion, to reach the blood in which it is transported to the liver for conversion into glucose (neoglucogenesis), before being returned to the blood. The accumulation of lactic acid is possible only if significantly more is produced than diffuses out of the muscle fibres. I therefore consider the molecular events to represent the major part of fatigue, but with the lactic acid modulating these processes within the body. A recent article by Fitts (2011) describes in detail the molecular, cellular and metabolic bases of muscle fatigue. Finally, Debold (2012) confirmed that fatigue occurs largely at the cross-bridge level, as suggested in this chapter, on the basis of the hybrid model.

CHAPTER EIGHT

Predictive and Explanatory Power of the Hybrid Model

ANALYSIS OF VARIOUS PROBLEMS

8.1 PRELIMINARY REMARKS

In the preceding chapters, many molecular and physiological properties of contracting muscles are explained on the basis of the hybrid model. In this chapter, I consider other salient problems that cannot easily be resolved on the basis of the swinging cross-bridge/lever-arm theories, but that can be accounted for by the hybrid model. It is impossible to know and to analyse all the properties and characteristics of contracting muscles under isometric or isotonic contraction conditions (steady states or transients). Moreover, some experimental results probably do not depend on the mode of force generation. In this context, it should be noted that most authors of unconventional models claim that their approach can explain many fundamental properties of contracting muscles. As an illustrative example, Jarosh (2008) wrote: 'This paper provides a comprehensive explanation of striated muscle mechanics and contraction on the basis of filament rotations…. The basic phenomena of muscle physiology… are explained and interpreted with the help of the model experiments'.

8.2 SOME TOO FREQUENTLY 'FORGOTTEN' EXPERIMENTAL DATA

I do not discuss, in this section, the problem of variations of the coefficient of shortening heat (see the brief historical account of this coefficient in Section 3.4.3.1.2) with sarcomere length (approximate linear relationship between this coefficient and the amount of overlap between the thick myosin and thin actin filaments in stretched fibres; Homsher et al. 1983; Lebacq 1980). Indeed, it is demonstrated, in Section 8.9, that there are many geometric and structural constraints in stretched fibres, regardless of the model of contraction. No clear conclusion can therefore be drawn in this area, as regards the advantage of the hybrid model over swinging cross-bridge/lever-arm theories, or vice versa. On the other hand, Irving and Woledge (1981) showed that the coefficient of shortening heat varies with the extent of shortening. This experimental result does not require 'particular molecular' explanations, and the 'phenomenological/semi-empirical' discussion proposed by Morel (1984a) in his Appendix I entitled 'Comments on the variations of the coefficient of shortening heat with the extent of shortening' is sufficient, because it is related solely to the cycling cross-bridges, regardless of their mechanical roles, and is therefore model independent.

An interesting problem concerns the slow stretching of isometrically contracting intact frog fibres, resulting in the stretch-induced activation studied by Edman (1975, 1980), Edman and Tsuchiya (1996) and Edman et al. (1978, 1979, 1982, 1993), who described the phenomenon as follows: 'force rose to a plateau value during stretch and failed to return within 2 s after stretch to the value expected from isometric contraction at the stretch length'. More recently, Rassier et al. (2003a,b) observed the same kind of behaviour with single isolated myofibrils from rabbit psoas muscle. The reverse phenomenon was observed by Maréchal and Plaghki (1979), with contracting whole frog muscles, showing a release-induced deactivation. Paul Edman and his colleagues, working on single intact frog fibres, confirmed the phenomenon described by Maréchal and Plaghki (1979), that is, a release-induced depression of isometric tetanic tension. These experimental observations were accounted for by Morel (1984a), on the basis of a phenomenological/semi-empirical approach to the mechanics of intact fibres independently of the mechanical roles of the cross-bridges (this previous approach is improved throughout Section 3.4). However, interpretation of the results obtained in Edman's group might depend on the model of contraction considered and explanations of a molecular nature are required. Amemiya et al. (1979, 1988) showed, by traditional x-ray diffraction, that no variations occur in the number of cross-bridges in the vicinity of the thin actin filaments during a slow stretch. Julian and Morgan (1981) measured muscle stiffness and found that stiffness remains constant throughout the response to slow stretch, from which it may be concluded that there is no significant change in the number of myosin heads attached to the thin actin filaments (see Section 5.1.4 for a critical analysis of the conclusions that can be drawn from measurements of stiffness). These three series of independent experiments are consistent and the origin of stretch-induced activation (and also, almost certainly, of release-induced deactivation) is therefore probably not related to the number of attached myosin heads. Matsubara and Yagi (1985) and Yagi and Matsubara (1984), working on whole frog muscles, studied the cross-bridge movements underlying the tension responses of active muscle to slow length changes, by time-resolved x-ray diffraction (traditional technique). The authors concluded that 'the tension responses to slow changes are due to shifts of the myosin heads along the thick filaments, and the elastic element responsible for tension production is located in the myosin molecules'. The second part of this sentence is in favour of the swinging cross-bridge/lever-arm theories and is also consistent with the swinging cross-bridge/lever-arm part of the hybrid model. The various experimental results presented above may also be connected to phenomena comparable to the slow return of a small proportion of myosin heads to the thick myosin filament backbone after the cessation of a short tetanus, in whole frog muscles. Indeed, according to Matsubara and Yagi (1978) and Yagi et al. (1977), ~80% of the heads attached during contraction rapidly return to the thick myosin filaments (<500 ms after cessation

of a tetanus), whereas ~20% of these heads very slowly return (>5 s after cessation of a tetanus) the technique used was time-resolved x-ray diffraction. This slow return was interpreted by Morel and Garrigos (1982b) on the basis of their model for the thick myosin filaments and the position of the myosin heads in these filaments (see also Figure 5.1, its legend and Appendix 5.I), with experimental confirmation by Morel et al. (1999). The hybrid model of contraction is based largely on this arrangement of the heads (see, in Section 5.1.1, pp. **170–172**, and Figure 5.1, its legend and Appendix 5.I). The stretch-induced activation and release-induced deactivation and the slow return, mentioned here, may therefore be accounted for by the hybrid model, itself based on the intimate structure of the thick myosin filaments. Nevertheless, it is unclear whether the experimental results recalled in the first part of this paragraph, on the one hand, and those obtained by Matsubara and Yagi (1978) and Yagi et al. (1977), on the other, have the same molecular basis. It is also unclear whether the short discussion presented here definitively supports the hybrid model or whether the phenomena observed are also consistent with the traditional swinging cross-bridge/lever-arm theories. This series of experimental results is therefore insufficient for comparison of the two types of model of muscle contraction.

The experimental results obtained by Horowitz et al. (1992) and Pollack et al. (1993), working on single intact frog fibres, appear to be at odds with the data presented and commented on in the preceding paragraph. For instance, Horowitz et al. (1992) found that a small decrease in sarcomere length induces a significant increase in force. However, this phenomenon can be detected only at sarcomere lengths >2.8 µm (see p. **91** in Section 3.7 for precise details). Thus, in the light of the results of Horowitz et al. (1992), the experimental results presented in the preceding paragraph would be questionable, because they were obtained at sarcomere lengths of ~2.0–2.2 µm, for which no increase (and no decrease) in force were observed by Horowitz et al. (1992). Moreover, at higher sarcomere lengths (>2.8 µm), a small release results in an increase in force level, which is apparently inconsistent with the results of Edman (1975, 1980), Edman and Tsuchiya (1996), and Edman et al. (1978, 1979, 1982, 1993) obtained with the same biological material. Nonetheless, on pp. **255–256** in Section 8.9, it is recalled that the technique used to record forces is a major parameter (fixed-end vs. sarcomere-isometric techniques; examples of the complexity of the various problems are also given by Horowitz et al. 1992). Thus, depending on the main features of the experiments, different results would be obtained and different conclusions could be drawn. The question therefore remains open, and I maintain the same line of reasoning as that highlighted in the last sentence of the preceding paragraph: on the basis of the experimental data available, it is not possible to distinguish between the swinging cross-bridge/lever-arm and hybrid model, at least concerning the phenomena described and discussed in this section. Fortunately, many other experimental, semi-empirical and phenomenological/semi-empirical results provide more conclusive evidence and arguments in favour of the hybrid model, as demonstrated throughout this book and in the following sections of this chapter.

The main conclusion that can be drawn concerning 'the force exerted by active striated muscle during and after change in length' (title of the paper by Abbott and Aubert 1952) is that the problem studied in this section is tricky and probably not closely related to the model of contraction. This is supported by more recent experimental results and discussions. For example, Leonard et al. (2010) studied experimentally force enhancement after stretch in single sarcomere from frog and attributed this phenomenon to actin–titin interactions or calcium binding to titin. This conclusion is consistent with the 'feeling' of Edman and Tsuchiya (1996), assuming that passive elements are involved in force enhancement by stretch. Nonetheless, a mixture of the various 'molecular phenomena' discussed in this section would correspond to the real behaviour of intact fibres/myofibrils/single sarcomeres.

8.3 THE NEGATIVE DELAYED HEAT

During a short tetanus, under steady-state conditions, MgATP is continuously hydrolysed. The splitting of MgATP releases a large amount of chemical energy and a limited proportion of this energy is instantaneously absorbed by Lohmann's reaction ($\sim$−2.4 × 10^{-20} J per MgATP molecule vs. ~9.0 × 10^{-20} J per MgATP molecule for MgATP splitting; see p. **195** in Section 6.2). It is shown, on pp. **211–212** in Section 6.3.5 and pp. **215–216** in

Section 6.4, that the subsequent 'enthalpy-related' and 'pressure-volume-related' phenomena result in increases in temperature within the fibre and the thick myosin filament core, respectively. These increases in temperature, particularly that in the thick myosin filament, induce partial dissociation of the internal B–B′-type head–head dimers (see, in Section 5.1.1, pp. **170–172** and **174–176** and Figure 5.1, for precise details on the B and B′ heads and their role, and pp. **180–181** in Section 5.1.2, Section 6.5 and Equation 6.4 concerning dimer dissociation). The small temperature excess, resulting from the 'pressure–volume-related' phenomenon, is transiently maintained, such that the reassociation of myosin internal B–B′-type head–head dimers is prevented for a few seconds, or less. This results in a balance between the dissociation and reassociation of these dimers. The mean number of dissociated internal B–B′-type head–head dimers therefore remains constant, at least for a short period, as does the elementary residual radial tethering force, F_{cb}, during this short tetanus (see pp. **180–181** in Section 5.1.2 for details about F_{cb}). When relaxation occurs, after a short tetanus, energy output from MgATP splitting stops and the heat excess within the thick myosin filament core dissipates: reassociation of the internal B–B′-type head–head dimers is no longer prevented. Tension therefore decreases, owing to both the swinging cross-bridges becoming gradually inactive and the radial tethering forces gradually reappearing (see p. **183** in Section 5.1.2). Thus, during relaxation, all the processes described for activation are reversed and the internal B–B′-type head–head dimers reassociate, with the term $(P\Delta V^*)_{t*}$ in Equation 6.1 becoming negative. As a straightforward consequence, heat must be absorbed during relaxation after short tetani. This phenomenon is familiar to specialists in muscle energetics and is called the 'negative delayed heat': 'it has been reported by several authors ... that the muscles absorb heat for a short time after the end of a short tetanus or a twitch' (Woledge et al. 1985). This 'immaterial' experimental observation strongly supports the lateral swelling part of the hybrid model. In my opinion, the swinging cross-bridge/lever-arm models cannot provide as clearcut an explanation to the negative delayed heat, unless extra ad hoc assumptions are introduced. To the best of my knowledge, no other simple explanation for the existence of the negative heat, other than that presented here, has ever been put forward.

8.4 THE UNEXPLAINED HEAT

The experimental data and calculations presented and discussed in Sections 6.3.4 and 6.4, for rabbit, demonstrate a maximal increase of ~1.56°C in temperature within the thick myosin filament core, during a short tetanus, regardless of temperature of the surrounding medium (see particularly p. **216** in Section 6.4). This maximal temperature increase results from full dissociation of the internal head–head dimers (B and B′ heads in Figure 5.1). As a first approximation, a similar maximal increase in temperature can reasonably be assumed to occur in frog muscle. Clearly, this maximal increase does not depend on 'unknown' chemical reactions (see Curtin and Woledge 1978, 1979; Gilbert et al. 1971; Homsher and Kean 1978; Homsher et al. 1979; Woledge et al. 1985, for precise details on this notion of unknown chemical reactions) and is intimately related to dissociation of the internal B–B′-type head–head dimers (see Section 6.4). This maximal increase in temperature ~ 1.56°C is probably, at least partly, responsible for the unexplained heat observed with frog fibres at ~0°C–5°C. The 'internal' volume of a thick myosin filament is ~1.49 × 10⁻¹⁶ cm³ (see p. **215** in Section 6.4) and that of an intact unit cell is ~3.44 × 10⁻¹⁵ cm³ (for a 'composite animal species'; see p. **206** in Section 6.3.5). The volume of a thick myosin filament therefore represents ~4.3% of the volume of a unit cell. Thus, the maximal increase in thick myosin filament core temperature of ~1.56°C would correspond to a maximal increase in fibre temperature of ~1.56°C × 4.3% ~0.0671°C, regardless of the bathing buffer and its temperature. This maximal value is consistent with the measured 'unexplained' increase in temperature observed in single frog fibres at ~0°C (Woledge et al. 1988). It should also be recalled that the apparent energy, resulting from MgATP splitting minus absorption in Lohmann's reaction, is $e_{app}^* \sim 6.4\,10^{-20}$ J per MgATP in vitro and e_{app} ~6.6 ± 1.0 × 10⁻²⁰ J per MgATP molecule in vivo (see p. **193** in Section 6.2). Thus, taking into account only the mean values, ~100 × (6.6 − 6.4) × 10⁻²⁰ J per MgATP molecule/6.4 × 10⁻²⁰ J per MgATP molecule, ~+ 3.1% of the unexplained heat, in intact fibres, could be accounted for in this way. However, taking into account the error bars (SEs) we obtain + 18.7% or ~ − 12.5%, and the value of ~ + 3.1% would be dubious, but cannot be rejected. The proportion of unexplained stable maintenance heat (steady state) is ~+15%–30%, as found in most species of frog (e.g. Curtin and Woledge 1978, 1979; Homsher and Kean 1978; Woledge et al. 1985; ~+15% in *Rana pipiens* and ~+30% in *Rana temporaria*). In any event, a large part of the unexplained heat is probably accounted for by the hybrid model, particularly its lateral swelling part, owing to the maximal increase in fibre temperature of ~0.0671°C, regardless of temperature of the medium surrounding the fibre, which is not related to any unknown chemical reaction (see above concerning the unknown reactions and the value of ~0.0671°C). The order of magnitude of the maximal increase of ~0.0671°C being consistent with the experimental results of Woledge et al. (1988), I think that neither the old cross-bridge models nor the recent swinging cross-bridge/lever-arm theories can provide such a simple explanation for this 'mysterious' phenomenon of unexplained heat/unexplained temperature increase.

Elliott and Worthington (1994, 1995), Jarosh (2008) and Ohno and Kodama (1991) provide other explanations for the unexplained heat. Thus, there are many possible explanations for this awkward phenomenon, which was previously considered to be difficult to account for and has been widely ignored by many specialists in the biochemistry, biophysics and physiology of muscle contraction, but not by specialists in muscle energetics. However, even these experts have been unable to provide entirely satisfactory explanations for this observation. There are thus now many complementary, possibly redundant, hypotheses. After writing this section, I read the paper by West et al. (2004), who worked on intact fibres from dogfish white muscle, at 12°C, and found that there was no detectable unexplained heat + work, at least within the wide range of experimental error. This confirms that the problem is complex and certainly depends on the animal species, as already found for *Rana pipiens* and *Rana temporaria* (see the preceding paragraph).

8.5 TEMPERATURE-INDUCED CONTRACTURE IN RESTING MUSCLES AND FIBRES (INTACT AND DEMEMBRANATED) FROM FROG AND RABBIT

As a preliminary remark, let us recall that there are processes for translating radial forces into axial forces (see pp. **102–103** in Section 3.10). The discussions applying to radial forces can therefore be applied to axial forces.

A major expected consequence of the hybrid model is that a substantial increase in temperature of the buffer surrounding a resting whole muscle or intact or demembranated fibre should induce contracture (non–Ca-dependent development of axial tension). In fact, sufficient increases in temperature result in the dissociation of all the internal B–B′-type head–head dimers (see Figure 5.1 concerning the arrangement of these dimers), leading to the disappearance of the mean elementary radial tethering, F_{cb}^0, per A-type head, resulting in a net radial expansive force and a subsequent axial force (see, in Section 5.1.1, pp. **170–172** and **174–176**, and Figure 5.1, its legend and Appendix 5.I, for precise details on the A-type heads and the generation of F_{cb}^0, respectively). Indeed, in synthetic thick myosin filaments from young adult rabbits, total dissociation of the ~100 internal B–B′-type head–head dimers (see p. **206** in Section 6.3.5 and p. **217** in Section 6.5 concerning this number of ~100) is observed at ~37°C–38°C (see, in Section 5.2, Figure 5.2, its legend and the corresponding comments [including Appendix 5.II]). This transition probably occurs at a higher temperature within the body, for example, ~41.3°C for humans or ~43.5°C for the rabbit (see legend to Figure 5.2 and Appendix 5.II). Other experimental results support this transition: temperature-induced dimer–monomer association–dissociation phenomena are systematically observed in vitro, with isolated myosin heads (S1), as recalled in Section 6.3.4. In situ/in vivo, regardless of the biological material used (whole muscle, intact or demembranated fibres, isolated myofibrils, intact unit cells) and under resting conditions, the mean elementary radial tethering force, F_{cb}^0, may steadily decrease, albeit probably very slowly, with increasing temperature of the surrounding buffer. Indeed, independently of the internal dimer dissociations, the force F_{cb}^0 is probably not very sensitive to temperature, unless there are minor and gradual temperature-induced changes in the conformation of the weakly bound cross-bridges, particularly for the A-type heads (see above for references on the A-heads). In this context, hypotheses are put forward on p. **248** in Section 8.8, concerning the possible dependence on temperature of the residual radial tethering force, NF_{cb}, in contracting intact unit cells (see Section 4.4.2.5.4 and pp. **180–181** in Section

5.1.2 for precise details about NF_{cb}). It is suggested that the mean elementary residual radial tethering force, F_{cb}, per 'contracting' A-type head, may decrease with increasing temperature. This hypothesis almost certainly applies to the force F_{cb}^0, which may also gradually decrease with increasing temperature, for the same reasons as for the force F_{cb}. As far as I know, no experimental evidence concerning the possible dependence on temperature of the number N_0 of A-type heads has been published (see p. **171** in Section 5.1.1 for the definition of N_0). Thus, possible small and gradual increases or decreases in the 'resting' radial tethering force, $N_0 F_{cb}^0$, per intact unit cell with temperature cannot be predicted. In any event, when the external temperature reaches its critical value (e.g. ~37°C–38°C for rabbit synthetic thick myosin filaments or ~43.5°C within the body; see above), all the internal B–B′-type head–head dimers dissociate and F_{cb}^0 completely disappears: a resting muscle should develop an axial force corresponding to a certain proportion of that developed by a fully contracting muscle (proportion of ~40% at 10°C, but possibly different at ~40°C), under given conditions, at the same temperature, and mostly related to the radial repulsive electrostatic forces between the myofilaments, minus a small proportion of radial attractive/compressive forces. The necessary references concerning these attractive/compressive forces are given in Chapter 4: see the third point in Section 4.4.2.4 (for general considerations) and pp. **155–156** in Section 4.4.2.5.3 (for more detailed considerations); it is demonstrated on p. **159** in Section 4.4.2.5.4 that the radial attractive/compressive forces Λ_r represent only ~1.11% of the radial electrostatic repulsive force $4(\gamma/2)F_{e,r}$ at rest, at 10°C, at the slack length of the fibre (the very small proportion of ~1.11% is reasonably assumed to remain very small at any temperature). As highlighted above, the proportion of ~40%, valid at 10°C, for the lateral swelling part of the mechanisms occurring in a fibre, according to the hybrid model, is not necessarily valid at any temperature, particularly those above 10°C. Nevertheless, a purely qualitative discussion is presented in this section and there is no need for accurate estimation, as a function of temperature, of the relative roles of the two mechanisms underlying the hybrid model.

Hill (1970b), working on whole frog muscles wrote: 'it has been long known that a resting muscle shows a reversible rubber-like response to a change of temperature: it develops tension when warmed and this reverses on cooling'. The critical temperature at which this 'contracture' occurs is ~28°C–35°C, at the slack length. This observation is consistent with the experimental results obtained in this monograph (see Figure 5.2, its legend, Appendix 5.II and the corresponding comments) and with the hybrid model (see the preceding and following paragraphs). Working with glycerinated fibres from rabbit psoas muscle, Ranatunga (1994) confirmed the findings of Hill (1970b) (the only difference was that the critical temperatures at which contractures occurred ranged between ~35°C and ~40°C, rather than between ~28°C and ~35°C). Hill (1970b) observed that the critical temperature was lower in stretched muscles. Ranatunga (1994) confirmed this finding: 'sharp tension rise occurs at a lower temperature when sarcomere length is higher'. Note that, in stretched fibres, complex phenomena occur, as recalled and discussed on pp. **251–254** in Section 8.9, for isometrically contracting fibres and resting fibres. Regardless of these problems, if resting muscles or fibres are moderately stretched, the myofilaments are brought closer together and the mean elementary radial tethering force, F_{cb}^0, decreases considerably (see Section 5.1.3, in particular p. **183**). Indeed, the stretching of muscles and intact fibres results in a decrease in myofilament spacing, owing to the isovolumic behaviour, as found by April et al. (1971), Brandt et al. (1967) and Elliott et al. (1963, 1965, 1967) (see pp. **4–5** in Chapter 1 for a brief discussion of this subject). In resting demembranated fibres, isovolumic behaviour is not maintained, but Matsubara and Elliott (1972), for example, have demonstrated that there is a linear decrease in myofilament spacing when the fibre is stretched. Thus, for both intact or demembranated fibres, in a stretched fibre, a temperature-induced contracture of the same amplitude as that observed in 'slack' fibres may be easier to induce. This is equivalent to lowering the critical temperature at which contracture occurs. At this point, we need to recall the experimental results obtained by Leake et al. (2004), showing that titin, which is assumed by most authors to be largely responsible for the passive elasticity of the sarcomere, shows temperature-dependent elasticity within the range 10°C–60°C. Thus, some of the dependence of passive elasticity on temperature would be related to titin. However, Leake et al. (2004) observed no break-

point in the elasticity–temperature relationship of titin. A sudden temperature-induced contracture cannot therefore be related to titin. In any event, the old results obtained by Hill (1970b) may be considered trivial, merely reflecting the putative release of calcium from the sarcoplasmic reticulum with increasing temperature (the sarcoplasmic reticulum is intact in whole muscles). More recent experiments, including those of Ranatunga (1994) recalled above, on chemically skinned fibres (sarcoplasmic reticulum largely removed) may answer this question, although the experimental data presented below, in this section, are frequently inconsistent and do not provide a clear solution. Many of these discrepancies between experimental studies from independent groups may be related to the use of traditionally skinned (old techniques) or permeabilised fibres (more recent techniques). Moreover, as pointed out in Sections 3.8, 4.4.2.1 and 8.7, skinned and permeabilised fibres do not behave like intact fibres and the specific methods used by different authors for demembranation (see Millman 1998 for a critical analysis of chemical skinning, including glycerination) may lead to different experimental results, particularly concerning the dependence of isometric tetanic tension on temperature (see Section 8.8) and the critical temperature at which contracture occurs in resting skinned or permeabilised fibres. This confusing situation is illustrated below, in this section.

Bershitsky and Tsaturyan (1992), working on chemically skinned fibres from rabbit psoas muscle, at sarcomere lengths of ~2.5–2.6 μm, claimed that 'in relaxed fibres the T-jump [temperature jump] did not induce tension changes'. This is possibly because the final temperature was below ~40°C, but the technique used to prepare skinned fibres may also have contributed to this conclusion, as already pointed out by Millman (1998; see the end of the preceding paragraph). Nonetheless, much of the paper by Bershitsky and Tsaturyan (1992) concerns isometrically contracting demembranated fibres (as highlighted by the authors, the active tensions had usual values for traditionally demembranated fibres: ~0.6–0.9 × 10^5 N m^{-2} at 10°C; see pp. **95–97** in Section 3.8.1 for comparison). The authors found a rapid biphasic transitory increase in tension, after the T-jump, followed by a slower decrease in tension on cooling (a typical result is given in their

Figure 5). This finding is not directly relevant to the problem of temperature-induced contracture in resting fibres studied in this section. However, I would like to comment briefly on this particular feature, because it seems to support the hybrid model. Indeed, within the framework of swinging cross-bridge/lever-arm concepts, let us assume that this biphasic increase in active tension results solely from increases in both swinging cross-bridge/lever-arm turnover rate and efficiency (i.e. increase in MgATPase activity), leading to a proportional increase in the axial contractile force (one-to-one relationship between isometric tension and MgATPase activity; see Section 3.4.6 for a discussion of the one-to-one/tight coupling process). To the best of my knowledge, no systematic experimental studies of MgATPase activity over the ~0°C–40°C temperature range have been performed (see, however, the complex experimental study by Hilber et al. 2001 discussed on p. **238**). By contrast, rates of MgATP cleavage up to room temperature (~20°C–22°C) have been published, with no signs of breakpoint observed within the ~0°C–22°C range. Assuming a steady increase in this enzymatic activity, up to ~39°C, results in self-consistent data, as demonstrated in Section 6.3.3 (see particularly p. **201**): the activation coefficient (i.e. MgATPase activity under contraction conditions, divided by MgATPase activity at rest) is independent of temperature, up to ~39°C. Thus, the swinging cross-bridge/lever-arm theories, based on the one-to-one process, predict solely a steady monophasic increase in isometric tetanic tension with temperature, directly related to the monophasic increase in the rate of MgATP breakdown. This conclusion is inconsistent with the experimental data of Bershitsky and Tsaturyan (1992), showing a biphasic increase in active tension, unless extra ad hoc hypotheses are introduced, and the pure swinging cross-bridge/lever-arm models are again called into question. In this context, Bershitsky and Tsaturyan (1992) proposed a 'simple three-state model of the cross-bridges kinetics … to explain the experimental data'. In my opinion, this is the traditional 'extra ad hoc' approach (see p. **39** in Section 3.4.2 and p. **263** in Section 8.11). Let us now take into account the hybrid model, in which ~60% of the axial contractile force is related to the swinging of the attached cross-bridges and ~40% is related to lateral swelling processes (these proportions

are valid at 10°C but are almost certainly comparable at other temperatures). In this ~40% indirect generation of the axial contractile force, the number, $n^*(\theta)$, of dissociations of the myosin B–B'-type head–head dimers within the thick myosin filament cores, during isometric tetanic contraction, increases with increasing temperature of the surrounding medium (see Equation 6.3). Otherwise, it should be noted that the radial electrostatic repulsive forces depend little on temperature (see particularly pp. **267–268** in Section 8.8 for a discussion). As a consequence, the indirectly generated axial tension also increases, essentially attributed to the decrease in the residual radial tethering force NF_{cb} (see pp. **158–159** in Section 4.4.2.5.4 and pp. **180–181** in Section 5.1.2 for precise details about this residual force), resulting in a limited increase in net radial repulsive force and, therefore, a small increase in the electrostatically related axial contractile force (see pp. **102–103** in Section 3.10 regarding the conversion processes). Finally, ~60% of the full axial contractile force results from a 'tightly coupled process' (axial force proportional to MgATPase activity), the remaining ~40% resulting from an 'extremely loosely coupled process' (radial expansive force extremely loosely coupled with MgATPase activity). Combining these two processes accounts for the very loose coupling between full axial contractile force and MgATPase activity (very loose coupling = combination of one-to-one coupling and extremely loose coupling). Moreover, the two mechanisms (directly and indirectly generated axial contractile force) depend on very different parameters, temperature being the only common determinant factor. They can therefore work practically independently, leading to the biphasic behaviour observed by Bershitsky and Tsaturyan (1992) (see also pp. **248–249** in Section 8.8 for other precise details).

Working with resting glycerinated fibres from rabbit psoas muscle fibres, at sarcomere lengths between ~2.3 μm and ~3.2 μm, Ranatunga (1994) found that the resting tensions (ranging between ~0.2 × 10^5 N m^{-2} and ~1.6 × 10^5 N m^{-2}) abruptly (and reversibly) increased with temperature (this increase is Ca independent) within the ~30°C–40°C range. The origin and some characteristics of this contracture at high temperature are described and discussed on pp. **234–236**. The results of Ranatunga (1994) are at odds with those of Bershitsky and Tsaturyan (1992) recalled at the start of the preceding paragraph. However, these contradictory results are not unexpected, given the complex behaviour of demembranated fibres (see Sections 3.8, 4.4.2.1 and 8.7 and below). In this context of complexity, Hilber et al. (2001) unwittingly confirmed (only qualitatively) the main features of the results obtained by Ranatunga (1994) and found, in permeabilised fibres from rabbit psoas muscle, that the resting tension was only ~0.05 × 10⁵ N m⁻² at ~35°C, but this very low value is approximately five and three times the values of the resting tension at ~25°C and ~30°C, respectively. These findings confirm that contracture occurs at high temperature. However, as pointed out in Section 8.8, many-faceted phenomena appear with increasing temperature and, moreover, the behaviour of chemically skinned or permeabilised fibres cannot be predicted at high temperature (e.g. ~35°C), probably because of the fragility and subsequent adulterations of demembranated fibres at such temperatures. The resting tensions measured by Hilber et al. (2001) are actually much lower than those obtained by Ranatunga (1994). These problems of fragility are pointed out at many places in this monograph (see p. **52** in Section 3.4.3.2, p. **200** in Section 6.3.2, p. **202** in Section 6.3.3 and p. **208** in Section 6.3.5). Thus, the very low resting tensions measured by Hilber et al. (2001) at ~25°C, ~30°C and ~35°C are probably related to the fragility of their chemically skinned fibres. Despite the qualifications regarding the experimental data of Hilber et al. (2001), let us assume that their results are, at least qualitatively, valid. According to their Figure 4A and B, under relaxing conditions, there is a simultaneous rapid increase in MgATPase activity and in resting tension between ~30°C and ~35°C. The authors suggested, therefore, that 'some active force may be generated under these conditions', possibly because 'the physiological regulatory mechanism has not been completely preserved in the demembranated rabbit psoas muscle fibre preparation'. In fact, Fuchs (1975) found that, in the absence of Ca^{2+}, heating regulated actomyosin in vitro (presence of troponin and tropomyosin in the F-actin filaments) reversibly inactivated the troponin–tropomyosin control mechanism and led to Ca-independent F-actin filament activation of myosin at high temperatures. However, the phenomena observed in vitro and in situ/in vivo are very different (see, in Chapter 9, p. **270**

for a quotation from Borejdo et al. 2006 and p. **272** for quotations from Grazi 2000 and the view of Xu et al. 2006a). The situation is therefore confusing: contracture at high temperature has not been definitively demonstrated in the case of demembranated fibres, although I think that it almost certainly occurs. In any event, the experimental findings of Hilber et al. (2001) and Ranatunga (1994) appear to be consistent with the onset of contracture at high temperature, at least qualitatively. For this reason, I refer, in this section, to the rapid increase in resting tension as 'temperature-induced contracture'. This viewpoint is, in my opinion, supported in the next paragraph.

Ranatunga (1994) demonstrated that the abrupt increase in resting tension at high temperature (contracture, recalled on p. **238**) depends on pH and the presence of inorganic phosphate (Pi). According to his experimental study, the critical temperature at which contracture occurs decreases with increasing pH (see Figure 4d in his paper), and, if 10 mM Pi is present, the abrupt increase in tension on warming is abolished (see Figure 5 in his paper). The isoelectric pH of myosin and thick myosin filaments was found to be ~5.0–5.4 for rabbit (e.g. Erdös and Snellman 1948; Millman 1998) and ~4.4 for crayfish (April 1978; April et al. 1972). The isoelectric pH of actin was found to be ~5.2 (A. Szent-Györgyi 1947) and the isoelectric pH of myofibrillar proteins, as a whole, was found to be ~4.0–4.5 (Elliott et al. 1978). Thus, any increase in pH to values above the isoelectric pH and, fortiori, above pH ~7.1 (at 20°C), used as a reference by Ranatunga (1994), must lead to an increase in the density of fixed negative electrical charges on the thick myosin and thin actin filaments (see also Millman 1986 and references therein) and, possibly, an increase or a decrease in the density of the bound anions (see Section 4.4.2.2 for some details on bound anions). As far as I know, there are no available experimental data on possible changes in the density of the bound anions with pH, and I assume that the increase in the density of fixed negative electrical charges largely prevails over hypothetical increase or decrease in the density of bound anions. Thus, increasing the pH results in an increase in the absolute values of the electrical surface potentials $\Psi_{m,r}$ (on the thick myosin filaments) and $\Psi_{a,r}$ (on the thin actin filaments) (see Section 4.4.2.2 concerning the relationship between the number of fixed electrical charges and the surface electrical potentials) and in a subsequent increase in the radial repulsive electrostatic forces between the myofilaments, via the parameter Q_r. This parameter is defined in the text below Equation 3.1 and depends on the electrical surface potentials through the mathematical function 'tanh', an asymptotic function that cannot exceed 1, resulting in variations smaller than those for the two electrical surface potentials, when the absolute values of these potentials are high (this is the case in the myofilament lattice, under physiological conditions, as recalled on pp. **161–162** in Section 4.4.2.6; the same is certainly true for pH values between 6.5 and 7.7, as used by Ranatunga 1994). However, increasing the pH should increase the radial repulsive electrostatic forces between the myofilaments. It is impossible to predict the dependence on pH of the radial tethering forces, $N_0 F_{cb}^0$ (described on pp. **174–176** in Section 5.1.1), and the radial attractive/compressive forces (see p. **235** for references on these forces), but it seems reasonable to assume that they are not very sensitive to pH. Increasing the pH should therefore increase the radial repulsive electrostatic forces and also the net radial repulsive (expansive) forces, and, consequently, the resting axial tension. Figure 4d in the paper by Ranatunga (1994) confirms that, at ~25°C (a temperature well below the critical temperatures inducing contracture), the resting axial force/tension increases when the pH increases from 6.5 to 7.1 and then to 7.7. I conclude that resting fibres are almost certainly closer to contracture at high pH, at a given temperature (~25°C for example). According to the hybrid model, the increase in pH should therefore lower the critical temperature at which contracture occurs, as experimentally demonstrated by Ranatunga (1994; see his Figure 4d).

Under otherwise identical conditions of pH (~7.1 at 20°C), we need to try to explain why, on warming, contracture is abolished if 10 mM Pi is added, as experimentally demonstrated by Ranatunga (1994; see his Figure 5). As shown by Bartels et al. (1993), ATP binds non-specifically to the thick myosin filaments (and obviously also to the enzymatic and dimerisation sites on the myosin heads). More generally, Elliott (1980) and Elliott and Hodson (1998) showed that various anions (especially Cl⁻) also bind to the protein filaments (e.g. actin thin and myosin thick filaments). Thus, the Pi anion almost certainly also binds non-specifically to the myofilaments, but the affinity of Pi for an actin thin and myosin thick filament is probably weaker than that of Cl⁻ or ATP (in a personal communication, the late Bill Harrington told me that Pi can bind non-specifically to synthetic thick myosin filaments). At a concentration of 10 mM, Pi can almost certainly compete with ATP (concentration of ATP generally ~2 mM in the experiments of Ranatunga 1994) and can replace it, at least partly, at the enzymatic and dimerisation sites (located on the myosin head) and also on the thick myosin filaments. As Pi is less electrically charged than ATP, the net negative electrical charge on the thick myosin filaments and heads almost certainly decreases, when 10 mM Pi is added. Thus, the radial repulsive electrostatic forces decrease through decreases in the absolute values of the electrical surface potentials $\Psi_{a,r}$ and, mostly, $\Psi_{m,r}$, via the parameter Q_r (see the preceding paragraph concerning Q_r). It is impossible to predict the dependence on the Pi concentration of the radial tethering and attractive/compressive forces. However, it seems reasonable to assume that they are not very sensitive to the concentration of Pi. Adding 10 mM Pi should therefore reduce the radial repulsive electrostatic forces and also the net radial repulsive (expansive) forces and, consequently, the resting axial tension. Thus, according to the hybrid model, in the presence of 10 mM Pi, it is certainly more difficult for fibres to display a contracture (at least within the ~30°C–40°C range; see the beginning of the preceding paragraph concerning this range of temperature). Note that the 'mechanical' role of Pi described here, under resting conditions, may also apply to contracting fibres. In this context, the experimental studies of Cooke and Pate (1985) and Tesi et al. (2000), for instance, performed on demembranated contracting fibres, and the interpretations given by these authors should be taken with caution, because Pi is considered essentially as a competitive inhibitor of MgATP at the enzymatic site, taking into account only the swinging cross-bridge/lever-arm theories and the one-to-one tight coupling between tension and MgATPase activity. From the analysis presented in this paragraph, the role of Pi is certainly more complex than most authors believe, with significant effects on both the 'enzymatic–mechanical effects' of the swinging cross-bridge/lever-arm part and the 'electrostatic–mechanical effects' of the lateral swelling part of the hybrid model.

In conclusion, temperature-induced contracture, predicted from the hybrid model, appears to be a genuine phenomenon, first studied experimentally and quantified by Hill (1970b) in whole frog muscles. This important feature was confirmed, to various extents, by experiments performed on intact, skinned and permeabilised fibres, recalled here. However, experimental observations of this temperature-induced contracture depend on various characteristics of the experiments, such as the maximal temperature reached, the origin of the fibre (e.g. rabbit or frog fibres) and the technique used to prepare demembranated fibres (see the critical analysis of the chemical techniques of skinning presented by Millman 1998). The particular question of the adulteration of demembranated fibres, mostly attributed to the age of the animals, is a major stumbling block, as highlighted at many places in this section; see also p. **52** in Section 3.4.3.2, p. **200** in Section 6.3.2, p. **202** in Section 6.3.3, p. **208** in Section 6.3.5 and p. **238**.

8.6 CONTRACTURE IN THE ABSENCE OF CALCIUM, AT MICROMOLAR LEVELS OF MgATP, IN THE CASE OF CRAYFISH MUSCLES

Reuben et al. (1971), working on mechanically skinned fibres from the walking leg of crayfish muscles (*Orconectes*), showed that tension is generated at micromolar levels of MgATP, in the absence of calcium (pCa > 9; resting conditions). The authors provided explanations based exclusively on the swinging cross-bridge/lever-arm theories, but their observation is also consistent with the hybrid model. Indeed, in the absence of MgATP, there are no myosin internal B–B′-type head–head dimers (see, in Section 5.1.1, Figure 5.1, its legend and Appendix 5.I for precise details on these dimers). There are no head–head dimeric interactions in the absence of MgATP, because soluble myosin and isolated heads (S1) do not form head–head dimers in vitro (Morel et al. 1998a,b). The same is true within the thick myosin filament core (Morel et al. 1999). Thus, in situ, when the concentration of MgATP in the buffers surrounding the skinned fibres is extremely low, the internal B–B′-type head–head dimers dissociate. The elementary radial tethering force, F_{cb}^0, therefore no longer exists (see, in Section 5.1.1, pp. **174–176**,

and Figure 5.1, its legend and Appendix 5.I concerning the mode of generation of F_{cb}^0; see also Sections 6.4 and 6.5 for discussions of the parameters controlling the association–dissociation processes and of the qualitative relationship between dissociation and radial tethering forces). The radial repulsive electrostatic forces automatically become dominant and are turned into ~40% of the full axial force (see pp. **102–103** in Section 3.10 concerning the mechanisms of translation of radial into axial forces). The hybrid model provides therefore a straightforward explanation of the experimental observation of Reuben et al. (1971).

8.7 WHY AND HOW DO ALL FIBRES SWELL LATERALLY UPON DEMEMBRANATION? SOME INSIGHTS INTO THE 'OSMOTIC COMPRESSION TECHNIQUE' USED TO COUNTERBALANCE THE LATERAL SWELLING

Andrews et al. (1991), April et al. (1971), Elzinga et al. (1989a,b), Ferenczi et al. (1984b), Ford and Podolsky (1972), Godt and Maughan (1977), Goldman and Simmons (1986), Linari et al. (1998), Matsubara and Elliott (1972), Matsubara et al. (1984, 1985), Maughan and Godt (1979), Reuben et al. (1971), Rome (1967, 1968, 1972) and many other groups studied various properties of mechanically or chemically skinned and permeabilised fibres from frog and rabbit. Other types of demembranated fibres have also been studied, including those from toadfish (e.g. Rome et al. 1999) and dogfish (e.g. West et al. 2004). All these authors reported systematic and unavoidable lateral swelling of the fibres, after removal of the sarcolemma in relaxing buffers. It should be noted that such a process of lateral swelling is also observed with other polyelectrolyte gels of filaments surrounded by membranes, particularly demembranated bovine corneal stroma (e.g. Elliott and Hodson 1998; Elliott et al. 1980; Maurice and Giardini 1951). Matsubara et al. (1984) considered the swelling of demembranated frog fibres to be caused by the 'abolition of the transmembrane osmotic constraints, which limits the fibre volume'. I present in this section an approach to this question different from the rather vague suggestion of Matsubara et al. (1984). I consider only the case of demembranation in relaxing buffers with physiological ionic strengths of ~150–200 mM and pH of ~7, at around the slack length of the fibres. These conditions correspond approximately to those used during most skinning or permeabilisation processes. Asayama et al. (1983) observed, on mechanically skinned frog fibres, an 'enormous swelling' of the sarcoplasmic reticulum that may contribute to the total lateral swelling of the fibres. However, as can be deduced from pp. **96–97** in Section 3.8.1, the isometric tetanic tensions recorded in chemically skinned and permeabilised fibres (sarcoplasmic reticulum largely absent, maybe 'partly' present and functional; see p. **92** in Section 3.7 for comments on this notion of 'partly' functional) are statistically similar to those recorded in mechanically skinned fibres (sarcoplasmic reticulum present and functional). Thus, the marked swelling of the sarcoplasmic reticulum, observed by Asayama et al. (1983), does not result in significantly different isometric tetanic tensions, within the broad limits of experimental error, and I take all types of demembranated fibres to be statistically similar.

Let us consider, as an illustrative example, the experimental studies of Goldman and Simmons (1986), performed on mechanically skinned frog fibres. The authors presumably measured fibre diameters, before skinning and just after removal of the sarcolemma, probably with a special light microscope that they had previously described (Goldman and Simmons 1984a). Goldman and Simmons (1986) observed that diameters increased by ~10%–30% upon skinning and attributed this to an increase in lattice spacing. However, Millman (1998) claimed that, regardless of the state of the fibre (intact or demembranated), there is a rather hazy relationship between fibre diameter and lattice spacing: 'the fibre diameter changes more than the filament lattice spacing' (see also p. **146** in Section 4.4.2.5.1 for a numerical example). On pp. **4–5** in the Introduction, I point out that there are no clear experimental data concerning the relationship between fibre diameter and lattice spacing in intact and demembranated fibres. From p. **5** in the Introduction, it may be deduced that, at rest, the relative decrease in fibre diameter (induced by osmotic compression) is approximately proportional to the relative decrease in myofilament spacing. Nonetheless, this conclusion, drawn from the experimental study of Kawai et al. (1993), is valid for rabbit psoas fibres chemically skinned by an unconventional technique,

not necessarily for mechanically skinned frog or rabbit fibres. At this point, I believe that there are several possible relationships, each valid under certain conditions (animal species, intact or demembranated fibres, technique of demembranation, temperature, etc.). Indeed, from p. **5** in the Introduction, the limited increase in width of intact frog fibres when contraction is triggered may depend on the cubic power of the increase in lattice spacing. Combining these conclusions and extending them to the demembranation of frog fibres in resting buffers, the values of ~10%–30%, corresponding to increases in fibre diameter after mechanical skinning (frog; see above), would give ~3%–9% (cubic power) and ~10%–30% (proportionality) for lattice spacing. At this stage, I take mean values of ~7%–20%, a range not too far from the old value of ~28% obtained by traditional x-ray diffraction on crayfish fibres (April et al. 1971). Several questions have been raised regarding the interpretation of results obtained by x-ray diffraction, for demembranated fibres (Morel 1985a; Morel and Merah 1997; see also p. **13** in Section 2.2). These reservations almost certainly do not concern frog half-fibres, as used in the experimental part of this monograph, in the presence of methanesulphonate (MS^-) as the major anion (see Sections 2.2 and 3.3, for instance), although no x-ray diffraction experiments have been performed on this biological material. On p. **98** in Section 3.8.1, multiplication factors of between ~1.18 and ~3.28 are proposed to pass from the cross-sectional area of an intact fibre to that of a demembranated fibre. These two factors correspond to factors of ~$(1.18)^{1/2}$ ~1.09 (corresponding to ~9%) and ~$(3.28)^{1/2}$ ~1.81 (corresponding to ~81%) to pass from the diameter of an intact fibre to that of a demembranated fibre. They correspond to increases in lattice spacing of ~$(1.09)^{1/3}$ ~1.03 (~3%, cubic power) or ~1.09 (~9%, proportionality), and ~$(1.81)^{1/3}$ ~1.22 (~22%, cubic power) or ~1.81 (~81%, proportionality). The notions of 'cubic power' and 'proportionality' are discussed on p. **5** in Chapter 1. The spectrum of estimates (~3%, ~9%, ~22% and ~81%) is obviously consistent with the ~7%–20% selected above.

The orders of magnitude of the kinetics and the amount of lateral swelling may be important. The experimental studies performed by Matsubara and Elliott (1972), with mechanically skinned frog fibres, in which the sarcoplasmic reticulum is a priori functional, are certainly the most detailed. However, the authors mostly used Cl⁻ as the major anion, raising questions about the problem of calcium release from the sarcoplasmic reticulum in the presence of Cl⁻ (see p. **14** in Section 2.2): were their skinned fibres actually in purely resting conditions? The authors checked that their fibres contracted in the presence of an excess of Ca^2, and their resting skinned fibres were certainly not fully activated. Nevertheless, they were possibly partly activated, owing to the Cl⁻-induced release of calcium from the sarcoplasmic reticulum. This phenomenon of Cl⁻-induced calcium release complicates interpretation of the 1972 results. Matsubara and Elliott (1972) also used propionate or sulphate as major anions, in comparative studies concerning the non-isovolumic behaviour of their skinned fibres, but they did not study the kinetics and amount of lateral swelling resulting from sarcolemma removal in the presence of these two anions. Thus, only the detailed results regarding the behaviour of mechanically skinned fibres in the presence of Cl⁻ as the major anion can be taken into account here. Despite the problems associated with this anion, we may consider, as a first approximation, that the possible partial activation, mentioned here, remained limited in the experiments performed in 1972 (e.g. the resting skinned fibres would have been partially activated, perhaps to between ~10% and ~30% of full activation).

Let us first consider the kinetics of lateral swelling. Matsubara and Elliott (1972) observed, under a binocular microscope, that the diameter of their skinned fibres, kept at the slack length (sarcomere length of ~2.0 µm), increased rather rapidly (over a few minutes) during removal of the sarcolemma, and, possibly, for a short period after full dissection, which lasted only a few minutes (this duration would have been overestimated by the authors, owing to their approximate technique of observation). The diameters of the skinned fibres then reached new values, remaining constant thereafter. Indeed, in their x-ray diffraction experiments (traditional technique), with an exposure time of 40 min, Matsubara and Elliott (1972) observed no subsequent lateral swelling. The question of the kinetics of lateral swelling is also considered on pp. **243–244**.

Moving on to the question of the amount of lateral swelling, Matsubara and Elliott (1972) observed rather rapid (see the preceding paragraph)

relative lateral swelling (increase in diameter) by ~13%, under a binocular microscope. As highlighted in the first column on this page, part of this ~13% relative swelling was possibly related to the partial activation of skinned fibres, resulting from the Cl⁻-induced release of Ca^{2+} from the intact and functional sarcoplasmic reticulum. Indeed, according to the hybrid model, activation must result in net radial repulsive (expansive) between myofilaments and, therefore, in lateral swelling of the myofilament lattice and an increase in fibre diameter. Thus, if the skinned fibres studied by Matsubara and Elliott (1972) had been in purely resting conditions, they would have been expected to display slightly less relative lateral swelling than actually observed experimentally (e.g. ~8%–10% rather than ~13%; the values of ~8%–10% are not given by Matsubara and Elliott 1972). However, as seen on p. **241**, for mechanically skinned frog fibres, the hypothetical initial increase of ~8%–10% in the diameter of a skinned fibre in the full resting state would have resulted in a smaller increase in lattice spacing, which would have been observed by x-ray diffraction between $\sim(1.08\text{–}1.10)^{1/3}$ ~1.03 (~3%) and ~8%–10% (see p. **241** concerning the cubic power and proportionality). The initial increase, just after removal of the whole sarcolemma, in the lattice spacing of the skinned fibres studied by Matsubara and Elliott (1972), which would have been observed by x-ray diffraction (between ~3% and ~10%), is slightly smaller than the ~7%–20% suggested on p. **241** from the experimental data of Goldman and Simmons (1984a). However, as pointed out by Millman (1998), so many parameters are involved in lateral swelling that it is impossible to anticipate the new balance between radial repulsive electrostatic, radial tethering and attractive/compressive forces, after sarcolemma removal and, therefore, to compare the quantitative results obtained by independent groups. Finally, partial activation in the experiments of Matsubara and Elliott (1972) (see above) can be neglected, as the crude value of ~13% (recalled at the beginning of this paragraph) for the diameter increase actually ranges between ~3% and ~81% (see p. **241**).

In resting fibres, before skinning, strong radial repulsive electrostatic forces are exerted between the myofilaments (see Sections 3.2 and 4.4.2.3), together with the radial tethering and attractive/compressive forces described in Sections 4.4.2.4

and 4.4.2.5.3 (including the small radial compressive forces exerted by the sarcolemma). There is a balance between these radial forces. Millman (1998) rightly pointed out that, in mechanically skinned fibre, the sarcolemmal components associated with the sarcolemma are removed, whereas they may be present in chemically skinned fibres. In any event, in demembranated fibres, the radial compressive forces exerted by the sarcolemma and associated non-sarcolemmal structures totally or partly disappear, leading to a tiny imbalance in favour of the radial repulsive electrostatic forces and a subsequent very small lateral swelling. As highlighted on p. **156** in Section 4.4.2.5.3, this initial limited lateral swelling is the starting point of the behaviour of demembranated fibres (massive lateral swelling upon demembranation), but no quantitative calculations can be proposed, owing to the profound differences between demembranated and intact fibres. In this context, it is recalled, on p. **156** in Section 4.4.2.5.3, that demembranated fibres are not equivalent to intact fibres. Nonetheless, in the following paragraphs, I use notations valid for intact unit cells belonging to intact fibres, for the sake of simplicity (this kind of approach is also used in Section 5.1.3).

According to the third point in Section 4.4.2.4 and p. **156** in Section 4.4.2.5.3, the radial attractive/compressive forces increase slightly upon lateral swelling. The radial tethering forces raise more complex problems. Indeed, the mean elementary radial tethering force, F_{cb}^0, increases considerably upon lateral swelling, even if this swelling is tiny (see Section 5.1.3). Moreover, the detachment of many A-type myosin heads (see, in Section 5.1.1, Figure 5.1, its legend, Appendix 5.I and pp. **170–172** for precise details on the A-type heads) is expected for significant lateral swelling, as demonstrated on pp. **165–167** in Section 4.4.2.8. Assuming that a rapidly increasing proportion of 'active weakly bound cross-bridges' (see pp. **170–172** in Section 5.1.1 concerning these structures) detach upon gradual lateral swelling, the mean radial tethering force, $N_0 F_{cb}^0$, may increase or decrease, or even remain unaltered (this notion of mean radial tethering force, corresponding to a mixture of unit cells in which the A-type myosin heads are detached and those in which they are not, is introduced in Section 5.1.3). Indeed, in case of significant increases in the surface-to-surface spacing, N_0 decreases

rapidly, whereas F_{cb}^0 increases rapidly. Thus, it is impossible to predict whether $N_0 F_{cb}^0$ increases or decreases and to anticipate the lateral spacing corresponding to the new balance between the radial repulsive electrostatic forces, tethering and attractive/compressive forces, as already pointed out by Millman (1998). However, Millman (1998) did not take into account the radial tethering forces, because their existence had not yet been experimentally demonstrated in 1998 (see Section 4.3 for experimental demonstrations and discussions of these forces). Millman (1998) therefore suggested that, after removal of the sarcolemma, the new lattice spacing results from a new balance between the radial repulsive electrostatic forces and the elastic attractive forces exerted by the Z discs and the M lines and a minor effect of the compressive forces exerted by part of the sarcolemma remaining in certain demembranated fibres and obviously in half-fibres (see the third point in Section 4.4.2.4 and p. **156** in Section 4.4.2.5.3). In the hybrid model, the major factor, making it difficult to predict the amount of lateral swelling, is the difficulty involved in estimating variations in the whole radial tethering force, owing to the 'unpredictable' variations of $N_0 F_{cb}^0$ when the myofilament lattice swells (see above). In any event, extremely interesting phenomena occur if there is too large a degree of lateral swelling, as pointed out in the next paragraph.

As experimentally demonstrated by Matsubara and Elliott (1972), moderately rapid lateral swelling (within a few minutes) occurs upon mechanical skinning, owing to removal of the sarcolemma (see p. **242** for a brief discussion). The duration of this swelling is of the same order of magnitude as the various durations observed by Tsuchiya (1988) for a different type of mechanically skinned fibre (from frog and using the old skinning technique described by Natori 1954a,b). The exact duration of the lateral swelling process is probably related to the procedure used for sarcolemma removal and certainly depends on the biological material and preparation technique (e.g. mechanical or chemical skinning, permeabilisation). Instantaneous lateral swelling would have been observed if the sarcolemma had been instantaneously removed. Indeed, in resting half-fibres, a rapid decrease in bulk ionic strength of the buffer (duration of ~300 ms; see p. **14** in Section 2.2) from ~40 mM to ~30–20 mM leads to a considerable increase in fibre diameter of ~20% (at $I \sim 20$–30 mM), in ~1–2 s or less (see pp. **165–167** in Section 4.4.2.8). At these low ionic strengths of ~20–30 mM, the active weakly bound cross-bridges (see the preceding paragraph for references) are all detached and radial tethering forces are no longer exerted (see again pp. **165–167** in Section 4.4.2.8), resulting in almost instantaneous lateral swelling. In this event ($N_0 \sim 0$), maximal lateral swelling results from the new balance between the forces analysed by Millman (1998) and here: radial repulsive electrostatic forces (see Sections 3.2 and 4.4.2.3) and the remaining attractive/compressive force exerted by the Z discs, the M lines and the half-sarcolemma (see the third point in Section 4.4.2.4 and p. **156** in Section 4.4.2.5.3). This new balance is established almost instantaneously, as recalled above. For the half-fibres, the processes operate rapidly to generate the 'enormous' lateral swelling (see above), whereas in the experiments of Matsubara and Elliott (1972), performed on mechanically skinned fibres, they were slower, even operating very slowly in the case of chemical skinning or permeabilisation (see the beginning of the preceding paragraph). The kinetics of lateral swelling is therefore highly dependent on the biological material and the technique of demembranation, for example.

As an intermediate conclusion, it should be noted that the lateral swelling part of the hybrid model explains the large degree of swelling of fibres upon demembranation, although it is impossible to predict quantitatively the amount of swelling in a given fibre, owing to the complex behaviour of demembranated fibres, recalled on pp. **243–244** (this accounts largely for the difference in the extent of lateral swelling between fibres).

The traditional technique used to reduce the initial lateral swelling in demembranated fibres involves lateral osmotic compression, using large impermeant polymers that do not penetrate the myofilament lattice (e.g. dextran or polyvinylpyrrolidone). Many experiments have been performed on demembranated fibres, and many of these experiments were reviewed by Millman (1998). If there is moderate compression, its major effect is to reduce the initial lateral swelling of resting demembranated fibres and also to increase the normalised axial isometric tetanic tension in contracting demembranated fibres (this notion of normalised tension

is frequently used in this monograph, to avoid the trivial geometric effect owing to the reduction in cross-sectional area in compressed fibres). Indeed, Goldman and Simmons (1984a), for instance, observed an increase in axial stiffness under moderate compression. Stiffness is roughly proportional to axial tension for a given biological material under similar conditions, and their experimental observations demonstrate that moderate compressions result in an increase in normalised axial isometric tetanic tension (see also p. **185** in Section 5.1.3, p. **187** in Section 5.1.4, and p. **252** in Section 8.9). On the other hand, if compression is strong, complex phenomena, mostly geometric and structural in nature, occur between the thick myosin and thin actin filaments (Bachouchi and Morel 1989a; see also pp. **251–254** in Section 8.9). Thus, only moderate compressions are taken into account here. In these conditions, the radial repulsive electrostatic forces exerted between myofilaments (at rest or under contraction conditions) increase with decreasing interfilament spacing (see Section 3.2 for comments on the radial repulsive electrostatic forces; on p. **31**, it is recalled that these forces increase exponentially with decreasing interfilament spacing). Moreover, under contraction conditions, the mean residual radial tethering force, NF_{cb}, decreases, even with moderate decreases in interfilament spacing (see Sections 5.1.2 for precise details on NF_{cb}, and 5.1.3, particularly p. **185**, concerning the decrease in NF_{cb}). The other radial elastic forces (related, for instance, to Z discs and M lines) also decrease slightly through decreases in lateral elastic strength as myofilament spacing decreases. The straightforward consequence of limited osmotic compressions is that the net radial repulsive (expansive) forces increase greatly, mostly caused by an increase in the repulsive electrostatic force, together with a decrease in the residual radial tethering force, NF_{cb}. According to the hybrid model, this increase in net radial repulsive force/normalised tension is turned into an increase in isometric axial contractile forces, as demonstrated logically and experimentally (see pp. **102–103** in Section 3.10 concerning conversion of radial forces into axial forces and above concerning the normalised tension and its properties). As illustrated on p. **158** in Section 4.4.2.5.4, it is much more difficult to propose such a simple explanation for the behaviour of actively contracting skinned fibres in the absence of both effective radial repulsive electrostatic forces and tethering forces (this is the case in pure swinging cross-bridge/lever-arm theories). For instance, it is imperative to introduce 'extra ad hoc assumptions' to these traditional models, related to the rotation of the cross-bridges only, which are thought to generate strong radial forces that can surprisingly be expansive, zero or compressive (see Section 3.8.2 for precise details and references).

Other physiological parameters that are generally ignored should also be taken into account. For instance, Wang and Fuchs (1995; the term force is used for normalised tension) confirmed that moderate osmotic compression affects isometric force development in skinned bundles from both cardiac and skeletal muscles, causing an increase in force, as described in the preceding paragraph. A large increase in osmotic compression leads to a decrease in isometric force, which is consistent with the analysis and discussion presented on pp. **251–254** in Section 8.9 and based on complex phenomena mostly resulting from steric hindrance of the cycling cross-bridges. However, for cardiac cells, as pointed out by Fabiato and Fabiato (1976, 1978) and Wang and Fuchs (1995), some of the change in isometric force is correlated with changes in calcium sensitivity (which varies with both sarcomere length and osmotic compression, i.e. lattice spacing), and a decrease in filament separation would result in a decrease in calcium sensitivity (Fuchs and Wang 1996). As recalled by Millman (1998), the 'length dependence [i.e. also lattice spacing dependence] of calcium activation may explain the Frank–Starling relationship', which was demonstrated by Lakatta (1987). More recently, Yagi et al. (2004), working on rat papillary muscle, demonstrated that sarcomere length and radial mass transfer of cross-bridges are intimately related, and this is the origin of the Frank–Starling law. All the same, I believe that variations in the radial forces (mostly repulsive electrostatic and tethering) are major factors, accounting for the behaviour of skinned fibres under osmotic compression, but calcium sensitivity would modulate the main features of these fibres, particularly at high levels of compression (see pp. **252–257** in Section 8.9 for precise details regarding the problem of strong compression). For moderate compression, at around the slack length of the fibres, as studied at many places in this section, calcium almost certainly plays a marginal role.

8.8 INCREASE IN ISOMETRIC TETANIC TENSION (SHORT TETANI–STEADY STATE) WITH TEMPERATURE. CASE OF RABBIT, WITH SOME INSIGHTS INTO THE CASE OF FROG

I should begin this section by drawing attention to the review by Kawai (2003) on the effect of temperature on isometric tension and tension transients in mammalian striated muscle fibres. The author analysed many experimental data, including his own and those obtained by other independent groups, in terms of the swinging cross-bridge/lever-arm models. A major assertion is that the 'tension per cross-bridge remains the same at different temperatures'. However, to reach this conclusion, the author introduced a 'six state cross-bridge model'. On p. **39** in Section 3.4.2, I express major reservations on the use of multistate models. In this section, I propose an entirely different approach, on the basis of the various results and discussions presented in this work, in terms of the hybrid model.

In intact and chemically skinned/permeabilised fibres, isometric tetanic tension increases with increasing temperature. On pp. **62–63** in Section 3.4.3.2, the Q_{10} for intact frog fibres is discussed and it is concluded that a uniform value of ~1.24, between 0°C and 20°C, can be retained. To the best of my knowledge, no data are available for temperatures above 20°C, because the frog is a cold-blooded animal and intact frog fibres are certainly very fragile at high temperatures and therefore rarely used at room temperature or above. In the absence of any extra phenomena, the Q_{10} certainly decreases at temperatures above 20°C (see p. **201** in Section 6.3.3 for the mode of calculation of Q_{10} as a function of temperature) and we can suggest an approximately constant value of ~1.20 between ~0°C and ~40°C–45°C. This value is rather low, indicating that, for intact frog fibres, the isometric tetanic tension depends little on temperature, unless the critical temperature inducing contracture is reached (see Section 8.5). For demembranated fibres from rabbit psoas muscle, the situation is more complex. Indeed, working on this biological material, Hilber et al. (2001) obtained a Q_{10} of ~3.5 at ~7°C–10°C and ~1.30 at ~20°C–25°C. Some of these differences in behaviour are probably related to the use of different biological materials (intact frog fibres vs. chemically skinned/permeabilised rabbit fibres). However, the major problem is probably, as repeatedly suggested in this monograph, that warm-blooded animals (e.g. rabbit) are frequently studied at too low a temperature (unfavourable conditions), whereas cold-blooded animals (e.g. frog) are generally studied at low temperature (favourable conditions). Indeed, as an illustrative example, the problem of the lability of cross-bridge order for rabbit muscle at low temperatures is described and discussed on p. **150** in Section 4.4.2.5.2 and p. **247**. When demembranated rabbit fibres are studied at high temperature (e.g. ~20°C–25°C), closer to the body temperature of 39°C (favourable conditions), the Q_{10} is ~1.30 (see above), comparable to ~1.20 for the frog within the range ~0–(40–45)°C (see above). Nonetheless, rabbit is frequently studied at temperatures between ~10°C (sometimes 5°C; see Table 1 in the paper by He et al. 1997) and ~15°C, and at room temperature (~20°C–22°C). For example, other experiments were performed by Hilber et al. (2001) and, from their various figures, we obtain, at 5°C, 10°C, 12°C and 15°C, values of Q_{10} of ~3.5, ~3.5, ~3.5 and ~2.4, respectively. Thus, for demembranated fibres from rabbit psoas muscle, Q_{10} is probably ~3.5 for the transition from ~5°C (also almost certainly ~0°C) to ~7°C–10°C (also up to 12°C, as suggested here), ~2.4 at 15°C and finally ~1.30 at ~20°C–25°C. As far as I know, there are no experimental data regarding the dependence on temperature of demembranated frog fibres. At this juncture, I suggest combining data for intact frog fibres and skinned rabbit fibres and I obtain the following values for a 'composite animal and biological material': Q_{10} ~ (1.20 + 3.5)/2 ~ 2.4 between ~0°C and ~12°C, Q_{10} ~ (1.20 + 2.4)/2 ~ 1.8 at ~15°C and Q_{10} ~ (1.20 + 1.30)/2 ~ 1.25 between 20°C and 25°C. Finally, I also recommend selecting Q_{10} ~ (2.4 + 1.8)/2 ~ 2.1 between ~12°C and ~15°C and Q_{10} ~ (1.8 + 1.25)/2 ~ 1.5 between ~15°C and ~20°C. One point of interest is the correction required to pass from room temperature (~20°C–22°C, mean value of ~21°C) to 10°C (reference temperature; see pp. **96–97** in Section 3.8.1). I select a barycentric mean value, given by Q_{10} ~ (2.4 × 2°C + 2.1 × 3°C + 1.5 × 5°C + 1.25 × 1°C)/11°C ~ 1.8. This estimate of ~1.8 is assumed to be valid up to ~30°C. These approximate values are used on pp. **96–97** in Section 3.8.1 to correct the many experimental values from independent groups working at temperatures between ~0°C and ~20°C–22°C (room temperature).

Given their dependence on the absolute temperature, the electrostatic forces are practically independent of temperature within the ~0–(40–45)°C range, provided that the surface electrical potentials, via $Q_x(I^*)$ (see text below Equation 3.1 for the definition of the parameter $Q_x(I^*)$), are also practically independent of temperature. Regini and Elliott (2001) studied small bundles of glycerinated fibres from rabbit psoas muscle between ~10°C and ~35°C, under resting or rigor conditions. They showed that, in both the rigor and relaxed states, the number of fixed negative electrical charges decreased with increasing temperature. They also showed that, in rigor, there was a dramatic step-function decrease in fixed negative electrical charges at ~28°C. There is also a decrease in these charges in relaxed muscle at around the same temperature, but the step-function is less distinct (see their Table 3). In relaxed muscle, the authors found fixed negative electrical charges to be essentially constant between ~10°C and ~25°C, decreasing significantly between ~25°C and ~30°C and becoming practically constant again between ~30°C and ~35°C (see their Table 3). As pointed out by the authors, the density of fixed negative electrical charges is not the only parameter involved. For example, when temperature is increased, bound anions may be released from the myofilament surfaces, but it is impossible to anticipate the number of bound anions released with increasing temperature. The various phenomena observed by Regini and Elliott (2001) are probably related to temperature-induced conformational changes in the thick myosin or thin actin filaments (see the following paragraphs).

Let us consider, as an illustrative example, the cases of the myosin heads (S1) and the synthetic thick myosin filaments in vitro, in the absence of F-actin. In the presence of 0.15 mM MgATP (relaxing conditions), the S1 monomer has been shown to undergo an abrupt conformational change at ~20°C–21°C (Morel and Guillo 2001). Furthermore, we found that the S1 dimer dissociates at ~21°C–22°C. It has also been observed, for synthetic thick myosin filaments, that the external head–head dimers (between interacting filaments) dissociate at ~ 27°C–29°C (see p. **18** in Section 2.4) and that the internal head–head dimers, buried within the filament core (see Figure 5.1), dissociate at ~37°C–38°C (see, in Section 5.2, Figure 5.2, its legend and Appendix 5.II). Barman and Travers (1985) and Barman et al. (1980), working in a special buffer (containing 40% ethylene glycol), observed temperature transitions for MgATPase activity within the range of ~12°C–14°C. I think that the series of experimental results obtained by Regini and Elliott (2001), Barman and Travers (1985), Barman et al. (1980) and my group are consistent, as they demonstrate that many phenomena occur as temperature increases. Other experimental data from various origins have also accumulated, concerning the myosin molecules included in thick myosin filaments and also thick myosin filaments in situ. Indeed, Ueno and Harrington (1981, 1986a,b) obtained experimental results that may be interpreted (e.g. Goldman et al. 1987) as showing that temperature changes lead to a reversible helix-to-coil transition in myosin subfragment 2 (S2; see Figure 5.1 regarding this part of the myosin molecule). Wakabayashi et al. (1988) also observed temperature-induced changes in thick myosin filaments in situ. Finally, several experiments have shown that S1 undergoes complex conformational changes with increasing temperature (e.g. Morita 1977; Morita and Ishigami 1977; Redowicz et al. 1987; Shriver and Sykes 1981).

Using traditional x-ray diffraction, Wray (1987) and Wray et al. (1988) studied helical order in the thick myosin filaments of demembranated fibres from rabbit psoas muscle and suggested that these structures would be 'disordered' at low temperature but would become increasingly ordered into a helical structure as the temperature increased (see also p. **150** in Section 4.4.2.5.2). In another experimental study, also based on x-ray diffraction from synchrotron radiation on demembranated fibres from rabbit psoas muscle, Malinchik et al. (1997) demonstrated that, under both resting and contracting conditions, complex structural temperature-induced changes occur in weakly bound cross-bridges (at rest) and in cross-bridges (under contraction conditions), and maybe also in thick myosin filament backbones. In demembranated fibre bundles stretched to non-overlap between the thick myosin and thin actin filaments, to prevent complications as a result of interactions with thin actin filaments, Xu et al. (1999), using x-ray diffraction from synchrotron radiation, suggested that the myosin heads (M) would have to be in the M.ADP.Pi state to display a good helical order, when observed at 20°C. By contrast, at 0°C, the myosin heads would be in the

M.ATP state and the thick myosin filaments would appear disordered. Xu et al. (2003) confirmed these findings and demonstrated a temperature dependence of conformation and helical order in thick myosin filaments from rabbit. Xu et al. (2006a,b) further confirmed these conclusions and showed that, under relaxing conditions, at ~25°C, in permeabilised fibres from rabbit psoas muscle, ~95% of the myosin heads were in the M.ADP.Pi state (good helical order; see above).

The experimental observations recalled in this section are consistent with the previous results of Regini and Elliott (2001) obtained with small bundles of glycerinated fibres from rabbit psoas muscle, those obtained by my group and the conclusions drawn from very different approaches. Regardless of the mechanisms involved in all these phenomena, they seem to result in a decrease in the fixed negative electrical charges and bound anions on the myofilaments, particularly the thick myosin filaments and possibly the thin actin filaments, with the surface electrical potentials on the myofilaments probably decreasing with increasing temperature, at least under resting conditions. Moreover, as recalled, on p. **245**, the Q_{10} for isometric tetanic tension in demembranated rabbit fibres is ~3.5 at ~7°C–10°C and only ~1.3 at ~20°C–25°C. This almost certainly confirms that, under contraction conditions, major and abrupt conformational changes in the myofilaments (particularly the thick myosin filaments) occur when the temperature increases, as also observed in vitro (see the two preceding paragraphs). There are therefore many concordant observations showing that, at the low or moderate temperatures used in the laboratory for rabbit fibres (between ~5°C–10°C and ~20°C–22°C), various conformations of the myofilaments exist that are highly temperature dependent, accounting for the two very different values of Q_{10} (~3.5 or ~1.3). From the experimental observations, recalled in this and the two preceding paragraphs, it may be inferred that the total electrical charge (fixed negative electrical charges plus bound anions) varies with increasing temperature, at least within the range ~5°C–30°C. However, even if a substantial variation in electrical surface potentials results from these phenomena, the corresponding variations in Q_r (at rest) must be smaller, because the mathematical function 'tanh' depends little on electrical surface potentials when these potentials are of the order of ~−100/131 mV (see pp. **161–162** in Section 4.4.2.6). Thus, only a limited variation in radial repulsive electrostatic forces would be expected. Regini and Elliott (2001) suggested that disorder would also gradually appear in the myofilament lattice with increasing temperature. This phenomenon may result in a decrease in the vectorial radial repulsive electrostatic forces, which would add to the possible limited variations in Q_r. Nevertheless, such disorders occur in resting conditions, under physiological conditions of ionic strength and pH, when there are only a few weakly bound cross-bridges (~9%; see p. **172** in Section 5.1.1). Under contraction conditions, many cross-bridges are attached (~82%; see Section 4.4.2.5.2, particularly pp. **152–153**), almost certainly stabilising the myofilament lattice and leading to a weaker 'disordering' effect than at rest. It can be concluded from this long discussion, that, during contraction, the radial repulsive electrostatic forces depend little on temperature.

According to the hybrid model, ~40% of the isometric tetanic tension should be observed as soon as some or all of the internal B–B′-type head–head myosin dimers dissociate in half-fibres, intact fibres and intact unit cells (see, in Section 5.1.1, pp. **170–172** and Figure 5.1, its legend and Appendix 5.I concerning these dimers; see Sections 5.1.2, 5.2, 6.3.5, 6.4 and 6.5 regarding dissociation). In isometrically contracting fibres, dimer dissociation is incomplete at low and moderate temperatures (e.g. ~0°C–15°C) and some internal B–B′-type head–head dimers remain bound, generating a residual radial tethering force, NF_{cb}, which is much weaker than the resting radial tethering force, $N_0F_{cb}^0$ (see p. **158** in Section 4.4.2.5.4, in which it is strongly suggested that, at 10°C, $NF_{cb} \sim 0.14N_0F_{cb}^0$). From Equation 5.2, it can be deduced that the residual radial tethering force, NF_{cb}, exerted under contraction conditions, should decrease with increasing temperature, because the number $n^*(\theta)$ of dissociations of internal B–B′-type head–head dimers increases, automatically resulting in a decrease in NF_{cb} (see Section 5.1.2, from which it can be inferred that there is a close, but unknown, correlation between $n^*(\theta)$ and NF_{cb}; i.e. when $n^*(\theta)$ increases, NF_{cb} should automatically decrease). According to Table II in the paper by Morel and Pinset-Härström (1975a), the rather loosely shaped myosin heads in vitro (S1) contain ~65% random coil and only ~35% α-helical structure (vs. up to ~90% for the rigid parts S2 and LMM; see Figure

5.1 for definitions). The heads are therefore probably rather 'compliant' and part of the decrease in NF_{cb} with increasing temperature may be related to a slight decrease in the stiffness or a slight increase in the length of S1 (i.e. a kind of dilatation) and a subsequent small additional decrease in the mean elementary radial residual tethering force F_{cb}. The 'stiffness' of the A-type heads (defined on pp. **170–172** in Section 5.1.1, see also Figure 5.1) therefore makes a contribution to the strengths of F_{cb} (and also F_{cb}^0), when the temperature is changed. Goldman et al. (1987) have already discussed the problem of the thermoelasticity of thin actin and thick myosin filaments and cross-bridges. The general notion of the radial elasticity of attached cross-bridges is not new (e.g. Xu et al. 1993), although interpretation of the experiments is not straightforward, by contrast to what was previously thought by Xu et al. (1987) (see also pp. **244–245** in Section 8.7 and Section 3.8.2 for other dubious inferences drawn by independent groups). Most of the elasticity is essentially associated with the myosin heads (also frequently called the myosin 'motors'; e.g. Chaudoir et al. 1999; Dominguez et al. 1998; Iwai et al. 2006; Piazzesi et al. 2007, and almost all recent papers), as emphasised above, on purely logical bases.

In the hybrid model, the small increase in isometric tension on increasing temperature should be regulated mostly by the decrease in residual radial tethering forces, NF_{cb}, resulting in an increase in net radial expansive forces, because the radial repulsive electrostatic forces are little dependent on temperature (see pp. **247–248** for a circumstantial discussion) and also by a small increase in the efficiency of the swinging of the extruding heads. This efficiency of swinging was discussed by Linari et al. (2005), working on intact frog fibres, but solely in terms of the traditional axial tilting of the myosin heads (swinging cross-bridge/lever-arm theories). They considered that their experimental data should be interpreted as meaning that this axial tilting is slightly higher at 17°C than at 0°C. To my mind, as demonstrated throughout this monograph, the overall increase in isometric tetanic tension with increasing temperature probably results from the two very different processes (lateral swelling mechanisms plus swinging cross-bridge/lever-arm processes) underlying the hybrid model. However, the dependences on temperature of the directly (swinging cross-bridge/lever-arm processes) and indirectly (lateral swelling process) generated axial contractile forces are very different. Although it is repeatedly pointed out in this work that it is difficult to compare demembranated fibres with intact fibres or half-fibres, the conclusion, concerning the 'double dependence' of axial contractile force on temperature, possibly provides an additional explanation for the biphasic increase in isometric active tension after a temperature jump in demembranated fibres from rabbit psoas muscle (see p. **237** in Section 8.5 for the final comment on the experiments of Bershitsky and Tsaturyan 1992).

8.9 SHORT TETANI AND STEADY-STATE CONDITIONS: THE TWO LENGTH–TENSION RELATIONSHIPS IN STRETCHED INTACT FROG FIBRES (LINEAR DESCENDING LIMB AND PARABOLA [FLAT CURVE]) AND RELATED PROBLEMS. COMPLEXITY OF THE SUBJECTS CONSIDERED

We must consider a major problem that I and my coworkers began to study approximately 40 years ago (Morel et al. 1976). What is the true length–tension relationship in stretched intact frog fibres? How can the hybrid model account for this relationship? At the end of the 1970s, a major controversy emerged concerning the true relationship, the protagonists of the debate each presenting their various experimental arguments and interpretations (e.g. Julian and Morgan 1979a,b; Julian et al. 1978a,b; Noble and Pollack 1977, 1978). In my group, the problem was also discussed, based on a simplified phenomenological/semi-empirical approach, deduced from the first illuminating biophysical model of muscle contraction presented by A. F. Huxley (1957). Our approach was independent of the mechanical roles of the cross-bridges, assuming only that they cyclically attach to and detach from the thin actin filaments and split MgATP during the cycle (e.g. Morel 1978, 1984a,b; Morel and Gingold 1977; Morel et al. 1976). We deduced, from the work of A. F. Huxley (1957), an 'energetic' expression of isometric tetanic tension, independent of the mechanical roles of the cross-bridges, which is again used in this monograph, in an improved form (see Section 3.4; see also this section and the general conclusion for additional precise details). I think that it is important to reconsider here this problem, in

the light of old and recent experimental, semi-empirical and structural results obtained by my group and several other independent groups. The analysis presented in this section includes a short historical account of the various problems.

Edman and Reggiani (1987), Granzier and Pollack (1990), Hashizume (1977) and ter Keurs et al. (1978) studied intact frog fibres (stimulated electrically). Fabiato and Fabiato (1978) studied skinned cardiac cells from frog (in this case, contraction was triggered by Ca^{2+}). All these authors found that the tension–length relationship did not follow the 'traditional' linear descending limb, in stretched fibres. Edman and Reggiani (1987) found that the 'descending limb' was slightly sigmoidal, whereas the other authors found it to be parabolic (also called 'flat shape'). This contrasts sharply with the approximate linear relationship obtained by Ramsey and Street (1940) (see also Gordon et al. 1966b for improvement of the results and discussion of the paper by Ramsey and Street 1940). This linear relationship was confirmed by Altringham and Bottinelli (1985), Bagni et al. (1988b), Edman (1966), Edman and Anderson (1968), Gordon et al. (1966b), Granzier and Pollack (1990) and Morgan et al. (1991), who studied isometric tetanic tension for sarcomere lengths from ~2.2 µm to ~3.6–3.7 µm (Edman 1966 and Edman and Anderson 1968 studied tension up to sarcomere lengths of only ~3.0 µm). Note that Altringham and Bottinelli (1985) and Granzier and Pollack (1990) obtained two possible length–tension relationships, that is, the linear descending limb and the parabolic (flat) shape (see below, in this section). Thus, at the end of the 1980s and the beginning of the 1990s, the situation remained unclear and few groups tried to resolve this problem. Most investigators ignored, and continue to ignore, the parabolic shape. The swinging cross-bridge/lever-arm models are mostly, and possibly even exclusively, based on the experimental observations of a purely linear descending limb, including the careful and detailed experimental study by Gordon et al. (1966b), which is generally considered beyond dispute and the essential basis of the swinging cross-bridge/lever-arm theories. However, as an illustrative example of the puzzle, ter Keurs et al. (1978), on the one hand, and Gordon et al. (1966a,b), on the other, working on the same type of biological material, at similar temperatures, used very different definitions of the 'true' isometric tetanic tension. These differences were at the heart of the dispute. Indeed, ter Keurs et al. (1978) took the true isometric tetanic tension to be that corresponding to the plateau in tension reached after a sufficient duration, but fulfilling the imperative condition of a short tetanus, lasting less than ~1.0–1.5 s (see Section 3.4.3.2 for precise details on this major question). By contrast, Gordon et al. (1966a,b) took the true tetanic tension earlier, just before the 'creep' they described and which they attributed solely to rearrangements in sarcomere lengths. These rearrangements were experimentally studied and confirmed by Cleworth and Edman (1972) and Edman and Reggiani (1984) and theoretically by Morgan et al. (1982). Thus, Gordon et al. (1966a,b) extrapolated the very rapid rise in tension after electrical stimulation and the part of the tension corresponding to the slow increase in tension, which they referred to as creep resulting from sarcomere rearrangements (see above and Figure 3 in the paper by Gordon et al. 1966b). The intersection of these two straight lines was assumed to give the true isometric tetanic tension. Their experiments lasted ~1.0–1.5 s and the conditions of short tetani, studied throughout this monograph and in Section 3.4.3.2 in particular, were fulfilled. In fibres at the slack length (sarcomere length of ~2.0–2.2 μm, corresponding to the plateau in the length–tension relationship; see the review by A. F. Huxley 2000) and at 3°C–4°C, there was no detectable creep (see, for instance, Figure 1 in the paper by Gordon et al. 1966b). By contrast, in stretched and highly stretched fibres, creep was clearly observed and its duration increased considerably with increasing sarcomere length (see Figure 9 in the paper by Gordon et al. 1966b). Working on white intact fibres from dogfish, at 12°C, West et al. (2004) observed exactly the same phenomenon of creep in stretched intact fibres. Independently of the mechanical roles of the cross-bridges and regardless of any rearrangements in sarcomere lengths (see above), Morel (1984a) demonstrated that the increase in creep duration with sarcomere length is an intrinsic property of an isometrically contracting intact fibre, in which the cross-bridges display the traditional cycle (attachment to and detachment from the thin actin filaments, with cleavage of one MgATP molecule per cycle). Taking the 'necessary drastic precautions' to control the sarcomere length of a fibre, some

of which were described by van Heuningen et al. (1982) and later improved, Altringham and Bottinelli (1985), Granzier and Pollack (1990) and Morgan et al. (1991) confirmed the linear length–tension relationship. For most investigators, the straightforward conclusion is that only the swinging cross-bridges generate the full axial contractile force, the cross-bridges acting independently, as has been claimed since 1966, based on the experiments and interpretation of Gordon et al. (1966a,b). At this point, the controversy seemed to have been resolved. However, Morel and Merah (1995) and Pollack et al. (1993) tried unsuccessfully to reopen the discussion, with entirely different arguments. Starting from the phenomenological/semi-empirical approach independent of the mechanical roles of the cross-bridges first introduced by Morel et al. (1976), Morel and Merah (1995) asked whether the cross-bridges were actually independent force generators. They identified several inconsistencies between this notion of independence, phenomenological/semi-empirical conclusions and structural data and concluded that this question remained unresolved. Let us examine this situation again.

Morel (1978, 1984a,b), Morel and Gingold (1977) and Morel et al. (1976) showed that the mechanics of intact unit cells and single intact fibres can be studied independently of the mechanical roles of the cross-bridges, provided only that they attach to and detach from the thin actin filaments and split MgATP during this cyclic process. This phenomenological/semi-empirical approach remains topical and powerful (see Section 3.4). The experimental part of this approach was and is still based on the long-standing observation that there is a shortening heat (Fenn effect; Fenn 1923, 1924; Fenn and Marsh 1935; quantified by A. V. Hill 1938 and then revisited by A. V. Hill 1964a,b). Curtin and Woledge (1978), Woledge (1971) and Woledge et al. (1985) claimed that the theory of A. F. Huxley (1957) is incompatible with the Fenn effect. Morel (1984a,b) introduced some apparently minor, but actually very important, modifications to A. F. Huxley's theory and demonstrated that the improved approach is fully compatible with the experiments analysed by Curtin and Woledge (1978), Woledge (1971) and Woledge et al. (1985) provided that the various experimental data are reinterpreted. At many places in Section 3.4.3, the correlation between the rate Π of MgATP splitting and the rate R of turnover of cycling cross-bridges is studied through the notion of inactive contacts, the proportion of which increases with shortening velocity (see p. **43** in Section 3.4.3.1.1). This notion makes it possible to avoid the problems raised by Curtin and Woledge (1978), Woledge (1971) and Woledge et al. (1985). Otherwise, Morel and Gingold (1977) and Morel et al. (1976) deduced, from the biophysical approach proposed by A. F. Huxley (1957), a purely energetic expression of the load lifted plus the coefficient of shortening heat, dependent solely on the characteristics of the attachment–detachment cycle of the cross-bridges (constants of attachment, f, and detachment, g) and on the velocity of shortening s_0V of the fibre. This approach, improved in Section 3.4 (Subsection 3.4.3 in particular), is totally independent of the mode of generation of the axial contractile force, as pointed out above and asserted throughout this work. The phenomenological/semi-empirical expression obtained is applicable when the velocity of shortening is zero (isometric contraction), and it is therefore possible to study the isometric tetanic tension under various conditions. For example, it is possible to study objectively the length–tension relationships and many other problems, regardless of the mode of generation of the contractile force. The phenomenological/semi-empirical approach proposed by Morel (1978), Morel (1984a,b), Morel and Gingold (1977) and Morel et al. (1976) was later improved and has been successfully used ever since by my group and in this book.

Morel (1984a) improved the previous phenomenological/semi-empirical approach proposed by Morel (1978), Morel and Gingold (1977) and Morel et al. (1976). In 1984, I suggested that some of the apparently biphasic rise in tension was related to the 'burst' in MgATPase activity (Pi and H^+ bursts), corresponding to the very rapid initial rise in tension (only a few tens of milliseconds; see pp. **210–211** in Section 6.3.5 for a circumstantial discussion; see also Figure 1 in the paper by Gordon et al. 1966b for purely experimental aspects), followed by lower transitory levels of MgATPase activity, corresponding to the slow increase in tension (creep). In Section 3.4.3.1.3, it is semi-empirically demonstrated that there is a rather loose correlation between the initial rise in tension and the changes in the rate of MgATP splitting (see Equation 3.22 and corresponding

comments, in which it is demonstrated that this correlation is complex). The estimate of the duration of the transitory period, Δt^*, for isometric tension at around the slack length of the fibre and intact unit cells (estimated at ~172–404 s, after the time lag of ~13–28 ms, at ~5.7°C; see p. **54** in Section 3.4.3.2) almost certainly includes the Pi and H^+ bursts and the creep, with a probable overlap between these phenomena. This adds further complexity to the problem of creep, mostly observed for long sarcomere lengths and described and discussed by Gordon et al. (1966a,b). Working on stretched intact fibres complicates the matter further (e.g. Bachouchi and Morel 1989a and pp. **251–254**). Thus, breaking up the rise in tension into three phases (rapid increase, followed by creep, and finally a plateau) is probably unwarranted, given the findings of phenomenological/semi-empirical studies (see Equation 3.22 and corresponding comments, demonstrating that the 'tension–time curve' is a continuous, not biphasic or triphasic, function of time t). Moreover, the plateau tension is almost certainly the true tension to be taken into account in intact fibres (this has been established for approximately 15 years; see Linari et al. 1998) and also in intact unit cells, at the microscopic level. This view is developed below, in this section. Concerning the phenomenon of creep at long sarcomere lengths, Pollack et al. (1993) claimed that the usual explanation, first suggested by Gordon et al. (1966a,b), is 'deceptively simple ... [and] has one serious fallacy' and they propose their own explanation, which is different from both that of Gordon et al. (1966a,b) and my own.

The plateau tension should be taken into account (see the preceding paragraph), for intact fibres at the slack length and stretched beyond the slack length. A parabola (also called the flat shape by Pollack and his colleagues) is more suitable than a straight line (descending limb), as demonstrated by Morel (1984a) in his phenomenological/semi-empirical approach. Furthermore, on the basis of simple geometric and structural considerations and referring to many experiments performed on osmotically compressed skinned fibres by several independent groups, Bachouchi and Morel (1989a) showed that, at sarcomere lengths exceeding ~2.4–2.5 μm, there are complex 'steric' problems concerning the external myosin heads making up a large part of the cross-bridges, for resting and contracting fibres (see Section 5.1.1, Figure 5.1, its legend, Appendix 5.I and the corresponding comments and pp. **170–172** and **174–176** concerning the external heads, the cross-bridges and the weakly bound cross-bridges). Such problems were also raised directly, based on purely experimental observations of isometrically contracting intact fibres by Granzier et al. (1989), who concluded that 'properties of the force generators change with sarcomere length'. Indeed, upon gradual stretching of the intact fibres, the interfilament spacing decreases (isovolumic behaviour of intact fibres; April et al. 1971; Brandt et al. 1967; Elliott et al. 1963, 1965, 1967; see pp. **4–5** in the Introduction for comments) and major distortions inevitably occur in the many myosin heads attached to the thin actin filaments, particularly during contraction (Bachouchi and Morel 1989a). Most heads are even crushed between the thin actin and thick myosin filaments at sarcomere lengths exceeding ~3.0 μm, regardless of the state of the fibres (at rest or under contraction conditions).

The conclusions drawn by Bachouchi and Morel (1989a) from geometric reasoning and structural data and those drawn by Granzier et al. (1989) from physiological observations are consistent with biochemical experiments performed on glycerinated fibres from rabbit psoas muscle (e.g. Morel 1984a; Tanaka et al. 1979) and with phenomenological/semi-empirical work on intact frog fibres (Morel 1984a). According to these independent and complementary studies, the specific rate of MgATP splitting per overlap cross-bridge sharply decreases for chemically demembranated fibres stretched beyond a sarcomere length of ~2.2 μm (Morel 1984a). For such fibres, stretched to a sarcomere length of ~3.68 μm (disappearance of the overlap between the thin actin and thick myosin filaments; see p. **25** in Section 3.2), the specific rate of MgATP breakdown per overlap cross-bridge (estimated by extrapolation) reaches values of only ~25% that recorded at ~2.2 μm (Morel 1984a). This value of ~25% represents only an order of magnitude, owing to the broad scattering of the experimental points (see Figure 5 in the paper by Morel 1984a). Moreover, Krasner and Maughan (1984) simultaneously measured isometric tension and MgATPase activity in osmotically compressed chemically skinned fibres from rabbit psoas muscle, at the slack length. Under osmotic compression conditions, sarcomere length and, therefore, the amount of overlap remain fixed,

but the surface-to-surface spacing between the myofilaments decreases with increasing osmotic pressure. This spacing also decreases when intact fibres are stretched (isovolumic behaviour; see the preceding paragraph for references). Thus, the decrease in interfilament spacing is common to both types of experiment, making it possible to compare these two situations. However, Ca^{2+} activation depends on sarcomere length and filament spacing (see pp. **258–259**), but I think that these phenomena are much less important than those described and discussed in this and the following paragraphs. Krasner and Maughan (1984), working on osmotically compressed skinned rabbit fibres, found that isometric force and MgATPase activity decreased when compression increased (with only a slight increase in force of ~4%, at very low compression similar experimental results have been reported by several independent groups - see p. **185** in Section 5.1.3, p. **187** in Section 5.1.4, and p. **244** in Section 8.7). Force and MgATPase activity are apparently coupled, but, in my opinion, these experimental results reveal only a loose coupling with gradual increases in compression. The authors also found that, for a centre-to-centre distance between the thick myosin and thin actin filaments corresponding to the disappearance of the overlap in stretched fibres, the isometric tension is zero, whereas the MgATPase activity is ~10%–15% that in non-stretched or non-compressed fibres (the experimental data of Krasner and Maughan 1984 were normalised by Morel 1984a). Note again the lack of close correlation between force/tension and MgATPase activity, for skinned rabbit fibres, consistent with the very loose coupling demonstrated on pp. **78–80** in Section 3.4.5 and pp. **83–84** in Section 3.4.6, for intact frog fibres, independently of the mode of force generation. This similar behaviour of the two animal species and the two types of fibres is discussed on p. **80** in Section 3.4.5. The residual value of ~10%–15% for MgATPase activity is close to the ~25% mentioned above, within the large range of experimental error. Krasner and Maughan (1984) suggested that decreasing the centre-to-centre distance between the thick myosin and thin actin filaments may hinder cross-bridge movement: 'Hindrance of cross movement is proposed to account for the inhibition of active force generation and (coupled) ATPase in compressed fibers'. This conclusion is consistent with the geometric

Morel

and structural conclusions drawn by Bachouchi and Morel (1989a). The number of cross-bridges able, in principle, to enter the cycling process, did not vary in the experiments performed by Krasner and Maughan (1984) (fixed amount of overlap). It can therefore be concluded that the specific MgATPase activity per overlap cycling cross-bridge decreases considerably when the myofilaments are brought closer together. Finally, it could be argued that the residual MgATPase activity measured at zero overlap is merely related to the existence of trace amounts of sarcoplasmic reticulum in the chemically demembranated fibres. However, this putative residual MgATPase activity is almost certainly marginal and cannot account for values of ~10%–15% and ~25%.

Kawai and Schulman (1985) and Zhao and Kawai (1993) provided definitive support for the previous experimental findings of Krasner and Maughan (1984) and for the steric hindrance assumption of Bachouchi and Morel (1989a). Similarly, Maughan and Godt (1981b) claimed that 'a likely explanation for the marked inhibition of force [at high osmotic pressure] is steric hindrance between myofilaments as they are compressed together'. More recently, Adhikari and Fajer (1996) studied external myosin head motions (see pp. **170–172** in Section 5.1.1 for comments on this notion of external heads) by EPR (electron paramagnetic resonance) in contracting, relaxed and rigor skinned fibres, from rabbit psoas muscle, subjected to various degrees of osmotic compression. They confirmed the experimental results of Kawai and Schulman (1985), Krasner and Maughan (1984), Maughan and Godt (1981b) and Zhao and Kawai (1993) and the semi-empirical, geometric and structural conclusions of Bachouchi and Morel (1989a). Indeed, for moderate osmotic compression, Adhikari and Fajer (1996) demonstrated that small increases in normalised isometric tension occur, with EPR rotational correlation times suggesting a modest change in average external myosin head motion. For high and very high levels of osmotic compression, normalised isometric tension gradually decreases and finally falls to zero and external myosin head motions are severely restricted (the initial increase in isometric tension followed by a decrease is a universal pattern of behaviour; see the review by Millman 1998; see also p. **185** in Section 5.1.3, p. **187** in Section 5.1.4, p. **244** in Section 8.7 and p. **252**). At a relative filament spacing corresponding to ~65% of the initial spacing (at zero external osmotic pressure), the normalised isometric tetanic tension is zero and the external myosin heads are entirely immobilised, as they are considered in 'rigor' conditions (Adhikari and Fajer 1996). However, to my mind, this is an involuntary abuse of language, because the external heads are merely 'crushed', and therefore immobilised, between the thick myosin and thin actin filaments, at very high osmotic pressures (Bachouchi and Morel 1989a) and this feature has nothing to do with the heads involved in the three-dimensional geometric arrowhead structure observed in rigor, at zero external osmotic pressure.

All the experimental and semi-empirical results, presented and commented on pp. **251–254**, are consistent with many myosin heads being strongly distorted in highly stretched fibres, even crushed between the thick myosin and thin actin filaments (Bachouchi and Morel 1989a), resulting in low and even very low specific MgATPase activities per overlap cross-bridge. However, a very small proportion of these highly distorted heads can cycle, with great difficulty, and split small amounts of MgATP during contraction (at approximately zero overlap, the specific rate of MgATP splitting per overlap cross-bridge is very low). As pointed out on pp. **252–253**, the experiments of Krasner and Maughan (1984) confirm the loose coupling, possibly very loose coupling, between the specific rate of MgATP splitting per overlap cross-bridge and the isometric tension recorded under these conditions (see also pp. **78–80** in Section 3.4.5 and pp. **83–84** in Section 3.4.6 concerning the very loose coupling in isometrically contracting fibres, at the slack length, for intact frog fibres, and also other animal species). Otherwise, the consistent experimental arguments and interpretations discussed on pp. **251–254** lead to the inevitable conclusion that most of the cross-bridges cannot rotate and work as traditionally described in swinging cross-bridge/lever-arm theories, which frequently ignore the major role of lattice spacing or provide hazy explanations of various experimental results obtained with stretched intact fibres or osmotically compressed demembranated fibres. It is therefore unreasonable to consider the cross-bridges as independent force generators, in pure swinging cross-bridge/lever-arm models and in the swinging cross-bridge/lever-arm part of the hybrid model (see also the paper by Morel and Merah 1995 in which the same conclusion is drawn, regardless of the mechanical roles of the cross-bridges). Thus, it is also unreasonable to consider the linear length–tension relationship to be beyond dispute. This view is commented on and supported below, in this section.

The geometric and structural considerations of Bachouchi and Morel (1989a), supported by experimental results obtained by independent groups (see pp. **251–254**), were reanalysed by Morel (1996), Morel and Merah (1997) and Morel et al. (1992). The conclusions drawn by Bachouchi and Morel (1989a) were borne out, except that the maximum chord of the myosin heads was found to be ~12–13 nm in these three series of papers, which is slightly higher than the ~11.2 nm, strongly suggested on p. **29** in Section 3.2, for the closest surface-to-surface spacing between the thin actin filaments and the backbone of the thick myosin filaments, at the reference sarcomere length of ~2.18 μm (see p. **24** in Section 3.2), at which the overlap between the thick myosin and thin actin filaments is maximal ($\gamma \sim 1.48$ μm; see p. **34** in Section 3.2). Morel and Merah (1997) reanalysed these problems of surface-to-surface spacing and drew some interesting conclusions. We cited the experimental work of Wakabayashi et al. (1988), suggesting that the actin-binding site on the myosin head is 'located from 2 nm to 6 nm from the tip of S1'. This conclusion was 'approximately' confirmed by Rayment et al. (1993a), working on S1 crystals. Thus, the problem of steric hindrance, occurring at sarcomere lengths somewhere between ~2.2 μm and ~2.4–2.5 μm and beyond (see pp. **251–254** concerning these sarcomere lengths), may be partly resolved, because the actin-binding site on a myosin head is located farther from the thin actin filaments (see discussion in the paper by Morel and Merah 1997). However, this partial 'reconciliation' is inconsistent with the decrease in specific MgATPase activity per overlap cross-bridge, recalled on pp. **251–254**. Furthermore, at higher sarcomere lengths, the problems of steric hindrance and related phenomena continue to apply.

As recalled by Millman (1998), the thin actin filaments are anchored to the Z discs in a square (tetragonal) structure, whereas they are arranged in a hexagonal structure in the A-band (in which the thin actin and thick myosin filaments coexist). Several experimental observations have been reported in this area, particularly on the Z discs and their behaviour at rest, during contraction, and in rigor (e.g. Edwards et al. 1989; Franzini-Armstrong and Porter 1964; Goldstein et al. 1986, 1987, 1988; Rowe 1971; Savel'ov 1986). Goldstein

et al. (1987) demonstrated a dependence of Z disc dynamics on sarcomere length. However, 'most recent experiments have avoided this problem' (Millman 1998). This is unfortunate, as Schiereck et al. (1992), using x-ray diffraction from synchrotron radiation on permeabilised fibres from rabbit *gracilis* muscle, at sarcomere lengths between 1.6 μm and 3.5 μm, demonstrated that the stretching of relaxed, partially and fully activated fibres leads to a gradual deformation of the hexagonal myofilament matrix: 'When sarcomere length increases the hexagonal lattice will change to a tetragonal lattice'. This largely ignored behaviour adds further complexity to the properties of stretched fibres at rest and under contraction. In this increasingly complex context, the notion of cross-bridges acting independently in a perfect double hexagonal lattice, with a subsequent linear descending limb in stretched fibres, appears again to be oversimplified (see also pp. **249–251** for a circumstantial discussion). It should, however, be recalled that the use of x-ray diffraction on demembranated fibres (e.g. Schiereck et al. 1992) also leads to major problems of interpretation (see Sections 3.8, 4.4.2.1 and 8.7), and the experimental results obtained by Schiereck et al. (1992) and their interpretation may be considered dubious. In any event, deformed unit cells in stretched to avoid confusion with the 'unit cells' in demembranated fibres and also 'unit cells' located at the periphery of the fibres in intact fibres (see Sections 3.7, 3.8, 4.4.2.1 and 8.7).

Some of the problems raised in this section and elsewhere in this monograph may be related to many of the arguments concerning intact unit cells, whereas all the experimental data concern whole fibres (intact, mechanically and chemically skinned, permeabilised) or isolated myofibrils. Throughout Chapter 3 (see particularly Section 3.7), explanations are proposed concerning the issue of intact fibres vs. intact unit cells, accounting for the gap between the 'microscopic' and 'macroscopic' tensions when an intact unit cell and an intact fibre are compared. Indeed, combining experimental results and a phenomenological/semi-empirical reasoning, it can be shown that the isometric tension, at 10°C, at the reference slack sarcomere length ($s_0 \sim 2.18$ μm; see p. **24** in Section 3.2), in an intact unit cell from the frog, is $T^* \sim 7.2 \times 10^5$ N m^{-2} (see p. **90** in Section 3.7), whereas it is only ~3.0×10^5 N m^{-2} in intact fibres and half-fibres (see Section 3.3). The simple and reasonable assumptions made throughout Chapter 3 can be retained in this section. However, it is possible that other unexpected parameters accounting for part of the 'drop' in tension when passing from intact unit cells to an intact fibre vary with increasing sarcomere length. These parameters would include, for instance, the width of the mitochondria in whole intact fibres, which take part in the 'apparent' isometric tetanic tension developed by a whole intact fibre (e.g. Merah and Morel 1993 for 'geometric' calculations) and the possible stretch-induced increase in radial and axial stiffness and elasticity of the sarcolemma and the sarcoplasmic reticulum (see p. **93** in Section 3.7 regarding the notion of stiffness/elasticity of these two structures).

Taking into account solely the energetic part of the biophysical approach of A. F. Huxley (1957), but rejecting the mechanical part (see Section 3.4 and the papers by Morel 1978; Morel 1984a,b; Morel and Gingold 1977; Morel et al. 1976; see also the general conclusion of this work), the conclusions drawn by my group over the last 40 years and the discussions presented in this book and by Bachouchi and Morel (1989a) typically concern microscopic intact unit cells (see pp. **254–255** concerning this notion of intact unit cells in stretched intact fibres and the corresponding abuse of language) whereas the linear length–tension relationship concerns macroscopic intact fibres. The problem of the rate of MgATP splitting and its very loose coupling with isometric tetanic tension concerns both intact unit cells and intact fibres, as recalled above, in this section. More generally, how should we reconcile the molecular data, valid for intact unit cells, with macroscopic data, valid for intact fibres? For example, the micromechanics of an intact unit cell are consistent with the parabolic (flat-shaped) length–tension 'micro-relationship' (see Morel 1984a). As recalled in the preceding paragraph, the behaviour of an intact fibre is different from that of an intact unit cell, at a given sarcomere length. Moreover, it is strongly suggested, in the two preceding paragraphs, that increasing sarcomere length may further increase the difference between an intact fibre and an intact unit cell. All these phenomena probably result in length–tension relationships being parabolic/flat-shaped for an intact unit cell (see above) and either linear or parabolic for the corresponding intact fibre, depending on the type of experiment performed. For instance,

the drastic mechanical constraints imposed on an intact fibre or the zone selected for observation and control of the behaviour of sarcomeres during contraction are probably extensively involved in this problem. To my mind, it is even possible to force an intact fibre to give different experimental results, that is, different length–tension relationships, with different techniques and devices, such as the use of sophisticated servo-controlled motors to keep sarcomere lengths strictly constant (this kind of problem was raised by Morel 1984a in his Appendix III entitled 'Can the cross-bridges be in a stable or metastable state?'). Thus, starting from the experimental and phenomenological/semi-empirical results recalled above, in this section, the situation is somewhat confusing. However, from Section 3.7, it appears that, under macroscopic conditions of isometric contraction, microscopic oscillations and movements occur in unit cells/intact unit cells (see Section 3.7, its title and particularly pp. **93–94** concerning the difference between unit cells and intact unit cells in an intact fibre) that are different from those in sarcomeres and whole fibres (with the possibility of cooperative phenomena when passing from unit cells to sarcomeres and then to whole fibres). On p. **67** in Section 3.4.3.3, the notion of 'purely' isometric tetanic contraction is studied and it is demonstrated that, under such strict conditions, it is not possible for any of the attached heads to rotate (zero step size distance). When no external mechanical constraints are imposed, it can reasonably be assumed that the attached heads can swing, with the occurrence of a working stroke, whereas, when drastic constraints are imposed, the notion of a working stroke becomes meaningless. This dichotomy almost certainly results in major differences and the possibility of there being two macroscopic length–tension relationships, depending on the mechanical constraints imposed on the whole fibre or on samples of sarcomeres in the fibre (see above). This view is commented on and confirmed in the next paragraph.

From the various discussions presented above, in this section, it may be inferred that both the linear and parabolic (flat) length–tension relationships are valid. For example, in the 1970s, Pollack and his group found the length–tension relationship to be parabolic (e.g. ter Keurs et al. 1978). However, Altringham and Bottinelli (1985), Bagni et al. (1988b) and Granzier and Pollack (1990), for instance, taking the 'necessary precautions' to control sarcomere length in a contracting fibre, found the traditional linear relationship. Granzier and Pollack (1990) presented a careful discussion of the 'fixed-end' technique (giving the parabolic shape) and the 'sarcomere-isometric' technique (giving the linear descending limb), and Horowitz et al. (1992) and Pollack et al. (1993) provided experimental arguments for the fixed-end technique resulting in the parabolic shape, also called the 'flat curve' by the authors, whereas the sarcomere-isometric technique results in the linear descending limb. From these observations, Pollack et al. (1993) posed four major questions: 'Why do the length–tension curves have two different shapes? Why is one flatter and higher [parabolic] than the other [linear]? Are these diverse results consistent with predictions of the cross-bridge theory? Or, is there another model that better explains the observations?' The authors provided experimental arguments to resolve these problems and suggested that the experimental data could be accounted for by the unconventional 'step-wise' theory proposed by Iwazumi (1970, 1979) and Pollack (1990). The authors developed their approach to the mechanisms of muscle contraction on the basis of 'quantal' and step-wise processes observed experimentally and largely described and commented on in the book by Pollack (1990). I agree with some inferences of Pollack et al. (1993), but I propose here and throughout this monograph an alternative approach to the interpretation of the various problems. I again confirm that it is almost certain that the drastic controls imposed by most authors (e.g. Altringham and Bottinelli 1985, Bagni et al. 1988b; Granzier and Pollack 1990) result in the parabolic relationship (which is compatible with many experimental, structural and semi-empirical results; see above, in this section) being artefactually converted into a linear length–tension relationship. Obviously, such controls on a sample of sarcomeres may have repercussions in intact unit cells. However, these macroscopic 'regulation effects', at the level of some sarcomeres, may be somewhat damped in intact unit cells, that is, at the microscopic level. The complex behaviour of an intact fibre may therefore be different from the simpler behaviour of intact unit cells. As suggested above, in this section, the parabolic relationship may be valid for intact unit cells, regardless of the techniques used in the intact fibre, whereas both the linear and parabolic (flat-shaped) relationships may be valid for intact fibres, much depending on the techniques used (see above). It is not yet possible to test this hypothesis in intact unit cells, and the problems raised here should be considered as open questions. Nevertheless, it is pointed out, on p. **206** in Section 6.3.5, that the cutting-edge techniques described and discussed by Borejdo et al. (2006) would be extremely useful for addressing the micromechanics of a single unit cell. Notwithstanding, the question of 'artefactual' results for intact fibres, owing to drastic controls of sarcomere length, is also clearly raised above. Finally, I note that ter Keurs and Elzinga (1981) reported, but did not further confirm, that, when sarcomere length is kept rigorously constant, the creep disappears and the isometric tetanic tension decreases linearly with increasing sarcomere length in stretched intact fibres. Clearly, this is consistent with the above discussions, according to which artefactual results may be obtained, depending on the technique used.

The hybrid model is entirely independent of the linear or parabolic nature of the length–tension relationship, unlike swinging cross-bridge/lever-arm theories, which are strongly linked to the linear descending limb. This statement does not rule out the possibility of some of the contractile force being related to the swinging cross-bridges, but the actual problem appears to be that of the relative proportions of the two mechanisms involved in the hybrid model, when sarcomere length is increased. At the slack sarcomere length and at 10°C, ~60% of isometric tetanic tension is related to the axial contractile forces directly generated by the swinging cross-bridges/lever-arms, and ~40% is indirectly related to the net radial repulsive forces, turned into axial contractile forces (see pp. **102–103** in Section 3.10). This conclusion is drawn from a combination of experimental results obtained and described in this work and elsewhere, and biophysical and phenomenological/semi-empirical calculations made in this monograph, for intact unit cells. Moreover, in the light of the discussions presented above, in this section, and in several other sections, it remains unclear whether the proportions of swinging cross-bridge/lever-arm and lateral swelling mechanisms are the same at the molecular and macroscopic levels. I therefore present, below, in this section, comments independent of this proportion.

In stretched intact fibres, the thick myosin and thin actin filaments are obviously closer together than at the slack length (isovolumic behaviour; see p. **252** for references). In resting stretched intact unit cells, the elementary mean tethering force, F_{cb}^{0*}, therefore decreases considerably and even reaches very low values, as for F_{cb}^{0} at the slack length; see Section 5.1.3). However, as recalled on pp. **251–254**, in resting and contracting intact unit cells, the proportion of distorted/crushed external myosin heads per unit length of overlap rapidly increases with decreasing interfilament distance and the corresponding number, N_0^*, of active weakly bound cross-bridges per unit length of overlap (see pp. **170–172** in Section 5.1.1 for the definition of N_0 and, therefore, N_0^*) is probably forced to increase (by 'crushing' between the thin actin and thick myosin filaments). This almost certainly leads to a very small residual resting radial tethering force $N_0^* F_{cb}^{0*}$ per unit length of overlap (despite the drastic decrease in F_{cb}^{0*} described above), contributing to the resting axial tension through the translation of the net radial expansive forces into axial forces (see the preceding paragraph). Another part of the resting axial force is almost certainly related to the 'radial counterforce', equivalent to a radial repulsive force, exerted by the myosin head crushed between the thick myosin and thin actin filaments. Under contraction conditions, in stretched fibres, the same 'universal' phenomenon is thought to occur as at the slack length and an extremely small residual radial tethering force $N^* F_{cb}^* \ll N_0^* F_{cb}^{0*}$ also exists and contributes to a residual active axial tension (see pp. **158–159** in Section 4.4.2.5.4 concerning the inequality between the residual radial tethering force under contraction conditions and the full radial tethering force, at rest, and at the slack length). As shown on pp. **251–254**, under contraction conditions, the specific MgATPase activity per overlap cross-bridge decreases sharply with the stretching of the intact unit cells. However, despite this decrease, the 'residual' specific MgATPase activity is effective and, in stretched intact unit cells and fibres, the 'net radial repulsive force' part of the hybrid model is functional during contraction, although much less efficient, as described above (decrease in the residual radial tethering forces from $N_0^* F_{cb}^{0*}$ to $N^* F_{cb}^* \ll N_0^* F_{cb}^{0*}$, resulting in weak net radial repulsive [expansive] forces and, therefore, in extremely small axial contractile forces, via the conversion process

recalled above). The swinging cross-bridge/lever-arm part of the hybrid model should also be strongly affected, as a result of both the sizeable decrease in specific MgATPase activity per overlap cross-bridge and steric hindrance, with rotation of the myosin heads becoming increasingly difficult as sarcomere length increases.

When the overlap between the thin actin and thick myosin filaments disappears, there are no longer cross-bridges attached to the thin actin filaments and no radial tethering force, either at rest or under contraction conditions. In principle, there are no detectable radial repulsive electrostatic forces and the axial electrostatically related tension does not exist (see the preceding paragraph, concerning the translation of radial into axial forces). Indeed, on p. **35** in Section 3.2, the 'active' radial repulsive electrostatic force between a thick myosin filament and its four neighbouring thin actin filaments in an intact unit cell, under physiological conditions of ionic strength (I = 180 mM; I* ~ 182 mM), is given by $4(\gamma/^*2)F_{e,a}$, where γ^* is the length of the overlap zone (the notation $\gamma^* < \gamma$ is the generalisation of the notation γ valid for full overlap). Thus, from a mathematical viewpoint, when γ^* becomes zero, the radial repulsive electrostatic force also becomes zero, but the problem is more complex, from a biophysical viewpoint. Indeed, there are residual radial repulsive electrostatic forces, related to the much reduced thick–thick filament spacing, with major consequences for the width of the envelope surrounding each thick myosin filament (see pp. **27–31** in Section 3.2). Moreover, these residual radial repulsive electrostatic forces are involved in resting and active tensions and in the very small active tension detected experimentally in highly stretched fibres (Gordon et al. 1966a). For instance, as demonstrated on pp. **27–28** in Section 3.2, the true radius R_m of a thick myosin filament depends on filament spacing (i.e. also on sarcomere length) and decreases slightly when the fibre is stretched, so that, at rest or under contraction conditions and at non-overlap (sarcomere length of ~3.68 μm; see p. **25** in Section 3.2), the residual radial repulsive electrostatic force would be ~5% that at maximal overlap (sarcomere length $s_0 \sim 2.18$ μm; see p. **24** in Section 3.2; see also p. **27** in Section 3.2 for precise details on the value of ~5%). From this value of ~5%, at zero overlap, it is strongly suggested, on p. **134** in Section 4.4.2.1, that when shifting from rest to contraction, there is a net radial expansive force of ~1.8 pN per intact unit cell. Otherwise, on p. **142** in Section 4.4.2.3, the radial repulsive electrostatic force at full overlap is ~ 598 pN per intact unit cell under contracting conditions and ~ 468 pN per intact unit cell at rest, whereas the net expansive force at full overlap is ~ 455 pN per intact unit cell (see p. **35** in Section 3.2). Thus, the residual expansive force of ~1.8 pN per intact unit cell represents ~0.30%, ~0.38% and ~ 0.40% of the three values at full overlap, respectively. Given this slight scattering, I choose a mean proportion of ~ t(0.30 + 0.38 + 0.40)/3 ~ 0.36%. The corrective factor for passing from radial to axial forces is B_a^* ~ 1 at full overlap (see text below Equation 3.45) and, assuming that B_a^* is not very sensitive to the amount of overlap, the relative residual axial contractile force would be ~1 × 0.36% ~ 0.36%. At full overlap, the structure of the myofilament lattice is hexagonal and clearly three-dimensional (see pp. **27–31** in Section 3.2). At zero overlap, the thick myosin and thin actin filaments do not interact (no attached cross-bridges), but the thick myosin filaments are cross-linked (see p. **17** in Section 2.4) and the myofilament lattice is somewhere between a two- and three-dimensional structure. Moreover, it is highlighted on p. **254** that the lattice is probably tetragonal. Consequently, the transmission of the radial into axial forces would be less efficient, leading to B_a^* ~ 0.8, for example, such that the value given above becomes ~0.8 × 0.36% ~ 0.29%. From Figures 12 and 13 in the paper by Gordon et al. (1966a), taking into account approximately 10 experimental points lying between sarcomere lengths of ~3.6 μm and ~3.7 μm, the estimate of the residual axial contractile force at zero overlap represents ~2.0–2.5% of the axial contractile force at full overlap (value not given by the authors). In the pure swinging cross-bridge/lever-arm models, there is no directly generated axial contractile force at zero overlap (no attached cross-bridges), provided that all sarcomeres are actually in the non-overlap state (see the perplexing discussion presented by Gordon et al. 1966a). At this stage, it should be pointed out that Gordon et al. (1966a) did not see this 'residual active axial contractile force' as very important and, thereafter, this property was ignored by the 'muscle community'. However, Pollack (1988) highlighted this property and cited several authors, who found that active forces could be recorded well beyond the disappearance of overlap (up to sarcomere lengths of ~4 μm; see Gordon et al. 1966b). Thus, Pollack (1988) deduced that 'for some theories, tension generation beyond overlap is tolerable', but that 'for the cross-bridge theory, tension beyond overlap is forbidden'. This is also my opinion and I consider that there is a genuine residual active axial contractile force at zero overlap, probably corresponding to ~ (0.36 + 0.29 + 2.0 + 2.5)/4 ~ 1.29% of the maximal axial contractile force at full overlap. According to the hybrid model, this residual axial contractile force at zero overlap is mostly related to the residual radial repulsive electrostatic forces, converted into axial contractile forces (see pp. **102–103** in Section 3.10). Finally, in the hybrid model, proportions of ~60% and ~40%, at the slack length, for the swinging cross-bridge/lever-arm and lateral swelling processes, respectively, are assumed throughout this work. It is impossible to estimate these proportions as functions of the sarcomere length in moderately and highly stretched fibres, owing to the complex phenomena occurring upon stretching, as discussed in this section.

In the 1970s, 1980s and 1990s, Allen et al. (1974), Endo (1972, 1973, 1975), Gordon and Ridgway (1975), Martyn and Gordon (1988), Güth et al. (1981), Stephenson and Wendt (1984) and Stephenson and Williams (1982), on the one hand, and Fabiato and Fabiato (1976, 1978), Hofman and Fuchs (1987) and Kuhn et al. (1990), on the other, working on skinned muscles and cardiac cells, respectively, observed that Ca^{2+} activation for isometric tetanic contraction depends on sarcomere length, that is, also on myofilament lattice spacing. Godt and Maughan (1981), for instance, suggested that, in skinned muscle fibres, Ca^{2+} activation depends on osmotic compression, that is, on filament spacing. In this context, Lakatta (1987), working on the heart, claimed that there is an intimate interaction of muscle length (i.e. also filament spacing) and calcium activation on the myofilaments. Wang and Fuchs (1995) studied the effects of osmotic compression (i.e. decrease in myofilament spacing) on skinned cardiac and skeletal muscle bundles. They demonstrated that both Ca^{2+} sensitivity and Ca^{2+} binding to myofilaments depend strongly on this spacing. This is equivalent to the findings of Allen et al. (1974), Endo (1972, 1973, 1975), Godt and Maughan (1981) and Gordon and Ridgway (1975), for example. Several other experimental studies have more recently been published on the subject of length-dependent (necessarily also

Morel

spacing-dependent) Ca^{2+} activation in cardiac muscle and its molecular basis, possibly in the myosin heads themselves, or in the whole cardiac cell (e.g. Fuchs and Martyn 2005; Konhilas et al. 2002; Martyn et al. 2004; Olsson et al. 2004; Xu et al. 2006a; Yagi et al. 2004). This adds complexity to tension–length relationships.

In this context of complexity, as repeatedly stressed in this monograph, a further problem appears in demembranated fibres. Indeed, as discussed in Section 8.7, Matsubara and Elliott (1972), for example, found, by traditional x-ray diffraction and stereomicroscopy, that the constant volume relationship does not hold in this biological material. Morel (1985a) and Morel and Merah (1997) carried out a critical analysis of this problem and came to a different conclusion, according to which certain interpretations of x-ray experiments were misleading. Even if we leave the reasoning of Morel (1985a) and Morel and Merah (1997) aside, it appears clear from all the figures and formulae presented by Matsubara and Elliott (1972) that filament spacing decreases linearly with increasing sarcomere length. Thus, the experiments concerning Ca^{2+} activation, recalled in the preceding paragraph, are almost certainly qualitatively similar in demembranated fibres. The stretching of these fibres also brings the myofilaments closer together; hence, Ca^{2+} sensitivity and Ca^{2+} binding are affected. As demonstrated by Bachouchi and Morel (1989a) and discussed above, in this section (see pp. **251–254**), stretching induces considerable distortions of the cross-bridges, but the thin actin filaments are also necessarily distorted. The affinity of TnC for Ca^{2+} is probably altered and the position of tropomyosin along the thin actin filaments may also be modified. This interpretation of the experimental findings is supported by the demonstration by Güth and Potter (1987), working on skinned rabbit psoas fibres, that the cycling movement of the cross-bridges during isometric contraction, at the slack length, itself modifies the structure of TnC and the Ca^{2+}-specific regulatory sites. We should also note that, when sarcomere length increases in skeletal fibres, the sarcoplasmic reticulum, which is present at the periphery of each myofibril, and, at least partly, in chemically skinned fibres and myofibrils (see pp. **92–93** in Section 3.7 for precise details concerning the residual sarcoplasmic reticulum in chemically skinned fibres), is stretched and may release additional Ca^{2+} into the sarcoplasm, as found by Stephenson and Wendt (1984) and Stephenson and Williams (1982). Thus, both the increase in sarcomere length and the subsequent decrease in filament spacing may contribute to the complexity of Ca^{2+}-related phenomena, via many intricate routes.

To conclude this many-faceted section, I would like to mention a paper that I published in the *Journal of Theoretical Biology* in 1978, entitled 'Force–velocity relationship in single muscle fibres'. In this paper, I analysed the experimental results of Edman and Hwang (1977) and Edman et al. (1976), showing a reversal of curvature in the P–V relationship at a relative load of ~0.80. I used the two-state model of Morel et al. (1976), derived from the theory of A. F. Huxley (1957). However, I ignored the mechanical expression of the force exerted by the attached cross-bridges (see throughout Section 3.4 for an updated demonstration of the validity of this approach; see also the general conclusion for an overview of this problem). Edman et al. (1976) observed that the reversal of curvature was weaker when force–velocity data were obtained from stretched fibres. Assuming that the sum (f + g) of the constants of attachment, f, and detachment, g, decrease significantly with increasing the sarcomere length, Morel (1978) was able to account for the experimental results of Edman et al. (1976) and suggested a simple formula: $(f + g)$ s^{-1} ~ 106 − 28s (s: sarcomere length in micrometres), independently of the model of contraction and of the many complex phenomena described in this section for stretched fibres. Unfortunately, f and g cannot be estimated individually, but I suggest here that f = μg, where μ is a coefficient of proportionality, assumed to be approximately independent of the sarcomere length s. We deduce, therefore, that g ~ $(106 − 28s)/(1 + μ)$. Otherwise, on p. **65** in Section 3.4.3.3, the duration of the cross-bridge cycle under isometric tetanic conditions is suggested to be $\tau_{c,\infty}$ ~ $1/R_{isom,\infty}$, where $R_{isom,\infty}$ is the turnover rate of the cycle. Combining Equations 3.5 and 3.19, we obtain $R_{isom,\infty} = (\Delta h/\sigma)fg/(f + g)$. Replacing f and g by the values proposed above, we deduce $R_{isom,\infty}$ ~ $(\Delta h/\sigma)[μ/1 + μ)^2](106 − 28s)$ s^{-1} (simple calculations not shown). Thus, the ratio of $R_{isom,\infty}(s)$ (at the sarcomere length s), to $R_{isom,\infty}(s_0)$ (at the reference sarcomere length s_0 ~ 2.18 μm; see p. **23** in Section 3.2) is given by $R_{isom,\infty}(s)/R_{isom,\infty}(s_0)$ ~ $(106 − 28s)$ $s^{-1}/45$ s^{-1}, clearly demonstrating that this ratio is a decreasing function of s. For example, when $s \sim 3.68$ μm (corresponding to the disappearance of overlap; see p. **24** in Section 2.2), the ratio is $R_{isom,\infty}(3.68$ μm$)/R_{isom,\infty}(s_0) \sim 0.06$–$0.07$ ($\sim 6\%$–7%, not too far from the experimental estimates of $\sim 10\%$–15% and $\sim 25\%$ for MgATPase activity, recalled on p. **252**). Thus, $R_{isom,\infty}$ is a steep decreasing function of s, and the duration, $\tau_{c,\infty}$, of the cross-bridge cycle (defined above) increases steeply with sarcomere length s. This conclusion can easily be interpreted in terms of steric hindrance of the cycling cross-bridges, rendering their turnover difficult. The various approaches presented in this section result, therefore, in a self-consistent description of stretched fibres.

8.10 MECHANICAL TRANSIENTS IN INTACT AND DEMEMBRANATED FROG FIBRES

In the 1980s, I investigated whether the lateral swelling model presented by my group (Morel 1975; Morel and Gingold 1979b; Morel and Pinset-Härström 1975b; Morel et al. 1976) could theoretically account for both the isometric and isotonic mechanical transients. I found that the responses were independent of the mode of generation of the axial contractile force, provided that the cross-bridges were assumed to be viscoelastic structures (unpublished calculations; this notion of viscoelasticity is also recalled on p. **261** in Section 8.11). Ford et al. (1977), Goldman and Simmons (1984a), Huxley and Simmons (1971, 1973), Huxley and Tideswell (1996), Mitsui and Chiba (1996), Siththanandan et al. (2006), Stienen and Blangé (1985), Sun et al. (2001), van den Hooff and Blangé (1984a,b) and Yagi et al. (2005) studied isometric transients experimentally and theoretically and suggested interpretations of their results based solely on the swinging cross-bridge/lever-arm approach. As $\sim 60\%$ of the hybrid model is conventional, the 'mechanical model' of Huxley and Simmons (1971, 1973) and the following improved approaches, accounting for isometric transients, remain valid for the hybrid model and can be applied here. Nonetheless, Davis and Harrington (1993) introduced new concepts into the approach suggested by Huxley and Simmons (1971, 1973). Otherwise, it has recently been demonstrated experimentally that there is an additional step in the complexity of the phenomena, regarding, for example, the type of fibre used to study isometric transients (Davis and Epstein 2003). Interpretations

of the experimental results of Huxley and Simmons (1971, 1973) should therefore be adjusted. To the best of my knowledge, isotonic transients were first studied by Podolsky (1960) and Podolsky and Nolan (1973). However, as for isometric transients, views concerning isotonic transients have changed considerably in the last 30 years (e.g. Caremani et al. 2006; Granzier et al. 1990; Lombardi et al. 1992; Sugi and Tsuchiya 1981a,b). Worthington and Elliott (2005) carried out a theoretical analysis of experimental data for isotonic transients obtained by Piazzesi et al. (2002a) and suggested that their interpretation of these experimental data results in a straightforward resolution of the puzzle of the working stroke. Nonetheless, I think that the situation remains puzzling, because the radial repulsive electrostatic forces and the residual radial tethering forces are demonstrated to be effective, throughout this monograph. It will therefore be important, in future discussions, to study the problem of mechanical transients based on the hybrid model.

8.11 ACTIVE SHORTENING AND FORCE–VELOCITY RELATIONSHIPS IN INTACT FROG FIBRES (ISOTONIC CONDITIONS), WITH SOME INSIGHTS INTO CARDIAC MUSCLES FROM RAT AND DEMEMBRANATED FIBRES FROM FROG, RABBIT AND RAT

In Section 3.4.4, active shortening is studied, mostly from a phenomenological/semi-empirical viewpoint, using the model-independent approach first described by Morel et al. (1976) and improved over approximately 40 years (e.g. Morel 1978, 1984a,b, 1990, 1991b; Morel and Gingold 1977; see Sections 3.4.2 and 3.4.3 for an updated phenomenological/semi-empirical approach). However, a major question remains: how can the hybrid model account for active shortening and how does the fibre shorten at a given velocity, for a given load lifted? Furthermore, is it possible to deduce a suitable force–velocity relationship from the hybrid model? The cross-bridges cyclically attach to and detach from the thin actin filaments and split MgATP, in pure lateral swelling models developed by Morel (1975), Morel and Gingold (1979b), Morel (1985a), Morel and Pinset-Härström (1975b) and Morel et al. (1976), and also in the hybrid model, as in traditional swinging cross-bridge/lever-arm models (this 'universal cycling behaviour' of the cross-bridges is demonstrated at many places in this monograph). As also demonstrated by Morel (1985a), pure lateral swelling models can account for active shortening, taking into account the negative forces exerted by the 'resisting' cross-bridges carried along the thin actin filaments by shortening, outside the interval of attachment ($-h_1$; $+h_2$) (see Equation 3.41 and corresponding comments for precise details on the proportion n' of attached cross-bridges outside the interval of attachment, below $-h_1$). The behaviour of the cross-bridges in pure swinging cross-bridge/lever-arm models is obviously similar, as deduced from the theoretical and biophysical approach of A. F. Huxley (1957). In the hybrid model, the mechanisms by which the net radial repulsive (expansive) forces are converted into axial contractile forces (see pp. **102–103** in Section 3.10), valid under isometric conditions, also apply under isotonic conditions. The negative forces exerted by the resisting cross-bridges also oppose the positive axial 'driving forces', induced both directly by the rotation of the cross-bridges and indirectly by the net radial repulsive (expansive) forces. This leads to a 'positive' difference between the positive axial driving forces of both origins and the resisting (negative) forces exerted by the cross-bridges that are carried along the thin actin filaments by shortening, resulting in a given shortening velocity for a given load.

Morel (1985a) considered the cross-bridges to be damped springs (viscoelastic structures) and the pure lateral swelling model provides a non-hyperbolic mechanical force–velocity relationship (reversal of curvature beyond a relative load of ~0.8; the notion mechanical is used here, as opposed to that of phenomenological/semi-empirical used by Morel 1978; see also discussion below, in this section). This mechanical relationship approximately fits the available experimental data obtained by Edman and Hwang (1977) and Edman et al. (1976) with intact frog fibres. The experimental non-hyperbolic force–velocity relationship, with reversal of curvature, was confirmed by two independent groups: on intact cardiac muscle from rat by van Heuningen et al. (1982) and on demembranated rabbit psoas muscle by Yamada et al. (1993). de Winkel et al. (1995), Grazi and Cintio (2001) and Zahalak (2000) rediscovered that cross-bridge force should be considered a non-linear function of cross-bridge strain, as for damped springs, as I suggested previously (see above). Applying the usual approach (simple

springs; see A. F. Huxley 1957) to cross-bridge properties during active shortening, the swinging cross-bridge/lever-arm models have been shown to be subject to various problems, including negative values for the square of one parameter (Morel 1985a). A. V. Hill (1922) and Gasser and Hill (1924) suggested that viscosity is relevant to muscle contraction (under isotonic conditions). Grazi (2007) and Grazi and Di Bona (2006) claimed that viscosity is 'an inseparable partner of muscle contraction'. The improvement of all the available theories will require the introduction into future mechanical theoretical studies of viscous-like frictional forces, originating from either the external medium (Chase et al. 1998; Endo et al. 1979) or the interfilament medium (Chase et al. 1998; Grazi 2007; Grazi and Di Bona 2006). These viscous forces were carefully described and analysed by Elliott and Worthington (2001, and references therein). A consideration of these viscous-like forces is obviously required in the hybrid model, to obtain the best fit of the theoretical mechanical force–velocity relationship to the experimental data for intact frog fibres. This should be addressed in future work.

Piazzesi et al. (2007) presented an experimental and semi-empirical study of the force–velocity relationship, based on the traditional approach, according to which the myosin heads (called the myosin motors by the authors) and their swinging alone are responsible for the generation of the axial contractile force. Comparing their own relationship with that obtained by A. F. Huxley (1957) from his mechanical expression of the force–velocity relationship, Piazzesi et al. (2007) wrote: 'the results do not support the conventional explanation of this relationship (A. F. Huxley 1957; Cooke 1997), in which the lower force exerted by shortening muscle is primarily a result of a lower force per motor'. In my opinion, these contradictory conclusions are related to the use of the mechanical expression of the force, which has been demonstrated to be inconsistent with the energetic expression (Morel and Gingold 1977; Morel 1984a; see also the general conclusion of this book). Finally, a weakness in the study of Piazzesi et al. (2007) is that the authors did not consider the experimental results of Edman and Hwang (1977) and Edman et al. (1976), concerning the non-hyperbolic relationship (see the preceding paragraph). Instead, like most specialists in muscle and muscle contraction, they took into account solely the 'old' hyperbolic relationship of A. V. Hill (1938), valid for whole frog muscle, but not for single intact or demembranated fibres.

A number of unanswered questions remain and are repeatedly posed in this monograph. They relate to various problems, such as the validity of combining experimental results for isometric tetanic tension and active shortening obtained with intact fibres (e.g. Edman and Hwang 1977; Edman et al. 1976), chemically skinned fibres (e.g. Potma and Stienen 1996; Potma et al. 1994), permeabilised fibres (e.g. He et al. 1997, 1998a,b, 1999) and unheld isolated myofibrils (e.g. Barman et al. 1998; Lionne et al. 1996) with phenomenological/semi-empirical results valid for intact unit cells (e.g. Merah and Morel 1993; Morel 1978, 1984a,b; Morel and Gingold 1977; Morel and Merah 1995; Morel et al. 1976), as extensively studied in this work, particularly throughout Section 3.4. This problem concerns quantitative analyses only. The intact fibres and intact unit cells, compared throughout this book, behave similarly in qualitative terms. A formal definition of intact unit cells is given in Section 3.1. Intact unit cells are located in the centre of a myofibril, itself lying in the centre of an intact fibre. In myofibrils from an intact fibre, the peripheral unit cells are in contact with the sarcoplasmic reticulum, leading to frictional (damping) phenomena during shortening (see p. **93** in Section 3.7). This results in the peripheral unit cells moving more slowly than the central unit cells and the two types of unit cells shortening at different velocities, with subsequent distortions in the myofibrils. There are also quantitative difficulties concerning the peripheral myofibrils in contact with the sarcolemma, for the same reasons (frictional phenomena). These 'mechanical differences' between intact (central) and peripheral unit cells can be extended to single fibres, as a function of their location in a whole muscle. This is probably the origin of the two different P–V relationships, obtained, for instance, by Edman and Hwang (1977) and Edman et al. (1976) with single fibres, and by A. V. Hill (1938, 1964a,b) with whole muscles. Moreover, as demonstrated on pp. **61–62** in Section 3.4.3.2, it is difficult to define the 'uniform' and 'steady' shortening velocity, and a simplified definition is used in this monograph and by all other authors (shortening velocity independent of time and location of the unit cell within the fibre).

Given the various biophysical problems described in the preceding paragraphs, it is impossible to predict whether ~60% of the shortening velocity is related to swinging cross-bridge/lever-arm mechanisms and ~40% to lateral swelling processes in an intact unit cell. In any event, there are no biophysical reasons for the radial repulsive electrostatic forces to disappear when the thin actin filaments slide over the thick myosin filaments. Nonetheless, as recalled on pp. **94–95** in Section 3.7, several experimental and phenomenological/semi-empirical results show that active sliding induces many distortions in the thin actin filaments and the myosin heads and, probably, in the thick myosin filaments (e.g. in the helical arrangement of the external myosin heads; see, in Section 5.1.1, pp. **170–172** and Figure 5.1, its legend and Appendix 5.I as concerns this notion of external heads). Such distortions almost certainly result in structural rearrangements in the actin molecules, the thin actin filaments, the myosin heads and the thick myosin filaments, leading to changes in the biophysical characteristics of the swinging cross-bridges and an increase or decrease in the number of fixed negative electrical charges and bound anions on the thin actin and thick myosin filaments. This probably also leads to changes in the biophysical behaviour of the resisting cross-bridges exerting negative forces (see the first paragraph of this section for precise details concerning the resisting cross-bridges). Both the directly generated axial contractile forces (resulting from the rotation of the cross-bridges) and indirectly generated axial contractile forces (resulting from net radial repulsive [expansive] forces between the myofilaments) can therefore be modified by sliding. Additional complex problems occur at the supramolecular level, when passing from central unit cells to peripheral unit cells and myofibrils located either in the centre or at the periphery of an intact fibre (see the preceding paragraph). Thus, as highlighted at the beginning of this paragraph, for a given load lifted by a fibre, it is impossible to determine whether ~40% of the shortening velocity is related to lateral swelling mechanisms and ~60% to swinging cross-bridge/lever-arm mechanisms, in any unit cell. All the same, it is well known that shortening is accompanied by a simultaneous lateral swelling of the fibre. This is clearly consistent with the lateral swelling part of this hybrid model.

As recalled on p. **261**, Edman and Hwang (1977) and Edman et al. (1976), working on intact frog fibres, at 2°C–3°C, were the first to find a non-hyperbolic force–velocity relationship. Working on trabeculae from the right ventricle of rat heart, presumably at room temperature (~20°C–22°C), van Heuningen et al. (1982) also obtained a non-hyberbolic shape, as did Yamada et al. (1993), who worked on chemically skinned fibres from rabbit psoas muscle, at room temperature (~20°C–22°C). These last two groups 'omitted' to point out that they worked on warm-blooded animals at room temperature (favourable temperature), whereas Edman and colleagues worked on cold-blooded animals at low temperature (favourable conditions). This is important, because both groups were working in 'ideal' temperature conditions, and their results are therefore consistent (Barman et al. 1998, for example, clearly posed the question of working with cold-blooded and warm-blooded animals at low temperatures). Morel (1978, 1984a), using the energetic expression of the load lifted $P(s_0V)$ (see Sections 3.4.2 and 3.4.3 and the general conclusion of this work), demonstrated that the non-hyperbolic relationship can easily be accounted for by the phenomenological/semi-empirical approach, regardless of the mode of force generation and the complex biophysical phenomena induced by active sliding (see the preceding paragraph). Using other experimental techniques, Edman (1988a,b) showed, again for intact frog fibres at ~2°C–3°C, that the force–velocity relationship has a more complex shape than previously found: a double hyperbola. This behaviour was confirmed on demembranated frog fibres by Lou and Sun (1993), working in collaboration with Edman. Taking into account solely the swinging cross-bridge/lever-arm models, Edman (1988a,b) claimed that the only possible explanation for his experimental results was changes in the biophysical characteristics of the cycling cross-bridges induced by active shortening, as apparently confirmed by Morel (1990), using the phenomenological/semi-empirical approach, independent of the mechanical roles of the cross-bridges (see above). However, I must acknowledge that, in 1990, I made several mistakes. I initially assumed that the constants of attachment, f, and detachment, g, were independent of the shortening velocity, but then concluded that they must depend on this velocity, which is obviously self-contradictory. This error probably resulted from my wish to account for the findings of Edman (1988a,b), particularly at low relative values of shortening velocity (<0.1),

corresponding to relative loads >0.8. In my opinion, it is difficult, if not impossible, to obtain clear experimental results in this region, because of the scattering of the experimental points and artefacts in the measurements of both P and V. It is also difficult to define the shortening velocity clearly, as recalled on p. **262**, and this is highly problematic at low values of this velocity. Thus, unlike Edman et al. (1997), I think that the double-hyperbolic P–V relationship is not valid and that the experimental studies of Edman (1988a,b) should be discounted, together with my own self-contradictory 'demonstration' in 1990.

I retain here only the non-hyperbolic force–velocity relationship with reversal of curvature. This position is supported by the works of van Heuningen et al. (1982) and Yamada et al. (1993) (see the preceding paragraph). In any event, the problem of the reversal of curvature or a double hyperbola does not interfere with the phenomenological/semi-empirical approach presented in this monograph (see p. **47** in Section 3.4.3.1.3 concerning the similar values of z and z_1). Edman et al. (1997) proposed a four-state model, based on the swinging cross-bridge/lever-arm concepts, accounting for the double-hyperbolic shape of the force–velocity relationship (see p. **39** in Section 3.4.2 for a critical opinion concerning the introduction of multistate models). Edman et al. (1997) found, for example, that their constant of attachment, $k_{30}(t,v)$, increased with shortening velocity. On experimental and semi-empirical bases, Piazzesi et al. (2007) and Yagi et al. (2006) reached similar conclusions, concerning the rate of attachment of the cycling cross-bridges (see pp. **261–262** for my doubts concerning the interpretation proposed by Piazzesi's group). All these conclusions depend strongly on the mechanical roles of the cross-bridges in the swinging cross-bridge/lever-arm processes. Moreover, the theoretical approach of Edman et al. (1997) 'leads to low average cycling rate (0.25 s^{-1}) in the model consistent with the low ATP splitting rate during isometric contraction of frog muscle (~1 s^{-1}; Curtin et al. 1974; Homsher 1987)'. As claimed on p. **10** in Chapter 1, the rate of MgATP breakdown is a major stumbling block and values of 0.25 s^{-1} and ~1 s^{-1} are unreasonable. Indeed, in their Table 1, He et al. (1997) gave values of ~13–15 s^{-1} for seven demembranated frog fibres at 5°C, corresponding to ~10–12 s^{-1} at ~2°C–3°C (this temperature is used in all the experiments of Edman and his group; $Q_{10} \sim 3.1$; see p. **202** in Section 6.3.3; multiplication factor of $\sim10-12$ s^{-1}/0.25 s^{-1} $\sim40-50$). This is one of the chief criticisms of the theory of Edman et al. (1997). By contrast, the 'theoretical' non-hyperbolic P–V relationship described by Morel (1978, 1984a), with a reversal of curvature, is a straightforward consequence of the phenomenological/semi-empirical approach, independent of the mechanical roles of the cycling cross-bridges, in which the constants of attachment, f, and detachment, g, are strictly independent of the shortening velocity. This conclusion is inconsistent with changes in the attachment–detachment characteristics of shortening fibres necessarily resulting from the mechanical role of the cross-bridges, in the swinging cross-bridge/lever-arm models (see above concerning the theory of Edman et al. 1997). This again casts doubt on the traditional 'mechanical approach', as denounced at many places in this monograph, whereas, in the hybrid model, considering the viscous-like frictional forces and taking into account the viscoelastic properties of the cross-bridges (see pp. **261–262** for brief discussions) could yield a mechanical approach consistent with phenomenological semi-empirical reasoning.

8.12 CARDIAC MUSCLE, CARDIOMYOPATHIES, HEART FAILURE

Many years ago, I showed, in light scattering studies (unpublished data), that cardiac myosin from young adult hamsters can form head–head dimers, in solution, in the presence of millimolar amounts of MgATP, as can myosin from the skeletal muscle of young adult rabbits (Morel et al. 1998a). This capacity, also demonstrated for young adult cardiomyopathic hamsters (BIO 14:6), is gradually lost upon ageing. Indeed, in $\sim$300-day-old cardiomyopathic animals, this capacity is much smaller than in healthy animals of the same age. Thus, if myosin head–head dimerisation is impossible, the thick myosin filaments are poorly structured (Morel et al. 1999; see also Pinset-Härström 1985; Pinset-Härström and Truffy 1978 concerning poorly structured filaments, but with no allusion to head–head dimers). According to the hybrid model, only $\sim$60% contraction can occur in $\sim$300-day-old BIO 14:6 cardiomyopathic hamsters (via the traditional rotation of the cross-bridges in the hybrid model), accounting for the untimely death of the animals (D'hahan et al.

1997). My group identified interesting pharmacological treatments that significantly increased survival ($\sim$+50%–60%) in cardiomyopathic hamsters (D'hahan 1999; D'hahan et al. 1997, 1998). However, we did not study the problem of myosin head–head dimerisation in the surviving animals. Heart failure is also a major problem. No relationship has yet been established between the absence of myosin head–head dimerisation and heart failure, although a caused relationship seems possible.

In the normal human heart, the contractile force is strong during systole (blood pressure $\sim$100–130 mm Hg). During diastole, the contractile force is weaker (blood pressure $\sim$60–80 mm Hg). For precise details, the reader should refer to some elaborate studies of the molecular and physiological properties of cardiac muscle, including heart failure, by Castro et al. (2010), Cazorla et al. (1999), Friel and Bean (1988), Gómez et al. (2001), Hunter et al. (1998), Joubert et al. (2008), Katz (1993, 2003, 2006), Kuzmin et al. (1998), Kuum et al. (2009), Legssyer et al. (1988), le Guennec et al. (2000), Métrich et al. (2010), Ventura-Clapier et al. (2004, 2011), Pucéat et al. (1991), Szigeti et al. (2007), Tuncay et al. (2011), Vassort (2001) and Vassort et al. (1976). Noble mostly studied computer models relating to the heart and its function (e.g. Noble 2002a,b, and references therein). Katz (2006) has also published a recent version of his book on the heart. Many experimental studies concerning parameters affecting isometric force have been performed (e.g. Allen et al. 1974; Brandt et al. 1998; Colomo et al. 1997; Fabiato and Fabiato 1976, 1978). On p. **176** in Section 5.1.1, I comment on the pioneering x-ray diffraction experiments performed on mammalian cardiac muscle by Matsubara and Millman (1974), in terms of the attachment of myosin heads to the thin actin filaments under diastolic and systolic conditions, and discuss other related experiments carried out by independent groups. Matsubara et al. (1975, 1979, 1980, 1989) studied the state of cardiac contractile proteins during the diastolic phase, the movement of myosin heads during a heartbeat, and the states of myosin heads during the systolic and diastolic phases. Matsubara (1980) has published a review on various properties of the heart studied by traditional x-ray diffraction and Matsubara et al. (1979, 1980) have studied, again by traditional x-ray diffraction, the heart muscle during contraction. On p. **245** in Section 8.7, plausible explanations of the Frank–Starling relationship are given. Gordon and Pollack (1980) discussed the Frank–Starling mechanism based on their experimental studies of the effect of calcium on the sarcomere length–tension relationship in rat cardiac muscle (further discussions of the roles of Ca^{2+} are also presented on pp. **258–259** in Section 8.9). Xu et al. (2006b), using x-ray diffraction from synchrotron radiation, analysed the implications of their findings concerning permeabilised rabbit cardiac trabeculae, for cardiac function, particularly the Frank–Starling relationship, from a molecular viewpoint. Fukuda and Granzier (2004) proposed that titin would play a central role in the Frank–Starling mechanisms. More recently, Smith et al. (2009) suggested that 'cooperative cross-bridge activation of thin filaments contributes to Frank–Starling mechanism'. Cardiac muscle has certain specific features (e.g. excitation–contraction mechanism) but displays several similarities to skeletal muscle. As illustrative examples, (i) the problem of 'living water' in skeletal muscle, studied on pp. **29–31** in Section 3.2, also applies to cardiac muscle (e.g. Hazlewood et al. 1971); (ii) the problems of fatigue or hypoxia in skinned rabbit and cardiac muscle were simultaneously studied by Godt and Nosek (1989) and their study gave similar results for both types of muscle; (iii) the problem of spontaneous oscillations in cardiac cells (e.g. Fabiato and Fabiato 1978; Linke et al. 1993) is similar to that of skeletal fibres, studied on pp. **91–94** in Section 3.7; (iv) the phenomena mediated by the osmotic compression of skeletal and cardiac skinned bundles of fibres/cells are similar (e.g. Wang and Fuchs 1995); (v) de Winkel et al. (1995) demonstrated that the cross-bridges in cardiac muscle have viscoelastic properties, like those of skeletal muscle (see pp. **261–262** in Section 8.11); and (vi) the controversy, discussed on pp. **91–92** in Section 3.7 and pp. **255–256** in Section 8.9, for skeletal muscle, also applies to cardiac muscle, particularly as concerns the quantal and step-wise mechanisms (A. F. Huxley 1986; Pollack 1986). More recent papers on both types of muscle have confirmed this quantal phenomenon (see Yakovenko et al. 2002, and references therein, for cardiac muscle).

The swinging cross-bridge/lever-arm theories, considered valid for skeletal muscles, are almost universally seen as also valid for cardiac muscle. However, as pointed out at the end of the preceding paragraph, other models are available, including that based on the experimental demonstration by Pollack (1986) of sarcomere shortening in discrete quantal steps. The hybrid model can be applied to cardiac muscle. In accordance with this model, I propose that the swinging cross-bridges/lever arms act in the traditional way and accounts for ~60% of contraction, corresponding to diastole. I suggest that lateral swelling mechanisms account for the remaining ~40% of contraction, resulting in full contraction (systole). If the hybrid model is accepted as the most appropriate approach, it can be concluded that, in cardiomyopathic hamsters, the gradual disappearance of lateral swelling mechanisms (see the first paragraph) would lead to the establishment of purely diastolic conditions in the heart, with no systole, making it impossible for the heart to work efficiently and inevitably leading to a gradual incapacity to work adequately and to the untimely death of the cardiomyopathic animals.

8.13 SMOOTH MUSCLE

It has been shown experimentally that myosin molecules from skeletal and smooth muscles 'show different strokes for different blocks' (Barsotti et al. 1996) and that 'ADP release produces a rotation of the neck region of smooth myosin but not skeletal myosin' (Gollub et al. 1996). The fully efficient swinging cross-bridge/lever-arm processes of the external heads (see Section 5.1.1 for precise details concerning the external heads and their roles) would therefore depend on the type of muscle. In other words, the rotation of the cross-bridges is likely to be more efficient in smooth muscles than in skeletal muscles. However, the role of the internal B–B′-type head–head dimers (see Figure 5.1 for the arrangement of these dimers) seems to be common to all types of muscle, as are the lateral swelling mechanisms. The experimental data of Barsotti et al. (1996) and Gollub et al. (1996), together with the interpretation presented here, may account for the specific force per myosin head developed in smooth muscles being higher than that in skeletal muscles (e.g. Warshaw and Fay 1983). However, the longer duration of the rise in tension in smooth muscles than in skeletal muscle (e.g. Jaworoski and Arno 1998, regarding chemically skinned smooth muscle fibres, and He et al. 1997, concerning chemically skinned skeletal muscle fibres) remains to be accounted for. Nonetheless, the most probable explanation may be structural in nature; that is, slow conformational changes occur in myosin and thick myosin filaments and thin actin filaments in the case of smooth muscle, whereas conformational changes may be rapid in striated muscle. Many other properties of smooth muscle are similar to those of striated and heart muscle. For example, a small proportion of the cross-bridges are attached to the thin actin filaments at rest and are viscoelastic in mammalian smooth muscles (Ratz and Speich 2010; Siegman et al. 1976; Speich et al. 2006), as in striated muscle (see pp. **170–172** in Section 5.1.1 concerning the weakly bound cross-bridges and their roles and pp. **261–262** in Section 8.11 regarding viscoelasticity of the cross-bridges). Moreover, Dantzig et al. (1999) suggested that, in stretched skinned smooth muscle, geometric and structural constraints in the filament lattice alter the motion of the myosin heads as predicted a decade previously from studies of stretched intact striated fibres from skeletal muscle by Bachouchi and Morel (1989a). This leads to the major consequences studied on pp. **251–254** in Section 8.9. Otherwise, under contraction conditions, the effect of sarcomere length on the sensitivity to calcium of striated and heart muscle, discussed on pp. **258–259** in Section 8.9, is qualitatively similar to that observed in vascular smooth muscle (e.g. van Heijst et al. 1999). In Section 8.9, it is demonstrated that there are two possible length–tension relationships in an intact frog fibre. In detrusor smooth muscles from rabbit, the situation is also complex, even more so than in striated muscle fibres. Indeed, Speich et al. (2009) found that 'the active length–tension relationship, or curve, does not have a unique peak tension value with single ascending and descending limbs, but instead reveals that multiple ascending and descending limbs can be exhibited in the same destrusor smooth muscle'.

In smooth muscle, concerning the location of the conformational changes in the external myosin heads (see above for necessary references on these heads in skeletal muscle), Warshaw et al. (2000) studied the domain at which the head can act as a mechanical lever and drew conclusions similar to those for skeletal muscles, regarding, for example, the swinging cross-bridge/lever-arm part of the hybrid model (see p. **173** in Section 5.1.1 for references on the 'domains' in skeletal muscle). Similarly, tension transients in single isolated smooth muscle cells, studied in the 1980s (e.g. Warshaw and Fay 1983), were found to be qualitatively similar to those found in single isolated striated muscle fibres in the 1970s (e.g. Ford et al. 1977; Huxley and Simmons 1971, 1973). Nonetheless, Warshaw and Fay (1983) suggested that 'The cross-bridge in smooth muscle is more compliant than the cross-bridge in striated muscle and transitions between several cross-bridge states occur more slowly in smooth muscle than in striated muscle' (see also the preceding paragraph, regarding possible slower conformational changes in smooth muscle than in striated muscle). Crystal structure experiments and their interpretation have generated more detailed information about the vertebrate smooth muscle myosin motor domain, that is, the myosin head (e.g. Dominguez et al. 1998). Working on HMM (heavy meromyosin; a myosin subfragment containing two S1 heads plus the S2 part; see Figure 5.1) from smooth muscle, Li and Ikebe (2003) demonstrated cooperativity between the two S1 heads, via interactions 'with each other', resulting in large increases in actin-activated MgATPase activity. Such cooperativity has also been suggested for myosin from striated muscle (see p. **170** in Section 5.1.1 for references). There currently seem to be no major qualitative differences between smooth and striated muscles, only quantitative differences. This conclusion is similar to that drawn in Section 8.12 for cardiac muscle.

CHAPTER NINE

In Vitro Motility

ANALYSIS OF EXPERIMENTAL DATA AND THEIR INTERPRETATIONS. MOVEMENT OF CELL ORGANELLES

About 25 years ago, forces of ~0.2 pN per myosin head were recorded between myosin or myosin heads (S1) and F-actin filaments in vitro, at ~25°C (Kishino and Yanagida 1988). In these experiments, a siliconed glass surface was coated with isolated myosin heads or whole myosin molecules (two heads per molecule) and F-actin filaments were moved over this surface. The authors suggested that the forces recorded during this 'sliding' should be multiplied by a factor of ~2–3, 'because the orientation of myosin and S-1 on the surface are random and therefore all heads are not expected to be able to interact with the actin', giving forces of ~0.2 × (2–3) ~ 0.4–0.6 pN per head. Kishino and Yanagida (1988) considered these values of ~0.4–0.6 pN per head to be 'comparable to force exerted in muscle during isometric contraction (about 1 pN)'. The estimate of ~1 pN per head was proposed more than 35 years ago by Oosawa (1977). Woledge (1988) criticised all these values and demonstrated that the isometric force in muscle is higher than ~3 pN per cross-bridge (presumably at ~0°C), corresponding to ~4.7–13.0 pN per cross-bridge at ~25°C (maximal interval; see below for the two possible values of Q_{10}). As a cross-bridge contains 1 or 2 heads and there is a wide margin of uncertainty, the values of ~4.7–13.0 pN also correspond to 1 head. The values of ~0.4–0.6 pN per head cannot therefore be considered good estimates for the value in muscle. Some months after his publication with Kishino, and after Woledge's demonstration

(1988), Yanagida claimed publicly that there was a calibration error in the experiments of Kishino and Yanagida (1988) and that their recorded forces should be multiplied by an additional factor of 2, giving forces of ~(0.4–0.6) × 2 ~ 0.8–1.2 pN per head at ~25°C. Averaging the four corrected values obtained at ~25°C by Kishino and Yanagida (1988) and suggested by Yanagida, we obtain the most probable mean value of ~0.8 pN per head. Taking Q_{10} ~ 1.20, as for the isometric tetanic tension recorded in intact frog fibres, between ~0°C and ~40°C–45°C (see p. **245** in Section 8.8), the mean force of ~0.8 pN per head at ~25°C corresponds to ~0.6 pN per head at 10°C (the reference temperature used in this monograph). Taking Q_{10} ~ 1.8 between ~25°C and 10°C, valid for demembranated fibres from rabbit psoas muscle (see p. **246** in Section 8.8), the mean force of ~0.8 pN at ~25°C becomes ~0.3 pN per head at 10°C. Many experiments were performed after those reported by Kishino and Yanagida (1988). For instance, Finer et al. (1994), working at 21°C, found that the force per HMM (two heads S1 + S2; see Figure 5.1 for definition of S1 and S2) 'covered a broad distribution, ranging from 1 pN to 7 pN'. The values of 1 pN to 7 pN, at 21°C, give values ranging between ~0.5 pN and ~5.6 pN per HMM at 10°C (see above for the two values of the Q_{10}; no correction performed to take into account the existence of two heads in HMM; see the next citation). Molloy et al. (1995a,b) found values of ~1.7–1.8 pN at 23°C, that is, a maximal interval of ~0.8 pN to ~1.4 pN at 10°C, and they stated that 'HMM and S1 produced approximately the same amount of force' (this conclusion is similar to that drawn in the next paragraph, concerning the discussion of the experimental results obtained for S1 and myosin by Kishino and Yanagida 1988, i.e. the impossibility of distinguishing between the forces generated by S1 and by a whole myosin molecule). The 'stepping forces' obtained, at room temperature (~22°C), on myosin V (a two-headed myosin) involved in the transport of vesicles organelles and proteins within cells, were of the order of magnitude of ~1–2 pN per head (e.g. Cappello 2008; Cappello et al. 2007; Dunn and Spudich 2007; Mehta 2001; Sellers and Veigel 2006; Yildiz et al. 2003), that is, ~0.5 pN or ~1.6 pN per head at 10°C (maximal interval). Averaging the eight available values, valid at 10°C, we obtain a mean estimate of ~1.4 pN per head. This value is significantly lower than the most probable force of ~7.4 pN per head at 10°C, in an intact unit cell (see p. **88** in Section 3.6). Most specialists in in vitro motility consider the strength of the forces they record in vitro to be comparable to that estimated in an intact fibre, providing support for swinging cross-bridge/lever-arm processes. However, neither A.F. Huxley (2000) nor Morel and D'hahan (2000) fully endorse this interpretation (see below, in this chapter). The views of A.F. Huxley (2000) and Morel and D'hahan (2000) are supported by the two different 'optimal' values (at 10°C) of ~1.4 pN per head, in vitro, and ~7.4 pN per head, in vivo (see above).

As recalled in the preceding paragraph, the forces recorded by most authors are of the same order of magnitude as those obtained by Kishino and Yanagida (1988). Only the results obtained by these authors are therefore considered in this and the following paragraphs. The authors found that the total force recorded was approximately proportional to the length of the F-actin filament in contact with the siliconed glass surface (see their Figure 4). Bachouchi and Morel (1989b) retreated the experimental points presented in this Figure 4 and found, for isolated myosin heads (S1), a mean value of ~5.5 pN µm^{-1}, whereas, for a whole myosin molecule (two heads), the mean value obtained was ~9.8 pN µm^{-1}, that is, a ratio of ~9.8 pN µm^{-1}/5.5 pN µm^{-1} ~ 1.8 (close to 2). The authors suggested that this ratio simply reflected the presence of two heads in the myosin molecule, a highly attractive conclusion. Scrutinising the

experimental results presented in Figure 4 of the paper by Kishino and Yanagida (1988), Bachouchi and Morel (1989b) found that the SD was ~1.3 pN µm^{-1} and SE ~ 0.4 pN µm^{-1} (n = 13) for isolated heads, and SD ~ 3.6 pN µm^{-1} and SE ~ 1.0 pN µm^{-1} (n = 14) for whole myosin. If the SEs are taken into account, the ratio of ~1.8 is of borderline statistical significance, as it may be as low as ~(9.8 − 1.0)/(5.5 + 0.4) ~ 8.8 pN µm^{-1}/5.9 pN µm^{-1} ~ 1.5 or as high as ~(9.8 + 1.0)/(5.5 − 0.4) ~ 10.8 pN µm^{-1}/5.1 pN µm^{-1} ~ 2.1 (these two values may be sufficiently close to the value of 2 expected for the two-headed myosin molecule). However, if the SDs (the most suitable tests; see below) are taken into account, the problem becomes much more difficult. Indeed, these two ratios become ~(9.8 − 3.6)/(5.5 + 1.3) ~ 6.2 pN µm^{-1}/6.8 pN µm^{-1} ~ 0.9 or ~(9.8 + 3.6)/(5.5 − 1.3) ~ 13.4 pN µm^{-1}/4.1 pN µm^{-1} ~ 3.3, which are not close to the ratio of ~2 (~0.9 corresponds to a single head per myosin molecule and ~3.3 corresponds to three heads). Thus, the conclusion drawn by Kishino and Yanagida (1988) is highly dependent on the statistical test employed and is therefore dubious, as the SD test is the most appropriate, as highlighted in Section 2.1. As the conclusion that can be drawn depends strongly on the statistical test used, the inference made by Kishino and Yanagida (1988), regarding isolated heads (S1) and two-headed myosin, should be taken with caution. On the basis of the 'SD test', Bachouchi and Morel (1989b) also considered the conclusion of Kishino and Yanagida (1988) that the forces recorded depend on the number of heads (1 or 2) to be dubious.

In this worrying context, Bachouchi and Morel (1989b) and Morel and D'hahan (2000), repeatedly pointed out, in vain, that the various proteins and protein assemblies must come into physical contact with air, solid or liquid surfaces, in all the experiments aiming to record the forces generated by the interaction of S1 or HMM, myosin molecules or thick myosin filaments with thin actin filaments in vitro (F-actin). Commenting on the experimental results obtained by Kishino and Yanagida (1988), Morel and Bachouchi (1988a) suggested that surface forces may interfere with the 'active' forces related to actin–myosin interactions. It is recalled, on p. **155** in Section 4.4.2.5.3, that the lowest surface force recorded at room temperature is ~20 × 10^3 pN µm^{-1} (olive oil–water interface). Very few experimental data are available for solid–water surface and solid–air forces, but, from specialist handbooks, it can be concluded that they are highly variable, because they depend principally on the characteristics of the various solid surfaces. At ~25°C, Kishino and Yanagida (1988) recorded a maximal active force of ~9.8 + 3.6 ~ 13.4 pN μm^{-1} for whole myosin molecules (see the preceding paragraph). As an illustrative example, this maximal value is considerably lower than the value of ~20 × 10^3 pN μm^{-1} (see above): 100 × 13.4 pN μm^{-1}/20 × 10^3 pN μm^{-1} ~ 0.07%. This proportion is negligible and I wonder whether trying to measure active forces related to actin–myosin interactions is illusory (see also below). In other words, Kishino and Yanagida (1988) and other authors who have recorded active forces in vitro would have recorded surface forces, as previously claimed by Bachouchi and Morel (1989b) and Morel and D'hahan (2000). In this context, it should be noted that the value of ~20 × 10^3 pN μm^{-1} used above is a crude value that is probably not entirely suitable (see the footnote on p. **98** in Section 3.8.2 and pp. **155–156** in Section 4.4.2.5.3 for a discussion): there is no straightforward relationship between the elastic moduli of the sarcolemma (of the order of ~10^6–10^7 pN μm^{-2}) and the pressure exerted by the curved sarcolemma on the myoplasm (~5–6 orders of magnitude lower), although both parameters have the same dimension. At this point, Bachouchi and Morel (1989b) applied a simple approach to the experiments of Kishino and Yanagida (1988). Using reasonable assumptions, we demonstrated that an active force of ~10 pN (corresponding to 53 heads, i.e. ~0.2 pN per myosin head; see Kishino and Yanagida 1988 regarding these numerical values) would correspond to a surface force between F-actin and the siliconed + myosin-coated surface (system bathed in an aqueous buffer) of ~3 dyn cm^{-1} ~ 3 × 10^3 pN μm^{-1}, lower than ~20 × 10^3 pN μm^{-1}, but not significantly so. As demonstrated, on pp. **267–268**, the force of ~0.2 pN per myosin head is underestimated and the most probable value is ~1.4 pN per myosin head (see p. **268**). Using the rule of three, the expected surface force equivalent to ~1.4 pN per myosin head would therefore be ~(1.4/0.2) × 3 × 10^3 pN μm^{-1} ~ 21 × 10^3 pN μm^{-1}, very close to ~20 × 10^3 pN μm^{-1}. I conclude that, in the study by Kishino and Yanagida (1988), the strength of the active and surface forces can be considered to be very similar.

It is difficult to extend this approach to the experiments of Kishino and Yanagida (1988) to other experimental studies of the active force, because a careful study of the techniques and methods used is required. Nonetheless, as a general rule, care is required in the interpretation of experimental values recorded in vitro and presented as representative of active forces related solely to actin–myosin interactions. We can pose the following three questions: (i) Do the forces recorded in vitro typically correspond to active forces or (ii) is it also essential to take into account surface forces? (iii) Is it too audacious to suggest that active forces as a result of cyclic interactions of myosin heads with actin would be partly or entirely concealed by passive surface forces? The severe criticism presented by Cyranoski (2000) mostly concerns the results obtained by Yanagida and his interpretation of the experimental data obtained by his group. However, the problem of recording surface forces is never addressed by Cyranoski (2000). I think that the groups that have recorded forces in vitro should be made aware of the 'surface force problem'.

Given the doubts concerning the validity of measurements of active force in vitro, it also remains unclear whether in vitro experimental studies can be extrapolated to muscle contraction (see also p. **267** concerning the calculation of Woledge 1988). In this context, Morel and D'hahan (2000) claimed: 'We believe there are mechanisms for muscle contraction (in this case considerable forces are developed, with small displacements) and other mechanisms for in vitro movement (giving large displacements, without necessarily generating substantial forces)'.

In the 1970s and 1980s, many experiments performed by Oplatka and coworkers showed that actin–myosin systems could generate movement in vitro (e.g. Hochberg et al. 1977; Oplatka and Tirosh 1973; Oplatka et al. 1974, 1977; Tirosh and Oplatka 1982). Unfortunately, these publications have been ignored, and, after the publication of the paper by Sheetz and Spudich (1983), Oplatka claimed publicly that he and his group were the first to observe in vitro motility. In the 1980s, many articles by other authors appeared on this subject of in vitro motility, but they systematically 'omitted' to cite the pioneering work of Oplatka and his group. More than 20 years after his 'public exasperation', Oplatka (2005) wrote: 'our discovery has been ignored for a long time until the so-called "in vitro essays" appeared. By using this artifact-laden technique the mechanochemical reactivity of the active myosin fragments was re-discovered without giving us any credit'. The authors ignoring the work of Oplatka's group include Higashi-Fujime (1985), who observed movements of thick myosin filaments along thin actin filaments in vitro, Sheetz and Spudich (1983), who demonstrated that myosin-coated beads move along actin cables, Toyoshima et al. (1987), who showed that myosin S1 (isolated head) is sufficient to move thin actin filaments in vitro, and Yanagida et al. (1984, 1985), who reported that thin actin filaments move along thick myosin filaments in vitro. It has thus been definitively demonstrated that directional movement can be generated upon interactions of myosin or its heads and F-actin in vitro, but, at this point, no attempts were made to try to record 'potential active' forces. Many experimental studies later confirmed such movements. I strongly suggest that the motility observed in vitro corresponds solely to displacements similar to those of cell organelles, as already claimed by Morel and Bachouchi (1988a). In this context, the major doubts raised above, in this chapter, concern the problems of quantifying the forces involved in generating in vitro motility. There are also problems relating to the choice as to the true rate of MgATP splitting to be taken into account when quantifying in vitro motility (Morel and D'hahan 2000; Morel et al. 1993). These doubts raise two key questions concerning muscle contraction. Do we actually need to measure the working stroke by carrying out in vitro experiments (Morel et al. 1993) and to detect the number of MgATP molecules split per attachment–detachment cycle, using the same 'universal' in vitro motility technique (e.g. Oiwa et al. 1991, concerning the 'molecular' work done by ATP-induced sliding between myosin and actin)? How important are these 'enzymatic–mechanical' parameters measured in vitro for the mechanical properties of the muscle sarcomere, in a context in which the swinging cross-bridge/lever-arm mechanisms of contraction are increasingly being superseded by other mechanisms, for example, the hybrid model presented and discussed in this book, the snap-back/impulsive model developed by Elliott and Worthington and the step-wise process suggested by Pollack and coworkers (many other models are recalled in the Introduction)? More recently, other crucial problems have been raised, concerning in vitro motility and the usual rotating cross-bridge models in vitro

and in muscle, by Bryant et al. (2007), Spudich and Sivaramakrishnan (2010) and Sweeney and Houdusse (2010), who expressed major doubts about the traditional swinging cross-bridge/lever-arm theories as the basic model for actomyosin motors (see the citations from these three independent groups in the Preface).

Realistically, it cannot be claimed that there is a consensus regarding the 'unconventional' molecular mechanisms underlying in vitro motility (e.g. Morel and Bachouchi 1988a,b, 1990; Morel and D'hahan 2000; Liu and Pollack 2004; Pollack 1990). In any event, as pointed out in the Introduction, conditions are very different in vivo/in situ and in vitro, raising questions about the link between in vitro motility and in vivo/in situ contraction. For example, as repeatedly recalled in this monograph, structural, biophysical and biochemical conditions are very different in vivo/in situ and in vitro. After writing this paragraph, I read the minireview by Borejdo et al. (2006). Regardless of differences in our approaches to muscle contraction and in vitro motility, Borejdo et al. (2006) and I follow the same train of thought, regarding conditions in vivo/in vitro and in vitro, as Borejdo et al. (2006) wrote: 'The lack of a crowded environment [in vitro compared to in situ] may be responsible for the fact that movement in solution may be generated differently than in muscle'.

The 'molecular jet' process, first suggested more than 25 years ago to account for in vitro motility (Morel and Bachouchi 1988a), has since been further discussed and improved (Morel and Bachouchi 1988b, 1990; Morel and D'hahan 2000; Morel et al. 1993). According to this model, the velocity of the motion of covaspheres (beads) coated with myosin along actin cables (see the experimental work of Sheetz and Spudich 1983) is proportional to the instantaneous initial rate of MgATP splitting (tight coupling). This rate corresponds principally to the bursts of H^+ and Pi and, to some extent, to the very rapid ejection of some MgADP molecules occurring during the first few milliseconds after contact between the myosin heads and F-actin (Morel and Bachouchi 1988b). The molecular jet hypothesis was criticised by Marin et al. (1990), but Morel and Bachouchi-Sahli (1992) demonstrated that the analysis of Marin et al. (1990) was misleading and erroneous. According to the molecular jet model, in vitro motility must obey the step-wise/stepping mechanisms, confirmed by several groups (e.g. Cappello 2008; Cappello et al.

2007; Mehta 2001, in the particular case of myosin V; see p. **268** for some details about myosin V). This step-wise/stepping mechanism is not new, because similar observations have already been reported by Uyeda et al. (1991a) for myosin II (muscle myosin). Indeed, studying the velocity of thin actin filaments in cyclic contacts with myosin II in vitro, the authors claimed that 'The progress of the filaments could be interpreted as a combination of intervals of "runs" of directed movement and "pauses" containing Brownian movements'. The authors therefore observed 'the presence of peaks in the velocity distribution'. Moreover, 'the peaks appeared to be regularly spaced'. These experimental observations are consistent with the molecular jet process (see Morel and Bachouchi 1988a,b), but this evidence is not sufficient to distinguish between the molecular jet process, the swinging cross-bridge lever-arm theories and the models developed by Elliott and Worthington (1994, 1997, 2001), Worthington and Elliott (1996a,b, 2003, 2005), Grazi and Di Bona (2006), Iwazumi (1970, 1979) and Pollack (1990), all of which are valid for muscle contraction and, supposedly, also for in vitro motility. For instance, Irving (1991) commented on the experimental findings of Uyeda et al. (1991b): 'Direct observation of force fluctuations at the piconewton level and of velocity quantization provide dramatic and powerful support for the conventional view of a motor which attaches to actin, executes a power stroke, and detaches'. Regardless of these views, Morel and Bachouchi (1988b) claimed, but did not demonstrate, that their reasoning valid for the molecular jet (tight coupling between velocity and initial rate of MgATP breakdown; see above) can be extended to swinging cross-bridge/lever-arm processes, which can themselves be systematically extended, by many specialists, to in vitro motility, and vice versa. In this context, the very loose coupling between tension and MgATPase activity occurring during isometric tetanic muscle contraction (see pp. **78–80** in Section 3.4.4.2 and pp. **83–84** in Section 3.4.4.3) and the tight coupling occurring during in vitro motility and movements of cell organelles (see above) cast other major doubts on the simple extrapolation of in vitro motility to in vivo/in situ contraction and vice versa.

Despite the remarkable techniques involved, it is unclear whether the experimental results obtained in vitro can be interpreted, in the case of muscle contraction, in terms of the ratchet model (e.g. Shimokawa et al. 2003; Wang and Oster 2002; Yanagida 2001) and the biased Brownian models (e.g. Esaki et al. 2003; Yanagida 2001; Yanagida et al. 2000a,b, 2007). The ratchet and biased Brownian models have the same molecular bases and I still consider, as I have claimed, with my coworkers, for more than 20 years, that these models cannot account for muscle contraction, contrary to the claim of Yanagida and coworkers, for instance. Probably as a response to the severe criticisms raised by Cyranoski (2000) concerning these concepts, Buonocore et al. (2004; in collaboration with Yanagida) suggested that the theoretical approach of Yanagida and coworkers can be combined with the swinging cross-bridge/lever-arm theories to account for muscle contraction. However, I do not think that this approach solves the major question posed above, in this section, concerning the difference between the molecular mechanisms underlying in vitro motility and muscle contraction.

Yanagida and his group, and an increasing number of other independent groups, have used a series of powerful techniques, all of which have been and are still being gradually improved (for precise details on some of these techniques applied to various species of myosin, see, for example, Cappello 2008; Cappello et al. 2007; Dunn and Spudich 2007; Finer et al. 1994; Funatsu et al. 1995; Ishijiama and Yanagida 2001; Kitamura et al. 1999, 2005; Kron and Spudich 1986; Mehta 2001; Mehta and Spudich 1998; Mehta et al. 1999; Molloy et al. 1995a,b; Saito et al. 1994; Sako and Yanagida 2003; Simmons et al. 1996; Spudich et al. 2011; Warshaw et al. 2000; Yildiz et al. 2003). Despite Yanagida's technological brilliance, it may not yet be the right time to consider, for instance, the question of the validity of loose coupling versus tight coupling (swinging cross-bridge/lever-arm theories) in actomyosin dynamics within the framework of the concepts put forward by Yanagida and his coworkers. The same is true for the question of the working stroke and the related question of the measurement of advancement down an actin filament by a myosin head caused by the splitting of a single MgATP molecule. Much more important and urgent problems should first be resolved to account for muscle contraction. For this reason, I study, in this monograph, a simple theory, based on sure and simple experimental data: the hybrid model, valid for a whole muscle, an intact fibre, a demembranated fibre, an isolated myofibril and an intact unit cell, but not for in vitro motility, which, in my opinion, obeys totally different biological and biophysical laws. Only after resolving all the elementary, but complex, problems of the exact mechanisms of muscle contraction should studies concerning the main features of the MgATPase cycle in muscle or in vitro be discussed in depth. In any event, all the in vitro experiments performed on various protein motors are extremely useful, for the comparison of these proteins. In this context, it has been shown that, by contrast to what has been reported for myosin II (found in skeletal muscle), the two heads of kinesin are essential to induce movement upon interaction with microtubule protofilaments (Berliner et al. 1995; Gilbert et al. 1995; Schnapp 1995). Another remarkable protein motor is myosin V, 'an unconventional myosin involved in transporting vesicles along actin cables in the cell. Like myosin II, myosin V is a two-headed molecule, but it differs in having a long lever arm … Myosin V appears to proceed along the actin filament by a "head-over-head" (Dunn and Spudich 2007; Yildiz et al. 2003) mechanism (Purcell et al. 2002)' (citation from Geeves and Holmes 2005). Laakso et al. (2008) used cutting-edge techniques to study the myosin I family, 'the widely expressed, single-headed and membrane-associated members of the myosin superfamily, that participate in regulating membrane dynamics and structure in nearly all eukaryotic cells'. The authors demonstrated that myosin I 'can act as a molecular force sensor', its actin-attachment kinetics depending strongly on the average inter-action force. Thus, kinesin, myosin I and myosin V are very different from myosin II. All these comparative results were obtained with the powerful techniques used by many authors cited at the beginning of this paragraph.

As there are serious problems associated with the quantitative measurements of active forces in vitro (see above, in this chapter), I think that the molecular jet hypothesis presented by Morel and Bachouchi (1988a,b) should be reconsidered. Indeed, this theory not only accounts for large displacements at velocities consistent with those measured in vitro but also predicts very small forces of only ~1.6–8.0 × 10^{-2} pN per myosin head

(optimal value of ~$[(1.6 + 8.0)/2] \times 10^{-2} \sim 4.8 \times 10^{-2}$ pN per head), upon the cyclic interaction of myosin and other protein motors (in the form of single molecules, fragments of these molecules or molecular assemblies, such as thick myosin filaments) with F-actin (Morel and Bachouchi 1988a). The molecular jet hypothesis may help to account for many phenomena concerning in vitro motility, the movement of cell organelles and the transport of vesicles along actin cables in the cell, but certainly not muscle contraction. It is clear to me that muscle contraction, in vitro motility and the movement of cell organelles obey different biological and biophysical laws, as already pointed out above, in this chapter. In this context, I am dubious about the concepts developed by Warshaw (1996), for example, according to which efforts to elucidate the mechanisms of muscle contraction have focused on measurements of single thin actin filaments over individual myosin molecules. This and similar approaches are still seen as topical and are widely accepted (e.g. Kitamura et al. 2005), but inappropriately, as demonstrated in this chapter.

To conclude and support my view concerning the questionable relationship between in vitro motility and muscle contraction, I would like to recall the words of Borejdo et al. (2006) (see p. **270**) and to cite two sentences written by Grazi (2000), who has presented, over the last 15 years or so, an unconventional approach to the mechanisms of muscle contraction. The author claimed that 'disrupting the sarcomere structure harms the functioning of the contractile apparatus' and that 'demonstration is still awaited that what is going on at the single molecule level bears any relationship with what is going on in muscle'. Xu et al. (2006), who are in favour of traditional swinging cross-bridge/lever-arm processes, noted that, in the various biochemical, biophysical, physiological and structural experimental results obtained with muscle fibres (intact or demembranated), there was frequently a gap between the experimental data obtained in vivo/in situ and in vitro. Nonetheless, some intimate properties of isolated actin filaments and myosin molecules from various origins can be studied and compared (in case of disease, for example), using in vitro motility as a working tool.

CHAPTER TEN

General Conclusion

FROM OLD CONCEPTS, OLD AND RECENT EXPERIMENTAL DATA AND REASONING AND NEW EXPERIMENTAL RESULTS, TO A HYBRID MODEL OF MUSCLE CONTRACTION

A hybrid mechanism of muscle contraction is proposed, based on many experimental results obtained here and elsewhere, by my group and other independent groups. This model combines swinging cross-bridge/lever-arm concepts and an improved lateral swelling process, deduced from the lateral swelling model previously proposed by my group (in the 1970s and 1980s). It is theoretically, semi empirically and experimentally demonstrated, as expected from the work of Morel et al. (1976), that the radial repulsive electrostatic forces exerted between the myofilaments are effective and are counterbalanced, at rest, by strong radial tethering forces exerted by some weakly bound cross-bridges between the thick myosin and thin actin filaments and some weak radial attractive/compressive forces. It is proposed that, when isometric contraction is triggered, there is a rapid drop in the strength of the radial tethering forces, which remain weak during short tetani and under steady-state conditions. Thus, the radial repulsive electrostatic forces between myofilaments automatically prevail over the radial residual tethering forces plus weak radial attractive/compressive forces, leading to ~40% of the axial isometric tetanic force, in intact frog fibres at around the slack length and under physiological conditions (e.g. surrounding buffer mimicking the extracellular medium within the body, at ~10°C). The remaining ~60% of the axial force results essentially from the rotation of the attached cross-bridges, according to the swinging cross-bridge/lever-arm theories. On relaxation, strong radial tethering forces reappear and the balance between the repulsive and tethering plus weak radial attractive/compressive forces is reestablished. As highlighted throughout this work, quantitative differences exist between an intact unit cell and an intact fibre. The proportions of ~40% and ~60% recalled above are valid in an intact fibre, but not necessarily in an intact unit cell (located in the centre of the intact fibre). However, the exact proportions are not a major issue and the proportions of ~40% and ~60% can be assumed to be identical in intact unit cells and intact fibres. Nonetheless, more obvious quantitative differences appear when a contracting 'microscopic' intact unit cell is compared to a contracting 'macroscopic' intact fibre. For example, in an isometrically contracting intact fibre, many biophysical phenomena are different from those existing in an intact unit cell, and, as a consequence, the isometric tetanic tension recorded in an intact fibre, at 10°C, is ~3.0×10^5 N m^{-2} (see Section 3.3), whereas this tension, deduced from several experimental data obtained by independent groups and estimated in this monograph semi empirically, by various methods, is found to be ~7.2×10^5 N m^{-2}, at 10°C, in an intact unit cell (see Section 3.7, particularly p. **90**).

The possible mechanisms by which the net radial repulsive (expansive) forces are translated into axial contractile forces were discussed in the 1970s (Dragomir et al. 1976; Elliott 1974; Morel et al. 1976). These processes are independent of the variations in fibre volume when contraction is triggered, and isovolumic behaviour, considered essential by most authors, is not a prerequisite for conversion of the radial repulsive forces into axial forces. Indeed, regardless of the exact mechanisms involved in this translation, it is shown, in the experimental part of this monograph, that this conversion is impressive, under both resting and contraction conditions, for intact fibres, half-fibres, demembranated fibres, isolated myofibrils and intact unit cells.

The hybrid model is the first approach taking into account major features of the structure of the thick myosin filaments and many properties of the myosin molecule and its heads. It was experimentally shown that the two heads of myosin are necessary for the formation of 'physiological' synthetic thick myosin filaments, and the arrangement of the myosin heads is also valid for natural myosin filaments (Morel et al. 1999). It is demonstrated in this book that this arrangement, with head–head dimers within the core of the thick myosin filaments, is required to induce ~40% contraction, via lateral swelling mechanisms, in an intact frog fibre, under conditions recalled in the first paragraph. The hybrid model has a strong predictive and explanatory power and accounts for many unexplained and somewhat 'mysterious' phenomena, clearly supporting this approach to muscle contraction. Within the body, many (but not all) skeletal muscles work around the slack length, and contributions of ~60% and ~40% for swinging cross-bridge/lever-arm and lateral swelling processes, respectively, probably best represent their actual behaviour.

Starting from experiments described in this monograph and many other experimental data accumulated by my group and many other independent groups, this work has provided me with the opportunity to analyse and account for many experimental data, some of which have remained unexplained or ignored for more than 30 or 40 years. I have paid particular attention to the problem of electrical phenomena in the myofilament lattice, the existence and properties of which have been ignored by most investigators, although several authors showed experimentally that many fixed negative electrical charges and bound anions are present on the myofilaments. Even when the

radial repulsive electrostatic forces are not discounted, they are mostly considered to play a passive role in the stability of the myofilament lattice, rarely being considered to play any active role in muscle contraction. It is shown in this book that the radial repulsive electrostatic forces are of the right order of magnitude and that they efficiently account for a large part of the axial contractile force.

As pointed out above and in several chapters and sections, many experimental results and their interpretations are valid for intact fibres, but quantitative comparison of intact fibres and intact unit cells located in the centre of myofibrils, themselves lying in the centre of an intact fibre, may be problematic. However, it is shown, throughout this work, that both qualitative and quantitative comparisons of these two biological materials are possible. Nonetheless, the observed differences between these biological materials should always be kept in mind and discussed point by point. Another major problem concerns the validity of studying molecular motors in vitro as a means of investigating muscle contraction. In this context, I agree with the opinions of Borejdo et al. (2006) and Grazi (2000) that muscle contraction and in vitro motility do not obey the same physical, physical–chemical and biological laws (see, in Chapter 9, pp. **270** and **272**, respectively). Nevertheless, in vitro studies of molecular motors and the development of advanced technologies should improve our knowledge of the MgATPase cycle and of the motility of cell organelles and properties of other cell structures (e.g. movements and exchanges in membranes, polar ascension of chromosomes during mitosis).

Most of the swinging cross-bridge/lever-arm models are based on the first cross-bridge (side-piece) theory of A.F. Huxley (1957). This pioneering approach was later adjusted by Huxley (1965, 1969) on the basis of more accurate experimental 'molecular' data published after 1957. Physiological results obtained by Huxley and Simmons (1971, 1973), working on isometric transients, added further improvements to the model of Huxley (1965, 1969). I demonstrate in this monograph that only part of the side-piece model can be retained (see also below). The phenomenological/semi empirical approach to any model, including the hybrid model presented and discussed in this book, is strongly connected to A.F. Huxley's theory (1957), but with a major difference in interpretation of his fundamental equation giving the total energy output in a contracting muscle, at any shortening velocity (see the next paragraph). With my coworkers, I indeed took into account the term proportional to the shortening velocity in this basic equation and compared it with the 'experimental' expression of total energy output (including the Fenn effect; Fenn 1923, 1924; Fenn and Marsh 1935; A.V. Hill 1938, 1964a,b). We demonstrated that the notion of cross-bridges cyclically attaching to and detaching from the thin actin filaments and splitting MgATP during the cycle should be considered beyond dispute and extremely powerful. Moreover, this phenomenological/semi empirical approach, derived from A.F. Huxley's theory (1957), is valid for any model of contraction, including the swinging cross-bridge/lever-arm models and the hybrid model (see above). From experimental data obtained for whole frog muscles, Curtin and Woledge (1978), Woledge (1971) and Woledge et al. (1985) claimed that A.F. Huxley's theory (1957) 'has now been experimentally disproved'. This assertion concerned mostly the expression of the energy output as a function of the velocity of shortening, deduced exclusively from the 'mechanical' expression of the force–velocity relationship proposed by A.F. Huxley (1957). To try to reconcile the contradictory views of A.F. Huxley (1957), on the one hand, and Curtin and Woledge (1978), Woledge (1971) and Woledge et al. (1985), on the other, Morel (1984a,b) introduced the simple notion of active contacts, with only some actin–myosin attachment–detachment cycles leading to the splitting of MgATP molecules. Moreover, Morel (1984a,b) suggested that an increase in the velocity of active shortening is associated with a decrease in the proportion of active contacts or, in other words, an increase in the proportion of inactive contacts (attachment–detachment cycle with no subsequent MgATP breakdown). Comparing the 'energetic' expression of tension (see the next paragraph) with experimental data obtained for isolated myofibrils and chemically skinned or permeabilised fibres, the notion of inactive contacts proves powerful, as demonstrated in this monograph (see Section 3.4.4.1). Finally, the new phenomenological/semi empirical approach presented in this book is consistent with old and recent experiments.

At this point, I would like to stress that Morel and Gingold (1977) and Morel et al. (1976) found that the energetic relationship between the load

lifted and shortening velocity, s_0V, is also valid for isometric tetanic contraction ($s_0V = 0$). This expression of the isometric tetanic tension, deduced from the energetic part of the theory of side-pieces (A.F. Huxley 1957), has frequently been used and improved by my group and is also used in his updated form in this work. More precisely, in his analytical expression of energy output, A.F. Huxley (1957) did not note that his fundamental equation contained, in a concealed form, the expression of the load lifted plus the coefficient of shortening heat as a function of the shortening velocity, entirely independent of the mechanical roles of the cross-bridges, provided only that they attach to and detach cyclically from the thin actin filaments and split MgATP during the cycle, as repeatedly recalled in this monograph. Approximately 20 years after the publication of the article by A.F. Huxley (1957), Morel and Gingold (1977) and Morel et al. (1976) discovered this 'concealed detail'. If A.F. Huxley (1957) had been aware of the potency of his equation, he would have deduced that he was in the presence of two contradictory expressions of the load lifted against the shortening velocity s_0V (isotonic contraction) and of the isometric tetanic tension ($s_0V = 0$): an energetic expression, deduced from the expression of energy output, and a totally independent mechanical expression that takes into account the linear distortion and elasticity of the attached cross-bridges. This would probably have modified the approach of A.F. Huxley (1957) to muscle contraction. Indeed, Morel and Gingold (1977) demonstrated that these two expressions are biophysically inconsistent. In many further publications, my group demonstrated that the energetic expression was extremely powerful to account for and predict many phenomena. A.F. Huxley's theory (1957) should therefore be considered extremely interesting and the cycling cross-bridge theory cannot be ruled out. From this phenomenological/semi empirical approach and many available experimental data by other independent groups, it is demonstrated, in this monograph, that there is a very loose coupling between the rate of 'isometric' MgATP splitting measured in vivo/in situ and isometric tetanic tension, at odds with the traditional view of tight (one-to-one) coupling, that is, proportionality between MgATP breakdown and isometric tension, in swinging cross-bridge/lever-arm theories. Thus, A.F. Huxley's theory (1957) cannot be used in its original form

ADDENDUM

As pointed out in the Preface, this monograph was completed between 2008 and 2011. Obviously, during this period, many papers concerning muscle proteins, whole muscle, muscle fibres (intact or demembranated) and muscle contraction were published. It would be impossible to write an updated book, taking into account all the publications appearing in each year, during the 2008–2011 period and thereafter. Thus, I suggest here a list of references corresponding to the period 2008–(2012–2013), with some references in 2014 and 2015. I have not read the corresponding papers (only the summaries, when available), but the reader may wish to select some papers, for comparison with the main body of this work. To my mind, when a new approach is proposed, it is imperative to try to attack it on solid experimental, semiempirical and theoretical bases. The reader will certainly find, in the following list of about 200–250 references, many new arguments for criticising (or for supporting, I hope) the analysis and the synthesis proposed in this 'treatise', particularly the hybrid model.

REFERENCES

Ait Mou, Y., J. Y. le Guennec, E. Mosca et al. 2008. Differential contribution of cardiac sarcomeric proteins in the myofibrillar force response to stretch. *Pflügers Arch. Eur. J. Physiol.* 457:25–36.

Al-Jumaily, A. M., P. Mbikou, and P. R. Redey. 2012. Effect of length oscillations on airway smooth muscle reactivity and cross-bridge cycling. *Am. J. Physiol.* 303:L286–94.

Allen, D. G., and S. Trajanovska. 2012. The multiple roles of phosphate in muscle fatigue. *J. Muscle Res. Cell Motil.* 30:199–207.

Aprodu, I., A. Redaelli, and M. Soncini. 2008. Actomyosin interaction: Mechanical and energetic properties in different nucleotide binding states. *Int. J. Mol. Sci.* 9:927–43.

Aprodu, I., M. Soncini, F. M. Montevecchi et al. 2010. Mechanical characterization of actomyosin complex by molecular mechanic simulations. *J. Appl. Biomater. Biomech.* 8:20–7.

Arjunan, S. P., and D. K. Kumar. 2013. Age-associated changes in muscle activity during isometric contraction. *Muscle Nerve* 47:545–9.

Arnold, D. B., and G. Gallo. 2014. Structure meets function: Actin filaments and myosin motors in the axon. *J. Neurochem.* 129:213–20.

Bang, M. L., M. Caremari, E. Brunello et al. 2009. Nebulin plays a direct role in promoting strong actin-myosin interactions. *Fed. Proc. FASEB J.* 23:4117–25.

Barclay, C. J., R. C. Woledge, and N. A. Curtin. 2010a. Inferring crossbridge properties from skeletal muscle energetics. *Prog. Biophys. Mol. Biol.* 102:3–71.

Barclay, C. J., R. C. Woledge, and N. A. Curtin. 2010b. Is the efficiency of mammalian (mouse) skeletal muscle temperature dependent? *J. Physiol. Lond.* 588:3819–31.

Bershitsky, S. Y., N. A. Koubassova, P. M. Bennett et al. 2010. Myosin heads contribute to the maintenance of filament order in relaxed rabbit muscle. *Biophys. J.* 99:1827–34.

Betters, C., C. Veigel, E. Homsher et al. 2014. To understand muscle you must take it apart. *Front. Physiol.* 5:90.

Bickham, D. C., T. G. West, M. R. Webb et al. 2011. Millisecond-scale biochemical response to change in strain. *Biophys. J.* 101:2445–54.

Billington, N., D. J. Revill, S. A. Burgess et al. 2014. Flexibility within the heads of muscle myosin-2 molecules. *J. Mol. Biol.* 426:894–907.

Böl, M., A. Schmitz, G. Nowak et al. 2012. A three-dimensional chemo-mechanical continuum model for smooth muscle contraction. *J. Mech. Behav. Biomed. Mater.* 13:215–29.

Borejdo, J., D. Szczesna-Cordary, P. Muthu et al. 2012. Single molecule detection approach to muscle study: Kinetics of a single cross-bridge during contraction of muscle. *Meth. Mol. Biol.* 875:311–34.

Boujemaa-Paterski, T., R. Galland, C. Suarez et al. 2014. Directed actin assembly and motility. *Meth. Enzymol.* 540:283–300.

Brunello, E., L. Fusi, M. Reconditi et al. 2009. Structural changes in myosin motors and filaments during relaxation of skeletal muscle. *J. Physiol. Lond.* 587:4509–21.

Burghardt, T. P., J. Li, and K. Ajtai. 2009. Single myosin lever arm orientation in a muscle fiber detected with photoactivable GFP. *Biochemistry* 48:754–65.

Campbell, K. S. 2010. Short-range mechanical properties of skeletal and cardiac muscles. *Adv. Exp. Med. Biol.* 682:223–46.

Capitano, M., and S. Franseco. 2013. Interrogating biology with force: single molecule high-resolution measurements with optical tweezers. *Biophys. J.* 105:1293–303.

Caremani, M., J. Dantzig, Y. E. Goldman et al. 2008. Effect of inorganic phosphate on the force and number of myosin cross-bridges during the isometric contraction of permeabilized muscle fibers from rabbit psoas. *Biophys. J.* 95:5798–808.

Caremani, M., L. Melli, M. Dolfi et al. 2013. The working stroke of the myosin motor in muscle is not tightly coupled to release of orthophosphate from its active site. *J. Physiol. Lond.* 591:5187–205.

Chandra, M., R. Mamidi, S. Ford et al. 2009. Nebulin alters cross-bridge cycling kinetics and increases thin filament activation: A novel mechanism for increasing tension and reducing tension cost. *J. Biol. Chem.* 284:30889–96.

Chen, X., K. Pavlish, and J. N. Benoit. 2008. Myosin phosphorylation triggers actin polymerisation in vascular smooth muscle. *Am. J. Physiol.* 295:H2172–7.

Cochran, J. C., M. E. Thomson, and F. J. Kull. 2013. Metal switch-controlled myosin II from *Dictyostelium discoideum* supports closure of nucleotides pocket during ATP binding coupled to detachment from actin filaments. *J. Biol. Chem.* 288:28312–23.

Coffee Castro-Zena, P. G., and D. D. Root. 2013. Asymmetric myosin binding to the thin filaments as revealed by a fluorescent nanocircuit. *Arch. Biochem. Biophys.* 535:14–21.

Colombini, B., M. Nocella, M. A. Bagni et al. 2010. Is the cross-bridge stiffness proportional to tension during muscle fiber activation? *Biophys. J.* 98:2582–90.

Cornachione, A. S., and D. E. Rassier. 2012. A non-cross-bridge, static tension is present in permeabilized skeletal muscle fibers after active force inhibition or actin extraction. *Am. J. Physiol.* 302:C566–74.

Davis, J. S., and N. D. Epstein. 2009. Mechanistic role of movement and strain sensitivity in muscle contraction. *Proc. Natl. Acad. Sci. U.S.A.* 106:6140–5.

Debold, E. P. 2012. Recent insights into muscle fatigue at the cross-bridge level. *Front. Physiol.* 3:141.

de Tombe, P. P., R. D. Mateja, K. Tachamba et al. 2010. Myofilament length dependent activity. *J. Mol. Cell. Cardiol.* 48:851–8.

Dominguez, R., and K. C. Holmes. 2011. Actin structure and function. *Annu. Rev. Biophys.* 40:169–86.

Duggal, D., J. Nagwekar, R. Rich et al. 2013. Phosphorylation of myosin regulatory light chain has minimal effect on kinetics and distribution of orientations of cross-bridges of rabbit skeletal muscle. *Am. J. Physiol.* 306:R222–32.

Dulyaninova, N. G., and A. R. Bresnick. 2013. The heavy chain has its day: Regulation of myosin-II assembly. *Bioarchitecture* 3:77–85.

Edman, K. A. P. 2010. Contractile performance of striated muscle. *Adv. Exp. Med. Biol.* 682:7–40.

Elangovan, R., M. Capitanio, L. Melli et al. 2012. An integrated in vivo and in situ study of myosin II from frog skeletal muscle. *J. Physiol. Lond.* 590:1227–42.

Elliott, G. F., and C. R. Worthington. 2012. Along the road not taken: How many myosin heads act on a single actin filament at any instant in working muscle? *Prog. Biophys. Mol. Biol.* 108:82–92.

Elting, M. W., and J. A. Spudich. 2012. Future challenges in single-molecule fluorescence and laser trap approaches to studies of molecular motors. *Dev. Cell* 23:1084–91.

Farman, G. P., D. Gore, E. Allen et al. 2011. Myosin orientation: A structural determinant for the Frank-Starling relationship. *Am. J. Physiol.* 300:H2155–60.

Fauler, M., K. Jurkat-Rott, and F. Lehman-Horn. 2012. Membrane excitability and excitation-contraction uncoupling in muscle fatigue. *Neuromuscul. Disord.* 22(Suppl. 3):S162–7.

Fitzsimmons, D. P., and R. L. Moss. 2008. Cooperativity in the regulation of force and the kinetics of force development in heart and skeletal muscles. *Adv. Exp. Med. Biol.* 592:177–89.

Fouchard, J., D. Mitrossilis, and A. Asnacios. 2011. Acto-myosin based response to stiffness and rigidity sensing. *Cell Adh. Migr.* 5:16–9.

Fukuda, N., S. Terui, K. Ishiwata et al. 2010. Titin-based regulations of diastolic and systolic functions of mammalian cardiac muscle. *J. Mol. Cell Cardiol.* 48:876–81.

Fusi, L., M. Reconditi, M. Linari et al. 2010. The mechanism of the resistance to stretch of isometrically contracting single muscle fibres. *J. Physiol. Lond.* 588:495–510.

Fusi, L., B. Brunello, M. Reconditi et al. 2014. The nonlinear elasticity of the muscle sarcomere and the compliance of myosin motors. *J. Physiol. Lond.* 592:1109–18.

George, N. T., T. C. Irving, C. D. Williams et al. 2013. The cross-bridge spring: Can cool muscles store elastic energy? *Science* 340:1217–20.

Gittings, W., J. Huang, and R. Vandenboom. 2012. Tetanic force potentiation of mouse fast muscle is shortening speed dependent. *J. Muscle Res. Cell Motil.* 33:359–68.

Gokhin, D. S., and V. M. Fowler. 2013. A two-segment model for thin filament architecture in skeletal muscle. *Nat. Rev. Mol. Cell Biol.* 14:113–9.

Gokhin, D. S., M. L. Bang, J. Zhang et al. 2009. Reduced thin filament length in nebulin-knockout skeletal muscle alters isometric contractile properties. *Am. J. Physiol.* 296:C1123–32.

Gomibuchi, Y., T. Q. P. Uyeda, and T. Wakabayashi. 2013. Bulkiness or aromatic nature of tyrosine-143 of actin is important for the weak binding between F-actin and myosin-ADP-phosphate. *Biochem. Biophys. Res. Commun.* 441:844–8.

Granzier, H. L. M. 2010. Activation and stretch-induced passive force enhancement—Are you pulling my chain? Focus on "Regulation of muscle force in the absence of actin-myosin-based cross-bridge interactions". *Am. J. Physiol.* 299:C11–3.

Grazi, E. 2010. Muscle mechanism. The acceleration of the load. *Arch. Biochem. Biophys.* 504:204–9.

Grazi, E. 2012. The Huxley–Simmons manoeuvre is still lacking the experimental evidence that the quick release is a pure elastic phenomenon. *Arch. Biochem. Biophys.* 522:121–4.

Grazi, E. 2013. The proposed mechanisms of skeletal muscle contraction. Possible uncertainties in the interpretation of the data. *OA Med. Hypothes.* 1(1):8.

Grazi, E., and S. Pozzati. 2009. The contribution of the elastic reaction is severely underestimated in studies on myofibril contraction. *Int. J. Mol. Sci.* 10:942–53.

Grazi, E., and S. Pozzati. 2010. Skeletal muscle contraction. The thorough definition of the contractile event requires both load acceleration and load mass to be known. *Theor. Biol. Med. Model.* 7:24.

Greenberg, M. J., and J. R. Moore. 2010. The molecular basis of frictional loads in the *in vitro* motility assay with applications to the study of the loaded mechanochemistry of molecular motors. *Cytoskeleton* 67:273–85.

Greenberg, M. J., T. R. Mealy, M. Jones et al. 2010. The direct molecular effects of fatigue and myosin regulatory light chain phosphorylation on the actomyosin contractile apparatus. *Am. J. Physiol.* 298:R969–96.

Günther, S., and K. Kruse. 2010. Spontaneous sarcomere dynamics. *Chaos* 20(4):045122.

Günther, M., and S. Schmitt. 2010. A macroscopic ansatz to deduce the Hill relation. *J. Theor. Biol.* 263:407–16.

Haldeman, B. D., R. K. Brizedine, K. C. Facemeyer et al. 2014. The kinetics underlying the velocity of smooth muscle myosin filament sliding on actin filaments in vitro. *J. Biol. Chem.* 289:21055–70.

Haraguchi, T., K. Honda, Y. Wanikawa et al. 2013. Function of the head-tail junction in the activity of myosin II. *Biochem. Biophys. Res. Commun.* 440:490–4.

Hasson, C. J., and G. E. Caldwell. 2012. Effects of age on mechanical properties of *dorsiflexor* and *plantarflexor* muscles. *Ann. Biomed. Eng.* 40:1088–101.

Herzog, W., V. Joumaa, and T. R. Leonard. 2010. The force–length relationship of mechanically isolated sarcomeres. *Adv. Exp. Med. Biol.* 682:141–61.

Herzog, W., T. Leonard, V. Joumaa et al. 2012. The three filament model of skeletal muscle stability and force production. *Mol. Cell Biomech.* 9:175–91.

Ibanez-Garcia, D., J. Requejo-Isidro, M. R. Webb et al. 2010. Fluorescence lifetime imaging reveals that the environment of the ATP binding site in muscle senses force. *Biophys. J.* 99:2163–9.

Iorga, B., L. Wang, R. Stehle et al. 2012. ATP binding and cross-bridge detachment steps during full Ca^{2+} activation: Comparison of myofibril and muscle fibre mechanics by sinusoidal analysis. *J. Physiol. Lond.* 590:3361–73.

Irving, T. C., Y. Wu, T. Bekyarova et al. 2011. Thick-filament strain and interfilament spacing in passive fiber bundles from rabbit skinned psoas muscles. *Biophys. J.* 100:1499–508.

Ishiwata, S., Y. Shimamoto, and M. Suzuki. 2010. Molecular motors as an auto-oscillator. *HFSP J.* 4:100–4.

Ishiwata, S., Y. Shimamoto, and N. Fukuda. 2011. Contractile system of muscle as an auto-oscillator. *Prog. Biophys. Mol. Biol.* 105:187–98.

James, R. S., J. Taillis, A. Herrel et al. 2012. Warmer is better: thermal sensitivity of both maximal and sustained power output in the *iliotibialis* muscle isolated from adult *Xenopus tropicalis*. *J. Exp. Biol.* 215:552–8.

Janssen, P. M. 2010. Kinetics of cardiac muscle contraction and relaxation are linked and determined by properties of the cardiac sarcomere. *Am. J. Physiol.* 299:H1092–9.

Jin, J. P. 2013. Myofilament and cytoskeleton proteins: Fine machineries of biological movements. *Arch. Biochem. Biophys.* 535:1–2.

Joumaa, V., D. E. Rassier, T. R. Leonard et al. 2008. The origin of passive force enhancement in skeletal muscle. *Am. J. Physiol.* 294:C74–8.

Joumaa, V., B. R. MacIntosh, and W. Herzog. 2012. New insights into force depression in skeletal muscle. *J. Exp. Biol.* 215:2135–40.

Kalganov, A., R. Novinger, and D. E. Rassier. 2010. A technique for simultaneous measurement of force and overlap between single muscle filaments of myosin and actin. *Biochem. Biophys. Res. Commun.* 403:351–6.

Karatzaferi, C., and P. B. Chase. 2013. Muscle fatigue and muscle weakness: What we know and what we wish we did. *Front. Physiol.* 4:125.

Kaya, M., and H. Higuchi. 2010. Nonlinear elasticity and an 8-nm working stroke of single myosin molecules in myofilaments. *Science* 329:686–9.

Kiani, F. A., and S. Fisher. 2013. Stabilization of the ADP/metaphosphate intermediate during ATP hydrolysis in pre-power stroke myosin: Quantitative anatomy of an enzyme. *J. Biol. Chem.* 288:35569–80.

Kosterina, N., H. Westerblad, and A. Eriksson. 2012. History effect and timing of force production introduced in a skeletal muscle model. *Biomech. Model Mechanobiol.* 11:947–57.

Kroon, M. 2010. A constitutive model for smooth muscle including active tone and passive visco-elastic behaviour. *Math. Med. Biol.* 27:129–55.

Kroon, M. 2011. Influence of dispersion in myosin filament orientation and anisotropic filament contractions in smooth muscle. *J. Theor. Biol.* 272:72–82.

Lee, J. Y., T. M. Iverson, and R. J. Dima. 2011. Molecular investigations into the mechanics of actin in different nucleotide states. *J. Phys. Chem. B* 115:186–95.

Lehman, W., A. Galinska-Rakóczy, V. Hatch et al. 2009. Structural basis for the activation of muscle contraction by troponin and tropomyosin. *J. Mol. Biol.* 388:673–81.

Lenz, M., T. Thoresen, M. L. Gardel et al. 2012. Contractile units in disordered actomyosin bundles arise from F-actin buckling. arXiv:1201.4110v1 [physics.bio-ph].

Leonard, T. R., and W. Herzog. 2010. Regulation of muscle force in the absence of actin–myosin-based cross-bridge interaction. *Am. J. Physiol.* 299:C14–20.

Linari, M., M. Caremari, and V. Lombardi. 2010. A kinetic model that explains the effect of inorganic phosphate on the mechanics and energetics of isometric contraction of fast skeletal muscle. *Proc. Biol. Sci.* 277:19–27.

Locher, M. R., M. V. Razumova, J. E. Stelzer et al. 2009. Determination of rate constants for turnover of myosin isoforms in rat myocardium: Implication for *in vivo* contractile kinetics. *Am. J. Physiol.* 297:H247–56.

Loiselle, D. S., K. Tran, E. J. Crampin et al. 2010. Why has reversal of the actin–myosin cross-bridge cycle not been observed experimentally? *J. Appl. Physiol.* 108:1465–71.

Lorenz, M., and K. C. Holmes. 2010. The acto–myosin interface. *Proc. Natl. Acad. Sci. U.S.A.* 107:12529–34.

Luther, P. K. 2009. The vertebrate muscle Z-disc: Sarcomere anchor for structure and signaling. *J. Muscle Res. Cell Motil.* 30:171–85.

MacIntosh, B. R. 2010. Cellular and whole muscle studies of activity dependent potentiation. *Adv. Exp. Med. Biol.* 682:315–42.

MacIntosh, B. R., R. J. Holash, and J. M. Renaud. 2012. Skeletal muscle fatigue: Regulation of excitation–contraction coupling to avoid metabolic catastrophe. *J. Cell Sci.* 125:2105–14.

Málnási-Csizmadia, A., and M. Kovács. 2010. Emerging complex pathways of the actomyosin power-stroke. *Trends Biochem. Sci.* 35:684–90.

Månson, A. R., D. Rassier, and G. Tsiavariliris. 2015. Poorly understood aspects of striated muscle contraction. *BioMed. Res. Intern.* 2015:245154.

Månsson, A. 2010. Actomyosin-ADP states, interhead cooperativity, and the force–velocity relation of skeletal muscle. *Biophys. J.* 98:1237–46.

Marcucci, L., and L. Truskinovsky. 2010a. Mechanics of the power stroke in myosin II. *Phys. Rev. E* 81:051915.

Marcucci, L., and L. Truskinovsky. 2010b. Muscle contraction: A mechanical perspective. *Eur. Phys. J. E Soft Matter* 32:411–8.

Marcucci, L., and T. Yanagida. 2012. From single molecule fluctuations to muscle contraction: A Brownian model of A. F. Huxley's hypotheses. *PLoS One* 7(7):pe40042.

Matsui, T. S., K. Ito, R. Kaunas et al. 2010. Actin stress fibers are at a tipping point between conventional shortening and rapid disassembly at physiological levels of MgATP. *Biochem. Biophys. Res. Commun.* 395:301–6.

Mettikolla, P., N. Callender, R. Luchkowski et al. 2010. Observing cycling of a few cross-bridges during isometric contraction of skeletal muscle. *Cytoskeleton* 67:400–11.

Meyer, G. A., A. D. McCulloch, and R. L. Lieber. 2011. A nonlinear model of passive muscle viscosity. *J. Biomech. Eng.* 133:091007.

Middle, K., R. Rich, P. Marandos et al. 2013. Comparison of orientation and rotational motion of skeletal muscle cross-bridges containing phosphorylated and dephosphorylated myosin regulatory light chain. *J. Biol. Chem.* 288:7012–23.

Mika, D., P. Bobin, H. Pomérance et al. 2013. Differential regulation of cardiac excitation-contraction coupling by cATP phosphodiesterase subtypes. *Cardiovasc. Res.* 100:336–46.

Miller, M. S., C. M. Dambacher, A. F. Knowles et al. 2009a. Alternative S2 hinge regions of the myosin rod affect myofibrillar structure and myosin kinetics. *Biophys. J.* 96:4132–43.

Miller, M. S., P. Vanburen, M. M. Lewinter et al. 2009b. Mechanisms underlying skeletal muscle weakness in human heart failure: Alterations in single fiber myosin content and function. *Circ. Heart Fail.* 2:700–6.

Miller, M. S., B. C. W. Tanner, L. R. Nyland et al. 2010a. Comparative biomechanics of thick filaments and thin filaments with functional consequences for muscle contraction. *BioMed. Res. Instru.* 2010:473423.

Miller, M. S., P. Vanburen, M. M. Lewinter et al. 2010b. Chronic heart failure decreases cross-bridge kinetics in single skeletal muscle fibres from humans. *J. Physiol. Lond.* 588:4039–53.

Minozzo, F. C., L. Hilbert, and D. E. Rassier. 2012. Pre-power-stroke cross-bridges contribute to force transients during imposed shortening in isolated muscle fibers. *PLoS One* 7(1):e29356.

Mitsui, T., and H. Ohshima. 2008. Remarks on muscle contraction mechanism. *Int. J. Mol. Sci.* 9:872–904.

Mitsui, T., N. Takai, and H. Ohshima. 2011. Remarks on muscle contraction mechanism. II. Isometric tension transient and isotonic velocity transient. *Int. J. Mol. Sci.* 12:1697–726.

Murrell, M., T. Thoresen, and M. Gardel. 2014. Reconstitution of contractile actomyosin arrays. *Meth. Enzymol.* 540:265–82.

Murtada, S. C., M. Kroon, and G. A. Holzapfel. 2010. A calcium-driven mechanochemical model for prediction of force generation in smooth muscle. *Biomech. Model. Mechanobiol.* 9:749–62.

Murtada, S. C., A. Arner, and G. A. Holzapfel. 2012. Experiments and mechanochemical modeling of smooth muscle contraction: Significance of filament overlap. *J. Theor. Biol.* 297:176–86.

Nishikawa, K. C., J. A. Monroy, T. E. Uyeno et al. 2012. Is titin a 'winding filament'? A new twist on muscle contraction. *Proc. Biol. Sci.* 279:981–90.

Nocella, M., B. Colombini, G. Benelli et al. 2011. Force decline during fatigue is due to both a decrease in the force per individual cross-bridge and the number of cross-bridges. *J. Physiol. Lond.* 589:3371–81.

Nocella, M., G. Cecchi, M. A. Bagni et al. 2013. Effect of temperature on crossbridge force changes during fatigue and recovery in intact mouse muscle fibres. *PLoS One* 8(10):e78918.

Nogueira, D. A., S. B. Silva, L. C. de Abreu et al. 2012. Effect of the rest interval duration between contractions on muscle fatigue. *BioMed. Eng. Online* 11:89.

Orzechowski, M., J. R. Moore, S. Fisher et al. 2014. Tropomyosin movement on F-actin during muscle activation explained by energy landscapes. *Arch. Biochem. Biophys.* 545:63–8.

Oshima, K., Y. Sugimoto, T. C. Irving et al. 2012. Head–head interactions of resting myosin cross-bridges in intact frog skeletal muscle, revealed by synchrotron X-ray fiber diffraction. *PLoS One* 7(12):e52421.

Pant, K. H., J. Watt, M. Greenberg et al. 2009. Removal of the cardiac myosin regulatory light chain increases isometric force production. *Fed. Proc. FASEB J.* 23:3571–80.

Park, K. H., L. Brotto, O. Lehoang et al. 2012. Ex vivo assessment of contractility, fatigability and alternans in isolated skeletal muscles. *J. Vis. Exp.* 2012(69):e4198.

Park-Holohan, S., M. Linari, M. Reconditi et al. 2012. Mechanics of myosin function in white muscle fibres of the dogfish *Scyliorhinus canicula*. *J. Physiol. Lond.* 590:1973–88.

Peterson, P., M. Kalda, and M. Vendelin. 2013. Real-time determination of sarcomere length of a single cardiomyocyte during contraction. *Am. J. Physiol.* 304:C519–31.

Piazzesi, G., M. Dolfi, E. Brunello et al. 2014. The myofilament elasticity and its effect on kinetics of force generation by myosin motor. *Arch. Biochem. Biophys.* 552–3:108–16.

Pinto, A., F. Sánchez, F. L. Alamo et al. 2012. The myosin interacting-heads motif is present in the relaxed thick filament of the striated muscle of scorpion. *J. Struct. Biol.* 180:469–78.

Plaçais, P. Y., M. Balland, T. Guérin et al. 2009. Spontaneous oscillations of a minimal actomyosin system under elastic loading. *Phys. Rev. Lett.* 103:158102.

Pollack, G. H. 2013. *The fourth phase of water: Beyond solid, liquid, and vapor.* Ebner and Sons, Seattle.

Preller, M., and K. C. Holmes. 2013a. The myosin start-of-power stroke state and how actin binding drives the power stroke. *Cytoskeleton* 70: 651–60.

Preller, M., and K. C. Holmes. 2013b. Myosin structure, allostery, and mechano-chemistry. *Structure* 21:1911–22.

Radocaj, A., T. Weiss, W. I. Helsby et al. 2009. Force-generating cross-bridges during ramp-shaped releases: Evidence for a new structural state. *Biophys. J.* 96:1430–46.

Rall, J. A. 2014. *Mechanism of muscular contraction.* Springer, New York.

Ranatunga, K. W. 2010. Force and power generating mechanism(s) in active muscle as revealed from temperature perturbation studies. *J. Physiol. Lond.* 588:657–70.

Ranatunga, K. W., and M. E. Coupland. 2010. Crossbridge mechanism(s) examined by temperature perturbation studies on muscle. *Adv. Exp. Med. Biol.* 682:247–66.

Ranatunga, K. W., H. Roots, G. J. Pinniger et al. 2010. Crossbridge and non-crossbridge contributions to force in shortening and lengthening muscle. *Adv. Exp. Med. Biol.* 682:207–21.

Rassier, D. E. 2009. Molecular basis of force development by skeletal muscles during and after stretch. *Mol. Cell Biomech.* 6:229–41.

Rassier, D. E. 2010. Striated muscles: From molecules to cells. *Adv. Exp. Med. Biol.* 682:1–6.

Rassier, D. E. 2012. The mechanisms of the residual force enhancement after stretch of skeletal muscle: Non-uniformity in half-sarcomeres and stiffness of titin. *Proc. Biol. Sci.* 279:2705–13.

Rassier, D. E., and I. Pavlov. 2010. Contractile characteristics of sarcomeres arranged in series or mechanically isolated from myofibrils. *Adv. Exp. Med. Biol.* 682:123–40.

Rassier, D. E., and C. Pun. 2010. Stretch and shortening of skeletal muscles activated along the ascending limb of the force–length relation. *Adv. Exp. Med. Biol.* 682:175–89.

Rassier, D. E., and I. Pavlov. 2012. Force produced by isolated sarcomeres and half-sarcomeres after an imposed stretch. *Am. J. Physiol.* 27:C240–8.

Reconditi, M., E. Brunello, L. Fusi et al. 2014. Sarcomere-length dependence of myosin filament structure in skeletal muscle fibres of the frog. *J. Physiol. Lond.* 592:1119–37.

Reymann, A. C., C. Guérin, M. Théry et al. 2014. Geometrical control of actin assembly and contractility. *Methods Cell Biol.* 120:19–38.

Romet-Lemonne, G., and A. Jégou. 2013. Mechanotransduction down to individual actin filaments. *Eur. J. Cell Biol.* 92:333–8.

Roots, H., G. Ball, J. Talbot-Ponsonby et al. 2009. Muscle fatigue examined at different temperatures in experiments on intact mammalian (rat) muscle fibers. *J. Appl. Phys.* 106:378–84.

Roots, H., G. J. Pinniger, G. W. Offer et al. 2012. Mechanism of force enhancement during and after lengthening of active muscle: A temperature dependence study. *J. Muscle Res. Cell Motil.* 33:313–25.

Rossier, O. M., N. Gauthier, N. Biais et al. 2010. Force generated by actomyosin contraction builds bridges between adhesive contacts. *EMBO J.* 29:1055–68.

Schnappacher-Tip, G., A. Jinha, and W. Herzog. 2011. Mapping the classical cross-bridge theory and backward steps in a three bead laser trap setup. *Math. Biosci.* 229:115–22.

Morel

Seebohn, B., F. Matinmehr, J. Köhler et al. 2009. Cardiomyopathy mutations reveal variable region of myosin converter as major element of cross-bridge compliance. *Biophys. J.* 97:806–24.

Seow, C. Y. 2013. Hill's equation of muscle performance and its hidden insight on molecular mechanisms. *J. Gen. Physiol.* 142:561–73.

Shabarchin, A. A., and A. K. Tsaturyan. 2010. Proposed role of the M-band in sarcomere mechanics and mechano-sensing: A model study. *Biomech. Model. Mechanobiol.* 9:163–75.

Shchepkin, D. V., A. M. Matyushenko, G. V. Kopylova et al. 2013. Stabilization of the central part of tropomyosin molecule alters the Ca^{2+}-sensitivity of actin–myosin interaction. *Acta Naturae* 5: 126–9.

Sich, N. M., T. J. O'Donnell, S. A. Coulter et al. 2010. Effects of actin–myosin kinetics on the calcium sensitivity of regulated thin filaments. *J. Biol. Chem.* 285:39150–9.

Smith, D. A., and D. G. Stephenson. 2011. An electrostatic model with weak actin–myosin attachment resolves problems with the stability of skeletal muscle. *Biophys. J.* 100:2688–97.

Smith, L. R., G. A. Meyer, and R. L. Lieber. 2013. Systems analysis of biological networks in skeletal muscle function. *Wiley Interdiscip. Rev. Syst. Biol. Med.* 5:55–71.

So, E., R. Stahlberg, and G. H. Pollack. 2012. Exclusion zone as an intermediate between ice and water. In *Water and society*, eds. D. W. Peper, and C. A. Brebbia, 1–11. WIT Press, Southampton, UK.

Spudich, J. A. 2012. One path to understanding energy transduction in biological systems. *Nat. Med.* 18:1478–82.

Stachowiak, M. R., and B. O'Shauhnessy. 2009. Recoil after severing reveals stress fiber contraction mechanisms. *Biophys. J.* 97:462–71.

Stehle, R., and B. Iorga. 2010. Kinetics of cardiac sarcomeric processes and rate-limiting steps in contraction and relaxation. *J. Mol. Cell Cardiol.* 48:843–50.

Stewart, M. A., K. Franks-Skiba, S. Chen et al. 2010. Myosin ATP turnover rate is a mechanism involved in thermogenesis in resting skeletal muscle fibres. *Proc. Natl. Acad. Sci. U.S.A.* 107:430–5.

Sugi, G., S. Chen, T. Kobayashi et al. 2014. Definite differences between *in vitro* actin-myosin sliding and muscle contraction as revealed using antibodies to myosin heads. *PloS One* 9(12):e93272.

Sugi, H. 2010. Electron microscopy visualization of the cross-bridge movement coupled with ATP hydrolysis in muscle thick filaments in aqueous solution, reminiscence and future prospects. *Adv. Exp. Med. Biol.* 282:77–103.

Sugi, H., H. Minoda, Y. Inayoshi et al. 2008. Direct demonstration of the cross-bridge recovery stroke in muscle thick filaments in aqueous solution by using the hydration chamber. *Proc. Natl. Acad. Sci. U.S.A.* 105:17396–401.

Swenson, A. M., D. V. Trivedi, A. A. Rausher et al. 2014. Magnesium modulates actin binding and ADP release in myosin motors. *J. Biol. Chem.* 289:10566–81.

Syamaladevi, D. P., J. A. Spudich, and R. Sowdhamini. 2012. Structural and functional insights on the myosin superfamily. *Bioinform. Biol. Insights* 6:11–21.

Takatsuhi, H., E. Bengtsson, and A. Månsson. 2014. Persistence length of fascin-cross-linked actin filament bundles in solution and the *in vitro* motility assay. *Biochim. Biophys. Acta* 1840: 1932–42.

Tamura, T., J. Wakayama, K. Inoue et al. 2009. Dynamics of thin filament activation in rabbit skeletal muscle fibers examined by time-resolved X-ray diffraction. *Biophys. J.* 96:1045–55.

Tanner, B. C. W., T. L. Daniel, and M. Regnier. 2012a. Filament compliance influences cooperative activation of thin filaments and the dynamics of force production in skeletal muscle. *PLoS Comput. Biol.* 8(5):e1002506.

Tanner, B. C. W., G. P. Farman, T. C. Irving et al. 2012b. Thick-to-thin filament surface distance modulates cross-bridges kinetics in *Drosophila* flight muscle. *Biophys. J.* 103:1275–84.

Tkachev, Y. V., J. Ge, I. V. Negrashov et al. 2013. Metal cation controls myosin and actomyosin kinetics. *Protein Sci.* 22:1766–74.

Várkuti, B. H., Z. Yang, B. Kintses et al. 2012. A novel actin binding site of myosin required for effective muscle contraction. *Nat. Struct. Mol. Biol.* 19:299–306.

von Wegner, F., S. Schurmann, R. H. A. Fink et al. 2009. Motor protein function in skeletal muscle—A multiple scale approach to contractility. *IEEE Trans Med. Imaging* 28:1632–42.

Walcott, S., P. M. Fagnant, K. M. Trybus et al. 2009. Smooth muscle heavy meromyosin phosphorylated on one of its two heads supports force and motion. *J. Biol. Chem.* 284:18244–51.

West, T. G., G. Hild, V. B. Siththanandran et al. 2009. Time course and strain dependence of ADP release during contraction of permeabilized skeletal muscle fibers. *Biophys. J.* 96:3281–94.

Williams, C. D., M. Regnier, and T. L. Daniel. 2010. Axial and radial forces of cross-bridges depend on lattice spacing. *PLoS Comput. Biol.* 6(12):e1001018.

Williams, C. D., M. Regnier, and T. L. Daniel. 2012. Elastic energy storage and radial forces in the myofilament lattice depend on sarcomere length. *PLoS Comput. Biol.* 8(11):e1002770.

Williams, C. D., M. K. Salcedo, T. C. Irving et al. 2013. The length–tension curve in muscle depends on lattice spacing. *Proc. Biol. Sci.* 280(1766): 20130697.

Zoghbi, M. E., J. L. Woodhead, R. L. Moss et al. 2008. Three-dimensional structure of vertebrate cardiac muscle myosin filaments. *Proc. Natl. Acad. Sci. U.S.A.* 105:2386–90.

Zot, H. G., J. E. Hasbun, and N. Van Minh. 2009. Striated muscle regulation of isometric tension by multiple equilibria. *PLoS One* 4(12):e8052.

REFERENCES

Abbott, B. C., and X. M. Aubert. 1952. The force exerted by active striated muscle during and after change in length. *J. Physiol. Lond.* 117:77–86.

Adhikari, B. B., and P. G. Fajer. 1996. Myosin head orientation and mobility during isometric contraction: Effects of osmotic compression. *Biophys. J.* 70:1872–80.

Adhikari, B. B., K. Hideg, and P. G. Fajer. 1997. Independent mobility of catalytic and regulatory domains of myosin heads. *Proc. Natl. Acad. Sci. U.S.A.* 94:9643–7.

Alberts, B., and R. Miake-Lye. 1992. Unscrambling the puzzle of biochemical machines: The importance of the details. *Cell* 68:415–20.

Alberts, B., A. Johnson, J. Lewis et al. 2007. *Molecular biology of the cell.* Garland Science, New York, USA.

Albet-Torres, N., M. J. Bloemink, T. Barman et al. 2009. Drug effect unveils inter-head cooperativity and strain-dependent ADP release in fast skeletal actomyosin. *J. Biol. Chem.* 284:22926–37.

Aldoroty, R. A., and E. W. April. 1984. Donnan potentials from striated muscle crystals: A-band and I-band measurements. *Biophys. J.* 46:769–79.

Aldoroty, R. A., N. A. Garty, and E. W. April. 1985. Donnan potentials from striated muscle liquid crystals: Sarcomere length dependence. *Biophys. J.* 47:89–96.

Aldoroty, R. A., N. A. Garty, and E. W. April. 1987. Donnan potentials from striated muscle liquid crystals. Lattice spacing dependence. *Biophys. J.* 51:371–81.

Allard, B., and O. Rougier. 1994. The effects of chloride ions in excitation-contraction coupling and sarcoplasmic reticulum calcium release in twitched muscle fibres. *J. Muscle Res. Cell Motil.* 15:563–71.

Allen, D. G., B. R. Jewell, and J. W. Murray. 1974. The contribution of activation processes to the length–tension relation of cardiac muscle. *Nature* 248:606.

Allen, D. G., J. Lännergren, and H. Westerblad. 1995. Muscle cell function during prolonged activity: Cellular mechanisms of fatigue. *Exp. Physiol.* 80:497–527.

Allen St. Claire, T., N. Ling, M. Irving et al. 1996. Orientation changes in myosin regulatory light chain following photorelease of ATP in skinned muscle fibers. *Biophys. J.* 70:1847–62.

Altringham, J. D., and R. Bottinelli. 1985. The descending limb of the sarcomere length-force relation in single muscle fibres from the frog. *J. Muscle Res. Cell Motil.* 6:585–600.

Altringham, J. D., R. Bottinelli, and J. Lacktis. 1984. Is stepwise sarcomere shortening an artefact? *Nature* 307:653–5.

Amemiya, Y., H. Sugi, and H. Hashizume. 1979. X-ray diffraction studies on the dynamic properties of crossbridges in skeletal muscle. In *Crossbridge mechanism in muscle contraction*, eds. H. Sugi, and G. H. Pollack, 425–43. Univ. Park Press, Baltimore, USA.

Amemiya, Y., H. Iwamoto, T. Kobayashi et al. 1988. Time-resolved X-ray diffraction studies of the effect of slow length changes on tetanized frog skeletal muscle. *J. Physiol. Lond.* 407:231–41.

Amos, L. A., H. E. Huxley, K. C. Holmes et al. 1982. Structural evidence that myosin heads may interact with two sites on F-actin. *Nature* 299:467–9.

Anazawa, T., K. Yasuda, and S. Ishiwata. 1992. Spontaneous oscillations of tension and sarcomere length in skeletal myofibrils. Microscopic measurements and analysis. *Biophys. J.* 61:1099–108.

Andreev, O. A., and J. Borejdo. 1991. Myosin heads can bind two actin monomers. *Biochim. Biophys. Res. Commun.* 177:350–6.

Andreev, O. A., A. L. Andreeva, V. S. Matkin et al. 1993. Two different rigor complexes of myosin subfragment-1 and actin. *Biochemistry* 32: 12035–46.

Andrews, M. A. W., D. W. Maughan, T. M. Nosek et al. 1991. Ion-specific and general ionic effects on contraction of skinned fast-twitch skeletal muscle from the rabbit. *J. Gen. Physiol.* 98:1105–25.

April, E. W. 1969. The effects of tonicity and ionic strength on tension and filament lattice volume in single muscle fibers. PhD diss., Columbia Univ., New York, USA.

April, E. W. 1978. Liquid crystalline contractile apparatus in striated muscle. *Am. Chem. Soc. Symp.* B 74:248–55.

April, E. W., and P. W. Brandt. 1973. The myofilament lattice: Studies on isolated fibers. III. The effect of myofilament spacing upon tension. *J. Gen. Physiol.* 61:490–508.

April, E. W., and D. Wong. 1976. Non-isovolumic behaviour of the unit cell of skinned muscle fibres. *J. Mol. Biol.* 101:107–14.

April, E. W., and R. A. Aldoroty. 1986. Donnan potentials generated by the surface charges on muscle filaments. In *Electrical double layers in biology*, ed. M. Blank, 287–300. Plenum Press, New York, USA.

April, E. W., P. W. Brandt, J. P. Reuben et al. 1968. Muscle contraction: The effect of ionic strength. *Nature* 220:182–3.

April, E. W., P. W. Brandt, and G. F. Elliott. 1971. The myofilament lattice: Studies on isolated fibers. I. The constancy of the unit-cell volume with variation in sarcomere length in a lattice in which the thin-to-thick myofilament ratio is 6:1. *J. Cell Biol.* 51:72–82.

April, E. W., P. W. Brandt, and G. F. Elliott. 1972. The myofilament lattice studies on isolated fibers. II. The effects of osmotic strength, ionic concentration, and pH on the unit cell volume. *J. Cell Biol.* 53:53–65.

Armstrong, C. F., A. F. Huxley, and F. J. Julian. 1966. Oscillatory responses in frog skeletal muscle fibre. *J. Physiol. Lond.* 186:26P–7P.

Asayama, J., L. E. Ford, and M. F. Surdyk-Droske. 1983. Relationship between sarcoplasmic reticulum volume and calcium capacity in skinned frog skeletal muscle fibres. *J. Muscle Res. Cell Motil.* 4: 307–19.

Ashley, R. 1972. A hybrid theory of muscle contraction. *J. Theor. Biol.* 36:339–54.

Auerbach, G., R. Huber, M. Grattinger et al. 1997. Closed structure of phosphoglycerate kinase from *Thermotoga maritima* reveals the catalytic mechanism and determinant of thermal stability. *Structure* 5:1475–83.

Baatsen, P. H., K. Trombitas, and G. H. Pollack. 1988. Thick filaments of striated muscles are laterally interconnected. *J. Ultrastruct. Mol. Struct. Res.* 98:267–80.

Bachouchi, N., and J. E. Morel. 1989a. Behaviour of the crossbridges in stretched or compressed muscle fibres. *J. Theor. Biol.* 141:143–57.

Bachouchi, N., and J. E. Morel. 1989b. Muscle contraction and *in vitro* movement. *J. Theor. Biol.* 141: 425–7.

Bachouchi, N., A. Gulik, M. Garrigos et al. 1985. Rabbit skeletal myosin heads in solution, as observed by ultracentrifugation and freeze-fracture electron microscopy: Dimerization and maximum chord. *Biochemistry* 24:6305–10.

Bachouchi, N., M. Garrigos, and J. E. Morel. 1986. MgATPase activity of myosin subfragment 1. The dimer is more active than the monomer. *J. Mol. Biol.* 191:247–54.

Bagni, M. A., G. Cecchi, and M. Schoenberg. 1988a. A model of force production that explains the lag between crossbridge attachment and force after electrical stimulation of striated muscle fibers. *Biophys. J.* 54:1105–14.

Bagni, M. A., G. Cecchi, F. Colomo et al. 1988b. Plateau and descending limb of the sarcomere length–tension relation in short length-clamped segments of frog muscle fibres. *J. Physiol. Lond.* 401:581–95.

Bagni, M. A., G. Cecchi, F. Colomo et al. 1990a. Myofilament spacing and force generation in intact frog muscle fibres. *J. Physiol. Lond.* 430: 61–75.

Bagni, M. A., G. Cecchi, F. Colomo et al. 1990b. Tension and stiffness of frog muscle fibres at full overlap. *J. Muscle. Res. Cell Motil.* 11:373–7.

Bagni, M. A., G. Cecchi, F. Colomo et al. 1992. Are weakly binding bridges present in resting intact muscle fibers? *Biophys. J.* 63:1412–5.

Bagni, M. A., G. Cecchi, F. Colomo et al. 1994a. Development of stiffness precedes cross-bridge attachment during the early tension rise in single frog muscle fibres. *J. Physiol. Lond.* 481:273–8.

Bagni, M. A., G. Cecchi, G. Griffiths et al. 1994b. Lattice spacing changes accompanying isometric tension development in intact single muscle fibers. *Biophys. J.* 67:1965–75.

Bagni, M. A., G. Cecchi, F. Colomo et al. 1995. Absence of mechanical evidence for attached weakly binding cross-bridges in frog relaxed muscle fibres. *J. Physiol. Lond.* 482:391–400.

Bagshaw, C. R. 1987. Are two heads better than one? *Nature* 326:746–7.

Bagshaw, C. R. 1993. *Muscle contraction*. Chapman and Hall, London, UK.

Bagshaw, C. R., and D. R. Trentham. 1973. The reversibility of adenosine triphosphatase cleavage by myosin. *Biochem. J.* 133:323–8.

Bagshaw, C. R., and D. R. Trentham. 1974. The characterization of myosin-product complexes and of product-release steps during the magnesium ion-dependent adenosine triphosphatase reaction. *Biochem. J.* 141:331–49.

Bagshaw, C. R., J. F. Eccleston, F. Eckstein et al. 1974. The magnesium ion-dependent adenosine triphosphatase of myosin: Two-step processes of adenosine triphosphatase association and adenosine diphosphate dissociation. *Biochem. J.* 141:351–64.

Bagshaw, C. R., D. R. Trentham, R. G. Wollcott et al. 1975. Oxygen exchange in the γ-phosphoryl group of protein-bound ATP during Mg^{2+} − dependent adenosine triphosphatase activity of myosin. *Proc. Natl. Acad. Sci. U.S.A.* 72:2592–5.

Baker, J. E., and D. D. Thomas. 2000. A thermodynamic muscle model and a chemical basis for A. V. Hill's muscle equation. *J. Muscle Res. Cell Motil.* 21:335–44.

Baker, J. E., L. Brust-Masher, S. Ramachandran et al. 1998. A large and distinct rotation of the myosin light chain domain occurs upon muscle contraction. *Proc. Natl. Acad. Sci. U.S.A.* 95:2944–9.

Ball, P. 2008. Water as an active constituent in cell biology. *Chem. Rev.* 108:74–108.

Banks, R. D., C. C. F. Blake, P. R. Evans et al. 1979. Sequence structure and activity of phosphoglycerate kinase: A possible hinge-bending enzyme. *Nature* 279:773–7.

Bárány, M. 1996. *Biochemistry of smooth muscle contraction*. Elsevier, New York, USA.

Barclay, C. J. 1992. Effects of fatigue on isometric force development in mouse fast- and slow-twitch muscle. *Am. J. Physiol.* 263:C1065–72.

Barclay, C. J. 1996. Mechanical efficiency and fatigue of fast and slow muscles of the mouse. *J. Physiol. Lond.* 497:781–94.

Barclay, C. J. 1999. A weakly coupled version of the Huxley crossbridge model can simulate energetics of amphibian and mammalian muscle. *J. Muscle Res. Cell Motil.* 20:163–76.

Barclay, C. J. 2005. Maximum contractile filament movement per ATP used in muscle contraction is approximately 1.3 nm not 13 nm. *Int. J. Biol. Macromol.* 37:154–5.

Barclay, C. J., N. A. Curtin, and R. C. Woledge. 1993. Changes in crossbridge and non-crossbridge energetics during moderate fatigue of muscle fibres. *J. Physiol. Lond.* 468:543–55.

Barclay, C. J., P. D. Arnold, and C. L. Gibbs. 1995. Fatigue and heat production in repeated contractions of mouse skeletal muscle. *J. Physiol. Lond.* 488:741–52.

Barclay, C. J., R. C. Woledge, and N. A. Curtin. 2010. Inferring crossbridge properties from skeletal muscle energetics. *Prog. Biophys. Mol. Biol.* 102:53–71.

Barman, T. E., and F. Travers. 1985. The rapid-flow-quench method in the study of fast reactions in biochemistry: Extension to subzero conditions. *Meth. Biochem. Anal.* 31:1–59.

Barman, T. E., A. Brun, and F. Travers. 1980. A flow-quench apparatus for cryoenzymic studies: Application to the creatine kinase reaction. *Eur. J. Biochem.* 110:397–403.

Barman, T., M. Brune, C. Lionne et al. 1998. ATPase and shortening rates in frog fast skeletal myofibrils by time-resolved measurements of protein-bound and free Pi. *Biophys. J.* 74:3120–30.

Barman, T., S. R. W. Bellamy, H. Gundfreund et al. 2006. The identification of chemical intermediate in enzyme catalysis by rapid quench-flow technique. *Cell Mol. Life Sci.* 63:2571–83.

Barrett, T. W., W. L. Petitcolas, and J. Robson. 1978. Laser Raman light-scattering observations of conformational changes in myosin induced by inorganic salts. *Biophys. J.* 23:349–58.

Barsotti, R. J., J. A. Dantzig, and Y. E. Goldman. 1996. Myosin isoforms show different strokes for different blocks. *Nat. Struct. Biol.* 3:737–9.

Bartels, E. M., and G. F. Elliott. 1980. Donnan potential measurements in the A- and I-bands of cross-striated muscles, and calculation of the fixed charges on the contractile proteins. *J. Muscle Res. Cell Motil.* 1:452–8.

Bartels, E. M., and G. F. Elliott. 1985. Donnan potentials from the A- and I-bands of glycerinated and chemically skinned muscle, relaxed and in rigor. *Biophys. J.* 48:61–76.

Bartels, E. M., P. Jensen, and O. Sten-Knudsen. 1976. The dependence of tension relaxation in skeletal muscle on the number of sarcomeres in series. *Acta Physiol. Scand.* 97:476–85.

Bartels, E. M., J. M. Skydsgaard, and O. Sten-Knudsen. 1979. The time course of the latency relaxation as a function of the sarcomere length in frog and mammalian muscle. *Acta Physiol. Scand.* 106:129–37.

Bartels, E. M., P. H. Cooke, G. F. Elliott et al. 1993. The myosin molecule charge response to nucleotide binding. *Biochim. Biophys. Acta* 1157:63–73.

Bartoo, M. L., V. I. Popov, L. Fearn et al. 1993. Active tension generation in isolated skeletal myofibrils. *J. Muscle Res. Cell Motil.* 14:498–510.

Bell, M. G., D. E. Dale, U. A. van den Heide et al. 2002. Polarized fluorescence depletion reports orientation distributions and rotational dynamics of muscle cross-bridges. *Biophys. J.* 83:1050–73.

Belton, P. S., and K. J. Paker. 1974. Pulsed NMR studies of water in striated muscle. III. The effects of water content. *Biochim. Biophys. Acta* 354:305–14.

Belton, P. S., R. R. Jackson, and K. J. Paker. 1972. Pulsed NMR studies of water in striated muscle. I. Transverse nuclear spin relaxation times and freezing effects. *Biochim. Biophys. Acta* 286:16–25.

Belton, P. S., K. J. Paker, and J. C. Sellwood. 1973. Pulsed NMR studies of water in striated muscle. II. Spin-lattice relaxation times and the dynamics of non-freezing fraction effects. *Biochim. Biophys. Acta* 304:56–64.

Benzinger, T. H. 1969. Ultrasensitive reaction calorimetry. In *Analysis methods of protein chemistry*, eds. P. Alexander and H. P. Lundgren, 85. Pergamon Press, Oxford.

Berger, C. E., P. M. Fagnant, S. Heizmann et al. 2001. ADP binding induces an asymmetry between the heads of unphosphorylated myosin. *J. Biol. Chem.* 276:23240–5.

Berliner, E., E. C. Young, K. Anderson et al. 1995. Failure of a single-headed kinesin to track parallel to microtubule protofilaments. *Nature* 373:718–21.

Bershitsky, S. Y., and A. K. Tsaturyan. 1992. Tension responses to joule temperature jump in skinned rabbit muscle fibres. *J. Physiol. Lond.* 447:425–48.

Bershitsky, S. Y., and A. K. Tsaturyan. 2002. The elementary force generation process probed by temperature and length perturbations in muscle fibres from the rabbit. *J. Physiol. Lond.* 540:971–88.

Bershitsky, S. Y., A. F. Tsaturyan, N. O. Bershistkaya et al. 1997. Muscle force is generated by myosin heads stereospecifically attached to actin. *Nature* 388:186–9.

Bianchi, C. P., and S. Narayau. 1982. Muscle fatigue and the role of the transverse tubules. *Science* 215:295–6.

Biosca, J. A., F. Travers, D. Hillaire et al. 1984. Cryoenzymatic studies on myosin subfragment 1: Perturbation of an enzyme reaction by temperature and solvent. *Biochemistry* 23:1967–76.

Blake, C. C. F., and P. R. Evans. 1974. Structure of horse muscle phosphoglycerate kinase. Some results on the chain conformation, substrate binding and evolution of the molecule from a 3 Ångström Fourier map. *J. Mol. Biol.* 84:585–601.

Blankenfeldt, W., N. H. Thomas, J. S. Wray et al. 2006. Crystal structures of human cardiac beta-myosin II S2 provide insight into the functional role of S2 subfragment. *Proc. Natl. Acad. Sci. U.S.A.* 103:17713–7.

Blinks, J. R. 1965. Influence of osmotic strength on cross-section and volume of isolated single muscle fibres. *J. Physiol. Lond.* 177:42–57.

Block, S. M. 1996. Fifty ways to love your lever: Myosin motors. *Cell* 87:151–7.

Blyakhman, F. A., T. Shklyar, and G. H. Pollack. 1999. Quantal length changes in single contracting sarcomeres. *J. Muscle Res. Cell Motil.* 20:529–38.

Blyakhman, F. A., A. Tourovskaya, and G. H. Pollack. 2001. Quantal sarcomere-length changes in relaxed single myofibrils. *Biophys. J.* 81:1093–100.

Bordas, J., G. P. Diakun, J. E. Harries et al. 1993. Two-dimensional X-ray diffraction of muscle: Recent results. *Adv. Biophys.* 27:15–33.

Bordas, J., A. Svenssen, M. Rothery et al. 1999. Extensibility and symmetry of actin filaments in contracting muscles. *Biophys. J.* 77:3197–207.

Borejdo, J. 1980. Tension fluctuations in contracting myofibrils and their interpretation. *Biophys. J.* 29:40–64.

Borejdo, J., and M. F. Morales. 1977. Fluctuations in tension during contraction of single muscle fibers. *Biophys. J.* 20:315–34.

Borejdo, J., A. A. Shepard, I. Akopova et al. 2004a. Rotation of the lever arm of myosin in contracting skeletal muscle fiber measured by two-photon anisotropy. *Biophys. J.* 87:3912–21.

Borejdo, J., A. A. Shepard, D. Dunka et al. 2004b. Changes in orientation of actin during contraction of muscle. *Biophys. J.* 86:2308–17.

Borejdo, J., J. Talent, and I. Akopova. 2006. Measuring rotation of a few cross-bridges in skeletal muscle. *Exp. Biol. Med.* 231:28–38.

Bottinelli, R., and C. Reggiani. 2000. Human skeletal muscle fibres: Molecular and functional diversity. *Prog. Biophys. Mol. Biol.* 73:195–262.

Brandt, P. W., and H. Grundfest. 1968. Sarcomere and myosin filament changes accompanying local contractile activation in crayfish muscle fibers. *Fed. Proc. FASEB J.* 27:375.

Brandt, P. W., E. Lopez, J. P. Reuben et al. 1967. The relationship between myofilament packing density and sarcomere length in frog striated muscle. *J. Gen. Physiol.* 33:255–64.

Brandt, P. W., F. Colomo, N. Piroddi et al. 1998. Force regulation by Ca^{2+} in skinned cardiac myocytes of frog. *Biophys. J.* 74:1994–2004.

Bray, D. 2000. *Cell movements – From molecules to motility*. Garland Publishing, New York, USA.

Bremel, R. D., and A. Weber. 1972. Cooperation within actin filament in vertebrate skeletal muscle. *Nat. New Biol.* 238:97–101.

Bremel, R. D., J. M. Murray, and A. Weber. 1973. Manifestation of cooperative behavior in the regulated actin filaments during actin-activated ATP hydrolysis in the presence of calcium. *Cold Spring Harb. Symp. Quant. Biol.* 37:267–75.

Brenner, B. 1980. Effects of free sarcoplasmic Ca^{2+} concentration on maximum unloaded shortening velocity: Measurements on single glycerinated rabbit psoas muscle fibres. *J. Muscle Res. Cell Motil.* 1:409–28.

Brenner, B. 1986. The cross-bridge cycle in muscle. Mechanical, biochemical, and structural studies on single skinned rabbit psoas fibers to characterize cross-bridge kinetics in muscle for correlation with the actomyosin ATPase in solution. *Basic Res. Cardiol.* 81:1–15.

Brenner, B. 1987. Mechanical and structural approaches to correlation of crossbridge action with actomyosin ATPase in solution. *Annu. Rev. Physiol.* 49:655–72.

Brenner, B. 1990. Muscle mechanics and biochemical kinetics. In *Molecular mechanism and biochemical kinetics*, ed. J. M. Squire, 76–149. MacMillan Press, London, UK.

Brenner, B. 1991. Rapid dissociation and reassociation of actomyosin cross-bridges during force generation: A newly observed facet of cross-bridge action in muscle. *Proc. Natl. Acad. Sci. U.S.A.* 88:10490–4.

Brenner, B., and L. C. Yu. 1985. Equatorial X-ray diffraction from single skinned psoas fibers during various degrees of activation. Changes in intensities and lattice spacing. *Biophys. J.* 48:829–34.

Brenner, B., and E. Eisenberg. 1987. The mechanism of muscle contraction. Biochemical, mechanical, and structural approaches to elucidate cross-bridge action in muscle. *Basic Res. Cardiol.* 82(Suppl. 2):3–16.

Brenner, B., and L. C. Yu. 1991. Characterization of radial force and radial stiffness in Ca^{2+}-activated skinned fibres of the rabbit psoas muscle. *J. Physiol. Lond.* 441:703–18.

Brenner, B., M. Schoenberg, J. M. Chalovich et al. 1982. Evidence for cross-bridge attachment in relaxed muscle at low ionic strength. *Proc. Natl. Acad. Sci. U.S.A.* 79:7288–91.

Brenner, B. M., L. C. Yu, and R. J. Podolsky. 1984. X-ray diffraction evidence for cross-bridge formation in relaxed muscle fibers at various ionic strengths. *Biophys. J.* 46:299–306.

Brenner, B., J. M. Chalovich, L. E. Greene et al. 1986. Stiffness of skinned rabbit psoas fibers in MgATP and MgPPi solutions. *Biophys. J.* 50:685–91.

Brenner, B., L. C. Yu, and J. M. Chalovich. 1991. Parallel inhibition of active force and relaxed fiber stiffness muscle by caldesmone: Implication for the pathway to force generation. *Proc. Natl. Acad. Sci. U.S.A.* 88:5739–43.

Brokaw, C. J. 1995. Weakly-coupled models for motor enzyme function. *Biophys. J.* 16:1013–27.

Brugman, C. T., H. van den Hooff, and T. Blangé. 1984. Some aspects of the role of quantum mechanics in the theory of muscle contraction. *J. Theor. Biol.* 107:173–7.

Brune, M., J. L. Hunter, J. E. T. Corrie et al. 1994. Direct, real-time measurement of rapid inorganic phosphate release using novel fluorescent probe and its application to actomyosin. subfragment 1 ATPase. *Biochemistry* 33:8262–71.

Brunello, E., P. Bianco, G. Piazzesi et al. 2006. Structural changes in the myosin filaments and cross-bridges during active force development in single intact frog muscle fibres: Stiffness and X-ray diffraction measurements. *J. Physiol. Lond.* 577:971–84.

Bryant, Z., D. Altman, and J. A. Spudich. 2007. The power stroke of myosin VI and the basis of reverse directionality. *Proc. Natl. Acad. Sci. U.S.A.* 104:772–7.

Buonocore, A., L. Caputo, Y. Ishii et al. 2004. A phenomenological model of myosin II dynamics in the presence of external loads. arXiv: q-bio/0411025v1 [q-bio. BM].

Burghardt, T. P., and K. Ajtai. 1985. Fraction of myosin cross-bridges bound to actin in active muscle fibers: Estimation by fluorescence anisotropy measurements. *Proc. Natl. Acad. Sci. U.S.A.* 82:8478–82.

Burghardt, T. P., T. Ando, and J. Borejdo. 1983. Evidence for crossbridge order in contraction of glycerinated skeletal muscle. *Proc. Natl. Acad. Sci. U.S.A.* 80:7515–9.

Burghardt, T. P., K. Ajtai, D. J. Chan et al. 2007. GFP-tagged regulatory light chain monitors single myosin lever-arm orientation in a muscle fiber. *Biophys. J.* 93:1–26.

Burke, M., E. Reisler, S. Himmelfarb et al. 1974. Myosin adenosine triphosphatase. Convergence of activation by actin and by SH modification at physiological ionic strength. *J. Biol. Chem.* 249:6361–73.

Burnell, E. E., M. E. Clarke, J. A. Hinke et al. 1981. Water in barnacle muscle. III. NMR studies of fresh fibers and membrane-damaged fibers equilibrated with selected solutes. *Biophys. J.* 33:1–26.

Burton, K. 1992. Myosin step size: Estimates from motility assays and shortening muscle. *J. Muscle Res. Cell Motil.* 13:590–607.

Burton, K., and A. F. Huxley. 1995. Identification of source of oscillations in apparent sarcomere length measured by laser diffraction. *Biophys. J.* 68: 2429–43.

Candau, R., B. Iorga, F. Travers et al. 2003. At physiological temperatures the ATPase rates of shortening *soleus* and psoas myofibrils are similar. *Biophys. J.* 85:3132–41.

Cantino, M. P., and J. M. Squire. 1986. Resting myosin cross-bridge configuration in frog muscle thick filaments. *J. Cell Biol.* 102:610–18.

Caplan, S. R. 1966. A characteristic of self-regulated linear energy converters. The Hill force-velocity relation for muscle. *J. Theor. Biol.* 13:63–86.

Cappello, G. 2008. Machines moléculaires, allumez vos feux de position! *Reflets Phys. (Paris)* 13:5–9.

Cappello, G., P. Pierobon, C. Symonts et al. 2007. Myosin V stepping mechanism. *Proc. Natl. Acad. Sci. U.S.A.* 104:15328–33.

Caremani, M., V. Lombardi, and M. Linari. 2006. The isotonic velocity transient in skinned muscle fibres from rabbit psoas. *J. Muscle Res. Cell Motil.* 27:401.

Carlson, F. D. 1975. Structural fluctuations in the steady-state of muscular contraction. *Biophys. J.* 15:633–49.

Carlson, F. D., and A. Sieger. 1960. The mechano-chemistry of muscular contraction. *J. Gen. Physiol.* 44:33–60.

Castro, L. R. V., J. Schittl, and R. Fishmeister. 2010. Feedback control through cAMP-dependent protein kinase contributes to differential regulation and compartmentation of cGMP in rat cardiac myocytes. *Circ. Res.* 107:1232–40.

Cazorla, O., G. Vassort, D. Garnier et al. 1999. Length modulation of active force in rat cardiac myocytes: Is titin the sensor? *J. Mol. Cell. Cardiol.* 31:1215–27.

Cecchi, G., M. A. Bagni, P. Griffiths et al. 1990. Detection of radial crossbridge force by lattice spacing changes in intact single muscle fibers. *Science* 250:1409–11.

Chai, B., and G. H. Pollack. 2010. Solute-free interfacial zones in polar liquids. *J. Phys. Chem. B.* 114:5371–5.

Chalovich, J. M. 1992. Actin mediated regulation of muscle contraction. *Pharmacol. Ther.* 55:95–148.

Chang, D. C., C. F. Hazlewood, B. L. Nichols et al. 1972. Spin echo studies on cellular water. *Nature* 235:170–1.

Chang, D. C., H. E. Rorschach, N. L. Nichols et al. 1973. Implications of diffusion coefficient measurements for the structure of cellular water. *Ann. N.Y. Acad. Sci. U.S.A.* 204:434–43.

Chase, P. B., T. M. Denkinger, and M. J. Kushmerick. 1998. Effects of viscosity on mechanics of single skinned fibres from rabbit psoas muscle. *Biophys. J.* 74:1428–38.

Chaudoir, B. M., P. A. Kowalczyk, and R. L. Chisholm. 1999. Regulatory light chain mutations affect myosin motor function and kinetics. *J. Cell Sci.* 122:1611–20.

Chichibu, S. 1961. Electrical properties of glycerinated crayfish muscle fiber. *Tohoku J. Exp. Med.* 73:170–9.

Civan, M. M., and M. Shporer. 1975. Pulsed nuclear magnetic resonance study of ^{17}O, ^2D, ^1H of water in frog striated muscle. *Biophys. J.* 15:299–306.

Claire, K., S. Highsmith, and R. Pecora. 1997. Skeletal muscle myosin subfragment 1 dimer. *Biophys. Chem.* 65:85–90.

Clark, M. E., E. E. Burnell, N. R. Chapman et al. 1982. Studies of water in barnacle muscle. IV. Factors contributing to reduced self-diffusion. *Biophys. J.* 39:289–99.

Clegg, J. S. 1984a. Properties and metabolism of the aqueous cytoplasm and its boundaries. *Am. J. Physiol.* 246:R133–51.

Clegg, J. S. 1984b. Intracellular water and the cytomatrix: Some methods of study and current views. *J. Cell Biol.* 99:167s–71s.

Clegg, J. S. 1986. Artemia cysts as a model for the study of water in biological systems. *Meth. Enzymol.* 127:230–9.

Cleveland, G. G., D. C. Chang, C. F. Hazlewood et al. 1976. Nuclear magnetic resonance measurement of skeletal muscle: Anisotropy of the diffusion coefficient of the intracellular water. *Biophys. J.* 16:1043–53.

Cleworth, D. R., and K. A. P. Edman. 1972. Changes in sarcomere length during isometric tension development in frog skeletal muscle. *J. Physiol. Lond.* 227:1–17.

Collins, E. W., and C. Edwards. 1971. Role of Donnan equilibrium in the resting potentials in glycerine-extracted muscle. *Am. J. Physiol.* 221:1130–3.

Colomo, F., N. Piroddi, C. Poggesi et al. 1997. Active and passive forces of isolated myofibrils from cardiac and fast skeletal muscle of the frog. *J. Physiol. Lond.* 500:535–48.

Conibear, P. B., and M. A. Geeves. 1998. Cooperativity between the two heads of rabbit skeletal heavy meromyosin in binding to actin. *Biophys. J.* 75:926–37.

Cook, C. S., H. Higuchi, and Y. E. Goldman. 1996. ATPase rate in contracting fibers from rabbit psoas muscle. *Biophys. J.* 70:A191.

Cooke, R. 1986. The mechanism of muscle contraction. *CRC Crit. Rev. Biochem.* 21:53–118.

Cooke, R. 1990. Force generation in muscle. *Curr. Opin. Cell Biol.* 2:62–6.

Cooke, R. 1995. The actomyosin engine. *Fed. Proc. FASEB J.* 19:636–42.

Cooke, R. 1997. The actomyosin interaction in striated muscle. *Physiol. Rev.* 77:671–97.

Cooke, R. 2004. The sliding filament model: 1972–2004. *J. Gen. Physiol.* 123:643–56.

Cooke, R. 2007. Modulation of the actomyosin interaction during fatigue of skeletal muscle. *Muscle Nerve* 36:756–77.

Cooke, R., and R. Wien. 1971. The state of water in muscle tissue as determined by proton nuclear magnetic resonance. *Biophys. J.* 11:1002–17.

Cooke, R., and I. D. Kuntz. 1974. The properties of water in biological systems. *Annu. Rev. Biophys. Bioeng.* 3:95–126.

Cooke, R., and K. E. Franks. 1978. Generation of force by single-headed myosin. *J. Mol. Biol.* 120:361–73.

Cooke, R., and W. Bialek. 1979. Contraction of glycerinated muscle fibers as a function of the ATP concentration. *Biophys. J.* 28:241–58.

Cooke, R., and K. E. Franks. 1980. All myosin heads form bonds with actin in rigor rabbit skeletal muscle. *Biochemistry* 19:2265–9.

Cooke, R., and E. Pate. 1985. The effects of ADP and phosphate on the contraction of muscle fibers. *Biophys. J.* 48:789–98.

Cooke, R., M. S. Crowder, and D. D. Thomas. 1982. Orientation of spin labels to cross-bridges in contracting muscle fibres. *Nature* 300:776–8.

Cooke, R., H. D. White, and E. Pate. 1994. A model of the release of myosin heads from actin in rapidly contracting muscle fibers. *Biophys. J.* 66:778–88.

Coomber, S. J., E. M. Bartels, and G. F. Elliott. 2011. Calcium-dependence of Donnan potentials in glycerinated rabbit psoas muscle in rigor, at and beyond filament overlap; a role for titin in the contractile process. *Cell Calcium* 50:91–7.

Cope, F. W. 1967. NMR evidence for complexing Na^+ in muscle, kidney, and brain, and by actomyosin. The relation of cellular complexing of Na^+ to water structure and to transport kinetics. *J. Gen. Physiol.* 50:1353–74.

Cope, F. W. 1969. Nuclear magnetic resonance evidence using D_2O for structured water in muscle and brain. *Biophys. J.* 9:303–19.

Corrie, J. E. T., B. D. Brandmeier, R. E. Fergusson et al. 1999. Dynamic measurement of myosin light-chain-domain tilt and twist in muscle contraction. *Nature* 400:425–30.

Coureux, P. D., A. L. Wells, J. Ménétrey et al. 2003. A structural state of the myosin V motor without bound nucleotide. *Nature* 408:764–6.

Craig, R. 1977. Structure of A-segments from frog and rabbit skeletal muscle. *J. Mol. Biol.* 109:69–81.

Craig, R. 1985. First sight of crossbridge crystals. *Nature* 316:16–7.

Craig, R., and G. Offer. 1976. Axial arrangement of crossbridges in thick filaments of vertebrate skeletal muscle. *J. Mol. Biol.* 102:325–32.

Craig, R., and J. L. Woodhead. 2006. Structure and function of myosin filaments. *Curr. Opin. Struct. Biol.* 16:204–12.

Craig, R., J. Trinick, and P. Knight. 1986. Discrepancies in length of myosin heads. *Nature* 320:688.

Crowther, R. A., R. Padrón, and R. Craig. 1985. Arrangement of the heads of myosin in relaxed thick filaments from tarantula muscle. *J. Mol. Biol.* 184:429–39.

Curmi, P. G. M., D. B. Stone, D. K. Schneider et al. 1988. Comparison of the structure of myosin subfragment 1 bound to actin and free in solution. A neutron scattering study using actin made 'invisible' by deuteration. *J. Mol. Biol.* 203:781–96.

Curtin, N. A., and R. C. Woledge. 1978. Energy changes and muscular contraction. *Physiol. Rev.* 48:690–761.

Curtin, N. A., and R. C. Woledge. 1979. Chemical changes during contraction of frog muscle: How are their time courses related? *J. Physiol. Lond.* 288:353–66.

Curtin, N. A., and R. C. Woledge. 1988. Power output and force–velocity relationship of live fibres white myotomal muscle of the dogfish *Scyliohinus canicula*. *J. Exp. Biol.* 140:188–97.

Curtin, N. A., C. Gilbert, K. M. Kretzschmar et al. 1974. The effect of the performance of work on total energy output and metabolism during muscular contraction. *J. Physiol. Lond.* 238:455–72.

Curtin, N. A., J. V. Howarth, J. A. Rall et al. 1986. Absolute values of myothermic measurements on single muscle fibres from frog. *J. Muscle Res. Cell Motil.* 7:327–32.

Cusanovich, M. A., and D. P. Flamig. 1980. Alterations in the mechanism of hydrolysis of ATP by cardiac myosin induced by thyrotoxicosis. *Fed. Proc. FASEB J.* 39:1934.

Cyranoski, D. 2000. Swinging against the tide. *Nature* 408:764–6.

D'hahan, N., K. Taouil, A. Dassouli et al. 1997. Long-therapy with trimetazidine in cardiomyopathic Syrian hamsters BIO 14:6. *Eur. J. Pharm.* 123:611–6.

D'hahan, N., K. Taouil, C. Janmot et al. 1998. Effect of trimetazindine and verapamil on the cardiomyopathic hamster myosin phenotype. *Brit. J. Pharm.* 123:611–6.

D'hahan, N. 1999. Trizetazidine potential mechanism of action in hypertrophic cardiomyopathy. *J. Cardiovasc. Pharmacol.* 33:500–6.

D'hahan. N., C. Moreau, A. L. Prost et al. 1999. Pharmacological plasticity of cardiac ATP-sensitive potassium channels toward diazoxide revealed by ADP. *Proc. Natl. Acad. Sci. U.S.A.* 96:12162–7.

Dahlstedt, A. J., A. Katz, R. Wieringa et al. 2000. Is creatine kinase responsible for fatigue? Studies of isolated skeletal muscle deficient in creatine kinase. *Fed. Proc. FASEB J.* 14:982–90.

Danker, P., and W. Hasselbach. 1971. Dependence of actomyosin ATPase activity on ionic strength and its modification by thiol group substitution. *FEBS Lett.* 16:273–4.

Dantzig, J. A., and Y. E. Goldman. 1985. Suppression of muscle contraction by vanadate. Mechanical and ligand binding studies in glycerol-extracted rabbit fibers. *J. Gen. Physiol.* 86:305–27.

Dantzig, J. A., R. J. Barsotti, S. Manz et al. 1999. The ADP release step of smooth muscle cross-bridge cycle is not directly associated with force generation. *Biophys. J.* 77:386–97.

Davies, R. E. 1963. A molecular theory of muscle contraction: Calcium-dependent contraction with hydrogen bond formation plus ATP-dependent extension of part of the myosin-actin cross-bridges. *Nature* 199:1068–74.

Davies, C. T. M., I. K. Mecrow, and J. M. White. 1982. Contractile properties of the human triceps surae with some observations of the effects of temperature and exercise. *Pflügers Arch. Eur. J. Physiol.* 49:255–69.

Davies, G. J., S. J. Gamblin, J. A. Littlechild et al. 1994. Structure of the ADP complex of the 3-phosphoglycerate kinase from *Bacillus stearothermophilus* at 1.65 Å. *Acta Crystallogr. Sect D. Biol. Crystallogr.* 50:202–9.

Davis, J. S. 1988. Myosin heads in muscle thick filament assembly. *Nature* 333:807.

Davis, J. S., and W. F. Harrington. 1993. A single order-disorder transition generates tension during the Huxley-Simmons phase 2 in muscle. *Biophys. J.* 65:1886–98.

Davis, J. S., and N. D. Epstein. 2003. Kinetic effects of fiber type on the two subcomponents of the Huxley-Simmons phase 2 in muscle. *Biophys. J.* 85:390–401.

Dawson, M. J., D. G. Gadian, and D. R. Wilkie. 1978. Muscular fatigue investigated by phosphorus nuclear magnetic resonance. *Nature* 274:861–6.

Dawson, M. J., D. G. Gadian, and D. R. Wilkie. 1980. Mechanical relaxation rate and metabolism studied in fatiguing muscle by phosphorus nuclear magnetic resonance. *J. Physiol. Lond.* 299:465–84.

de Winkel, M. E., T. Blangé, and B. W. Teijtel. 1995. Viscoelastic properties of cross-bridges in cardiac muscle. *Am. J. Physiol.* 268:H987–98.

Debold, E. P. 2012. Recent insights into muscle fatigue at the cross-bridge level. *Front. Physiol.* 3:151.

Decostre, V., P. Bianco, V. Lombardi et al. 2005. Effect of temperature on the working stroke of muscle myosin. *Proc. Natl. Acad. Sci. U.S.A.* 102:13927–32.

Delay, M. J., N. Ishide, R. C. Jacobson et al. 1981. Stepwise sarcomere shortening analysis by high-speed cinematography. *Science* 213:1523–5.

Deshayes, C., G. F. Elliott, K. Jennison et al. 1993. Electric charge measurements on F- and G-actin from rabbit muscle at physiological ionic strength. *J. Physiol. Lond.* 459:275P.

Dickinson, M., G. Farman, M. Frye et al. 2005. Molecular dynamics of cyclically contracting insect flight muscle *in vivo*. *Nature* 433:330–4.

Diegel, J. G., and M. M. Pintar. 1975. Origin of the non-exponentiality of the water proton spin relaxation in tissues. *Biophys. J.* 15:855–60.

Dijkstra, S., J. J. van der Gon, T. Blangé et al. 1973. A simplified sliding-filament muscle model for simulation purposes. *Kybernetik* 12:94–101.

Dominguez, R., Y. Freyzon, K. M. Trybus et al. 1998. Crystal structure of a vertebrate smooth muscle myosin motor domain and its complex with the essential light chain: Visualization of the pre-power stroke state. Cell 94:559–71.

Dragomir, C. T., A. Barbier, and D. Ungureanu. 1976. A new hybrid theory of muscle contraction. J. Theor. Biol. 61:221–44.

Drost-Hansen, W. 1969. Structure of water near solid interfaces. Ind. Eng. Chem. 61:10–47.

Drost-Hansen, W. 1971a. Role of water in cell-wall interactions. Fed. Proc. FASEB J. 30:1539–50.

Drost-Hansen, W. 1971b. Structure and properties of water near biological interfaces. In Chemistry of cell interface, ed. H. D. Brown, 2–184. Academic Press, New York, USA.

Drost-Hansen, W. 2001. Temperature effects on cell-functioning. A critical role for vicinal water. Cell Mol. Biol. 47:865–84.

Drury, A. W., and A. Szent-Györgyi. 1929. The physiological activity of adenine compounds with especial reference to their action upon the mammalian heart. J. Physiol. Lond. 68:213–37.

Dubuisson, M. 1954. Muscular contraction. Thomas. Springfield, USA.

Duke, T. A. J. 1999. Molecular model of muscle contraction. Proc. Natl. Acad. Sci. U.S.A. 96:2770–5.

Duke, T. A. J. 2000. Cooperativity of myosin molecules through strain-dependent chemistry. Phil. Trans. R. Soc. Lond. Ser. B Biol. Sci. 355:529–38.

Dunn, A. R., and J. A. Spudich. 2007. Dynamics of the unbound head during myosin V processive translocation. Nat. Struct. Mol. Biol. 14:246–8.

Eaton, B. L., and F. A. Pepe. 1974. Myosin filaments showing a 430 Å axial repeat. J. Mol. Biol. 82: 421–3.

Edelman, E. H., and R. Padrón. 1984. X-ray evidence that actin is 100 Å. Nature 307:56–8.

Edman, K. A. P. 1966. The relation between sarcomere length and active tension in isolated semitendinosus fibres of the frog. J. Physiol. Lond. 183:407–17.

Edman, K. A. P. 1975. Mechanical deactivation induced by active shortening in isolated muscle fibres of the frog. J. Physiol. Lond. 246:255–75.

Edman, K. A. P. 1979. The velocity of unloaded shortening and its relation to sarcomere length and isometric forces in vertebrate muscle fibres. J. Physiol. Lond. 291:143–59.

Edman, K. A. P. 1980. Depression of mechanical performance by active shortening during twitch and tetanus of vertebrate muscle fibres. Acta Physiol. Scand. 109:15–26.

Edman, K. A. P. 1988a. Double-hyperbolic nature of the force–velocity relation in frog skeletal muscle. Adv. Exp. Med. Biol. 226:643–52.

Edman, K. A. P. 1988b. Double-hyperbolic force–velocity relation in frog muscle fibres. J. Physiol. Lond. 404:301–21.

Edman, K. A. P., and K. E. Anderson. 1968. The variation in active tension with sarcomere length in vertebrate skeletal muscle and its relation to fibre width. Experientia 24:134–6.

Edman, K. A. P., and J. C. Hwang. 1977. The force velocity relationship in vertebrate muscle fibres at varied tonicity of the extracellular medium. J. Physiol. Lond. 269:255–72.

Edman, K. A. P., and A. R. Mattiazzi. 1981. Effect of fatigue and altered pH on isometric force and velocity of shortening at zero load in frog muscle fibres. J. Muscle Res. Cell Motil. 2:321–34.

Edman, K. A. P., and C. Reggiani. 1984. Redistribution of sarcomere length during isometric contraction of frog muscle fibres and its relation to tension creep. J. Physiol. Lond. 351:169–98.

Edman, K. A. P., and C. Reggiani. 1987. The sarcomere length–tension relation determined in short segments of intact fibres of frog. J. Physiol. Lond. 385:709–32.

Edman, K. A. P., and F. Lou. 1990. Changes in force and stiffness induced by fatigue and intracellular acidification in frog muscle fibres. J. Physiol. Lond. 424:133–49.

Edman, K. A. P., and T. Tsuchiya. 1996. Force enhancement following stretch in single sarcomere. J. Physiol. Lond. 490:191–205.

Edman, K. A. P., L. A. Mulieri, and B. Scubon-Mulieri. 1976. Non-hyperbolic force–velocity relationship in single muscle fibres. Acta Physiol. Scand. 98:255–72.

Edman, K. A. P., and N. A. Curtin. 2001. Synchronous oscillations of length and stiffness during loaded shortening if muscle fibres. J. Physiol. Lond. 534:553–63.

Edman, K. A. P., G. Elzinga, and M. I. M. Noble. 1978. Enhancement of mechanical performance by stretch during tetanic contractions of vertebrate skeletal muscle fibres. J. Physiol. Lond. 281:139–55.

Edman, K. A. P., G. Elzinga, and M. I. M. Noble. 1979. The effects of stretch on contracting skeletal muscle fibres. In Cross-bridge mechanism of muscle contraction, eds. H. Sugi, and G. H. Pollack, 297–310. Univ. Park Press, Baltimore, USA.

Edman, K. A. P., G. Elzinga, and M. I. M. Noble. 1982. Residual force enhancement after stretch of contracting frog single muscle fibers. *J. Gen. Physiol.* 80:769–84.

Edman, K. A. P., C. Caputo, and F. Lou. 1993. Depression of tetanic force induced by loaded shortening of frog muscle fibres. *J. Physiol. Lond.* 466:535–52.

Edman, K. A. P., A. Månsson, and C. Caputo. 1997. The biphasic force–velocity relationship in frog muscle fibres and its evaluation in terms of crossbridge function. *J. Physiol. Lond.* 503:141–56

Edwards, R. J., M. A. Goldstein, J. P. Schroeter et al. 1989. The Z-band lattice in skeletal muscle in rigor. *J. Ultrastruct. Mol. Struct. Res.* 102:59–65.

Eisenberg, E., and C. Moos. 1970. Actin activation of heavy meromyosin adenosine triphosphatase. *J. Biol. Chem.* 245:2451–6.

Eisenberg, E., and T. L. Hill. 1978. A crossbridge model of muscle contraction. *Prog. Biophys. Mol. Biol.* 33: 55–82.

Eisenberg, E., and L. E. Greene. 1980. The relation between muscle physiology and muscle biochemistry. *Annu. Rev. Physiol.* 42:293–309.

Eisenberg, E., and T. L. Hill. 1985. Muscle contraction and free energy transduction in biological systems. *Science* 227:999–1006.

Eisenberg, E., T. L. Hill, and Y. D. Chen. 1980. Crossbridge model of muscle contraction. Quantitative analysis. *Biophys. J.* 29:195–227.

Elliott, G. F. 1968. Force-balances and stability in hexagonally-packed polyelectrolyte systems. *J. Theor. Biol.* 21:71–87.

Elliott, G. F. 1973. Donnan and osmotic effects in muscle fibers without membranes. *J. Mechanochem. Cell Motil.* 2:83–9.

Elliott, G. F. 1974. Comments presented in *The physiology of cell and organisms*, ed. R. M. Holmes (S 321), 45–6. Unit 6, The Open Univ. Press, Oxford, UK.

Elliott, G. F. 1980. Measurements of the electric charge and ion-binding of the protein filaments in intact muscle and cornea, with implication for filament assembly. *Biophys. J.* 32:95–7.

Elliott, G. F. 2007. X-rays, twitching muscles, and burning anodes. *Physiol. News* 67:6–10.

Elliott, A., and G. Offer. 1978. Shape and flexibility of the myosin molecule. *J. Mol. Biol.* 123:505–19.

Elliott, G. F., and E. M. Bartels. 1982. Donnan potential measurements in extended hexagonal polyelectrolyte gels such as muscle. *Biophys. J.* 38:195–9.

Elliott, G. F., and C. R. Worthington. 1994. How muscle may contract. *Biochim. Biophys. Acta* 1200:109–16.

Elliott, G. F., and C. R. Worthington. 1995. Electrical forces in muscle contraction. *Biophys. J.* 68:327s.

Elliott, G. F., and C. R. Worthington. 1997. The muscle motor: 'Simultaneous' lever or sequential impulses? *Int. J. Biol. Macromol.* 21:271–5.

Elliott, G. F., and S. A. Hodson. 1998. Cornea, and the swelling of polyelectrolyte gels of biological interest. *Rep. Prog. Phys.* 61:1325–65.

Elliott, G. F., and C. R. Worthington. 2001. Muscle contraction: Viscous-like frictional forces and the impulsive model. *Int. J. Biol. Macromol.* 29:213–8.

Elliott, G. F., and C. R. Worthington. 2006. Crossbridges act sequentially in muscle contraction. *J. Muscle Res. Cell Motil.* 27:491.

Elliott, G. F., J. Lowy, and C. R. Worthington. 1963. An X-ray and light diffraction study of the filament lattice of striated muscle in the living state and in rigor. *J. Mol. Biol.* 6:295–305.

Elliott, G. F., J. Lowy, and B. M. Millman. 1965. X-ray diffraction from living striated muscle during contraction. *Nature* 206:1357–8.

Elliott, G. F., J. Lowy, and B. M. Millman. 1967. Low-angle X-ray diffraction studies of living striated muscle during contraction. *J. Mol. Biol.* 25:31–45.

Elliott, G. F., E. Rome, and M. Spencer. 1970. A type of contraction hypothesis applicable to all muscles. *Nature* 226:417–20.

Elliott, G. F., J. M. Goodfellow, and A. E. Woolgar. 1980. Swelling studies of bovine corneal stroma without bounding membrane. *J. Physiol. Lond.* 298:453–79.

Elliott, G. F., E. M. Bartels, P. H. Cooke et al. 1984. Evidence for a simple Donnan equilibrium under physiological conditions. *Biophys. J.* 45:487–88.

Elliott, G. F., E. M. Bartels, and R. A. Hughes. 1986. The myosin filament: Charge amplification and charge condensation. In *Electrical double layers in biology*, ed. M. Blank, 277–85. Plenum Press, New York, USA.

Elzinga, G., G. J. M. Stienen, and M. G. A. Wilson. 1989a. Isometric force production before and after chemical skinning in isolated muscle fibres of the frog *Rana temporaria*. *J. Physiol. Lond.* 410: 171–85.

Elzinga, G., J. V. Howarth, J. A. Rall et al. 1989b. Variation in the normalized tetanic force of single frog muscle fibres. *J. Physiol. Lond.* 410:157–70.

Endo, M. 1973. Length dependence of activation of skinned muscle fibers by calcium. *Cold Spring Harb. Symp. Quant. Biol.* 37:505–10.

Endo, M. 1972. Stretch-induced increase in activation of skinned muscle fibres by calcium. *Nature* 237:211–3.

Endo, M. 1975. Conditions required for calcium-induced release of calcium from the sarcoplasmic reticulum. *Proc. Jpn. Acad. B* 51:467–72.

Endo, M. 1977. Calcium release from the sarcoplasmic reticulum. *Physiol. Rev.* 57:71–108.

Endo, M., M. Tanaka, and Y. Ogawa. 1970. Calcium induced release of calcium from the sarcoplasmic reticulum of skinned muscle fibres. *Nature* 228:34–6.

Endo, M., T. Kitazawa, and Y. Kakuta. 1979. Effect of 'viscosity' of the medium on mechanical properties of skinned skeletal fibers. In *Crossbridge mechanisms in muscle contraction*, eds. H. Sugi, and G. H. Pollack, 365–74. Univ. Tokyo Press, Japan.

Engelhardt, W. A., and M. N. Ljubimova. 1939. The adenosine triphosphatase splitting of myosin. *Nature* 144:668–70.

Erdös, T., and O. Snellman. 1948. Electrophoretic investigations of crystal myosin. *Biochim. Biophys. Acta* 2:642–9.

Esaki, S., Y. Ishii, and T. Yanagida. 2003. Model describing the biased Brownian movement of myosin. *Proc. Jpn. Acad. Ser. B* 79:9–14.

Esaki, S., Y. Ishii, M. Nishikawa et al. 2007. Cooperative actions between myosin heads bring effective functions. *Biosystems* 88:293–300.

Fabiato, A., and F. Fabiato. 1976. Dependence of calcium release, tension generation and restoring forces on sarcomere length in skinned cardiac cells. *Eur. J. Cardiol (Suppl.)* 4:13–27.

Fabiato, A., and F. Fabiato. 1978. Myofilament-generated tension oscillations during partial calcium activation and activation dependence of the sarcomere length–tension in skinned cardiac cells. *J. Gen. Biol.* 72:667–99.

Fenn, W. O. 1923. A quantitative comparison between the energy liberated and the work performed by the isolated *sartorius* muscle of the frog. *J. Physiol. Lond.* 58:175–203.

Fenn, W. O. 1924. The relation between the work performed and the energy liberated by isolated *sartorius* muscle of the frog. *J. Physiol. Lond.* 58:373–95.

Fenn, W. O., and B. S. Marsh. 1935. Muscular force at different speeds of shortening. *J. Physiol. Lond.* 85:277–97.

Ferenczi, M. A. 1986. Phosphate burst in permeable muscle fibers of the rabbit. *Biophys. J.* 50:471–7.

Ferenczi, M. A., Y. E. Goldman, and R. M. Simmons. 1984a. The dependence of force and shortening velocity on substrate concentration in skinned muscle fibres from *Rana temporaria*. *J. Physiol. Lond.* 350:519–43.

Ferenczi, M. A., E. Homsher, and D. R. Trentham. 1984b. The kinetics of magnesium adenosine triphosphate cleavage in skinned muscle fibres of the rabbit. *J. Physiol. Lond.* 352:575–99.

Ferenczi, M. A., S. Y. Bershistky, N. A. Koubassova et al. 2005. The 'roll and lock' mechanism of force generation in muscle. *Structure* 13:131–41.

Finch, E. D., and L. D. Homer. 1974. Proton nuclear magnetic resonance relaxation measurements in frog muscle. *Biophys. J.* 14:907–21.

Finch, E. D., J. F. Harmon, and B. H. Muller. 1971. Pulsed NMR measurements of the diffusion constant of water in muscle. *Arch. Biochem. Biophys.* 147:299–310.

Finer, J. T., R. M. Simmons, and J. A. Spudich. 1994. Single molecule mechanics: Piconewton forces and nanometer steps. *Nature* 368:113–9.

Fink, F. H. A., D. G. Stephenson, and D. A. Williams. 1986. Potassium and ionic strength effects on the isometric force of skinned twitch muscle fibre of the rat and the toad. *J. Physiol. Lond.* 370:317–37.

Fisher, A. J., C. A. Smith, J. B. Thorson et al. 1995a. Structural studies of myosin-nucleotide complexes. A revised model for the molecular basis of muscle contraction. *Biophys. J.* 68:19s–28s.

Fisher, A. J., C. A. Smith, J. B. Thorson et al. 1995b. X-ray structures of the myosin motor domain of *Dictyotelium discoidium* complexed with $MgADP*BeF_x$ and $MgADP*AlF_4^-$. *Biochemistry* 34:8960–72.

Fitts, R. H. 1994. Cellular mechanism of muscle fatigue. *Physiol. Rev.* 74:49–94.

Fitts, R. H. 2011. Cellular, molecular and metabolic basis of muscle fatigue. Handbook of Physiology, Exercise: Regulation and integration of multiple systems. *Compr. Physiol. Suppl.* 29:1151–83.

Flachner, B., Z. Kováry, A. Varga et al. 2004. Role of phosphate chain mobility of MgATP in completing the 3-phosphoglycerate kinase catalytic site: Binding, kinetic, and crystallographic studies with ATP and MgATP. *Biochemistry* 43:3436–49.

Flamig, D. P., and M. A. Cusanovich. 1981. Aggregation-linked kinetics heterogeneity in bovine cardiac myosin subfragment 1. *Biochemistry* 20:6780–7.

Flicker, P. F., T. Walliman, and P. Vibert. 1983. Electron microscopy of scallop myosin. Location of regulatory light chain. *J. Mol. Biol.* 169:723–41.

Ford, L. E., and R. J. Podolsky. 1970. Regeneration calcium release within muscle cells. *Science* 167:58–9.

Ford, L. E., and R. J. Podolsky. 1972. Intracellular calcium movements in skinned muscle fibres. *J. Physiol. Lond.* 223:21–33.

Ford, L. E., and M. F. Surdyk. 1978. Electron microscopy on skinned muscle cells. *J. Gen. Physiol.* 72:5a.

Ford, L. E., A. F. Huxley, and R. M. Simmons. 1977. Tension responses to sudden length change in stimulated frog muscle fibres near slack length. *J. Physiol. Lond.* 269:441–515.

Ford, L. E., A. F. Huxley, and R. M. Simmons. 1981. The relation between stiffness and filament overlap in stimulated frog muscle fibres. *J. Physiol. Lond.* 311:219–49.

Ford, L. E., K. Nakagawa, J. Desper et al. 1991. Effect of osmotic compression on the force–velocity properties of glycerinated rabbit skeletal muscle cells. *J. Gen. Physiol.* 97:73–88.

Franzini-Armstrong, C., and K. P. Porter. 1964. The Z discs of skeletal muscle fibres. *Z. Zellforsch.* 61:661–72.

Friel, D. D., and B. P. Bean. 1988. Two ATP-activated conductances in bullfrog atrial cells. *J. Gen. Physiol.* 81:1–27.

Fuchs, F. 1975. Thermal inactivation of calcium regulatory mechanism of human skeletal muscle actomyosin. *Anesthesiology* 42:585–9.

Fuchs, F., and D. A. Martyn. 2005. Length-dependent Ca^{2+} activation in cardiac muscle: Some remaining questions. *J. Muscle Res. Cell Motil.* 26:199–212.

Fuchs, F., and Y. P. Wang. 1996. Sarcomere length versus interfilament spacing as determinants of cardiac myofilament. *J. Mol. Cell. Cardiol.* 28:1375–83.

Fukuda, N., and H. L. M. Granzier. 2004. Role of the giant elastic protein titin in the Frank–Starling mechanism of the heart. *Curr. Vasc. Pharmacol.* 2:135–9.

Funatsu, T., Y. Harada, M. Tokunaga et al. 1995. Imaging of single fluorescent molecules and individual ATP turnovers by single myosin molecules in aqueous solutions. *Nature* 374:555–9.

Fung, B. M. 1975. Orientation of water in striated frog muscle. *Science* 190:800–2.

Fung, B. M., and T. W. McGaughy. 1974. The state of water in muscle as studied by pulsed NMR. *Biochim. Biophys. Acta* 343:663–73.

Fung, B. M., D. L. Durham, and D. A. Wassil. 1975. The state of water in biological systems as studied by proton and deuterium relaxation. *Biochim. Biophys. Acta* 399:191–202.

Galinska-Rakoczy, A., P. Engel, C. Xu et al. 2008. Structural basis for the regulation of muscle contraction by troponin and tropomyosin. *J. Mol. Biol.* 379:929–35.

Garrigos, M., J. E. Morel, and J. Garcia de la Torre. 1983. Reinvestigation of the shape and state of hydration of the skeletal myosin subfragment 1 monomer in solution. *Biochemistry* 22:4961–9.

Garrigos, M., S. Mallam, P. Vachette et al. 1992. Structure of the myosin heads in solution and the effect of light chain 2 removal. *Biophys. J.* 63:1462–70.

Gasser, H. S., and A. V. Hill. 1924. The dynamics of muscular contraction. *Proc. R. Soc. Lond. Ser. B* 96:398–437.

Geerlof, A., P. S. Schmidt, F. Travers et al. 1997. Cryoenzymic studies on yeast 3-phosphoglycerate kinase. Attempt to obtain the kinetics of the hinge-bending motion. *Biochemistry* 36:5538–45.

Geerlof, A., F. Travers, T. Barman et al. 2005. Perturbation of yeast 3-phosphoglycerate kinase reaction mixtures with ADP: Transient kinetics of formation of ATP from bound 1,3-bisphosphoglycerate. *Biochemistry* 44:14948–55.

Geeves, M. A. 1991. The dynamics of actin and myosin association and the crossbridge model of muscle contraction. *Biochem. J.* 274:1–14.

Geeves, M. A., and T. R. Trentham. 1982. Protein-bound adenosine 5′-triphosphate properties of a key intermediate of the magnesium-dependent subfragment 1 adenosine phosphatase from rabbit skeletal muscle. *Biochemistry* 21:2782–9.

Geeves, M. A., and K. C. Holmes. 1999. Structural mechanism of muscle contraction. *Annu. Rev. Biochem.* 68:687–728.

Geeves, M. A., and K. C. Holmes. 2005. The molecular mechanism of muscle contraction. *Adv. Protein Chem.* 71:161–93.

Geeves, M. A., R. Fedorov, and D. J. Manstein. 2005. Molecular mechanism of actomyosin-based motility. *Cell. Mol. Life Sci.* 62:1462–77.

Gergely, J. 1964. *Biochemistry of muscle contraction.* J. & A. Churchill, London, UK.

Gergely, J., and J. C. Seidel. 1983. *Conformational changes and dynamics of myosin.* Wiley-Blackman, Malden, USA.

Gilbert, C., K. M. Kretzschmar, D. R. Wilkie et al. 1971. Chemical change and energy output during muscular contraction. *J. Physiol. Lond.* 218:163–93.

Gilbert, S. P., M. R. Webb, M. Brune et al. 1995. Pathway of processive ATP hydrolysis by kinesin. *Nature* 373:671–6.

Godt, R. E., and D. W. Maughan. 1977. Swelling of skinned muscle fibers of the frog. Experimental observation. *Biophys. J.* 19:103–16.

Godt, R. E., and D. W. Maughan. 1981. Influence of osmotic compression on calcium activation and tension in skinned muscle fibers from the rabbit. *Pflügers Arch. Eur. J. Physiol.* 391:334–7.

Godt, R. E., and D. W. Maughan. 1988. On the composition of the cytosol of relaxed skeletal muscle of the frog. *Am. J. Physiol.* 254:C591–604.

Godt, R. E., and T. M. Nosek. 1989. Changes of intracellular milieu with fatigue or hypoxia depress contraction of skinned rabbit and cardiac muscle. *J. Physiol. Lond.* 412:155–80.

Goldman, Y. E., and R. M. Simmons. 1977. Active and rigor muscle stiffness. *J. Physiol. Lond.* 269:55P–7P.

Goldman, Y. E., and R. M. Simmons. 1984a. Control of sarcomere length in skinned muscle fibres of *Rana temporaria* during mechanical transients. *J. Physiol. Lond.* 350:497–518.

Goldman, Y. E., and R. M. Simmons. 1984b. Control of sarcomere length in skinned muscle fibres at altered filament spacing. *J. Physiol. Lond.* 378:175–94.

Goldman, Y. E., and R. M. Simmons. 1986. The stiffness of frog skinned muscle fibres at altered lateral filament spacing. *J. Physiol. Lond.* 378:175–94.

Goldman, Y. E., and A. F. Huxley. 1994. Actin compliance: Are you pulling my chain? *Biophys. J.* 67:2131–6.

Goldman, Y. E., J. A. McCray, and K. W. Ranatunga. 1987. Transient tension changes initiated by laser temperature jumps in rabbit psoas muscle fibres. *J. Physiol. Lond.* 392:71–95.

Goldstein, M. A., L. H. Michael, J. P. Schroeter et al. 1986. The Z band lattice in skeletal muscle before, during and after tetanic contraction. *J. Muscle Res. Cell Motil.* 7:527–36.

Goldstein, M. A., L. H. Michael, and J. P. Schroeter. 1987. Z band dynamics as a function of sarcomere length and the contractile state of muscle. *Fed. Proc. FASEB J.* 1:133–42.

Goldstein, M. A., L. H. Michael, J. P. Schroeter et al. 1988. Structural states in the Z band of skeletal muscle correlates with states of active and passive tension. *J. Gen. Physiol.* 92:113–9.

Gollub, J. C., R. Cremo, and R. Cooke. 1996. ADP release produces a rotation of the neck region of smooth muscle myosin and HMM. *Nat. Struct. Biol.* 3:796–801.

Gómez, A., S. Guatimosim, K. Dilly et al. 2001. Heart failure after myocardial infarction. Altered excitation–contraction couplage. *Circulation* 104:688–93.

Goodall, M. C. 1956. Auto-oscillations in extracted muscle fibres. *Nature* 177:1238–9.

Goody, R. S. 2003. The missing link in the muscle cross-bridge cycle. *Nat. Struct. Biol.* 10:773–5.

Gordon, A. M., and E. B. Ridgway. 1975. Muscle activation: Effects of small length changes on calcium release in single fibres. *Science* 189:881–4.

Gordon, A. M., and G. H. Pollack. 1980. Effects of calcium on the sarcomere length–tension relation in rat cardiac muscle. Implications for the Frank–Starling mechanism. *Circ. Res.* 47:610–9.

Gordon, A. M., A. F. Huxley, and F. J. Julian. 1966a. Tension development in highly stretched vertebrate muscle fibres. *J. Physiol. Lond.* 184:143–69.

Gordon, A. M., A. F. Huxley, and F. J. Julian. 1966b. The variation in isometric tension with sarcomere length in vertebrate muscle fibres. *J. Physiol. Lond.* 184:170–92.

Gordon, A. M., R. E. Godt, S. K. Donaldson et al. 1973. Tension in skinned frog muscle fibers in solutions of varying ionic strength and neutral salt composition. *J. Gen. Physiol.* 62:550–74.

Gordon, A. M., E. Homsher, and M. Regnier. 2000. Regulation of contraction in striated muscle. *Physiol. Rev.* 80:853–924.

Grabareck, Z. J., P. C. Grabareck, P. C. Laevis et al. 1983. Cooperative binding to the Ca^{2+}-specific sites on troponin C in regulated actin and actomyosin. *J. Biol. Chem.* 258:14098–102.

Granzier, H. L. M., and G. H. Pollack. 1990. The descending limb of the force-sarcomere length relation of the frog revisited. *J. Physiol. Lond.* 421:595–615.

Granzier, H. L. M., J. Meyers, and G. H. Pollack. 1987. Stepwise shortening of muscle fibre segments. *J. Muscle Res. Cell Motil.* 8:242–51.

Granzier, H. L. M., D. H. Burns, and G. H. Pollack. 1989. Sarcomere length dependence of the force–velocity relation in single frog muscle fibers. *Biophys. J.* 55:499–507.

Granzier, H. L. M., A. Mattiazzi, and G. H. Pollack. 1990. Sarcomere dynamics during isotonic velocity transients in single frog muscle fibers. *Am. J. Physiol.* 259:C266–78.

Gray, B. F., and I. Gonda. 1977a. The sliding filament model of muscle contraction. I. The quantum mechanical formalism. *J. Theor. Biol.* 69:167–85.

Gray, B. F., and I. Gonda. 1977b. The sliding filament model of muscle contraction. II. The energetic and dynamical predictions of a quantum mechanical transducer model. *J. Theor. Biol.* 68:187–230.

Grazi, E. 1997. What is the diameter of the actin filament? *FEBS Lett.* 405:249–52.

Grazi, E. 2000. A highly non-ideal solution: The contractile system of skeletal muscle. *Eur. Biophys. J.* 29:535–41.

Grazi, E. 2007. Protein osmotic pressure and viscosity are the inseparable partners of muscle contraction. *Curr. Topics Biochem. Res.* 9(1):39–52.

Grazi, E. 2008. Water and muscle contraction. *Int. J. Mol. Biol.* 9:1435–52.

Grazi, E., and O. Cintio. 2001. Thermodynamic features of myosin filament suspensions: Implication for modelling of muscle contraction. *Biophys. J.* 81:313–20.

Grazi, E., and C. Di Bona. 2006. Viscosity as an inseparable partner of muscle contraction. *J. Theor. Biol.* 242:853–61.

Gregorio, C. C., H. L. M. Granzier, H. Sorimachi et al. 1999. Muscle assembly: A titanic achievement. *Curr. Opin. Cell Biol.* 11:18–25.

Griffiths, P. J., C. C. Ashley, M. A. Bagni et al. 1993. Time-resolved equatorial X-ray diffraction measurements in single intact muscle fibres. In *Mechanisms of myofilament sliding in muscle contraction*, eds. H. Sugi, and G. H. Pollack, 409–22. Plenum Press, New York, USA.

Griffiths, P. J., M. A. Bagni, B. Colombini et al. 2006. Effects of the number of actin-bound S1 and axial force on X-ray patterns in intact skeletal muscle. *Biophys. J.* 91:3370–82.

Grussaute, H., F. Ollagnon, and J. E. Morel. 1995. F-actin–myosin subfragment-1 (S1) interactions: Identification of the refractory state of S1 with the S1 dimer. *Eur. J. Biochem.* 228:524–9.

Grynkiewicz, G., M. Poenie, and R. Y. Tsien. 1985. A new generation of Ca^{2+} indicators with greatly improved fluorescence properties. *J. Biol. Chem.* 260:3440–50.

Gu, J., S. Xu, and L. C. Yu. 2002. A model of cross-bridge attachment to actin in the A*M*ATP state based on X-ray diffraction from permeabilised rabbit psoas muscle. *Biophys. J.* 82:2123–33.

Guinier, A. 1939. La diffraction des rayons X aux très petits angles: Application à l'étude de phénomènes ultramicroscopiques. *Ann. Phys. (Paris)* 11:161–237.

Gulati, J. 1983. Magnesium ion-dependent contraction of skinned frog muscle fibers in calcium-free solution. *Biophys. J.* 441:113–21.

Gulati, J., and A. Babu. 1982. Tonicity effects on intact single muscle fibers: Relation between force and cell volume. *Science* 215:1109–12

Güth, K., and J. D. Potter. 1987. Effect of rigor and cycling cross-bridges on the structure of troponin C and the Ca^{2+} affinity of the Ca^{2+}-specific regulatory sites in skinned rabbit psoas fibres. *J. Biol. Chem.* 262:13627–35.

Güth, K., H. J. Kuhn, T. Tsuchyia et al. 1981. Length dependent state of activation – Length change dependent kinetics of cross-bridges in skinned insect flight muscle. *Biophys. Struct. Mech.* 7: 139–69.

Hansen, J. 1971. Pulsed NMR study of water in muscle and brain tissue. *Biochim. Biophys. Acta* 230:482–6.

Hanson, J., E. J. O'Brien, and P. M. Bennett. 1971. Structure of the myosin-containing filament assembly. A-segment separated from frog skeletal muscle. *J. Mol. Biol.* 58:865–71.

Harada, Y., A. Noguchi, A. Kishino et al. 1987. Sliding movement of single actin filaments on one-headed myosin filaments. *Nature* 326:805–8.

Harada, Y., K. Sakurada, T. Aoki et al. 1990. Mechanochemical coupling in actomyosin energy transduction studied by in vitro movement assay. *J. Mol. Biol.* 77:549–68.

Harafuji, Y., and Y. Ogawa. 1980. Re-examination of the apparent binding constant of ethylene glycol bis (β-aminoethyl)–N,N,N′,N′-tetraacetic acid with calcium around neutral pH. *J. Biochem. Tokyo* 87:1305–12.

Harford, J. J., and J. M. Squire. 1992. Evidence for structurally different attached states of myosin cross-bridges on actin during contraction of fish muscle. *Biophys. J.* 63:387–96.

Harlos, K., M. Vas, and C. C. F. Blake. 1992. Crystal structure of the binary complex of pig muscle phophoglycerate kinase and its substrate 3-phospho-D-glycerate. *Proteins* 12:133–44.

Harrington, W. F. 1971. A mechanochemical mechanism for muscle contraction. *Proc. Natl. Acad. Sci. U.S.A.* 68:685–89.

Harrington, W. F. 1979. On the origin of the contractile force in skeletal muscle. *Proc. Natl. Acad. Sci. U.S.A.* 76:5066–70.

Harrington, W. F., and M. E. Rodgers. 1986. Myosin. *Annu. Rev. Biochem.* 53:35–73.

Harris, S. P., W. T. Heller, M. L. Greaser et al. 2003. Solution structure of heavy meromyosin by small-angle scattering. *J. Biol. Chem.* 278:6034–40.

Harry, J. D., A. W. Ward, N. C. Heglund et al. 1990. Cross-bridge cycling theories cannot explain high-speed lengthening behavior in frog muscles. *Biophys. J.* 57:201–8.

Hasan, H., and J. Unsworth. 1985. Hydration effects on muscle response. *Physiol. Chem. Physics Med. NMR* 17:131–4.

Morel

Haselgrove, J. C. 1973. X-ray evidence for conformational changes in the actin-containing filaments of vertebrate striated muscle. *Cold Spring Harb. Symp. Quant. Biol.* 37:341–52.

Haselgrove, J. C. 1975. X-ray evidence for conformational changes in the myosin filaments of vertebrate striated muscle. *J. Mol. Biol.* 92:113–43.

Haselgrove, J. C. 1980. A model of myosin crossbridge structure consistent with the low-angle X-ray diffraction pattern of vertebrate striated muscle. *J. Muscle Res. Cell Motil.* 1:117–91.

Haselgrove, J. C., and H. E. Huxley. 1973. X-ray evidence for radial crossbridge movement and for the sliding filament model in actively contracting skeletal muscle. *J. Mol. Biol.* 77:549–68.

Haselgrove, J. C., and C. D. Rodger. 1980. The interpretation of X-ray diffraction patterns from vertebrate striated muscle. *J. Muscle Res. Cell Motil.* 1:371–90.

Hashizume, H. 1977. Length dependence of tension in a frog toe muscle activated by caffeine rapid cooling. *Jikiekai Med.* 24:145–54.

Haugen, P., and O. Sten-Knudsen. 1976. Sarcomere lengthening and tension drop in the latent period of isolated frog skeletal muscle fibers. *J. Gen. Physiol.* 68:247–65.

Hazlewood, C. F. 1972. Pumps or no pumps? *Science* 82:815–6.

Hazlewood, C. F. 1973. Physicochemical state of the ions and water in living tissues and model systems. *Ann. N.Y. Acad. Sci. U.S.A.* 204:1–631.

Hazlewood, C. F. 1975. A role for water in the exclusion of cellular medium. Is a sodium pump needed? *Cardiovasc. Dis.* 2:83–104.

Hazlewood, C. F. 1979. A view of the significance and understanding of the physical properties of cell-associated water. In *Cell-associated water*, eds. W. Drost-Hansen, and J. S. Clegg, 165–259. Academic Press, New York, USA.

Hazlewood, C. F., B. L. Nichols, and N. F. Chamberlain. 1969. Evidence for the existence of a minimum of two phases of ordered water in skeletal muscle. *Nature* 222:747–50.

Hazlewood, C. F., D. C. Chang, B. L. Nichols et al. 1971. Interactions of water molecules with macromolecular structures in cardiac muscle. *J. Mol. Cell Cardiol.* 2:117–31.

Hazlewood, C. F., D. C. Chang, B. L. Nichols et al. 1974. Nuclear magnetic resonance transverse relaxation times of water protons in skeletal muscle. *Biophys. J.* 25:583–606.

Hazzard, J. H., and M. A. Cusanovich. 1986. Binding and hydrolysis of ATP by cardiac myosin subfragment-1: Effect of solution parameters on transient kinetics. *Biochemistry* 25:8141–9.

He, Z. H., R. K. Chillingworth, M. Brune et al. 1997. ATPase kinetics on activation of rabbit and frog permeabilised isometric muscle fibres. *J. Physiol. Lond.* 501:125–48.

He, Z. H., R. K. Chillingworth, M. Brune et al. 1998a. The ATPase activity in isometric and shortening skeletal muscle fibres. *Adv. Exp. Med. Biol.* 453:331–41.

He, Z. H., G. J. M. Stienen, J. P. F. Barents et al. 1998b. Rate of phosphate release after photoliberation of ATP in slow and fast skinned skeletal muscle fibres. *Biophys. J.* 75:2389–401.

He, Z. H., R. K. Chillingworth, M. Brune et al. 1999. The efficiency of contraction in rabbit skeletal muscle fibres, determined from the rate of release of inorganic phosphate. *J. Physiol. Lond.* 517:839–54.

Hellam, D. C., and R. J. Podolsky. 1969. Force measurements in skinned muscle fibres. *J. Physiol. Lond.* 204:207–30.

Herbst, M. 1976. Studies on the relation between latency relaxation and resting cross-bridges of frog skeletal muscle. *Pflügers Arch. Eur. J. Physiol.* 364:71–6.

Herrmann, C., F. Houadjeto, F. Travers et al. 1992. Early steps of the Mg^{2+}-ATPase of relaxed myofibrils. A comparison with Ca^{2+}-activated myofibrils and myosin subfragment-1. *Biochemistry* 31:8036–42.

Herrmann, C., C. Lionne, F. Travers et al. 1994. Correlation of actoS1, myofibrillar and muscle fibre ATPases. *Biochemistry* 33:4148–54.

Herzog, W., T. R. Leonardi, V. Joumaa et al. 2008. Mysteries of muscle contraction. *J. Appl. Biomech.* 24:1–13.

Heuser, J. E. 1983. Procedure of freeze-drying molecules adsorbed to mica flakes. *J. Mol. Biol.* 169:155–95.

Heuser, J. E., and R. Cooke. 1983. Actin-myosin interactions visualized by the quick-freeze, deep-etch replica technique. *J. Mol. Biol.* 169:97–122.

Hibberd, M. G., and D. R. Trentham. 1986. Relationship between chemical and mechanical events during muscle contraction. *Annu. Rev. Biophys. Biophys. Chem.* 15:119–61.

Higashi-Fujime, S. 1985. Unidirectional sliding of myosin filaments along the bundles of F-actin filaments spontaneously formed during super-precipitation. *J. Cell Biol.* 101:2335–44.

Highsmith, S., and A. J. Murphy. 1992. Electrostatic changes at the actomyosin-subfragment 1 interface during force generating reaction. *Biochemistry* 31:385–9.

Highsmith, S., K. Duigan, R. Cooke et al. 1996. Osmotic pressure probe of actin-myosin hydration changes during ATP hydrolysis. *Biophys. J.* 70:2830–7.

Higuchi, H. 1987. Lattice swelling with the selective digestion of elastic components in single skinned fibers of frog muscle. *Biophys. J.* 52:29–32.

Higuchi, H., and Y. E. Goldman. 1991. Sliding distance between actin and myosin filaments per ATP molecule hydrolysed in skinned muscle fibers. *Nature* 352:352–4.

Higuchi, H., and Y. E. Goldman. 1995. Sliding distance per ATP molecule hydrolysed by myosin heads during isotonic shortening of skinned muscle fibers. *Biophys. J.* 69:1491–507.

Hilber, K., Y. B. Sun, and M. Irving. 2001. Effects of sarcomere length and temperature on the rate of ATP utilisation by rabbit psoas muscle fibres. *J. Physiol. Lond.* 531:771–80.

Hill, A. V. 1911. The position occupied by the production of heat in the chain of processes constituting a muscular contraction. *J. Physiol. Lond.* 42:1–43.

Hill, A. V. 1913a. The energy degraded in the recovery processes of stimulated muscles. *J. Physiol. Lond.* 46:28–80.

Hill, A. V. 1913b. The absolute mechanical efficiency of the contraction of an isolated muscle. *J. Physiol. Lond.* 46:435–69.

Hill, A. V. 1922. The maximum work and efficiency of human muscles and their most economical speed. *J. Physiol. Lond.* 56:19–41.

Hill, A. V. 1925. Length of muscle and the heat and tension developed in an isometric contraction. *J. Physiol. Lond.* 60:237–63.

Hill, A. V. 1932. The revolution in muscle physiology. *Physiol. Rev.* 12:56–67.

Hill, A. V. 1938. The heat of shortening and the dynamic constants of muscle. *Proc. R. Soc. Lond. Ser. B* 126:136–95.

Hill, A. V. 1939. The mechanical efficiency of frog's muscle. *Proc. R. Soc. Lond. Ser. B.* 127:434–51.

Hill, A. V. 1948. On the time required for diffusion and its relation to processes in muscle. *Proc. R. Soc. Lond. Ser. B.* 136:446–53.

Hill, A. V. 1949a. The heat of activation and the heat of shortening in a muscle twitch. *Proc. R. Soc. Lond. Ser. B.* 136:195–211.

Hill, A. V. 1949b. Adenosone triphosphate and muscular contraction. *Nature* 163:320.

Hill, A. V. 1949. The onset of contraction. *Proc. R. Soc. Lond. Ser. B* 136:242–54.

Hill, A. V. 1951. The transition from rest to full activity in muscle: The velocity of shortening. *Proc. R. Soc. Lond. Ser. B* 138:329–38.

Hill, A. V. 1953. The mechanics of active muscle. *Proc. R. Soc. Lond. Ser. B* 141:104–17.

Hill, A. V. 1964a. The effect of load on the heat of shortening of muscle. *Proc. R. Soc. Lond. Ser. B* 159:297–318.

Hill, A. V. 1964b. The ratio of mechanical power developed to total power expanded during muscular contraction. *Proc. R. Soc. Lond. Ser. B* 159:319–24.

Hill, D. K. 1968a. Tension due to interactions between the sliding filaments in resting striated muscle. The effect of stimulation. *J. Physiol. Lond.* 199:637–84.

Hill, T. L. 1968b. On the sliding filament model of muscular contraction. II. *Proc. Natl. Acad. Sci. U.S.A.* 61:98–105.

Hill, T. L. 1968c. Phase transition in the sliding filament model of muscular contraction. *Proc. Natl. Acad. Sci. U.S.A.* 61:1194–200.

Hill, A. V. 1970a. *First and last experiments in muscle mechanics.* Cambridge Univ. Press, UK.

Hill, D. K. 1970b. The effect of temperature in the range 0–35°C on the resting tension of frog's muscle. *J. Physiol. Lond.* 208:725–39.

Hill, T. L. 1970c. Sliding filament model of muscular contraction. V. Isometric force and interfilament spacing. *J. Theor. Biol.* 29:395–410.

Hill, T. L. 1974. Theoretical formalism for the sliding filament model of contraction of striated muscle. Part I. *Prog. Biophys. Mol. Biol.* 28:267–340.

Hill, T. L. 1975. Theoretical formalism for the sliding filament model of contraction of striated muscle. Part II. *Prog. Biophys. Mol. Biol.* 29:105–59.

Hill, T. L. 1977. *Free energy transduction in biology.* Academic Press, New York, USA.

Hill, T. L. 1978. Binding of monovalent and divalent myosin fragments onto sites on actin. *Nature* 274:825–6.

Hill, A. V., and W. Hartree. 1920. The four phases of heat production of muscle. *J. Physiol. Lond.* 54:84–128.

Hill, A. V., and J. V. Howarth. 1959. The reverse of chemical reactions in contracting muscle during an applied stretch. *Proc. R. Soc. Lond. Ser. B* 151:169–93.

Hill, T. L., and G. M. White. 1968a. On the sliding filament model of muscular contraction. III. Kinetics of crossbridge fluctuations in configuration. *Proc. Natl. Acad. Sci. U.S.A.* 61:514–21.

Hill, T. L., and G. M. White. 1968b. On the sliding filament model of muscular contraction. IV. Calculation of force-velocity curves. *Proc. Natl. Acad. Sci. U.S.A.* 61:889–96.

Hill, T. L., E. Eisenberg, Y. D. Chen et al. 1975. Some self-consistent two-state sliding filament models of muscle contraction. *Biophys. J.* 15:335–72.

Hill, T. L., E. Eisenberg, and L. E. Greene. 1980. Theoretical model for the cooperative equilibrium binding of myosin subfragment 1 to the actin-troponin-tropomyosin complex. *Proc. Natl. Acad. Sci. U.S.A.* 77:3186–90.

Himmel, D. M., S. Gourinath, L. Reshetnikova et al. 2002. Crystallographic findings on the internally uncoupled and near-rigor states of myosin: Further insights into the mechanics of the motor. *Proc. Natl. Acad. Sci. U.S.A.* 99:12645–50.

Hinke, J. A. M. 1980. Water and electrolyte content of the myofilament phase in chemically skinned barnacle fiber. *J. Gen. Physiol.* 75:532–51.

Hirose, K., C. Franzini-Armstrong, Y. E. Goldman et al. 1994. Structural changes in muscle crossbridges accompanying force generation. *J. Cell Biol.* 127:763–79.

Hochberg, A., W. Low, R. Tirosh et al. 1977. A study of the dynamic properties of actomyosin systems by quasi-elastic light scattering. *Biochim. Biophys. Acta* 460:308–17.

Hofman, P. A., and F. Fuchs. 1987. Effect of length and cross-bridge attachment on Ca^{2+} binding to cardiac troponin C. *Am. J. Physiol.* 253:C90–6.

Holmes, K. C. 1996. Muscle proteins – Their actions and interactions. *Curr. Opin. Struct. Biol.* 6:781–9.

Holmes, K. C. 1997. The swinging lever-arm hypothesis of muscle contraction. *Curr. Biol.* 7:R112–8.

Holmes, K. C., and M. A. Geeves. 2000. The structural basis of muscle contraction. *Phil. Trans. R. Soc. Ser. B Biol. Sci.* 355:419–31.

Homsher, E. 1987. Muscle enthalpy production and its relationship to actomyosin ATPase. *Annu. Rev. Physiol.* 49:673–90.

Homsher, E., and C. J. Kean. 1978. Skeletal muscle energetics and metabolism. *Annu. Rev. Physiol.* 40:93–131.

Homsher, E., C. J. Kean, A. Wallner et al. 1979. The time course of energy balance in an isometric tetanus. *J. Gen. Physiol.* 73:553–67.

Homsher, E., M. Irving, and A. Walner. 1981. High-energy phosphate metabolism and energy liberation associated with rapid shortening in frog skeletal muscle. *J. Physiol. Lond.* 321:423–36.

Homsher, E., M. Irving, and J. Lebacq. 1983. The variation of the shortening heat with sarcomere length in frog muscle. *J. Physiol. Lond.* 345:107–21.

Homsher, E., F. Wang, and J. R. Sellers. 1992. Factors affecting movement of F-actin propelled by skeletal muscle heavy meromyosin. *Am. J. Physiol.* 262:C714–23.

Hopkins, S. C., C. Sabido-David, J. E. T. Corrie et al. 1998. Fluorescence polarization transients from rhodamine isomers on the myosin regulatory light chain in skeletal muscle fibers. *Biophys. J.* 74:3093–110.

Horiuti, K. 1986. Some properties of the contractile system and sarcoplasmic reticulum of skinned slow fibres from *Xenopus*. *J. Physiol. Lond.* 373:1–23.

Horowitz, A., and G. H. Pollack. 1993. Force-length relation in isometric sarcomeres in fixed-end tetani. *Am. J. Physiol.* 264:C19–25.

Horowitz, A., H. P. Wussling, and G. H. Pollack. 1992. Effect of small release on force during sarcomere-isometric tetani in frog muscle fibers. *Biophys. J.* 63:3–17.

Houadjeto, M., T. Barman, and F. Travers. 1991. What is the true ATPase activity of contracting myofibrils. *FEBS Lett.* 281:105–7.

Houdusse, A., and H. L. Sweeney. 2001. Myosin motors: Missing structures and hidden springs. *Curr. Opin. Struct. Biol.* 11:182–94.

Houdusse, A., V. N. Kalabokis, D. M. Himmel et al. 1999. Atomic structure of scallop myosin subfragment S1 complexed with MgADP: A novel conformation of the myosin head. *Cell* 97:459–70.

Houdusse, A., A. G. Szent-Györgyi, and C. Cohen. 2000. Three conformational states of scallop myosin S1. *Proc. Natl. Acad. Sci. U.S.A.* 97:11238–43.

Howard, J. 1997. Molecular motors: Structural adaptations to cellular functions. *Science* 389:561–6.

Howard, J. 2001. *Mechanics of motor proteins and the cytoskeleton*. Sinauer Associates, Sunderland, USA.

Hoyle, G. 1969. Comparative aspects of muscle. *Annu. Rev. Physiol.* 31:43–82.

Hoyle, G. 1970. How is muscle turned on and off? *Sci. Am.* 222:84–93.

Hoyle, G. 1983. *Muscles and their neuronal control.* John Wiley and Sons, Hoboken, USA.

Huang, X., H. M. Holden, and F. M. Rausel. 2005. Channeling of substrates and intermediates in enzyme-catalyzed reactions. *Annu. Rev. Biochem.* 79:149–80.

Hubley, M. J., B. R. Locke, and T. S. Moerland. 1996. The effects of temperature, pH, and magnesium on the diffusion coefficient of ATP in solutions of physiological ionic strength. *Biochim. Biophys. Acta* 1291:115–21.

Hunter, P. J., A. D. McCulloch, and H. E. D. J. ter Keurs. 1998. Modelling the mechanical properties of cardiac muscle. *Prog. Biophys. Mol. Biol.* 69:425–43.

Huxley, H. E. 1953. X-ray analysis and the problem of muscle. *Proc. R. Soc. Lond. B* 141:59–62.

Huxley, A. F. 1957. Muscle structure and theories of contraction. *Prog. Biophys. Biophys. Chem.* 7:255–318.

Huxley, H. E. 1963. Electron microscope studies on the structure of natural and synthetic protein filaments from skeletal muscle. *J. Mol. Biol.* 7:281–308.

Huxley, A. F. 1964. Muscle. *Annu. Rev. Physiol.* 26:131–52.

Huxley, H. E. 1965. The mechanism of muscular contraction. *Sci. Am.* 213:18–27.

Huxley, H. E. 1969. The mechanism of muscular contraction. *Science* 164:1356–66.

Huxley, H. E. 1971. The structural basis of muscular contraction. *Proc. R. Soc. Lond. Ser. B* 178:131–49.

Huxley, A. F. 1973a. A note suggesting that the cross-bridge attachment during muscle contraction may take place in two stages. *Proc. R. Soc. Lond. B* 183:83–6.

Huxley, H. E. 1973b. Molecular basis of contraction in cross-striated muscle. In *The structure and function of muscle,* ed. G. H. Bourne, 301–87. Academic Press.

Huxley, H. E. 1973c. Muscular contraction and cell motility. *Nature* 243:445–9.

Huxley, H. E. 1973d. Structural changes in the actin and myosin-containing filaments during contraction. *Cold Spring Harb. Symp. Quant. Biol.* 37:361–76.

Huxley, A. F. 1974. Muscular contraction. *J. Physiol. Lond.* 243:1–44.

Huxley, H. E. 1975. The structural basis of contraction and regulation in skeletal muscle. *Acta Anat. Nippon* 50:310–25.

Huxley, H. E. 1979. Time-resolved X-ray diffraction studies on muscle. In *Cross-bridge mechanism in muscle contraction,* eds. H. Sugi, and G. H. Pollack, 391–405. Tokyo Univ. Press, Japan.

Huxley, A. F. 1980a. *Reflections on muscle.* Liverpool Univ. Press, UK.

Huxley, H. E. 1980b. The movement of myosin cross-bridges during contraction. In *Muscle contraction: Its regulatory mechanisms,* eds. S. Ebashi, K. Maruyama, and M. Endo, 391–405. Jpn. Sci. Soc. Press, Japan.

Huxley, H. E. 1980c. The movement of myosin cross-bridges during contraction. In *Muscle contraction: Its regulatory mechanisms,* eds. S. Ebashi, K. Maruyama, and M. Endo, 33–43. Jpn. Sci. Soc. Press, Japan.

Huxley, A. F. 1986. Comments on 'Quantal mechanisms in cardiac contraction'. *Circ. Res.* 59:9–14.

Huxley, A. F. 1988. Prefatory chapter: Muscular contraction. *Annu. Rev. Physiol.* 59:1–16.

Huxley, H. E. 1990. Sliding filaments and molecular motile systems. *J. Biol. Chem.* 265:8347–50.

Huxley, H. E. 1996. A personal view of muscle and motility mechanisms. *Annu. Rev. Physiol.* 58:1–19.

Huxley, A. F. 1998. Muscle: Support for the lever arm. *Nature* 396:317–8.

Huxley, A. F. 2000. Mechanics and models of the myosin motor. *Phil. Trans. R. Soc. Lond. Ser. B Biol. Sci.* 355:433–40.

Huxley, H. E. 2004. Fifty years of muscle and the sliding filament hypothesis. *Eur. J. Biochem.* 271:1403–5.

Huxley, A. F., and R. Niedergerke. 1954. Structural changes in muscle during contraction: Interference microscopy of living muscle fibres. *Nature* 173:971–3.

Huxley, H. E., and J. Hanson. 1954. Changes in the cross-striations of muscle during contraction and stretch and their structural interpretation. *Nature* 173:973–6.

Huxley, H. E., and W. Brown. 1967. The low-angle X-ray diagram of vertebrate striated muscle and its behaviour during contraction and rigor. *J. Mol. Biol.* 30:384–434.

Huxley, A. F., and R. M. Simmons. 1971. Proposed mechanism of force generation in striated muscle. *Nature* 233:533–8.

Huxley, A. F., and R. M. Simmons. 1973. Mechanical transients and the origin of muscle force. *Cold Spring Harb. Symp. Quant. Biol.* 37:669–80.

Huxley, H. E., and M. Kress. 1985. Crossbridge behaviour during muscle contraction. *J. Muscle Res. Cell Motil.* 6:153–61.

Huxley, A. F., and S. Tideswell. 1996. Filament compliance and tension transients in muscle. *J. Muscle Res. Cell Motil.* 17:507–11.

Huxley, A. F., and S. Tideswell. 1997. Rapid regeneration of power stroke in contracting muscle by attachment of the second myosin head. *J. Muscle Res. Cell Motil.* 18:111–4.

Huxley, H. E., A. Stewart, H. Sosa et al. 1994. X-ray diffraction measurements of the extensibility of acin and myosin filaments in contracting muscles. *Biophys. J.* 67:2411–21.

Huxley, H. E., R. M. Simmons, A. R. Faruqi et al. 1981. Millisecond time-resolved changes in X-ray reflections from contracting muscles during rapid mechanical transients, recorded using synchrotron radiation. *Proc. Natl. Acad. Sci. U.S.A.* 78:2297–301.

Huxley, H. E., M. Kress, A. R. Faruqi et al. 1988. X-ray diffraction studies on muscle during rapid shortening and their implication concerning cross-bridge behaviour. In *Molecular mechanism of muscle contraction*, eds. H. Sugi, and G. H. Pollack, 347–52. Plenum Press, New York, USA.

Huxley, H. E., M. Reconditi, A. Stewart et al. 2006. X-ray interference studies of crossbridges action in muscle contraction: Evidence from muscles during steady shortening. *J. Mol. Biol.* 363:762–72.

Inoue, A., T. Arata, H. Takenaka et al. 1979. Functional implication of two-headed structure of myosin. *Adv. Biophys.* 13:1–143.

Iorga, B., R. Candau, F. Travers et al. 2004. Does phosphate release limit the ATPases of *soleus* myofibrils? Evidence that (A)M.ADP.Pi states predominate on the cross-bridge cycle. *J. Muscle Res. Cell Motil.* 25:367–78.

Ip, W., and J. E. Heuser. 1983. Direct visualization of the myosin cross-bridge helices on relaxed psoas thick filaments. *J. Mol. Biol.* 205:677–83.

Irving, M. 1991. Biomechanics goes quantum. *Nature* 352:284–6.

Irving, M. 1985. Muscle contraction: Weak and strong crossbridges. *Nature* 316:292–3.

Irving, M. 1987. Muscle mechanics and probes of the crossbridge cycle. In *Fibrous protein structure*, eds. J. M. Squire, and P. J. Vibert, 496–526. Academic Press, London, UK.

Irving, M. 1991. Biomechanics goes quantum. *Nature* 352:284–6.

Irving, M. 1995. Give in the filaments. *Nature* 374:14–5.

Irving, M., and R. C. Woledge. 1981. The dependence on extent of shortening of extra energy liberation by rapidly shortening frog skeletal muscle. *J. Physiol. Lond.* 321:411–22.

Irving, M., and Y. E. Goldman. 1999. Motor proteins: Another step ahead for myosin. *Nature* 398:463–5.

Irving, M., V. Lombardi, G. Piazzesi et al. 1992. Myosin head movements are synchronous with the elementary force-generating process in muscle. *Nature* 357:688–91.

Irving, M., T. Allen St. Claire, C. Sabido-David et al. 1995. Tilting of the light-chain region of myosin during step length changes and active force generation in skeletal muscle. *Nature* 357:688–91.

Ishii, Y., and T. Yanagida. 2000. Single molecule detection in life science. *Single Mol.* 1:5–16.

Ishijiama, A., and T. Yanagida. 2001. Single molecule nanobioscience. *Trends Biochem. Sci.* 26:438–44.

Ishijiama, A., T. Doi, K. Sakudara et al. 1991. Subpiconewton force fluctuations of actomyosin in vitro. *Nature* 352:301–6.

Israelachvili, J. N. 1974. van der Waals forces in biological systems. *Q. Rev. Biophys.* 6:341–67.

Ito, K., T. P. Q. Uyeda, Y. Suzuki et al. 2003. Requirement of domain-domain interaction for conformational changes and functional ATP hydrolysis in myosin. *J. Biol. Chem.* 276:6034–40.

Iwai, S., D. Hanamoto, and S. Chaen. 2006. A point mutation in the SH1 helix alters elasticity and thermal stability of myosin II. *J. Biol. Chem.* 281:30736–44.

Iwazumi, T. 1970. A new theory of muscle contraction. PhD diss., Univ. Pennsylvania, USA.

Iwazumi, T. 1979. A new field theory of muscle contraction. In *Cross-bridge mechanism in muscle*, eds. H. Sugi, and G. H. Pollack, 611–23. Univ. Tokyo Press, Japan.

Iwazumi, T. 1989. Molecular mechanisms of muscle contraction. *Physiol. Chem. Phys. Med. NMR* 21:187–219.

Iwazumi, T., and G. H. Pollack. 1981. The effect of sarcomere non-uniformity on the sarcomere shortening steps. *J. Cell Physiol.* 106:321–47.

Iwazumi, T., and M. I. M. Noble. 1989. An electrostatic mechanism of muscular contraction. *Int. J. Cardiol.* 24:267–75.

Jacobson, R. C., R. Tirosh, M. Delay et al. 1983. Quantized nature of sarcomere shortening steps. *J. Muscle Res. Cell Motil.* 4:529–42.

Jarosh, R. 2000. Muscle force arises by actin filament rotation and torque in the Z-filaments. *Biochem. Biophys. Res. Commun.* 279:677–82.

Jarosh, R. 2008. Large-scale models reveal the two component mechanics of striated muscle. *Int. J. Mol. Sci.* 9:2658–723.

Jaworoski, A., and A. Arno. 1998. Temperature sensitivity of force and shortening velocity in maximally activated skinned smooth muscle. *J. Muscle Res. Cell Motil.* 19:247–55.

Johnston, I. A. 1983. Dynamic properties of fish muscle. In *Fish biomechanics*, eds. P. W. Webb, and D. Weiths, 36–57. Praeger, Westpoint, USA.

Jontes, J. D. 1995. Theories of muscle contraction. J. Struct. Biol. 115:116–43.

Joubert, F., J. R. Wilding, D. Fortin et al. 2008. Local energetic regulation of sarcoplasmic reticulum and myosin ATPase is differently impaired in rats with heart failure. J. Physiol. Lond. 586:5182–92.

Julian, F. J., and D. L. Morgan. 1979a. Intersarcomere dynamics during fixed-end tetanic contraction of frog muscle fibres. J. Physiol. Lond. 293:365–78.

Julian, F. J., and D. L. Morgan. 1979b. The effect of tension on non-uniform distribution of length changes applied to frog muscle fibres. J. Physiol. Lond. 293:379–92.

Julian, F. J., and R. L. Morgan. 1981. Tension, stiffness, unloaded shortening speed and potentiation of frog muscle fibres at sarcomere lengths below optimum. J. Physiol. Lond. 319:205–17.

Julian, F. J., R. L. Moss, and M. R. Sollins. 1978a. The mechanism for vertebrate striated muscle contraction. Circ. Res. 42:2–14.

Julian, F. J., M. R. Sollins, and R. L. Moss. 1978b. Sarcomere length non-uniformity in relation to tetanic responses of stretched muscle fibres. Proc. R. Soc. Lond. B 200:109–16.

Jung, H. S., S. Komatsu, M. Ikebe et al. 2008. Head–head and head–tail interactions. A general mechanism for switching off myosin II activity in cells. Mol. Biol. Cell 19:3234–42.

Kasturi, S. R., D. C. Chang, and C. F. Hazlewood. 1980. Study of anisotropy in nuclear magnetic resonance relaxation times of water protons in skeletal muscle. Biophys. J. 30:369–81.

Katayama, E. 1989. The effects of various nucleotides on the structure of actin-attached myosin subfragment 1 studied by quick-freeze etch electron microscopy. J. Biochem. Tokyo 106:751–70.

Katz, A. M. 1993. Metabolism of the failing heart. Cardioscience 4:199–203.

Katz, A. M. 2003. Heart failure: A hemodynamic disorder complicated by maladaptive proliferative responses. J. Cell. Mol. Med. 7:1–10.

Katz, A. M. 2006. Physiology of the heart. Lippincott, Williams and Wilkins, New York, USA.

Katzir-Katchalsky, A. 1949. Rapid swelling and deswelling of reversible gels of polymeric acids by ionisation. Experientia 5:319–20.

Katzir-Katchalsky, A., and P. F. Curran. 1965. Non-equilibrium thermodynamics in biophysics. Harvard. Univ. Press, USA.

Kawai, M. 2003. What do we learn by studying the temperature effect on isometric tension and tension transients in mammalian striated muscle fibres? J. Muscle Res. Cell Motil. 24:127–38.

Kawai, M., and M. I. Schulman. 1985. Crossbridge kinetics in chemically skinned rabbit psoas fibres when the actin-myosin lattice spacing is altered by dextran T-500. J. Muscle Res. Cell Motil. 6:313–32.

Kawai, M., J. S. Wray, and T. Zhao. 1993. The effect of the lattice spacing change on cross-bridge kinetics in chemically skinned rabbit psoas rabbit fibers. I. Proportionality between the lattice spacing and the fiber width. Biophys. J. 64:187–96.

Kedem, O., and S. R. Caplan. 1965. Degree of coupling and its relation to energy conversion. Trans. Faraday Soc. 61:1897–911.

Kensler, R. W. 2002. Mammalian cardiac muscle thick filaments: Their periodicity and interactions with actin. Biophys. J. 82:1407–508.

Kensler, R. W. 2005. The mammalian cardiac muscle thick filament: Crossbridge arrangement. J. Struct. Biol. 149:303–12.

Kensler, R. W., and M. Stewart. 1983. Frog skeletal muscle thick filaments are three-stranded. J. Cell Biol. 96:1797–802.

Kensler, R. W., and M. Stewart. 1993. The relaxed crossbridge pattern in isolated rabbit psoas muscle thick filaments. J. Cell Biol. 105:841–8.

Kensler, R. W., S. Peterson, and M. Norberg. 1994. The effects of changes in temperature or ionic strength on isolated rabbit and fish skeletal muscle thick filaments. J. Muscle Res. Cell Motil. 15:69–79.

Kinosita, H., S. Ishiwata, H. Yoshimara et al. 1984. Submicrosecond and microsecond rotational motions of myosin heads in solution and in myosin synthetic filaments as revealed by time-resolved optical anisotropy decay measurements. Biochemistry 23:5963–75.

Kishino, A., and T. Yanagida. 1988. Force measurement by micromanipulation of a single actin filament by glass needles. Nature 334:746–7.

Kitamura, K., M. Tokunaga, A. H. Iwane et al. 1999. A single myosin head moves along an actin filament with regular steps of 5.3 nanometers. Nature 397:129–34.

Kitamura, K., A. Ishijiama, M. Tokunaga et al. 2001. Single-molecule nanobiotechnology. JSAP Int. 4:4–9.

Kitamura, K., M. Tokunaga, S. Esaki et al. 2005. Mechanism of muscle contraction based on stochastic properties of single actomyosin motors observed in vitro. Biophysics 1:1–19.

Klein, M. P., and D. E. Phelps. 1969. Evidence against orientation of water in rat phrenic nerve. *Nature* 224:70–1.

Knappeis, G. G., and F. Carlsen. 1968. The ultrastructure of the M line in skeletal muscle. *J. Cell Biol.* 38:202–11.

Konhilas, J. P., T. C. Irving, and P. P. de Tombe. 2002. Frank–Starling law of the heart and the cellular mechanisms of length-dependent activation. *Pflüfers Arch. Eur. J. Physiol.* 445:305–10.

Koretz, J. F. 1979. Structural studies of synthetic filaments prepared from column-purified myosin. *Biophys. J.* 27:423–32.

Koubassova, N. A., and A. K. Tsaturyan. 2011. Molecular mechanisms of actin-myosin motor in muscle. *Biochemistry (Moscow)* 76:1484–506.

Kováry, Z., and M. Vas. 2004. Protein conformer selection by sequence-dependent packing contacts in crystals of 3-phosphoglycerate kinase. *Proteins* 55:198–209.

Kováry, Z., B. Flachner, G. Naray-Szabó et al. 2002. Crytallographic and thiol-reactivity studies on the complex of pig muscle phosphoglycerate kinase with ATP analogues: Correlation between nucleotide binding mode and helix flexibility. *Biochemistry* 41:8796–808.

Kraft, T., J. M. Chalovich, L. C. Yu et al. 1995. Parallel inhibition of active force and relaxed fiber stiffness by caldesmon fragments at physiological ionic strength and temperature conditions: Additional evidence that weak cross-bridge binding to actin is an essential intermediate for force generation. *Biophys. J.* 68:2404–18.

Krasner, B., and D. W. Maughan. 1984. The relationship between ATP hydrolysis and active forces in compressed and swollen skinned muscle fibres of the rabbit. *Pflügers Arch. Eur. J. Physiol.* 400: 160–5.

Kron, S. J., and J. A. Spudich. 1986. Fluorescent actin filaments move on myosin fixed to glass surface. *Proc. Natl. Acad. Sci. U.S.A.* 83:8272–6.

Kuhn, H. J. 1981. The mechanochemistry of force production in muscle. *J. Muscle Res. Cell Motil.* 2: 7–44.

Kuhn, W., B. Hargitay, A. Katachalsky et al. 1950. Reversible dilatation and contraction by changing the state of ionization of high polymer acid networks. *Nature* 165:514–7.

Kuhn, H. J., C. Bletz, and J. C. Rüegg. 1990. Stretch induced increase in the Ca^{2+} sensitivity of myofibrillar ATPase activity in skinned fibres from pig ventricles. *Pflügers Arch. Eur. J. Physiol.* 415:741–6.

Kuntz, P. A., K. Loth, J. G. Watterson et al. 1980. Nucleotide induced head-head interactions in myosin. *J. Muscle Res. Cell Motil.* 1:15–30.

Kuschmerick, M. J., and R. J. Podosky. 1969. Ionic mobility in muscle cells. *Science* 166:1297–8.

Kushmerick, M. J., and R. E. Davies. 1969. The chemical energetics of muscle contraction. II. The chemistry, efficiency and power of maximally working sartorius muscle. *Proc. R. Soc. Lond. B.* 174:315–53.

Kushmerick, M. J., R. E. Larson, and R. E. Davies. 1969. The chemical energetics of muscle contraction. I. Activation heat, heat of shortening and ATP utilization for activation-relaxation processes. *Proc. R. Soc. Lond. B* 174:293–313.

Kuum, M., A. Kaasik, F. Joubert et al. 2009. Energetic state is a strong regulator of sarcoplasmic reticulum Ca^{2+} loss in cardiac muscle: Different efficiencies of different energy sources. *Cardiovasc. Res.* 83:89–96.

Kuzmin, A. I., V. L. Lakomkin, V. I. Kapelko et al. 1998. Interstitial ATP level and degradation in control and postmyocardial infracted rats. *Am. J. Physiol.* 275:C766–71.

Laakso, J. M., J. H. Lewis, and E. M. Ostap. 2008. Myosin I can act as a molecular force sensor. *Science* 321:133–6.

Labbé, J. P., R. Bertrand, E. Audemard et al. 1984. The interaction of skeletal myosin subfragment 1 with the polyanion, heparin. *Eur. J. Biochem.* 143:315–22.

Lakatta, E. G. 1987. Starling's law of the heart is explained by an intimate interaction of muscle length and myofilament calcium activation. *J. Am. Cell. Cardiol.* 10:1157–64.

Lampinen, M. J., and T. Noponen. 2005. Electric dipole theory and thermodynamics of acto-myosin molecular motor in muscle contraction. *J. Theor. Biol.* 236:397–421.

le Guennec, J. Y., O. Cazorla, A. Lacampagne et al. 2000. Is titin the length sensor in cardiac muscle? Physiological and physiopathological perspectives. *Adv. Exp. Med. Biol.* 481:337–48.

Leake, M. C., D. Wilson, M. Gautel et al. 2004. The elasticity of single titin molecules using a two-bead optical tweezer assay. *Biophys. J.* 87:1112–35.

Lebacq, J. 1980. Origine de la chaleur de raccourcissement dans un muscle strié. PhD diss., Univ. Cathol. Louvain, Belgium.

Legssyer, A., J. Poggioli, D. Renard et al. 1988. ATP and other adenine compounds increase mechanical activity and inositol triphosphate production in rat heart. *J. Physiol. Lond.* 401:186–99.

Lehman, W. 1978. Thick filament-linked calcium regulation in vertebrate striated muscle. *Nature* 274:80–1.

Lehninger, A. L. 2008. *Principles of biochemistry.* W. H. Freeman, New York, USA.

Lehrer, S. S., and M. A. Geeves. 1998. The muscle thin filament as a classical cooperative/allosteric regulatory system. *J. Mol. Biol.* 277:1081–9.

Lenart, T. D., J. M. Murray, C. Franzini-Armstrong et al. 1996. Structure and periodicities of crossbridges in relaxation and during contraction initiated by photolysis of caged calcium. *Biophys. J.* 71:2289–306.

Leonard, T. R., M. Duvall, and W. Herzog. 2010. Force enhancement following stretch in single sarcomere. *Am. J. Physiol.* C1398–401.

Levy, H. M., F. Ramirez, and K. K. Shukla. 1979. Muscle contraction: A mechanism of energy conversion. *J. Theor. Biol.* 81:327–33.

Levy, C., H. E. D. J. ter Keurs, Y. Yaniy et al. 2005. The sarcomeric control of energy conversion. *Ann. N.Y. Acad. Sci. U.S.A.* 1047:345–65.

Li, X. D., and M. Ikebe. 2003. Two functional heads are required for full activation of smooth muscle myosin. *J. Biol. Chem.* 278:29435–41.

Linari, M., and R. C. Woledge. 1995. Comparison of energy output during ramp and staircase shortening in frog muscle fibres. *J. Physiol. Lond.* 487:699–710.

Linari, M., J. Dobbie, M. Reconditi et al. 1998. The stiffness of skeletal muscle in isometric contraction and rigor: The fraction of myosin heads bound to actin. *Biophys. J.* 74:2459–73.

Linari, M., R. C. Woledge, and N. A. Curtin. 2003. Energy storage during stretch of active single fibres from frog skeletal muscle. *J. Physiol. Lond.* 548:461–74.

Linari, M., E. Brunello, M. Reconditi et al. 2005. The structural basis of the increase in isometric force production with temperature in frog skeletal muscle. *J. Physiol. Lond.* 567:459–69.

Linari, M., M. Caremari, C. Piperio et al. 2007. Stiffness and fraction of myosin motors responsible for active force in permeabilised muscle fibers from rabbit psoas. *Biophys. J.* 92:2476–90.

Linari, M., G. Piazzesi, and V. Lombardi. 2009. The effect of myofilament compliance on force generation by myosin motors in muscle. *Biophys. J.* 96:583–92.

Ling, G. N. 1965. The physical state of water in living cells and model systems. *Ann. N.Y. Acad. Sci. U.S.A.* 125:401–17.

Ling, G. N. 1970. The physical state of water in living cells and its physiological significance. *Int. J. Neurosci.* 1:129–52.

Ling, G. N. 1976. Structured water or pumps? *Science* 193:530–2.

Ling, G. N. 1977. The physical state of water and ions in living cells and a new theory of the energization of biological work performance by ATP. *Mol. Cell Biochem.* 15:159–72.

Ling, G. N. 1980. The role of multilayer polarization of cell water in the swelling or shrinking of living cells. *Physiol. Chem. Phys.* 12:383–4.

Ling, G. N. 1984. *Search of the physical basis of life.* Plenum Press, New York, USA.

Ling, G. N. 1988a. Solute exclusion by polymer and protein-dominated water: Correlation with results of nuclear magnetic resonance (NMR) and calorimetric studies and their significance for the understanding of the physical state of water in living cells. *Scanning Microsc.* 2:871–84.

Ling, G. N. 1988b. A physical theory of the living state: Application to water and solute distribution. *Scanning Microsc.* 2:899–913.

Ling, G. N. 1993. A quantitative theory of solute distribution in cell water according to molecular weight. *Physiol. Chem. Phys. Med. NMR* 25:145–75.

Ling, G. N. 2003. A new theoretical foundation for the polarized-oriented multilayer theory of cell water and for inanimate systems demonstrating long-range dynamics structuring of water molecules. *Physiol. Chem. Phys. Med. NMR* 35:91–130.

Ling, G. N., and W. X. Hu. 1988. Studies of the physical state of water in living cells and model systems. X. The dependence of the equilibrium distribution coefficient of a solute in polarized water on the molecular weights of the solute: Experimental confirmation of the 'size rule' in model studies. *Physiol. Chem. Phys. Med. NMR* 20:293–307.

Ling, G. N., and W. X. Hu. 2004. How much water is made 'non-free' by 36% native hemoglobin? *Physiol. Chem. Phys. Med NMR* 36:143–8.

Ling, G. N., C. Miller, and M. M. Ochsenfeld. 1973. The physical state of solute and water according to the association-induction hypothesis. *Ann. N.Y. Acad. Sci. U.S.A.* 204:6–50.

Ling, G. N., Z. Niu, and M. M. Ochsenfeld. 1993. Prediction of polarized multiplayer theory of solute distribution confirmed from a study of the equilibrium distribution in frog muscle of twenty-one electrolytes including five cryoprotectants. *Physiol. Chem. Phys. Med. NMR* 25:177–208.

Ling, N. C., C. Shrimpton, J. Sleep et al. 1996. Fluorescence probes of the orientation of myosin regulatory light chains in relaxed, rigor and contracting muscle. *Biophys. J.* 70:1836–46.

Linke, W. A., M. L. Bartoo, and G. H. Pollack. 1993. Spontaneous sarcomeric oscillations at intermediate activation levels in single isolated cardiac myofibrils. *Circ. Res.* 73:724–34.

Lionne, C., M. Brune, M. R. Webb et al. 1995. Time resolved measurements show that phosphate release is the rate limiting step on myofibrillar ATPase. *FEBS Lett.* 364:59–62.

Lionne, C., F. Travers, and T. Barman. 1996. Mechanochemical coupling in muscle: Attempts to measure simultaneously shortening and ATPase rates in myofibrils. *Biophys. J.* 70:887–95.

Lionne, C., B. Iorga, R. Candau et al. 2002. Evidence that phosphate release is the rate-limiting step on the overall ATPase of psoas myofibrils prevented from shortening by chemical cross-linking. *Biochemistry* 41:13297–308.

Lionne, C., B. Iorga, R. Candau et al. 2003. Why choose myofibrils to study muscle myosin ATPase? *J. Muscle Res. Cell Motil.* 24:139–48.

Liu, X., and G. H. Pollack. 2002. Mechanics of F-actin characterized with microfabricated cantilevers. *Biophys. J.* 85:2705–15.

Liu, X., and G. H. Pollack. 2004. Stepwise sliding of single actin and myosin filaments. *Biophys. J.* 86:353–8.

Lombardi, V., G. Piazzesi, and M. Linari. 1992. Rapid regeneration of the acto-myosin power stroke in contracting muscle. *Nature* 355:638–41.

Lorand, L., and C. Moos. 1956. Auto-oscillations in extracted muscle fibres systems. *Nature* 177:1239.

Lou, F., and Y. B. Sun. 1993. The high-force region of the force–velocity relation in frog skinned muscle fibres. *Acta Physiol. Scand.* 148:243–52.

Lowey, S., H. S. Slayter, A. G. Weeds et al. 1969. Substructure of the myosin molecule. I. Subfragments of myosin by enzymic digestion. *J. Mol. Biol.* 42:1–6.

Lovell, S. J., and W. F. Harrington. 1981. Measurement of the fraction of myosin heads bound to actin in rabbit skeletal myofibril in rigor *J. Mol. Biol.* 149:659–74.

Lovell, S. J., P. J. Knight, and W. F. Harrington. 1981. Fraction of myosin heads bound to thin filaments in rigor fibrils from insect flight and vertebrate muscles. *Nature* 293:664–6.

Luther, P., and J. M. Squire. 1978. Three dimensional structure of the vertebrate muscle M-region. *J. Mol. Biol.* 125:313–24.

Lymn, R. W. 1978. Myosin subfragment-1 attachment to actin. *Biophys. J.* 21:93–8.

Lymn, R. W., and E. W. Taylor. 1970. Transient state phosphate production in the hydrolysis of nucleotide triphosphates by myosin. *Biochemistry* 9:2975–91.

Lymn, R. W., and E. W. Taylor. 1971. Mechanism of adenosine triphosphate hydrolysis by actomyosin. *Biochemistry* 10:4617–24.

Ma, S., and G. I. Zahalak. 1991. A distribution-moment model of energetics in skeletal muscle. *J. Biomech.* 24:21–35.

Ma, Y. Z., and E. W. Taylor. 1994. Kinetic mechanism of myofibril ATPase. *Biophys. J.* 66:1542–53.

MacIntosh, B. R., R. R. Neptune, and A. J. van den Bogert. 2000. Intensity of cycling and cycle ergonomy: Power output and energy cost. In *Biomechanics and Biology of movement*, eds. B. M. Nigg, B. R. MacIntosh, and J. Mester, 129–547. Human Kinetics, Leeds.

Maciver, S. K. 1996. Myosin II function in non-muscle cells. *Bioassays* 18:179–82.

Magid, A., and M. K. Reedy. 1980. X-ray diffraction observations of chemically skinned frog skeletal muscle processed by an improved method. *Biophys. J.* 30:27–40.

Magid, A., H. P. Ting-Ball, M. Carvell et al. 1984. Connecting filaments, core filaments, and side-struts: A proposal to add three new load-bearing structures to the sliding filament model. *Adv. Exp. Med. Biol.* 170:307–28.

Malinchik, S., and V. V. Lednev. 1992. Interpretation of the X-ray diffraction pattern from relaxed skeletal muscle and modelling of the thick filament structure. *J. Muscle Res. Cell Motil.* 13:406–19.

Malinchik, S., S. Xu, and L. C. Yu. 1997. Temperature-induced structural changes in the myosin thick filament of skinned rabbit psoas muscle. *Biophys. J.* 73:2304–12.

Mannherz, H. G. 1968. ATP-spaltung und ATP-diffusion in oscillierenden extra hierten Müskelfasern. *Pflügers Arch.* 303:230–48.

Maréchal, G., and L. Plaghki. 1979. The deficit of the isometric tetanic tension redeveloped after a release of frog muscle at a constant velocity. *J. Gen. Physiol.* 73:453–67.

Margossian, S. S., and S. Lowey. 1973. Substructure of the myosin molecule. IV. Interactions of myosin and its subfragments with adenosine triphosphate and F-actin. *J Mol. Biol.* 74:13–30.

Margossian, S. S., and S. Lowey. 1982. Preparation of myosin and its subfragments from rabbit skeletal muscle. *Meth. Enzymol.* 85:55–71.

Margossian, S. S., and H. S. Slayter. 1987. Electron microscopy of cardiac myosin: Its shape and properties as determined by the regulatory light chain. *J. Muscle Res. Cell Motil.* 8:437–47.

Margossian, S. S., H. D. White, J. Lefford et al. 1993. Functional effects of LC1-reassociation with cardiac papain Mg.S1. *J. Muscle Res. Cell Motil.* 14: 3–14.

Marin, J. L., J. Muñiz, and M. Huerta. 1990. Further analysis of the molecular jet hypothesis during muscle contraction. *J. Theor. Biol.* 147:373–6.

Martin-Fernandez, M. L., J. Bordas, G. Diakun et al. 1994. Time-resolved X-ray diffraction studies of myosin head movements in live frog *sartorius* muscle during isometric and isotonic contractions. *J. Muscle Res. Cell Motil.* 15:319–48.

Martson, S. B. 1973. The nucleotide complexes of myosin in glycerol-extracted muscle fibres. *Biochim. Biophys. Acta* 305:397–412.

Martyn, D. A., and A. M. Gordon. 1988. Length and myofilament spacing-dependent changes in calcium sensitivity of skeletal fibres: Effects of pH and ionic strength. *J. Muscle. Res. Cell Motil.* 9:424–45.

Martyn, D. A., and A. M. Gordon. 1992. Force and stiffness in glycerinated rabbit psoas fibers. *J. Gen. Physiol.* 99:795–816.

Martyn, D. A., P. B. Chase, M. Regnier et al. 2002. A simple model with myofilament compliance predicts activation-dependent cross-bridge kinetics in skeletal skinned fibres. *Biophys. J.* 83:3425–34.

Martyn, D. A., B. B. Adhikari, M. Regnier et al. 2004. Response of equatorial X-ray reflections and stiffness to altered sarcomere length and myofilament lattice in relaxed skinned cardiac muscle. *Biophys. J.* 86:1002–11.

Matsubara, I. 1980. X-ray diffraction studies on the heart. *Annu. Rev. Biophys. Bioeng.* 9:81–105.

Matsubara, I., and G. F. Elliott. 1972. X-ray diffraction studies on skinned single fibres of frog skeletal muscle. *J. Mol. Biol.* 72:657–69.

Matsubara, I., and B. M. Millman. 1974. X-ray diffraction patterns from mammalian heart muscle. *J. Mol. Biol.* 82:527–36.

Matsubara, I., and N. Yagi. 1978. A time-resolved diffraction study of muscle during twitch. *J. Physiol. Lond.* 278:297–307.

Matsubara, I., and N. Yagi. 1985. Movements of cross-bridges during and after slow length changes in active frog skeletal muscle. *J. Physiol. Lond.* 361:151–63.

Matsubara, I., N. Yagi, and H. Hashizume. 1975. Use of an X-ray television for diffraction of the frog striated muscle. *Nature* 255:728–9.

Matsubara, I., N. Yagi, and M. Endoh. 1979. Movement of myosin heads during a heart beat. *Nature* 278:474–6.

Matsubara, I., N. Yagi, and M. Endoh. 1980. The states of myosin heads in heart muscle during systolic and diastolic phases. *Eur. Heat J. Suppl. A* 17–20.

Matsubara, I., Y. E. Goldman, and R. M. Simmons. 1984. Changes in the lateral filament spacing of skinned muscle fibres when cross-bridges attach. *J. Mol. Biol.* 173:15–33.

Matsubara, I., Y. Umazume, and N. Yagi. 1985. Lateral filamentary spacing in chemically-skinned murine muscles during contraction. *J. Physiol. Lond.* 360:135–48.

Matsubara, I., N. Yagi, D. W. Maughan et al. 1989. X-ray diffraction study on heart muscle during contraction. *Prog. Clin. Biol. Res.* 315:481–6.

Maughan, D. W., and R. E. Godt. 1979. Stretch and radial compression studies on relaxed skinned muscle fibres of the frog. *Biophys. J.* 28: 391–402.

Maughan, D. W., and R. E. Godt. 1980. A quantitative analysis of elastic, entropic, electrostatic, and osmotic forces within relaxed skinned muscle fibres. *Biophys. Struct. Mech.* 7:17–40.

Maughan, D. W., and R. E. Godt. 1981a. Inhibition of force production in compressed skinned muscle fibers of the frog. *Pflügers Arch. Eur. J. Physiol.* 390:161–3.

Maughan, D. W., and R. E. Godt. 1981b. Radial forces within muscle fibers in rigor. *J. Gen. Physiol.* 77:49–64.

Maurice, D. M., and A. A. Giardini. 1951. Swelling of the cornea *in vivo* after destruction of the limiting layers. *Brit. J. Ophthal.* 35:791–7.

Maw, M. C., and A. J. Rowe. 1980. Fraying of A-filaments into three subfilaments. *Nature* 286:412–4.

McClare, C. W. 1972a. A 'molecular energy' muscle model. *J. Theor. Biol.* 35:569–96.

McClare, C. W. 1972b. A quantum mechanical muscle model. *Nature* 240:88–90.

McLaughling, S. G. A., and J. A. M. Hinke. 1966. Sodium and water binding in single striated muscle fibers of the giant barnacle. *Can. J. Physiol. Pharmacol.* 44:837–48.

Mehta, A. D. 2001. Myosin learns to walk. *J. Cell Sci.* 141:1981–98.

Mehta, A. D., and J. A. Spudich. 1998. Single myosin molecule mechanics. *Adv. Struct. Biol.* 5:229–70.

Mehta, A. D., R. S. Rief, J. A. Spudich et al. 1999. Single molecule biomechanics with optical methods. *Science* 283:1689–95.

Mendelson, R. A. 1982. X-ray scattering by myosin S1: Implication for the steric blocking model of muscle control. *Nature* 298:665–7.

Mendelson, R. A. 1985. Length of myosin subfragment 1. *Nature* 318:20.

Mendelson, R. A., and K. M. Kretzshmar. 1980. Structure of myosin subfragment 1 from low-angle X-ray scattering. *Biochemistry* 9:4103–8.

Mendelson, R. A., and P. D. Wagner. 1984. X-ray scattering by a single-headed heavy meromyosin cleavage of the myosin head from the rod does not change its shape. *J. Mol. Biol.* 177:153–71.

Ménétret, J. F., W. Hofman, R. R. Schröder et al. 1991. Time-resolved cryo-electron microscopic study of the dissociation of actomyosin induced by photolysis of photolabile nucleotides. *J. Mol. Biol.* 219:139–44.

Merah, Z., and J. E. Morel. 1993. Isometric tension exerted by a myofibril of the frog at 0°C: Geometrical considerations. *J. Muscle Res. Cell Motil.* 14:552–3.

Merli, M., A. N. Szilágyi, N. Flatcher et al. 2002. Nucleotide binding to pig muscle 3-phosphoglycerate kinase in the crystal and in solution: Relationship between substrate antagonism and interdomain communication. *Biochemistry* 41:111–9.

Métrich, M., M. Berthouze, E. Morel et al. 2010. Role of the cyclic cAMP-binding protein Epac in cardiovascular physiology and pathophysiology. *Pflügers Arch. Eur. J. Physiol.* 459:435–46.

Mijailovich, S. M., J. J. Fredberg, and J. P. Butler. 1996. On the theory of muscle contraction: Filament extensibility and the development of isometric force and stiffness. *Biophys. J.* 71:1475–84.

Mills, R. 1973. Self-diffusion in normal and heavy water in the range 1-45 deg. *J. Phys. Chem.* 77:685–8.

Miller, A., and R. T. Tregear. 1970. Evidence concerning crossbridge attachment during muscle contraction. *Nature* 226:1060–1.

Miller, A., and J. Woodhead-Galloway. 1971. Long-range forces in muscle. *Nature* 229:470–3.

Milligan, R. A., and P. F. Flicker. 1987. Structural relationship of actin, myosin and tropomyosin revealed by cryo-electron microscopy. *J. Cell Biol.* 105:29–39.

Millman, B. M. 1986. Long range forces in cylindrical systems: Muscle and virus gels. In *Electrical double layers in biology*, ed. M. Blank, 301–12. Plenum Press. New York, USA.

Millman, B. M. 1998. The filament lattice of striated muscle. *Physiol. Rev.* 78:359–91.

Millman, B. M., and B. G. Nickel. 1980. Electrostatic forces in muscle and cylindrical gel systems. *Biophys. J.* 32:49–63.

Millman, B. M., and T. C. Irving. 1988. Filament lattice of frog striated muscle: Radial forces, lattice stability, and filament compression in the A-band of relaxed and rigor muscle. *Biophys. J.* 54:437–47.

Millman, B. M., T. J. Racey, and I. Matsubara. 1981. Lateral forces in the filament lattice of intact frog skeletal muscle. *Biophys. J.* 33:189–202.

Mitchell, P. 1974. A chemiosmotic molecular mechanism for proton-translocation adenosine triphosphatase. *FEBS Lett.* 43:189–94.

Mitsui, T. 1999. Induced potential model of muscular contraction mechanism and myosin molecule. *Adv. Biophys.* 36:107–58.

Mitsui, T., and H. Chiba. 1996. Proposed modification of the Huxley-Simmons model for myosin head motion along an actin filament. *J. Theor. Biol.* 182:147–59.

Mobley, B. A., and B. R. Eisenberg. 1975. Sizes of components in frog skeletal muscle measured by methods of stereology. *J. Gen. Physiol.* 66:31–45.

Moisescu, D. G., and R. Thielececk. 1978. Calcium and strontium concentration changes within skinned muscle preparation following a change in the external bathing solution. *J. Physiol. Lond.* 275:241–62.

Molloy, J. E. 2005. Muscle contraction: Actin filament enter the fray. *Biophys. J.* 89:1–2.

Molloy, J. E., J. E. Burns, J. Kendrick-Jones et al. 1995a. Movement and force produced by a single myosin head. *Nature* 378:209–12.

Molloy, J. E., J. E. Burns, J. C. Sparrow et al. 1995b. Single-molecule mechanics of heavy meromyosin and S1 interacting with rabbit or *Drosophila* actins using optical tweezers. *Biophys. J.* 68:298s–305s.

Moore, P. B., H. E. Huxley, and D. J. Derosier. 1970. Three-dimensional reconstruction of F-actin, thin filaments and decorated thin filaments. *J. Mol. Biol.* 50:279–95.

Moos, C., G. Offer, R. Starr et al. 1975. Interaction of C-protein with myosin, myosin rod and light meromyosin. *J. Mol. Biol.* 97:1–9.

Morales, M. F., and J. Botts. 1979. On the molecular basis for chemomechanical energy transduction in muscle. *Proc. Natl. Acad. Sci. U.S.A.* 76:3857–9.

Morel, J. E. 1975. Un modèle universel de la motilité. *La Recherche (Paris)* 58:676–80.

Morel, J. E. 1976. Structure of interfacial water. Application to muscle contraction. *Colloques Intern. CNRS* 246:209–14.

Morel, J. E. 1978. Force–velocity relationship in single muscle fibres. *J. Theor. Biol.* 73:445–51.

Morel, J. E. 1984a. A theoretical and semi-empirical approach to muscle mechanics and energetics, independent of the mechanical role of the crossbridges. I. Mechanics of a single fibre. *Prog. Biophys. Mol. Biol.* 44:47–71.

Morel, J. E. 1984b. A theoretical and semi-empirical approach to muscle mechanics and energetics, independent of the mechanical role of the crossbridges. II. Energetics of a whole muscle. *Prog. Biophys. Mol. Biol.* 44:72–96.

Morel, J. E. 1985a. Models of muscle contraction and cell motility: A comparative study of the usual concepts and the swelling theories. *Prog. Biophys. Mol. Biol.* 46:97–126.

Morel, J. E. 1985b. Discussion of the state of water in the myofilament lattice and other biological systems, based on the fact that the usual concepts of colloid stability cannot explain the stability of the myofilament lattice. *J. Theor. Biol.* 112:847–58.

Morel, J. E. 1990. Velocity-induced modifications in the cross-bridges and/or the actin filaments during shortening of muscle fibres. *J. Theor. Biol.* 146:347–54.

Morel, J. E. 1991a. The isometric force exerted per myosin head in a muscle fibre is 8 pN. Consequence on the validity of the traditional concepts of force generation. *J. Theor. Biol.* 151:285–8.

Morel, J. E. 1991b. Cross-bridge cycling theories and high-speed lengthening behavior in frog muscle. *Biophys. J.* 60:290–1.

Morel, J. E. 1996. Myosin heads in solution: How to interpret the data obtained by means of X-ray and neutron scattering? *J. Theor. Biol.* 179:87–90.

Morel, J. E., and I. Pinset-Härström. 1975a. Ultrastructure of the contractile system of striated skeletal muscle and the processes of muscular contraction. I. Ultrastructure of the myofibril and source of energy. *Biomedicine* 22:88–96.

Morel, J. E., and I. Pinset-Härström. 1975b. Ultrastructure of the contractile system of striated skeletal muscle

and the processes of muscular contraction. II. Releasing system and mechanisms of muscular contraction. *Biomedicine* 22:186–94.

Morel, J. E., and M. P. Gingold. 1977. On a simple way to verify the experimental agreement of any two-state mathematical model of muscle contraction. *J. Theor. Biol.* 68:437–47.

Morel, J. E., and M. P. Gingold. 1979a. Does water play a role in the stability of the myofilament lattice and other filament arrays? In *Cell-associated water*, eds. W. Drost-Hansen, and J. S. Clegg, 53–67. Academic Press, New York, USA.

Morel, J. E., and M. P. Gingold. 1979b. Stability of a resting muscle, mechanism of muscular contraction, and a possible role of the two myosin heads. *Acta Protozool. (Poland)* 18:179.

Morel, J. E., and M. Garrigos. 1982a. Dimerization of the myosin heads in solution. *Biochemistry* 21:2679–86.

Morel, J. E., and M. Garrigos. 1982b. The possible roles of the two myosin heads. *FEBS Lett.* 149:8–16.

Morel, J. E., and N. Bachouchi. 1988a. Muscle contraction and movement of cellular organelles: Are there two different types of mechanisms for their generation? *J. Theor. Biol.* 132:83–96.

Morel, J. E., and N. Bachouchi. 1988b. Comments on the molecular jet process for the in vitro movements: Is it valid to use steady-state kinetics to analyse actin–myosin motility? *J. Theor. Biol.* 135:119–21.

Morel, J. E., and N. Bachouchi. 1988c. Myosin heads and assembly of muscle thick filaments. *Nature* 332:591.

Morel, J. E., and N. Bachouchi. 1990. Movement of actin filaments along tracks of myosin heads. *J. Theor. Biol.* 145:135–6.

Morel, J. E., and N. Bachouchi-Sahli. 1992. Comments on 'further analysis of the molecular jet hypothesis during muscle contraction'. *J. Theor. Biol.* 154:403–4.

Morel, J. E., and Z. Merah. 1992. Muscle contraction and in vitro movement: Role of actin? *J. Muscle Res. Cell Motil.* 13:5–6.

Morel, J. E., and Z. Merah. 1995. Muscle contraction: Are the cross-bridges independent force generators? *J. Theor. Biol.* 176:431–2.

Morel, J. E., and Z. Merah. 1997. Shape and length of myosin heads and radial compressive forces in muscle. *J. Theor. Biol.* 184:133–8.

Morel, J. E., and N. D'hahan. 2000. The myosin motor: Muscle contraction and in vitro movement. *Biochem. Biophys. Acta* 1474:128–32.

Morel, J. E., and N. Guillo. 2001. Steady-state kinetics of MgATP splitting by native myosin RLC-free subfragment 1. *Biochim. Biophys. Acta* 1526:115–8.

Morel, J. E., I. Pinset-Härström, and M. P. Gingold. 1976. Muscular contraction and cytoplasmic streaming: A new general hypothesis. *J. Theor. Biol.* 62:17–51.

Morel, J. E., I. Pinset-Härström, and A. M. Bardin. 1979. Internal structure of myosin synthetic filaments. *Biol. Cell. (Paris)* 34:9–16.

Morel, J. E., N. Bachouchi-Salhi, and Z. Merah. 1992. Shape and length of the myosin heads. *J. Theor. Biol.* 156:73–90.

Morel, J. E., N. Bachouchi-Sahli, and Z. Merah. 1993. In vitro movement of actin filaments over myosin: Role of actin and related problems. *J. Theor. Biol.* 160:265–9.

Morel, J. E., N. D'hahan, K. Taouil et al. 1998a. Native myosin from adult rabbit skeletal muscle: Isoenzymes and states of aggregation. *Biochemistry* 37:5457–63.

Morel, J. E., K. Taouil, N. D'hahan et al. 1998b. Dimerization of native myosin LC2(RLC)-free subfragment 1 from adult rabbit skeletal muscle. *Biochemistry* 37:15129–36.

Morel, J. E., N. D'hahan, P. Bayol et al. 1999. Myosin thick filaments from adult rabbit skeletal muscles. *Biochim. Biophys. Acta* 1472:413–30.

Morgan, D. L. 1990. New insights into the behaviour of muscle during active lengthening. *Biophys. J.* 57:347–54.

Morgan, D. L., D. R. Claflin, and F. J. Julian. 1991. Tension as a function of sarcomere length and velocity of shortening in single skeletal fibres of the frog. *J. Physiol. Lond.* 441:719–32.

Morgan, D. L., S. Mochon, and F. J. Julian. 1982. A quantitative model of inter-sarcomere dynamics during fixed-end contractions of single frog muscle fibers. *Biophys. J.* 39:189–96.

Morimoto, K., and W. F. Harrington. 1974. Evidence for structural changes in vertebrate thick filament induced by calcium. *J. Mol. Biol.* 88:693–709.

Morita, F. 1977. Temperature induced analog reaction of adenylyl imidodiphosphate to an intermediate step of heavy meromyosin adenosine triphosphatase. *J. Biochem. Tokyo* 81:313–20.

Morita, F., and F. Ishigami. 1977. Temperature dependence of the decay of the UV absorption difference spectrum of heavy meromyosin induced by adenosine triphosphate and inosine triphosphate. *J. Biochem. Tokyo* 81:305–12.

Mornet, D., R. Bertrand, P. Pantel et al. 1981. Structure of the acto-myosin interface. *Nature* 292:301–6.

Mulieri, L. A. 1972. The dependence of the latency relaxation on sarcomere length and other characteristics of isolated muscle fibres. *J. Physiol. Lond.* 223:333–54.

Munson, K., M. J. Smerdon, and R. G. Yount. 1986. Crosslinking of myosin subfragment 1 and heavy meromyosin by use of vanadate and bis-(adenosine 5' triphosphate) analogue. *Biochemistry* 25:7640–50.

Muñiz, J., J. L. Marin, L. Yeomans et al. 1996. Electrostatic forces as a possible mechanism underlying skeletal muscle contraction. *Gen. Physiol. Biophys.* 15:441–9.

Murai, S., T. Arata, and A. Inoue. 1995. Binding of myosin and its subfragment-1 with antibodies specific to the two heads of the myosin molecule. *J. Biochem. Tokyo* 137:914–79.

Myburgh, K. H., K. Franks-Skiba, and R. Cooke. 1995. Nucleotide turnover rate measured in fully relaxed skeletal muscle myofibrils. *J. Gen. Physiol.* 106:957–73.

Nagornyak, E., N. E. Blyakhman, and G. H. Pollack. 2004. Effect of sarcomere length on step size in relaxed rabbit psoas muscle. *J. Muscle Res. Cell Motil.* 25:37–45.

Nassar-Gentina, V., J. V. Passoneau, J. L. Vergera et al. 1978. Metabolic correlates of fatigue and recovery from fatigue in single frog muscle fibers. *J. Gen. Physiol.* 72:593–606.

Natori, R. 1954a. The role of myofibrils, sarcoplasm and sarcolemma in muscle contraction. *Jikiekai Med. J.* 1:18–28.

Natori, R. 1954b. The property and contraction process of isolated myofibrils. *Jikiekai Med. J.* 1:119–26.

Naylor, G. R. S. 1982. On the average electrostatic potential between the filaments in striated muscle and its relation to a simple Donnan potential. *Biophys. J.* 38:201–4.

Naylor, G. R. S., E. M. Bartels, T. D. Bridgman et al. 1985. Donnan potentials in rabbit psoas muscle in rigor. *Biophys. J.* 48:47–59.

Neering, I. R., L. A. Quesenberry, V. A. Morris et al. 1991. Non-uniform volume changes during muscle contraction. *Biophys. J.* 59:926–32.

Nielsen, B. G. 2002. Entropic elasticity in the generation of muscle force: A theoretical model. *J. Theor. Biol.* 218:99–119.

Noble, D. 2002a. Modelling the heart: From genes to cells to the whole organism. *Science* 358:1678–82.

Noble, D. 2002b. Modelling the heart: Insights, failures and progress. *Bioessays* 24:1156–63.

Noble, M. I. M., and G. H. Pollack. 1977. Molecular mechanism of contraction. *Circ. Res.* 40:333–42.

Noble, M. I. M., and G. H. Pollack. 1978. Response to 'the mechanism for vertebrate striated muscle contraction'. *Circ. Res.* 42:15–6.

Nyitrai, M., R. Rossi, N. Adamek et al. 2006. What limits the velocity of fast skeletal muscle contraction in mammals? *J. Mol. Biol.* 355:432–42.

Offer, G. 1974. The molecular basis of muscular contraction. In *Companion of biochemistry*, eds. A. T. Bull, J. R. Lagnado, J. G. Thomas, and K. F. Tipton, 633–71. Longmans, New York, USA.

Offer, G. 1987. Myosin filaments. In *Fibrous protein structure*, eds. J. M. Squire, and P. J. Vibert, 307–56. Academic Press, London, UK.

Offer, G., and A. Elliott. 1978. Can a myosin molecule bind to two actin filaments? *Nature* 271:325–9.

Offer, G., and K. Ranatunga. 2010. Cross-bridge and filament compliance in muscle: Implication for tension generation and lever arm swing. *J. Muscle Res. Cell Motil.* 31:245–65.

Offer, G., P. J. Knight, S. A. Burgess et al. 2000. A new model for the surface arrangement of myosin molecules in tarantula thick filaments. *J. Mol. Biol.* 298:239–60.

Ohno, F., and T. Kodama. 1991. Kinetics of adenosine triphosphate hydrolysis by shortening myofibrils from rabbit psoas muscle. *J. Physiol. Lond.* 441:685–702.

Oiwa, K., S. Chaen, and H. Sugi. 1991. Measurements of work done by ATP-induced sliding between rabbit muscle myosin and algal cell actin cable in vitro. *J. Physiol. Lond.* 437:751–63.

Okamura, N., and S. Ishiwata. 1988. Spontaneous oscillatory contraction of sarcomeres in skeletal myofibrils. *J. Muscle Res. Cell Motil.* 9:111–9.

Olsson, M. C., J. R. Patel, D. P. Fitzsimons et al. 2004. Basal myosin light chain phosphorylation is a determinant of Ca^{2+} sensitivity of force and activation dependence of the kinetics of myocardial force development. *Am. J. Physiol.* 287:H2712–8.

Oosawa, F. 1977. Actin-actin bond strength and the conformational change of F-actin. *Biorheology* 14:11–9.

Oosawa, F. 2000. The loose coupling mechanism in molecular machines of living cells. *Genes Cells* 5:9–16.

Oosawa, F., and S. Hayashi. 1986. The loose coupling mechanism in molecular machines of living cells. *Adv. Biophys.* 22:151–83.

Oplatka, A. 1972. On the mechanochemistry of muscular contraction. *J. Theor. Biol.* 34:378–403.

Oplatka, A. 1989. Changes in the hydration shell of actomyosin are obligatory for tension generation and movement. *Prog. Clin. Biol. Res.* 315:45–9.

Oplatka, A. 1994. The role of water in the mechanism of muscular contraction. *FEBS Lett.* 355:1–3.

Oplatka, A. 1997. Critical review of the swinging crossbridge theory and the cardinal active role of water in muscle contraction. *Crit. Rev. Biochem. Mol. Biol.* 32:307–60.

Oplatka, A. 2005. The simultaneous collapse of both the swinging crossbridge theory of muscle and the in vitro motility essays. *Cell Mol. Biol.* 51:753–66.

Oplatka, A., and R. Tirosh. 1973. Active streaming in actomyosin solutions. *Biochim. Biophys. Acta* 305:684–8.

Oplatka, A., H. Gadasi, and J. Borejdo. 1974. The contraction of 'ghost' myofibrils and glycerinated muscle fibres irrigated with heavy meromyosin subfragment-1. *Biochem. Biophys. Res. Commun.* 58:905–12.

Oplatka, A., H. Gadasi, R. Tirosh et al. 1977. Demonstration of mechanochemical coupling in systems containing actin, ATP and non-aggregating active myosin derivatives. *J. Mechanochem. Cell Motil.* 2:295–306.

Ostap, E. M., V. A. Barnett, and D. D. Thomas. 1995. Resolution of three structural states of spin-labeled myosin in contracting muscle. *Biophys. J.* 69:177–88.

Overbeek, J. Th. G. 1956. The Donnan equilibrium. *Prog. Biophys. Biophys. Chem.* 6:57–88.

Page, S. G., and H. E. Huxley. 1963. Filament lengths in striated muscle. *J. Cell Biol.* 19:369–90.

Palmer, B. M., T. Suzuki, Y. Wang et al. 2007. Two-state model of actin-myosin attachment-detachment predicts C-process of sinusoidal analysis. *Biophys. J.* 93:760–9.

Palmer, L. S., A. Cunliff, and J. M. Hough. 1952. Dielectric constant of water film. *Nature* 170:796.

Parsegian, V. A. 1973. Long-range physical forces in the biological milieu. *Annu. Rev. Biophys. Bioeng.* 2:221–5.

Parsegian, V. A. 1975. Long range van der Waals forces. In *Physical chemistry: Enriching topics from colloid and surface science*, eds. H. van Olphen, and K. J. Mysels, 27–72. UPAC Commission 1.6, Colloid and Surface Chemistry, Theorex.

Pate, E., and R. Cooke. 1989. A model of crossbridge action: The effects of ATP, ADP and Pi. *J. Muscle Res. Cell Motil.* 10:181–96.

Pate, E., H. D. White, and R. Cooke. 1993. Determination of the myosin step size from mechanical and kinetic data. *Proc. Natl. Acad. Sci. U.S.A.* 90:2451–5.

Pate, E., N. Naber, M. Matsuka et al. 1997. Opening of the myosin nucleoside triphosphate binding domain during the ATPase cycle. *Biochemistry* 36:551–67.

Pemrick, S. M., and C. Edwards. 1974. Differences in the charge distribution of glycerol-extracted muscle fibers in rigor, relaxation, and contraction. *J. Gen. Physiol.* 64:551–67.

Pepe, F. A. 1971. Structure of the myosin filament of striated muscle. *Prog. Biophys. Mol. Biol.* 22:75–96.

Perry, S. V. 1956. Relation between chemical and contractile function and structure of skeletal muscle cell. *Physiol. Rev.* 36:1–76.

Perry, S. V. 1967. The structure and interactions of myosin. *Prog. Biophys. Mol. Biol.* 17:325–81.

Pézolet, M., M. Pigeon-Gosselin, R. Savoie et al. 1978. Laser Raman investigation of single muscle fibres on the state of water in muscle tissue. *Biochim. Biophys. Acta* 544:394–406.

Phan, B., P. Cheung, C. J. Miller et al. 1994. Extensively methylated myosin subfragment-1: Examination of local structure, interactions with nucleotides and actin, and ligand-ligand conformational changes. *Biochemistry* 33:11286–95.

Piazzesi, G., and V. Lombardi. 1995. A cross-bridge model that is able to explain mechanical and energetic properties of shortening muscle. *Biophys. J.* 68:1966–79.

Piazzesi, G., M. Reconditi, I. Dobbie et al. 1999. Changes in conformation of myosin heads during the development of isometric contraction and rapid shortening in single frog muscle fibres. *J. Physiol. Lond.* 514:305–12.

Piazzesi, G., L. Lucii, and V. Lombardi. 2002a. The size and speed of the working stroke of muscle myosin and its dependence on the force. *J. Physiol. Lond.* 545:145–51.

Piazzesi, G., M. Reconditi, M. Linari et al. 2002b. Mechanism of force generation by myosin heads in skeletal muscle. *Nature* 415:659–62.

Piazzesi, G., M. Reconditi, N. Koubassova et al. 2003. Temperature dependence of the force-generating process in single fibres from frog skeletal muscle. *J. Physiol. Lond.* 549:93–106.

Piazzesi, G., M. Reconditi, M. Linari et al. 2007. Skeletal muscle performance determined by modulation of number of myosin motors rather than motor force or stroke size. *Cell* 131:784–95.

Pinset-Härström, I. 1985. MgATP specifically controls in vitro self-assembly of vertebrate skeletal myosin in the physiological pH range. *J. Mol. Biol.* 182:159–72.

Pinset-Härström, I., and J. Truffy. 1979. Effect of adenosine triphosphate, inorganic phosphate and divalent cations on the size and structure of synthetic myosin filaments. An electron microscope study. *J. Mol. Biol.* 134:179–88.

Pinset-Härström, I., J. E. Morel, and A. M. Bardin. 1975. Observation par microscopie électronique des filaments de myosine 'in vitro'. *Bull. Inf. Sci. Techn. CEA (Saclay)* 207:45–8.

Podlubnaya, Z. A., D. I. Levitsky, I. A. Shuvakova et al. 1987. Ordered assemblies of myosin minifilaments. *J. Mol. Biol.* 196:729–32.

Podolsky, R. J. 1960. Kinetics of muscular contraction: The approach to the steady-state. *Nature* 188:666–8.

Podolsky, R. J., and A. C. Nolan. 1973. Muscle contraction transients, cross-bridge kinetics, and the Fenn effect. *Cold Spring Harb. Symp. Quant. Biol.* 37:661–8.

Podolsky, R. J., T. St. Onge, L. C. Yu et al. 1976. X-ray diffraction of actively shortening muscle. *Proc. Natl. Acad. Sci. U.S.A.* 73:813–7.

Pollack, G. H. 1983. The cross-bridge theory. *Physiol. Rev.* 63:1049–113.

Pollack, G. H. 1984. A proposed mechanism of contraction in which stepwise shortening is a basic feature. In *Cross-bridge mechanisms in muscle contraction*, eds. G. H. Pollack, and H. Sugi, 787–92. Plenum Press, New York, USA.

Pollack, G. H. 1986. Quantal mechanism in cardiac contraction. *Circ. Res.* 59:1–8.

Pollack, G. H. 1988. Cross-bridge theory: Do not disturb! *J. Mol. Cell Cardiol.* 20:563–70.

Pollack, G. H. 1990. *Muscles and molecules: Uncovering the principles of biological motion.* Ebner and Sons, Seattle.

Pollack, G. H. 1995. Muscle contraction mechanism: Are alternative engines gathering steam? *Cardiovasc. Res.* 29:737–46.

Pollack, G. H. 1996. Phase transition and the molecular mechanism of contraction. *Biophys. Chem.* 59:315–28.

Pollack, G. H. 2001. *Cells, gels and the engines of life. A new unifying approach to cell function.* Ebner and Sons, Seattle, USA.

Pollack, G. H. 2003. The role of aqueous interfaces in the cell. *Adv. Colloid Interface Sci.* 103:173–96.

Pollack, G. H., T. Iwazumi, H. E. D. J. ter Keurs et al. 1977. Sarcomere shortening in striated muscle occurs in stepwise fashion. *Nature* 168:757–9.

Pollack, G. H., H. L. M. Granzier, A. Mattiazzi et al. 1988. Pauses, steps, and the mechanism of contraction. *Adv. Exp. Med. Biol.* 226:617–42.

Pollack, G. H., A. Horowitz, M. Wussling et al. 1993. Shortening-induced tension enhancement: Implication for length–tension relations. *Adv. Exp. Med. Biol.* 332:679–88.

Pollack, G. H., F. A. Blyahkman, X. Liu et al. 2005. Sarcomere dynamics, stepwise shortening and the nature of contraction. *Adv. Exp. Med. Biol.* 332:679–88.

Pollard, T. D. 2000. Reflections on a quarter century of research on contractile systems. *Trends Biochem. Sci.* 22:607–11.

Pollard, T. D., D. Bhandari, P. Maupin et al. 1993. Direct visualization by electron microscopy of the weakly bound intermediates in the actomyosin adenosine triphosphate cycle. *Biophys. J.* 64:454–71.

Portzehl, H., P. Zaoralek, and J. Gaudin. 1969a. Association and dissociation of the thick and thin filaments within the myofibrils in conditions of contraction and relaxation. *Biochim. Biophys. Acta* 189:429–39.

Portzehl, H., P. Zaoralek, and J. Gaudin. 1969b. The activation by Ca²⁺ of the ATPase of extracted muscle fibrils with variation of ionic strength, pH and concentration of MgATP. *Biochim. Biophys. Acta* 189:440–8.

Potma, E. J., and G. J. M. Stienen. 1996. Increase in ATP consumption during shortening in skinned fibres from rabbit psoas muscle: Effects of inorganic phosphate. *J. Physiol. Lond.* 496:1–12.

Potma, E. J., G. J. M. Stienen, J. P. F. Barends et al. 1994. Myofibrillar ATPase activity and mechanical performance of skinned fibres from rabbit psoas muscle. *J. Physiol. Lond.* 474:303–17.

Poulsen, F. R., J. Lowy, P. M. Cooke et al. 1987. Diffuse X-ray scatter from myosin heads in oriented synthetic filaments. *Biophys. J.* 51:959–67.

Prochniewicz, E., Q. Zhung, P. A. Janmey et al. 1996. Cooperativity in F-actin: Binding of gelsolin at the barbed end affects structure and dynamics of the whole filament. *J. Mol. Biol.* 260:756–66.

Pucéat, M., O. Clément, F. Scamps et al. 1991. Extracellular ATP-induced acidification leads to cytosolic calcium transient rise in single rat cardiac myocytes. *Biochem. J.* 274:761–72.

Purcell, T. J., C. Morris, J. A. Spudich et al. 2002. Role of the lever arm in the processive stepping of myosin V. *Proc. Natl. Acad. Sci. U.S.A.* 99:14159–64.

Rall, J. A. 1982. Sense and non-sense about the Fenn effect. *Am. J. Physiol.* 242:H1–5.

Rall, J. A., and R. C. Woledge. 1990. Influence of temperature on mechanics and energetics of muscle contraction. *Am. J. Physiol.* 259:R197–203.

Rall, J. A., E. Homsher, A. Wallner et al. 1976. A temporal dissociation of energy liberation and high energy phosphate splitting during shortening in frog skeletal muscle. *J. Gen. Physiol.* 18:13–27.

Ramaswamy, K. S., M. L. Palmer, van der Meulen et al. 2011. Lateral transmission of force is impaired in skeletal muscles of dystrophic mice and very old rabbits. *J. Physiol. Lond.* 589:1195–208.

Ramsey, R. W., and S. F. Street. 1940. The isometric length–tension diagram of isolated skeletal muscle fibres of the frog. *J. Cell Comp. Physiol.* 15:11–34.

Ranatunga, R. W. 1994. Thermal stress and Ca-independent contractile activation of mammalian skeletal muscle fibers at high temperature. *Biophys. J.* 66:559–85.

Rapoport, S. I. 1972. Mechanical properties of the sarcolemma and myoplasm in frog muscle as a function of sarcomere length. *J. Gen. Physiol.* 59:550–85.

Rapoport, S. I. 1973. The anisotropic elastic properties of the sarcolemma of the frog *semitendinosus* muscle fiber. *Biophys. J.* 13:14–36.

Rassier, D. E., W. Herzog, and G. H. Pollack. 2003a. Dynamics of individual sarcomeres during and after stretch in activated single myofibrils. *Proc. Biol. Sci.* 270:1735–40.

Rassier, D. E., W. Herzog, and G. H. Pollack. 2003b. Stretch-induced force enhancement and stability of skeletal muscle myofibrils. *Adv. Exp. Med. Biol.* 558:501–15.

Ratz, P. H., and J. E. Speich. 2010. Evidence that actomyosin crossbridges contribute the 'passive' tension in detrusor smooth muscle. *Am. J. Physiol.* 298:F1424–35.

Rayment, I., H. M. Holden, M. Whitaker et al. 1993a. Structure of the actin-myosin complex and its implications for muscle contraction. *Science* 261:58–65.

Rayment, I., W. Rypniewski, K. Schmidt-Bäse et al. 1993b. Three-dimensional structure of myosin subfragment-1: A molecular motor. *Science* 261:50–8.

Reconditi, M., M. Linari, C. Lucii et al. 2004. The myosin motor in muscle generates a smaller and shorter working stroke at higher load. *Nature* 428:578–81.

Redowicz, M. J., L. Szilágyi, H. Strzeleca-Golaszeska. 1987. Conformational transition in the myosin head induced by temperature, nucleotide and actin: Studies on subfragment-1 of myosin from rabbit and frog fast skeletal muscle with a limited proteolytic method. *Eur. J. Biochem.* 165:353–62.

Reedy, M. C. 2000. Visualizing myosin's power stroke in muscle contraction. *J. Cell Sci.* 113:3551–62.

Reedy, M. K., K. C. Holmes, and R. T. Tregear. 1965. Induced changes in orientation of the cross-bridges of glycerinated insect flight muscle. *Nature* 207:1276–80.

Reggiani, C., R. Bottinelli, and G. J. Stienen. 2000. Sarcomeric myosin isoforms: Fine tuning of a molecular motor. *News Physiol. Sci.* 15:26–33.

Regini, J. W., and G. F. Elliott. 2001. The effect of temperature on the Donnan potentials in biological polyelectrolyte gels: Cornea and striated muscle. *Int. J. Biol. Macromol.* 28:245–55.

Regnier, M., C. Morris, and E. Homsher. 1995. Regulation of cross-bridge transition from a weakly to strongly bound state in skinned rabbit muscle fibers. *Am. J. Physiol.* 38:C1532–9.

Reuben, J. P., P. W. Brandt, M. Berman et al. 1971. Regulation of tension in the skinned crayfish muscle fiber. I. Contraction and relaxation in the absence of Ca (pCa> 9). *J. Gen. Physiol.* 57:385–407.

Rimm, D. L., J. H. Sinard, and T. D. Pollard. 1989. Location of the head–tail junction of myosin. *J. Cell Biol.* 108:1783–9.

Rizzino, A. A., W. W. Barouch, D. Eisenberg et al. 1970. Actin-heavy meromyosin binding. Determination of binding stoichiometry from adenosine triphosphatase kinetic measurements. *Biochemistry* 9:2402–8.

Robinson, J. D. 1989. Solvent effects on substrate and phosphate interactions with the $(Na^+ + K^+)$-ATPase. *Biochim. Biophys. Acta* 994:95–103.

Rome, E. 1967. Light and X-ray diffraction studies of the filament lattice of glycerol-extracted rabbit psoas muscle. *J. Mol. Biol.* 27:591–602.

Rome, E. 1968. X-ray diffraction studies of the filament lattice of striated muscle in various bathing media. *J. Mol. Biol.* 32:331–44.

Rome, E. 1972. Relaxation of glycerinated muscle: X-ray diffraction studies. *J. Mol. Biol.* 65:331–45.

Rome, L. C., C. Cook, D. A. Syme et al. 1999. Trading force for speed: Why superfast crossbridge kinetics leads to superlow forces? *Proc. Natl. Acad. Sci. U.S.A.* 96:5828–31.

Rorschach, H. E., D. C. Cheung, C. F. Hazlewood et al. 1973. The diffusion of water in striated muscle. *Ann. N.Y. Acad. Sci. U.S.A.* 204:444–52.

Rowe, R. W. 1971. Ultrastructure of the Z line of skeletal muscle fibers. *J. Cell Biol.* 51:674–85.

Rüdel, R., and F. Zite-Ferenczi. 1979. Do laser diffraction studies on striated muscles indicate stepwise shortening? *Nature* 278:573–5.

Sabido-David, C., S. C. Hopkins, L. D. Saraswat et al. 1998. Orientation changes of fluorescence probes at five sites on the myosin regulatory light chain during contraction of single skeletal muscle fibres. *J. Mol. Biol.* 279:387–402.

Saito, K., T. Aoki, and T. Yanagida. 1994. Movement of single myosin filaments and myosin step size on actin filament suspended in solution by a laser trap. *Biophys. J* 66:769–77.

Sako, Y., and T. Yanagida. 2003. Single-molecule visualization in cell biology. *Nat. Rev. Mol. Cell Biol.* Supp:SS1–5.

Sandow, A. 1944. Studies of the latent period of muscular contraction. Method. General properties of latency relaxation. *J. Cell Comp. Physiol.* 24:221–56.

Sandow, A. 1966. Latency relaxation: A brief analytical review. *Med. Coll. Va. Q.* 2:82–9.

Sandow, A. 1970. Skeletal muscle. *Annu. Rev. Physiol.* 32:87–138.

Sato, K., M. Ohtaki, Y. Shimamoto et al. 2011. A theory on auto-oscillations and contraction in striated muscle. *Prog. Biophys. Mol. Biol.* 105:199–207.

Savel'ov, V. B. 1996. Effect of storage and stimulation of a muscle on the intensity of the Z-reflection of its equatorial diffraction pattern. *Biophysics* 31:785–7. (translated from *Biofizika (Russian)* 31:720–1).

Schaub, M. C., J. G. Watterson, and P. G. Wasser. 1977. Evidence for head-to-head interactions in myosin from cardiac and skeletal muscle. *Basic Res. Cardiol.* 72:124–32.

Schiereck, P. E. 1982. Stretch-induced activation and deactivation in intact left ventricle. In *Cardiovascular system dynamics: Models and measurements*, ed. T. Kenner, 69–76. Springer, New York, USA.

Schiereck, P. E., E. L. Debeer, R. L. Gruneman et al. 1992. Tetragonal deformation of the hexagonal myofilament matrix in single skinned skeletal muscle fibres owing to change in sarcomere length. *J. Muscle Res. Cell Motil.* 13:573–80.

Schnapp, B. J. 1995. Two heads are better than one. *Nature* 373:655–6.

Schoenberg, M. 1980a. Geometrical factors influencing muscle force development. I. The effect of filament spacing upon axial forces. *Biophys. J.* 30:51–68.

Schoenberg, M. 1980b. Geometrical factors influencing muscle force development. II. Radial forces. *Biophys. J.* 30:69–78.

Schoenberg, M. 1985. Equilibrium muscle cross-bridge behaviour. Theoretical considerations. *Biophys. J.* 48:467–75.

Schoenberg, M. 1988. Characterization of the myosin adenosine triphosphate (M.ATP) cross-bridge in rabbit and frog skeletal muscle fibers. *Biophys. J.* 54:135–48.

Schutt, C. E., and U. Lindberg. 1992. Actin as the generator of tension during muscle contraction. *Proc. Natl. Acad. Sci. U.S.A.* 89:319–23.

Schutt, C. R., and U. Lindberg. 1993. A new perspective on muscle contraction. *FEBS Lett.* 325:59–62.

Schwienbacher, C., F. Magri, G. Trombetta et al. 1995. Osmotic properties of the calcium regulated actin filaments. *Biochemistry* 34:1090–5.

Scordilis, S. P., H. Tedeschi, and C. Edwards. 1975. Donnan potential of rabbit skeletal muscle myofibrils. I. Electrofluorometric detection of potential. *Proc. Natl. Acad. Sci. U.S.A.* 72:1325–9.

Sellers, J. R. 2004. Fifty years of contractility research post sliding filament hypothesis. *J. Muscle Res. Cell Motil.* 25:475–82.

Sellers, J. R., and C. Veigel. 2006. Walking with myosin V. *Curr. Opin. Cell Biol.* 18:68–73.

Seymour, J., and E. J. O'Brien. 1985. Structure of myosin decorated actin filaments and natural thin filaments. *J. Muscle Res. Cell Motil.* 6:725–55.

Shapiro, P. J., K. Tawada, and R. J. Podolsky. 1979. X-ray diffraction of skinned muscle fibers. *Biophys. J.* 25:18a.

Shear, D. B. 1970. Electrostatic forces in muscle contraction. *J. Theor. Biol.* 28:531–46.

Sheetz, M. P., and J. A. Spudich. 1983. Movement of myosin-coated fluorecent beads on actin cables in vitro. *Nature* 303:31–5.

Shepard, A. A., D. Dumka, I. Akopova et al. 2004. Simultaneous measurement of rotation of myosin, actin and ADP in contracting skeletal muscle fiber. *J. Muscle Res. Cell Motil.* 25:549–57.

Shimokawa, T., S. Sato, A. Buonocore et al. 2003. A chemically driven fluctuating ratchet model for actomyosin interaction. *Biosystems* 71:179–87.

Shiner, J. S., and J. Solaro. 1982. Activation of the thin-filament-regulated by calcium: Considerations based on nearest-neighbor lattice statistics. *Proc. Natl. Acad. Sci. U.S.A.* 79:4637–41.

Shriver, J. W., and B. D. Sykes. 1981. Energetics and kinetics of the interconversion of two myosin subfragment-1-adenosine 5′diphopsphate complexes as viewed by phosphorus-31 nuclear magnetic resonance. *Biochemistry* 20:6357–62.

Shu, Y. G., and H. L. Shi. 2006. Mechanochemical coupling of molecular motor. *AAPPS Bull.* (China) 6:8–10.

Siegman, M. J., T. M. Butler, S. V. Mooers et al. 1976. Crossbridge attachment, resistance to stretch, and viscoelasticity in resting mammalian smooth muscle. *Science* 191:383–5.

Simmons, R. M. 1983. What can the muscle biochemist tell to the muscle physiologist? *Biochem. Soc. Trans.* 11(Pt. 2):149–50.

Simmons, R. M. 1991. Moving story. Muscle and molecules. Uncovering the principles of biological motion. By Gerald Pollack (Book Review). *Nature* 351:452.

Simmons, R. M. 1992a. Testing time for muscle. *Curr. Biol.* 2:373–5.

Simmons, R. M. 1992b. *Muscular contraction.* Cambridge Univ. Press, UK.

Simmons, R. M. 1996. Molecular motors: Single-molecule mechanics. *Curr. Biol.* 6:392–4.

Simmons, R. M., and B. R. Jewell. 1974. Mechanics and models of muscular contraction. *Recent Adv. Physiol.* 9:87–147.

Simmons, R. M., and T. L. Hill. 1976. Definition of free energy levels in biochemical reactions. *Nature* 263:615–8.

Simmons, R. M., J. T. Chen, S. Chen et al. 1996. Quantitative measurements of force and displacement using an optical trap. *Biophys. J.* 70:1813–22.

Simmons, R. M., J. T. Finer, S. Chu et al. 1993. Optical tweezers: Glasperlenspiel II. *Curr. Biol.* 3:303–11.

Siththanandan, V. B., J. L. Donnelly, and M. A. Ferenczi. 2006. Effect of strain on actomyosin kinetics in isometric muscle fibers. *Biophys. J.* 90:3653–65.

Sjöström, M., and J. M. Squire. 1977. Fine structure of the A-band in cryosection. The structure of the A-band of human skeletal muscle fibres from ultrathin cryosections negatively stained. *J. Mol. Biol.* 102:49–68.

Sleep, J. A., and S. J. Smith. 1981. Actomyosin ATPase and muscle contraction. *Curr. Top. Bioenerg.* 11:239–86.

Small, J. V. 1988. Myosin filaments on the move. *Nature* 331:568–9.

Smith, D. A., and M. A. Geeves. 1995. Strain-dependent cross-bridge cycle for muscle. *Biophys. J.* 69:523–37.

Smith, D. A., and S. M. Mijailovich. 2008. Toward a unified theory of muscle contraction. II. Predictions with the main-field approximation. *Ann. Biomed. Eng.* 36:1353–73.

Smith, D. A., and D. G. Stephenson. 2009. The mechanism of spontaneous oscillatory contractions in skeletal muscle. *Biophys. J.* 96:3682–91.

Smith, N. P., C. J. Barclay, and D. S. Loiselle. 2005. The efficiency of muscle contraction. *Prog. Biophys. Mol. Biol.* 88:1–58.

Smith, D. A., M. A. Geeves, J. Sleep et al. 2008. Towards a unified theory of muscle contraction. I. Foundations. *Ann. Biomed. Eng.* 36:1624–40.

Smith, L. C., M. Tainter, M. Regnier et al. 2009. Cooperative crossbridge activation of thin filaments contributes to Frank–Starling mechanism in cardiac muscle. *Biophys. J.* 96:3692–702.

Speich, J. E., K. Quintero, C. Dosier et al. 2006. A mechanical model for adjustable passive stiffness in rabbit detrusor. *J. Appl. Physiol.* 101:1645–55.

Speich, J. E., A. M. Almari, H. Bhatia et al. 2009. Adaptation of the length-active tension relationship in rabbit destrusor. *Am. J. Physiol.* 297:F1119–28.

Spencer, M., and C. R. Worthington. 1960. A hypothesis of contraction in striated muscle. *Nature* 187:388–91.

Spudich, J. A. 1994. How molecular motors work. *Nature* 372:515–8.

Spudich, J. A. 2001. The myosin swinging cross-bridge model. *Nat. Rev. Mol. Cell Biol.* 2:387–92.

Spudich, J. A. 2011a. Molecular motors. Beauty in complexity. *Science* 331:1143–4.

Spudich, J. A. 2011b. Molecular motors: Fifty years of interdisciplinary research. *Mol. Biol. Cell* 22:3436–9.

Spudich, J. A., and S. Sivaramakrishnan. 2010. Myosin VI: An innovative motor that challenged the swinging lever arm hypothesis. *Nat. Rev. Med. Cell Biol.* 11:128–37.

Spudich, J. A., J. Finer, R. M. Simmons et al. 1995. Myosin structure and function. *Cold Spring Harb. Symp. Quant. Biol.* 60:783–91.

Spudich, J. A., S. E. Rice, R. S. Rock et al. 2011. Optical traps to study properties of molecular motors. *Cold Spring Harb. Protoc.*

Squire, J. M. 1971. General model for the structure of all myosin-containing filaments. *Nature* 233:457–62.

Squire, J. M. 1972. General model of myosin filament structure. II. Myosin filaments and cross-bridges interactions in vertebrate striated and insect flight muscles. *J. Mol. Biol.* 72:125–38.

Squire, J. M. 1973. General model of myosin filament structure. III. Molecular packing arrangements in myosin filaments. *J. Mol. Biol.* 77:291–323.

Squire, J. M. 1975. Muscle filament structure and muscle contraction. *Annu. Rev. Biophys. Bioeng.* 4:137–63.

Squire, J. M. 1981. *The structural basis of muscular contraction.* Plenum Press.

Squire, J. M. 1983. Molecular mechanism in muscular contraction. *Trends Neurosci.* 6:409–13.

Squire, J. M. 1989. In pursuit of myosin function. *Cell Regul.* 1:1–11.

Squire, J. M. 1994. The actomyosin interaction – shedding light on structural events: 'Plus ça change, plus c'est la même chose'. *J. Muscle Res. Cell Motil.* 15:227–31.

Squire, J. M. 1997. Architecture and function in the muscle sarcomere. *Curr. Opin. Struct. Biol.* 7:247–57.

Squire, J. M. 2011. *Muscle contraction.* eLS. Wiley & Sons Ltd., Chichester, USA.

Squire, J. M., R. J. Podolsky, J. S. Barry et al. 1991. X-ray diffraction testing for weak-binding cross-bridges in relaxed bony fish muscle fibers at low ionic strength. *J. Struct. Biol.* 107:221–6.

Squire, J. M., H. A. Al-Khayat, C. Knupp et al. 2005. Molecular architecture in muscle contractile assemblies. *Adv. Protein Chem.* 71:17–87.

Steffen, W., D. Smith, R. M. Simmons et al. 2001. Mapping the actin filament with myosin. *Proc. Natl. Acad. Sci. U.S.A.* 98:14949–54.

Stehle, R., and B. Brenner. 2000. Cross-bridge attachment during high-speed shortening of skinned fibers of the rabbit psoas muscle: Implications for cross-bridge action during maximum velocity of filament sliding. *Biophys. J.* 78:1458–73.

Stehle, R., C. Lionne, F. Travers et al. 2000. Kinetics of the initial steps of rabbit psoas myofibrillar ATPases studied by tryptophan and pyrene fluorescence stepped-flow and rapid flow-quench. Evidence that cross-bridges detachment is slower than ATP binding. *Biochemistry* 39:7508–20.

Stein, L. A., and M. P. White. 1987. Biochemical kinetics of porcine cardiac subfragment 1. *Circ. Res.* 60:39–49.

Stephenson, D. G., and D. A. Williams. 1982. Effects of sarcomere length on the force–pCa relation in fast- and slow-twitch skinned muscle fibres from the rat. *J. Physiol. Lond.* 333:637–53.

Stephenson, D. G., and I. R. Wendt. 1984. Length-dependence of changes in sarcoplasmic calcium concentration and myofibrillar calcium sensitivity in skinned muscle fibres. *J. Muscle Res. Cell Motil.* 5:243–72.

Stephenson, D. G., and D. A. Williams. 1985. Temperature-dependent calcium sensitivity changes in skinned muscle fibres of rat and toad. *J. Physiol. Lond.* 360:1–12.

Stephenson, D. G., A. W. Stewart, and G. J. Wilson. 1989. Dissociation of force from myofibrillar MgATPase and stiffness at short sarcomere length in rat and toad skeletal muscle. *J. Physiol. Lond.* 410:351–66.

Stephenson, D. G., I. R. Wendt, and Q. G. Forrest. 1981. Non-uniform ion distribution and electrical potentials in sarcoplasmic region of skeletal muscle fibres. *Nature* 34:690–2.

Stewart, M., and R. W. Kensler. 1986a. Arrangement of myosin heads in relaxed thick filaments from *Limulus* and scorpin muscle. *J. Cell Biol.* 101:402–11.

Stewart, M., and R. W. Kensler. 1986b. Arrangement of myosin heads in relaxed thick filaments from frog skeletal muscle. *J. Mol. Biol.* 192:831–51.

Stewart, M., R. W. Kensler, and R. J. C. Levine. 1985. Three-dimensional reconstruction of thick filaments from *Limulus* and scorpion muscle. *J. Cell Biol.* 101:402–11.

Stienen, G. J. M., and J. Blangé. 1985. Tension responses to rapid length changes in skinned muscle fibres of the frog. *Pflügers Arch. Eur. J. Physiol.* 405:5–11.

Stienen, G. J. M., J. L. Kiers, R. Bottinelli et al. 1996. Myofibrillar ATPase activity in skinned human skeletal muscle fibres: Fibre type and temperature dependence. *J. Physiol. Lond.* 493:299–307.

Street, S. F. 1983. Lateral transmission of tension in frog myofibers: A myofibrillar network and transverse cytoskeletal connections are possible transmitters. *J. Cell. Physiol.* 114:346–64.

Sugi, H., and Tsuchiya. 1981a. Isotonic velocity transients in frog muscle fibres following quick changes in load. *J. Physiol. Lond.* 319:219–38.

Sugi, H., and Tsuchiya. 1981b. Enhancement of mechanical performance in frog muscle fibres after quick increases in load. *J. Physiol. Lond.* 319:239–52.

Sugi, H., H. Iwamoto, T. Akimoto et al. 2003. High mechanical efficiency of the cross-bridge power-stroke in skeletal muscle. *J. Exp. Biol.* 206:1201–6.

Sukharova, M., J. Morrissette, and R. Coronado. 1994. Mechanism of chloride-dependent release of

Ca^{2+} in the sarcoplasmic reticulum of rabbit skeletal muscle. *Biophys. J.* 67:751–65.

Sun, Y. B., K. Hilber, and M. Irving. 2001. Effect of active shortening on the rate of ATP utilisation by rabbit psoas muscle fibres. *J. Physiol. Lond.* 531:781–91.

Sussman, M. V., and L. Chin. 1966. Liquid water in frozen tissue: Study by nuclear magnetic resonance. *Science* 151:324–5.

Sutoh, K. 1983. Mapping of actin-binding sites on the heavy chain of myosin subfragment 1. *Biochemistry* 22:1579–85.

Suzuki, S., and G. H. Pollack. 1986. Bridgelike interconnections between thick filaments in stretched skeleton muscle fibers observed by the freeze-fracture method. *J. Cell Biol.* 102:1093–8.

Sweeney, H. L., and A. Houdusse. 2010a. Structural and functional insights into the myosin motor mechanisms. *Annu. Rev. Biophys.* 39:539–57.

Sweeney, H. L., and A. Houdusse. 2010b. Myosin VI rewrites the rules for myosin motors. *Cell* 141:573–82.

Szabó, J., A. Varga, B. Flatcher et al. 2008. Communication between the nucleotide site and the main molecular hinge of 3-phosphoglycerate kinase. *Biochemistry* 47:6735–44.

Szent-Györgyi, A. 1947. *Chemistry of muscle contraction.* Academic Press, New York, USA.

Szent-Györgyi, A. 1949. Free energy reactions and the contraction of actomyosin. *Biol. Bull. Woods Hole* 96:141–61.

Szent-Györgyi, A. 1951. *Chemistry of muscular contraction.* Academic Press, New York, USA.

Szent-Györgyi, A. 1953. *Chemistry and physiology of contraction in body and heart muscle.* Academic Press, New York, USA.

Szigeti, G. P., J. Almássy, M. Sztretye et al. 2007. Alterations in the calcium homeostasis of skeletal muscle from post-myocardial infarcted rats. *Pflügers Arch. Eur. J. Physiol.* 455:541–53.

Szilágyi, A. N., M. Ghosh, E. Garman et al. 2001. A 1.8 Å resolution structure of pig 3-phosphoglycerate kinase with bound MgADP and 3-phophoglycerate in open conformation: New insight into the role of the nucleotide in domain closure. *J. Mol. Biol.* 306:400–11.

Takahashi, K. 1978. Topography of the myosin molecule as visualized by an improved negative staining method. *J. Biochem. Tokyo* 83:905–8.

Takezawa, Y., Y. Sugimoto, and K. Wakabayashi. 1998. Extensibility of the actin and myosin filaments in various states of skeletal muscles as studied by X-ray diffraction. *Adv. Exp. Med. Biol.* 453:309–17.

Tameyasu, T. 1994. Oscillatory contraction of single sarcomeres in single myofibrils of glycerinated striated adductor muscle of scallop. *Jpn. J. Physiol.* 44:295–318.

Tameyasu, T., T. Toyoki, and H. Sugi. 1985. Non-steady motion in unloaded contraction of single frog cardiac cells. *Biophys. J.* 48:461–5.

Tanaka, H., M. Tanaka, and H. Sugi. 1979. The effect of sarcomere length and stretching on the rate of ATP splitting in glycerinated rabbit psoas muscle fibres. *J. Biochem. Tokyo* 86:1587–93.

Tanford, C. 1967. *Physical chemistry of macromolecules.* John Wiley & Sons, Chischester, USA.

Taylor, E. W. 1972. Chemistry of muscle contraction. *Annu. Rev. Biochem.* 41:577–617.

Taylor, E. W. 1979. Mechanism of actomyosin ATPase and the problem of muscle contraction. CRC Crit. *Rev. Biochem.* 6:103–64.

Taylor, E. W. 1989. Actomyosin ATPase mechanism and muscle contraction. In *Muscle energetics*, eds. R. J. Paul, G. Elzinga, and K. Yamada, 9–14. Alan R. Liss, USA.

Taylor, E. W. 1993. Molecular muscle. *Science* 261:35–6.

Taylor, K. A., and L. A. Amos. 1970. A new model for the geometry of the binding of myosin cross-bridges to muscle thin filaments. *J. Mol. Biol.* 147:297–324.

Taylor, E. W., and A. G. Weeds. 1977. Transient phase of ATP hydrolysis by myosin sub-fragment-1. *FEBS Lett.* 75:55–60.

Taylor, E. W., R. W. Lymn, and G. Moll. 1970. Myosin-product complex and its effect on the steady-state rate of nucleotide triphosphate hydrolysis. *Biochemistry* 9:2984–91.

ter Keurs, H. E. D. J., and Elzinga, G. 1981. The sarcomere length–tension relation of frog muscle: Effects of sarcomere motion and species. Presented at the VII Intern. Biophys. Congress, Mexico, 275.

ter Keurs, H. E. D. J., T. Iwazumi, and G. H. Pollack. 1978. The sarcomere length–tension relation in skeletal muscle. *J. Gen. Physiol.* 72:565–92.

Tesi, C., N. Bachouchi, T. Barman et al. 1989. Cryoenzymic studies on myosin: Transient evidence of two types of head with different ATP binding properties. *Biochimie (Paris)* 71:187–99.

Tesi, C., F. Colomo, S. Nencini et al. 2000. The effect of inorganic phosphate on force generation in single myofibrils from rabbit skeletal muscle. *Biophys. J.* 78:3081–92.

Tesi, C., F. Colomo, N. Piroddi et al. 2002. Characterization of the cross-bridge force-generating step using inorganic phosphate and BDM in myofibrils from rabbit skeletal muscle. *J. Physiol. Lond.* 541:187–99.

Thames, M. D., L. E. Teichholz, and R. J. Podolsky. 1974. Ionic strength and the contraction kinetics of skinned muscle fibers. *J. Gen. Physiol.* 63:509–30.

Thomas, D. D. 1987. Spectroscopic probes of muscle cross-bridge rotation. *Annu. Rev. Physiol.* 49:691–709.

Thomas, D. D., and R. Cooke. 1980. Orientation of spin-labeled myosin heads in glycerinated muscle fibers. *Biophys. J.* 32:891–906.

Thomas, D. D., S. Ramachandran, O. Roopnarine et al. 1995. The mechanism of force generation in myosin: A disorder-to-order transition, coupled to internal structural changes. *Biophys. J.* 68:135s–41s.

Tigyi, J., N. Kallay, A. Tigyi-Sebes et al. 1981. Distribution and function of water and ions in resting and contracted muscle. In *International cell biology*, ed. H. D. Schweiger, 925–49. Springer-Verlag, Berlin.

Tirosh, R. 1984. 1 kf/cm²: The isometric tension of muscle contraction: Implication to cross-bridge and hydraulic mechanisms. *Adv. Exp. Med. Biol.* 170:531–9.

Tirosh, R., and A. Oplatka. 1982. Active streaming against gravity in glass microcapillaries of solutions containing acto-heavy meromyosin and native tropomyosin. *J. Biochem.* 91:1435–40.

Titus, M. A. 1993. Myosins. *Curr. Opin. Cell Biol.* 5:77–81.

Toride, M., and H. Sugi. 1989. Stepwise sarcomere shortening in locally activated frog muscle fibers. *Proc. Jpn. Acad. Ser. B* 85:49–52.

Tourovskaya, A., and G. H. Pollack. 1998. Stepwise length changes during stretch in single sarcomere of single myofibrils. *Biophys. J.* 78:A153.

Toyoshima, C., and T. Wakabayashi. 1979. Three-dimensional image analysis of the complex of thin filaments and myosin molecules from skeletal muscle. I. Tilt angle of myosin subfragment 1 in the rigor complex. *J. Biochem. Tokyo* 86:1886–90.

Toyoshima, C., and T. Wakabayashi. 1985. Three-dimensional image analysis of the complex of thin filaments and myosin molecules from skeletal muscle. IV. Reconstruction from minimal- and high-dose images on the actin-tropomyosin-myosin subfragment 1 complex. *J. Biochem. Tokyo* 97:219–43.

Toyoshima, Y. Y., S. J. Kron, and J. A. Spudich. 1990. The myosin step size measurement of the unit displacement per ATP hydrolysis in an *in vitro* assay. *Proc. Natl. Acad. Sci. U.S.A.* 87:7130–4.

Toyoshima, Y. Y., S. J. Kron, E. M. McNally et al. 1987. Myosin subfragment-1 is sufficient to move actin filament in vitro. *Nature* 328:536–9.

Trentham, D. R. 1977. The adenosine triphosphatase reactions of myosin and actomyosin and their relation to energy transduction in muscle. *Biochem. Soc. Trans.* 5:5–22.

Trentham, D. R., R. G. Bardsley, and J. F. Eccleston. 1972. Elementary processes of the magnesium ion-dependent adenosine triphosphatase activity of heavy meromyosin. A transient kinetic approach to the study of kinases and adenosine triphophatases and a colorimetric inorganic phosphate assay *in situ*. *Biochem. J.* 126:635–44.

Trentham, D. R., J. F. Eccleston, and C. R. Bagshaw. 1976. Kinetic analysis of ATPase mechanisms. *Q. Rev. Biophys.* 9:217–81.

Trinick, J., and A. Elliott. 1979. Electron microscope studies of thick filaments from vertebrate skeletal muscle. *J. Mol. Biol.* 131:133–6.

Trinick, J., P. Knight, and A. Whiting. 1984. Purification and properties of native titin. *J. Mol. Biol.* 180:331–56.

Trombitás, K., L. Frey, and G. H. Pollack. 1993. Filament lengths in frog *semitendinosus and tibialis anterior* muscle fibres. *J. Muscle Res. Cell Motil.* 14:167–72.

Truong, X. T. 1974. Viscoelasticity wave propagation and rheologic properties of skeletal muscle. *Am. J. Physiol.* 236:256–65.

Tsaturyan, A. K., N. A. Koubassova, M. A. Ferenczi et al. 2005. Strong binding of myosin heads stretches and twists the actin helix. *Biophys. J.* 88:1902–10.

Tsaturyan, A. K., S. Y. Bershitsky, N. A. Koubassova et al. 2011. The fraction of myosin motors that participate in isometric contraction of rabbit muscle fibers at near-physiological temperature. *Biophys. J.* 101:404–10.

Tskhovrekova, L., J. Trinick, J. A. Sleep et al. 1997. Elasticity and unfolding of single molecules of the giant muscle protein titin. *Nature* 387:308–12.

Tsong, T. Y., T. Karr, and W. F. Harrington. 1979. Rapid helix-coil transition in the S2-region of myosin. *Proc. Natl. Acad. Sci. U.S.A.* 76:1109–13.

Tsuchiya, T. 1988. Passive interaction between sliding filaments in the osmotically compressed skinned muscle fibers of the frog. *Biophys. J.* 53:415–23.

Tuncay, E., A. Bilginoglu, N. N. Sozmen et al. 2011. Intracellular free zinc during cardiac excitation–contraction cycle: Calcium and redox dependencies. *Cardiovasc. Res.* 89:634–42.

Tyska, M. J., D. F. Dupuis, W. H. Guilford et al. 1999. Two heads of myosin are better than one for generating force and motion. *Proc. Natl. Acad. Sci. U.S.A.* 96:4402–7.

Ueno, H., and W. F. Harrington. 1981. Conformational transition in the myosin-hinge upon activation of muscle. *Proc. Natl. Acad. Sci. U.S.A.* 78:6101–5.

Ueno, H., and W. F. Harrington. 1986a. Temperature-dependence of local melting in the myosin subfragment-2 region in the rigor cross-bridges. *J. Mol. Biol.* 190:59–68.

Ueno, H., and W. F. Harrington. 1986b. Local melting in the subfragment-2 region of myosin in activated muscle and its correlation with contractile force. *J. Mol. Biol.* 190:69–82.

Ullrick, W. C. 1967. A theory of contraction for striated muscle. *J. Theor. Biol.* 190:53–69.

Umazume, Y., and N. Kasuga. 1984. Radial stiffness of frog muscle fibres in relaxed and rigor conditions. *Biophys. J.* 45:783–8.

Uyeda, T. Q. P., S. Kron, and J. A. Spudich. 1990. Myosin step size estimation from slow sliding movement of actin over low densities of heavy meromyosin. *J. Mol. Biol.* 214:699–710.

Uyeda, T. Q. P., H. M. Warrick, S. J. Kron et al. 1991a. Velocities of actin filaments occur in discrete populations when observed in an in vitro myosin motility assay. *Biophys. J.* 59(2 Part 2):186a.

Uyeda, T. Q. P., H. M. Warrick, S. J. Kron et al. 1991b. Quantized velocities at low myosin densities in an in vitro motility assay. *Nature* 352:307–11.

Uyeda, T. Q. P., P. D. Abramson, and J. A. Spudich. 1996. The neck region of the myosin motor domain acts as a lever arm to generate movement. *Proc. Natl. Acad. Sci. U.S.A.* 93:4459–64.

Vale, R. D., and R. A. Milligan. 2000. The way things move: Looking under the hood of molecular proteins. *Science* 288:88–95.

van den Hooff, H., and T. Blangé. 1984a. A 10 ms component in the tension transients of isolated intact skeletal muscle fibres of the frog. *Pflügers Arch. Eur. Physiol.* 400:137–43.

van den Hooff, H., and T. Blangé. 1984b. Superfast tension transients from intact muscle fibres. *Pflügers Arch. Eur. J. Physiol.* 400:280–5.

van Heijst, B. G., E. de Widt, U. A. van der Heide et al. 1999. The effect of length on the sensitivity to phenylephrine and calcium in intact and skinned vascular smooth muscle. *J. Muscle Res. Cell Motil.* 20:11–8.

van Heuningen, R., W. H. Rijnsberg, H. E. D. J. ter Keurs. 1982. Sarcomere length control in striated muscle. *Am. J. Physiol.* 242:H411–20.

Varga, A., C. Lionne, P. Lallemand et al. 2001. Direct kinetic evidence that lysine 215 is involved in the phospho-transfer step of human 3-phosphoglycerate kinase. *Biochemistry* 48:6998–7008.

Varga, A., B. Flatcher, E. Gráczer et al. 2005. Correlation between conformational stability of the ternary enzyme-substrate complex and domain closure of 3-phosphoglycerate kinase. *FEBS Lett.* 272:1867–85.

Varga, A., B. Flatcher, P. Konarev et al. 2006. Substrate-induced double side H-bond network as a mean of domain closure in 3-phosphoglycerate kinase. *FEBS Lett.* 580:2698–706.

Vassort, G. 2001. Adenosine 5′-triphosphate. A P_2-prurinergic agonist in the myocardium. *Physiol. Rev.* 81:767–806.

Vassort, G., M. Roulet, K. Mongo et al. 1976. Relaxation of frog myocardium. *Rec. Adv. Stud. Cardiac. Struct. Metab.* 11:143–7.

Ventura-Clapier, R., A. Garnier, and V. Veksler. 2004. Energy metabolism in heart failure. *J. Physiol. Lond.* 555:1–13.

Ventura-Clapier, R., A. Garnier, V. Veksler et al. 2011. Bioenergetics of the failing heart. *Biochim. Biophys. Acta.* 1813:1360–72.

Verwey, E. J. W., and J. Th. G. Overbeek. 1948. *Theory of stability of lyophobic colloids.* Elsevier, Amsterdam, Netherlands.

Vibert, P. J., and R. Craig. 1982. Three-dimensional reconstruction of thin filaments decorated with a Ca^{2+}-regulated myosin. *J. Mol. Biol.* 157:299–320.

Vibert, P. J., and C. Cohen. 1988. Domains, motions and regulation in the myosin head. *J. Muscle Res. Cell Motil.* 9:296–305.

Vibert, P. J., J. Lowy, J. C. Haselgrove et al. 1972. Structural changes in the actin-containing filaments of muscle. *Nat. New Biol.* 236:182–3.

Vigoreaux, J. O. 1994. The muscle Z band: Lessons in stress management. *J. Muscle Res. Cell Motil.* 15:237–55.

Villaz, M., M. Ronjat, M. Garrigos et al. 1987. The remotor muscle of the lobster antenna: Sarcoplasmic reticulum and skinned fiber experiments. *Tissue Cell* 19:135–43.

Villaz, M., M. Robert, L. Carrier et al. 1989. G-protein dependent potentiation of calcium release from sarcoplasmic reticulum of skeletal muscle. *Cell. Sign.* 1:493–506.

Vivaudou, M. B., C. Arnoult, and M. Villaz. 1991. Skeletal muscle ATP-sensitive K^+-channels recorded from sarcolemmal blebs of split fibers: ATP inhibition is reduced by magnesium and ADP. *J. Membrane Biol.* 122:165–75.

Vol'kenstein, M. V. 1970. Physics of muscle contraction (translated from Russian). *Sov. Phys. Uspekhi* 13:269–88.

Wakabayashi, K., and C. Toyoshima. 1981. Three-dimensional image analysis of the complex of thin filaments and myosin molecules. II. The multi-domain structure of actin–myosin S1 complex. *J. Biochem. Tokyo* 90:269–88.

Wakabayashi, T., T. Akiba, K. Hirose et al. 1988. Temperature-induced change of thick filament and location of the functional sites on myosin. *Adv. Exp. Med. Biol.* 226:39–48.

Wakabayashi, K., H. Tanaka, H. Saito et al. 1991. Dynamic X-ray diffraction of skeletal muscle contraction: Structural changes of actin filaments. *Adv. Biophys.* 27:3–13.

Wakabayashi, K., M. Tokunaga, L. Kohno et al. 1992. Small-angle synchrotron X-ray scattering reveals distinct shape changes of the myosin heads during hydrolysis of ATP. *Science* 258:443–7.

Wakabayashi, T., Y. Sugimoto, H. Tanaka et al. 1994. X-ray diffraction evidence for the extensibility of actin and myosin filaments during muscle contraction. *Biophys. J.* 67:2422–35.

Wakabayashi et al. 2001. Muscle contraction mechanism: Use of X-ray synchrotron radiation. Exact reference: eLS, John Wiley & Sons Ltd., USA.

Walker, M. L., H. D. White, B. Belknap et al. 1994. Electron cryomicroscopy of actin–myosin–S1 during steady state ATP hydrolysis. *Biophys. J.* 66:1563–72.

Walker, M. L., P. Knight, and J. Trinick. 1985. Negative staining of myosin molecules. *J. Mol. Biol.* 184:535–42.

Walker, M. L., S. A. Burgess, J. K. Sellers et al. 2000. Two-headed binding of a processive myosin to F-actin. *Nature* 405:804–7.

Walliman, T., D. Turner, and H. M. Eppenberger. 1977. Localization of creatine kinase isoenzymes in myofibrils. I. Chicken skeletal muscle. *J. Cell Biol.* 75:297–317.

Wang, G., and M. Kawai. 2001. Effect of temperature on elementary steps of the cross-bridge cycle in rabbit *soleus* slow-twitch muscle fibres. *J. Physiol. Lond.* 531:219–34.

Wang, Y. P., and F. Fuchs. 1995. Osmotic compression of skinned cardiac and skeletal muscle bundles: Effects on force generation, Ca^{2+} sensitivity and Ca^2 binding. *J. Mol. Cell. Cardiol.* 27:1235–44.

Wang, H., and G. Oster. 2002. Ratchets, power strokes, and molecular motors. *App. Phys. A* 75:315–23.

Wang, G., W. Ding, and M. Kawai. 1999. Does thin filament compliance diminish the cross-bridge kinetics? A study in rabbit psoas fibres. *Biophys. J.* 76:978–84.

Warshaw, D. M. 1996. The *in vitro* motility assay: A window into the myosin molecular motor. *Science* 219:1438–41.

Warshaw, D. M., and F. S. Fay. 1983. Tension transients in single isolated smooth muscle cells. *Science* 219:1438–41.

Warshaw, D. M., W. H. Guilford, Y. Freyzon et al. 2000. The light chain binding domain of expressed smooth muscle heavy meromyosin acts as a mechanical lever. *J. Biol. Chem.* 275:37167–72.

Webb, M. R., and D. R. Trentham. 1981. The mechanism of ATP hydrolysis catalyzed by myosin and actomyosin, using rapid reaction techniques to study oxygen exchange. *J. Biol. Chem.* 256:10910–16.

Weber, A., and R. Herz. 1961. Requirement for calcium in the syneresis of myofibrils. *Biochem. Biophys. Res. Commun.* 6:364–8.

Weber, A., and R. Herz. 1963. The binding of calcium to actomyosin systems in relation to their biological activity. *J. Biol. Chem.* 238:599–605.

West, T. G., N. A. Curtin, M. A. Ferenczi et al. 2004. Actomyosin energy turnover declines while force remains constant during isometric muscle contraction. *J. Physiol. Lond.* 555:27–43.

Westerblad, H., and D. G. Allen. 1991. Changes in myoplasmic calcium concentration during fatigue in single mouse muscle fibres. *J. Gen. Physiol.* 98:615–35.

Westerblad, H., and D. G. Allen. 1992. Myoplasmic free Mg^{2+} concentration during repetitive stimulation of single fibres from mouse skeletal muscle. *J. Physiol. Lond.* 453:413–34.

Westerblad, H., and J. Lännergren. 1994. Changes of the force–velocity, isometric tension and relaxation rate during fatigue in intact single fibres of *Xenopus* skeletal muscle. *J. Muscle Res. Cell Motil.* 15:287–98.

Westerblad, H., J. A. Lee, J. Lännergren et al. 1991. Cellular mechanisms of fatigue in skeletal muscle. *Am. J. Physiol.* 261:C195–209.

White, D. C. S., and J. Thorson. 1973. The kinetics of muscle contraction. *Prog. Biophys. Mol. Biol.* 27:173–255.

White, H. D., B. Belknap, M. Walker et al. 1995. Polyethylene glycol produces large increases in the amount of myosin-S1 bound to actin during steady state ATP hydrolysis. *Biophys. J.* 68:17a.

Wiggins, P. M. 1971. Water structure as a determinant of ion distribution in living tissue. *J. Theor. Biol.* 37:131–46.

Wiggins, P. M. 1972. Intracellular pH and the structure of cell water. *J. Theor. Biol.* 37:363–71.

Wiggins, P. M. 1973. Ionic partition between surface and bulk water in a silica gel. A biological model. *Biophys. J.* 13:385–98.

Wiggins, P. M. 1990. Role of water in some biological processes. *Microbiol. Rev.* 54:432–49.

Wiggins, P. M. 1996. High and low density water in resting, active and transformed cells. *Cell Biol. Int.* 20:429–35.

Wiggins, P. M. 2001. High and low density intracellular water. *Cell Mol. Biol.* 47:735–44.

Wiggins, P. M. 2008. Life depends upon two kinds of water. *PLoS One* 3(1):e1406.

Wiggins, P. M., and B. A. McClement. 1987. Two states of water found in hydrophobic clefts: Their possible contribution to mechanisms of cation pumps and other enzymes. *Int. Rev. Cytol.* 108:240–303.

Wiggins, P. M., R. van Ryn, and D. Ormrod. 1991. Donnan membrane equilibrium is not directly applicable to distribution of ions and water in gels or cells. *Biophys. J.* 60:8–14.

Winkelman, D. A., H. Mekeel, and I. Rayment. 1985. Packing analysis of crystalline myosin subfragment 1. Implication for the size and shape of the myosin head. *J. Mol. Biol.* 181:487–501.

Woessner, D. E., and B. S. Snowden Jr. 1973. A pulsed NMR study of dynamics and ordering of water molecules in interfacial systems. *Ann. N.Y. Acad. Sci. U.S.A.* 204:113–24.

Woledge, R. C. 1971. Heat production and chemical change in muscle. *Prog. Biophys. Mol. Biol.* 22:37–74.

Woledge, R. C. 1988. Force dilemma to come in muscle contraction. *Nature* 334:655.

Woledge, R. C., and S. P. Canfield. 1971. Heat of splitting of phosphocreatine *in vivo* and *in vitro*. Paper presented at the first European Biophysis Congress, Verlag der Wiener Medizinishen Akademie, Vienna, Austria, eds. E. Broda, A. Locker, and H. Springer-Lederer, 355–9.

Woledge, R. C., and P. J. Reilly. 1988. The molar enthalpy change for the hydrolysis of phosporylcreatine under the conditions in muscle cells. *Biophys. J.* 54:97–104.

Woledge, R. C., N. A. Curtin, and E. Homsher. 1985. *Energetic aspects of muscle contraction.* Academic Press, London, UK.

Woledge, R. C., M. G. A. Wilson, J. V. Howarth et al. 1988. The energetics of work and heat production by single muscle fibres from the frog. *Adv. Exp. Med. Biol.* 226:677–88.

Woledge, R. C., C. J. Barclay, and N. A. Curtin. 2009. Temperature change as a probe of muscle cross-bridge kinetics: A review and discussion. *Proc. R. Soc. Lond. Ser. B* 276:2685–95.

Woodhead, J. L., F. Q. Zhao, R. Craig et al. 2005. Atomic model of a myosin filament in the relaxed state. *Nature* 436:1195–9.

Worthington, C. R. 1962. Conceptual model for a force–velocity relation of muscle (Hill's equation). *Nature* 193:1283–4.

Worthington, C. R. 1964. Impulsive (electrical) forces in muscle. In *Biochemistry of muscle contraction*, ed. J. Gergely, 511–9. Little, Brown, and Company, New York, USA.

Worthington, C. R., and G. F. Elliott. 1996a. Muscle contraction: The step-size distance and the impulse-time per ATP. *Int. J. Biol. Macromol.* 18:123–31.

Worthington, C. R., and G. F. Elliott. 1996b. The step-size distance in muscle contraction: Properties and estimates. *Int. J. Biol. Macromol.* 19:287–94.

Worthington, R. C., and G. F. Elliott. 2003. Muscle contraction: Energy rate equations in relation to efficiency and step-size distance. *Int. J. Biol. Macromol.* 32:149–58.

Worthington, C. R., and G. F. Elliott. 2005. Muscle contraction: A new interpretation of the transient behaviour of muscle. *Int. J. Biol. Macromol.* 32:149–58.

Wray, J. S. 1987. Structure of relaxed myosin filaments in relation to nucleotide state in vertebrate skeletal muscle. *J. Muscle Res. Cell Motil.* 8:62.

Wray, J. S., R. S. Goody, and K. C. Holmes. 1988. Towards a molecular mechanism for the cross-bridge cycle. *Adv. Exp. Med. Biol.* 226:49–59.

Xu, S., M. Kress, and H. E. Huxley. 1987. X-ray diffraction studies of the structural state of cross-bridges in skinned frog *sartorius* muscle at low ionic strength. *J. Muscle Res. Cell Motil.* 8:39–54.

Xu, S., B. Brenner, and L. C. Yu. 1993. State-dependent radial elasticity of attached cross-bridges in single skinned fibres from rabbit psoas muscle. *J. Physiol. Lond.* 465:749–65.

Xu, S., S. Malinchick, D. Gilroy et al. 1997. X-ray diffraction studies of cross-bridges weakly bound to actin in relaxed skinned fibres of rabbit psoas muscle. *Biophys. J.* 73:2292–303.

Xu, S., J. Gu, T. Rhodes et al. 1999. The M.ADP.Pi state is required for helical order in the thick filaments of skeletal muscle. *Biophys. J.* 77:2665–76.

Xu, S., J. Gu, G. Melvin et al. 2002. Structural characterization of weakly attached cross-bridges in the A*M*ATP state in permeabilised rabbit psoas muscle. *Biophys. J.* 82:2111–22.

Xu, S., G. Offer, J. Gu et al. 2003. Temperature and ligand dependence of conformation and helical order in myosin filaments. *Biochemistry* 42:390–401.

Xu, S., J. Gu, B. Belknap et al. 2006a. Structural characterization of the binding of myosin.ADP.Pi to actin in permeabilised rabbit psoas muscle. *Biophys. J.* 92:3370–82.

Xu, S., D. Martyn, J. Zaman et al. 2006b. X-ray diffraction studies of the thick filament in thin permeabilised myocardium from rabbit. *Biophys. J.* 91:3768–75.

Yagi, N. 2007. A structural origin of the latency relaxation in frog skeletal muscle. *Biophys. J.* 92:162–71.

Yagi, N., and I. Matsubara. 1984. Cross-bridge movements during slow length change of active muscle. *Biophys. J.* 45:611–4.

Yagi, N., and I. Matsubara. 1989. Structural changes in the thin filament during activation studied by X-ray diffraction of highly stretched skeletal muscle. *J. Mol. Biol.* 208:359–63.

Yagi, N., and S. Takemori. 1995. Structural changes in myosin cross-bridges during shortening of frog skeletal muscle. *J. Muscle Res. Cell Motil.* 16:57–63.

Yagi, N., M. H. Itho, H. Nakajima et al. 1977. Return of myosin heads to thick filaments after muscle contraction. *Science* 197:685–9.

Yagi, N., E. J. O'Brien, and I. Matsubara. 1981a. Behaviour of myosin projections in frog striated muscle during isometric contraction. *Adv. Physiol. Sci.* 5:137–40.

Yagi, N., E. J. O'Brien, and I. Matsubara. 1981b. Changes of thick filament structure during contraction of frog striated muscle. *Biophys. J.* 33:121–38.

Yagi, N., H. Okuyama, H. Toyota et al. 2004. Sarcomere length dependence of lattice volume and radial mass transfer of myosin cross-bridges in rat papillary muscle. *Pflügers Arch. Eur. J. Physiol.* 448:153–6.

Yagi, N., H. Iwamoto, J. Wakayama et al. 2005. Structural changes of actin-bound myosin heads after a quick length change in frog skeletal muscle. *Biophys. J.* 89:1150–64.

Yagi, N., H. Iwamoto, and K. Inoue. 2006. Structural changes of cross-bridges on transition from isometric to shortening state in frog skeletal muscle. *Biophys. J.* 91:4110–20.

Yakovenko, O., T. Blyakhman, and G. H. Pollack. 2002. Fundamental step size in single cardiac and skeletal sarcomeres. *Am. J. Physiol.* 283:C735–42.

Yamada, T., O. Abe, T. Kobayashi et al. 1993. Myofilament sliding per ATP molecule in rabbit muscle fibres using laser flash photolysis of caged ATP. *J. Physiol. Lond.* 466:229–43.

Yamamoto, K., M. Tokunaga, K. Sutoh et al. 1985. Location of the SH group of the alkali light chain on the myosin heads as revealed by electron microscopy. *J. Mol. Biol.* 183:287–90.

Yanagida, T. 1990. Loose coupling between chemical and mechanical reactions in actomyosin energy transductions. *Adv. Biophys.* 26:75–95.

Yanagida, T. 2001. Muscle in. In the article by R. D. Astumian entitled 'Making molecules into motors'. *Sci. Am.* 285:64.

Yanagida, T. 2007. Muscle contraction mechanism based on actin filament rotation. *Adv. Exp. Med. Biol.* 592:359–67.

Yanagida, T., K. Nakase, K. Nishiyama et al. 1984. Direct observation of motion of single F-actin filaments in the presence of myosin. *Nature* 307:58–60.

Yanagida, T., T. Arata, and F. Oosawa. 1985. Sliding distance of actin filament induced by a myosin crossbridge during one ATP hydrolysis cycle. *Nature* 316:366–9.

Yanagida, T., S. Esaki, A. H. Iwane et al. 2000a. Single-motor mechanics and models of myosin motor. *Phil. Trans. R. Soc. Lond. Ser. B Biol. Sci.* 355:441–7.

Yanagida, T., K. Kitamura, H. Tanaka et al. 2000b. Single molecule analysis of the actomyosin motor. *Curr. Opin. Cell Biol.* 12:20–5.

Yanagida, T., M. Ueda, T. Murata et al. 2007. Brownian motion, fluctuation and life. *Biosystems* 88:228–42.

Yang, J. T., and C. C. Wu. 1977. The shape of myosin subfragment 1. An equivalent oblate ellipsoidal model based on hydrodynamic properties. *Biochemistry* 16:5785–9.

Yang, P., T. Tameyasu, and G. H. Pollack. 1998. Stepwise dynamics of connecting-filaments in single myofibrillar sarcomeres. *Biophys. J.* 74:1673–83.

Yildiz, A., J. N. Forkey, S. A. McKinney et al. 2003. Myosin V walks hand-over-hand: Single fluorophore imaging with 1.5 nm localization. *Science* 300:2061–5.

Yoo, H., D. R. Baker, C. M. Pirie et al. 2011. Characteristics of water adjacent to hydrophilic interfaces. In *Water: The forgotten biological molecule*, eds. D. le Bihan, and H. Fukuyama, 123–36. Pan Stanford Publishing, Singapore.

Yoshizaki, K., Y. Seo, H. Nishikawa et al. 1982. Application of pulsed-gradient ^{31}P NMR on frog muscle to measure the diffusion rates of phosphorus compounds in cells. *Biophys. J.* 38:209–11.

Yount, R. G., D. Lawson, and I. Rayment. 1995. Is myosin a 'back door' enzyme? *Biophys. J.* 68:44S–7S.

Yu, L. C., and B. Brenner. 1989. Structures of actomyosin cross-bridges in relaxed and rigor muscle fibers. *Biophys. J.* 55:441–53.

Yu, L. C., R. M. Dowben, and K. Kornacker. 1970. The molecular mechanisms of force generation in striated muscle. *Proc. Natl. Acad. Sci. U.S.A.* 66:1199–205.

Zahalak, G. J. 2000. The two-state cross-bridge model of muscle is one asymptotic limit of multistate models. *J. Theor. Biol.* 204:67–42.

Zhao, T., and M. Kawai. 1993. The effect of the lattice spacing change on cross-bridge kinetics in chemically skinned rabbit psoas muscle fibers. II. Elementary steps affected by the spacing change. *Biophys. J.* 64:197–216.

Zhao, F. Q., R. Craig, and J. L. Woodhead. 2009. Head–head interaction characterizes the relaxed state of Limulus muscle myosin filaments. *J. Mol. Biol.* 385:423–31.

Zheng, J. M., and G. H. Pollack. 2003. Long range forces extending from polymer surfaces. *Phys. Rev. E. Stat. Nonlin. Soft Matter Phys.* 68:031408.

Zheng, J. M., W. C. Chin, E. Khijniak et al. 2006. Surfaces and interfacial water: Evidence that hydrophilic surfaces have long-range impact. *Adv. Colloid Interface Sci.* 127:19–27.

Zoghbi, M. E., J. L. Woodhead, R. Craig et al. 2004. Helical order in tarantula thick filaments requires the 'closed' conformation of the myosin heads. *J. Mol. Biol.* 342:1223–6.

INDEX